CONSERVATION BIOLOGY IN SUB-SAHARAN AFRICA

Conservation Biology in Sub-Saharan Africa

John W. Wilson and Richard B. Primack

OpenBook Publishers

https://www.openbookpublishers.com

Updated digital material and resources associated with this volume are available at https://doi.org/10.11647/OBP.0177#resources

The Lounsbery Foundation has generously contributed to the publication of this volume.

ISBN Paperback: 978-1-78374-750-4
ISBN Hardback: 978-1-78374-751-1
ISBN Digital (PDF): 978-1-78374-752-8
ISBN Digital ebook (epub): 978-1-78374-753-5
ISBN Digital ebook (mobi): 978-1-78374-754-2
ISBN XML: 978-1-78374-755-9
DOI: 10.11647/ OBP.0177

Cover photo and design by Anna Gatti.

This book is dedicated to our spouses Lesley Starke and Margaret Primack, who supported our efforts to write this book and our decision to provide it for free to students and researchers across Africa

Contents

List of Boxes

Foreword

Even for those of us who are one step removed from nature in our present-day lives, looking back on places we called home as children can reveal disturbing truths. This is certainly true for me. I grew up in a small mining town in Zimbabwe called Kadoma, and as soon as school was out, my parents would bundle me off to Mhondoro, a rural area to the east.

I loved it there and happily spent hours herding cattle across the quintessential untamed African savannah. I have many fond memories of swimming and catching fish in Mhondoro too. I recall how the surrounding forests were lush, teeming with wildlife; rivers were bountiful, full of fish and fowl in the pristine fresh water. City folk went home with their hands full of gifts from nature and agricultural fields.

Today, when I return to my village in Mhondoro, my heart breaks. The lush forests and wildlife are gone, replaced with barren fields and a whimpering stream where the river once ran. I now bring my own food and bottled water when I visit. And worst of all, the people of Mhondoro who I had always associated with nature's abundance are today poor and disenfranchised and have few if any options for bettering their lives.

Tragically, the story of my village is shared by thousands of villages across Africa that are suffering the worst impacts of climate change, population growth and harmful development choices. Faced with the challenge of feeding their families and generating cash incomes, farmers, like those of Mhondoro, end up expanding their crops increasingly deeper into wild lands and forests. These encroachments not only bring their families into dangerous conflict with wildlife, they simultaneously endanger and destroy the forests and fertile soils that would otherwise support their agricultural bounty.

But this outcome is not inevitable.

Africa suffers the greatest burden of global heating and deteriorating nature. As such, there is recognition that a "*new deal for nature*" is needed if we are to avert the worst climate and nature crisis. A new deal that transforms the way we produce our food and choose what to consume, the way we develop infrastructure, including our cities, roads, housing and dams, produce our energy and the way we value nature in our economic systems. The search is on for solutions.

But Africa risks being left behind or having to acquiesce to solutions that are not fit nor ideal for the continent. If Africa is to meaningfully define solutions for a new deal for nature, we must support research capacity and skills building within its populations, including investments in faculty and research leaders, facilities and infrastructure, and expanding career opportunities for budding researchers to apply their findings in real world settings.

Fortunately, there is now widespread acknowledgement that African researchers are best placed to ask questions and find solutions to the challenges facing Africa. Hence, we are witnessing a slow move away from the notion that researchers from high income countries must be parachuted in to identify and address the continent's problems.

This textbook, *Conservation Biology in Sub-Saharan Africa*, is a critical first step as it goes a long way towards focusing attention to the urgent need to define the nature of the problem and develop practical, context specific solutions to Africa's many environmental challenges. It discusses how our lives are inextricably linked to a healthy environment, explores how inclusive conservation can provide greater benefits to all, and illustrates how grassroots action can ensure that nature's many beneficial contributions will remain available for generations to come.

By being distributed for free, this textbook ensures that its readership can include those who stand to benefit the greatest, including African researchers and practitioners. Open access textbooks like it are critical to expanding access to African research and improving intra-African research collaboration and capacity.

If we are to "leave no one behind," as agreed by global leaders in 2015 and encapsulated in Sustainable Development Goals, farmers like those from Mhondoro will also need access to the best available information, science and solutions. That is a deal that we owe them—and the many people who call Africa their home. This book makes an important contribution to the challenge. I hope that, thanks to the efforts of the experts featured in this textbook and others, one day when I return to Mhondoro, it will more closely resemble its prior self that I loved as a child.

Maxwell Gomera

Director: Biodiversity and Ecosystem Services Branch
UN Environment

Preface

We are excited to present the very first conservation biology textbook dedicated entirely to an African audience. The need for this work has never been more pressing than now. Africa has some of the fastest growing human populations on Earth. This growth, together with a much-needed push for development to ensure that all Africans can live healthy and prosperous lives, exerts unsustainable pressure on the region's rich and unique biological treasures. Consequently, Africa is losing its natural heritage faster than ever before. It is sobering to consider that there is a very real risk that our children may never have the opportunity to see gorillas, rhinoceros, or elephants in the wild.

To address this alarming loss of Africa's natural heritage, there is an urgent need to produce the next cohort of well-trained conservation leaders, able to confront conservation challenges head-on, and to secure a sustainable future for all. This effort starts early, by exposing children from a young age to the wonders of the natural world. But it is also important to ensure that those children who later choose a career in biodiversity conservation are well-prepared for the road ahead. To facilitate this capacity building, we have compiled this textbook, designed for use in conservation biology courses, and as a supplemental text for other courses in the natural sciences and environmental policy. While the main target audience for this book is early-career conservationists, we strived for a balance between theory, empirical data, and practical guidelines to also make the book a valuable resource for mid- and late-career professionals. To further remove obstacles to training, we made every effort to ensure that this work is accessible to as wide an audience as possible. For that reason, we are making this textbook available for free, under a Creative Commons (CC BY) license, to guarantee the rights for anyone to use and spread this work to whoever wishes to make a difference in the future of Africa's biodiversity and its people.

Scope

This textbook focuses on the Afrotropics, one of Earth's eight major terrestrial ecozones. This area includes continental Africa south of the Sahara Desert, continental islands (e.g. the Seychelles) that drifted away from Africa millions of years ago, and oceanic islands with a volcanic origin (e.g. the Comoros archipelago, São Tomé, and

Príncipe) which share many biological characteristics with the Afrotropics. This area is also generally known as Sub-Saharan Africa, which we use throughout this book as a convenient and acceptable way to designate this ecologically (African parts of Afrotropics) and geographically (Africa south of the Sahara) distinct region.

Composite satellite image of Africa, with the Sahara Desert (sand-coloured) in the north and the Afrotropical ecoregion's tropical ecosystems (in green) further south featuring prominently. The area between the sand-coloured and green regions is the Sahel, which marks the northern boundary of the Afrotropics. Photograph by NASA, https://commons.wikimedia.org/wiki/File:Africa_satellite_orthographic.jpg, CC0.

Deviating slightly from the typical scope of some other books focussed on the Afrotropics, this textbook does not cover south-west Arabia—these areas, together with North Africa, are covered in a sister textbook to this one, published in Arabic (Primack and El-Demerdash, 2003). Similarly, this book excludes Madagascar and the Mascarene Islands, which are covered in two sister textbooks published in French (Primack and Ratsirarson, 2005; Primack et al., 2012). The text does cover a few areas not usually considered part of the Afrotropics, but which share several affinities with the Afrotropical region. These additional areas include oceans within 200 nautical miles from Sub-Saharan Africa, and oceanic islands in the Atlantic that are usually treated as part of the Palearctic realm, namely Cabo Verde, St. Helena, Ascension, and the Tristan da Cunha archipelago.

Taxonomy and the IUCN Red List categories

The International Union for Conservation of Nature and Natural Resources (IUCN) maintains a comprehensive online database (https://www.iucnredlist.org) that summarises the threat status of many species on Earth. The classification system used to compile this database is discussed in detail in Section 8.5. We foreshadow this discussion by alerting readers that the threat status of each assessed species that is mentioned in the text is indicated with one of the seven acronyms mentioned below, right after its scientific name. To facilitate this indication, we generally follow the IUCN's taxonomy in this textbook. We are fully aware that the lag time between taxonomic updates and IUCN assessment updates may create the appearance that this textbook's taxonomy might sometimes be outdated. We made every effort to highlight important taxonomic discrepancies when relevant to the text. For common names, we tried to use the most-widely used terms across Africa.

EX	Extinct
EW	Extinct in the Wild
EN	Endangered
VU	Vulnerable
NT	Near Threatened
LC	Least Concern
DD	Data Deficient

Organisation of the book

This book contains 15 chapters and four appendices. While there is broad overlap in the topics covered in each chapter, the first three chapters are meant to be introductory, while chapter four provides an overview on the importance of biodiversity for our own wellbeing. Chapters five to seven outline the most important threats to biodiversity, while chapters eight to fifteen suggest overarching solutions to the current biodiversity crisis. We ensured that the main body text, which is nearly entirely comprised of examples from Sub-Saharan Africa, covers examples from a range of organisms living in terrestrial, freshwater, and oceanic environments. Within the main body text several words are written in bold-face—these represent the first mention and/or explanation of specialist terms listed in the glossary. Also included are over 50 Boxes, authored by conservation researchers and practitioners from governments, universities, and nongovernmental organisations in West, Central, East, and Southern Africa. These case studies cover interdisciplinary topics such as public health, sacred spaces, energy, agriculture, law, sustainable development, and leadership. They expose readers to the voices of conservationists in the region, provide compelling examples of on-the-ground work, and offer insights into real-life conservation issues readers may face in their careers.

In each chapter, following the main body of text, there is a brief summary of main take-home messages, a list of discussion questions, and a list of suggested readings.

The discussion questions are formulated in such a way that there are no definitive right or wrong answers—rather, they are meant to stimulate discussion among readers so that they can develop their own conservation philosophies. We tried to restrict the suggested reading lists to works that are freely available either on publisher websites or online depositories accessible through Google Scholar (https://scholar.google.com). Note that the suggested readings are not meant to be absolute; in fact, they should be adapted to meet local contexts and syllabus requirements. Each chapter concludes with an extensive bibliography, which serves to provide a starting point for readers interested in specific conservation topics, and lecturers interested in adapting the suggested reading list. The textbook concludes with four appendices, meant to encourage readers to take the field's activist spirit to heart.

While we have made every attempt to ensure that the content is current and comprehensive, we recognise that mistakes do creep in; the field of conservation biology is currently also rapidly evolving, including being more inclusive of previously marginalised communities. We thus plan on continue updating this text, and welcome comments and suggestions from readers who share our interest in protecting Sub-Saharan Africa's natural heritage.

Bibliography

Primack, R.B., and J. Ratsirarson. 2005. *Principe de Base de la Conservation de la Biodiversité* (in French) (Antananarivo: University of Antananarivo).

Primack, R.B., F. Sarrazin, and J. Lecompte. 2012. *Biologie de la Conservation* (in French) (Paris: Dunod).

Primack R.B., and M. El-Demerdash. 2003. *Essentials of Conservation Biology* (in Arabic) (Cairo: Mars Publications).

Acknowledgements

We sincerely appreciate the contributions of everyone who helped ensure that this project was a success. In particular, we would like to thank Peter Ryan for his guidance and support throughout the project, Abe Miller-Rushing for writing several boxes and providing editorial input, and the Richard Lounsbery Foundation for funding support. Peter Ryan, David Cumming, and Adrian Shrader provided invaluable reviews to ensure the content is accurate and pitched at the right level for conservation students. We would also like to recognise the support from Nigel Bennett, Duan Biggs, Meg Boeni, Israel Borokini, and Mike Scantlebury at various stages of the project. Additional input on ex situ conservation was provided by Dalia Amor Conde, Johanna Staerk, and Simeon Smeele at Species360, while Peter Maniloff provided input on environmental economics. We thank all our photographers and box authors for their willingness to contribute to this work. Alessandra Tosi, Rupert Gatti, and the rest of the staff at Open Book Publishers helped to transform the manuscript into a finished product. Lastly, a special thanks to Lesley Starke and Margaret Primack for their encouragement, support, advice, and input throughout the project.

List of Acronyms

CAR	Central African Republic
CBD	Convention on Biodiversity Diversity
CBNRM	community-based natural resource management
CITES	Convention on International Trade in Endangered Species of Fauna and Flora
DRC	Democratic Republic of the Congo, formerly Zaire
EIA	environmental impact assessment
FFI	Fauna & Flora International
FSC	Forest Stewardship Council
GDP	gross domestic product
GEF	Global Environmental Facility
GIS	geographic information system
GPS	Global Positioning System
FZS	Frankfurt Zoological Society
HIV	human immunodeficiency virus
IBA	Important Bird and Biodiversity Area
IPBES	Intergovernmental Science-Policy Platform on Biodiversity and Ecosystem Services
IUCN	International Union for Conservation of Nature and Natural Resources
IPCC	Intergovernmental Panel on Climate Change
MPA	marine protected area
MSC	Marine Stewardship Council
MVP	minimum viable population
NGO	non-governmental organisation
PADD	protected area downgrading, downsizing, and degazettement
PAAZA	Pan-African Association for Zoos and Aquaria
PVA	population viability analysis
RLE	Red List for Ecosystems
RSPO	Roundtable on Sustainable Palm Oil

SCB	Society for Conservation Biology
SDM	species distribution modelling
sp.	species (plural: spp.)
TEK	traditional ecological knowledge
TFCA	transfrontier conservation area
TNC	The Nature Conservancy
UK	United Kingdom
UN	United Nations
UNEP	United Nations Environment Programme
UNESCO	United Nations Educational Scientific and Cultural Organisation
USA	United States of America
USAID	United States Agency for International Development
WCMC	World Conservation Monitoring Centre, a centre within UNEP
WCS	Wildlife Conservation Society
WWF	World Wide Fund For Nature
ZSL	Zoological Society of London

1. What is Conservation Biology?

East Africa's great migration is one of the most famous wildlife spectacles on Earth. Each year, tens of thousands of tourists from around the world flock to the region to see the 1.7 million common wildebeest (*Connochaetes taurinus*, LC) and hundreds of thousands of other plains mammals make their way from Tanzania's Ngorongoro Conservation Area, through the Serengeti Plains, to Kenya's Maasai Mara National Reserve. Photograph by Daniel Rosengren, https://commons.wikimedia.org/wiki/File:Wildebeest_Migration_in_Serengeti_National_Park,_Tanzania.jpg, CC BY 4.0.

https://doi.org/10.11647/OBP.0177.01

Popular interest in protecting **biological diversity**—which describes the amazing range of species, **genetic diversity** within each species, and the multitude of Earth's complex **biological communities** with their associated **ecosystem processes**—has intensified during the past few decades. During this time, scientists and the public have recognised that biological diversity (often shortened to **biodiversity**) is being lost at increasing rates. Across the world, human activities are destroying ecological communities that have developed over millions of years. Over the next several decades, thousands of species and millions of **populations** will likely go **extinct**.

The fundamental driver of all the biodiversity losses we are currently witnessing is a rapidly expanding human population coupled with increased consumptive needs. In 1850, after roughly 300,000 years of *Homo sapiens* on the planet, there were around 1 billion people on Earth. By 1987, not even 140 years later, the world's human population had surpassed 5 billion. By 2017, there were 7.5 billion humans globally, of which over 1 billion lived in Sub-Saharan Africa (World Bank, 2019). With this many people, the human population grows by tens of millions of people each year, even with modest population growth (Figure 1.1). To make matters worse, Sub-Saharan Africa has the fastest population growth rate in the world, with a projected human population estimate of over 4 billion people by the year 2100—a number that is well beyond the ecological capacity of the region to support.

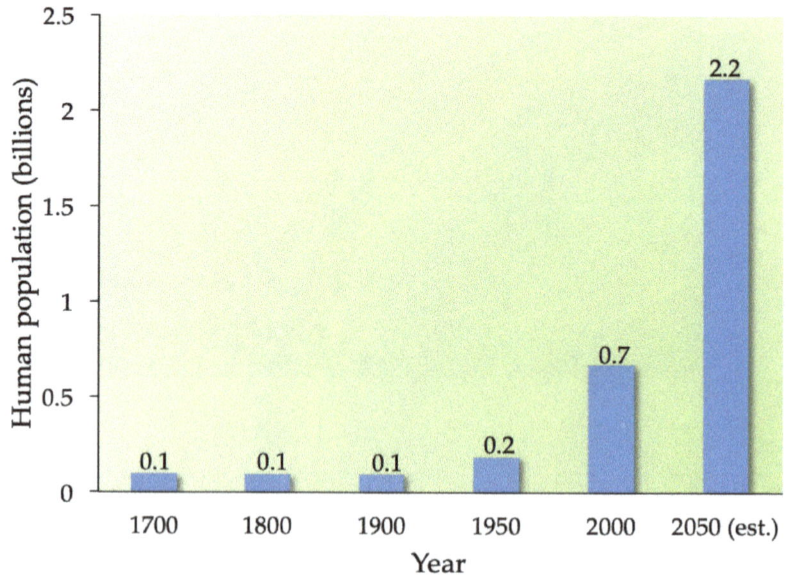

Figure 1.1 Sub-Saharan Africa's human population crossed the 1 billion mark in 2015. At the current annual population growth rate of 2.7%, more than 28 million people will be added to the region in 2019. This number will escalate each subsequent year as increases are compounded. Sources: Biraben, 2003; World Bank, 2019, CC BY 4.0.

To survive and prosper, people use **natural resources**. They harvest and use oil, water, and wildlife products, and convert natural ecosystems for agriculture, cities, roads, and industrial activities. This consumption, which reduces natural **habitat** and the associated **wildlife** populations, is intensifying because of the demands of a rapidly increasing human population. Consumption of resources also increases as countries develop and industrialise: the average citizen of the USA uses five times more resources than the average global citizen, 11 times more than the average Chinese citizen, and 32 times more than the average Kenyan citizen (Worldwatch Institute, 2015). This growth in the number of humans, together with their ever-more-intensive use of natural resources, is the fundamental driver behind most current species extinctions.

For conservation biologists and other nature lovers, the widespread extinction of species and destruction of natural ecosystems are incredibly discouraging.

For conservation biologists and other nature lovers, the widespread extinctions of species and destruction of natural ecosystems are incredibly discouraging. Perhaps nowhere in the world is this issue as dramatic as in Africa with its rich and spectacular wildlife, but also its significant socio-economic challenges, such as a rapidly increasing human population, persistent poverty, weak **governance** structures, and many people's near-obligate dependence on natural resources. Many Africans are also confused by the importance and need for conservation actions, pointing to the romanticised but inaccurate notion that humans have been living in relative harmony with nature since humans first made an appearance on Earth (see Box 8.1). But it is possible, and indeed necessary, to find ways to ensure the persistence of biodiversity. Actions taken, or not taken, during the next few decades will determine how many species and natural areas will continue to survive. Someday, people will likely look back and say that this time—the first half of the 21st century—was an important and exciting time when people worked together, and acted locally and globally, to prevent the extinction of many species and ecosystems. Examples of successful conservation efforts are described throughout this textbook.

1.1. Conservation Biology is Still Evolving

As a distinct scientific field, **conservation biology** is an integrated, multidisciplinary subject that developed in response to the challenge of preserving populations, species, ecosystems, and **biological interactions**. The main aim of conservation biology is to ensure the long-term preservation of biodiversity. To achieve its aim, conservation biology has set three goals:

- To document Earth's biological diversity.
- To investigate how humans influence species, **evolution**, and ecosystem processes.
- To investigate practical approaches to protect and restore biological communities, maintain genetic diversity, and prevent the extinction of species.

The first two goals describe typical scientific research investigating objective facts. The third goal, however, is a part of what makes conservation biology a **normative discipline**; that is, conservation biology incorporates human values, not just facts, to understand and achieve its value-laden goals (Lindenmayer and Hunter, 2010). In this sense, conservation biology is related to **environmentalism**, in which people aim to protect the natural environment for its own sake (see Section 4.3.2). However, conservation biology is at its core a **scientific discipline**; it is founded on scientific principles. This is not to say you must be a scientist to practice conservation biology; there are many people who are not scientists who apply the principles of conservation biology in their professional and personal lives.

The emergence of conservation biology as a distinct scientific field in the 1970s has given rise to the formation of various formal societies representing the field in a united voice. Most notable among these is the **Society for Conservation Biology** (**SCB**, Figure 1.2), which is a non-profit international professional organisation with a mission to advance "the science and practice of conserving the Earth's biological diversity". To facilitate opportunities where like-minded people can share ideas locally, the SCB has regional branches, including an active Africa Section (http://conbio.org/groups/sections/africa) which hosts regular conferences. In addition to the SCB, a great number of other local, national, and regional conservation organisations also exist and act as mouthpieces for grassroots movements and as custodians of nature. Many of these groups focus on specific animals or local **protected areas**. Others organically adapt their missions and visions in response to a specific need or threat. For example, established in 1913 as an exchange forum between collectors of rare plants, the Botanical Society of South Africa now actively works toward protecting those rare plants in their natural habitats.

Figure 1.2 The logo of the Society for Conservation Biology (SCB) has several layers of symbolism. Enclosed in the circle of life are ocean waves, representing change. The bird symbolises beauty, and the leaves (the bird's wings) remind us of nature's productivity. Image courtesy of SCB, all rights reserved.

Conservation biology also has a history of adapting to new challenges. The very first conservation activities, in Africa and beyond, were geared towards securing the rights to valuable natural resources for people in powerful positions, such as kings and tribal chiefs, enforced through a strictly adherence to cultural norms and **customary laws** (Section 2.2). But as a growing human population expanded its influence on the environment, and wildlife started to decline, earliest conservation models gradually shifted towards **fortress conservation** approaches (Wilshusen et al., 2002) which aimed to shield wildlife from people by setting aside protected areas where human activities were strictly controlled.

Today, however, as human populations are exploding, and consumption is increasing, even protected areas are increasingly unable to withstand the multitude of threats to biodiversity that ignore property boundaries and political borders. In response, fortress conservation approaches are beginning to make way for large-scale integrated activities that highlight the social and economic benefits of biodiversity conservation. To do this, new alliances are being formed and new agendas are being established, such as those that directly link human health with environmental health (Box 1.1). These **integrated conservation** philosophies that pursue strategies that benefit both humans and biodiversity show much promise because they focus on fundamental extinction drivers, and advocate for more inclusive sustainable development. In this way, the practice of conservation has evolved from just a plan to save the environment to a vision that accomplishes its goals through sustainable development and social justice.

In recent years, conservation practice has evolved from just a plan to save the environment to a vision that includes sustainable development and social justice.

Box 1.1 Conservation Through Public Health: A Case Study

Gladys Kalema-Zikusoka

Conservation Through Public Health, Kampala, Uganda.

http://www.ctph.org

Conservation Through Public Health (CTPH) is a grassroots non-governmental organisation (NGO) and non-profit that promotes biodiversity conservation by enabling people, wildlife, and livestock to coexist. The organisation was founded in 2003 after fatal scabies skin disease outbreaks in mountain gorillas (*Gorilla beringei beringei*, EN) were traced to people living around Bwindi Impenetrable National Park, Uganda, who had limited access to basic health services (Kalema-Zikusoka et al., 2002). Since then, CTPH has contributed to conservation and sustainable development in Africa by improving human and animal health and welfare in and around protected areas.

One of the main goals of CTPH is to reduce disease transfer between humans and gorillas. We accomplish this through an integrated **population, health, and environment (PHE)** programme that was established in 2007 with funding from the US Agency for International Development (USAID). As a first step, piloted around Bwindi, CTPH held consultative meetings with local leaders, during which at least one Village Health and Conservation Team (VHCT) volunteer was selected from each village and two from each parish (consisting of 11 villages) to oversee distribution of family planning supplies. This initiative rapidly expanded into a sustainable social service delivery network that promotes family planning, hygiene, and sanitation. The network resulted in a 20% to 60% (national average is 30%) increase in new users to modern family planning, and a 10% to 60% increase in adoption of hand washing facilities at homes visited by VHCTs. VHCT volunteers also refer people suffering from infectious diseases and malnutrition to local health centres and promote more sustainable alternative livelihoods. Another group of community volunteers, the "Human and Gorilla Conflict Resolution" (HUGO) team, in turn collect gorilla faecal samples left on communal land to monitor their health (Figure 1.A), and visually monitor gorillas for clinical signs of disease inside and outside protected areas (Gaffikin and Kalema-Zikusoka, 2010). In the process, we have seen reduced disease incidences in the gorillas, reduced conflict between people and gorillas, and improved attitudes toward conservation. One unintended outcome has been increased gender equality: men are now more involved in family planning, and women are more involved in natural resource management.

Figure 1.A A park ranger from the Uganda Wildlife Authority teaching HUGO community volunteers how to collect faecal samples from gorilla night nests during a CTPH training workshop. Photograph by CTPH, CC BY 4.0.

Our experience in initiating and managing PHE programmes for the past 10 years has taught us several lessons. One of the most important lessons to ensure project sustainability is to regularly engage with local leaders and the government. The Uganda Wildlife Authority, Uganda's Ministry of Health, and local health centres all attend CTPH meetings with VHCTs. Attendance by and representation of these groups not only informs them of our activities, but also provides a platform to inform or train the VHCTs in what they would like them to disseminate to the local communities.

We have also learnt that PHE-implementing partners and projects need to be well-suited to each other and each site; this remains true even though health needs are often the same, regardless of the location. For example, at Mount Elgon National Park in Uganda, we found that training VHCTs in reducing conflict with park management played a key role in changing community attitudes toward conservation. In contrast, at Virunga National Park, Democratic Republic of the Congo (DRC), we found that VHCTs needed to work more closely with local health centres to prevent disease transmission between people and gorillas, and to promote family planning in a largely Catholic country.

Lastly, we found that establishing income-generating projects for groups rather than individuals was key to sustaining VHCT networks and programme goals beyond donor funding cycles where we have had no volunteer dropouts in the first 10 years of initiating the PHE programme. These key components were accomplished by initiating livestock group enterprises and by encouraging VHCT volunteers to invest generated income into Village Saving and Loan Associations (see http://www.care.org/vsla).

Yet, as we consider how to best invest limited conservation resources, some difficult questions arise. With seemingly more work to be done than can be accomplished, should we let some species go extinct (Bottrill et al., 2008)? Which species? Who decides? How can we even dare to think that we can play god? Such questions predictably bring about strongly opinionated and emotional debate (Soulé, 2013 vs. Marvier, 2014; Tallis and Lubchenco, 2014). Given the successful track record of fortress conservation initiatives in preventing extinctions despite limited budgets (Young et al., 2014), as well as the promising progress of more complex people-centred initiatives (Pooley et al., 2014), it seems clear that conservation relies on some balance between these two conservation philosophies (Sodhi et al., 2011). Conservation biologists of tomorrow will be able to fine-tune the balance between these strategies by closely inspecting the successes and failures of our actions today.

1.2 The Role of Conservation Biologists

While there are a few extinctions that have only one cause, more generally, extinctions occur because several factors acted simultaneously and/or sequentially. Blaming a certain industry or specific group of people for an extinction (or other biodiversity loss) is thus simplistic, ineffective, and often counter-productive. Though challenging, a better approach would be to better understand how local, national, and international links led to those losses, and to find viable alternatives to prevent it from happening again. To succeed in this challenge, conservation biologists should strongly consider taking on one or more of several roles:

- Conservation biologists should be *curious*. The world around us is full of natural wonders waiting to be discovered. These discoveries underpin conservation action, by allowing us to define all the different components of biodiversity, enabling us to better understand the needs of different species, and providing us with opportunities to celebrate our conservation successes.

- Conservation biologists must be *good listeners*. Sometimes, the only difference between attracting a new ally and making an enemy, or between developing a landscape and saving a species from extinction, is the way we communicate. Conservationists must be careful and respectful listeners, especially to opposing perspectives. Careful listening is particularly important in rural areas, where villagers often have practical concerns related to their daily contact with wildlife, such as staying safe and preventing crop damage and livestock loss. Quite often, those villagers may also have unique insights into wildlife ecology that could prove valuable in local conservation measures.

- Conservation biologists must be *law-abiding citizens*. Activities that involve wildlife and ecosystems are regulated by laws and regulations. These laws are important because ethical boundaries differ from person to person—activities acceptable to one group of people may be considered harmful by another. As conservation biologists, abiding by environmental laws is especially important if we want others to take those laws seriously.

Laws are important because ethical boundaries differ from person to person—activities acceptable to one person may be immensely harmful to another.

- Conservation biologists should become *effective communicators*. They should be able to discuss the problems facing biodiversity in depth, as well as the consequences of losing biodiversity, to as broad a range of people as possible. Groups like hunters, community leaders and organisers, and church leaders may be interested in participating in conservation efforts once they recognise that their activities, health, and emotional well-being depend on conservation action.

- Conservation biologists could become *politically active leaders*, so that they can influence public opinion and policy. As a starting point, those interested in this role can join a conservation organisation to learn more about broader issues. They could also use their personal networks to form alliances with lawyers, citizen groups, and politicians.

- Conservation biologists could become *pro-active* land managers. Those taking on this task must be willing to walk on the land and go out on the water to find out what is really happening. They should also talk with local people to communicate their knowledge to others in ways that are clear and easily understood.

- Above all, a conservation biologist must be *honest*. To encourage effective action, both from the public and through policy, conservationists must present arguments backed by reliable evidence. To do otherwise, conservation biologists could lose credibility, which would very likely delay or even compromise conservation efforts.

It is worth taking a moment to distinguish between two important pillars of conservation action, namely **conservation advocacy** and **conservation science**. Conservation advocacy describes the roles that conservation biologists adopt to guide social, political, and economical systems towards a personally-preferred outcome—adopting environmentally-friendly practices; incorporating these activities makes conservation biology a normative discipline. Conservation science, in contrast, describes activities that conservation biologists undertake to generate knowledge, like objectively describing biodiversity and measuring biodiversity's response to stressors and safeguards. While conservation advocacy and conservation science often support and inform each other as to the next steps required for "doing conservation", it is important to distinguish between these two pillars to ensure that policy makers and other stakeholders in the environment understand when we advocate for personal preferences and when we offer objective findings (Rykiel, 2001; Lackey, 2007; Nelson and Vucetich, 2009). The next section will further expand on the importance of science in conservation biology.

1.3 The Value of Scientific Methods

The field of conservation biology applies **scientific methods** to achieving its goals. Like the medical sciences, which apply principles from physiology, anatomy, and genetics to problems of human health, conservation biologists solve biodiversity problems using principles from fields, such as mathematics, veterinary medicine, social sciences, and several natural sciences (Figure 1.3). Conservation biology differs from these and other component disciplines in that its primary goal is the long-term preservation of biodiversity. Unlike many other scientific fields, conservation biology can also be described as a **crisis discipline** (Soulé, 1985; Kareiva and Marvier, 2012). That is, conservation biologists are often required to take creative steps to respond to imminent threats, typically without a complete knowledge of the systems requiring attention. Conservation scientists must also articulate long-term visions for conservation beyond solving immediate problems.

To be effective, conservation biologists must demonstrate the relevance of their findings to a range of stakeholders. To be successful in this task, the importance of sound scientific principles cannot be over-emphasised. Nature is a complex network of many interdependent connections and **feedback loops**. Science is underpinned by principles that provide conservationists the necessary quantitative and qualitative tools to better measure and control for all these different aspects of biodiversity. Such

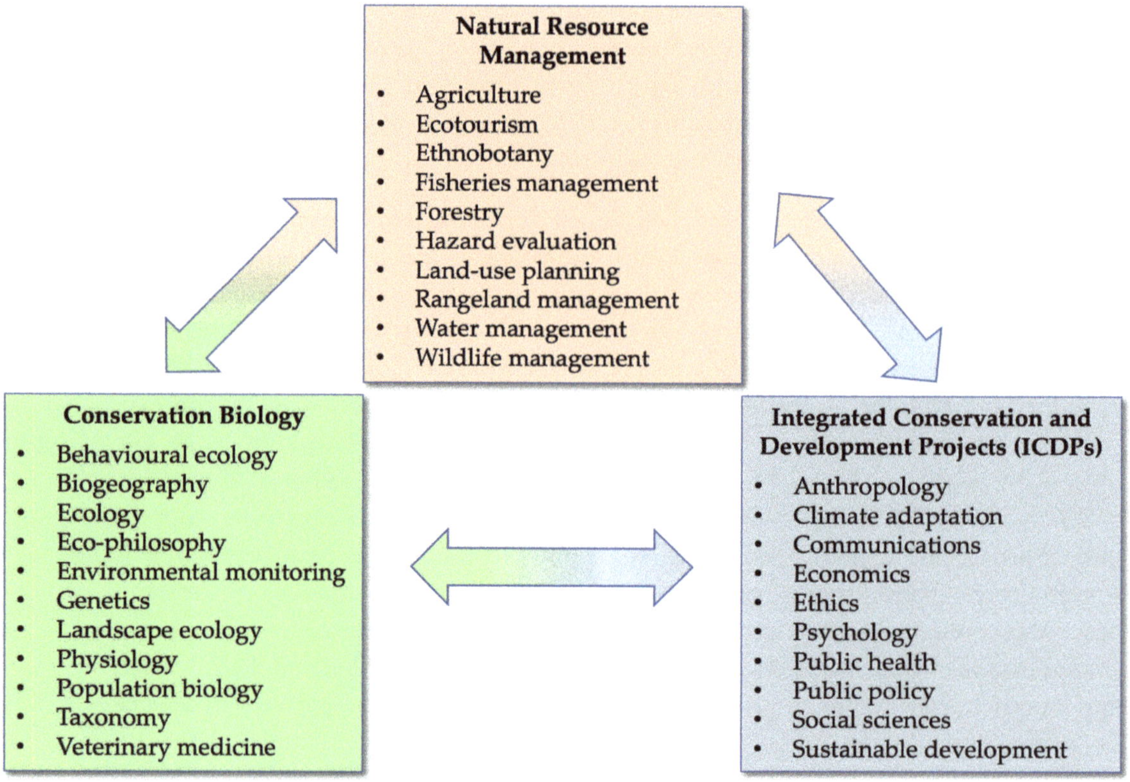

Figure 1.3 Conservation biology draws from many other sciences to protect biodiversity. It is closely related to natural resource management, which aims to manage biodiversity primarily for the benefit of humans. Integrated conservation and development projects (ICDP) are projects that manage nature for the benefit of both humans and biodiversity. After Kareiva and Marvier, 2012; Temple, 1991, CC BY 4.0.

measurements allow us to gain a better understanding of complex natural systems, and the consequences of human activities. Reliable, unbiased data obtained from sound and transparent scientific methods also facilitate policy making that is too often based on value judgments by non-experts who must balance many needs and different sources of information (Ntshotsho et al., 2015).

One of the cornerstones of modern science is to identify a hypothesis (a proposed explanation for a specific observation) to evaluate. The best hypotheses, often expressed as goals or objectives, are usually those that are SMART:

Specific: not overly general;
Measurable: has both units and a method of measurement;
Attainable: realistic to achieve;
Relevant: related to what needs to be accomplished;
Time-bound: achievable within a specific timeframe.

Identifying SMART goals and objectives is an essential aspect of conservation biology. Without such benchmarks, practitioners cannot know whether their tasks were successful, or when management actions should be adjusted to achieve success. While this may seem obvious, many previous conservation projects have failed because biologists neglected to set SMART goals and objectives (Tear et al., 2005). While lofty, "We're going to save all species" is not a SMART conservation goal because it is overly general, hard to measure, unrealistic, and not time bound. In contrast, "We want to protect 25% of our country's wetlands within the next 10 years" is a SMART goal because it sets a very clear and measurable objective. In general, it is wise to set smaller short-term (e.g. quarterly), and medium-term (e.g. annual) goals as one works towards long-term (e.g. 5–10 years) objectives; this allows one to constantly assess progress, which in turn provides opportunities for celebrations and strategic adjustments as and when needed.

Setting specific, measurable, realistic and timebound goals and objectives is essential for effective conservation.

Another scientific standard that conservation practitioners must adopt at a larger scale is the transfer of knowledge gained from unique and specialist experiences. Conservation activities are too often hampered by the lack of guidance from credible and available sources. This forces conservationist managers to base important decisions on biased anecdotes, personal intuition, and even myths (Sutherland et al., 2004). Successful conservation actions on the other hand often rely on results and guidelines that were disseminated to the broader community by practitioners who faced similar challenges earlier. To maximise this learning from each other's successes and mistakes, it is crucial for conservation scientists and managers to make every effort to ensure knowledge transfer, by carefully tracking their activities, and publishing their results and experiences in scientific journals and reports.

Biodiversity conservation, however, is not accomplished by simply setting SMART goals, measuring outcomes, and publishing results in scientific journals and reports. It is also important for conservation biologists to engage in **public outreach** activities, during which they can build on the public's existing connection to nature, help them better understand the value of biodiversity in their local area, and enable them to actively contribute in conservation projects. When interacting with the public, conservation practitioners must be sensitive to the complicated emotions and diverging interests of different groups of people (Milfont et al., 2017), especially vulnerable peoples who may be negatively impacted—hopefully only in the short term—by conservation actions. This requires a sense of emotional awareness, because the words we choose matter when we encourage others to care for and reduce

Public outreach builds on the public's existing connection to nature and helps them better understand the value of local biodiversity.

their impact on nature. Equally important, conservation biologists, as with any field of science, should be sceptical of their results. The process of generating data is not equal to generating facts, because data can be fraught with bias, imprecision, and uncertainty. This is perhaps even more important when sharing findings with lay people, as scientists have rigorous training in understanding uncertainty and connecting cause and effect. Putting scientific findings in context with adequate and clear explanation is a challenge to all scientists, but it is necessary, especially when partnering with conservationists not specifically trained as scientists.

1.4 Environmental Ethics

Most human societies today aim to protect biodiversity through rules and regulations (Chapter 12). An alternative approach is to change the fundamental materialistic values of modern society to values that prioritise genuine and lasting human well-being. This is the goal of **environmental ethics**, a discipline within philosophy that emphasises the ethical values of biodiversity. The foundation of environmental ethics lies in the philosophical principle that every organism of Earth has a right to exist, regardless of its usefulness to humans, so any action that negatively impacts biodiversity would be considered unethical.

Because human quality of life is intricately linked to the ability of the natural world to prosper (Chapter 4), the ethical arguments for biodiversity conservation hold even for people who value only human life. Or, in other words, respect for human life—even our instincts for self-preservation—should compel us to preserve biodiversity. In contrast, if we neglect our assumed responsibility to act as guardians of life on Earth, future generations will suffer with a lower quality of life. We can already see signs of this today: as species are lost and natural ecosystems replaced with sprawling cities, children are increasingly deprived of the wonderful experience of seeing a 'new' animal (Figure 1.4) or pretty flower. We can imagine that we are borrowing Earth from future generations, and that it is our responsibility to ensure that they receive it in good condition.

Because of this close link between nature and human well-being, the concept of nature preservation has permeated through the value systems of most human cultures, philosophies, and religions throughout history. This is especially relevant in Africa, where most (if not all) traditional societies have a deep connection with nature that is woven into their spiritual beliefs and customs (Figure 1.5). Our responsibility to protect animals is also explicitly described in Jewish, Christian, and Islamic traditions. Other major religions, including Hinduism, Buddhism, and Taoism also strongly support the preservation of non-human life. In light of accelerated biodiversity losses, faith-based groups have recently started playing a more active role in conservation, particularly among urbanised people. They do this by informing adherents that it is wrong to allow the destruction of nature, and that such destructive activities can have negative consequences for all people on Earth. These links between faith-based

Figure 1.4 A ragged-tooth shark (*Carcharias taurus*, VU) fascinates two kids at the Two Oceans Aquarium, South Africa. People—and especially kids—enjoy seeing wildlife, as shown by the increased popularity of protected areas, zoos, and other institutions where biodiversity can be seen. Photograph by Karen Schermbrucker, courtesy of Two Oceans Aquarium, CC BY 4.0.

organisations and conservation have given rise to consortiums such as the Forum on Religion and Ecology (http://fore.yale.edu), the Alliance of Religions and Conservation (http://www.arcworld.org), and the SCB's Religion and Conservation Working Group (https://twitter.com/ReligionConBio), as well as the emerging field of spiritual ecology (Vaughan-Lee, 2016).

Figure 1.5 The spiritual connection between people and nature features strongly in ancient rock art made by Bushmen (also known as San, or First People) of Southern Africa, believed to be the oldest human population on Earth. Pictured here is a common eland (*Tragelaphus oryx*, LC), drawn by shamans to open the portals of the spiritual world. Photograph by Alan Manson, https://commons.wikimedia.org/wiki/File:Shaman_at_work_Game_Pass_2004_0522_121946AA.jpg, CC BY 4.0.

Environmental ethics has strong links to the **environmental justice** movement and has recently established strong ties to the social justice movement. Some of the most exciting developments in this direction involve initiatives that combine protection activities with community upliftment programmes that improve the well-being of local peoples (Box 1.2; see also Section 14.3). These developments have shown that when poor and marginalised people are empowered to protect the environment, they may act as strong local guardians of forests, coastal areas, and other ecosystems that may have been destroyed otherwise.

Box 1.2 The Okapi Wildlife Reserve: Protecting Nature and Providing for People

Rosmarie Ruf & Marcel Enckoto

Okapi Conservation Project,
Okapi Wildlife Reserve,
Epulu, DRC.
https://www.okapiconservation.org

The Okapi Wildlife Reserve, a World Heritage Site in peril, is located within the dense, tropical Ituri Forest in north-eastern DRC. The reserve was created to protect the okapi (*Okapia johnstoni*, EN) (Stephenson and Newby, 1997). Researchers estimate that there are between 10,000 and 25,000 okapi (Figure 1.B) left in the wild, but with populations appearing to have declined by more than 50% over the last 15 years (Kümpel et al., 2015). The reserve also protects charismatic species like forest elephants (*Loxodonta cyclotis*), 14 species of primates, including chimpanzees (*Pan troglodytes*, EN), leopards, forest buffalo (*Syncerus caffer nanus*), and bongo antelope (*Tragelaphus eurycerus*, NT).

Since 1987, the Okapi Conservation Project has partnered with the Congo Institute for Conservation of Nature, the government agency responsible for the Okapi Wildlife Reserve's management, to provide financial and technical support for the operation of the reserve and preservation of the surrounding Ituri Forest. The project is partially managed and funded by Wildlife Conservation Global, a non-profit NGO based in Florida, USA.

Despite the support, conservation managers in this region face various challenges due to political instability since the 1990s. This has led to the deaths and displacement of millions of residents and rampant poaching in the area and beyond. Epulu Station, the reserve's headquarters, was tragically attacked in 2012, resulting in hostage taking, destruction and looting of the headquarters' infrastructure, and the deaths of staff and families, as well as 14 okapis at the captive breeding station.

To meet its goals in this difficult environment, the Okapi Conservation Project has seven objectives aimed at enhancing conservation, safety, and community:

- Financially supporting the operation of the reserve, paying warden and guard bonuses, building informer and monitoring networks, and providing necessities, such as food rations for patrols, fuel and spare parts for travel and field equipment.

Figure 1.B The okapi, a relative of the giraffe, at ZooPark de Beauval, France. This species, one of the DRC's natural treasures, survives only because of the dedication of a devoted group of conservation biologists. Photograph by Daniel Jolivet, https://www.flickr.com/photos/sybarite48/7973333500, CC BY 2.0.

- Maintaining and building infrastructure in the region that includes an airstrip, okapi pens, patrol posts for rangers, and high-quality tourism facilities. The project has also outfitted 20 health centres with necessities and has overseen the refurbishment and setup of territory offices in Mambasa and Wamba, the construction of a primary school in Epulu, and rehabilitation of medical dispensaries in Sondo and Koki.
- Breeding okapi in captivity for release into their natural habitat to boost non-captive population numbers and genetic diversity. Between 1987 and 2012, the project succeeded in producing 11 calves.
- Promoting environmental education, public engagement, and public awareness. Project staff achieves this by developing and implementing school programmes in and around the reserve, presenting seminars in primary and secondary schools, and producing radio broadcasts. They also facilitate focus group meetings with women and farmers, public meetings in villages, work with local committees, and produce outreach materials such as conservation films, calendars, leaflets, and posters. The Okapi Conservation Project has already supplied 112 schools with educational materials.
- Facilitating tourism activities, including visits around the Epulu Station and zoo, forest walks, and participation in traditional hunting.

- Promoting food security in local communities by providing seeds and farming tools to more than 900 farmers in and around the reserve. In addition, more than 142 women have benefited from sewing and embroidery materials, and participation in community farm fields.
- Offering medical care to more than 300 families, a total of more than 1,500 family members, who work at the reserve.

The work of staff at the Okapi Conservation Project and Congo Institute for Conservation of Nature is not easy due to political instability, breakdown of law and order, and lack of financial security. The Reserve and surrounding area also face increasing pressures from mining, poaching, and logging interests. But the project is necessary to help preserve the unique biodiversity of this Global Biodiversity Hotspot. In coming years, we hope that governance of the region will continue to improve and restore peace, justice, and the socio-economic status of local people. In such a situation, local communities and ecosystems, including the okapi, will benefit.

1.4.1 Conservation biology's ethical principles

Conservation biology rests on a set of underlying ethical principles that is generally agreed upon (Soulé, 1985) and can be summarised as follows:

- *The diversity of species and biological communities should be preserved*: Most people appreciate biodiversity. Hundreds of millions of people visit national parks, game reserves, zoos, botanical gardens, and aquaria each year. They spend money and take actions to protect these places and species. People also recognise that biodiversity has economic value, whether through tourism, consumption, or other services.
- *The untimely extinction of populations and species should be prevented*: Throughout history, species have occasionally died off as a result of natural, non-human causes. The loss of a local population was generally temporary, until a new population established itself through dispersal. However, human activities have increased the rate at which species are going extinct by more than a hundredfold (Box 1.3). Meanwhile, there is no similar increase in the rate at which new populations and species are being created.
- *Ecological complexity should be maintained*: In complex natural environments, biodiversity expresses many of its most valuable features and interactions. Although the biodiversity of species may be partially preserved in captivity, maintaining ecological complexity requires that natural areas be preserved.
- *Evolution should continue*: Evolution creates new species, increases biodiversity over time, and facilitates adaptation to changing environmental conditions.

People can help preserve these evolutionary processes by maintaining genetic diversity in wild populations and allowing populations to exchange genetic material. In captivity, many natural evolutionary processes do not occur, which can hamper survival when species are **reintroduced** in the wild.

- *Biodiversity has intrinsic value*: The value of species, communities, and ecosystems does not depend on their utility to people. They are **intrinsically valuable** on their own, with unique evolutionary histories and ecological roles. There are certain iconic species that people simply want to have around, but other, lesser-known species or species seen as problematic to people are not less valuable.

Box 1.3 Biodiversity: Can Humanity be Saved?

Nkengifor Nkeshia Valery

Regina International Cameroon,
Member of Union Farms of Africa,
Yaoundé, Cameroon.

vnkeh@yahoo.com

What happened over the past 200 years that we have arrived where we are? How did we get to this modern paradox? A society where we cherish comfort at the cost of the ever-increasing destruction of our planet. Never in the history of humanity has the environment been degraded to the point that even the air we breathe has become cancerous. Animals are exploited by industries at an alarming rate and those remaining are killed to enrich a privileged few. And all this evil happens with our complicity as indirect consumers. Our inheritance from God, the source of all our nourishment, does not belong to us. Yet it has been bought and exploited by multinational corporations and financial markets that hinder us from cultivating sustainably. We are pushed to feed ourselves and our crops with chemical products that are dangerous to our long-term health. We are also experiencing the start of the sixth mass extinction episode of biodiversity (Ceballos et al., 2017). As a result, the natural world has declared World War III against humanity. This is a war fought not by nation against nation, but that the environment has declared against the whole human race.

This war condemns us to live in an illusion of freedom; we are, in fact, destroyed at an increasing rate by different dangerous diseases and rendered slaves of the polluted environments that we blindly accept. The question we need to ask is not whether we should act to save our planet, but what future and meaning we are going to give the word "HUMANITY". We are all actors in a civilisation that we are constructing; to quote the Indian leader Mahatma Gandhi: "If we could change ourselves, the tendencies in the world would also change. As a man changes his own nature, so does the attitude of the world change towards him. [...] We need not wait to see what others do". Let us pause and ask ourselves what we want the future to say of us. Are we a destructive generation, or a generation that is ready to sustainably preserve its biodiversity? It is a question every reader needs to ponder. The future is judging no one and blaming no one, but it needs us to change our habits towards protecting the world's biodiversity.

To change our attitude and make the world a better place, I drafted the following poem with passion to see my words become action for every lover of biodiversity

WORLD CHANGERS

WE ARE A PEOPLE OF PEACE
CALLED FORTH OUT OF HUMANITY INTO RESTORING LIFE TO OUR NATURAL HABITAT.
WE ARE GOVERNED AND GUIDED BY A SENSE OF SUSTAINABILITY.
CONSERVATION AND PROTECTION IS OUR PRIORITY IN ALL THINGS AT ALL TIMES
WE ARE LED AND DRIVEN BY THE SPIRIT OF AN ENVIRONMENT FREE OF POLLUTION
WE ARE CALLED TO EFFECT AND AFFECT EVERY LIFE THAT WE COME IN CONTACT WITH
TOWARDS THE SUSTAINABLE DEVELOPMENT OF THE ENVIRONMENT

WE ARE CALLED BY HUMANITY TO BE WORLD CHANGERS
WE REFUSE TO CONFORM WITH THE THINKING PATTERN OF THE WORLD SYSTEM BECAUSE WE ARE WORLD CHANGERS

These principles are not absolute, nor are conservation biologists required to agree with them—they are actively discussed and debated. But many individuals and organisations agree with two, three, or all the principles, and support conservation efforts.

1.5 Summary

1. Conservation biology has three goals: (a) to document Earth's biological diversity; (b) to investigate how humans influence species, evolution, and ecosystem processes; and (c) to investigate approaches to protect and restore biological communities, maintain genetic diversity, and prevent the extinction of species.

2. Because conservation is multidisciplinary and requires a deep understanding of natural processes and human society, conservation biologists must take on multiple roles. Specifically, conservation biologists, as a group, must be curious, good communicators, effective educators, law-abiding citizens, and effective managers and practitioners of conservation projects.
3. Conservation biology relies on scientific evidence and ethical principles that underpins the preservation of biodiversity. Conservation biologists generally agree that biodiversity should be preserved, untimely extinctions should be prevented, ecological complexity should be maintained, evolution should continue, and biodiversity has intrinsic value.
4. Science provides conservation biologists with useful tools that guide the setting of clear, achievable, and measurable goals, to monitor conservation actions to assess whether goals have been met, and to communicate in a clear and unbiased manner.
5. Environmental ethics appeals to people of different walks to life to preserve biodiversity. It holds that biodiversity must also be protected because human well-being and economic opportunities are linked to a healthy environment.

1.6 Topics for Discussion

1. Is conservation biology substantially different in its approach from other fields of science, such as physics, chemistry, or medicine? If so, how? How is it linked to but different from environmentalism and social justice?
2. Looking at the titles of Chapters 5–7, what do you think are the biggest threats to biodiversity near where you live? Explain your answers.
3. There have been three broad conservation approaches throughout history: early models that secured natural resources for powerful people, fortress conservation, and integrated conservation. How do these approaches differ and how do they complement each other? How do you think each of these approaches can contribute to conservation today?
4. Which statements about the ethical principles of conserving biodiversity in this chapter do you agree with? Which do you disagree with? Explain your answers.

1.7 Suggested Readings

Bottrill, M.C., L.N. Joseph, J. Carwardine, et al. 2008. Is conservation triage just smart decision making? *Trends in Ecology and Evolution* 23: 649–54. https://doi.org/10.1016/j.tree.2008.07.007 Should we let some species go extinct?

Elosegi, A., M.O. Gessner, and R.G. Young. 2017. River doctors: Learning from medicine to improve ecosystem management. *Science of the Total Environment* 595: 294–302. https://doi.org/10.1016/j.scitotenv.2017.03.188 Comparing conservation to the health sciences

Illingworth, S. 2017. Delivering effective science communication: Advice from a professional science communicator. *Seminars in Cell and Developmental Biology* 70: 10–16. https://doi.org/10.1016/j.semcdb.2017.04.002 Tips to help you better communicate with public audiences.

Kareiva, P., and M. Marvier. 2012. What is conservation science? *BioScience* 62: 962–69. https://doi.org/10.1525/bio.2012.62.11.5 A review of an article published in 1985, offering a revised set of core principles that should guide conservation biology today.

Meyer, J.L., P.C. Frumhoff, S.P. Hamburg, et al. 2010. Above the din but in the fray: environmental scientists as effective advocates. *Frontiers in Ecology and the Environment* 8: 299–305. https://doi.org/10.1890/090143 Scientists have a responsibility to be advocates as well as researchers.

Sutherland, W.J., A.S. Pullin, P.M. Dolman, et al. 2004. The need for evidence-based conservation. *Trends in Ecology and Evolution* 19: 305–08. https://doi.org/10.1016/j.tree.2004.03.018 How do conservationists make decisions?

Tallis, H., and J. Lubchenco. 2014. Working together: A call for inclusive conservation. *Nature* 515: 27–28. https://doi.org/10.1038/515027a There is a risk that conservation can be polarising. We should rather be working together.

Tear, T.H., P. Kareiva, P.L. Angermeier, et al. 2005. How much is enough? The recurrent problem of setting measurable objectives in conservation. *BioScience* 55: 835–49. https://doi.org/10.1641/0006-3568(2005)055[0835:HMIETR]2.0.CO;2 Guidelines for setting goals in conservation.

Bibliography

Biraben J.-N. 2003. The rising numbers of humankind. *Populations and Societies* 394: 1–4. https://www.ined.fr/en/publications/population-and-societies/the-rising-numbers-of-humankind-en

Bottrill, M.C., L.N. Joseph, J. Carwardine, et al. 2008. Is conservation triage just smart decision making? *Trends in Ecology and Evolution* 23: 649–54. https://doi.org/10.1016/j.tree.2008.07.007

Ceballos, G., P.R. Ehrlich, and R. Dirzo. 2017. Biological annihilation via the ongoing sixth mass extinction signalled by vertebrate population losses and declines. *Proceedings of the National Academy of Sciences* 114: E6089–E6096. https://doi.org/10.1073/pnas.1704949114

Gaffikin, L., and G. Kalema-Zikusoka. 2010. *Integrating human and animal health for conservation and development: Findings from a program evaluation in southwest Uganda* (Kampala: CTPH; Stanford: EARTH, Inc.; Boston: John Snow, Inc). https://www.jsi.com/JSIInternet/Inc/Common/_download_pub.cfm?id=11196&lid=3

Kalema-Zikusoka, G., R.A. Kock, and E.J. Macfie. 2002. Scabies in free-ranging mountain gorillas (*Gorilla beringei beringei*) in Bwindi Impenetrable National Park, Uganda. *Veterinary Record* 150: 12–15. http://doi.org/10.1136/vr.150.1.12

Kareiva, P., and M. Marvier. 2012. What is conservation science? *BioScience* 62: 962–69. https://doi.org/10.1525/bio.2012.62.11.5

Lackey, R.T. 2007. Science, scientists, and policy advocacy. *Conservation Biology* 21: 12–17. https://doi.org/10.1111/j.1523-1739.2006.00639.x

Lindenmayer, D., and M. Hunter. 2010. Some guiding concepts for conservation biology. *Conservation Biology* 24: 1459–68. https://doi.org/10.1111/j.1523-1739.2010.01544.x

Marvier, M. 2014. New conservation is true conservation. *Conservation Biology* 28: 1–3. https://doi.org/10.1111/cobi.12206

Milfont, T.L., P.G. Bain, Y. Kashima, et al. 2017. On the relation between social dominance orientation and environmentalism: A 25-nation study. *Social Psychological and Personality Science* 2017: 1948550617722832. https://doi.org/10.1177/1948550617722832

Nelson, M.P., and J.A. Vucetich. 2009. On advocacy by environmental scientists: what, whether, why, and how. *Conservation Biology* 23: 1090–101. https://doi.org/10.1111/j.1523-1739.2009.01250.x

Ntshotsho, P., H.E. Prozesky, K.J. Esler, et al. 2015. What drives the use of scientific evidence in decision making? The case of the South African Working for Water program. *Biological Conservation* 184: 136–44. https://doi.org/10.1016/j.biocon.2015.01.021

Pooley, S., J.A. Mendelsohn, and E.J. Milner-Gulland. 2014. Hunting down the chimera of multiple disciplinarity in conservation science. *Conservation Biology* 28: 22–32. http://doi.org/10.1111/cobi.12183

Rykiel, E.J. 2001. Scientific objectivity, value systems, and policymaking. *BioScience* 51: 433–36. https://doi.org/10.1641/0006-3568(2001)051[0433:SOVSAP]2.0.CO;2

Sodhi, S.N., R. Butler, W.F. Laurance, et al. 2011. Conservation successes at micro-, meso- and macroscales. *Trends in Ecology and Evolution* 26: 585–94. https://doi.org/10.1016/j.tree.2011.07.002

Soulé, M.E. 1985. What is conservation biology?: A new synthetic discipline addresses the dynamics and problems of perturbed species, communities, and ecosystems. *BioScience* 35: 727–34. https://doi.org/10.2307/1310054

Soulé, M.E. 2013. The "new conservation." *Conservation Biology* 27: 895–97. https://doi.org/10.1111/cobi.12147

Stephenson, P.J., and J.E. Newby. 1997. Conservation of the Okapi Wildlife Reserve, Zaïre. *Oryx* 31: 49–58. https://doi.org/10.1046/j.1365-3008.1997.d01-3.x

Sutherland, W.J., A.S. Pullin, P.M. Dolman, et al. 2004. The need for evidence-based conservation. *Trends in Ecology and Evolution* 19: 305–08. https://doi.org/10.1016/j.tree.2004.03.018

Tallis, H., and J. Lubchenco. 2014. Working together: A call for inclusive conservation. *Nature* 515: 27–28. https://doi.org/10.1038/515027a

Tear, T.H., P. Kareiva, P.L. Angermeier, et al. 2005. How much is enough? The recurrent problem of setting measurable objectives in conservation. *BioScience* 55: 835–49. https://doi.org/10.1641/0006-3568(2005)055[0835:HMIETR]2.0.CO;2

Temple, S.A. 1991. Conservation biology: New goals and new partners for managers of biological resources. In: *Challenges in the Conservation of Biological Resources: A Practitioner's Guide*, ed. by D.J. Decker, et al. (Boulder: Westview Press).

Vaughan-Lee, L. 2016. *Spiritual Ecology: The Cry of the Earth* (Point Reyes: Golden Sufi Center).

Wilshusen, P.R., S.R. Brechin, C.L. Fortwangler, et al. 2002. Reinventing a square wheel: Critique of a resurgent "protection paradigm" in international biodiversity conservation. *Society and Natural Resources* 15: 17–40. https://doi.org/10.1080/089419202317174002

World Bank. 2019. *World Bank Open Data: Sub-Saharan Africa*. http://data.worldbank.org/region/sub-saharan-africa

Worldwatch Institute. 2015. *Vital Signs, v. 22: The Trends that are Shaping our Future* (Washington: The Worldwatch Institute). http://www.worldwatch.org/Vital-Signs-22

Young, R.P., M.A. Hudson, A.M.R. Terry, et al. 2014. Accounting for conservation: Using the IUCN Red List Index to evaluate the impact of a conservation organization. *Biological Conservation* 180: 84–96. https://doi.org/10.1016/j.biocon.2014.09.039

2. Introduction to Sub-Saharan Africa

The forested hills of Bwindi Impenetrable National Park, Uganda, are home to Africa's iconic mountain gorillas and over 350 bird species. The park, a World Heritage Site, is situated in the Albertine Rift, a Biodiversity Hotspot. Photograph by Jason Houston/USAID, https://www.flickr.com/photos/usaid-biodiversity-forestry/25422552517, CC0.

https://doi.org/10.11647/OBP.0177.02

Not only is Africa the second most-populous continent in the world, its human population is also incredibly diverse. Consider, for example, that over 2,000 native languages are spoken across the continent (Lewis et al., 2014). (Interestingly, there are strong positive correlations between linguistic diversity and biodiversity, as well as between the loss of species and languages; Gorenflo et al., 2012). Africa is also economically diverse; the continent contains some of the poorest nations in the world but also some of the fastest growing economies (World Bank, 2017). Herein also lies a major challenge: Africa's diverse human population—already over 1 billion people—is expected to double over the next 25 years (World Bank, 2019). To stimulate **economic growth** and provide resources for a growing and upwardly mobile human population, once unending wildernesses are constantly being cleared for agriculture, timber, expanding cities, and other human activities. In the process, the remaining natural areas are being polluted, **overharvested**, and fragmented, particularly in areas of outstanding conservation value (Balmford et al., 2011).

This environmental destruction we are witnessing across Africa holds negative consequences for all people on the continent. Among the most vulnerable are **traditional peoples** who rely on natural products such as firewood, wild animals, and wild edible fruits and roots to maintain their way of life. The destruction of the environment also makes it more challenging for city dwellers to access basic needs such as clean drinking water, clean air, and **wilderness areas** where they can fulfil their spiritual and emotional needs. With Africa's human population and consumption expected to grow substantially for many years to come, there is an urgent need to find ways to ensure that the region's unique environmental treasures are preserved, for the benefit of current and future generations.

2.1 Sub-Saharan Africa's Natural Environment

Much of the African continent encompasses the Afrotropical **ecoregion**, which is separated from other ecoregions by the Indian Ocean to the East, the Atlantic Ocean to the West, and the Saharan Desert to the North. These major geographic features have acted as barriers to movement since the African continent first took its current shape, enabling species and ecosystems characteristic of the region to evolve in relative isolation from those of other ecoregions. The Afrotropical ecoregion can be further subdivided into eight terrestrial **biomes** (Figure 2.1), each with its own distinct climate, geology, and **biota** (Burgess et al., 2004):

- *Tropical and subtropical savannahs and grasslands*: Sub-Saharan Africa's largest biome is a mosaic of grasslands, woodlands, bushlands, thickets, and semi-arid **drylands** that are maintained by fire and grazing. East and Southern Africa's miombo and mopane savannah-woodland ecosystems are included in this ecosystem.

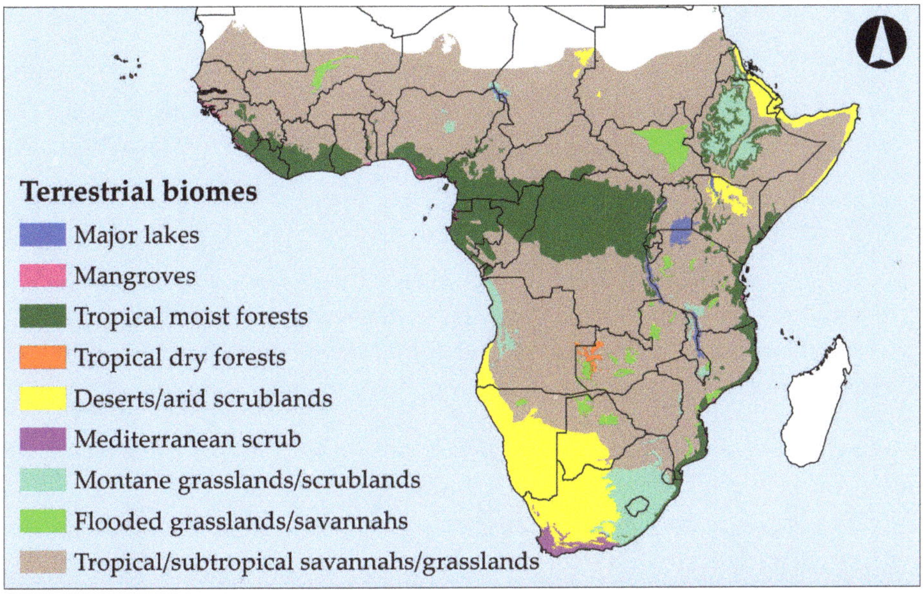

Figure 2.1 Simplified map of Sub-Saharan Africa's eight terrestrial biomes. The region's topographic complexity, the diversity of biomes, and the multiple ecological transition zones between the different biomes have given rise to a rich biodiversity. After Olson et al., 2001. Map by Johnny Wilson, CC BY 4.0.

- *Deserts and arid scrublands*: A biome of areas where evaporation exceeds precipitation, generally with rainfall < 250 mm/year. Generally associated with searing daytime temperatures and wind-swept sand dunes, this biome contains scrub deserts rich in succulent plants, rocky mountain deserts, and arid grassland-savannah mosaics, such as the **Sahel** region located just south of the Sahara.

- *Tropical moist forests*: Lowland broadleaf ecosystems with near-continuous canopies that run as a broad band across equatorial Africa. This biome is characterised by high rainfall (> 2 m/year), low variability in temperatures, and very high **species diversity**.

- *Montane grasslands and scrublands*: A patchily distributed biome that occurs at altitudes > 800 m and has enough rainfall that a variety of grasses can thrive. Generally lacking trees except along some rivers and streams, it includes high altitude heathlands and other Afro-alpine areas.

- *Mediterranean scrub*: A scrubland ecosystem of limited extent, better known as the Fynbos or Cape Floristic Region, that is situated at Africa's southwestern tip. Characterised by hot dry summers and cool moist winters, it contains one of Earth's richest concentrations of **endemic** plant species.

- *Flooded grasslands and savannahs*: Grasslands, marshes, and shallow lakes that are periodically flooded by water that can be fresh, brackish, or hypersaline.

When flooded, these areas host some of the largest water bird congregations in the region.

- *Tropical dry forests*: A highly restricted forest type that can be found in western Zambia and adjacent Angola, as well as on Cabo Verde. While these areas may receive high rainfall, they are characterised by seasonal droughts that can last several months.
- *Mangroves*: Coastal wetlands of tropical climates characterised by distinctive woody plants with aerial roots that can tolerate saltwater. Typically associated with intertidal zones and muddy bottoms, mangroves provide nursery grounds for many aquatic animal species.

In addition to these terrestrial biomes, Sub-Saharan Africa also contains several aquatic biomes. Prominent freshwater biomes include several large rivers along with their headwaters and deltas, numerous small rivers, multiple large and small lakes, as well as a variety of wetland ecosystems such as swamps, bogs, and salt marshes (WWF/TNC, 2013). Prominent marine biomes include tropical coral reefs along Africa's east coast, as well as temperate continental shelves and seas along South Africa and Namibia (Spalding et al., 2007). There are also several important oceanic upwellings—areas of high productivity where surface waters are fertilised by nutrient-rich waters that "wells up" from below; these include the tropical Gulf of Guinea upwelling along West Africa, and the Benguela upwelling ecosystem along Africa's southwest coast.

The variety of biomes present in Sub-Saharan Africa is the result of variable geology and a long history of changes in climate and ecological communities. For example, when Earth's climate was warmer, tropical moist forests were more widely distributed. As the planet cooled during **glacial periods**, forests contracted and became fragmented while grasslands expanded; some new biomes developed as the climate changed and species moved around. Even today, biome boundaries are still shifting: for example, over the last few decades the boundary between the Sahara Desert and Sahel has shifted by hundreds of kilometres southward (Foley et al., 2003). The development, fragmentation, and movement of these and other biomes, as well as the influence of major dispersal barriers, such as large rivers and mountain ranges, have stimulated **speciation**, as different populations became specialised to conditions that were restricted to their particular elevations or on certain sides (wet or dry, sunny or shady) of mountain ranges.

> Sub-Saharan Africa boasts tremendous species richness, the result of a complex geological and environmental history.

Due to this dynamic geological, climatic, and environmental history, as well as all the factors that have promoted speciation, Sub-Saharan Africa boasts tremendous **species richness**. The region is particularly well known for its mammals, particularly its charismatic terrestrial **megafauna** and other large mammals that attract millions of tourists from all around the world each year (Figure 2.2). Among the most famous are the **Big**

Five animals—lions (*Panthera leo*, VU), savannah elephants (*Loxodonta africana*, VU), African buffalo (*Syncerus caffer*, NT), African leopards (*P. pardus*, VU), and black rhinoceros (*Diceros bicornis*, CR). Other notable mammals include cheetahs (*Acinonyx jubatus*, VU), the fastest mammal on Earth; Maasai giraffes (*Giraffa camelopardalis tippelskirchii*, VU), the world's tallest mammal; the giant eland (*Tragelaphus derbianus*, VU), the world's largest antelope; and Africa's four species of great apes. Many small mammals are also noteworthy. For example, East Africa's naked mole-rat (*Heterocephalus glaber*, LC) is the world's only mammalian thermoconformer—meaning it is almost entirely cold-blooded; like reptiles their body temperature tracks ambient temperatures (Buffenstein and Yahav, 1991). The naked mole-rat and Southern Africa's Damaraland mole-rat (*Fukomys damarensis*, LC) are the only known eusocial mammals; like some ants and bees, only one female (the queen) reproduces with one to three breeding males, while all the other colony members are sterile workers (Jarvis et al., 1994).

Figure 2.2 The thrill to go on a guided safari walk with the Big Five in a protected area, such as Zambia's South Luangwa National Park pictured here, is a major drawcard to foreigners visiting Africa. Dangerous mammals calm down significantly when they are not persecuted. Photograph by Time + Tide, CC BY 4.0.

While Africa's large mammals are a major tourist drawcard, the region hosts many other rich and noteworthy wildlife assemblages. With more than 2,100 bird species, 1,400 of them found nowhere else on Earth (Sinclair and Ryan, 2011), the Afrotropics may be the most taxonomically diverse bird region on Earth (Lotz et al., 2013). Among the many bird species that call Africa home is the world's largest **extant species** of bird, the red-necked ostrich (*Struthio camelus camelus*); standing up to 2.74 m tall, it is in dire need of conservation attention (Miller et al., 2011). Africa is also home to the world's heaviest extant flying animal, the kori bustard (*Ardeotis kori*, NT), which can weigh over 20 kg (Dunning, 2008). Over 100,000 insects have been described in Sub-Saharan Africa (Miller and Rogo, 2001), which include the world's smallest butterfly, the dwarf blue (*Oraidium barberae*, LC) of Southern Africa, and the aptly named goliath beetles (*Goliathus* spp.), which can be found throughout much of tropical Africa. The region also hosts a great number of noteworthy endemic amphibians and reptiles, which include the world's largest frog, the Goliath frog (*Conraua goliath*, EN) of Cameroon and Equatorial Guinea, and the black mamba (*Dendroaspis polylepis*, LC), arguably the world's most feared snake, which is widespread across Africa's savannahs. Lastly, Africa is home to Jonathan the Aldabra giant tortoise (*Aldabrachelys gigantea*, VU); having hatched in 1832, he is considered the oldest living terrestrial animal in the world.

The region's plant richness, estimated at over 45,000 species (Klopper et al., 2007), is also important from a global perspective. Many plant species have high economic value, particularly those that have been domesticated in the region, and are now important crops across the world. Primary among these are coffee—second only to tea in worldwide popularity as a beverage—which is native to West and Central Africa (*Coffea robusta*) and Ethiopia (*Coffea arabica*). Other important crops that originated in in the Afrotropics include okra, black-eyed peas, watermelon, and African oil palm. Conserving the wild genetic diversity of these domesticated plants in their native ranges is important because they may serve as "insurance" for today's crops that may be less productive in future due to **anthropogenic climate change** (Davis et al., 2012). Others, such as the wide variety of plants utilised in traditional medicine to treat malaria, may one day lead to new antimalarial drugs (Chinsembu, 2015). Similarly, many plant species also have high evolutionary value. These include **relict species** that survived previous **mass extinction events**, such as cycads (*Encephalartos* spp.) (unfortunately several cycad species are now *Extinct in the Wild*), and **Lazarus species** that were once believed to be extinct, such as the unique jellyfish tree (*Medusagyne oppositifolia*, CR) of the Seychelles.

> Species that have survived previous mass extinction events are unable to withstand the current onslaught of human activities.

A few small and isolated African ecosystems are particularly rich in species. Particularly noteworthy is the Rift Valley lakes, such as Lake Victoria, Lake Malawi, and Lake Tanganyika, which hold the richest freshwater fish diversity in the world. For example, nearly 14% of the world's freshwater fish species occur in Lake Malawi (also known as Lake Nyasa). Moreover, over 90% of Lake Malawi's 500–1,000 (numbers vary by source) fish species (Figure 2.3) are endemic, and thus found nowhere else on Earth. The Cape Floristic Region is home to the greatest concentration of non-tropical endemic species in the world, including speciose well-known plant **genera** like *Protea* and *Erica*. The Succulent Karoo, directly north of the Cape Floristic Region, may be the most floristically rich desert in the world (Mittermeier et al., 2004). Africa has deservedly received international acclaim for these and many other natural wonders. Prominently, more than 37 sites in Sub-Saharan Africa have already been recognised as natural **World Heritage Sites**. One such site is also Africa's oldest national park, Virunga National Park in eastern DRC, which contains at least 218 mammal and 706 bird species (WHC, 2007).

Figure 2.3 Lakes in Central Africa's Rift Valley hold the richest freshwater fish communities on Earth. Many species, such as these brightly coloured cichlids from Lake Malawi, face extinction because of overfishing, pollution, and invasive species. Photograph by OakleyOriginals, https://www.flickr.com/photos/oakleyoriginals/8589738572, CC BY 2.0.

2.2 History of Conservation in Sub-Saharan Africa

Traditional communities have long held a belief that humans are physically and spiritually connected to nature, and that communal needs outweighed individual desires. This also extended to natural resources, which were considered communal property that must also be shared with the spirits of the ancestors and future generations. Managing natural resources this way required strict adherence to **customary law** systems that imposed controls on the collection of animal and plant products. Some animals and plants were also worshiped, which leads to mythical superstitions and taboos that prohibited the killing of culturally and spiritually important animals, as well as totem species that bond families and villages together. Customary laws also created Africa's first protected areas, such as royal hunting grounds (areas where kings and traditional chiefs had exclusive hunting rights) and areas of spiritual significance (Box 2.1), where access and harvesting of natural resources were restricted.

Traditional African communities have long shared the belief that humans are physically and spiritually connected to nature, and that communal needs outweigh individual desires.

Box 2.1 Sacred Spaces: A Tradition of Forest Conservation in Benin

Emile N. Houngbo

School of Agribusiness and Agricultural Policies, National University of Agriculture, Cotonou, Republic of Benin.

✉ *enomh2@yahoo.fr*

The importance of forests for human life has been recognised for millennia. That is why public approaches have historically been adopted for their protection. Today, some of the most effective programmes are those that integrate local communities and their traditional knowledge with scientific forest management. The *Convention on Biological Diversity* (CBD) recognises the value of the cultural practices of traditional peoples for (a) practicing conservation and maintaining biodiversity and (b) promoting sustainable use. The sacred forests of Benin are recognised as a tangible heritage, both natural and cultural; their management by the local community is a major achievement in modern conservation.

Forest protection, an ancient reality in Benin

The life of traditional communities of Benin is closely linked to conservation of its forests, also known as *Zoun* in the local Goun language. Many social practices of Beninese communities rely on leaves, animals, water, stones, and other resources; the areas that provide these natural resources are called sacred forests because they are inhabited by deities or spirits, serve as spaces for rituals, or represent the seat of past kings. Monitoring of sacred forests is often entrusted to members of a certain lineage. For example, custody of the forest of the city of Abomey is the responsibility of traditional chief Dah Djagba, whose ancestors were installed near the sacred spring Didonou by King Houegbadja of Abomey in the 17th century. A sacred forest is a point of contact between a community and a spirit or deity, and between the visible and the invisible. The value and protection of the sacred forest is passed down from generation to generation, as are the rules and regulations. Typically hunting and setting fires in sacred forests are prohibited, while logging for timber and gathering plants for food and medicine are strictly regulated, with these products shared between priests and caretakers of the site (Juhe-Beaulaton and Roussel, 2002). The *Aloe vera* plant, for example, has long been used by vodun (spirit) priests during religious ceremonies to heal the wounds of new initiates.

Sacred forests today

Sacred forests have significant spiritual capital, or the power to influence the communities that revere them. They influence the collective consciousness regarding experiences as basic as rain, health, and the collection of spring water (in the case of the Abomey forest), or as complex as religious ceremonies, fertility, and overall happiness. Sacred sites also play an important role in cult practices (Roussel, 1994): funeral rites, ceremonies for dead infants, rites for accidental deaths (Laine, 1990; Sokpon et al., 1998), and healing ceremonies with medicinal plants. Meetings of secret societies such as the *Zangbeto*, *Kuvito*, and *Oro*, and religious or social ceremonies and ordeals are held in sacred forests. They also play an important role in the exercise of justice and social cohesion; disobeying the traditional rules and damaging the sacred forest can cause harm to the whole community (bad harvests, epidemics, drought, and mosquito infestations) or the person responsible (accidents, illness, or misfortune). The wrongdoer may need to perform a rite of reparation, such as an animal sacrifice or offering to repair the damage that they have caused.

Figure 2.A Tomato farm on the periphery of the Gbevozoun sacred forest, Benin. Farming encroachment has reduced this sacred forest from 1.6 km^2 to 0.5 km^2 in recent years; other sacred forests face a similar fate. Photograph by Emile N. Houngbo, CC BY 4.0.

Resistance to human pressures

One difficulty of managing sacred forests today is that they are often not well delineated. Under the influence of population growth, the area occupied by a sacred forest sometimes diminishes to a minimum size under communal protection. Some sacred forests in Benin, such as the 32 km^2 Birni forest, 11 km^2 Tanekas forest, and 2 km^2 Natitingou forest, have vanished due to human pressure on the land. The peripheral zone of Gbevozoun sacred forest in which the Gbevo deity is believed to dwell is currently encroached by agriculture (Figure 2.A), and only a central core of 0.5 km^2 of the forest's original 1.6 km^2 is still protected. The Honhoue sacred forest, meanwhile, still retains an area of 0.04 km^2 that has not shrunk over time. This is due the local community's belief in the power of the Honhoue divinity and 40 other deities that dwell in the forest.

Sacred forests are based on traditions of safeguarding religious ceremonies and nature for the future, and they continue to be a means of protecting biodiversity. They may be a resource for conservation of rare plant species for medicinal purposes, and even future improvement of agro-biodiversity. The preservation of sacred forests is crucial to community involvement in conservation.

This culturally driven system of checks and balances was greatly disrupted with the arrival of European settlers in the 17th century. Armed with guns, and little thought given to sustainability, the earliest colonists killed thousands of animals for food, trophies, sport, and profit. Following concerns about declining wildlife populations, particularly at the southern tip of South Africa, Sub-Saharan Africa's first formal environmental legislation was introduced in 1657, followed by the region's first formal environmental law in 1684 (MacKenzie, 1997). Significantly, this first law separated protected species, such as the common hippopotamus (*Hippopotamus amphibious*, VU), from pest species (which at the time included lions). Unfortunately, these early laws and regulations were of little consequence as an increasing number of colonists, lured by the promise of unlimited hunting on unexplored lands, arrived in the region. Consequently, by 1700, populations of every animal over 50 kg within 200 km from Cape Town were **extirpated** (Rebelo, 1992). These developments also led to Africa's first modern human-caused mammal extinctions. First to disappear was the bluebuck (*Hippotragus leucophaeus*, EX) around 1798. Nearly a century later, in 1871, the Cape warthog (*Phacochoerus aethiopicus aethiopicus*, EX)—more closely related to East Africa's desert warthog (*Phacochoerus aethiopicus delamerei* LC) than the widespread common warthog (*Phacochoerus africanus* LC)—disappeared, followed by the quagga (*Equus quagga quagga*, EX) around 1878 (the last captive individual died in 1883). Elsewhere, bontebok (*Damaliscus pygargus pygargus*, NT), Cape mountain zebra (*Equus zebra zebra*, VU), southern white rhinoceros (*Ceratotherium simum simum*, NT), and black wildebeest (*Connochaetes gnou*, LC) were all reduced to about a dozen individuals at one or two locations.

Ecosystems—forests in particular—near early European settlements similarly suffered as early colonists perceived them as an "inexhaustible" supply of fuel and timber. This widespread overharvesting prompted the Cape Colony's Governor in

1778 to appoint its first professional nature conservator, Johann Fredrick Meeding, to exercise some control over **deforestation**. But, like controls on hunting large mammals, these efforts generally only had a local and temporary impact.

2.2.1 The 1800s and launching of formal conservation efforts

Interest in the formal protection of Africa's biodiversity started to intensify during the 19th century. Most of the initial steps were taken in South Africa, which had the largest early colonial settlements and, hence, the most species threatened by human activities. First, in 1822, the *Game Law Proclamation* introduced hunting licence fees and closed seasons for selected species, followed by regulations to protect 'open spaces' in 1846 and forests in 1859. A major step towards ecosystem protection was taken in 1876 with the creation of the Cape Colony's Department of Forests and Plantations, while the appointment of a Superintendent of Woods and Forests in 1881 led to initial efforts towards the scientific management of ecosystems. Then, in 1886, the British government passed the *Cape Act for the Preservation of Game* (in 1891 extended to other British South African Territories), followed by the *Cape Forest Act* of 1888. The *Cape Forest Act* played an instrumental role in the proclamation of the Cape Colony's first formally protected areas, namely the Tsitsikamma and Knysna Forest Reserves, in 1888; today these lands are incorporated into South Africa's Garden Route National Park (Figure 2.4). These were followed by the appointment of Southern Africa's first formal game warden, H. F. van Oordt, in 1893, to manage Pongola Nature Reserve, proclaimed in 1894. (Pongola was degazetted and converted into agriculture land in 1921 but re-established in 1979). Thereafter, protected areas were established at regular intervals across South Africa, starting with Groenkloof Nature Reserve in February 1895, then Hluhluwe Valley and Umfolozi Junction Game Sanctuaries (today the Hluhluwe-iMfolozi Park) in April 1895. (St Lucia Game Reserve, today part of iSimangaliso Wetland Park, was also established sometime in 1895.)

Figure 2.4 Sub-Saharan Africa's first formally protected area was established to stop logging of the Tsitsikamma coastal forests, South Africa. Photograph by Androstachys, https://commons.wikimedia.org/wiki/File:Knysna_Forest00.jpg, CC BY 4.0.

West and Central Africa saw its first steps towards formal conservation efforts in 1885, with the establishment of forest reserves to protect valuable timber products (Brugiere and Kormos, 2009). The region's first game reserves were gazetted as early as 1889 in the DRC to protect elephants. Unfortunately, these efforts were of little consequence as ivory hunters continued to slaughter the region's elephant populations. It was only after colonial governments raised concerns about declining ivory revenues that the region passed its first formal environmental law in 1892, with the ratification of the *Congo Basin Convention* to regulate the ivory trade in French, Portuguese, and Belgian territories (Cioc, 2009).

In East Africa, colonial authorities passed its first formal environmental legislations in 1888. These initial laws called for game reserve establishment, hunting quotas for common species, strict protection for breeding females and immature animals, and hunting bans for rare species (Prendergast and Adams, 2003). While protected area establishment was initially slow, a circular from Lord Salisbury (the UK's Prime Minister at the time) in which he called for protected areas and hunting restrictions to prevent large mammal extinctions, prompted the passing of the *German East African Game Ordinance* of 1896. That same year, East Africa saw the proclamation of its first modern protected areas, both in Tanzania: one along the Rufiji river (today included in Selous Game Reserve), and one west of Mount Kilimanjaro.

Initial laws and regulations to protect Africa's environment were greatly expanded in 1900, with the signing of the *Convention on the Preservation of Wild Animals, Birds, and Fish in Africa*, during the *International Conference of the African Colonial Powers* held in London, UK. The most innovative agreement of this treaty was the establishment of Schedules that afforded different species different levels of protection. Species on Schedule 1 included rare and valuable species for which all hunting was prohibited; Schedule 2 and 3 included species for which hunting of young animals and accompanying females was prohibited; Schedule 4 included species for which hunting was allowed 'in limited numbers'; and Schedule 5 included 'harmful' species whose populations needed to be reduced. While this **convention** never went into force (because not enough parties ratified it), several signatories continued to follow the convention's agreements by establishing wildlife reserves. Among the first to act were Ghana and Sierra Leone, which took their first formal steps towards conserving the environment in 1901. Soon afterwards, in 1903, Africa's first conservation **non-governmental organisation (NGO)** was established, namely the Society for the Preservation of Wild Fauna of the Empire (today known as Fauna & Flora International, or FFI).

In 1925, Africa's first national park, the Albertine Rift's Albert National Park (today divided into the DRC's Virunga and Rwanda's Volcanoes National Parks) was proclaimed. The following year, South Africa's Sabie Game Reserve (which was originally gazetted in 1898) was renamed and expanded as Kruger National Park. Although most early laws focused on protecting rare and 'valuable' mammals, birds, tortoises, and timber forests, the welwitschia (*Welwitschia mirabilis*) (Figure 2.5) was the first African plant to enjoy formal protection after colonial powers ratified the

Figure 2.5 The welwitschia, a primitive gymnosperm found only in the Namib Desert of Namibia and Angola, was the first African plant to enjoy formal protection. It is adapted to collect coastal fog on its single pair of leaves which appear as many, having been torn apart by harsh desert conditions. Considered a living fossil, some welwitschias may be over 2,000 years old. Photograph by nhelia, https://pixabay.com/photos/welwitschia-mirabillis-namibia-49479, CC0.

1933 *Convention Relative to the Preservation of Fauna and Flora in the Natural State* (often referred to as the *London Convention*).

From the outset however, colonial governments managed Africa's earliest protected areas with policies more representative of Western values, which emphasised the need for nature to be shielded from human activities, and conservation management to be centralised. This top-down, protectionist "fines and fences" strategy, also known as "fortress conservation", showed little regard for the rights and cultural practices of local communities. In fact, local peoples were more likely seen as a threat to the environment. Consequently, many of Africa's first formally protected areas were established on land forcibly taken from communal ownership, and access to

Figure 2.6 A photo from East Africa in the late 1800s, illustrating typical African conservation of the time: restricting hunting privileges and wildlife trade to rich colonists who shipped their bounties to Europe, with little if any benefit to Africa. From Wikipedia, https://en.wikipedia.org/wiki/File:Ivory_1880s.jpg, CC0.

natural resources on which the local peoples previously relied upon was prohibited. Paradoxically, hunting privileges were reserved for wealthy elites on protected areas set aside for colonists' enjoyment (Figure 2.6). These practices, termed **eco-colonialism** for the similarity to the abuses of native rights by colonial powers, caused a growing rift between conservation authorities and deeply offended local peoples.

2.2.2 Conservation efforts after colonialism

Following World War II (1939–1945), after which many African countries regained independence, there was an urgent need for new conservation treaties that also addressed the needs of local peoples. Tanzania's first president, Julius Nyerere, most vividly expressed this at the 1961 *Pan-African Symposium on the Conservation of Nature and Natural Resources in Modern African States* (Watterson, 1963), in a speech that became known as the *Arusha Manifesto*:

> "The survival of our wildlife is a matter of grave concern to all of us in Africa. These wild creatures amid the wild places they inhabit are not only important as a source of wonder and inspiration, but are an integral part of our natural resources and our future livelihood and well-being. In accepting the trusteeship of our wildlife we solemnly declare that we will do everything in our power to make sure that our children's grand-children will be able to enjoy this rich and precious inheritance. The conservation of wildlife and wild places calls for specialist knowledge, trained manpower, and money, and we look to other nations to cooperate with us in this important task – the success or failure of which not only affects the continent of Africa but the rest of the world as well.

Soon after the *Arusha Manifesto*, the *African Charter for the Protection and Conservation of Nature* was established in 1963. This was followed by the *African Convention on the Conservation of Nature and Natural Resources* (*Algiers Convention* in short), which was adopted by member states of the Organisation of African Unity (which preceded the African Union) in 1968. The *Algiers Convention* provided a major break from colonial conservation models by acknowledging the principle that environmental management is a common responsibility among all Africans, while it also called for conservation of soil and water, and for environmental research and conservation (IUCN, 2004).

Despite the progress and extended scope of the *Algiers Convention*, conservation policies implemented by early post-colonial governments unfortunately continued to resemble those of colonial governments, notably the centralised and authoritarian style of decision-making. Similarly, the visions of well-funded international conservation organisations operating in the region generally reflected the perceptions and policies of developed nations, and thus lacked adequate consideration of local cultures (Abrams et al., 2009). Consequently, in the years following Africa's decolonisation, conservation largely remained a polarising endeavour that continued to uproot the lives of tens of millions of **conservation refugees** over time (Geisler and de Sousa, 2001).

2.3 Conservation in Sub-Saharan Africa Today

Building on the environmental laws and protected areas system Africans have inherited from the tumultuous past has not been easy. The scars left in the collective psyche by forced relocations and exclusions have been difficult to mend, with many conservation initiatives still struggling to shake the unfortunate association. Nevertheless, Africa's passionate conservation biologists and the broader public have shown tremendous fortitude and initiative in advancing the biodiversity conservation agenda over the last few decades. Much of this progress can be attributed to a growing realisation that conserving biodiversity is best achieved when combined with the social and economic upliftment of local people.

Perhaps the first true step to conservation reform came at the 1975 **World Parks Congress** hosted in the DRC, when the **International Union for Conservation of Nature and Natural Resources (IUCN)** adopted its first resolution that recognised the rights and needs of traditional peoples. Over the next few decades, conservation policies of national governments followed, many of which included local people in very explicit terms. One example is Namibia's Constitution, passed in 1990, stating that:

> "The State shall actively promote and maintain the welfare of the people by adopting, inter alia, policies aimed at the following: maintenance of ecosystems, essential ecological processes and biological diversity of Namibia and utilisation of living natural resources on a sustainable basis for the benefit of all Namibians, both present and future".

As the previous centralised and authoritarian style of conservation policy making has made way for more inclusive conservation activities (Abrams et al., 2009), an increasing number of local communities have become active participants in environmental programmes and policy development inside and on the periphery of protected areas. Two notable examples are **biosphere reserves** (Section 13.5.2) and **transfrontier conservation areas** (**TFCA**, Box 2.2), both pioneering strategies in promoting human-wildlife coexistence. Several governments are also expanding their protected areas networks by experimenting with private ownership of protected areas (Box 2.3) and **co-management** partnerships (Section 13.1.4), a land tenure model in which local people share the decision-making and other responsibilities of protected areas management with public institutions (Borrini-Feyerabend et al., 2004). In recent years, **integrated conservation and development projects** (**ICDPs**, Section 14.3) have also emerged as viable options to link conservation and socio-economic development.

Box 2.2 Why Go Transfrontier? (And Why Not?)

Tamar Ron

Biodiversity Conservation consultant.

✉ *tamarron@bezeqint.net*

The past two decades have brought high praise and gaining momentum for TFCAs in Southern Africa, as in other parts of the world (e.g. Vasilijevic et al., 2015; Zunckel, 2014). While Africa's first TFCA, the W National Park, was established already in 1954 by the governments of Benin, Burkina-Faso, and Niger, it was only after the Kgalagadi Transfrontier Park was established in 1999 (between the governments of South Africa and Botswana) that TFCAs have become a prominent component of the concepts driving biodiversity conservation and tourism development in Southern Africa, and across the continent.

TFCAs can support biodiversity conservation in several ways. They help protect large conservation areas and ecological corridors, facilitate cross-border knowledge exchange and cooperation in conservation and enforcement efforts, and promote mainstreaming conservation considerations into land-use planning. These benefits, in turn, offer socio-economic advantages through eco-tourism, sustainable use of natural resources, increased attraction for investors and donors, and, in some cases, supporting peace-building efforts.

Establishing a TFCA, however, entails challenges and risks (Vasilijevic et al., 2015; Zunckel, 2014; Ron, 2007). These processes are often top-down in nature, involving long and costly high-level negotiations between governments with critical conservation funds being spent on multiple cross-border meetings of senior officials and coordination efforts. Due to financial and political considerations, too often the focus remains at the central governments' level, with limited engagement with local stakeholders and on-the-ground impact. At times, many residents in the concerned area are not even aware that they live in a TFCA, or how this can change their lives.

Political and financial challenges at the local, national and regional levels may hinder the establishment of TFCAs. National inter-agency competition, disagreements within and between local communities, and conflict between international agencies, NGOs, and supporting donors may all have negative consequences. Facilitated cross-border movement of people and goods can cause security challenges and other risks, such as disease transfer, spread of invasive species, increased human-wildlife conflict, and increased illegal wildlife traffic and other criminal activities. In establishing a TFCA, it is thus essential to consult and engage all key stakeholders, and especially local communities, beginning in the planning phase, as well as to prioritise investment in on-the-ground impact-generating activities, to achieve conservation, social and development goals.

My experience in developing the Mayombe Transfrontier Initiative, between Angola, Republic of the Congo, DRC, and Gabon was most revealing (Ron, 2011a). In 2000, we initiated conservation efforts in the Angolan component of the Mayombe forest. From the start, it became clear that the striking difference in the level of degradation between the countries that share the Mayombe forest (Figure 2.B) could not be sustainable. Moreover, uncontrolled logging for timber and poaching of primates, elephants, parrots, pangolins, and other threatened species were driven, to a large extent, by illegal cross-border wildlife traffickers. It was evident that cooperation between the four countries that shared the forest was essential (Ron, 2003), so we solicited financial support from several international organisations. Initial support focused on high-level meetings and negotiations (Ijang et al. 2012). Unfortunately, local stakeholders perceived these mediation attempts as unbalanced. Finally, through governmental leadership, a Memorandum of Understanding was signed between the first three countries in 2009, with Gabon joining in 2013. A study was implemented through extensive consultation with stakeholders, and a strategic plan focusing on the most needed on-the-ground activities was adopted (Ron, 2011b). While conservation efforts have progressed at the national level, the same originally identified threats are still prominent throughout the TFCA, so it is now critical that substantial funding be allocated to the strategy's actual on-the-ground implementation.

Figure 2.B The Mayombe Forest; the tree line in the photo marks the border between Angola (top) and the Republic of the Congo. Efforts are currently underway to protect forest and surrounding area as a TFCA. Photograph by Tamar Ron, CC BY 4.0.

So, what is the conclusion? Go transfrontier? The answer is yes—but not in every case—and very carefully. Perspective must be kept through long term planning, while keeping the focus on local-level priorities.

Box 2.3 Privately Owned Lands for African Conservation

Graeme Cumming

ARC Centre of Excellence in Coral Reef Studies,
James Cook University,
Townsville, Australia.

✉ *gscumming@gmail.com*

With rates of species loss increasing and natural communities under pressure worldwide from human demands, the creation and maintenance of protected areas continues to be a vitally important conservation strategy. At the 2014 World Parks Congress in Sydney, Australia, there was widespread recognition of the need to increase the total amount of land and ocean under protection. However, this cannot be achieved by governments simply setting aside more land. Protected areas are ultimately created by people for people, and if they are to be successful, they must be created and managed in a way that is socially acceptable and sustainable.

Committing more land to biodiversity conservation means achieving a consensus between political, economic, societal, and ecological forces. This is particularly important in heavily populated landscapes, especially in Africa where local communities still bear the scars of a history of colonialism and top-down decision-making. One possible solution is to provide incentives that encourage private landowners to engage voluntarily in conservation. The area of land in private nature reserves in South Africa (both individually- and community-owned) is already estimated to be nearly twice the extent of government-owned protected areas (de Vos et al., 2019). The dynamics of privately protected areas and their overall contributions to biodiversity are, however, largely undocumented and poorly understood.

The number of privately protected areas in South Africa has increased rapidly since the end of apartheid in 1994 (de Vos et al., 2019). This increase can be partly attributed to increased tourism in South Africa and partly to the removal of perverse subsidies that kept marginal agricultural land in production (see also Section 4.5.3). Unlike statutory reserves, privately protected areas receive little or no financial support from the government and must ensure their own survival by generating revenue. They can be economically self-sufficient only if they can generate enough income from tourism. Two models appear to be particularly effective: either offering a high-cost, high-investment Big Five game viewing experience (i.e. staying in a comfortable bungalow, being guided by knowledgeable individuals), or providing a cheaper, lower-investment experience that focuses on affordable accommodation with access to hiking trails, striking scenery, and outdoor recreational opportunities (Clements et al., 2016). These models may be particularly effective in areas adjacent to national parks. For example, Shamwari Private Game Reserve, one of the more successful upper-end privately protected areas (Figure 2.C), is adjacent to Addo Elephant National Park in the Eastern Cape.

Figure 2.C A group of tourists watching two young giraffes (*Giraffa camelopardalis*, VU) play-fighting on Shamwari Private Game Reserve, South Africa. Shamwari successfully linked luxurious accommodation with wildlife safari activities to tap into the profitable conservation industry on private lands. Photograph by Iky's Photographic, https://commons.wikimedia.org/wiki/File:Shamwari_Private_Game_Reserve.jpg, CC BY-SA 4.0.

The conservation value of private lands, and particularly those that stock large herbivores, has been questioned in South Africa because of concerns about economic influences on their management. For example, tourist demand for wildlife viewing experiences can drive the overstocking of large animals, such as elephants, in small Southern African protected areas, even though higher densities of elephants do not necessarily provide a better tourism experience (Maciejewski and Kerley, 2014). Overstocking of large mammals can also lead to the conversion of woodlands to thickets, decreasing both conservation and tourism value (Cumming et al., 1997). Conversely, many private lands in the Western and Eastern Cape of South Africa have high conservation potential;

many private lands in the Cape sit lower in the landscape than parks, which, in water-scarce South Africa, have been focused on mountainous water catchment areas, and many harbours threatened lowland vegetation (Winter et al., 2007). Lowland ecosystems with their richer soils are under higher pressure from agriculture and settlement, meaning that well-managed private areas may make a disproportionately large contribution to the conservation of globally rare and endemic fynbos plants and animals (e.g. proteas, heathers, reptiles, and birds). Several governmentally supported programmes, such as the stewardship programme of the South African National Biodiversity Institute (SANBI), have been created to foster biodiversity conservation on private lands by providing information and encouraging good management practices (Rouget et al., 2014).

The owners and managers of privately protected areas could potentially interact with one another, and with the leadership of provincial and national parks, on a wide range of issues. But the managers of private lands are often poorly connected in these networks and may not benefit from knowledge sharing in the same way as managers of established reserves (Maciejewski and Cumming, 2015). In addition, many privately protected areas are not profitable, with the result that financial demands may push managers to make short-term decisions that attract revenue (e.g. overstocking large herbivores or suppressing wildfires) but have harmful long-term ecological consequences. Possible measures to ensure that private conservation efforts are both sustainable and effective include governmental interventions through tax breaks and support, and improved integration of private lands with national and provincial parks and their managers. Private conservation has considerable promise as a strategy for Africa, but its full potential will only be realised if it is achieved equitably with secure land tenure and supportive governments.

Through these different conservation partnerships models (see also Chapter 13), Africans have surpassed expectations in how rapidly they have expanded their conservation areas network. Illustrating the progress, Cameroon has augmented its existing protected areas system with nine new national parks between 2000 and 2015, with an additional nine in the proposal phase (UNEP-WCMC, 2019). The new parks include Takamanda National Park, which connects with Nigeria's Cross River National Park to form one of West Africa's largest continuous formally protected areas; it also plays a critical role in protecting the world's last remaining Cross River gorillas (*Gorilla gorilla diehli*, CR), of which fewer than 300 remain. As of mid-2019, protected areas covered over 38% of Tanzania's land area (more than 361,000 km^2, an area larger than Germany or Côte d'Ivoire [UNEP-WCMC, 2019])!

Sub-Saharan Africa's **marine protected areas** (**MPA**, Section 13.4.1) are similarly also expanding. For example, in 2017, Gabon declared 26% of its territorial waters protected, offering a haven to at least 20 species of whales and dolphins, and 20 species of sharks and rays (Parker, 2017). More recently, the Seychelles created two new MPAs that cover an area of 210,000 km^2—an area the size of Great Britain. The South African government, in collaboration with World Wide Fund For Nature (WWF), has taken the addition step by creating a forum (http://mpaforum.org.za) to improve MPA governance, and a website (https://www.marineprotectedareas.org.za) to teach the public more about South Africa's rapidly expanding MPA system.

It is important, however, to keep in mind that protecting a certain area of land and water should not in itself be the only goal in conservation. Even when a country has numerous protected areas, certain unique ecosystems may remain unprotected. Being safeguarded in name is not enough, protected areas must also be maintained and managed to achieve meaningful conservation success. Too many protected areas are nothing more than **paper parks**, areas that are protected on paper but not in reality. Two of the most important causes of protected area failure are lack of buy-in from local people, and lack of investment, financially or otherwise, from local and national governments (Watson et al., 2014; McClanahan et al., 2016; Gill et al., 2017).

Fortunately, African conservation biologists regularly employ a can-do attitude, shown in a long history of resourcefulness in the face of resource constraints. For example, conservationists from all over the region have established, and are partnering with, non-profit NGOs to facilitate a variety of innovative mechanisms to advance biodiversity conservation (see also Section 15.3). One notable example is the African Parks Network; as of mid-2019, African Parks, in partnership with its host governments, are managing 15 national parks in nine countries, covering 10.5 million hectares. Through this collaboration, which includes extensive community engagement and law enforcement, several once-declining parks are now seeing their wildlife prospering. For example, lions were reintroduced to Rwanda in 2016 after a 20-year absence, elephant strongholds in Chad and the DRC are been secured, and populations of threatened large mammals on Zambia's Liuwa Plains have increased by 50% to over 100% in just a few years (African Parks, 2016). Not only do recovering wildlife populations here and elsewhere attract more tourists, they also provide opportunities to attract new people to conservation, through **environmental education** (Figure 2.7), public health services, and other community upliftment programmes that improve the well-being of local peoples (see Box 1.2). These benefits then provide additional positive feedback towards wildlife conservation, for example by encouraging an increasing number of poachers to transition into new fulfilling lives as conservation professionals (Cooney et al., 2017).

By seeing and being exposed to all the social and economic benefits biodiversity conservation efforts offer, many local communities have been inspired to take the lead in protecting wildlife on their own lands. For example, community efforts have successfully safeguarded Mali's savannah elephants (Canney and Ganamé, 2015) and Rwanda's mountain gorillas (*Gorilla beringei beringei*, EN) (Kalpers et al., 2003) through periods of conflict. Locally managed forest reserves now protect more than 36,000 km^2 of land in Tanzania (Roe et al., 2009), while conservation efforts on **community**

Figure 2.7 Environmental education plays an important role in teaching people about the importance of their natural heritage and conservation. Here a group of school children releases a ringed yellow-fronted tinkerbird (*Pogoniulus chrysoconus*, LC) in Wondo Genet, Ethiopia, as part of a project that combines citizen science with long-term wildlife monitoring. Photograph by Çağan Şekercioğlu, CC BY 4.0.

conserved areas in Kenya have renewed hope for the future of the world's rarest antelope, the hirola (*Beatragus hunter*, CR) (King et al., 2016). These examples have set a positive, enterprising tone that has enabled conservation to play in increasingly prominent role in multiple economies through the creation of job opportunities while also improving Africans' overall quality of life.

2.4 Ongoing Conservation Challenges

Despite many examples of progress, conservation challenges and conflicts persist across Africa. As a result, the region lags in several aspects with regards to safeguarding our natural heritage (Table 2.1). The causes are many and vary by region. Below is a discussion of some of the more prominent impediments to effective conservation action in Africa.

2.4.1 Persistent poverty

There is a direct link between poverty and conservation failure (Oldekop et al., 2016; Hauenstein et al., 2019). This is a problem particularly in Africa, where millions of people live in extreme poverty that is difficult to escape. Faced with hard choices to ensure there is food on the table, poverty can drive desperate people to illegally collect natural products from protected areas, even though they likely understand the detriment these actions may have on society at large and their own futures. Other vulnerable peoples that live close to the land, such as traditional hunter-gatherers and **pastoralists**, are increasingly pushed into wildlife sanctuaries by mining, deforestation, agricultural

> Poverty can drive desperate people to illegal actions, even though they understand the detriment these actions may have on society at large and their own futures.

expansion, and development that encroach on their traditional lands. Lacking the resources to defend their land and/or support to transition to new lifestyles, these marginalised communities are often left desolate, with few if any legal options to support their livelihoods.

Table 2.1 A comparison between the number of species and number of threatened species for several major groups of animals and plants present in Sub-Saharan Africa.

Group	Species assessed by IUCN[a]	Species threatened with extinction		Data deficient species
		Number[b]	Percentage	
Vertebrate animals	10,463	1,464	14	1,427
Mammals	1,226	203	17	196
Primates	97	39	40	1
Carnivores	85	13	15	3
Bats	248	23	9	55
Birds	2,265	233	10	18
Birds of prey[c]	141	32	23	0
Vultures	10	7	70	0
Amphibians	840	213	25	152
Reptiles	736	109	15	123
Ray-finned fishes	5,650	637	11	846
Cichlids	1,026	232	23	146
Arthropods	2,368	637	27	334
Arachnids	186	142	76	2
Insects	1,796	396	22	246
Ants	8	6	100	0
Butterflies	305	72	24	32
Dragonflies	737	70	10	67
Plants	4,916	2,165	44	294
Cycads	68	48	71	0
Ferns	115	47	41	3

Source: IUCN, 2019, current as of April-2019

[a] *Low species richness generally reflects inadequate data because only a few species were evaluated. For example, 100% of ants are listed as threatened, but only eight species have been evaluated; there are more ant species in many African towns and villages.*

[b] *Categories included: Extinct in the Wild, Critically Endangered, Endangered, Vulnerable*

[c] *Includes raptors, falcons, and owls*

Further complicating matters, many well-intentioned citizens and organisations from western countries continue to have overly simplistic views of Africa. By imposing their outsider views on rare species management in Africa, these groups exacerbate the impacts of poverty, by cutting off funding sources of well-functioning conservation programmes. A good example comes from regulating trophy hunting of rare animals. Some African mammals, such as lions and elephants, are globally rare, but locally common in well-managed private game reserves and community conserved areas. Due to their global rarity, land managers of such well-managed populations can earn large fees from foreign hunters targeting these sought-after trophy species; the money earned supports local communities by boosting the local economy and conservation efforts (Lindsey et al., 2007; IUCN/PACO, 2009; Cooney et al., 2017). Unfortunately, the hunting of rare species remains controversial because many people dislike seeing charismatic animals killed. Consequently, campaigns from western countries (e.g. Hance, 2018) have significantly impeding the African trophy hunting industry, with no exemption for effective self-supporting land managers. By limiting and threatening the benefits regulated trophy hunting can bring to well-managed conservation areas and poor communities (Mbaiwa, 2018), there is fear that these campaigns will achieve the opposite of their intended purposes, by removing the incentive to protect those rare and/or charismatic species. Conservation requires all parties involved to weigh the benefits as well as unintended consequences of wildlife trade—i.e. overharvesting and black markets (Lenzen et al., 2012; Hsiang and Sekar, 2016), **land grabbing** (see Section 5.2), corruption, and terrorism (Christy and Stirton, 2015) and adapt as and when needed. Section 14.3 provides some solutions on how to link conservation with development.

2.4.2 Obstructive mindsets

Colonial Africa has provided many examples showing that conservation activities implemented in an authoritarian manner are bound to fail. Yet, authoritarian mindsets continue to impede conservation efforts throughout the region. Work from Guinea-Bissau has shown that authoritarian conservation actions that disempower or displace local communities are more likely to worsen than overcome conservation challenges in post-colonial Africa (Cross, 2015). Conservation in Africa is as much about people as it is about wildlife; this book provides many examples to show how human welfare and conservation are tied to one another.

At the same time, integrating diverging cultural beliefs about the natural environment into conservation practices also remains an obstacle (Dickman et al., 2015). Many Africans continue to fixate on cultural justifications ("We have been hunting for many generations", Figure 2.8) without acknowledging that human population growth, more sophisticated weapons, and increased levels of consumption are putting unsustainable pressure on natural landscapes. Others believe that the destruction of nature is simply not possible because their ancestors will intervene before this happens, effectively removing individual or community responsibility from conservation management and planning. Breaking down such barriers is hard, frustrating, and takes a long time to achieve. It requires an interdisciplinary approach (Section 1.1) bringing together aspects of conservation science and the social sciences to find common ground. Despite the challenges to putting effective conservation into practice, it is important to remember that fortress conservation models—telling people how they should act, with little to no local input—are more likely to produce enduring counter-productive results.

Figure 2.8 A group of hunters carry a western lowland gorilla (*Gorilla gorilla gorilla*, CR) that was shot while raiding crops in southern Cameroon. While retaliatory killings is the traditional method for dealing with problem animals (but see Section 14.4), killing rare species such as gorillas is generally forbidden by customary and statutory laws. Photograph by Edmond Dounias/CIFOR, CC BY 4.0.

2.4.3 Weak governance/institutional structures

Africa's natural environment and its people often fall victim to weak governance and institutional structures. It is well-known that weak policies, failing governments, and civil conflict hamper conservation efforts and drive biodiversity declines (Nackoney et al., 2014; Brito et al., 2018; Daskin and Pringle, 2018). But even in well-functioning countries, government officials turning a blind eye (either willingly, or because they lack capacity) may enable corporations to cut corners for increased profits at the cost of the environment. Corruption and greed also fuel land grabbing (Section 5.2), black markets (Hauenstein et al., 2019), and unwarranted protected area degazettement (Section 13.7.3). There is broad interest to challenge these behaviours which benefit only a handful of people at the cost of thousands of others (Box 2.4). Fixing these issues will rely on strengthening institutional capacity on multiple levels (Amano et al., 2018).

Weak policies, failing governments, and civil conflict hamper conservation efforts and drive biodiversity declines.

Box 2.4 Malawi: No Longer a Weak Link in the Elephant Ivory Trafficking Chain?

Jonathan Vaughan

Lilongwe Wildlife Trust,
Lilongwe, Malawi.
http://www.lilongwewildlife.org

International efforts to combat illegal wildlife trade—now the fourth largest transnational crime in the world (Nellemann et al., 2016)—have intensified in recent years, but Malawi has escaped public scrutiny due to its small size and relatively small wildlife numbers. Despite these factors, Malawi's wildlife populations have been decimated by poaching in the last few decades. For example, Kasungu National Park's wildlife was so abundant in the 1980s that animals were translocated to the Kruger National Park in South Africa. Back then, elephants numbered as high as 2,000. Today, there are no more than 60.

Southern Africa's principle transit hub for wildlife trafficking

In 2016, CITES identified Malawi as a "country of primary concern", and Southern Africa's principle transit hub for ivory trafficking. Malawi's own *Illegal Wildlife Trade Review* (Waterland et al., 2015), published a year earlier, had come to similar conclusions, uncovering evidence of large-scale international trafficking of bushmeat, carnivore pelts, tortoises, pangolins, orchids, ivory, and rhino horn. The revelations served a wake-up call for urgent action to protect not just Malawi's own wildlife but also wild populations throughout Southern Africa.

Central to region's poaching hotspots

Why is Malawi such a significant link in the trafficking chain? The first clue is geography. Malawi is surrounded by Africa's biggest elephant poaching hotspots. Selous Game Reserve in Tanzania reportedly lost 25,000 elephants between 2009 and 2013, while 1,000 elephants were killed in Mozambique's Niassa Province in 2011, alone (Booth and Dunham, 2016). Poaching in Zambia's Luangwa Valley is well above the CITES average (Nyirenda et al., 2015). Wasser et al. (2015) found that all the study samples of ivory seized from consignments weighing more than half a tonne between 2006 and 2014 originated from ecosystems immediately bordering Malawi.

Malawi has already been implicated in some of the largest ivory seizures in the world. The biggest impoundment ever—at 7.5 tonnes, equivalent to over 1,500 elephants—was made in Singapore in 2002 and had been shipped from

Malawi's capital, Lilongwe (Wasser et al., 2007, 2015). In 2013, 2.6 tonnes of ivory were confiscated from a container within Malawi's borders at Mzuzu. Fifty cases were recorded between 2010 and 2014, including numerous smaller examples of ivory trafficking. With an estimated 10% interception rate, the true scale of ivory trafficking was evidently much larger than previously thought (Waterland et al., 2015).

Risk-reward ratio in favour of criminals

Malawi's weak wildlife legislation was another significant factor. Coupled with under-resourced law enforcement and high levels of corruption, this offered an attractive risk-to-reward ratio for wildlife criminals. The individuals convicted of trafficking in the 2013 Mzuzu case escaped with a fine of just US $5,000 for a 2.6-tonne haul. This paled in comparison to the penalties handed out in other countries. For example, during the same period, a Zambian man was sent to prison for five years for trafficking 12.5 kg of ivory, a South African man received 10 years and a US $392,000 fine for trafficking one tonne of ivory and, in Kenya, a man was fined US $233,000 for trafficking a single tusk weighing 3.4 kg.

While sentencing in the Mzuzu case was hampered to some extent by limitations of the law, it was also indicative of the fact that, historically, trafficking was not treated as a serious crime in Malawi. Most wildlife prosecutions had taken place in lower courts and have been prosecuted by lower-ranked officials. The average fine for ivory trafficking was found to be just US $40 between 2011 and 2014. This is an extremely low amount given the potential profits from the trade of ivory and, thus, provides virtually no deterrent to traffickers. Awareness, motivation, and cooperation within and between departments like the police and border forces were found to be severely lacking. Government resources to combat wildlife crime are also limited, with many other causes competing for funding and attention.

Management of wildlife crime data also made life easier for wildlife criminals. Take the case of a Chinese national who was arrested and prosecuted for an ivory trafficking offence under one name, deported under a second name, and reported by the INTERPOL country office to INTERPOL headquarters under a third. This shows the ease with which criminals are circumventing the weak systems currently in place.

Turning the Tide

Today, however, things are changing. Recommendations from the 2015 *Illegal Wildlife Trade Review* were swiftly executed, strengthening the process from investigations and arrest right through to prosecution and sentencing. As a result, in just four years, over 1.5 tonnes of ivory were confiscated, average monthly arrests for wildlife crime jumped from 0.7 to 9.5, and custodial sentence rates rose to over 90%, with judgments passed of up to 18 years. Remember that, in comparison, no-one convicted of a wildlife crime between 2010 and 2015 had been put behind bars and the average fine was just $40.

Other initiatives included improving protected area management, launching the country's first wildlife crime investigations unit, and establishing an Inter-Agency Committee to Combat Wildlife Crime to improve cooperation and information sharing. Critical amendments to wildlife legislation were also passed in record time, and technical expertise from partners was harnessed to maximise impact. Lilongwe Wildlife Trust is currently the only NGO sanctioned to prosecute wildlife crime cases in partnership with an African government. In short, Malawi has strengthened each stage of the enforcement chain.

These successes came about largely as a result of a collaborative, innovative, and holistic approach that moved beyond traditional wildlife conservation to incorporate practices used in combatting serious organised crime.

Support from the very top

Strengthened legislation and enforcement will continue to be a critical deterrent, but advocacy has also been a critical tool for securing high-level political will and turning it into action. The President of Malawi, himself, His Excellency Peter Mutharika, backed the nation's "Stop Wildlife Crime" campaign (Figure 2.D) and the Malawi Parliamentary Conservation Caucus continues to raise awareness through the media, essentially holding stakeholders such as the police or judiciary to account by highlighting both successes and questionable outcomes.

Focus on trafficking

Poaching has been a major focus of conservation efforts elsewhere in Africa, and local poachers can still expect to feel the full weight of the Malawian law. However, bringing traffickers to justice is proving a more effective use of limited resources. After all, it is members of organised international crime syndicates that ultimately exploit local communities, incite corruption, threaten our national security, and provide the routes to overseas markets.

What's next for Malawi?

Sustaining Malawi's astounding turnaround will be no easy feat. But with continued determination, as well as local and international cooperation and support, we believe that these criminal networks can be disrupted enough to halt the impending extinction of one of Africa's most iconic species.

The same tenacity and high-level commitment we have witnessed in the last five years must now be applied to other conservation challenges, as attention is being turned to the protection of Malawi's wider biodiversity. In 2018, a further

Figure 2.D Campaigners taking to the streets in support of Malawi's "Stop Wildlife Crime" campaign, which the President of Malawi also supports. Photograph by Lilongwe Wildlife Trust, CC BY 4.0.

216 species of animals, plants and trees were placed under legal protection, and lessons from combatting wildlife crime can now be applied to other illegal or unsustainable practices, such as trades in timber, charcoal, and fish.

When it comes to pioneering conservation, Malawi is one to watch. Let's hope that there are more achievements to celebrate in another five years' time.

2.4.4 Skills shortages

Scientific advances depend on increased or updated knowledge. That is also true for conservation biology—effective conservation depends on local experts who can design and implement monitoring and research projects, apply **adaptive management** (Section 10.2.3) when needed, act as managers and advocates for conservation activities, and increase awareness of the importance of the environment (Laurance, 2013). It is thus of great concern that conservation in Africa continues to face an enduring skills shortage (Wilson et al., 2016). Illustrating the problem, a recent review found that, over the past three decades, only 129 of the scientific articles focussed on West African birds were produced in international journals by local authors. This productivity contrasts strongly with Europe, where 12,380 ornithological articles were produced over the same time (Cresswell, 2018). Another review, covering all of Africa, found that less than 30% of the continent's birds received attention in international journals (Beale, 2018). While high-impact publications are not the only metric to estimate conservation success, they provide an accurate accounting of persistent knowledge gaps, as well as skills shortages further down the hierarchy, from researchers and teachers to rangers and other fieldworkers down to **citizen scientists**.

There are many reasons for these skills shortages. Some of the most prominent foundational issues include a fragmented communication network that limits skills transfer, financial and other resources limitations, a shortage of quality training institutes, and overburdened teachers at existing educational facilities. Fortunately, many of these shortfalls are currently being addressed. For example, new people are being involved in conservation activities through citizen science projects (see Box 15.3), innovative funding mechanisms are being developed (Section 15.3), legal and organisational structures are being adapted to foster increased collaboration (Section 15.4) and freely-accessible resources such as this textbook are being made available.. It is important to continue to build on this progress by supporting such initiatives, and continuously highlighting to others the importance of nature to their own well-being.

2.4.5 Competing interests

Because of competing interests (for land, natural resources, etc.), there is always a risk that a wealthy business will threaten a conservation initiative with competing offers that typically include promises of jobs and development (Koohafkan et al., 2011). Local peoples, especially those in poverty, may find it hard to turn down such attractive counteroffers, even if they recognise that those offers rarely live up to the promises made. Conservation biologists should carefully consider what such offers on the table might look like and factor in how their conservation programmes compete and bring better results for all.

Like stock market investments, the benefits to be gained from conservation may take years to materialise.

People concerned with the environment have worked hard to better highlight that conservation has the potential to be profitable and to spur sustainable development. These activities have seen the emergence of fields such as **environmental economics**, and methods to put a market value on **ecosystem services** (Section 4.5). Unfortunately, some conservation biologists have fallen into a trap of (over)emphasising the economic benefits that conservation can bring, without a realistic representation of the upfront investment required or the length of time required for a tangible return on investment. Like stock market investments, the benefits to be gained from conservation may take years to materialise, sometimes with very little to show for it in the meantime. Given that all investments require either expendable capital or credit, willing stakeholders with neither are essentially being asked to maintain a more restrictive livelihood over an unsustainable (and often undisclosed or unknown) period of time. It is crucial for conservation biologists to set realistic expectations and to offer a balanced approach that provides interim funding/credit options. Such options could perhaps include **microloans,** village savings and loan associations (http://www.care.org/vsla), or community conservation banks (https://sema.fzs.org/en/conservation-banks) such as those established by Frankfurt Zoological Society (FZS) in Tanzania. It is also important to incorporate benefits beyond immediate financial gain when starting or expanding conservation programmes. Conservation actions should also aim to provide concrete benefits, whether financial or otherwise, to local communities from an early stage. In that way, if a project comes to a premature end, one can still point to the progress made, which will make it easier to engage with that community when future opportunities arise.

2.5 Conclusion

Because of the many challenges that conservation projects continue to face, the list of Sub-Saharan African species and ecosystems that are threatened with extinction and destruction continues to grow every year. In a recent assessment, BirdLife International identified 51 **Important Bird and Biodiversity Areas (IBA)** in Sub-Saharan Africa—many of them national parks—in danger of ecosystem degradation (BirdLife International, 2019). A **United Nations (UN)** assessment similarly found that the outlook of 12 natural World Heritage Sites situated in Sub-Saharan African "in danger" (http://whc.unesco.org/en/danger). These are substantial and challenging problems that will keep conservation biologists very busy in the future. These problems need to be faced head-on to ensure that future generations will also be able to enjoy the natural treasures and resources the region has to offer.

2.6 Summary

1. Sub-Saharan Africa supports extremely diverse ecological communities across its eight terrestrial biomes (which include forests, savannahs, woodlands, grasslands, scrublands, deserts, and mangroves) as well as multiple freshwater and marine ecosystems. The region's complex climate, geology, and history have contributed to the development of its exceptional biodiversity.
2. Conservation in Africa has gone through major changes over the past few centuries including traditional relationships with nature; exploitation of wildlife and natural resources by European settlers in the 17th and 18th centuries; western practices of setting aside land shielded from human influences; and more recently integrated conservation and development practices.
3. Africa's conservation biologists and the broader public have shown tremendous fortitude and initiative to overcome the various challenges facing biodiversity over the last few decades. This includes greatly expanding the protected areas network, passing laws protecting the environment, and establishing productive partnerships.

4. By reaping the benefits from conservation activities in and around protected areas, many private individuals and local communities have been inspired to take the lead in protecting biodiversity on their own lands.
5. Historical legacies, poverty, greed, weak governance, consumptive needs by an increasing human population, and competing interests remain challenges to conservation in Africa. Many of these challenges lead to threats to the future persistence of many species and ecosystems, including environmental degradation and overharvesting.

2.7 Topics for Discussion

1. The human population of Sub-Saharan Africa is predicted to increase dramatically in coming decades. How do you think this growth will affect the region's biodiversity? Do you think that the increase in human population will increase consumptive needs?
2. What are the main international and national organisations contributing to conservation in your region? What projects are they working on? What are the most important goals of those projects? What do you think are the biggest challenges facing those projects?
3. Conservation in Africa has gone through several stages through history. Can you summarise each of these stages in two or three sentences? What do you think are the strengths and weaknesses of each stage?

2.8 Suggested Readings

Abrams, R.W., E.D. Anwana, A. Ormsby, et al. 2009. Integrating top-down with bottom-up conservation policy in Africa. *Conservation Biology* 23: 799–804. https://doi.org/10.1111/j.1523-1739.2009.01285.x Africa requires locally-adapted conservation policies.

African Parks. 2019. *Unlocking the value of protected areas*. African Parks Annual Report 2018 (Johannesburg: African Parks). https://www.africanparks.org/unlocking-value-protected-areas An overview of activities undertaken by a successful conservation NGO.

Balmford, A., J.L. Moore, T. Brooks, et al. 2011. Conservation conflicts across Africa. *Science* 291: 2616–19. https://doi.org/10.1126/science.291.5513.2616 Many of the challenges to conservation in Africa are rooted in population growth.

Beale, C.M., S. van Rensberg, W.J. Bond, et al. 2013. Ten lessons for the conservation of African savannah ecosystems. *Biological Conservation* 167: 224–32. https://doi.org/10.1016/j.biocon.2013.08.025 Learning from past efforts to guide future actions.

Cooney, R., D. Roe, H. Dublin, et al. 2017. From poachers to protectors: Engaging local communities in solutions to illegal wildlife trade. *Conservation Letters* 10: 367–74. https://doi.org/10.1111/conl.12294 Guidelines for engaging the local community in conservation activities.

Cross, H. 2015. Displacement, disempowerment and corruption: Challenges at the interface of fisheries, management and conservation in the Bijagós Archipelago, Guinea-Bissau. *Oryx* 50:

693–701. https://doi.org/10.1017/S003060531500040X Authoritative top-down strategies will likely worsen rather than solve conservation conflicts.

Hauenstein, S., M. Kshatriya, J. Blanc, et al. 2019. African elephant poaching rates correlate with local poverty, national corruption and global ivory price. *Nature Communications* 10: 2242. https://doi.org/10.1038/s41467-019-09993-2 Linking poverty, corruption, and wildlife declines.

Kalpers, J., E.A. Williamson, M.M. Robbins, et al. 2003. Gorillas in the crossfire: Population dynamics of the Virunga mountain gorillas over the past three decades. *Oryx* 37: 326–37. https://doi.org/10.1017/S0030605303000589 African conservationists have had to deal with a very dynamic, and sometimes dangerous environment.

Prendergast, D.K., and W.M. Adams. 2003. Colonial wildlife conservation and the origins of the Society for the Preservation of the Wild Fauna of the Empire (1903–1914). *Oryx* 37: 251–60. https://doi.org/10.1017/S0030605303000425 Consider how conservation biologists today face similar challenges to those a century ago.

Bibliography

Abrams, R.W., E.D. Anwana, A. Ormsby, et al. 2009. Integrating top-down with bottom-up conservation policy in Africa. *Conservation Biology* 23: 799–804. https://doi.org/10.1111/j.1523-1739.2009.01285.x

African Parks. 2016. *African Parks annual report 2015: Conservation at scale* (Johannesburg: African Parks). https://www.africanparks.org/sites/default/files/uploads/resources/2017-05/APN_AnnualReport_2015.pdf

Amano, T., T. Székely, B. Sandel, et al. 2018. Successful conservation of global waterbird populations depends on effective governance. *Nature* 553: 199–202. https://doi.org/10.1038/nature25139

Balmford, A., J.L. Moore, T. Brooks, et al. 2011. Conservation conflicts across Africa. *Science* 291: 2616–19. https://doi.org/10.1126/science.291.5513.2616

Beale, C.M. 2018. Trends and themes in African ornithology. *Ostrich* 89: 99–108. https://doi.org/10.2989/00306525.2017.1407834

BirdLife International. 2019. *IBA's in Danger-Site Summary*. http://datazone.birdlife.org/site/ibaidsites

Booth, V.R., and K.M. Dunham. 2016. Elephant poaching in Niassa Reserve, Mozambique: Population impact revealed by combined survey trends for live elephants and carcasses. *Oryx* 50: 94–103. https://doi.org/10.1017/S0030605314000568

Borrini-Feyerabend, G., M. Pimbert, M.T. Farvar, et al. 2004. *Sharing power. Learning by doing in co-management of natural resources throughout the world* (Cenesta: IIED and IUCN/CEESP/CMWG). http://pubs.iied.org/G01089

Brito, J.C., S.M. Durant, N. Pettorelli, et al., 2018. Armed conflicts and wildlife decline: Challenges and recommendations for effective conservation policy in the Sahara-Sahel. *Conservation Letters* 11: e12446. https://doi.org/10.1111/conl.12446

Brugiere, D., and R. Kormos. 2009. Review of the protected area network in Guinea, West Africa, and recommendations for new sites for biodiversity conservation. *Biodiversity and Conservation* 18: 847–68. https://doi.org/10.1007/s10531-008-9508-z

Buffenstein, R., and S. Yahav. 1991. Is the naked mole-rat *Heterocephalus glaber* an endothermic yet poikilothermic? *Journal of Thermal Biology* 16: 227–32. https://doi.org/10.1016/0306-4565(91)90030-6

Burgess, N., J.D. Hales, E. Underwood, et al. 2004. *Terrestrial Ecoregions of Africa and Madagascar: A Conservation Assessment* (Washington: Island Press).

Canney, S., and N. Ganamé. 2015. The Mali elephant project, Mali. In: *Conservation, Crime and Communities: Case Studies of Efforts to Engage Local Communities in Tackling Illegal Wildlife Trade*, ed. by D. Roe (London: IIED). http://pubs.iied.org/14648IIED/

Chinsembu, K.C. 2015. Plants as antimalarial agents in Sub-Saharan Africa. *Acta Tropica* 152: 32–48. https://doi.org/10.1016/j.actatropica.2015.08.009

Christy, B., and B. Stirton. 2015. How killing elephants finances terror in Africa. *National Geographic*. http://on.natgeo.com/1I5N2aO.

Cioc, M. 2009. *The Game of Conservation: International Treaties to Protect the World's Migratory Animals* (Columbus: Ohio University Press).

Clements, H., J. Baum, and G.S. Cumming. 2016. Money and motives: An organizational ecology perspective on private land conservation. *Biological Conservation* 197: 108–15. https://doi.org/10.1016/j.biocon.2016.03.002

Cooney, R., C. Freese, H. Dublin, et al. 2017. The baby and the bathwater: Trophy hunting, conservation and rural livelihoods. *Unasylva* 249: 3–16. http://www.fao.org/3/i6855en/I6855EN.pdf

Cooney, R., D. Roe, H. Dublin, et al. 2017. From poachers to protectors: Engaging local communities in solutions to illegal wildlife trade. *Conservation Letters* 10: 368–74. https://doi.org/10.1111/conl.12294

Cresswell, W. 2018. The continuing lack of ornithological research capacity in almost all of West Africa. *Ostrich* 89: 123–29. https://doi.org/10.2989/00306525.2017.1388301

Cross, H. 2015. Displacement, disempowerment and corruption: challenges at the interface of fisheries, management and conservation in the Bijagós Archipelago, Guinea-Bissau. *Oryx* 50: 693–701. https://doi.org/10.1017/S003060531500040X

Cumming, D.H.M., M.B. Fenton, I.L. Rautenbach, et al. 1997. Elephants, woodlands and biodiversity in southern Africa. *South African Journal of Science* 93: 231–36

Daskin, J.H., and R.M. Pringle. 2018. Warfare and wildlife declines in Africa's protected areas. *Nature* 553: 328–32. https://doi.org/10.1038/nature25194

Davis A.P., T.W. Gole, S. Baena, et al. 2012. The impact of climate change on indigenous Arabica coffee (*Coffea arabica*): Predicting future trends and identifying priorities. *PLoS ONE* 7: e47981. https://doi.org/10.1371/journal.pone.0047981

de Vos, A., H.S. Clements, D. Biggs, et al. 2019. The dynamics of proclaimed privately protected areas in South Africa over 83 years. *Conservation Letters* 12: e12644. https://doi.org/10.1111/conl.12644

Dickman, A., P.J. Johnson, F. van Kesteren, et al. 2015. The moral basis for conservation: How is it affected by culture? *Frontiers in Ecology and the Environment* 13: 325–31. https://doi.org/10.1890/140056

Dunning, J.B. Jr. 2008. *CRC Handbook of Avian Body Masses* (Boca Raton: CRC Press).

Foley, J.A., M.T. Coe, M. Scheffer, et al. 2003. Regime shifts in the Sahara and Sahel: Interactions between ecological and climatic systems in Northern Africa. *Ecosystems* 6: 524–32. https://doi.org/10.1007/s10021-002-0227-0

Geisler, C., and R. de Sousa. 2001. From refuge to refugee: the African case. *Public Administration and Development* 21: 159–70. https://doi.org/10.1002/pad.158

Gill, D.A., M.B. Mascia, G.N. Ahmadia, et al. 2017. Capacity shortfalls hinder the performance of marine protected areas globally. *Nature* 543: 665–69. https://doi.org/10.1038/nature21708

Gorenflo, L.J., S. Romaine, R.A. Mittermeier, et al. 2012. Co-occurrence of linguistic and biological diversity in biodiversity hotspots and high biodiversity wilderness areas. *Proceedings of the National Academy of Sciences* 109: 8032–37. https://doi.org/10.1073/pnas.1117511109

Hance, J. 2018. Trump's elephant, lion trophy hunting policy hit with double lawsuits. *Mongabay* https://news.mongabay.com/2018/03/trumps-elephant-lion-trophy-hunting-policy-hit-with-double-lawsuits.

Hauenstein, S., M. Kshatriya, J. Blanc, et al. 2019. African elephant poaching rates correlate with local poverty, national corruption and global ivory price. *Nature Communications* 10: 2242. https://doi.org/10.1038/s41467-019-09993-2

Hsiang, S., and N. Sekar. 2016. Does legalization reduce black market activity? Evidence from a global ivory experiment and elephant poaching data. *NBER Working Paper* 22314 (Cambridge: National Bureau of Economic Research). https://doi.org/10.3386/w22314

Ijang, T.P., N. Cleto, N.W. Ewane, et al. 2012. Transboundary dialogue and cooperation: First lessons from igniting negotiations on joint management of the Mayombe Forest in the Congo Basin. *International Journal of Agriculture and Forestry* 2: 121–31. https://doi.org/10.5923/j.ijaf.20120203.08

IUCN. 2004. *An introduction to the African Convention on the Conservation of Nature and Natural Resources* (Gland: IUCN). https://portals.iucn.org/library/efiles/documents/EPLP-056-rev.pdf

IUCN. 2019. *The IUCN Red List of Threatened Species.* http://www.iucnredlist.org

IUCN/PACO. 2009. *Big game hunting in West Africa. What is its contribution to conservation?* (Ouagadougou: IUCN/PACO). https://portals.iucn.org/library/sites/library/files/documents/2009-074-En.pdf

Jarvis, J.U.M., M.J. O'Riain, N.C. Bennett, et al. 1994. Mammalian eusociality: A family affair. *Trends in Ecology and Evolution* 9: 47–51. https://doi.org/10.1016/0169-5347(94)90267-4

Juhé-Beaulaton, D., and B. Roussel. 2002. Les sites religieux vodun, des patrimoines en permanente evolution. In: *Patrimonialiser la nature tropicale: Dynamiques locales, enjeux internationaux,* ed. by M.-C. Cormier-Salem et al. (Paris: IRD). http://horizon.documentation.ird.fr/exl-doc/pleins_textes/divers09-03/010028405.pdf

Kalpers, J., E.A. Williamson, M.M. Robbins, et al. 2003. Gorillas in the crossfire: Population dynamics of the Virunga mountain gorillas over the past three decades. *Oryx* 37: 326–37. https://doi.org/10.1017/S0030605303000589

King, J., A. Wandera, M.I. Sheikh, et al. 2016. *Status of Hirola in Ishaqbini Community Conservancy* (Masalani: Ishaqbini Hirola Community Conservancy; Northern Rangelands Trust). https://nrt-kenya.squarespace.com/s/Hirola-sanctuary-status-report-May2016-mjwy.pdf

Klopper, R.R., L. Gautier, C. Chatelain, et al. 2007. Floristics of the angiosperm flora of Sub-Saharan Africa: An analysis of the African Plant Checklist and Database. *Taxon* 56: 201–08.

Koohafkan, P., M. Salman, and C. Casarotto. 2011. Investments in land and water. *SOLAW Background Thematic Report* TR17 (London: FAO). http://www.fao.org/fileadmin/templates/solaw/files/thematic_reports/TR_17_web.pdf

Kümpel, N.F., A. Quinn, and S. Grange. 2015. The distribution and population status of the elusive okapi, *Okapia johnstoni*. *African Journal of Ecology* 53: 242–45. https://doi.org/10.1111/aje.12221

Laine, A. 1990. Mythe et Réalités: Les Enfants nés Pour Mourir en Afrique de l'Ouest. *Episteme* 1: 87–97.

Laurance W.F. 2013. Does research help to safeguard protected areas? *Trends in Ecology and Evolution* 28: 261–66. https://doi.org/10.1016/j.tree.2013.01.017

Lenzen, M., D. Moran, K. Kanemoto, et al. 2012. International trade drives biodiversity threats in developing nations. *Nature* 486: 109–12. https://doi.org/10.1038/nature11145

Lewis, M.P., G.F. Simons, and C.D. Fennig. 2014. *Ethnologue: Languages of the World* (Dallas: SIL International). https://www.ethnologue.com

Lindsey, P.A., P.A. Roulet, and S.S. Romanach. 2007. Economic and conservation significance of the trophy hunting industry in sub-Saharan Africa. *Biological Conservation* 134: 455–69. https://doi.org/10.1016/j.biocon.2006.09.005

Lotz, C.N., J.A. Caddick, M. Forner, et al. 2013. Beyond just species: Is Africa the most taxonomically diverse bird continent? *South African Journal of Science* 109(5/6): 1–4. https://doi.org/10.1590/sajs.2013/20120002

Maciejewski, K., and G. Cumming. 2015. The relevance of socioeconomic interactions for the resilience of protected area networks. *Ecosphere* 6: 1–14. https://doi.org/10.1890/ES15-00022.1

Maciejewski, K., and G.I. Kerley. 2014. Elevated elephant density does not improve ecotourism opportunities: Convergence in social and ecological objectives. *Ecological Applications* 24: 920–26. https://doi.org/10.1890/13-0935.1

MacKenzie, J.M. 1997. *The Empire of Nature: Hunting, Conservation and British Imperialism* (Manchester: Manchester University Press).

Mbaiwa, J.E. 2018. Effects of the safari hunting tourism ban on rural livelihoods and wildlife conservation in Northern Botswana. *South African Geographical Journal* 100: 41–61. https://doi.org/10.1080/03736245.2017.1299639

McClanahan, T.R., and P.S. Rankin. 2016. Geography of conservation spending, biodiversity, and culture. *Conservation Biology* 30: 1089–101. https://doi.org/10.1111/cobi.12720

Miller, J.M., S. Hallager, S.L. Monfort, et al. 2011. Phylogeographic analysis of nuclear and mtDNA supports subspecies designations in the ostrich (*Struthio camelus*). *Conservation Genetics* 12: 423–31. https://doi.org/10.1007/s10592-010-0149-x

Miller, S.E., and L.M. Rogo. 2011. Challenges and opportunities in understanding and utilisation of African insect diversity. *Cimbebasia* 17: 197–218

Mittermeier, R.A., P. Robles-Gil, M. Hoffman, et al. 2004. *Hotspots Revisited: Earth's Biologically Richest and Most Endangered Terrestrial Ecoregions* (Chicago: University of Chicago Press).

Nackoney, J., G. Molinario, P. Potapov, et al. 2014. Impacts of civil conflict on primary forest habitat in northern Democratic Republic of the Congo, 1990–2010. *Biological Conservation* 170: 321–28. https://doi.org/10.1016/j.biocon.2013.12.033.

Nellemann, C., R. Henriksen, A. Kreilhuber, et al. 2016. *The rise of environmental crime—A growing threat to natural resources, peace, development and security* (Cambridge: UNEP). http://wedocs.unep.org/handle/20.500.11822/7662

Nyirenda, V.R., P.A. Lindsey, E. Phiri, et al. 2015. Trends in the illegal killing of African elephants (*Loxodonta africana*) in the Luangwa and Zambezi Ecosystems of Zambia. *Environment and Natural Resources Research* 5: 24–36. https://doi.org/10.5539/enrr.v5n2p24

Oldekop, J.A., G. Holmes, W.E. Harris, et al. 2016. A global assessment of the social and conservation outcomes of protected areas. *Conservation Biology* 30: 133–41. https://doi.org/10.1111/cobi.12568

Olson, D.M., E. Dinerstein, E.D. Wikramanayake, et al. 2001. Terrestrial ecoregions of the world: A new map of life on Earth. *BioScience* 51: 933–38. https://doi.org/10.1641/0006-3568(2001)051[0933:TEOTWA]2.0.CO;2

Parker, L. 2017. New ocean reserve, largest in Africa, protects whales and turtles. *National Geographic*. http://on.natgeo.com/2samx3a

Prendergast, D.K., and W.M. Adams. 2003. Colonial wildlife conservation and the origins of the Society for the Preservation of the Wild Fauna of the Empire (1903–1914). *Oryx* 37: 251–60. https://doi.org/10.1017/S0030605303000425

Rebelo, A.G. 1992. Red Data Book species in the Cape Floristic Region: Threats, priorities and target species. *Transactions of the Royal Society of South Africa* 48: 55–86. https://doi.org/10.1080/00359199209520256

Roe D., F. Nelson, and C. Sandbrook. 2009. Community management of natural resources in Africa: Impacts, experiences and future directions. *Natural Resource Issues* 18 (London: IIED). http://pubs.iied.org/pdfs/17503IIED.pdf

Ron, T. 2003. The conservation of the Maiombe Forest, Cabinda, Angola, within the framework of a transfrontier conservation initiative. *The World Parks Congress*, September 2003, Durban, South Africa.

Ron, T. 2007. *Southern Africa Development Community (SADC) proposed framework for transfrontier conservation areas (TFCAs)*. Issues and options report. Approved by the SADC Integrated Committee of Ministers (ICM). http://www.tbpa.net/docs/SDC_SADC%20final%20report%20draft_Tamar%20Ron.pdf

Ron, T. 2011a. *Potential for designating protected areas for conservation and for identifying conservation corridors as part of the planning process of the Mayombe Forest Ecosystems Transfrontier Conservation Area*. Report for the Governments of Angola, Congo, and DRC, UNEP and IUCN.

Ron, T. 2011b. *Towards a transboundary protected area complex in the Mayombe Forest Ecosystems. Five years strategic plan and roadmap*. With inputs from Angola, Congo, DRC, UNEP and IUCN. Adopted by the Mayombe Transfrontier Initiative's Governments on March 2013.

Rouget, M., M. Barnett, R.M. Cowling, et al. 2014. Conserving the Cape Floristic Region. In: *Fynbos: Ecology, Evolution, and Conservation of a Megadiverse Region*, ed. by N. Allsopp et al. (Oxford: Oxford University Press).

Roussel, B. 1994. Des dieux à l'homme. Les plantes des vaudous. *Hommes et Plantes* 7–8: 46–49.

Sinclair, I., and P. Ryan. 2011. *Birds of Africa South of the Sahara*. Penguin Random House, Johannesburg, South Africa.

Sokpon, N., A. Ametepe, and V. Agbo. 1998. Forêts sacrées et conservation de la biodiversité au Bénin: cas du pays Adja au Sud-Ouest du Bénin, *Annales des Sciences Agronomiques* 1: 47–64.

Spalding, M.D., H.E. Fox, G.R. Allen, et al. 2007. Marine ecoregions of the world: A bioregionalization of coastal and shelf areas. *BioScience* 57: 573–83. https://doi.org/10.1641/B570707

UNEP-WCMC. 2019. *World Database on Protected Areas*. http://www.protectedplanet.net

Vasilijevic, M., K. Zunckel, M. McKinney, et al. 2015. Transboundary Conservation: A systematic and integrated approach. *Best Practice Protected Area Guidelines*, Series 23 (Gland: IUCN). https://doi.org/10.2305/IUCN.CH.2015.PAG.23.en

Wasser, S.K., C. Mailand, R. Booth, et al. 2007. Using DNA to track the largest ivory seizure since the 1989 trade ban. *Proceedings of the National Academy of Sciences* 104: 4228–33. https://doi.org/10.1073/pnas.0609714104

Wasser, S.K., L. Brown, C. Mailand, et al. 2015. Genetic assignment of large seizures of elephant ivory reveals Africa's major poaching hotspots. *Science* 349: 84–87. https://doi.org/10.1126/science.aaa2457

Waterland, S, J. Vaughan, E. Lyman, et al. 2015. *Illegal wildlife trade review of Malawi* (Lilongwe: DNPW). https://www.lilongwewildlife.org/wp-content/uploads/IWT-Review-Malawi.pdf

Watson, J.E., N. Dudley, D.B. Segan, et al. 2014. The performance and potential of protected areas. *Nature* 515: 67–73. https://doi.org/10.1038/nature13947

Watterson, G.G. 1963. Conservation of nature and natural resources in modern African states. *IUCN Publications New Series* 1 (Morges: IUCN). https://portals.iucn.org/library/sites/library/files/documents/NS-001.pdf

WHC (World Heritage Committee). 2007. *Nomination of natural, mixed and cultural properties to the World Heritage List - Virunga National Park.* Decision 31COM 8B.74 (Christchurch: UNESCO. https://whc.unesco.org/en/decisions/1377

Wilson, K.A., N.A. Auerbach, K. Sam, et al. 2016. Conservation research is not happening where it is most needed. *PLoS Biology* 14: e1002413. https://doi.org/10.1371/journal.pbio.1002413

Winter, S.J., H. Prozesky, and K.J Esler. 2007. A case study of landholder attitudes and behaviour toward the conservation of renosterveld, a critically endangered vegetation type in Cape Floral Kingdom, South Africa. *Environmental Management* 40: 46–61. https://doi.org/10.1007/s00267-006-0086-0

World Bank. 2017. *Global economic prospects, June 2017: A fragile recovery* (Washington: World Bank). https://openknowledge.worldbank.org/bitstream/handle/10986/26800/9781464810244.pdf

World Bank. 2019. *World Bank Open Data: Sub-Saharan Africa.* http://data.worldbank.org/region/sub-saharan-africa

WWF/TNC. 2013. *Freshwater Ecoregions of the World.* http://www.feow.org

Zunckel, K. 2014. *Southern African Development Community transfrontier conservation guidelines: The establishment and development of TFCA initiatives between SADC Member States* (Gabarone: SADC TFCA).

3. What is Biodiversity?

A "silverback" mountain gorilla (*Gorilla beringei beringei,* EN) in Mgahinga Gorilla National Park, Uganda. Once thought to be one species, genetic analyses have shown that there are two gorilla species, each with two subspecies. Thanks to the efforts of dedicated conservationists and local communities in the Albertine Rift, the IUCN downlisted the mountain gorilla from *Critically Endangered* to *Endangered* in 2018. The three other subspecies are still considered *Critically Endangered*. Photograph by Nina R, https://www.flickr.com/photos/150102727@N06/31467129021, CC BY 2.0.

https://doi.org/10.11647/OBP.0177.03

Conservation biology aims to improve the protection of biodiversity—that is, all the species, genetic diversity, and ecosystems on Earth. By this definition, the process of documenting life on Earth requires us to consider biodiversity on three different levels (Figure 3.1):

- **Species diversity**: The full variety of species, from single-celled organisms like bacteria to larger multicellular organisms like animals and everything in between.
- **Genetic diversity**: The full range of variability in genetic material within a species. This variation can occur spatially as differences between populations or as differences between individuals of the same population.
- **Ecosystem diversity**: The full variety of ecosystems—i.e., assemblages of species and the physical environments in which they live.

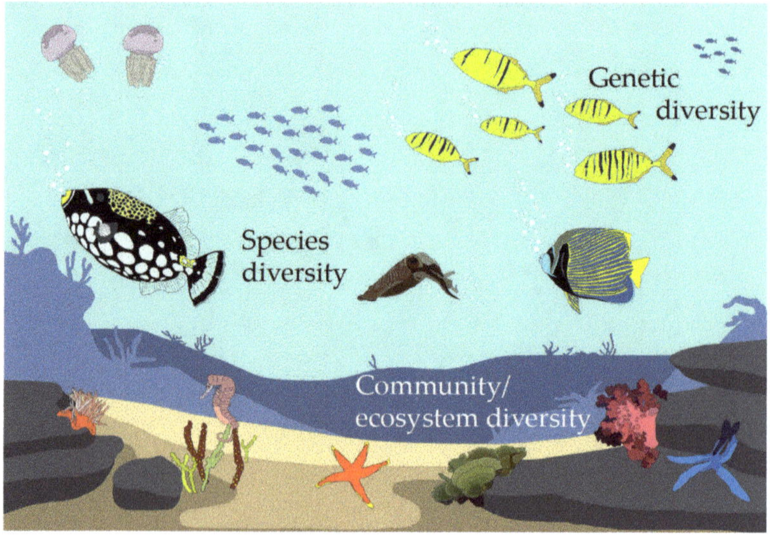

Figure 3.1 A region's biodiversity includes the full complement of that area's species diversity (all the area's species), genetic diversity (the full range of genetic variation found within each of those species), and ecosystem diversity (the variety of ecosystems and ecological processes). CC BY 4.0.

The relationship between species, genetic, and ecosystem diversities is complex and interdependent. That is, a species cannot exist without genetic diversity or ecosystem diversity, and vice versa. For that reason, it is virtually impossible to affect one aspect of diversity without affecting the other. We can therefore think of species, genetic, and ecosystem diversities simply as different ways to measure the variety of life.

3.1 Species Diversity

In general, the first step in responding to the conservation need of a species or population is to know its identity. For this reason, one of the three main goals of conservation biology is to document all life on Earth or, in plain language, to give each species a name. The task of giving each species a (formal) name falls on specialist scientists known as **taxonomists.** Taxonomists (and the people assisting them) explore nature, collect specimens of plants, animals, and other organisms, describe/name those specimens, and store the specimens in permanent collections, such as natural history museums and **herbaria** (there are currently over 6,500 natural history museums in the world). These permanent collections, affectionally called "Libraries of Life", provide the material and locations that taxonomists use to describe species and to develop systems for biodiversity classifications.

When a species is formally described, it is given a unique two-part name, known as a binomial name. For example, the **binomial** name for the lion is *Panthera leo*. The first part of the name, *Panthera*, identifies the generic epithet (or simply **genus**); in this case, the panthers or big cats. The second part of the name, *leo*, identifies a subset within the genus known as the specific epithet (or simply species); in this case, the lion. This binomial system thereby both identifies a lion as its own species and connects it to other closely-related species: Africa and Asia's leopards (*P. pardus*, VU); Asia's snow leopard (*P. uncia*, VU); Asia's tigers (*P. tigris*, EN); and South America's jaguars (*P. onca*, NT) (Figure 3.2).

Figure 3.2 The world's large predatory cats: (A) tiger, (B) slow leopard, (C) leopard, (D) jaguar, and (E) lion. By looking at their binomial names one can immediately see the five species are closely related. CC BY 4.0.

Binomial species names, as well as the taxonomic relationships between different species, form the backbone of taxonomic databases, as compiled and organised by biodiversity informatics projects. Some biodiversity informatics projects focus on one group of species, while others focus on certain regions. For example, all known marine species are listed in the World Register of Marine Species (http://www.marinespecies.org), while the Catalogue of Afrotropical Bees (https://doi.org/10.15468/u9ezbh) collates information of only African bees. In some cases, multiple projects—each using different assumptions to suit different user groups better—may catalogue the same group of species. For example, the world's fungi, are listed both in Index Fungorum (http://www.indexfungorum.org) and MycoBank (http://www.mycobank.org), while bird names are indexed by at least seven different projects, each a little different from the other. There are even some biodiversity informatics projects that attempt to catalogue all life on Earth; examples include Catalogue of Life (http://www.catalogueoflife.org), Encyclopaedia of Life (http://eol.org), and Wikispecies (https://species.wikimedia.org).

3.1.1 What is a species?

There are three rules of thumb that taxonomists use to describe a species:

- **Morphological definition of species**: Individuals that are distinct from other groups in their **morphology**, physiology, or biochemistry.
- **Biological definition of species**: Individuals that breed (or could breed) with each other in the wild, but do not breed with members of other groups.
- **Evolutionary definition of a species**: Individuals that share a common evolutionary past, usually indicated by genetic similarities.

In practice, conservation biologists generally rely on the morphological definition to identify species. The ability to recognise physical or morphological differences between organisms is handy even when the actual identity of specimens is unknown. In such cases, field biologists may refer to the unknown species as **morphospecies** (Figure 3.3), at least until an expert identifies the unknown individuals or a taxonomist gives them an official scientific name. In contrast, the biological definition of species relies on information that is difficult to obtain and thus not readily available. The biological definition also fails to recognise recent speciation, which can cause closely related but distinct species to interbreed. Similarly, it is generally impractical for fieldworkers to measure differences in genetic sequences to distinguish one species from another because these procedures currently require expensive, immovable laboratory equipment.

> Taxonomists can use morphological, biological, and genetic information to identify species.

Despite the practical difficulties of applying the biological and evolutionary definitions in the field, both provide important guidelines for conservation efforts.

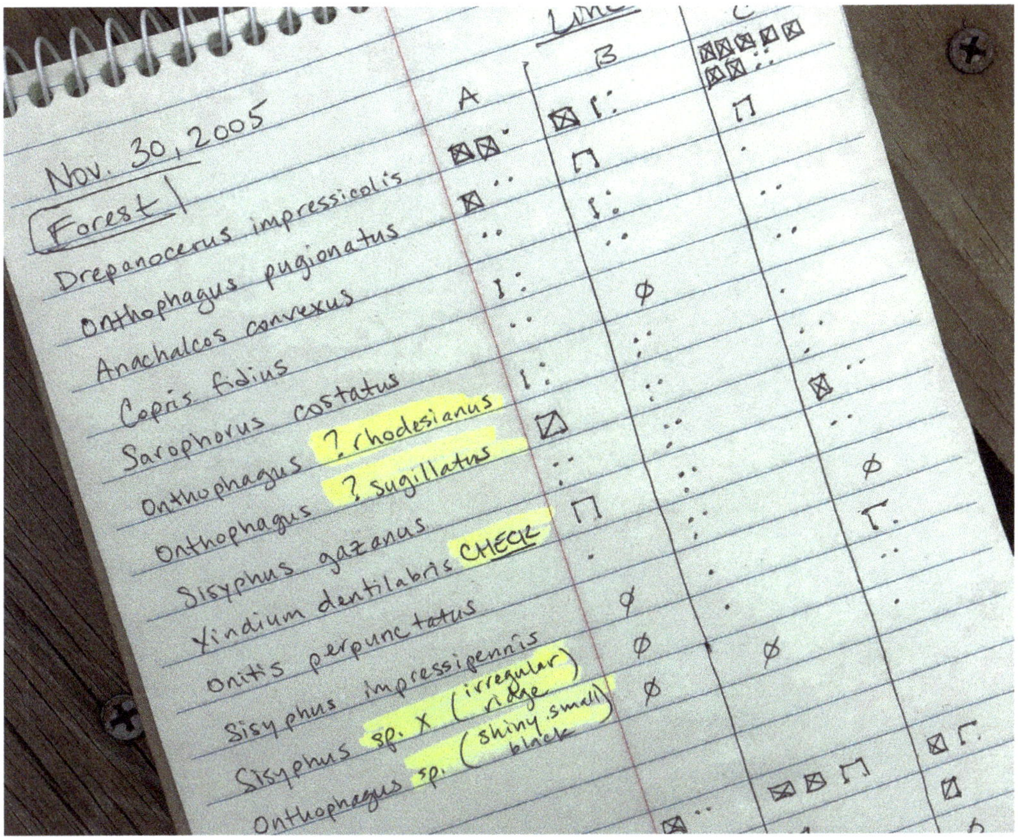

Figure 3.3 Field sheet showing application of the morphospecies concept during a dung beetle surveys in South Africa. Unidentified (and potentially misidentified) specimens are noted with a variety of descriptors (highlighted) and collected for identification at a later stage. Photograph by Lesley Starke, CC BY 4.0.

The biological species definition allows us to better understand species **biogeography** and the mechanisms that prevent two closely-related species to interbreed. The evolutionary species definition in turn allows us to better understand how and why the genetic makeup of populations change over time, through processes such as random **mutations**, **natural selection**, **emigration**, and **immigration**. It is thus important for conservation biologists to acknowledge the importance of maintaining these dynamic processes in protecting natural systems, and where possible, include them in their fieldwork (Box 3.1).

3.2 Genetic Diversity

Every extant species on Earth consists of at least one **population**, a group of individuals at a certain place that generally look alike and can potentially breed with each other to produce offspring. These population(s) can be very small (just a few individuals), very large (billions of individuals), or anything in-between. The individuals within

Box 3.1 Finding a Needle in a Haystack: Monitoring Species Using eDNA

Tammy Robinson and Clova Mabin

Centre for Invasion Biology, Stellenbosch University, Stellenbosch, South Africa.

✉ *trobins@sun.ac.za, clovamabin@gmail.com*

Trying to find threatened species in aquatic systems can be like trying to find a needle in a haystack. Traditionally, researchers have set off with nets, buckets, and even snorkels and scuba gear to painstakingly search for threatened species in ecosystems, ranging from streams to coral reefs. While searching in a small system, such as a pond, might not seem too difficult, it can be a real challenge to find tiny, inconspicuous organisms in and amongst the mud, stones, and plants, especially when they are trying their best to remain hidden. Things get even trickier when combing through large ecosystems like lakes or bays. These difficulties make it hard to reliably monitor the status or distribution of threatened aquatic species.

However, scientists have recently developed a new search tool called environmental DNA, (eDNA in short), where researchers collect and search water samples for the DNA of the species they are interested in. The eDNA technique was first developed by a biologist trying to detect organisms in sediment (Willerslev et al., 2003) but is now being used by conservationists working in all kinds of aquatic ecosystems. Organisms continually release small amounts of DNA into the water by sloughing off skin or other cells and releasing bodily wastes. This DNA then mixes in the surrounding environment, allowing those organisms to be detected through genetic analyses without actually sampling them directly.

Researchers have been testing just how useful eDNA is for finding threatened species in ponds and streams (Thomsen et al., 2012). They detected the eDNA of fish, shrimp, dragonflies, and amphibians in most ponds where the species were known to occur and found no trace of the eDNA of these species where they were absent. The most exciting development was their ability to detect eDNA evidence of threatened species in places where they had previously occurred but not been recently recorded by traditional search methods. Field observations and experiments also showed that eDNA can persist for up to two weeks in fresh water, and that concentrations can correspond to population sizes; this suggests that scientists may be able to monitor the abundance of rare aquatic species to a high degree of accuracy using this approach. For example, Lake Victoria could be searched for rare cichlid fish species that may still be present at low numbers even though researchers have not seen them for several years.

eDNA technology also holds considerable promise for the management of aquatic invasive species, if they could be detected as new arrivals before their numbers grow enough to be detected by conventional methods (Takahara et al., 2013). Early detection will give conservation managers a head-start and enable them to react quickly to invasions and increase their chances of preventing the environmental damage associated with invasive species. In a local twist to the tale, ongoing work in South Africa is applying eDNA as a tool for measuring the success of management efforts aimed at removing the invasive marine European shore crab (*Carcinus maenas*) (Figure 3.A) that could outcompete or threaten native African marine species. It is hoped that eDNA will be able to track the decline in crab numbers as the species is removed and then be used to monitor for any new arrivals should the crabs re-invade.

Figure 3.A Researchers in South Africa have been using eDNA to monitor the success of a control programme aimed at invasive European shore crabs. Photograph by Clova Mabin, CC BY 4.0.

This exciting new approach in detecting species is rapidly developing and improving our efficiency at monitoring threatened and invasive species. This makes the process less like looking for a needle in a haystack, and more like finding the millions of needles right under your nose.

each population generally differ genetically from one another to some degree. This **genetic variation**, a component of genetic diversity **(Figure 3.4)**, exists because the **genes**—the functional units of hereditary information that provide the blueprint of an organism—in different individuals are made up of slightly different **DNA** sequences. Different forms of a gene, which arise through **mutations** that change DNA sequences, are known as **alleles**. The **gene pool**, in turn, consists of the total diversity of genes and alleles in a population or species. The particular mix of genes and alleles in an individual is its **genotype**. The expression of an individual's genotype, as determined by the environment where an organism has developed, is its **phenotype**—that is, the organism's morphology, anatomy, physiology, and biochemistry. Common characteristics to describe a person include height, hair colour, and blood type, which taken together begin to describe that person's phenotype.

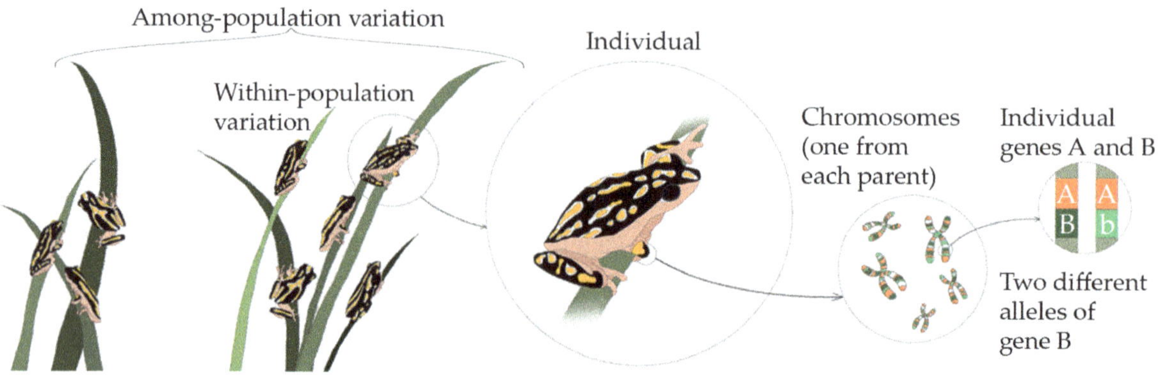

Figure 3.4 Genetic diversity arises due to variation in the alleles of individual genes and variation in chromosomes from different parents, which give rise to genetic variation between individuals, both within the same population and between different populations. CC BY 4.0.

In species which reproduce asexually, the potential for increased genetic diversity is limited to DNA mutations. However, sexual reproduction creates new genetic combinations by bringing together **chromosomes** from each parent. This process, called recombination, results in offspring that are genetically unique from their parents. Genetic mutations provide the foundation of genetic variation, but sexual reproduction dramatically increases genetic diversity by randomly mixing alleles in different combinations.

Two factors determine a species' genetic diversity: the number of genes that have multiple alleles (**polymorphic genes**) and the number of alleles present in a population for each polymorphic gene. If a gene is polymorphic, some individuals will have two different forms of the gene—that is, they will be **heterozygous** because they received different alleles of the same gene from their parents. Some individuals will have two of the same forms of the

Genetic diversity enables species to adapt to environmental change.

gene—they will be **homozygous** because each parent gave them the same allele. In general, the greater the genetic diversity in a population, especially the greater number of alleles present, the more capable a species will be to adapt to changing circumstances in their environment. Genetics also affect an individual organism's development, physiology, and **fitness**—the relative ability of individuals to survive and reproduce. This same principle gives humans the ability to select and breed crops and domestic animals with characteristics that benefit the production and quality of food (Davis et al., 2012). Many rare species have relatively low genetic diversity, especially in populations which have dwindled to small sizes. Low genetic diversity limits small populations' ability to adapt to changes in environmental conditions and leaves them at risk of extinction when conditions do change. Section 8.7 discusses the importance of maintaining genetic diversity in greater detail.

3.3 Ecosystem Diversity

Those who have climbed Africa's highest mountains have likely noticed how the plants and animals present gradually change, as one moves from tall lowland forest to moist, mid-elevation forest with a low canopy, then into grassy alpine meadows, and lastly, onto cold, windy, and rocky mountain peaks. We see these changes because, as we move across the landscape, physical conditions (e.g. geology, soil type, temperature, precipitation) change, and so also the species adapted to different environmental **niches**, as determined by the varying conditions. Thus, one by one, the species present at one location are replaced by new species better suited to the new conditions. We can see how the whole landscape changes in response to dynamic biotic and abiotic components of the environment (Figure 3.5). The variety of life resulting from these environmental changes is what gives rise to ecosystem diversity.

Figure 3.5 Climate plays an important role in the distribution of biodiversity. That is why we see a gradual decline in species diversity as one moves from warm and humid lowlands towards cold and windy peaks of high mountains. This photo was taken on Mount Kilimanjaro in Tanzania around 3,800 m above sea level. Photograph by Andreas Ensslin, CC BY 4.0.

Ecosystem diversity describes the full variety of ecosystems of an area, while the term "ecosystem" describes all the organisms in an area, as well as the physical and chemical environment with which those organisms interact. An important component of any ecosystem is its **biological community (or ecological community)**, defined as all the living individuals, populations, and species of a place, as well as all the biological interactions among those organisms. The abiotic (or physical) environment, especially climate, energy, and nutrients availability, greatly affects the structure, composition, and characteristics of an area's biological community (or biotic environment), and ultimately the type of ecosystem present (Figure 3.6). For example, water that evaporates from leaves, the ground, and other surfaces may later become rain or snow that provides drinking water that sustains life. Sunlight energy, in turn, enables **photosynthetic** plants (or **primary producers**) to grow; the energy from the sun is later transferred to animals that eat the plants (**herbivores**, or **primary consumers**), and then to animals that eat other animals (**carnivores**, or **secondary consumers**). The physical environment similarly affects aquatic ecosystems. For example, in freshwater stream, the biological community present is determined in large part by the physical characteristics of the stream, including water chemistry, temperature, flow rate, and substrate.

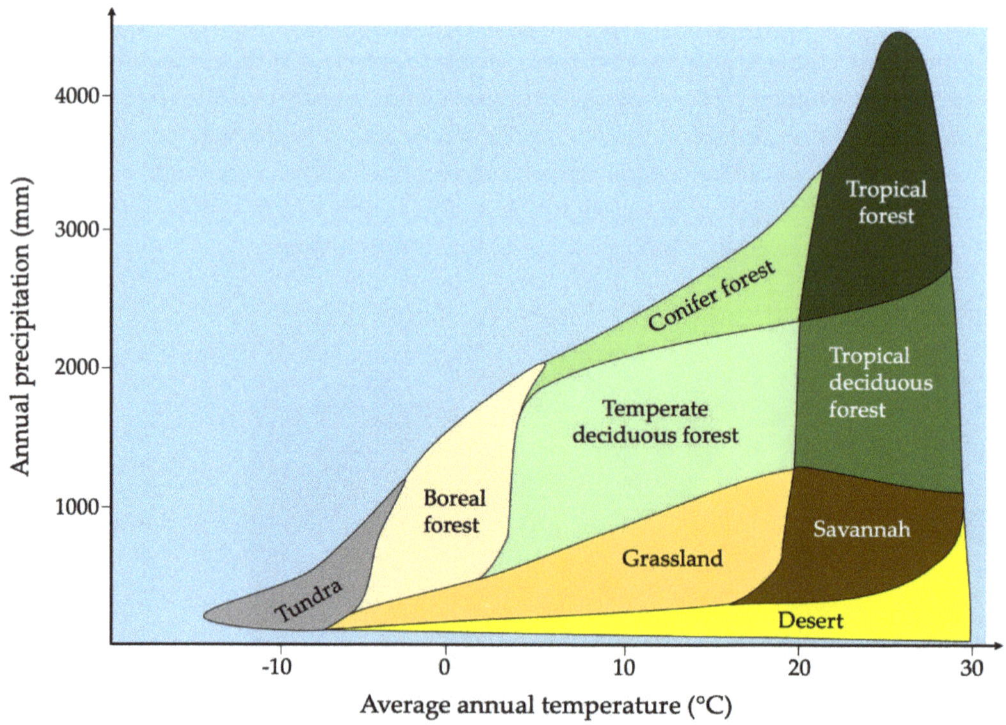

Figure 3.6 An area's abiotic components strong influence its biotic environment. For example, average temperature and precipitation determine which biome will dominate, which in turn influences which species will be present. After Whittaker, 1975, CC BY 4.0

At local scales, biological communities themselves can play prominent roles in altering the physical environment. For example, the trees present in a forest ecosystem can influence wind speed, light, humidity, soil chemistry, and temperature. Likewise, marine biological communities, such as kelp forests, seagrass beds, and coral reefs, can affect water temperature, water chemistry, sunlight penetration, and wave energy.

Within a biological community, individual species have specific ecological roles and have different requirements for survival. These roles and requirements enable different species to coexist, and in cases of interdependency, necessitate that they do so. For example, a given plant species may grow only in one type of soil, be pollinated by one type of insect, or have its seeds dispersed by only one type of animal. If any one of these requirements restricts the population size or distribution of that plant, it is considered a **limiting resource**. Even animal dung, usually considered a waste product, may become a limiting resource to species that rely on it for feeding and breeding. For example, studies from Côte d'Ivoire and Southern Africa have linked dung beetle population declines to the extirpation of large herbivores such as elephants and buffaloes (Nichols et al., 2009).

Environmental conditions that regulate the abundance of limiting resources may change over time. Consequently, many ecological communities can undergo major shifts in their composition over time. This is particularly prominent during ecological **succession**, which describes the gradual process during which ecosystems change after a disturbance. Consider, for example, an **old-growth forest** that is cleared by a logging operation. Shortly after clearing and abandonment, the soil absorbs more sunlight, resulting in high temperatures and low humidity during the day. These early stages present an ideal environment for **pioneer species,** such as sun-loving butterflies, annual herbs, and grasses, with wind-dispersed seeds. In a few years' time, the early successional herb-field or grassland transition to a scrubland, which accommodates a new suite of species. As the shrubs mature, forest trees germinate in the shade provided by the shrubs. Over the course of decades, as the forest trees mature, the forest canopy is gradually re-established which, in turn, provide opportunities for species characteristic of mid- and late-successional stages, such as shade-tolerant wildflowers of moist soils. Eventually, after many decades, **climax species** representative of mature forests, such as birds that nest in the holes of dead trees, start **colonising** the area.

3.4 Patterns of Biodiversity

Developing a strategy to conserve biodiversity requires a firm understanding of where **threatened species** and populations occur, why they are threatened, what their needs are, and what role they play in their respective ecosystems. By obtaining an understanding of species' distributions, biologists simultaneously gain an initial rough "estimate" of genetic diversity and ecosystem diversity. While addressing these questions is a critical task, finding appropriate answers can be complex, expensive, and take a long time to solve. This is in no small part because identifying species can, at times, be a very challenging endeavour.

3.4.1 Challenging species identifications

Before biologists can determine a species' distribution, needs, and population status, it is important to know the identity of the individuals being studied. While this may sound like a straightforward task, the process of identifying (and naming) a species can be deceptively hard, even for professional taxonomist. For example, a recent study found that 58% of 4,500 wild African ginger (*Aframomum* spp.) specimens that were deposited by professional biologists across 40 herbaria in 21 countries were given the wrong name (Goodwin et al., 2015)!

Identifying species can be hard, in part because the three tests biologists use to separate different species—morphology, biology, and evolution—do not always give the same results. That is because the methods and assumptions of each test are different. For example, some species have several varieties with easily observed morphological differences but are biologically and genetically similar enough that all those varieties are still considered a single species. A well-known example is the single species *Canis familiaris*, or domestic dog, whose wildly variable and numerous breeds can interbreed despite their large morphological differences. In contrast, some butterflies are considered distinct species because they cannot interbreed and have a characteristic genetic makeup, even though they cannot be separated by the naked eye.

> Describing species can be difficult, in part, because the multiple methods used by biologists to separate species do not always give the same results.

Another important aspect complicating species identifications is that **speciation**—whereby one species evolves into another—is a slow and gradual process; for some species, it may take many thousands of years. Consequently, much controversy exists about where to draw the "new species" line; in other words, when is a species distinct enough to be considered a separate species? Africa's iconic giraffes (*Giraffa camelopardalis*, VU) are a case in point. Taxonomists recently suggested that the region's giraffes—previously considered a single species—may, in fact, consist of four (Fennessy et al., 2016) or even six (Brown et al., 2007) species. Unfortunately, the final number of giraffe species is still disputed because of the different assumptions made by each study and how that impacts the number of species (Bercovitch et al., 2017). Similarly, biologists often struggle to **split** and identify **cryptic species**—undescribed species that are wrongly grouped with other similar-appearing species. A recent study estimated that 60% of newly discovered species are cryptic (Ceballos and Ehrlich, 2009). Even well-known groups may suffer from this problem: there is a reasonable chance that the bushbuck (*Tragelaphus scriptus*. LC) and klipspringer (*Oreotragus oreotragus*, LC) may in fact consist of several cryptic species yet to be described (Plumptre and Wronski, 2013; Groves et al., 2017).

To complicate matters even further, some species are closely related enough that they sometimes mate and produce **hybrids.** These hybrids blur the distinction between species, particularly those that may be early in the process of speciation. For some **taxa**, hybridisation naturally occurs in areas where the distribution ranges of related species overlap (e.g. de Jong and Butynski, 2010). Such natural hybridisation plays an important role in speciation (the evolution of new species); for example, it may have contributed to the high diversity of cichlid fishes in Africa's Rift Valley lakes (Salzburger et al., 2002). But hybridisation can also be detrimental to conservation efforts, particularly when it involves rare species and/or human disturbance. For example, when humans reduce one species' populations so much that they struggle to find reproductive partners of their own kind (e.g. vaz Pinto et al., 2016), when humans remove dispersal barriers that kept related species apart (e.g. Mondol et al., 2015), or when humans force related species that naturally occupy separate distributions to live together through **translocations** (e.g. Grobler et al., 2011; Benjamin-Fink and Reilly, 2017; van Wyk et al., 2017). While some hybrids may be sterile and thus unable to reproduce, at other times the resulting offspring can be quite strong in an evolutionary sense—a condition known as **hybrid vigour** (or **heterosis**)—and may outcompete their parent species. Such is the case with a land snail from the Seychelles (*Pachnodus velutinus*, EX), which was recently driven to extinction by hybridisation with a closely-related species (Gerlach, 2009)—hybrid individuals can still be found where *P. valutinus* used to occur.

Conversely, there may also be times when conservation biologists get it wrong and prioritise a species that does not warrant specific status. The Liberian greenbul (*Phyllastrephus leucolepis*) is one such example. Known from only a handful of records, this species was considered *Critically Endangered* until 2016, when geneticists discovered that the Liberian greenbul was the same species as the common icterine greenbul (*Phyllastrephus icterinus*, LC), but with an unusual coloration due to nutrient deficiencies (Collinson et al., 2018).

3.4.2 Implications of challenging species identifications

The difficulties in distinguishing between species have several practical conservation implications. First, when it is hard to identify a species, it may also be hard to determine that species' true population size and distribution which, in turn, impacts its conservation status. This was illustrated in a study on **bushmeat** markets in Guinea-Bissau, which showed how primate misidentifications hide the true impact of hunting on some of the region's most impacted species (Minhós et al., 2013). It also hampers captive breeding projects, by making the captive populations susceptible to **outbreeding depression**, which occurs when individuals that are not closely related (i.e. from different populations) breed and produce offspring (Conservation genetics is discussed in more detail in Section 8.7). Lastly, identification challenges with cryptic species can also cause delays in the formal description process, a necessary step in writing effective laws to protect them. The recent controversy among biologists arguing whether Africa's elephants are one or two species is a case in point. African elephants were already considered threatened when biologists thought they were a single species. This all changed in 2005, when taxonomic authorities officially recognised two elephant species, in effect dividing a single threatened species into two (thus even more imperilled) species (CBD, 2015). Yet, to avoid leaving hybrid elephants (e.g. Mondol et al., 2015) with an uncertain conservation status, the IUCN continues to assess elephants as one single species (Blanc, 2008); thus, their current *Vulnerable* assessment may not be an accurate reflection of each species' true conservation status.

Despite these challenges, conservationist biologists need to make every effort to obtain correct identifications. For most studies, morphological methods may be adequate. But when there is doubt, it is important for researchers to confirm their identifications with additional methods. Recent progress in making genetic technology more widely accessible through hand-held devices (Pennisi, 2016; Parker et al., 2017) and techniques such as **DNA barcoding** has also greatly enhanced our ability to correctly classify cryptic species, allowing us to give those species the conservation attention they deserve (Box 3.2).

Box 3.2 Golden Mole Conservation Requires a Sound Taxonomy

Sarita Maree[1,2] and Samantha Mynhardt[2]

[1]Department of Genetics, &
[2]Department of Zoology and Entomology,
University of Pretoria, South Africa.

smaree@zoology.up.ac.za;
samantha.mynhardt@up.ac.za

Golden moles (*Chrysochloridae*) are small, subterranean insectivores that rank among Africa's most unique, most threatened, and yet poorly studied mammals thanks to their secretive burrowing lifestyle. Ten of the 21 known species are currently threatened with extinction (IUCN, 2019) as their highly restricted and naturally fragmented sandy soil habitats are under threat from human activities. Current conservation efforts are severely jeopardised by taxonomic uncertainties and ambiguous evolutionary relationships, thus far based on morphological and limited genetic data, which suggest that many distinct but cryptic species remain undescribed (Taylor et al., 2018).

To remedy the dearth in knowledge on two endemic South African golden mole species, we analysed molecular data of individuals collected across the entire distribution range of both Juliana's golden mole (*Neamblysomus julianae*, EN) and Hottentot golden mole (*Amblysomus hottentotus*, LC) (Figure 3.B). In contrast to the widespread Hottentot golden mole, the Juliana's golden mole counts among South Africa's most imperilled mammals and is known from only three **range-restricted**, geographically isolated populations (Maree, 2015, Maree et al., 2016; Maree, 2017; Taylor et al., 2018). These three populations, together covering less than 160 km^2 occur in southeastern Pretoria (Gauteng population), the district of Modimolle (Limpopo, ~ 120 km north of Pretoria), and in southwestern Kruger National Park (Mpumalanga, ~ 400 km east of Pretoria) (Figure 3.C).

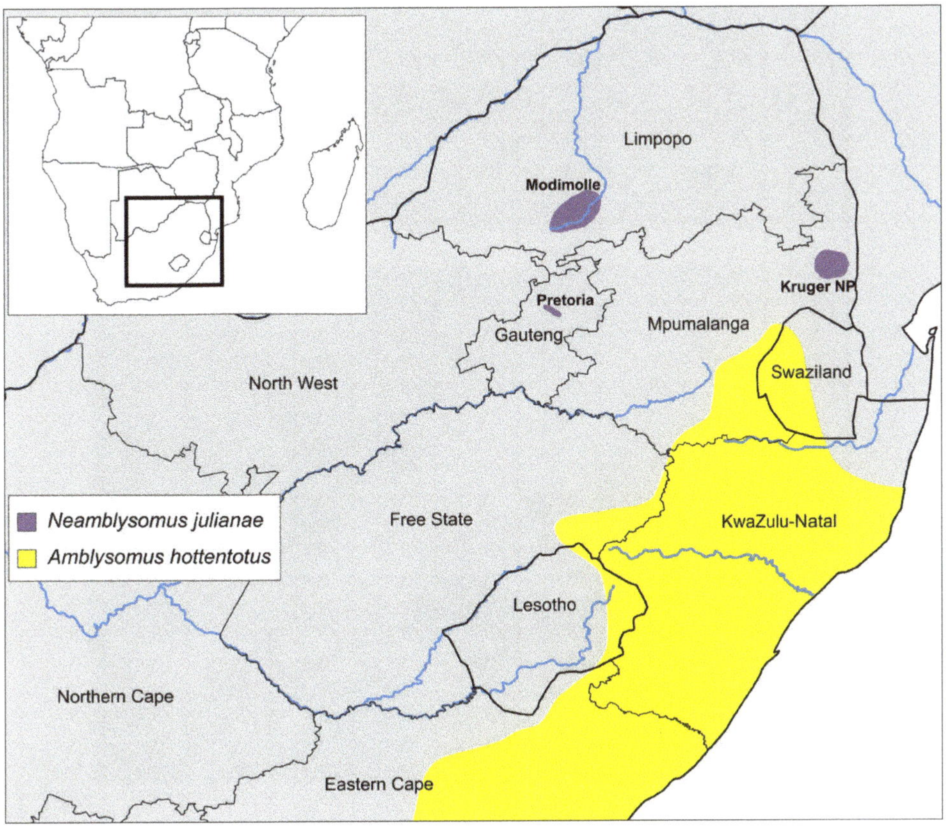

Figure 3.B The known geographic distribution of the widespread Hottentot golden mole and range-restricted Juliana's golden mole. Map by Arrie Klopper, after IUCN, 2019, CC BY 4.0.

Using molecular and other genetic methods, we have gained insights about the evolutionary relationships and gene flow between these two golden mole species, which have several conservation implications. First, preliminary findings suggest that the Hottentot golden mole contains several morphologically similar, but evolutionary distinct and genetically divergent lineages, some of which would represent undescribed cryptic species (Mynhardt et al., 2015; Taylor et al., 2018). Similarly, preliminary evidence suggests the Juliana's golden mole contains pronounced genetic separation between the Mpumalanga population and the Gauteng and Limpopo populations. This also corresponds

Figure 3.C (Top) The Juliana's golden mole is one of Africa's most threatened mammals. Photograph by Craig R. Jackson, CC BY 4.0. (Bottom) The Hottentot golden mole is generally thought of as widespread across southeastern South Africa but may in fact consist of several undescribed cryptic and potentially threatened species. Photograph by Samantha Mynhardt, CC BY 4.0.

to morphological differences observed between these populations, which collectively suggest that the Mpumalanga population of Juliana's golden mole might in fact be a cryptic species (Maree, 2015; Maree et al., 2016; Maree, 2017; Taylor et al., 2018). Unfortunately, in each of these cases the knowledge gaps remaining precluded definitive conclusions. Rigorous geographic sampling and additional molecular/genomic analyses will be needed to confirm the taxonomic status and geographic boundaries of putative new species within these and other golden mole taxa (Taylor et al., 2018).

Our results show that genetic frameworks contribute substantially to informed conservation decision-making. For golden moles and other taxa, some newly described species will undoubtedly be considered more threatened than in their previous species designations. Threat assessments on the Juliana's golden mole has already identified the Gauteng population as *Critically Endangered* due to severe habitat loss and transformation within its highly restricted and already fragmented range (~ 22 km^2 in extent) caused by rapid urbanisation and opencast sand mining. This pressure is exacerbated by this species' extreme habitat specificity and poor dispersal capabilities (Jackson and Robertson, 2011; Maree, 2015; Maree et al., 2016; Maree, 2017; Taylor et al., 2018). **Species distribution modelling** (SDM, discussed in Section 10.1.1) predicted several regions throughout Gauteng, Mpumalanga, and Limpopo Provinces where the species could potentially occur, but subsequent surveys led to the discovery of only two new localities around Modimolle (Jackson and Robertson, 2011). This finding emphasises that the protection of all suitable habitats remaining for the species and the Pretoria population, in particular, would be key to its persistence. Strategies to achieve this ought to be incorporated into current conservation planning (Maree, 2015; Maree et al., 2016; Maree, 2017; Taylor et al., 2018).

We also illustrated the importance of maintaining the integrity of geographically isolated and/or genetically unique populations, lest yet undescribed species be lost to extinction before they could be fully recognised. A sound taxonomy, obtained through genetic analyses, thus contributes substantially to informed conservation decision-making. Even in the absence of such information, it is still crucial that isolated populations be managed as distinct units to conserve the evolutionary history of different species and populations.

Because the demand for expert taxonomists outstrips their availability, there is also a need to train and employ more taxonomists, particularly in the tropics and other species-rich areas. The public can help in this endeavour. In 2015, **citizen scientists**—volunteers participating in scientific projects—discovered 51 of 60 new dragonfly species from Africa that were described that year (Dijkstra et al., 2015). For conservation biologists, it is also important to not become despondent about the lagging efforts to describe species. They should instead take an example from motivated parrot lovers who were motivated to work even harder to get their study species recognised as distinct (Box 3.3). It is also important to keep in mind that species are never fixed; evolve all the time, albeit at different rates, due to challenges and opportunities presented by their environment.

Box 3.3 Does Tardy Recognition of a Species Hamper its Conservation?

Colleen T. Downs

School of Life Sciences, University of KwaZulu-Natal, Pietermaritzburg, South Africa.

✉ *downs@ukzn.ac.za*

The usefulness of subspecies in conservation has long been a subject of controversy (Coetzer et al., 2015). Accurately drawing the line between an individual species and other similar animals is important for effective studies of biodiversity, and for planning and implementing official conservation strategies. Across Africa, there are many species with very broad historical distributions that are thought to contain locally adapted varieties. However, the distributions of many of these species are now fragmented and disjointed, mainly because of changes in available habitat. Examples include reptiles, such as the Nile crocodile (*Crocodylus niloticus*, LC), mammals, such as the common hippopotamus (*Hippopotamus amphibious*, VU), and a wide range of bird species. As a result of this fragmentation, various subspecies, recognised by morphology and habitat distribution, are now recognised as individual species. Modern DNA technology allows these discoveries to be supported with genetic evidence.

Protecting a newly recognised species can be difficult; genetic testing takes time and funding, and if an animal or plant is threatened before it has full species status, conservation success is that much more difficult. An example is the Cape parrot (*Poicephalus robustus*, EN), a forest species which was first suggested in 1997 to be a separate species and distinct from the more widespread grey-headed parrot (*Poicephalus fuscicollis*, LC) of Africa's savannah ecosystems. Additional support for the Cape parrot (Figure 3.D) being a separate species came from ecological and morphological data in 2002 (Wirminghaus et al., 2002) and separate genetic evidence in 2015 (Coetzer et al., 2015). Although many published bird guides reflect the change, the species was recently recognised as a species by authorities (e.g. BirdLife International, 2017), which affected its ability to receive legal protection. The Cape parrot is endemic to South Africa, with a distribution primarily restricted to southern mist-belt Afromontane forests in the Eastern Cape and southern KwaZulu-Natal plus a relict population in Limpopo Province. Cape parrots are restricted in their distribution by their specialised habitat and dietary requirements for particular fruits. A decrease in this species' abundance over the past 50 years is a consequence of several factors, including habitat fragmentation and degradation, food and nest site shortages, illegal trade of the birds for pets and aviculture, and disease.

Figure 3.D Juvenile Cape parrots feeding on pecan nuts near Creighton, KwaZulu-Natal, South Africa. Until recently the conservation efforts targeting this species was hampered by lack of international recognition. Photograph by C.T. Downs, CC BY 4.0.

Dedicated researchers have recognised the importance of determining population size and raising the awareness of the plight of the Cape parrot and the forests for which it is a flagship species. Current abundance of the Cape parrot is relatively low but stable, with an estimate of fewer than 1,600 birds in the wild (Downs et al., 2014). Estimates are based on an annual census held since 1998, organized by citizen scientists. For the Cape parrot, tardy genetic recognition of full species status was overcome by conservationists' perseverance. We must be vigilant if we want to protect other still-hidden species from future extinction.

3.4.3 Measuring species diversity

Biologists have developed three quantitative measures of species diversity as a means of measuring and comparing species diversity (Figure 3.7):

- **Alpha diversity** (or **species richness**), the most commonly referenced measure of species diversity, refers to the total number of species found in a particular biological community, such as a lake or a forest. Bwindi Forest in Uganda, with an estimated 350 bird species, has one of the highest alpha diversities of all African ecosystems.
- **Gamma diversity** describes the total number of species that occur across an entire region, such as a mountain range or continent, that includes many

ecosystems. The Albertine Rift, which includes Bwindi Forest, has more than 1,074 species of birds, a very high gamma diversity for such a small region.

- **Beta diversity** connects alpha and gamma diversity. It describes the rate at which species composition changes across a region. For example, if every wetland in a region was inhabited by a similar suite of plant species, then the region would have low beta diversity; in contrast, if several wetlands in a region had plants communities that were distinct and had little overlap with one another, the region would have high beta diversity. Beta diversity is calculated as gamma diversity divided by alpha diversity. The beta diversity for forest birds of the Albertine Rift is about 3.0, if each ecosystem in the area has about the same number of species as Bwindi Forest.

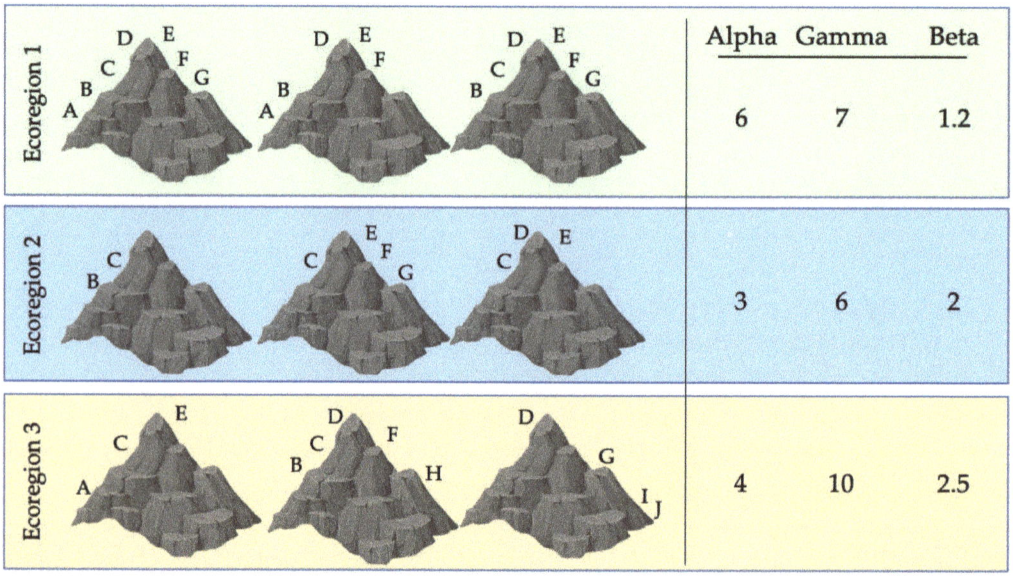

Figure 3.7 Biodiversity indices for nine mountain peaks across three ecoregions. Each symbol represents a different species; some species have populations on only one peak, while others are found on two or more peaks. The variation in species richness on each peak results in different alpha, gamma, and beta diversity values for each ecoregion. This variation has implications for how we divide limited resources to maximise protection. If only one ecoregion can be protected, ecoregion 3 may be a good choice because it has high gamma (total) diversity. However, if only one peak can be protected, should a peak in ecoregion 1 (with many widespread species) or ecoregion 3 (with several unique, range-restricted species) be protected? After Primack, 2012, CC BY 4.0.

It is important to note that alpha, beta, and gamma diversity describe only part of what is meant by biodiversity. For example, none of these three terms completely account for genetic diversity, which allows species to adapt as conditions change (Section 8.7.1). It also neglects the importance of ecosystem diversity, which results from the collective response of species to their dynamic environment. However, these diversity measures are useful for comparing different regions, and identifying locations with high concentrations of **native species** that should be protected.

3.4.4 How many species exist?

To date, taxonomists have described about 1.5 million species that share this planet with us (Costello et al., 2012). While this total may seem impressive, available evidence suggests that this estimate vastly underestimates the true extent of Earth's biodiversity. In fact, even now, after all the exploration in years gone by, several thousand new species are being described each year. Many new discoveries are made by skilled researchers recognising new species by being able to discern variation in morphological characters; that includes the discoveries of a new small forest antelope from West Africa (Colyn et al., 2010) and a new species of shark off Mozambique (Ebert and Cailliet, 2011). Such discoveries can also be rather surprising and unexpected. For example, an amateur botanist recently discovered two new flowering plants in the heavily studied Cape Floristic Region (Bello et al., 2015). Similarly, the lesula (*Cercopithecus lomamiensis*)—a species of monkey long known to local hunters—was only formally described after biologists discovered this "different" monkey on a leash in a remote village of the DRC (Hart et al., 2012). Some recent discoveries even include entire new communities in unexpected places. For example, in 2007, grassland **surveys** by citizen scientists in an area starting 5 km from South Africa's Johannesburg metropolitan area found previously unknown populations of five threatened bird species, as well as a number of regionally threatened birds and mammals; these discoveries were instrumental in recognising this area as the Devon Grasslands Important Birding Area (Marnewick et al., 2015).

> New genetic technologies have highlighted that there are many thousands of species yet to be described.

The most exciting and newsworthy discoveries of new species generally involve higher-level taxa, especially living fossils. For example, in 1938, biologists across the world were stunned by the report of a strange fish caught in the Indian Ocean off South Africa. This fish, subsequently named coelacanth (*Latimeria chalumnae*, CR), belongs to a group of marine fishes that were common in ancient seas but were thought to have gone extinct 65 million years ago. Coelacanths are of interest to evolutionary biologists because they show certain features of muscles and bones in their fins that are comparable to the limbs of the first vertebrates that crawled onto land. Following the initial discovery, coelacanths have been found along Africa's Indian Ocean coast from South Africa to the Comoros and through to Kenya. Unfortunately, the entire coelacanth population, estimated at fewer than 500 individuals, is currently highly threatened because of ongoing fishing pressures (Musick, 2000).

Although field surveys have proven to be of great importance for discovering new species and populations, perhaps the greatest taxonomic progress has come from advances in genetic analyses which help to separate cryptic species previously **lumped** under more widespread species. For example, advances in genetic research recently highlighted that the African clawed frog (*Xenopus laevis*, LC)—a popular model organism in biomedical research—consists of seven distinct species (Evans et al., 2015). Similarly, using new genetic methods, scientists recently confirmed that the slender-snouted crocodile (*Mecistops cataphractus*, CR) consists of two different species, one endemic to West Africa and the other to Central Africa (Shirley et al., 2018).

The presence of so many undiscovered species and communities makes precise estimates of species diversity incredibly difficult, especially in Africa where so many areas remain scientifically unexplored. Our most recent estimates, combining genetic analysis of well-known groups with mathematical patterns, suggests there are between 1–6 billion distinct species on Earth (Table 3.1) of which there are only about 163 million animals and 340 thousand plants (Larsen et al., 2017)—this is obviously much greater than the current catalogue of 1.5 million species! Given the amount of new species that continue to hide in plain sight, so to speak, there is no doubt that a great number of species and communities are waiting to be discovered by eager African adventurers over the next several decades.

Estimates suggest there are somewhere between 1–6 billion distinct species on Earth. The most diverse group of species is bacteria.

Table 3.1 Estimated living biomass and number of species for each kingdom of life, following the seven-kingdom system (Ruggiero et al., 2015). Note how plants weigh the most, but bacteria have the most species.

Kingdom	Weight (Gt)[a]	Number of species (in million)	% of all species[b]	Number of described species[c]	% of described species
Animals	2	163	7	1,205,336	< 1
Fungi	12	165	7	135,110	< 0.1
Plants	450	0.382[c]	< 0.5	364,009	95
Chromista	Unknown	0.025[c]	< 0.5	23,428	94
Protozoans	4	163	7	2,686	0.1
Archaea	7	0.0005	< 0.5	377	75
Bacteria	70	1,746	78	9,982	0.1

[a] *As gigatonnes of carbon, from Bar-On et al., 2018*

[b] *From Larsen et al. (2017)'s Table 1, Scenario 1*

[c] *From http://www.catalogueoflife.org*

3.4.5 Where are most species found?

Because it is so hard to obtain accurate estimates of species numbers, many conservation biologists have recently started to focus their efforts on understanding and planning around patterns of species diversity. This makes sense: regions with many species of one taxon tend to also have many species of other taxa, so protecting one diverse group of species will likely also protect many other species, even if those other species are not well understood. Consequently, many conservation biologists see the forests of the Congo Basin, Albertine Rift, and West Africa as critical conservation priorities because these areas hold Africa's greatest species concentrations, particularly birds, mammals, and butterflies. But there are very important outliers. For example, due to factors that include the geology and soil characteristics, size and variability of the environment, historical circumstances, or climatic conditions, none of these tropical forest areas have as many plant species as the Cape Floristic Region of South Africa—an area of unparalleled importance for plant diversity. Species diversity relationships may also break down at the local scale; for example, amphibians are likely more diverse in wet, shady riverbeds, whereas reptiles may be more diverse in drier, open habitats even if only tens of metres of space separate the reptiles from a riverbed full of amphibians.

Table 3.2 Number of endemic and native mammal species as a function of the environment, for a range of African countries.

Country	Dominant ecoregion	Area (× 1000 km²)	Number of endemic mammals	Number of native mammals	Mammals per 1000 km²
Seychelles	Oceanic island	0.45	2	24	53.3
Cabo Verde	Oceanic island	4.03	0	29	7.2
Rwanda	Montane forest	26.8	1	189	7.05
Eq. Guinea	Lowland forest	28.1	3	184	6.55
Burundi	Montane forest	27.8	1	144	5.18
Sierra Leone	Varied Forest	71.7	0	197	2.75
Zimbabwe	Savannah	391	0	204	0.52
Zambia	Savannah	753	5	242	0.32
Namibia	Desert	825	3	206	0.25
South Africa	Varied	1,221	31	307	0.25
South Sudan	Sahel	644	1	151	0.24
DRC	Varied	2,345	26	438	0.19
Niger	Sahel	1,267	0	134	0.11

Source: IUCN, 2019.

By examining all these patterns of species diversity across the world, biologists have discovered at least two general frameworks governing species richness. The first framework is that stable ecosystems usually have many species, while ecosystems that were subjected to more recent glaciation usually have fewer species. This observation explains why tropical ecosystems are generally considered the world's most species-rich environments (Table 3.2). While tropical grasslands, wetlands, and other ecosystems all hold relatively high species diversity, species richness of tropical forests are particularly noteworthy; even though these areas occupy only about 7% of Earth's land surface, they contain over half of the world's species (Corlett and Primack, 2010). This is, in a large part, due to the relatively large global distribution of the tropical forests and the diversity of geological history between these areas of South and Central America, Africa, Asia, and Australia, which has resulted in unique assemblages of species that have evolved in isolation from each other.

Tropical forests are not the only species-rich tropical ecosystem. Tropical coral reefs, colonies of tiny aquatic invertebrates that form entire ecosystems (Figure 3.8), are the marine equivalent of tropical forests both in terms of species richness and complexity. These areas not only provide homes for corals, but also for huge numbers of fish, molluscs, and marine mammals that find shelter in these highly productive and sheltered ecosystems. In Africa, tropical coral reefs are most widespread and diverse in coastal East Africa, but unique tropical coral reef communities can also be found along Mozambique and South Africa's north-eastern coast.

Figure 3.8 Coral reefs such as this one at Zanzibar's Mnemba Atoll, off the north coast of Tanzania, are highly diverse underwater ecosystems composed of the accumulated skeletons of billions of tiny marine invertebrates. These underwater landscapes provide habitat for at least 25% of all marine species. Photograph by Kamal Karim, https://www.flickr.com/photos/118534047@N06/22449100152, CC BY 2.0.

High levels of species diversity, especially among plants, can also be found in ecosystems with a Mediterranean climate, such as southwestern Africa, as well as southwestern Australia, California, central Chile, and the Mediterranean Basin of southern Europe and North Africa. The climate of a Mediterranean-type ecosystem is characterised by cool, moist winters, hot, dry summers, resulting in distinctive plant adaptations such as short twigs and stiff leaves. A combination of special environmental factors, including a considerably old geological age, complex site characteristics (such as varied topography and soils), and frequent fires facilitated rapid speciation and helped to prevent any one species from dominating. Today, although regions with a Mediterranean climate cover only 2% of Earth's surface, 20% of all plant species are found here (Underwood et al., 2009). The Cape Floristic Region—the only Mediterranean climate in Sub-Saharan Africa—is particularly important to conservationists as it has the highest concentration of higher plant diversity (over 9,000 species) in the world.

The second framework governing pattern of species diversity is that locations with high numbers of species usually hold many endemic species. The Cape Floristic Region, for example, boasts more than 6,200 endemic plant species, which include 12 endemic families and 160 endemic genera. Similarly, Lake Malawi holds nearly 14% of the world's freshwater fishes (500–1,000 species, totals vary by source), with more than 90% of those being endemic.

Biogeographic transition zones—also known as **ecotones**—regions where different ecosystems meet and overlap, are a special case of areas that contain great species diversity and high levels of endemicity. These areas share environmental factors of two or more environments, allowing for the mixture of biodiversity from those component environments, while unique features within these areas often also give rise to unique species. A case in point is the Maputaland Centre of Endemism, situated in far southern Mozambique. Here, biological communities from northern tropical and southern temperate ecosystems overlap, resulting in surprisingly high levels of species richness as well as endemism (van Wyk, 1996).

Today is an exciting time of biological exploration. Methods and technologies for exploration are improving rapidly, and we are learning more about the value and function the diversity of life on Earth. As genetic techniques advance and become more accessible, an increasing number of people are participating in recording the presence of species in locations around the world; this includes amateur **naturalists** and citizen scientists who contribute to bird surveys, plant walks, and other **natural history** activities. With this increased knowledge of biodiversity also comes an acute awareness that human activities damage ecosystems and reduce diversity. Hopefully this broader awareness will spur more people to take responsibility to protect and restore that biodiversity.

3.5 Summary

1. Earth's biodiversity includes the entire range of living species (species diversity), the genetic variation that occurs among individuals within a

species (genetic diversity), and, at a higher level, the biological communities in which species live and their associations with the physical and chemical environment (ecosystem diversity).

2. For practical purposes, most ecologists and conservationists identify species in the field according to their morphology, although improvements in genetic techniques are allowing more species to be identified according to their evolutionary past, revealing many cryptic species that people did not realise were there.
3. There are several ways to measure and compare biodiversity. The most popular measurement is species richness in a particular community, such as a forest or grassland (alpha diversity), species richness across a larger landscape, such as a mountain range (gamma diversity), and the rate of change of species composition as one crosses a large region (beta diversity).
4. It is estimated that there may be as many as 2 billion species on Earth. Most species already described are insects, while the best-known species include birds and mammals. The majority of species still need to be discovered.
5. Variation in climate, topography, and geological age are all factors that affect patterns of species richness. Geological age and complexity provide environmental variation, which in turn allows opportunities for genetic isolation, local adaptation, and speciation, given enough time. Tropical forests, coral reefs, and Mediterranean-type ecosystems host a disproportionately large amount of the world's biodiversity.

3.6 Topics for Discussion

1. Think of any group of species (birds, trees, or maybe insects) that can be found in the area where you live. Do you think it is important to be able to identify these species? Why? How many species can you personally identify at this moment? What steps would you take to learn to identify more species?
2. Which ecosystems in your country are particularly species-rich, and which are species-poor? Describe some factors that make ecosystems species-rich or species-poor.
3. Where in Africa do you think most undescribed species are lurking? Explain your answer.

3.7 Suggested Readings

Ceballos, G., and P.R. Ehrlich. 2009. Discoveries of new mammal species and their implications for conservation and ecosystem services. *Proceedings of the National Academy of Sciences*

106: 3841–46. https://doi.org/10.1073/pnas.0812419106 Even familiar taxa contain many undescribed species.

Bar-On, Y.M., R. Phillips, and R. Milo. 2018. The biomass distribution on Earth. *Proceedings of the National Academy of Sciences* 25: 6505–11. https://doi.org/10.1073/pnas.1711842115 Note the impact of humans on the composition of life on Earth.

Ebach, M.C., J.J. Morrone, L.R. Parenti, et al. 2007. International code of area nomenclature. *Journal of Biogeography* 35: 1153–57. https://doi.org/10.1111/j.1365-2699.2008.01920.x Scientists also grapple with confusing terminology when describing biodiversity.

Gippoliti, S., F.P.D. Cotterill, D. Zinner, et al. 2018. Impacts of taxonomic inertia for the conservation of African ungulate diversity: An overview. *Biological Reviews* 93: 115–30. https://doi.org/10.1111/brv.12335 Taxonomic inertia, or the delay in recognising distinct species, slows conservation efforts.

Hart, J.A., K.M. Detwiler, C.C. Gilbert, et al. 2012. Lesula: A new species of *Cercopithecus* monkey endemic to the Democratic Republic of Congo and implications for conservation of Congo's Central Basin. *PLoS ONE* 7: e44271. https://doi.org/10.1371/journal.pone.0044271 Many species known to local people still need to be formally described.

Joppa, L.N., D.L. Roberts, and S.L. Pimm. 2011. The population ecology and social behavior of taxonomists. *Trends in Ecology and Evolution* 26: 551–53. https://doi.org/10.1016/j.tree.2011.07.010 The number of taxonomists and the number of species described per year are steadily increasing.

Laikre, L., F.W. Allendorf, L.C. Aroner, et al. 2010. Neglect of genetic diversity in implementation of the Convention on Biological Diversity. *Conservation Biology* 24: 86–88. https://doi.org/10.1111/j.1523-1739.2009.01425.x A greater emphasis on genetic diversity needs to be part of conservation efforts.

Larsen, B.B., E.C. Miller, M.K. Rhodes, et al. 2017. Inordinate fondness multiplied and redistributed: The number of species on Earth and the new pie of life. *Quarterly Review of Biology* 92: 229–65. https://doi.org/10.1086/693564 We have much to learn about life on Earth.

Minhós, T., E. Wallace, M.J.F. da Silva, et al. 2013. DNA identification of primate bushmeat from urban markets in Guinea-Bissau and its implications for conservation. *Biological Conservation* 167: 43–49. https://doi.org/10.1016/j.biocon.2013.07.018 Misidentifications could have significant conservation implications.

Bibliography

Bar-On, Y.M., R. Phillips, and R. Milo. 2018. The biomass distribution on Earth. *Proceedings of the National Academy of Sciences* 25: 6505–11. https://doi.org/10.1073/pnas.1711842115

Bello A., C.H. Stirton, S.B.M. Chimphango, et al. 2015. *Psoralea diturnerae* and *P. vanberkelae* (Psoraleeae, Fabaceae): Two new species restricted to the core Cape Region of South Africa. *PhytoKeys* 44: 97–107. https://doi.org/10.3897/phytokeys.44.8999

Benjamin-Fink, N., and B.K. Reilly. 2017. Conservation implications of wildlife translocations: The state's ability to act as conservation units for wildebeest populations in South Africa. *Global Ecology and Conservation* 12: 46–58. https://doi.org/10.1016/j.gecco.2017.08.008

Bercovitch, F.B., P.S.M. Berry, A. Dagg, et al. 2017. How many species of giraffe are there? *Current Biology* 27: R136–R137. https://doi.org/10.1016/j.cub.2016.12.039

BirdLife International. 2017. *Poicephalus robustus. The IUCN Red List of Threatened Species* 2017: e.T119194858A119196714. http://doi.org/10.2305/IUCN.UK.2017-3.RLTS.T119194858A119196714.en

Blanc, J. 2008. *Loxodonta africana. The IUCN Red List of Threatened Species* 2008: e.T12392A3339343. http://doi.org/10.2305/IUCN.UK.2008.RLTS.T12392A3339343.en

Brown, D.M., R.A. Brenneman, K.P. Koepfli, et al. 2007. Extensive population genetic structure in the giraffe. *BCM Biology* 5: 57. https://doi.org/10.1186/1741-7007-5-57

CBD (Center for Biological Diversity). 2015. *Petition to reclassify and uplist African elephants from Threatened to Endangered under the Endangered Species Act as two separate species: forest elephants* (Loxodonta cyclotis) *and savannah elephants* (Loxodonta africana) (Tucson: CBD). https://www.biologicaldiversity.org/species/mammals/pdfs/African_Elephant_Uplisting_Petition.pdf

Ceballos, G., and P.R. Ehrlich. 2009. Discoveries of new mammal species and their implications for conservation and ecosystem services. *Proceedings of the National Academy of Sciences* 106: 3841–46. https://doi.org/10.1073/pnas.0812419106

Coetzer, W.R., C.T. Downs, M.R. Perrin, et al. 2015. Molecular systematics of the Cape Parrot (*Poicephalus robustus*): Implications for taxonomy and conservation. *PLoS ONE* 10: e0133376. https://doi.org/10.1371/journal.pone.0133376

Collinson, J.M., M. Päckert, Y. Lawrie, et al. 2018. Taxonomic status of the Liberian Greenbul *Phyllastrephus leucolepis* and the conservation importance of the Cavalla Forest, Liberia. *Journal of Ornithology* 159: 19–27. https://doi.org/10.1007/s10336-017-1477-0

Colyn, M., J. Hulselmans, G. Sonet, et al. 2010. Discovery of a new duiker species (Bovidae: Cephalophinae) from the Dahomey Gap, West Africa. *Zootaxa* 2637: 1–30. http://doi.org/10.11646/zootaxa.2637.1.1

Corlett, R., and R.B. Primack. 2010. *Tropical Rainforests: An Ecological and Biogeographical Comparison* (Malden: Wiley-Blackwell). https://doi.org/10.1002/9781444392296

Costello, M.J., S. Wilson, and B. Houlding. 2012. Predicting total global species richness using rates of species description and estimates of taxonomic effort. *Systematic Biology* 61: 871–83. http://doi.org/10.1093/sysbio/syr080

Davis A.P., T.W. Gole, S. Baena, et al. 2012. The impact of climate change on indigenous Arabica coffee (*Coffea arabica*): Predicting future trends and identifying priorities. *PLoS ONE* 7: e47981. http://doi.org/10.1371/journal.pone.0047981

de Jong, Y.A., and T.M. Butynski. 2010. Three sykes's monkey *Cercopithecus mitis* × vervet monkey *Chlorocebus pygerythrus* hybrids in Kenya. *Primate Conservation* 25: 43–56. https://doi.org/10.1896/052.025.0109

Dijkstra, K.-D.B., J. Kipping, and M. Nicolas. 2015. Sixty new dragonfly and damselfly species from Africa (Odonata). *Odonatologica* 44: 447–678.

Downs, C.T., M. Pfeiffer, and L. Hart. 2014. Fifteen years of annual Cape Parrots (*Poicephalus robustus*) censuses: Current population trends and conservation contributions. *Ostrich* 85: 273–80. https://doi.org/10.2989/00306525.2014.959088

Ebert, D.A., and G.M. Cailliet. 2011. *Pristiophorus nancyae,* a new species of sawshark (Chondrichthyes: Pristiophoridae) from southern Africa. *Bulletin of Marine Science* 87: 501–12. https://doi.org/10.5343/bms.2010.1108

Evans B.J., T.F. Carter, E. Greenbaum, et al. 2015. Genetics, morphology, advertisement calls, and historical records distinguish six new polyploid species of African clawed frog (*Xenopus,* Pipidae) from West and Central Africa. *PLoS ONE* 10: e0142823. https://doi.org/10.1371/journal.pone.0142823

Fennessy, J., T. Bidon, F. Reuss, et al. 2016. Multi-locus analysis reveals four giraffe species instead of one. *Current Biology* 26: 2543–49. https://doi.org/10.1016/j.cub.2016.07.036

Gerlach, J. 2009. *Pachnodus velutinus. The IUCN Red List of Threatened Species* 2009: e.T40091A10304648. http://doi.org/10.2305/IUCN.UK.2009-2.RLTS.T40091A10304648.en

Goodwin, Z.A., D.J. Harris, D. Filer, et al. 2015. Widespread mistaken identity in tropical plant collections. *Current Biology* 25: R1066–R1067. https://doi.org/10.1016/j.cub.2015.10.002

Grobler, J.P., I. Rushworth, J.S. Brink, et al. 2011. Management of hybridization in an endemic species: decision making in the face of imperfect information in the case of the black wildebeest—*Connochaetes gnou*. *European Journal of Wildlife Research* 57: 997–1006. http://doi.org/10.1007/s10344-011-0567-1

Groves, C.P., F.P.D. Cotterill, S. Gippoliti, et al. 2017. Species definitions and conservation: A review and case studies from African mammals. *Conservation Genetics* 18: 1247–`156. https://doi.org/10.1007/s10592-017-0976-0

Hart, J.A., K.M. Detwiler, C.C. Gilbert, et al. 2012. Lesula: A new species of *Cercopithecus* monkey endemic to the Democratic Republic of Congo and implications for conservation of Congo's Central Basin. *PLoS ONE* 7: e44271. http://doi.org/10.1371/journal.pone.0044271

IUCN, 2019. *The IUCN Red List of Threatened Species.* http://www.iucnredlist.org

Jackson, C.R., and M.P. Robertson. 2011. Predicting the potential distribution of an endangered cryptic subterranean mammal from few occurrence records. *Journal for Nature Conservation* 19: 87–94. http://doi.org/10.1016/j.jnc.2010.06.006

Larsen, B.B., E.C. Miller, M.K. Rhodes, et al. 2017. Inordinate fondness multiplied and redistributed: The number of species on Earth and the new pie of life. *Quarterly Review of Biology* 92: 229–65. https://doi.org/10.1086/693564

Maree S., N.C. Bennett, and G.N. Bronner. 2016. A conservation assessment of *Neamblysomus julianae*. In: *The Red List of Mammals of South Africa, Swaziland and Lesotho*, ed. by M.F. Child, et al. (Cape Town and Johannesburg: SANBI and EWT).

Maree S. 2017. Planning for persistence of a Juliana's Golden Mole (*Neamblysomus julliane*) subpopulation threatened by urban development on the Bronberg Ridge of Pretoria (Tshwane), South Africa. *IUCN/SSC Afrotheria Specialist Group Newsletter* 13: 24–33. https://www.afrotheria.net/newsletter.php

Maree, S. 2015. *Neamblysomus julianae. The IUCN Red List of Threatened Species* 2015: e.T1089A21285354. http://doi.org/10.2305/IUCN.UK.2015-2.RLTS.T1089A21285354.en

Marnewick, M.D., E.R. Retief, N.T. Theron, et al. 2015. *Important Bird and Biodiversity Areas of South Africa* (Johannesburg: BirdLife South Africa). https://www.birdlife.org.za/images/IBA/Documents/IBA%20Version%202.pdf

Minhós, T., E. Wallace, M.J.F. da Silva, et al. 2013. DNA identification of primate bushmeat from urban markets in Guinea-Bissau and its implications for conservation. *Biological Conservation 167*: 43–49. https://doi.org/10.1016/j.biocon.2013.07.018

Mondol, S., I. Moltke, J. Hart, et al. 2015. New evidence for hybrid zones of forest and savanna elephants in Central and West Africa. *Molecular Ecology* 24: 6134–47. https://doi.org/10.1111/mec.13472

Musick, J.A. 2000. *Latimeria chalumnae. The IUCN Red List of Threatened Species* 2000: e.T11375A3274618. http://doi.org/10.2305/IUCN.UK.2000.RLTS.T11375A3274618.en

Mynhardt, S., S. Maree, I. Pelser, Net al. 2015. Phylogeography of a morphologically cryptic golden mole assemblage from South-Eastern Africa. *PLoS ONE* 10: e0144995. https://doi.org/10.1371/journal.pone.0144995

Nichols, E., T.A. Gardner, C.A. Peres, et al. 2009. Co-declining mammals and dung beetles: An impending ecological cascade. *Oikos* 118: 481–87. https://doi.org/10.1111/j.1600-0706.2009.17268.x

Parker, J., A.J. Helmstetter, D. Devey, et al. 2017. Field-based species identification of closely-related plants using real-time nanopore sequencing. *Scientific Reports* 7: 8345. https://doi.org/10.1038/s41598-017-08461-5

Pennisi, E. 2016. Pocket DNA sequencers make real-time diagnostics a reality. *Science* 351: 800–01. https://doi.org/10.1126/science.351.6275.800

Plumptre, A.J., and T. Wronski. 2013. *Tragelaphus scriptus* Bushbuck. In: *Mammals of Africa: v. VI: Pigs, Hippopotamuses, Chevrotain, Giraffes, Deer and Bovids*, ed. by J. Kingdon and M. Hoffmann (London: Bloomsbury Publishing).

Primack, R.B. 2012. *A Primer for Conservation Biology* (Sunderland:Sinauer).

Ruggiero, M.A., D.P. Gordon, T.M. Orrell, et al. 2015. A higher-level classification of all living organisms. *PLoS ONE* 10: e0119248. http://doi.org/10.1371/journal.pone.0119248

Salzburger, W., S. Baric, and C. Sturmbauer. 2002. Speciation via introgressive hybridization in East African cichlids? *Molecular Ecology* 11: 619–25. http://doi.org/10.1046/j.0962-1083.2001.01438.x

Shirley, M.H., A.N. Carr, J.H. Nestler, et al. 2018. Systematic revision of the living African slender-snouted crocodiles (Mecistops Gray, 1844). *Zootaxa* 4504: 151–93. http://doi.org/10.11646/zootaxa.4504.2.1

Takahara, T., T. Minamoto, and H. Doi. 2013. Using environmental DNA to estimate the distribution of an invasive fish species in ponds. *PLoS ONE* 8: e56584. https://doi.org/10.1371/journal.pone.0056584

Taylor, W.A., S. Mynhardt, and S. Maree. 2018. Family *Chrysochloridae* (Golden Moles). In: *The Handbook of the Mammals of the World*, v. 8, ed. by. D.E. Wilson, and R.A. Mittermeier (Barcelona: Lynx Ediciones).

Thomsen, P.F., J. Kielgast, L.L. Iversen, et al. 2012. Monitoring endangered freshwater biodiversity using environmental DNA. *Molecular Ecology* 21: 2565–73. https://doi.org/10.1111/j.1365-294X.2011.05418.x

Underwood, E.C., J.H. Viers, K.R. Klausmeyer, et al. 2009. Threats and biodiversity in the Mediterranean biome. *Diversity and Distributions* 15: 188–97. https://doi.org/10.1111/j.1472-4642.2008.00518.x

van Wyk, A.E. 1996. Biodiversity of the Maputaland Centre. In: *The Biodiversity of African Plants*, ed. by L.J.G. van der Maesen et al. (Dordrecht: Kluwer Academic).

van Wyk, A.M., D.L. Dalton, S. Hoban, et al. 2017. Quantitative evaluation of hybridization and the impact on biodiversity conservation. *Ecology and Evolution* 7: 320–30. http://doi.org/10.1002/ece3.2595

vaz Pinto, P., P. Beja, N. Ferrand, et al. 2016. Hybridization following population collapse in a critically endangered antelope. *Scientific Reports* 6: 18788. https://doi.org/10.1038/srep18788

Whittaker, R.H. 1975. *Communities and Ecosystems* (New York: Macmillan).

Willerslev, E., A.J. Hansen, J. Binladen, et al. 2003. Diverse plant and animal genetic records from holocene and pleistocene sediments. *Science* 300: 791–95. https://doi.org/10.1126/science.1084114

Wirminghaus, J.O., C.T. Downs, M.R. Perrin, et al. 2002. Taxonomic relationships of the subspecies of the Cape Parrot *Poicephalus robustus* (Gmelin). *Journal of Natural History* 36: 361–78. https://doi.org/10.1080/00222930010004250

4. Why Should We Protect Biodiversity?

The charismatic baobab (*Adansonia digitata*) standing tall outside Dakar, Senegal. Sometimes called the "Tree of Life" for its enormous value to humans, the baobab is also a keystone species. In addition to providing food for a great number of species, baobabs also provide an important refuge for several bat and bird species that exclusively use these trees for roosting, nesting, and breeding. Photograph by Mattia Menchetti, https://www.inaturalist.org/photos/24373230, CC BY-SA 4.0.

 https://doi.org/10.11647/OBP.0177.04

All of us depend on nature for survival, whether we live off the land, or in a city where we can buy natural resources, transported to us from a distance, at the market. When we do not take care of nature, our quality of life suffers. To illustrate this point, in the book, *Collapse* (2011), prize-winning author Jared Diamond describes how, throughout history, ineffective responses to ailing environments have contributed to human conflicts. In one case study, Diamond examines how overpopulation contributed to Rwanda's collapse into genocide in the early 1990s. Prior to the genocide, Rwanda had one of the highest human population densities in the world, putting enormous strain on its natural resources. Widespread deforestation led to erosion, which in turn contributed to famine, further escalating conflict over what resources remain.

Rwanda—situated in the Albertine Rift Biodiversity Hotspot—is not the only country in which environmental degradation has led to human pain and suffering. Between 1950 and 2000, 80% of the world's armed conflicts occurred within the boundaries of the world's 36 **Global Biodiversity Hotspots** (Hanson et al., 2009)—areas with high levels of biodiversity that also suffer from substantial environmental degradation. Even today, environmental degradation continues to play a major role in fuelling ongoing conflicts, such as those of the Middle East (Gleick, 2014), West Africa's Sahel region (Benjaminsen, 2008), and the Horn of Africa (Markakis, 1995). Preventing these conflicts, which also impact biodiversity negatively (Nackoney et al., 2014; Brito et al., 2018; Daskin et al., 2018), from escalating and new conflicts from developing requires political and societal changes. People in government and local communities must recognize the value of **healthy ecosystems** and become their champions. After all, complex and adaptive ecosystems provide jobs, food, and other resources, thereby contributing to our overall well-being.

But what exactly are we losing when we fail to protect biodiversity? Why should we care if a species goes extinct, or an ecosystem becomes degraded? What evidence do we have that the natural world is our life support system? To better understand the importance of biodiversity for human well-being and quality of life, and the variety of benefits people freely gain from biodiversity, the UN brought together a group of leading scientists to study **nature's contributions to people** (**NCP**, Díaz et al., 2018), more commonly referred to as **ecosystem services**. This group, called the **Intergovernmental Science-Policy Platform on Biodiversity and Ecosystem Services** (**IPBES**), recognises three categories of ecosystem services, namely material contributions, regulation services, and nonmaterial contributions. Note that there are broad overlaps and interdependence among the three categories; consequently, some contributions and services can easily fit under more than one category.

> A healthy environment improves our overall wellbeing by enabling us to live healthy and prosperous lives. In other words, it is our life support system.

4.1 Material Contributions

Nature's material contributions to people, also called provisioning services, commodity values or **direct use values**, represent contributions derived from the direct extraction and physical consumption of natural resources (Figure 4.1). This category is often the most visible and marketed of all ecosystem services. Also, because of their important contribution to the economy, economists are often interested in calculating the values of material contributions and associated services, which they do by monitoring the cost of each product at several points along its life cycle, as well as the behaviours of target groups of people.

Figure 4.1 (Top) A Mandari fisherman from South Sudan carrying smoked fish to the local market. Photograph by Leonard Tedd/DFID, https://www.flickr.com/photos/dfid/8379215187, CC BY-SA 2.0. (Bottom) A lady from Burkina Faso returns to her village with a bundle of firewood for cooking. Photograph by Jose Navarro, CC BY 4.0.

Material contributions can be subdivided into four subcategories. The first subcategory is energy resources, such as firewood and biofuels. The second is food resources, such as drinking water, bushmeat, and edible fruit. The third is materials, companionship, and labour, which include natural products used to make clothes, ornamental resources used for decorations, and animals used for biomedical research, as pets, and for labour. The fourth is medicinal, biochemical, and genetic resources, which include medicinal plants used to cure ailments, psychoactive fungi used in spiritual ceremonies, and genetic stocks used to improve crops.

Many people, especially those in rural areas, obtain many of the material contributions they need for survival from the surrounding environment. These products, which include bushmeat, perfumes from aromatic plants, and firewood, are often assigned to **consumptive use values**. In contrast, material contributions that are sold at commercial markets, whether locally or internationally, are assigned to **productive use values**. Because of material contributions' importance in sustaining people's material assets and health, it is important to ensure that these products are sustainably harvested (Box 4.1).

Box 4.1 Research on Hunting Underpins Conservation in Central Africa

Katharine Abernethy[1,2] *and Lauren M. Coad*[3]

[1]*Biological and Environmental Sciences, Faculty of Natural Sciences, University of Stirling, UK.*

[2]*Institut de Recherches en Ecologie Tropicale, CENAREST, Gros Bouquet, Libreville, Gabon.*

[3]*CIFOR, Jalan CIFOR Situ Gede, Sindang Barang Bogor (Barat) 16115, Indonesia.*

✉ *k.a.abernethy@stir.ac.uk*

A major threat to wildlife in Africa is hunting. Subsistence hunting has been practiced for thousands of years, but new technologies allow hunters to have higher impacts than they had in the past. Improved access routes and vehicles allow hunters to cover more ground and sell to a greater client base, while habitat encroachment from logging or agriculture squeezes wildlife into smaller areas. Growing human populations are pushing the overall demand for wildlife products to a level that the remaining fauna simply cannot support. Yet wild meat is a critically important resource in rural Central Africa, so managing hunting is an important issue for conservation and human welfare (Coad et al., 2010).

Our 20-year research programme looked at hunting in Central Africa to determine how conservation may be most effective. We studied how human communities rely on hunting, impacts of hunting on wildlife and ecosystems, law enforcement challenges, and alternative practices. We found that across Central Africa hunters are in the poorer sections of society and hunt for very similar reasons: food and income. In rural villages, most able-bodied men hunt, but usually < 10% of men make most kills and have disproportionately important impacts on wildlife. These hunters have invested most in equipment and local assets; thus, they have the most to lose and are resistant to regulations or alternatives. The more successful a hunter, the more meat he sells (Coad et al., 2013). Only around 40–60% of hunted meat is consumed directly within the community; smoked or frozen meat can be traded up to 1,000 km away. Even remote villages now trade meat as a commodity to buy supplies such as fuel and medicines.

Under subsistence-driven hunting, studies show that larger-bodied species (> 20 kg) are targeted first. As these decline, smaller species are hunted (Ingram et al., 2015). During this process, the wildlife community changes and, as large predators, browsers, and seed dispersers are lost, ecosystem functioning is compromised (Abernethy et al., 2013).

Commercial hunting often targets illegal trophies, which is only lucrative if hunters have access to clients. These illegal hunters are often recruited directly by the buyer and local people may not necessarily participate, or even benefit at all. If profits are high, hunters can access better weaponry and surveillance than law enforcers, making them difficult and dangerous to apprehend. In the past 20 years, species such as elephants, rhinoceros, lions, and gorillas, have suffered drastic declines that authorities have not been able to combat.

Although wildlife protection laws are generally strong in the region, law enforcement is underfunded and complex. Commercial hunting is regulated but subsistence hunting is allowed, making the identification of illegal hunting difficult as most hunters sell only part of their catch. Alternative livelihood projects have been promoted in the hope of reducing hunting without complex enforcement. However, our review of these projects shows negligible impact, as they have generally been on a small scale and were often unreliable in generating better revenues than hunting (Wicander and Coad, 2015).

Our research shows that the effective regulation of hunting is desperately needed to preserve Central Africa's ecosystems and the sustainability of rural communities. This will require balancing law enforcement and long-term community outreach with policy interventions—such as lobbying—to change laws or awareness campaigns. A conservation practitioner tasked with trying to manage hunting should ask who hunts, why they hunt, where hunting pressure is greatest, and how hunting affects the local ecosystem in order to determine whether they are tackling a subsistence issue or a commercially-driven one, and from there to decide which strategies could be used and who the interventions will affect. This will help to ensure planning for fair, long-term solutions, which have broad local support and the best chance of success.

4.2 Regulating Services

Regulating services maintain nature's ability to provide material contributions, usually in indirect ways. For example, our ability to obtain food indirectly depends on a variety of subtle yet important ecosystem processes (e.g. energy cycling), as well as more observable services (e.g. pollination). For that reason, regulation services are sometimes called **indirect use values**. Regulating services (together with nonmaterial contributions, discussed below) are also sometimes referred to as **non-consumptive use values** because they provide economic benefits without needing to be collected, harvested, consumed, converted, or destroyed during use.

The economic benefits we gain from regulating services are estimated to be larger than all the different kinds of material contributions together, especially in areas where ecosystems are intact (Costanza et al., 2014). Even so, these benefits do not always appear in descriptions of national economies because those statistics generally focus on material contributions. Nonetheless, maintaining regulating services is very important. When damaged ecosystems cannot provide these benefits, substitute resources must be found—often at great expense—to avoid economic collapse. In Section 4.2.4, we discuss one such example, by considering the value and replacement costs of water maintenance services obtained from forests.

The economic benefits we gain from regulating services are estimated to be larger than all the different kinds of material contributions together.

Regulating services can be subdivided into many different subcategories depending on context, each overlapping to varying degrees with one another. Following is a discussion of some prominent subcategories of regulating services.

4.2.1 Maintaining ecosystem stability

Perhaps the most important indirect contribution we gain from biodiversity is its ability to maintain conditions that enable life on Earth to persist. This principle complements the **Gaia hypothesis**, which proposes that all the biological, physical, and chemical properties on Earth interact to form a complex, self-regulating superorganism, and that these interactions maintain the conditions and processes necessary for life to persist (Lovelock, 1988).

There are two complementary theories that explain the importance of maintaining a variety of different species if one is to conserve this superorganism (Ehrlich and Walker, 1998). Originally proposed by American ecologist Paul Ehrlich, the **rivet-popper hypothesis** compares biodiversity to the rivets (some of which may be redundant) that hold an aeroplane together. Just as an aeroplane can only lose so many rivets before it falls apart, so will the progressive loss of species systematically weaken an ecosystem until the entire system collapses. A well-known example of the rivet-popper hypothesis is the **mutualistic relationships** many plants have with all the various pollinators and **seed dispersers** (Section 4.2.5), in this context representing the rivets holding the system together. We might not immediately notice the systematic loss of pollinators we are currently experiencing (Gallai et al., 2008; Dirzo et al, 2014), but eventually these losses will catch up with us, perhaps in the form of food insecurity.

The **species redundancy hypothesis**, proposed by African ecologist Brian Walker, holds that biodiversity and ecosystem stability is best maintained not by focussing on preserving individual species, but by preserving redundancy in ecosystem functioning, by ensuring that each ecosystem is composed of a variety of (seemingly redundant) species performing similar roles. In other words, we should not focus our efforts on protecting just one or two important pollinating species, but a variety of them, to ensure that a variety of plants (and hence entire ecosystems) can also continue to survive. In this way, if one pollinator is lost due to an environmental disturbance or disease, the system will not collapse because other pollinating species will be able to compensate for the loss of that one species.

Keystone species provide such an outsized contribution to ecosystem functioning that their loss will greatly alter ecosystem composition and functioning.

It is important to note that there are some individual species that provide such an outsized contribution to ecosystem functioning that their loss will greatly alter ecosystem composition and functioning. These "pilots" of natural ecosystems are generally known as **keystone species** (Figure 4.2). The keystone species concept was originally proposed after scientists observed that removing sea stars from intertidal zones allowed their **prey** (mussels) to increase uncontrollably which, in turn, pushed species, such as sea urchins and other shellfish, away, leaving an overall poorer ecosystem (Paine, 1969). Apex **predators,** such as lions (*Panthera leo*, VU) and cheetahs (*Acinonyx jubatus*, VU), are also keystone species because of their role in keeping herbivore populations under control. If these apex predators were to disappear, increasing herbivore populations would lead to overgrazing, and ultimately also herbivore declines. This top-down control predators exert on herbivores also answers one of modern ecology's oldest questions: "why is the world green?" (Hairston et al., 1960).

An **ecosystem engineer** is a special type of keystone species that extensively modifies the physical environment, thereby creating and maintaining habitats for other species. Mount-building termites are important ecosystem engineers in many African ecosystems because their activities alter physical, chemical, and biological soil properties (Jouquet et al., 2011), and their massive mounts (some mounts are 10 m high, 20 m across, and may be over 2,000 years old) support distinctive ecological communities and serve as refuges for a large variety of animals and even plants (Loveridge and Moe, 2004; van der Plas et al, 2013). Elephants are also ecosystem engineers; their dramatic foraging habit of pushing over trees provides suitable habitats to countless small animals (Pringle, 2008). Elephants also open up dense vegetation, which allows grasses to thrive, in turn providing food for grazing antelope (Valeix et al., 2011). Holes dug by elephants sometimes make water more accessible,

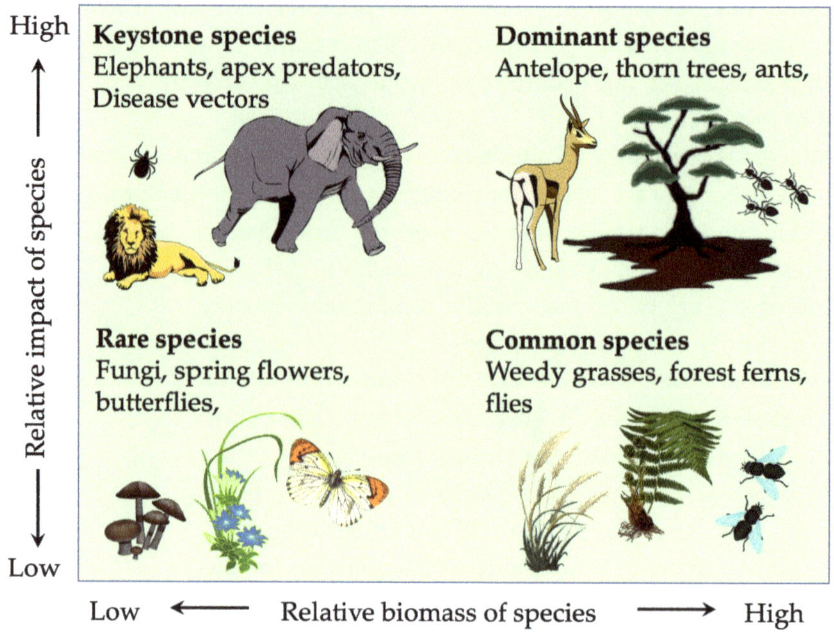

Figure 4.2 Although keystone species constitute only a small proportion of their ecosystem's overall living biomass, they have such disproportionately important roles that their disappearance would lead to drastic environmental changes. This contrasts with rare species that constitute a low proportion of overall biomass and have a minimal impact on their ecosystems' organisation. Like keystone species, dominant species have a significant impact on their environment; however, they also make up a large proportion of an area's living biomass. Common species, in turn, have a relatively minimal impact on their communities despite making up a large proportion of the living biomass. After Power et al., 1996, CC BY 4.0.

while elephant dung provides food for butterflies and dung beetles and creates an important germination environment for seeds and fungi. But too many elephants can also damage ecosystems by reducing the number of large trees on which other species depend (Cumming et al., 1997). It is important to remember that water is an important limiting resource for elephants (Chamaillé-Jammes et al., 2008), so there is a greater risk for elephants to become overly destructive in areas where humans artificially increase aboveground water availability.

> The loss of keystone species from an ecosystem may create an extinction cascade—a series of linked extinction events following one another.

Because so many species depend on ecosystem engineers and other keystone species for survival, their disappearance from an ecosystem can create an **extinction cascade**—a series of linked extinction events following one another. A related phenomenon known as a **trophic cascade** describes the situation where one keystone species' loss has rippling effects at other **trophic levels**. Some of the best-studied trophic cascades involve apex predators and their role in suppressing prey populations (Estes et al., 2011), but disease pathogens can also be a keystone species that leads to trophic cascades. For example, the **introduction** of rinderpest from Asia to Africa in the late 1800s caused catastrophic ungulate population declines in East Africa through the early 1900s. With no primary consumers, grasslands were encroached by woody plants; these changes in the primary producer community also increased the intensity and frequency of wildfires, leading to cascading impacts throughout these savannah communities. An extensive vaccination programme finally saw the disease eradicated in the 1960s, allowing ungulate population and grasslands to recover; and wildfires to become less destructive (Holdo et al., 2009).

4.2.2 Maintaining ecosystem productivity

Plants and algae—in this context known as primary producers—use photosynthesis to capture and store energy from sunlight in their living tissue. This ability of ecosystems to generate living biomass, starting with plants trapping the sun's energy, is known as **ecosystem productivity**. Primary consumers (i.e. herbivores) can then harvest this captured energy by eating plant material. The energy (and nutrition) obtained from plants enable herbivores to generate their own living biomass, in the form of growth and reproduction, before they, themselves, are consumed by secondary consumers (e.g. carnivores, predators, **omnivores**). This cycle ends (or starts, depending on one's perspective) when **decomposers** and **detritivores** (e.g. fungi, earthworms, and millipedes) that break down complex plant and animal tissues into simple compounds such as nitrates, and phosphates. These simple compounds are then released into the soil and water, from where primary producers can take them up again.

4.2.3 Climate regulation

Many of us were taught from a young age that plants are the "lungs of the planet" (Figure 4.3) because they convert carbon dioxide (CO_2) into breathable oxygen (O_2) during photosynthesis. This contribution, whereby plants regulate the atmosphere's CO_2/O_2 balance through carbon absorption and storage (termed **carbon sequestration**) forms part of the atmospheric carbon cycle and plays a major role in regulating global climate patterns. The reduction in plant life through deforestation or other human activities is thus of major concern because of the reduced capacity of plants to sequester atmospheric carbon dioxide, a **greenhouse gas** that contributes to climate change (Chapter 6). The important role of plant communities in the atmospheric carbon cycle is now even being recognised by global markets. For example, the carbon-storing capacity of the Congo Basin's forests has an estimated value at over US $2.5 billion per year (Hughes, 2011). As part of the worldwide effort to reduce carbon dioxide emissions and address climate change, industrial countries and corporations have started paying some landowners to preserve and restore ecosystems that store significant amounts of carbon (Section 10.4).

Plants are also important in regulating regional climate conditions by influencing both the water cycle via transpiration, and local heating and cooling via solar radiation absorption. For example, forests and other vegetation often absorb more

Figure 4.3 A very visual advertisement campaign used emotion and guilt to raise awareness of deforestation. It shows a forest as a pair of lungs, rivers symbolising veins and arteries, and water representing blood. The left lung is healthy, but the right one is partially cut down, symbolic of a cancer, to remind us that ongoing deforestation will increase our own personal discomfort. Image by TBWA\Paris, CC BY 4.0.

heat than bare soil due to their respective **albedos**. Because heat rises, heat absorbed by vegetation enables water vapor released by plants via transpiration to rise higher into the atmosphere, where it subsequently condenses and falls as rain. In contrast, the loss of vegetation is often associated with reduced rainfall (Garcia-Carreras and Parker, 2011), which can in turn reduce agricultural productivity and biodiversity (Lawrence and Vandecar, 2015).

Lastly, trees keep local areas cool by providing shade and releasing water vapor into the atmosphere (Morakinyo et al., 2013; Kardan et al., 2015). This cooling effect increases people's comfort and work efficiency, and reduces the need for fans or air conditioners, leading to higher productivity and cost savings (Balogun et al., 2014; Ogueke et al., 2017). Trees also act as windbreaks, thereby reducing evaporation and erosion in agricultural areas, and reducing the loss of heat from homes and other buildings in cold weather. The value of shade trees is also recognised in agro-ecosystems, as a strategy for coffee and cacao farmers to increase crop yields (Section 14.1.1) and to adapt to increasing temperatures due to climate change (Jaramillo et al., 2011).

4.2.4 Conserving soil and water quality

> Wetlands play a prominent role in flood control. They are also very effective in immobilising pathogens and toxic pollutants released into the environment by human activities.

Wetlands play a prominent role in regulating soil and water quality, as well as flood control. During heavy rains, wetlands slow the speed of rushing floodwater, which lowers flood height and reduces erosion. Wetlands also act as natural sponges: they absorb vast amounts of floodwater during heavy rains, which is then released more slowly and evenly afterwards, thereby maintaining water sources used for drinking, irrigation, hydropower generation, and transport. Wetlands are also very effective in breaking down and immobilising pathogens, toxic pollutants, and excess nutrients released into the environment from agricultural activities, sewage, industrial wastes, and pesticides. One study from South Africa found that wetlands were almost 100% effective in preventing further spread of highly toxic organophosphorus pesticides (Schulz and Peall, 2001).

Wetlands are, however, not the only ecosystem that maintain soil and water quality and quality. In fact, maintaining complex and adaptive ecological communities of all kinds are of vital importance in buffering ecosystems against flooding and drought, protecting fertile soils, and maintaining water quality (see also Section 10.2.1). In intact ecosystems, plant foliage and dead leaves intercept rain, which slows the flow of water from upper reaches of **catchment areas** into streams and rivers; this allows for a slow release of water for days or even weeks after rains have ceased. Soil is anchored in place by plant roots and aerated by soil organisms; this combination increases the soil's capacity to absorb water and hold nutrients. All these aspects together reduce flooding and limit erosion of fertile topsoil which, in turn, limits loss of essential nutrients that would otherwise occur after heavy rains.

The economic benefits of water quality maintenance services provided by intact plant communities are enormous. In the late 1980s, the New York City administration paid US $1.5 billion to local authorities in rural New York State to protect their water supplies by maintaining forests in the catchment area that surrounded the city's reservoirs, and by improving agricultural practices in the catchment area. While US $1.5 billion may seem like a lot of money, at the time it was considered a pittance compared to the US $9 billion that the man-made water filtration systems—doing the same job—would have cost over just the first 10 years in operation (NRC, 2000).

A situation very similar to the one in New York is currently playing out in Kenya. The Mau Forest Complex is one of East Africa's largest montane forests and serves as the principle catchment area for waters that flow into the famed Mara River and Lake Victoria. But large-scale deforestation in the Mau Forest Complex over the past few decades (Figure 4.4) has resulted in reduced water storage, flow regulation, **groundwater** discharge, and water purification, causing annual economic losses of over US $65 billion to Kenya's energy, tourism, and agricultural sectors (UNEP, 2012). The situation in Kenya was so severe that the 2008 inauguration of a hydropower station was postponed due to low water levels; this station later achieved only 50% of its production capacity as a result of deforestation in the Mau Complex. To avoid further losses, the Kenyan government initiated a multi-stakeholder taskforce to investigate options to restore the Mau complex's degraded forests (Prime Minister's Task Force, 2009). Since then, tens of thousands of trees have been planted to reverse deforestation in the area.

4.2.5 Pollination and seed dispersal

Pollination describes the transfer of pollen grains from male parts of a flower to female parts to allow fertilization and production of offspring. Some plants can be pollinated

Figure 4.4 Logging, fire, and agriculture reduced the Mau Forest Complex, Kenya's most important catchment, to a quarter of its original size, in the process damaging the region's hydroelectric, tea, and tourism industries. Restoration plans are currently underway to reverse the destruction through extensive reforestation projects. Photograph by Patrick Shepherd/CIFOR, https://www.flickr.com/photos/cifor/36978973483, CC BY 4.0.

by wind, but others require animals to pollinate their flowers; examples include birds, bats, bees, flies, butterflies, and other insects (Figure 4.5). These pollination services are important for the persistence of many wild plants, as well as for many fruit, seed, and vegetable crops that we utilise as food (Box 4.2). Research from The Gambia has shown that management practices that increase the abundance of bats and bees to contribute to increased yields and sweetness of African locust bean (*Parkia biglobosa*) crops (Lassen et al., 2012). In contrast, work done in Zambia, Mozambique, and Uganda showed that pollinator collapse could increase malnutrition rates by over 50% which, in turn, could increase death rates among children and mothers during childbirth (Ellis et al., 2015). Luckily, many agricultural systems in Africa are still friendly to pollinators (see Box 7.4). Given the dependency on animal-assisted pollination in many agricultural systems, it is critical to maintain or expand pollinator-friendly practices. Our ability to continue benefitting from these services will depend on our ability to maintain and expand on those pollinator-friendly activities.

Many fruit and seed-bearing plants also depend on a process called **seed dispersal** to reproduce, colonise vacant habitats, and avoid competing with parent plants for limiting resources. Seed dispersal describes the physical movement of seeds by fruit-eating and seed-eating birds, large herbivores, primates, and a range of other animals away from the parent plant. Due to specialised features, some seeds can stick to animals' fur, allowing them to be carried along much further distances than wind could, and different directions than water could. Many animals also

Box 4.2 Are Wild Pollinators Important in African Agriculture?

Abraham J. Miller-Rushing

*Acadia National Park, US National Park Service,
Bar Harbor, ME, USA.*

Pollinators and food security are so closely tied to one another they should almost be considered synonymous terms. But when people think of pollination, they often only think of honeybees, which people domesticated more than 8,500 years ago for honey production. However, wild pollinators, which include a variety of insects, birds, and mammals, are often more effective at pollinating than honeybees are. One estimate suggests wild pollinators can double fruit production compared to honeybees (Garibaldi et al., 2013). This is most likely because the morphological and behavioural diversity of wild pollinators allow for more specialised pollination relationships with plants. For example, some wild pollinators have longer proboscis (i.e. insect tongues) that enable them to pollinate deeper flowers (Figure 4.A), something honeybees cannot do. African crops rely even more on wild pollinators than do crops in other areas of the world because it can be difficult to maintain aggressive African honeybee hives and prevent them from being damaged by wild animals (African Pollinators Initiative, 2007).

Figure 4.A With their long proboscis, wild pollinators, such as this white barred hawk moth (*Leucostrophus alterhirundo*) from Mozambique, are highly efficient pollinators. Photograph by Celesta von Chamier, https://www.inaturalist.org/observations/1124702, CC BY 4.0.

Eggplant, papaya, coffee, and palm oil—crops of huge economic and cultural importance—highlight the value of wild pollinators to local and global economies. Eggplants are hermaphroditic; in other words, they can self-pollinate. Even so, pollination from two wild bee species, namely the doubleband carpenter bee (*Xylocopa caffra*) and a type of sweat bee (*Lipotriches rufipes*), increase fruit production far beyond that of self-pollination (Gemmil-Herren and Ochieng, 2008). In contrast, papaya trees are dioecious (i.e. they have separate male and female trees) and thus depend on crosspollination (i.e. pollinators take pollen grains from male flowers on one tree to female flowers on another tree) to produce fruit. While a wide variety of wild bees and butterflies visit papaya flowers, only some hawkmoths and skipper butterflies are effective papaya pollinators, probably because they have long proboscises that can penetrate the deep papaya flowers (African Pollinators Initiative, 2007). A healthy and diverse pollinator community also help coffee plants (which relies on a variety of pollinators, Samnegård et al., 2014) and oil palm (which requires cross pollination by specialist oil palm weevils, African Pollinators Initiative, 2007) produce more fruits, thereby increasing their economic value.

Despite their value to natural ecosystems and food security, wild pollinator populations are declining worldwide (Gallai et al., 2008; Dirzo et al., 2014). To avoid losing them forever, it is important to preserve wild pollinators through the conservation and restoration of native ecosystems (Chapter 10), sustainable agricultural practices, such as the reduced use of pesticides and herbicides (Section 14.1.1), and by communicating the value of pollinators to the general public, land managers, and politicians. Additionally, monitoring and research programmes aimed at pollinators could enhance our understanding of their value, ecology, and conservation.

consume seeds and fruits, providing opportunities for dispersal when the consumer moves off looking for more food, a resting spot, or mates to interact with. For some plants, seed dispersal involves a critical step required for germination, namely **seed scarification**. One method of scarification involves an animal breaking the seed's hard coat by biting it. Alternatively, stomach acids may weaken the consumed seed's hard coat while it passes through the animal's digestive tract. Without this step, seeds requiring scarification may not be able to germinate; those plants' persistence thus depends upon the animals that consume them. While the importance of pollination for food security is well known, the importance of seed dispersal should not be underestimated. A study from Côte d'Ivoire found that primates provided necessary seed dispersal services for at least 25 fruiting plant species important to humans (Koné et al., 2008).

Figure 4.5 (Left) A Cape sugarbird (*Promerops cafer*, LC) feeding on a pincushion (*Leucospermum* sp.), and in the process pollinating the plant. Photograph by Rafael Tosi, https://macaulaylibrary.org/asset/118353841, CC BY 4.0. (Right) A scoliid wasp (Scoliidae) pollinating a creeping foxglove (*Asystasia gangetica*) flower. Photograph by Peter Vos, https://www.inaturalist.org/observations/10965989, CC BY 4.0.

4.2.6 Hazard detection and mitigation

When intact, nature is our first line of defence against many natural disasters. Consider, for example, the contribution of **mangrove swamps** in protecting us from cyclones/hurricanes (van Bochove et al., 2014), or the contribution of wetlands in flood control (Section 5.3.3). In contrast, degrading the natural environment can have severe consequences. For example, a 2010 landslide in Uganda that buried three villages, killing over 300 people and displacing 8,000 more, was attributed to deforestation activities three years earlier (Gorokhovich et al., 2013). To prevent such disasters, and harness all the other contributions of forests, there are numerous projects across Africa working to reverse deforestation (Section 10.3). Unfortunately, Africa's tropical forests regenerate very slowly—sometimes requiring more than 100 years (Bonnell et al., 2011). It is thus critical to prevent ecosystem degradation in the first place, rather than having to resort to costly restoration projects.

In addition to keeping us safe, biodiversity can also be used to help track environmental changes. Species used for this purpose, called **indicator species** or environmental monitors are, by definition, associated with unique environmental conditions or sets of ecosystem processes. Tracking changes in their population sizes, distributions, and behaviour of can thus serve as a substitute for expensive detection equipment (Section 10.1). Aquatic filter feeders, such as mussels and clams are particularly useful in this regard because their tissues accumulate chemical pollutants. A study from Senegal's mangrove swamps detected **heavy metal** pollution using clams, mussels, and snails after tests barely detected those pollutants in the area's sediments (Bodin et al., 2013). But even common everyday species can serve as indicator species: for example, conservation authorities around the world are using bird abundances and behaviours to better understand the impact of climate change (http://climatechange.birdlife.org).

Sentinel species are a special type of indicator species that can act as an early warning system for environmental hazards because they are more sensitive to certain conditions than humans are. Lichens are particularly well-known sentinel species. Being sensitive to air pollution and chemicals in rainwater, some lichens cannot survive in polluted areas. Thus, their presence is generally a sign of good air quality, while their absence may signal air pollution (Bako et al., 2008). Another example is seabirds, whose declining populations can serve as an early-warning system for overfishing (Paiva et al., 2015). Some sentinel species can even be used directly for human health purposes. For example, the non-profit NGO APOPO has been taking advantage of the incredibly fine sense of smell of southern giant pouched rats (*Cricetomys ansorgei* LC)—affectionately called HeroRATs—to detect landmines (Figure 4.6), tuberculosis (Reither et al., 2015), salmonella infections (Mahoney et al., 2014), and even people trapped under collapsed structures (LaLonde et al., 2015).

Figure 4.6 With the help of specially trained southern giant pouched rats, over 80,000 landmines and other unexploded remnants of war were found and destroyed in Mozambique starting in 2008; the country was declared landmine free in 2015. Photograph by APOPO, CC BY 4.0.

Lastly, some species can be used to mitigate various sources of pollution. For example, through a process called **biosorption**, the superior absorption capabilities of some lichens, plants, fungi, and microorganisms offer some of the cheapest and most effective methods for removing toxic heavy metals (Fosso-Kankeu and Mulaba-Bafubiandi, 2014) from the environment. Scientists also recently discovered a plastic-eating fungus (Khan et al., 2017) that may provide a potential solution to plastic pollution.

4.2.7 Pest and disease control

Every day, predators, such as owls and bats, keep us healthy by controlling populations of disease vectors, such as rats and mosquitoes. This process, where predatory (and

parasitic) species regulate populations of pests and other nuisance species, is known as **biological control**, or biocontrol in short (Box 4.3). The use of insectivores (i.e. insect-eating species), such as bats and birds, to control crop pests is well established in traditional farming systems (Abate et al., 2000). But even on commercial crop farms, natural enemies, such as bats and birds, play an important role in keeping pests under control (Taylor et al., 2018). Some plants also play a part in biocontrol efforts: recent research found that some native plants used for intercropping in traditional agricultural systems emit chemical signals that kill and drive pest species away from crops (Khan et al., 2010). With an increasing number of studies illustrating the significant benefits gained from natural pest control systems, enhanced farming practices that facilitate greater ecosystem complexity (Section 14.1.1) will hopefully play a bigger role in food security in future.

Box 4.3 Biological Control Saves the Cassava Crop

Meg Boeni and Richard Primack

Biology Department,
Boston University,
Boston, MA, USA.

As it stands along the farm-plot boundary,
its base appears beautiful like a bride's feet...
O cassava to whom the bembe drum beats a salute
that never reaches an end...
It is no small service the cassava renders us in this our land

Yoruba Poem
(Babalola, 1966)

This traditional song from Nigeria praises the cassava, a South American crop brought to tropical Africa in the 16th century, and upon which millions of Africans have since relied for food and income.

Disaster struck in the 1970s, when an agricultural scientist that brought a new variety of cassava from South America to Africa also accidentally introduced a new pest: the cassava mealybug (*Phenacoccus manihoti*) (Neuenschwander, 2001). Previously unknown to science, the bug attacked the new shoots of cassava plants, laying its eggs at their tips and stripping them of their leaves. As it spread through Central and West Africa, the mealybug wiped out 80–90% of the productivity of most cassava fields, threatening large parts of Africa with famine.

With so many Africans relying on the cassava as a primary food source, scientists had to find a solution, and quickly. The bug's waxy coating that protected it from pesticides complicated this effort. With conventional pest-control methods failing, scientists turned to biological control, hoping that introducing a natural predator would counteract the spread of the invasive insect. Researchers searching for the source of the mealybug finally found a candidate in the fields of Paraguay, where cassava, known locally as mandioca, was also an important food staple. Here, investigators discovered that mealybug numbers were kept low by a tiny wasp called *Anagyrus lopezi* that attack the mealybugs' eggs and larvae (Figure 4.B). *A. lopezi* passed laboratory tests for host specificity—it fed and bred exclusively on cassava mealybugs and would not attack other African insects. And so, the International Institute of Tropical Agriculture (IITA) began field tests using the wasp as a biological control agent.

Figure 4.B (Top) A vial containing the parasitic wasp *Anagyrus lopezi* at a biocontrol release site. Photograph by Rod Lefroy/CIAT, https://www.flickr.com/photos/ciat/4809242082, CC BY 2.0. (Bottom) A cassava farmer from Tanzania smiles broadly, very happy with her crop. Photograph by Holly Holmes/CGIAR RTB, https://www.flickr.com/photos/129099219@N03/33324350781, CC BY 2.0.

Results were astounding; the quick-spreading Paraguayan wasp reduced crop losses by an impressive 95% (Neuenschwander, 2001), all without the danger of pollution and poisoning associated with traditional pesticides. While identifying and introducing the biocontrol agent required significant resources, estimates suggest gains of 370–740 times the original investment, depending on the region considered (Zeddies et al., 2009), making it well worth the cost. Today, *A. lopezi* is found everywhere where the cassava mealybug survives in Africa. Bolstered by this success, the IITA has subsequently expanded its biological control programmes to fight tropical pests on crops, such as cowpeas, maize, and bananas.

In 2008, the cassava mealybug was discovered in Southeast Asia, where it repeated the damage inflicted in Africa (Graziosi et al., 2016). Scientists are now replicating Africa's biocontrol efforts to reduce crop failure in Vietnam, Thailand, Cambodia, and China. In conjunction with a number of local parasites, they hope that *A. lopezi* will halt the spread of the mealybug before it reaches even larger fields in India (Parsa et al., 2012). The control of the cassava mealybug is certainly one case where biological control was able to achieve great success.

Most Africans are familiar with **scavengers,** such as jackals and vultures, that work as nature's clean-up crew, picking at food scraps left in the field by large predators. Together with the range of flesh-eating insects, detritivores, and decomposers, scavengers play a crucial role in keeping us healthy by sanitising the environment (O'Bryan et al., 2018). While it is all too easy to take these activities for granted, some people actively welcome these services. For example, in northern Ethiopia, spotted hyenas (*Crocuta crocuta*, LC) are tolerated in urban settlements because they consume livestock carcasses and sometimes even human corpses, which pose a disease risk (Yirga et al., 2015).

Recent experiences have shown that without proper care, the sanitary services provided by wildlife can collapse over a very short time. For example, during what is known as the Asian vulture crisis of the 1990s, vulture populations in India, Pakistan, and Nepal declined precipitously in a matter of years from **secondary poisoning** after eating carcasses of dead animals treated with the anti-inflammatory drug diclofenac. With nothing else available to remove carcasses of dead animals as efficiently as vultures, rotting flesh contaminated drinking water and allowed populations of rats and **feral** dogs (*Canis familiaris*) to proliferate. While vultures have stomach acids which kill pathogens, dogs and rats do not and thus became major pathogen vectors, spreading deadly diseases such as rabies, anthrax, and plague. The estimated healthcare costs in the face of Asia's vulture crisis amounted to over US $1 billion per year (Markandya et al.,

Scavengers such as vultures and jackals are nature's clean-up crew; they keep us healthy by sanitising the environment.

2008). Today, Africa is facing its own vulture crisis. But instead of one threat, Africa's vultures face a multitude of human-made threats, making solving this crisis much more complex (Box 4.4).

Box 4.4 Conservation Lessons from the Asian and African Vulture Crises

Ara Monadjem

Department of Biological Sciences, University of Eswatini, Kwaluseni, Eswatini.

✉ *ara@uniswa.sz*

A common perception among laypeople and conservationists alike is the idea of safety in numbers for wildlife species. After all, is a widely distributed and abundant species not safe from the threats of extinction? The answer is a firm no! As the collapse of central Asia's vulture populations (Oaks et al., 2004) demonstrates, species numbering in the millions can disappear in the space of just a few years.

The Asian vulture crisis shares some similarities with the demise of the passenger pigeon (*Ectopistes migratorius*, EX) in North America a century ago. This pigeon was once the most abundant bird on Earth; yet, despite numbering in the billions, it was driven to extinction in a short span of time, primarily due to hunting over a 20-year period in the late 1800s. In the case of Asian vultures, the threat was not hunting, but rather a nonsteroidal anti-inflammatory drug (NSAID)—diclofenac—which is fed to sick cattle and then ingested by vultures when they feed on dead livestock. As diclofenac is deadly toxic to vultures, the widespread use of this treatment on the Indian subcontinent (which includes India, Nepal, and Pakistan) has seen catastrophic vulture population declines. With one of Asia's major natural trash disposal systems gone, the area experienced a human health crisis from widespread drinking water contamination and increased incidence of diseases carried by ubiquitous and increasing rat and feral dog populations (Markandya et al., 2008).

The Asian vulture crisis is instructive on several grounds. First, it took a long time to detect and confirm the vulture declines because regular and standardised monitoring of the three affected vulture species had not been conducted. Second, the extent of the decline was extreme, with vulture numbers declining by over 95% within a decade. Third, the declines were due to a single threat—contamination from diclofenac, which were subsequently found to be deadly-poisonous to vultures (Oaks et al., 2004). Thanks to the concerted efforts of conservationists and politicians, and the rapid reactions of the governments of India, Pakistan, and Nepal, diclofenac was removed from the market in 2006. Vulture populations in Asia have since stabilised, with even a cautious suggestion of an increase.

Now, Africa faces its own vulture crisis (Ogada et al., 2015). However, in contrast to the Asian crisis, Africa's crisis involves a greater number of species, and spans a larger geographical area. Importantly, it also includes a greater number of threats, including poisoning, harvesting for traditional medicine and for food, and electrocution following contact with power lines. Many vultures also die when they scavenge on poisoned carcasses meant to kill problem predators (Figure 4.C). To this list of lethal causes, one should also add the universal threats of habitat loss and persecution of birds of prey.

Figure 4.C Conservation biologists inspect several white-backed vultures that were poisoned at South Africa's Kruger National Park. Poachers across Africa intentionally kill vultures for traditional medicine, and unintentionally kill them while setting traps for large predators. Photograph by Andre Botha, CC BY 4.0.

Thanks to long-term monitoring, the collapse of African vulture populations has been well documented. Of the 95 vulture populations being monitored,

89% were either extirpated or experienced severe declines. Across eight study species, the mean rate of decline is estimated at 4.6% per year (i.e. one out of 20 birds that are dying per year are not being replaced). The charismatic Rüppell's vulture (*Gyps rueppellii*, CR) has declined by 85% across its range; consequently, this species is now considered highly threatened by the IUCN, as are the hooded vulture (*Necrosyrtes monachus*, CR), white-headed vulture (*Trigonoceps occipitalis*, CR), and African white-backed vulture (*Gyps africanus*, CR). Only slightly better off, at least for now, are the lappet-faced vulture (*Torgos tracheliotos*, EN) and Egyptian vulture (*Neophron percnopterus*, EN), the Cape vulture (*Gyps coprotheres*, VU), and the bearded vulture (*Gypaetus barbatus*, NT).

The collapse of Africa's vulture populations is cause for serious concern among conservation biologists, wildlife and livestock managers, and public health officials. Unlike in Asia, however, workable solutions to Africa's vulture crisis have not yet been found. This may be due to the multitude of threats, and the complexity of the problem exacerbated by the involvement of individual poachers, local communities, and government structures across more than 40 countries. If conservationists and governments can work together, as they did in Asia, then perhaps Africa's vultures and the ecosystem services that they provide can still be saved.

4.3 Nonmaterial Contributions

Nonmaterial contributions from nature, also called cultural services, include the subjective and psychological aspects of nature that influence our perceptions about quality of life. These contributions can be divided into three subcategories: inspiration and learning support, supporting psychological and physical experiences, and supporting individual and group identities.

4.3.1 Inspiration and learning support

Nature has inspired artists and writers throughout history. Consequently, many books, television programmes, movies, and websites produced for entertainment purposes are based on natural themes. This infusion of nature into popular culture is worth billions of dollars per year. To take one example, the 1994 Disney blockbuster *The Lion King*, based on the lives of a variety of African savannah animals, generated revenues estimated at just under US $1 billion from theatre attendances alone. It was so successful that three movie sequels, an animated television series, and several video games and books followed. A musical based on *The Lion King* movie plot continues to be a top-earning title in box-office history for both stage productions and films.

Movies featuring stunning natural landscapes and charismatic wildlife often increase the desire of moviegoers to visit natural areas where they can see these landscapes and animals first-hand. But it can also raise awareness of environmental issues in new audiences. While many documentaries are created with this purpose in mind, such benefits can also extend to blockbuster movies meant for broader audiences (Silk et al., 2018). For example, Disney's *Happy Feet* (2006) highlighted the threat of overfishing and plastic pollution to penguins; *Avatar* (2009) raised awareness of **habitat loss** and overharvesting; and *The Jungle Book* (2016) exposed audiences to the plight of pangolins. Such exposure can even lead to environmentally conscious behavioural changes. For example, moviegoers were willing to donate 50% more money to climate mitigation after watching the apocalyptic movie *The Day After Tomorrow* (2004) (Balmford et al., 2004). Perhaps, in part, due to the influence of environmentally-orientated movies, an increasing number of movie stars (and other celebrities) have started using their stardom as a platform from where they promote biodiversity conservation efforts in Africa (Duthie et al. 2017; see also https://wildfor.life).

Scientists and engineers also sometimes turn to nature to seek inspiration for new technologies or to solve innovation challenges. For example, the water-vapor collecting capacity of the racing stripe darkling beetle (*Stenocara gracilipes*) from the Namib Desert in Namibia (Figure 4.7) inspired engineers who developed self-filling water bottles (Clark, 2012), irrigation systems to overcome drought conditions (Scott, 2011), fog-free windows and mirrors (Parker and Lawrence, 2011), and methods for controlling condensation and frost on aircraft surfaces (Boreyko et al., 2016). While these and other scientific endeavours, collectively known as **biomimicry**, provide many social and economic benefits, their primary value comes from new knowledge, improved education, and enriched human experiences.

Scientists and engineers often seek inspiration from nature for new technologies or to solve innovation challenges.

Figure 4.7 The racing stripe darkling beetle is endemic to one of the world's most arid regions, Namibia's Namib Desert. To survive, it collects water from early-morning fog with the bumps on its back. In a classic case of biomimicry, creative entrepreneurs are copying these features to create self-filling water bottles and fog-free windows. Photograph by Alex Rebelo, https://www.inaturalist.org/observations/11086737, CC BY 4.0.

4.3.2 Supporting psychological and physical experiences

While the economic benefits gained from nature incentivises biodiversity protection, many people believe that the aesthetic values of nature provide an even greater incentive for conservation. This principle rests on the fact that nearly everyone enjoys wildlife and landscapes aesthetically. Even city-dwellers who are superficially removed from nature find a sense of relief and well-being when they have opportunities to come in close contact with the natural world. But what if dragonflies and butterflies disappeared? What if our favourite sports team's mascot ceased to exist in nature? What if there were no more forests filled with bird flocks or monkey troops?

The intangible but desirable aesthetic values people attach to certain aspects of nature are known as **amenity values**. Amenity values are becoming increasingly important in many local and national economies throughout Africa, in the form **ecotourism**. At any one time, there are millions of tourists traveling and spending money across Africa to see particular species or to experience unique ecosystems. This includes scuba divers approaching a coral reef (Figure 4.8), birdwatchers visiting a rare species' stakeout, and people on a safari to view the many **flagship species** for which Africa's savannahs are so well known. Ecotourism has long been a major industry in southern and East Africa. For example, ecotourism generated over US $1 billion in annual revenue in the Cape Floristic Region more than a decade ago (Turpie et al., 2003), and has accounted for over 15% of Kenya's **gross domestic product (GDP)** at times (WWF and BSI, 2006). Ecotourism is also becoming increasingly important in other parts of Africa. For example, since overcoming periods of social unrest, Burundi, Rwanda, and Uganda have created profitable local industries charging tourists high fees to visit **habituated** populations of mountain gorillas (*Gorilla beringei beringei*, EN). Also, in South Africa, some bird guides earn an average of US $362 per month by showing tourists the unique birds their local area has to offer (Biggs et al., 2011).

Figure 4.8 Scuba divers on vacation at Ponta do Ouro, Mozambique, appreciating a large potato bass (*Epinephelus tukula*, LC). The income to be gained from ecotourism activities often outweighs the profits from unsustainable harvesting, and thus provides a strong economic justification for biodiversity conservation. Photograph by Derek Keats, https://www.flickr.com/photos/dkeats/36684179721, CC BY 2.0.

In recent years, volunteer-based ecotourism has emerged as a lucrative industry that combines ecotourism with learning opportunities. These organisations offer aspiring conservationists and citizen scientists hands-on experience while bringing financial and other logistical support to rural and protected areas. Many wildlife sanctuaries and conservation NGOs also offer volunteer opportunities and field courses that combine conservation action with local community outreach and education programmes. The research done by professional scientists and citizen science volunteers can be used in locally-relevant educational materials. Biological field stations (Section 13.1.5) often host these activities; the stations can also provide training and jobs for local community members.

The revenue and jobs generated by ecotourism provides a strong and immediate justification to protect areas rich with biodiversity or to restore areas that have been degraded. Ecotourism can even be integrated directly in plans for future development, protection, and restoration. One such example is integrated conservation and development projects (ICDPs, Section 14.3), which provide models for how empowered rural communities can successfully establish accommodation, develop expertise in nature guiding, and sell local handicrafts at curio stores to obtain multiple stable income streams. The revenue obtained from ecotourism also allows local people to move away from unsustainable hunting, fishing, or grazing practices towards lifestyles that can be maintained in the long term.

> The revenue and jobs generated by ecotourism provides a strong and immediate justification to protect biodiversity and restore areas that have been degraded.

Still, many of Africa's ecotourism resources remain under-utilised. To use one example, only a few locations in Africa cater to people who enjoy the thrill of swimming with sharks in their natural habitat. Beyond removing fear and instilling a healthy respect for sharks, this industry also plays an important role in conservation by showing how living sharks bring greater economic benefit than a once-off catch. For example, shark diving at just one location in South Africa is estimated at US $4.4 million annually (Hara et al., 2003); similar industries in the Maldives (Cagua et al., 2014) and Palau (Vianna et al., 2012) generate even more revenue. Presenting unique recreational experiences and a growing global ecotourism sector, more and more African countries will hopefully explore these and other opportunities soon. It is worth noting that the long-term effects of shark diving operations are largely unknown, particularly as it relates to possible behavioural changes from using bait to attract sharks to people, and an active area of current research (Gallagher and Huveneers, 2018).

Although ecotourism can provide many valuable conservation and economic benefits (Thiel et al., 2014), care must always be taken that these activities abide by accepted ethical standards (Hayward et al., 2012). It is also worth remembering that wildlife ecotourism is often geared towards wealthy western markets, making it prohibitively costly to the people who live near facilities, and are most vulnerable to factors such as **human-wildlife conflict**. As such, it is important to consider what portion of the generated funds are invested locally versus reserved to enrich well-compensated shareholders in the far-away capital. Are local people given opportunities to further their training and education, and to advance their careers within ecotourism organisations? Unfortunately, in many areas of Africa, local people continue to receive only the smallest percentage of money spent by tourists. Similarly, even though national parks themselves may receive large numbers of foreign visitors, governments continue to use only a small percentage of the generated funds on park management (Lindsey et al., 2014; Balmford et al., 2015).

4.3.3 Supporting individual and group identities

Many people care deeply about biodiversity. The thought of a charismatic animal or a special landscape (Figure 4.9) may elicit a strong emotional response, which leads to a desire to protect plants, animals, and natural places. For some people, this desire is associated with a hope to someday see those unique species or landscapes in person. Others do not expect or even desire to see these species and landscape themselves, yet they value their existence. In either case, these individuals recognise the **existence values** of wildlife and nature—the benefit people receive from simply knowing that an ecosystem or species exists. **Bequest values** (also known as beneficiary values) is a component of existence values, defined as the perceived benefit people receive from preserving a natural resource or species for future generations.

Figure 4.9 Each year, after the first spring rains, South Africa's Namaqua National Park comes alive with a rich tapestry of colour, attracting wildflower enthusiasts from all over the world to this otherwise barren semi-desert landscape. Photograph by LBM1948, https://commons.wikimedia.org/wiki/File:Sur%C3%A1frica,_Namaqualand_02.jpg, CC BY-SA 4.0.

The desire to ensure the protection of biodiversity has prompted a wide range of people to establish, join, or otherwise contribute to conservation organisations. For many people involved in these organisations, their participation stems from the ethical premise that wildlife are equal to human life, and that biodiversity conservation offers genuine and long-lasting well-being. This environmental philosophy is often described as **deep ecology**, the ethical premise that species and biodiversity have a right to exist independent of their possible benefits to humans, and that humans have an inherent responsibility to protect species and biodiversity (see also Section 1.4). Deep ecology holds that social structures (including politics, economics, technology, and ideology) must change radically to reduce the destruction of Earth's biodiversity and to enhance people's quality of life. It emphasises and prioritises the natural environment, aesthetics, religion, and culture, rather than material consumption. Although the ethical appreciation of biodiversity is similar in deep ecology and conservation biology, deep ecology includes broader goals for personal, social, and political change.

Biodiversity also forms the basis of spiritual, celebratory, and other social-cohesion experiences for many people. It ensures people experience a sense of place and belonging, reminds them of childhood experiences, and gives a sense of connection when they experience natural sights, sounds and smells. This is especially true for Africans, many of whom attach deeply-held spiritual, cultural, and symbolic values to the environment. Even the money of most, perhaps all, African countries features aspects of nature, as if to add a little extra (if only symbolic) value to those coins and bills. All these factors play a major role in people's sense of who we are—our identity.

> Biodiversity forms the basis of spiritual, celebratory, and other social-cohesion experiences for many people.

4.4 The Long-Term View: Option Values

The **option values** of biodiversity describe nature's potential to provide currently unknown or unrealised benefits at some point in the future. For example, while many species may not currently have any realised material contributions, a small number of taxa may have enormous potential to support new industries or prevent major agricultural crops from collapsing. For this reason, scientists continuously search for species with hidden uses: entomologists search for insects that can control pest species, microbiologists search for bacteria useful in biochemical manufacturing, and agricultural scientists search for genetic varieties of plants that can produce more food to feed a growing human population. As fears of antibiotic resistance become reality, archaea (widespread single-celled microorganisms with no nucleus which are also thought to be the oldest life forms on Earth) may be used to develop new classes of antibiotic medicine (Metcalf et al., 2014). Some researchers also hope that studying primates—the likely original source of diseases such as HIV/AIDS, Ebola, and malaria

(Martin et al., 2005)—may allow us to one day find cures for these diseases. It is worth noting that the effectiveness of using animals to study human diseases remains highly controversial (Archibald and Clotworthy, 2007; Festing and Wilkinson, 2007; Rollin, 2007), and that many people believe that the suffering and death of animals during biomedical research is unethical.

This continuous search for valuable or useful natural products, called **bioprospecting**, has already contributed a great amount to global **economic development**, and is expected to become even more important in the coming decades. This is particularly true in the rush to find replacements for climate sensitive crops that may be threatened by climate change. For example, researchers hope that the genetic diversity in wild coffee populations can act as an insurance policy in case our warming planet damages currently popular commercial strains (Davis et al., 2012). There is also much hope that plants from Africa will lead to new medical treatments, for diseases such as malaria, cancer, and high blood pressure (Gurib-Fakim, 2017). It is for reasons such as these that losing even small portions of expansive ecosystems concerns scientists. The extinction of even one valuable species or gene can represent a tremendous loss to humanity, even if many other species are preserved.

> Bioprospecting, the search for valuable or useful natural products, has already contributed greatly to global economic development.

4.5 Environmental Economics

It should be clear from reading this chapter that the well-being of people around the world is fundamentally linked to opportunities for biodiversity to survive and prosper. That means that when we destroy an ecosystem or let a species go extinct, we also put at risk our own ability to survive and prosper. To fully account for these risks, decisions that negatively affect biodiversity must account for *all* the costs and benefits (hidden or otherwise), including the impacts on ecosystem services, before the decision is implemented.

One of the most popular methods for accounting for potential harm to biodiversity, especially when weighing public policy and commercial decisions, is to attach market (or monetary) values to the ecosystem services. For some ecosystem services, it is rather straightforward to estimate a market value. For example, how much would it cost to replace a natural pollination service with hand pollination by farm workers? But for many services, estimating a monetary value is much more difficult. For example, how do we calculate the value of the Congo Basin's carbon stocks? Where do we even start to estimate the value of breathing clean air, or knowing that dolphins exists?

To examine these kinds of complex questions, conservation biologists look to a sub-discipline within economics called **environmental economics**. Environmental economics broadly examines the contribution of ecosystem services to global economies. An important component of this examination involves estimating the market value of all the different ecosystem services we benefit from, but it also includes studying the environmental costs of economic transactions, environmental policies, and other decisions that impact the environment.

Environmental economics strengthen arguments for biodiversity protection by examining the contribution of ecosystem services to global economies.

4.5.1 Placing a price on the natural world

Approximating the market values of ecosystem services is no small feat, in part because nature's contributions to people vary by location and perspective (Díaz et al., 2018). There also continue to be technical (e.g. Kling et al., 2012) and ethical (e.g. McCauley, 2006; Silvertown, 2015) disagreements about the need and methods used to translate nature's services into monetary terms. Nevertheless, including such estimates has become a widely accepted norm in economics models and conservation activities (Guerry et al., 2015). To accomplish this task, economists rely on seven main methods to estimate the market values of ecosystem services (Farber et al., 2002):

- *Market value*: The price a person is willing to pay for a specific product or service. For example, how much is a person willing to pay for a bundle of firewood at a local market?
- *Avoidance cost*: The cost society avoids paying because a specific ecosystem service exists. For example, how much does society avoid paying for water filtration service otherwise provided by a region's forests and wetlands?
- *Replacement cost*: The cost society would have incurred if a specific ecosystem service had to be replaced. For example, how much would society have to pay in extra healthcare costs and in clean-up costs for diseased carcasses to replace the sanitation services provided by vultures?
- *Factor income*: The additional income generated by the enhancement of an ecosystem service. For example, how much would a reduction in water pollution increase the income of fishermen through healthier fish populations?
- *Travel cost*: The additional travel cost a person is willing to pay to experience an ecosystem service otherwise not available to them. For example, how much extra is a person willing to pay for transport to participate in recreational activities at a cleaner lake?
- *Hedonic pricing*: The additional expense a person is willing to pay to experience an ecosystem service. For example, how much extra is someone willing to pay for a house with an ocean view, compared to an inland house?
- *Contingent valuation*: The additional expenses a person is willing to pay for an alternative hypothetical scenario. For example, how much is someone willing to pay for cleaner air, or the right to catch more fish?

Using a combination of these methods, a range of ecosystem services have been valued in recent years. For example, the services offered by pollinating insects around the world have been valued at US $153 billion per year (Gallai et al., 2008). In just South Africa's Western Cape Province, free pollination services provided by wild insects to the local fruit industry, valued at US $500 million, has been estimated at nearly US $360 million per year (Allsopp et al., 2008). The replacement cost of tropical forests is also increasingly appreciated in carbon sequestration markets, where heavy greenhouse gas emitters pay huge sums of money to conserve forests to become more **carbon neutral** (Section 10.4). For example, the **United Nations Environmental Programme (UNEP)** has estimated that their forests are worth 4.2 times more intact than the value that could be earned through logging; the value of just Kenya's remaining Mau forest, if left intact, is estimated at US $1.3 billion *per year* (UNEP, 2012). One ambitious study estimated the value of all of Earth's ecosystem services at US $145 trillion annually (Costanza et al., 2014), which is almost double the current US $78 trillion value of the global economy. By comparing the value of ecosystem services over time, Costanza et al. (2014) also estimated that we are losing US $4.3–20.2 trillion per year in ecosystem services through land degradation.

> The combined value of all of Earth's ecosystem services may be double the current value of the global economy.

4.5.2 Environmental economics' biggest contributions

Since its development, environmental economics has contributed to conservation biology in several very important ways. Perhaps the most important contribution is that it has enabled conservation biologists to better communicate the value of ecosystem goods and services to audiences like government officials and business leaders, who often base decisions on economic considerations. By doing this, environmental economics has also focussed our attention on the wide range of goods and services that biodiversity provides and has elevated these topics into corridors where they were not previously discussed. These efforts have already paid dividends; in 2012, several Africa countries signed the *Gaborone Declaration*, a pledge to integrate the value of ecosystem services into their economies.

Environmental economics also enabled conservation biologists to better account for environmental impacts of environmental damaging activities. In doing so, the field highlighted how activities that appear profitable are running at an economic loss when properly accounting for otherwise ignored environmental (and social) damages. While such calculations have traditionally focussed on imbalances in overharvesting of material contributions (see negative **externalities**, Section 4.5.3), recent developments have also started accounting for damages inflicted on regulating services and nonmaterial

> Environmental economics enable us to better communicate the value of biodiversity to those who base decisions on economic considerations.

contributions, such as the loss of nature's contribution to climate regulation (Auffhammer et al., 2017; Hsiang et al., 2017).

4.5.3 Environmental economics' biggest challenges

Despite all the direct and indirect contributions of environmental economics to biodiversity conservation, there are also several challenges facing the field. Some of these challenges relate to methodological complexities of valuing ecosystem services, but many challenges also have their roots in governance failures. Following is a discussion of the most important challenges facing environmental economics.

Accounting for negative externalities

Modern economics is built on the principle of **voluntary transactions**—that is, a transaction occurs only when it benefits all the stakeholders involved. However, environmental (and social) harm often arises when some hidden costs are passed on to people not directly involved in the transactions. The unregulated use of **open-access resources**—resources such as water, air, and fish populations that are freely used by many different groups of people—provide many opportunities for this kind of abuse. Consider a company that dumps chemical waste into a river instead of properly disposing of it. While the company may benefit from this cost-cutting measure, people further downstream bear the environmental and social costs of the company's "free" waste disposal by having to contend with polluted drinking water, loss of swimming and other recreational opportunities, and loss of fish as a safe food source. Damage inflicted on rivers and other open-access resources also represent a classic example of the **tragedy of the commons**—while some people initially benefit from abusing the "free" ecosystem services, those values are gradually lost to all of society, including those who abused it (NRC, 2002).

The hidden costs of economic transactions that are passed on to people not directly involved are generally known as negative **externalities** (Figure 4.10). Because negative externalities allow a small number of people to benefit at the expense of the rest of society, they often lead to **market failures**, characterised by transactions that do not lead to optimal outcomes for all stakeholders. Governments may correct for these kinds of market failures by imposing taxes on activities that are harmful to the environment. Carbon taxes imposed on greenhouse gas emitters (see climate change, Chapter 6) is a common example. But many times, governance structures fail, or even exacerbate, the impact of negative externalities, by artificially maintaining destructive activities with tax incentives, direct payments, and price regulations. For example, subsidies give foreign fishing fleets operating off Africa a competitive advantage over local fisherman and artificially inflate their profitability despite declining fish populations (Brashares et

> Because negative externalities allow a small number of people to benefit at the expense of the rest of society, they often lead to market failures.

al., 2004; Sumaila and Pauly, 2006; Mallory, 2013). The financial incentives governments provide to maintain destructive activities are more often referred to as **perverse subsidies** (Myers and Kent, 2001). The size of perverse subsidies is often very large, regularly dwarfing conservation spending. For example, US $26 billion in subsidies were provided to the Africa's **fossil fuel** industry just in 2015 (Whitley and van der Burg, 2015), compared to just US $381 million spent annually to secure Africa's protected areas with lions (Lindsey et al., 2018).

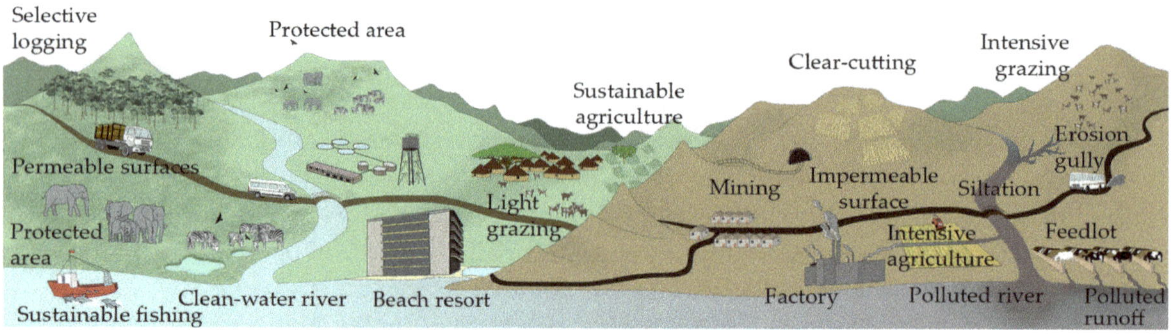

Figure 4.10 Politicians, developers, and industries all too often fail to account for negative externalities (right side of figure) when they consider the contribution of destructive economic activities to society. Accounting for these negative externalities—and redistributing perverse subsidies to activities that provide public benefits (left side of figure)—will help us transition to more sustainable lifestyles. CC BY 4.0.

There are many reasons why governance structures continue to fail nature and allow market failures to occur. For example, due to the prevailing mindset of pursuing economic growth at all costs, politicians, developers, and industries often skew their cost-benefit analyses by prioritising the short-term benefits gained from destructive sectors over long-term societal well-being and sustainability. Another factor is intense lobbying by industries benefitting from perverse subsidies, which leads to corruption and other questionable decisions. Solving these challenges will rely on a society that prioritises economic development (Section 15.1) and establishes structures (i.e. passing and enforcing environmental laws, Chapter 12) that fully account for negative externalities.

Determining ownership

Another problem that plagues environmental economists and other stakeholders is deciding who owns the commercial rights to biodiversity. Imagine a biochemist from a wealthy country traveling in a rural part of West Africa. The biochemist falls sick, but luckily local villagers help the chemist get better with the aid of a traditional healing plant. Once back home, the biochemist scientifically demonstrates that this plant can be used to synthesise a new effective medicine. Do the profits from this new medicine belong to the biochemist, the organisation that sponsored his/her trip, or the local people in the area who helped the biochemist?

In the past, corporations and scientists (generally from wealthier countries) travelled extensively (often to poorer countries in the tropics) to collect species from which commercially valuable products might be obtained. These new products were then sold, but all profits were kept by the corporations while the people in the poorer source countries received little to no financial benefit. One such example is the production of palm oil, of which Malaysia and Indonesia currently contribute 85% of the global vegetable oil supply. This industry is entirely dependent on the oil palm (*Elaeis guineensis*, LC), and its specialist pollinator, the oil palm weevil (*Elaeidobius kamerunicus*), both imported from West Africa. Yet, West Africa have seen little benefit from the profits palm oil generated in Southeast Asia (Mbugua, 2017). (Note this exploitation goes multiple ways; for example, South America has also seen little benefit from profits generated from cacao production in West Africa.)

> Scientists, economists, politicians, and others are currently debating who owns the commercial rights to the world's biodiversity.

To combat this unfair exploitation, called **biopiracy**, many developing countries now require scientists and corporations to obtain permits before they can collect biological material for commercial or research purposes. Also, at the international level, nearly 100 countries have agreed to the fair sharing of benefits arising from the use of biological resources, through the ***Nagoya Protocol*** (see Section 12.2.1 for further discussion on international laws). Through these and similar laws and agreements, the hope is that a greater portion of the profits gained from biodiversity will be allocated to people who protect biodiversity and who live in the areas from where it is extracted.

A more inclusive approach

The valuation of ecosystem services has traditionally relied on generalised principles of economics and natural sciences. While this focus enabled scientists to develop broadly applicable themes and metrics in ecosystem evaluation, it also neglected the role of context and culture in understanding nature's role in people's lives. Many people have also remained uneasy about commodifying nature (i.e. giving it a market value), because some of the most important contributions of biodiversity are not easily converted into monetary metrics. Consequently, many feared that the transactional approach to ecosystem services would lead to social inequity concerns and alienate people offended by the idea that nature's metaphysical properties must compete against commercial interests.

To address these concerns, the valuation and classification of ecosystem services are currently undergoing several major transformations. Prominently, the UN's most recent classification scheme (Díaz et al., 2018) has given a more prominent voice to a wider range of stakeholders, including the social sciences, and recognises the importance of culture and context in nature's contributions to people. This exciting area of research is actively developing, and readers are encouraged to track developments and reactions associated with 2019 IPBES Global Assessment at https://www.ipbes.net/news/ipbes-global-assessment-preview.

4.6 Summary

1. People value biodiversity in many ways. The reasons vary from person to person, and from region to region. But generally, nature's contributions to people, also called ecosystem services, are divided into three overlapping and interdependent categories, namely material contributions, regulating services, and nonmaterial contributions.
2. Material contributions include benefits people get from consuming natural resources (e.g. drinking water or burning wood for cooking) or using natural resources in production and trade (e.g. timber to build homes or other structures).
3. Biodiversity also provides a large variety of regulating services that enable people to benefit from nature's material contributions. Some of these contributions include ecosystem productivity, water and soil protection, climate regulation, pollination, seed dispersal, and disaster prevention and detection.
4. People attach nonmaterial values to biodiversity which are difficult to quantify, and thus to account for, in modern economic systems. These values include support for inspiration and learning, support for psychological and physical experiences, and support for personal and group identities.
5. Environmental economics studies the implications of economic transactions, environmental policies, and other decisions that impact the environment. This field has highlighted how damage to the environment, such as pollution caused by industry, are not always fully considered when making political and development decisions, leading to unsustainable economic practices and market failure. Accounting for negative externalities and perverse subsidies can help policymakers design incentives that promote sustainable practices.

4.7 Topics for Discussion

1. Think of a recent infrastructure development near where you live, such as a recently-built road or dam. Try to come up with a list of ecosystem services that were damaged by this development. Who carries the costs of these lost services? Do you think the benefits from the development were worth the costs? Explain your answer.
2. Do individual organisms, populations, species, and biological communities have rights? What about physical features such as lakes, rivers, and mountains? While explaining your answer, also think about where we should draw the line of moral responsibility in how we care for nature.
3. A European botanist on holiday visits your area. During a short hike, you show this botanist a plant used as a traditional treatment for malaria. The

botanist takes samples of this plant back to Europe, where subsequent testing shows that it can be used to develop an effective anti-malarial drug. Who do you think should receive the profits from this new drug? The botanist who undertook the trip, and you because you showed the botanist the plant? What about the organisation that funded the drug's development, and the scientists who synthesised the new drug? What about all the people who educated you and your family in the plant's value? If the profits belong to multiple entities, how should it be divided?

4. More than a decade ago, the shark ecotourism industry at Gansbaai, South Africa, was estimated at US $4.4 million annually (Hara et al., 2003)—it has been increasing ever since. There are an estimated 900 great white sharks (*Carcharodon carcharias*, VU) living in Gansbaai (Towner et al., 2013). Assuming the average white shark lives for 70 years (Hamady et al., 2014), what is the value of each shark at Gansbaai? Can you find (or estimate) the price that a single shark sold for food would obtain on the world market? How do these values compare? What do you think is the best use of the sharks?

4.8 Suggested Readings

Allsopp, M.H., W.J. de Lange, and R. Veldtman. 2008. Valuing insect pollination services with cost of replacement. *PLoS ONE* 3: e3128. https://doi.org/10.1371/journal.pone.0003128 A study estimating the economic value of local pollination services.

Costanza, R., R. de Groot, P. Sutton, et al. 2014. Changes in the global value of ecosystem services. *Global Environmental Change* 26: 152–58. https://doi.org/10.1016/j.gloenvcha.2014.04.002 An attempt to value all ecosystem services.

Farber, S.C., R. Costanza, and A.M. Wilson. 2002. Economic and ecological concepts for valuing ecosystem services. *Ecological Economics* 41: 375–92. https://doi.org/10.1016/S0921-8009(02)00088-5 Methods for estimating the value of ecosystem services.

Isbell, F., V. Calcagno, A. Hector, et al. 2011. High plant diversity is needed to maintain ecosystem services. *Nature* 477: 199–202. https://doi.org/10.1038/nature10282 Maintaining ecosystem services requires protecting a diversity of species.

Koné, I., J.E. Lambert, J. Refisch, et al. 2008. Primate seed dispersal and its potential role in maintaining useful tree species in the Taï region, Côte d'Ivoire: Implications for the conservation of forest fragments. *Tropical Conservation Science* 1: 293–306. https://doi.org/10.1177%2F194008290800100309 Maintaining primate populations is important also for humans who rely on forest resources.

Markandya, A., T. Taylor, A. Longo, et al. 2008. Counting the cost of vulture decline—an appraisal of the human health and other benefits of vultures in India. *Ecological Economics* 67: 194-204. https://doi.org/10.1016/j.ecolecon.2008.04.020 A study illustrating the value of vultures.

Naidoo, R., B. Fisher, A. Manica, et al. 2016. Estimating economic losses to tourism in Africa from the illegal killing of elephants. *Nature Communications* 7: 13379. https://doi.org/10.1038/ncomms13379 Africa loses US $25 million annually from elephant poaching.

Peterson, G.D., Z.V. Harmackova, M. Meacham, et al. 2018. Welcoming different perspectives in IPBES: "Nature's contributions to people" and "Ecosystem services". *Ecology and Society* 23: 39. https://doi.org/10.5751/ES-10134-230139 Addressing shortcomings of the ecosystem services concept

Schleicher, J., M. Schaafsma, N.D. Burgess, et al. 2018. Poorer without it? The neglected role of the natural environment in poverty and wellbeing. *Sustainable Development* 25: 83–98. https://doi.org/10.1002/sd.1692 The environment and human well-being are intricately linked.

Bibliography

Abate, T., A. van Huis, and J.K.O. Ampofo. 2000. Pest management strategies in traditional agriculture: An African perspective. *Annual Review of Entomology* 45: 631–59. https://doi.org/10.1146/annurev.ento.45.1.631

Abernethy, K.A., L. Coad, G. Taylor, et al. 2013. Extent and ecological consequences of hunting in Central African rainforests in the twenty-first century. *Philosophical Transactions of the Royal Society B*: 368: 20120303. https://doi.org/10.1098/rstb.2012.0303

African Pollinators Initiative. 2007. *Crops, Browse and Pollinators in Africa: An Initial Stock-taking* (Rome: FAO). http://www.fao.org/3/a-a1504e.pdf

Allsopp, M.H., W.J. de Lange, and R. Veldtman. 2008. Valuing insect pollination services with cost of replacement. *PloS ONE* 3: e3128. https://doi.org/10.1371/journal.pone.0003128

Archibald, K., and M. Clotworthy. 2007. Comment on 'The ethics of animal research' by Festing and Wilkinson. *EMBO Reports* 8: 794–96. https://doi.org/10.1038/sj.embor.7401057

Auffhammer, M., P. Baylis, and C.H. Hausman. 2017. Climate change is projected to have severe impacts on the frequency and intensity of peak electricity demand across the United States. *Proceedings of the National Academy of Sciences* 114: 1886–91. https://doi.org/10.1073/pnas.1613193114

Babalola, S.A. 1966. *The Content and Form of Yoruba Ijala*. Oxford Library of African Literature (Oxford: Oxford University Press).

Bako, S.P., S. Afolabi, and I.I. Funtua. 2008. Spatial distribution and heavy metal content of some bryophytes and lichens in relation to air pollution in Nigeria's Guinea Savanna. *International Journal of Environment and Pollution* 33: 195–206. https://doi.org/10.1504/IJEP.2008.019393

Balmford, A., A. Manica, L. Airey, et al. 2004. Hollywood, climate change, and the public. *Science* 305: 1713. https://doi.org/10.1126/science.305.5691.1713b

Balmford, A., J.M. Green, M. Anderson, et al. 2015. Walk on the wild side: Estimating the global magnitude of visits to protected areas. *PloS Biology* 13: e1002074. https://doi.org/10.1371/journal.pbio.1002074

Balogun, A.A., T.E. Morakinyo, and O.B. Adegun. 2014. Effect of tree-shading on energy demand of two similar buildings. *Energy and Buildings* 81: 305–15. https://doi.org/10.1016/j.enbuild.2014.05.046

Benjaminsen, T. 2008. Does supply-induced scarcity drive violent conflicts in the African Sahel? The case of the Tuareg rebellion in northern Mali. *Journal of Peace Research* 45: 819–36. https://doi.org/10.1177/0022343308096158

Biggs, D., J. Turpie, C. Fabricius, et al. 2011. The value of avitourism for conservation and job creation—an analysis from South Africa. *Conservation and Society* 9: 80–90. https://doi.org/10.4103/0972-4923.79198

Bodin, N., R. N'Gom-Kâ, S. Kâ, et al. 2013. Assessment of trace metal contamination in mangrove ecosystems from Senegal, West Africa. *Chemosphere 90*: 150–57. https://doi.org/10.1016/j.chemosphere.2012.06.019

Bonnell, T.R., R. Reyna-Hurtado, and C.A. Chapman. 2011. Post-logging recovery time is longer than expected in an East African tropical forest. *Forest Ecology and Management* 261: 855–64. https://doi.org/10.1016/j.foreco.2010.12.016

Boreyko, J.B., R.R. Hansen, K.R. Murphy, et al. 2016. Controlling condensation and frost growth with chemical micropatterns. *Scientific Reports* 6: 19131. https://doi.org/10.1038/srep19131

Brashares J.S., C.W. Epps, and C.J. Stoner. 2010. Ecological and conservation implications of mesopredator release. In: *Trophic Cascades,* ed. By J. Terborgh and J. Estes (Washington: Island Press).

Brashares, J.S., P. Arcese, M.K. Sam, et al. 2004. Bushmeat hunting, wildlife declines, and fish supply in West Africa. *Science* 306: 1180–83. https://doi.org/10.1126/science.1102425

Brito, J.C., S.M. Durant, N. Pettorelli, et al. 2018. Armed conflicts and wildlife decline: Challenges and recommendations for effective conservation policy in the Sahara-Sahel. *Conservation Letters* 11: e12446. http://doi.org/10.1111/conl.12446

Cagua E.F., N. Collins, J. Hancock, et al. 2014. Whale shark economics: A valuation of wildlife tourism in South Ari Atoll, Maldives. *PeerJ* 2: e515. https://doi.org/10.7717/peerj.515

Chamaillé-Jammes, S., H. Fritz, M. Valeix, et al. 2008. Resource variability, aggregation and direct density dependence in an open context: The local regulation of an African elephant population. *Journal of Animal Ecology* 77: 135–44. https://doi.org/10.1111/j.1365-2656.2007.01307.x

Clark, L. 2012. This self-filling water bottle mimics a desert beetle. *Wired Magazine.* http://wrd.cm/1OcZRnF

Coad, L., J. Schleicher, E.J. Milner-Gulland, et al. 2013. Social and ecological change over a decade in a village hunting system, central Gabon. *Conservation Biology* 27: 270–80. https://doi.org/10.1111/cobi.12012

Coad, L., K. Abernethy, A. Balmford, et al. 2010. Distribution and use of income from bushmeat in a rural village, Central Gabon: Bushmeat income in Gabon. *Conservation Biology* 24: 1510–18. https://doi.org/10.1111/j.1523-1739.2010.01525.x

Costanza, R., R. de Groot, P. Sutton, et al. 2014. Changes in the global value of ecosystem services. *Global Environmental Change* 26: 152–58. https://doi.org/10.1016/j.gloenvcha.2014.04.002

Cumming, D.H.M., M.B. Fenton, I.L. Rautenbach, et al. 1997. Elephants, woodlands and biodiversity in southern Africa. *South African Journal of Science* 93: 231–36.

Daskin, J.H., and R.M. Pringle. 2018. Warfare and wildlife declines in Africa's protected areas. *Nature* 553: 328–32. https://doi.org/10.1038/nature25194

Davis A.P., T.W. Gole, S. Baena, et al. 2012. The impact of climate change on indigenous arabica coffee (*Coffea arabica*): Predicting future trends and identifying priorities. *PloS ONE* 7: e47981. https://doi.org/10.1371/journal.pone.0047981

Diamond, J. 2011. *Collapse: How Societies Choose to Fail or Succeed* (London: Penguin Books).

Díaz, S., U. Pascual, M. Stenseke, et al. 2018. Assessing nature's contributions to people. *Science* 359: 270–72. https://doi.org/10.1126/science.aap8826

Dirzo, R., H.S. Young, M. Galetti, et al., 2014. Defaunation in the Anthropocene. *Science* 345: 401–06. https://doi.org/10.1126/science.1251817

Duthie, E., D. Veríssimo, A. Keane, et al. 2017. The effectiveness of celebrities in conservation marketing. *PloS ONE* 12: e0180027. https://doi.org/10.1371/journal.pone.0180027

Ehrlich, P., and B. Walker. 1998. Rivets and redundancy. *BioScience* 48: 387–88. https://doi.org/10.2307/1313377

Ellis A.M., S.S. Myers, and T.H. Ricketts. 2015. Do pollinators contribute to nutritional health? *PloS ONE* 10: e114805. https://doi.org/10.1371/journal.pone.0114805

Estes, J.A., J. Terborgh, J.S. Brashares, et al. 2011. Trophic downgrading of planet Earth. *Science* 333: 301–06 https://doi.org/10.1126/science.1205106

Farber, S.C., R. Costanza, and A.M. Wilson. 2002. Economic and ecological concepts for valuing ecosystem services. *Ecological Economics* 41: 375–92. https://doi.org/10.1016/S0921-8009(02)00088-5

Festing, S., and R. Wilkinson. 2007. The ethics of animal research. *EMBO Reports* 8: 526–30. https://doi.org/10.1038/sj.embor.7400993

Fosso-Kankeu, E., and A.F. Mulaba-Bafubiandi. 2014. Implication of plants and microbial metalloproteins in the bioremediation of polluted waters: A review. *Physics and Chemistry of the Earth* 67: 242–52. https://doi.org/10.1016/j.pce.2013.09.018

Gallagher, A.J., and C.P.M. Huveneers. 2018. Emerging challenges to shark-diving tourism. *Marine Policy* 96: 9–12. https://doi.org/10.1016/j.marpol.2018.07.009

Gallai, N., J.M. Salles, J. Settele, et al. 2009. Economic valuation of the vulnerability of world agriculture confronted with pollinator decline. *Ecological Economics* 68: 810–21. https://doi.org/10.1016/j.ecolecon.2008.06.014

Garcia-Carreras, L., and D.J. Parker. 2011. How does local tropical deforestation affect rainfall? *Geophysical Research Letters* 38: L19802. https://doi.org/10.1029/2011GL049099

Garibaldi, L.A., I. Steffan-Dewenter, R. Winfree, et al. 2013. Wild pollinators enhance fruit set of crops regardless of honey bee abundance. *Science* 339: 1608–11. https://doi.org/10.1126/science.1230200

Gemmill-Herren, B., and A.O. Ochieng. 2008. Role of native bees and natural habitats in eggplant (*Solanum melongena*) pollination in Kenya. *Agriculture, Ecosystems and Environment* 127: 31–36. https://doi.org/10.1016/j.agee.2008.02.002

Gleick, P.H. 2014. Water, drought, climate change, and conflict in Syria. *Weather, Climate, and Society* 6: 331–40. https://doi.org/10.1175/WCAS-D-13-00059.1

Gorokhovich, Y., S. Doocy, F. Walyawula, et al. 2013. Landslides in Bududa, eastern Uganda: Preliminary assessment and proposed solutions. In: *Landslide Science and Practice,* ed. by C. Margottini et al. (Berlin: Springer). https://doi.org/10.1007/978-3-642-31325-7

Graziosi, I., N. Minato, E. Alvarez, et al. 2016. Emerging pests and diseases of South-east Asian cassava: A comprehensive evaluation of geographic priorities, management options and research needs. *Pest Management Science* 72: 1071–89. https://doi.org/10.1002/ps.4250

Guerry, A.D., S. Polasky, J. Lubchenco, et al. 2015. Natural capital and ecosystem services informing decisions: From promise to practice. *Proceedings of the National Academy of Sciences* 112: 7348–55. https://doi.org/10.1073/pnas.1503751112

Gurib-Fakim, A. 2017. Capitalize on African biodiversity. *Nature* 548: 7. https://doi.org/10.1038/548007a

Hairston, N.G., F.E. Smith, and L.B. Slobodkin. 1960. Community structure, population control, and competition. *American Naturalist* 94: 421–25. https://doi.org/10.1086/282146

Hamady, L.L., L.J. Natanson, G.B. Skomal, et al. 2014. Vertebral bomb radiocarbon suggests extreme longevity in white sharks. *PloS ONE* 9: e84006. https://doi.org/10.1371/journal.pone.0084006

Hanson, T., T.M. Brooks, G.A. Da Fonseca, et al. 2009. Warfare in biodiversity hotspots. *Conservation Biology* 23: 578–87. https://doi.org/10.1111/j.1523-1739.2009.01166.x

Hara, M., I. Maharaj, and L. Pithers. 2003. *Marine-based tourism in Gansbaai: A socio-economic* study (Cape Town: DEAT).

Hayward, M.W., M.J. Somers, G.I. Kerley, et al. 2012. Animal ethics and ecotourism. *South African Journal of Wildlife Research* 42: iii–v. https://doi.org/10.3957/056.042.0207

Holdo, R.M., A.R.E. Sinclair, A.P. Dobson, et al. 2009. A disease-mediated trophic cascade in the Serengeti and its implications for ecosystem C. *PloS Biology* 7: e1000210. https://doi.org/10.1371/journal.pbio.1000210

Hsiang, S., R. Kopp, A. Jina, et al. 2017. Estimating economic damage from climate change in the United States. *Science* 356: 1362–69. http://doi.org/10.1126/science.aal4369

Hughes, N.J. 2011. The economic value of Congo Basin protected areas goods and services. *Journal of Sustainable Development* 4: 130–42

Ingram, D., L.M. Coad, B. Collen, et al. 2015. Indicators for wild animal offtake: Methods and case study for African mammals and birds. *Ecology and Society* 20: 40. http://doi.org/10.5751/ES-07823-200340

Jaramillo J., E. Muchugu, F.E. Vega, et al. 2011. Some like it hot: The influence and implications of climate change on coffee berry borer (*Hypothenemus hampei*) and coffee production in East Africa. *PloS ONE* 6: e24528. https://doi.org/10.1371/journal.pone.0024528

Jouquet, P., S. Traoré, C. Choosai, et al. 2011. Influence of termites on ecosystem functioning. Ecosystem services provided by termites. *European Journal of Soil Biology* 47: 215–22. https://doi.org/10.1016/j.ejsobi.2011.05.005

Kardan, O., P. Gozdyra, B. Misic, et al. 2015. Neighborhood greenspace and health in a large urban center. *Scientific Reports* 5: 11610. https://doi.org/10.1038/srep11610

Khan, S., S. Nadir, Z.U. Shah, et al. 2017. Biodegradation of polyester polyurethane by *Aspergillus tubingensis*. *Environmental Pollution* 225: 469–80. https://doi.org/10.1016/j.envpol.2017.03.012

Khan, Z.R., C.A. Midega, T.J. Bruce, et al. 2010. Exploiting phytochemicals for developing a 'push–pull'crop protection strategy for cereal farmers in Africa. *Journal of Experimental Botany 61*: 4185–96. https://doi.org/10.1093/jxb/erq229

Kling, C.L., D J. Phaneuf, and J. Zhao. 2012. From Exxon to BP: Has some number become better than no number? *Journal of Economic Perspectives* 26: 3–26. https://doi.org/10.1257/jep.26.4.3

Koné, I., J.E. Lambert, J. Refisch, et al. 2008. Primate seed dispersal and its potential role in maintaining useful tree species in the Taï region, Côte-d'Ivoire: Implications for the conservation of forest fragments. *Tropical Conservation Science* 1: 293–306. https://doi.org/10.1177/194008290800100309

LaLonde, K.B., A. Mahoney, T.L. Edwards, et al. 2015. Training pouched rats to find people. *Journal of Applied Behavior Analysis* 48: 1–10. https://doi.org/10.1002/jaba.181

Lassen, K.M., A. Ræbild, H. Hansen, et al. 2012. Bats and bees are pollinating *Parkia biglobosa* in The Gambia. *Agroforestry Systems* 85: 465–75. https://doi.org/10.1007/s10457-011-9409-0

Lawrence, D., and K. Vandecar. 2015. Effects of tropical deforestation on climate and agriculture. *Nature Climate Change* 5: 27–36. https://doi.org/10.1038/nclimate2430

Lindsey P.A., V.R. Nyirenda, J.L. Barnes, et al. 2014. Underperformance of African protected area networks and the case for new conservation models: Insights from Zambia. *PloS ONE* 9: e94109. https://doi.org/10.1371/journal.pone.0094109

Lindsey, P.A., J.R.B. Miller, L.S. Petracca, et al. 2018. More than $1 billion needed annually to secure Africa's protected areas with lions. *Proceedings of the National Academy of Sciences*: E10788-E10796. https://doi.org/10.1073/pnas.1805048115

Lovelock, J. 1988. *The Ages of Gaia: A Bibliography of Our Living Earth* (Oxford: Oxford University Press).

Loveridge, J.P., and S.R. Moe. 2004. Termitaria as browsing hotspots for African megaherbivores in miombo woodland. *Journal of Tropical Ecology* 20: 337–43. https://doi.org/10.1017/S0266467403001202

Mahoney, A., T.L. Edwards, K.B. LaLonde, et al. 2014a. Pouched rats' (*Cricetomys gambianus*) detection of Salmonella in horse feces. *Journal of Veterinary Behavior* 9: 124–26. https://doi.org/10.1016/j.jveb.2014.02.001

Mallory, T.G. 2013. China's distant water fishing industry: Evolving policies and implications. *Marine Policy* 38: 99–108. https://doi.org/10.1016/j.marpol.2012.05.024

Markakis, J. 1995. Environmental Degradation and Social Conflict in the Horn of Africa. In: *Environmental Crisis: Regional Conflicts and Ways of Cooperation*, ed. by K.R. Spillman and G. Bächler (Zurich: Center for Security Studies; Berne: Swiss Peace Foundation). http://www.css.ethz.ch/content/dam/ethz/special-interest/gess/cis/center-for-securities-studies/pdfs/Environmental_Crisis_1995.pdf

Markandya, A., T. Taylor, A. Longo, et al. 2008. Counting the cost of vulture decline—an appraisal of the human health and other benefits of vultures in India. *Ecological Economics* 67: 194–204. https://doi.org/10.1016/j.ecolecon.2008.04.020

Martin, M.J., J.C. Rayner, P. Gagneux, et al. 2005. Evolution of human-chimpanzee differences in malaria susceptibility: Relationship to human genetic loss of N-glycolylneuraminic acid. *Proceedings of the National Academy of Sciences* 102: 12819–24. https://doi.org/10.1073/pnas.0503819102

Mbugua, S. 2017. Conservation leaders in Africa call for a crackdown on biopiracy. *Mongabay* https://news.mongabay.com/2017/10/conservation-leaders-in-africa-call-for-a-crackdown-on-biopiracy

McCauley, D.J. 2006. Selling out on nature. *Nature* 443: 27–28. https://doi.org/10.1038/443027a

Metcalf, J.A., L.J. Funkhouser-Jones, K. Brileya, et al. 2014. Antibacterial gene transfer across the tree of life. *Elife* 3: e04266. https://doi.org/10.7554/eLife.04266

Morakinyo, T.E., A.A. Balogun, and O.B. Adegun. 2013. Comparing the effect of trees on thermal conditions of two typical urban buildings. *Urban Climate* 3: 76–93. http://doi.org/10.1016/j.uclim.2013.04.002

Myers, N., and J. Kent. 2001 *Perverse Subsidies: How Tax Dollars Can Undercut the Environment and the Economy* (Washington: Island Press).

Nackoney, J., G. Molinario, P. Potapov, et al. 2014. Impacts of civil conflict on primary forest habitat in northern Democratic Republic of the Congo, 1990–2010. *Biological Conservation* 170: 321–28. https://doi.org/10.1016/j.biocon.2013.12.033.

Neuenschwander, P. 2001. Biological control of the cassava mealybug in Africa: A review. *Biological Control* 21: 214–29. https://doi.org/10.1006/bcon.2001.0937

NRC (National Research Council). 2000. *Watershed Management for Potable Water Supply: Assessing the New York City Strategy* (Washington: National Academy Press). https://doi.org/10.17226/9677

NRC (National Research Council). 2002. *The drama of the commons* (Washington: National Academies Press). https://doi.org/10.17226/10287

O'Bryan, C.J., A.R. Braczkowski, H.L. Beyer, et al. 2018. The contribution of predators and scavengers to human well-being. *Nature Ecology and Evolution* 2: 229–36. https://doi.org/10.1038/s41559-017-0421-2

Oaks, J.L., M. Gilbert, M.Z. Virani, et al. 2004. Diclofenac residues as the cause of vulture population decline in Pakistan. *Nature* 427: 630–33. https://doi.org/10.1038/nature02317

Ogada, D., P. Shaw, R.L. Beyers, et al. 2015. Another continental vulture crisis: Africa's vultures collapsing toward extinction. *Conservation Letters* 9: 98–97. https://doi.org/10.1111/conl.12182

Ogueke, N.V., A.F. Nwakanma, T. Ngharamike, et al. 2017. Energy-saving potentials of some local trees. *Energy Efficiency* 10: 171–81.

Paine, R.T. 1969. A note on trophic complexity and community stability. *American Naturalist* 103: 91–93. https://doi.org/10.1086/282586

Paiva, V.H., P. Geraldes, I. Rodrigues, T. Melo, et al. 2015. The foraging ecology of the endangered Cape Verde shearwater, a sentinel species for marine conservation off West Africa. *PloS ONE* 10: e0139390. https://doi.org/10.1371/journal.pone.0139390

Parker, A.R., and C.R. Lawrence. 2001. Water capture by a desert beetle. *Nature* 414: 33–34. https://doi.org/10.1038/35102108

Parsa, S., T. Kondo, and A. Winotai. 2012. The cassava mealybug (*Phenacoccus manihoti*) in Asia: First records, potential distribution, and an identification key. *PloS ONE* 7: e47675. https://doi.org/10.1371/journal.pone.0047675

Prime Minister's Task Force. 2009. *Report of the Prime Minister's Task Force on the conservation of the Mau Forests Complex* (Nairobi: Prime Minister's Task Force). https://www.kws.go.ke/file/1448/download?token=vy_ga198

Pringle, R.M. 2008. Elephants as agents of habitat creation for small vertebrates at the patch scale. *Ecology* 89: 26–33. https://doi.org/10.1890/07-0776.1

Reither K., L. Jugheli, T.R. Glass, et al. 2015. Evaluation of giant African pouched rats for detection of pulmonary tuberculosis in patients from a high-endemic setting. *PloS ONE* 10: e0135877. https://doi.org/10.1371/journal.pone.0135877

Rollin, B.E. 2007. Animal research: A moral science. *EMBO Reports* 8: 521–25. https://doi.org/10.1038/sj.embor.7400996

Samnegård, U., P.A. Hambäck, S. Nemomissa, et al. 2014. Dominance of the semi-wild honeybee as coffee pollinator across a gradient of shade-tree structure in Ethiopia. *Journal of Tropical Ecology* 30: 401–08. https://doi.org/10.1017/S0266467414000327

Schulz, R., and S.K.C. Peall. 2001. Effectiveness of a constructed wetland for retention of nonpoint-source pesticide pollution in the Lourens River catchment, South Africa. *Environmental Science and Technology* 35: 422–26. https://doi.org/10.1021/es0001198

Scott, K. 2011. Airdrop water harvester wins 2011 James Dyson Award. *Wired Magazine*. http://www.wired.co.uk/article/dyson-award-2011-winners

Silk, M.J., S.L. Crowley, A.J. Woodhead, et al. 2018. Considering connections between Hollywood and biodiversity conservation. *Conservation Biology* 31: 597–606. https://doi.org/10.1111/cobi.13030

Silvertown, J. 2015 Have ecosystem services been oversold? *Trends in Ecology and Evolution* 30: 641–48. http://doi.org/10.1016/j.tree.2015.08.007

Sumaila, U.R., A.S. Khan, A.J. Dyck, et al. 2010. A bottom-up re-estimation of global fisheries subsidies. *Journal of Bioeconomics* 12: 201–25. https://doi.org/10.1007/s10818-010-9091-8

Taylor, P.J., I. Grass, A.J. Alberts, et al. 2018. Economic value of bat predation services—a review and new estimates from macadamia orchards. *Ecosystem Services* 30C: 372–81. https://doi.org/10.1016/j.ecoser.2017.11.015

Thiel, M., M.A. Penna-Díaz, G. Luna-Jorquera, et al. 2014. Citizen scientists and marine research: Volunteer participants, their contributions, and projection for the future. *Oceanography and Marine Biology* 52: 257–314. https://doi.org/10.1201/b17143

Towner, A.V., M.A. Wcisel, R.R. Reisinger, et al. 2013. Gauging the threat: The first population estimate for white sharks in South Africa using photo identification and automated software. *PloS ONE* 8: e66035. https://doi.org/10.1371/journal.pone.0066035

Turpie, J.K., B.J. Heydenrych, and S.J. Lamberth. 2003. Economic value of terrestrial and marine biodiversity in the Cape Floristic Region: Implications for defining effective and socially optimal conservation strategies. *Biological Conservation* 112: 233–51. https://doi.org/10.1016/S0006-3207(02)00398-1

UNEP. 2012. *The role and contribution of montane forests and related ecosystem services to the Kenyan economy* (Nairobi: UNEP). http://wedocs.unep.org/handle/20.500.11822/8513

Valeix, M., H. Fritz, R. Sabatier, et al. 2011. Elephant-induced structural changes in the vegetation and habitat selection by large herbivores in an African savanna. *Biological Conservation* 144: 902–12. http://doi.org/10.1016/j.biocon.2010.10.029

van Bochove, J., E. Sullivan, and T. Nakamura. 2014. *The Importance of Mangroves to People: A Call to Action* (Cambridge: UNEP). http://wedocs.unep.org/handle/20.500.11822/9300

van der Plas, F., R. Howison, J. Reinders, et al. 2013. Functional traits of trees on and off termite mounds: understanding the origin of biotically-driven heterogeneity in savannas. *Journal of Vegetation Science* 24: 227–38. https://doi.org/10.1111/j.1654-1103.2012.01459.x

Vianna, G.M.S., M.G. Meekan, D.J. Pannell, et al. 2012. Socio-economic value and community benefits from shark-diving tourism in Palau: A sustainable use of reef shark populations. *Biological Conservation* 145: 267–77. https://doi.org/10.1016/j.biocon.2011.11.022

Whitley S., and L. van der Burg, 2015. *Fossil fuel subsidy reform in sub-Saharan Africa: from rhetoric to reality* (London and Washington: New Climate Economy). https://newclimateeconomy.report/workingpapers/workingpaper/fossil-fuel-subsidy-reform-in-sub-saharan-africa-from-rhetoric-to-reality-2

Wicander, S., and L.M. Coad. 2015. *Learning our Lessons: A Review of Alternative Livelihood Projects in Central Africa* (Gland: IUCN). https://www.iucn.org/content/learning-our-lessons-review-alternative-livelihood-projects-central-africa

WWF, and BSI. 2006. *Poverty and environmental issues: Governance institutions, institutional frameworks and opportunities for communities* (Nairobi: UNEP). https://www.unpei.org/sites/default/files/e_library_documents/kenya-poverty-environment-issues.pdf

Yirga, G., H. Leirs, H.H. de Iongh, et al. 2015. Spotted hyena (*Crocuta crocuta*) concentrate around urban waste dumps across Tigray, northern Ethiopia. *Wildlife Research* 42: 563–69. https://doi.org/10.1071/WR14228

Zeddies, J., R.P. Schaab, P. Neuenschwander, et al. 2001. Economics of biological control of cassava mealybug in Africa. *Agricultural Economics* 24: 209–19. https://doi.org/10.1111/j.1574-0862.2001.tb00024.x

5. The Scramble for Space

Mount Nimba Strict Nature Reserve, which straddles the border region of Guinea and Côte d'Ivoire, is a World Heritage Site situated in the Guinean Forest of West Africa Biodiversity Hotspot. The ecosystem, which supports chimpanzees that use stone tools, is threatened by iron mining, agriculture, and deforestation. Photograph by Guy Debonnet, http://whc.unesco.org/en/documents/123989, CC BY-SA 3.0 IGO.

https://doi.org/10.11647/OBP.0177.05

The right to shelter, food, and association are basic human needs recognised in many international charters and country constitutions. Like humans, wildlife also needs areas where they can find protection, nourishment, and mates to have any hope of survival. The area where a species can survive and meet their basic needs is known as its **habitat**. It is often useful to think of a habitat as a multi-dimensional space, characterised by suitable levels of many different environmental variables. Some species, including humans, are highly tolerant of changes in their environmental conditions; consequently, such **generalist species** find it relatively easy to move to a new area in the unfortunate event that their "home" is destroyed. In contrast, **specialist species**—those that can only survive within a narrow range of environmental conditions—often do not have anywhere else to go when their habitat is lost, and consequently they go extinct.

In a world where intact natural ecosystems are increasingly being altered by the activities of an ever-increasing human population and its consumptive needs, habitat loss has emerged as the number one threat facing biodiversity today. The expansion of human activity causes massive disturbances to natural ecosystems by altering, degrading, and outright destroying wildlife habitats. A number of specialist species have already been pushed to extinction. But even generalist species are increasingly falling victim to habitat loss: pushed out of their shrinking habitats, they come into conflict with humans while trying to meet their needs near urban centres and on agricultural land. Eventually our own lives will suffer, whether through lost ecosystem services, or sorrow for all the wonderful landscapes and species that have disappeared under our watch. In this chapter we delve into the causes and consequences of this increased competition for space between man and wildlife.

> The primary threat to Africa's biodiversity today is habitat loss and degradation.

5.1 What is Habitat Loss?

Habitat loss is defined as the outright destruction of natural ecosystems, an inevitable consequence of expanding human populations and human activities. The theory of **island biogeography** (MacArthur and Wilson, 1967) offers a good explanation for why habitat loss drives species extinctions. Using oceanic islands as a model system, one of the theory's main predictions is that large islands have more species than small ones because they can accommodate more individuals, which causes those species to be better buffered against extinctions (Section 8.7). Empirical evidence offers strong support for this observation, also known as the **species-area relationship**. For example, large African islands generally hold more bird species than small islands (Figure 5.1). In addition, 62 of the 79 (63%) Sub-Saharan Africa's species that went extinct over the past few centuries (IUCN, 2019) have been confined to oceanic islands, rather than the continental mainland which in effect functions like one very big island.

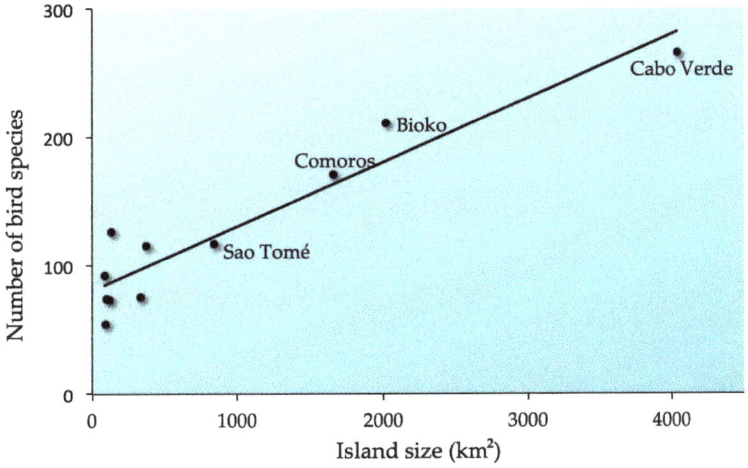

Figure 5.1 Area size greatly influences species richness, as evidenced by the bird species richness on several prominent volcanic islands around Africa. This observation, known as the species-area relationship, explains why habitat loss is so devastating to biodiversity—the more we reduce the amount of habitat left for species to live in, the more extinctions we will see in the coming years. Source: Avibase (https://avibase.bsc-eoc.org), following BirdLife International 2018 taxonomy, CC BY 4.0

The species-area relationship underpins much of conservation biology today. By applying the relationship's principles to "islands" of suitable habitat surrounded by a "sea" of damaged or unsuitable habitat (the "matrix"), conservation biologists know that conserving large areas of suitable habitat is much more effective in protecting biodiversity (Box 5.1). This is especially true when trying to protect species that have large home ranges, and/or occur in low densities: they can only live in habitat patches that are large enough to maintain viable populations (Chapter 9 discusses the relationship between population size and population viability in more detail). Observations of extirpations in differently sized habitat patches support this application. For example, researchers have found that nearly 50% of Ghana's forest bird species are sensitive to habitat size, with 25% of species never found in forest patches smaller than 0.1 km^2 (Beier et al., 2002). One Ghanaian species that seems particularly sensitive to habitat patch size is the icterine greenbul (*Phyllastrephus icterinus*, LC); due to habitat loss, this once-common species decreased by 90% during one study's 15-year period (Arcilla et al., 2015).

It is important to understand that species living in ecosystems that are not conspicuously destroyed may also experience the effects of habitat loss, and hence suffer population declines. This is because habitat loss often manifests itself, at least initially, through less visible but equally threatening **habitat degradation**. For example, disturbances such as overgrazing do not immediately change the organisation of dominant plants and other structural features of an ecological community. First, barely noticeable, a few sensitive habitat specialists disappear, being unable to cope with high levels of grazing. Soon, **invasive species** that can tolerate trampling start

Box 5.1 The Importance of Liberia's Forest Network to the Survival of the Pygmy Hippopotamus

Mary Molokwu-Odozi[1] *and Kathryn Phillips*[2]

[1]*Fauna & Flora International, Harmon Compound, Congo Town, Monrovia, Liberia.*

[2]*Fauna & Flora International, Cambridge, UK.*

✉ *mary.molokwu@fauna-flora.org; kathryn.phillips@fauna-flora.org*

Remaining populations of the pygmy hippopotamus (*Choeropsis liberiensis*, EN) are found predominately within transboundary West African rainforests spanning Côte d'Ivoire, Guinea, Liberia, and Sierra Leone (Ransom et al., 2015). Liberia contains the largest intact blocks (over 40%) of this Upper Guinean rainforest, a Global 200 ecoregion (Olson et al., 2002). An elusive animal, little is known about the pygmy hippo's distribution, population status, and ecology. Pygmy hippo numbers are currently estimated at fewer than 2,500 individuals across its range, with the expectation of further decline as a result of agricultural expansion, logging, development, and hunting (Ransom et al., 2015).

Within Liberia, pygmy hippopotamus populations are found in the major forest blocks of the southeast and northwest (Figure 5.A), which are separated by an area of degraded land with high human densities (FFI and FDA, 2013). The southeastern forest block is made up of several large chunks of national, communal, and protected forests fragmented by logging routes and concessions. Although populations are well documented within protected areas, recent reports indicate that populations also exist outside formally protected forests (Hillers et al., 2017). The establishment and management of forest corridors linking key habitats is therefore a conservation priority.

In the last few years, huge investment in agriculture, logging, and mining has increased pressure on forests for conversion and from increased human settlements and access roads. Weak law enforcement in Liberia's protected areas and limited operational capacity has led to increased incursion of illegal activities, such as poaching and mining in these critical habitats. Sapo National Park, Liberia's only national park and second largest in West Africa after Taï National Park in Côte d'Ivoire, is believed to be a stronghold for the species. However, pygmy hippo numbers remain low at an average encounter rate of 0.12 individuals/km from 2007–2009 (Vogt, 2011) to 0.15 individuals/km (2014–2016 data), i.e. one per 7–8 km, much lower than records from Taï (Vogt, 2011). Sapo National Park has historically suffered from—and continues to suffer from—mining and hunting pressures; hundreds of illegal miners who were

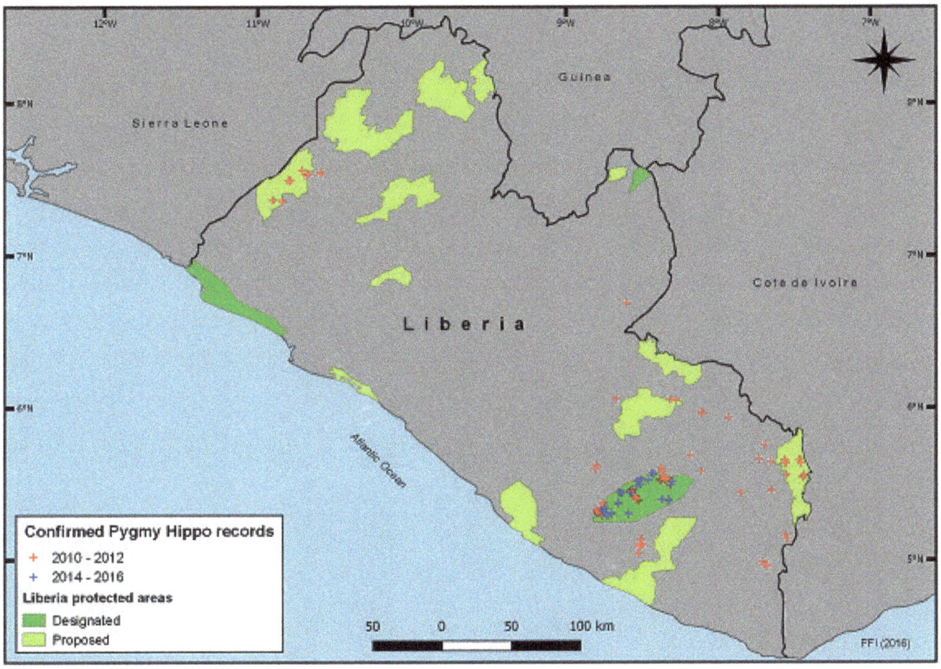

Figure 5.A (Top) Night-time camera trap image of the pygmy hippopotamus taken in Sapo National Park, Liberia. Photograph by FFI, CC BY 4.0. (Bottom) Distribution of pygmy hippopotamus in Liberia based on confirmed records from 2010-2016. Map by Benedictus Freeman/FFI, CC BY 4.0.

evacuated after the civil crises of 2002–2007 and in 2010–2011 and again during the 2014-2015 Ebola crisis, reoccupied a large section of the park, where hunting signs (trails, camps, gun shells) were encountered almost every kilometre of walking within the park's 1,804 km^2 area. The miners have once again been removed from the park, this time with local community support.

Recent efforts to save Liberia's declining forests have brought about an increase in activities by national and international NGOs and the formation of transboundary collaboration initiatives. For example, an agreement has been established between the Liberian and Sierra Leonean governments, creating the Gola Transboundary Peace Park and significant progress has been made in the development of the Taï-Grebo-Krahn-Sapo Transboundary Forest Complex with Côte d'Ivoire. The Liberian and Guinean government have also commenced a bilateral agreement for the conservation and sustainable management of the Ziama-Wonegizi-Wologizi Transboundary Forest Landscape. Agricultural investments have also evolved to promote public-private partnerships in green-growth and community-based forest protection initiatives. Notable support has also come from the Government of Norway to help Liberia fully halt deforestation by 2020.

Fauna & Flora International (FFI) has worked in Liberia since 1997 focusing on the pygmy hippopotamus as a flagship species. These efforts have contributed to increased knowledge of the species in Liberia, including recording (in collaboration with the Zoological Society of London, ZSL) the first footage of the species in Liberia. FFI also developed a pygmy hippopotamus national action plan and will be revising the regional conservation strategy for the pygmy hippo in 2019. FFI has also established monitoring programmes and a training and research centre at Sapo National Park. FFI's capacity building programme saw the development of the first conservation biology curriculum for Liberia's premier university and engagement of close to 1,000 children in a conservation education programme focused on the pygmy hippopotamus. Effective transboundary and protected area law enforcement will be key towards safeguarding and increasing remaining pygmy hippopotamus numbers, whilst awareness raising, collaborative forest management and national/regional policies to reduce deforestation will be needed to secure habitats for pygmy hippopotamus populations to thrive.

occupying the niches left open by the extirpated sensitive species. Eventually, when livestock eat the last remaining edible morsels of palatable plants not choked out by invasive species, all that is left of the once productive grassland is a field full of dense, unpalatable, invasive shrubbery.

5.1.1 What is habitat fragmentation?

As governments and industries implement measures to accelerate economic growth, ecosystems that formerly covered large, continuous swathes of land are being increasingly subdivided into smaller parcels by roads, farm fields, towns, and other human constructs. A recent study estimated that roads have divided the African continent into more than 50,000 individual units of land; the median unit size was an alarming 6.75 km^2 (Ibisch et al., 2016). This process, known as **habitat fragmentation**, divides once large and widespread wildlife populations—many already suffering from habitat loss—into several increasingly smaller **subpopulations**. Habitat fragmentation thereby hastens extinctions, as each of these fragmented subpopulations are more exposed to a range of deleterious genetic effects (Section 8.7) than the previously large and connected population.

Habitat fragmentation creates small and isolated subpopulations that have fewer opportunities to find food, water, shelter, and mates.

As if they are victims of double jeopardy, habitat fragmentation also impedes these smaller subpopulations' dispersal and colonisation abilities. Most species, especially those that occur in low densities, have large home ranges and/or live in ephemeral habitats, and must be able to move freely across the landscape to find shelter, food, water, and mates. A recent global review found that habitat fragmentation has already reduced the average distance of animal movements by two-thirds—from 22 km to 7 km—over the past few decades (Tucker et al., 2018). If they cannot move freely, these individuals cannot fulfil their needs and are at risk of extinction. **Habitat interior** specialists are particularly vulnerable to habitat fragmentation, as they are often reluctant to disperse over degraded or cleared areas, even if only a few metres wide (Blake et al., 2008; van der Hoeven et al., 2010). And yet, many habitat specialists face barriers much larger than a few metres. This includes Cameroon's few remaining drill (*Mandrillus leucophaeus*, EN) populations, which are facing extinction because individuals are reluctant or unable to disperse over agricultural land that stretches over hundreds of metres (Morgan et al., 2013).

Physical barriers that impede the ability of wildlife to move freely across the landscape also represent a form of habitat fragmentation. Dispersal impeded by human-constructed barriers, such as railways; dams; water-filled ditches; roads; and fences (Figure 5.2), can have disastrous consequences for biodiversity. Consider, for example, Africa's seasonal drylands. These areas were historically characterised by vast herds of **migratory** herbivores constantly moving from one area to another after fresh pasture. But as land management systems changed over time, the construction of roads and erection of fences to mark property boundaries impeded the ability of these herds to move freely after the resources they needed to stay alive (Durant et al., 2015; Hopcraft et al., 2015; Stabach et al., 2016). Restricted to only small parts of their range, these once-migratory animals were forced to overgraze the areas they already exploited, leading to extensive population declines. Through this process, Africa has already lost seven mass migrations, each involving millions of animals (Harris et al., 2009). Considering the economic stimulus provided by tourists visiting East Africa's famous Serengeti-Mara herbivore migration each year, the loss of these seven mass migrations have come at a huge cost to economies elsewhere. Luckily, through diligent conservation efforts, all of Africa's once-migratory herbivores have managed to persist in small and scattered populations throughout their range (Hoffmann et al., 2015). Section 11.3.1 discusses how some herbivore populations are reverting to their old migration routes after fence removals.

Figure 5.2 Common wildebeest (*Connochaetes taurinus*, LC) at Kenya's Maasai Mara that died after a fence stopped them from continuing their migration. Photograph by Teklehaymanot G. Weldemichel, CC BY 4.0.

Habitat loss and habitat fragmentation may even threaten the survival of species that are not as obviously dependent on large-scale movements for survival. As discussed in Section 4.2.5, many plants cannot persist without seed dispersal. Unfortunately, many seed dispersers, including forest primates (Estrada et al., 2017) as well as frugivorous birds, such as parrots, orioles, turacos, and hornbills (Lehouck et al., 2009), are sensitive to habitat fragmentation. In one of the few studies looking at this issue in Africa, researchers found that valuable timber trees in Tanzania's East Usambara Mountains are being extirpated as forest fragments become too small to support viable populations of fruit-eating birds (Cordeiro et al., 2009). The loss of these important seed dispersers will therefore have knock-on effects on the plants that depend on them for survival. Eventually, if enough seed dispersers, or perhaps even a single keystone species, disappear because of habitat fragmentation, entire ecosystems may eventually collapse.

5.1.2 What are edge effects?

> Edge effects exacerbate the impact of habitat fragmentation by reducing the functional size of habitat patches.

Edge effects are closely associated with, and exacerbate, the negative effects of habitat loss and fragmentation by altering environmental conditions in the habitat interiors. Dense woodlands, thickets, and forests are especially vulnerable to edge effects. Imagine a tropical forest, especially its large trees forming a continuous leafy canopy. These continuous canopies regulate the **microclimate** of a forest's understory by blocking sunlight and wind and maintaining humidity during the day, but also trapping heat rising from the forest floor at night. When the forest's trees are felled, the continuous canopy is fragmented, which in turn compromises the canopy's ability to regulate the forest's microclimate. Cleared areas, as well as forested areas directly adjacent to the cleared areas, will consequently be sunnier, warmer, windier, and dryer during the day, and cooler at night; these climatic changes also disturb nutrient cycles and biomass balances (Haddad et al., 2015). All of these changes further reduce the size of the forest patch to be smaller than the remaining canopy might indicate (Figure 5.3) as the new conditions prevent forest specialists such as shade-loving mosses, seedlings of late-successional trees, and humidity-sensitive amphibians from living in forest edges, leaving them with less interior forest habitat for which they must compete. Importantly, these microclimatic changes can penetrate a forest patch over much greater distances than one might expect. For instance, some forest birds in Uganda are sensitive to edge effects as far as 500 m from cleared areas (Dale et al., 2000).

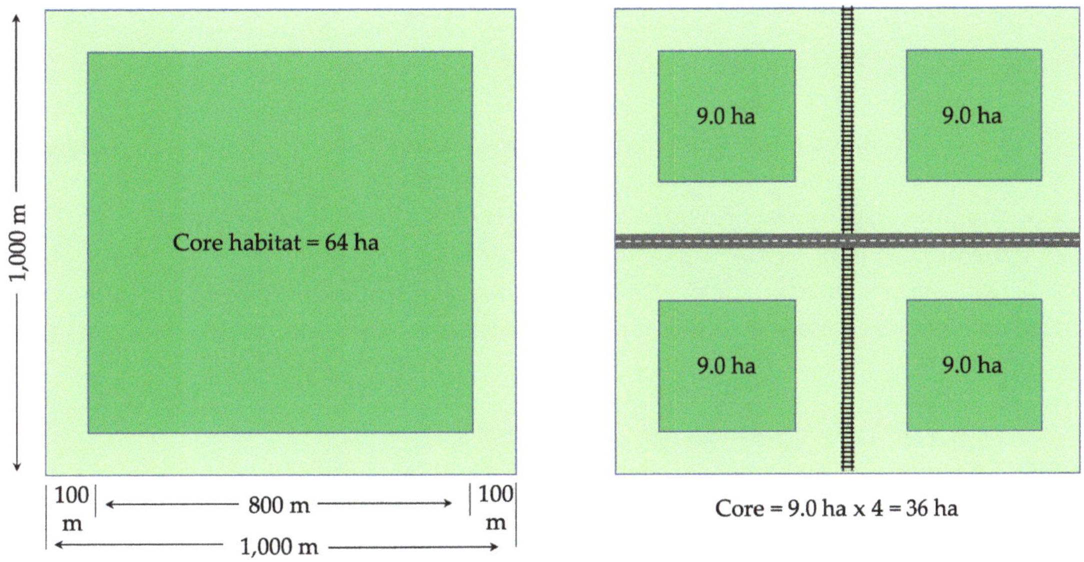

Figure 5.3 An illustration showing how habitat fragmentation and edge effects reduce habitat area. (A) A 100-ha forest patch, where edge effects (grey) penetrate 100 m into the forest: approximately 64 ha of the forest is still core habitat suitable for forest interior species. (B) The same 100-ha forest patch now bisected by a road and a railway. Although the road and railway take up very little area, it increases the patch's perimeter: area ratio. The resulting edge effects leave more than half of the forest unsuitable for interior species. After Primack, 2012, CC BY 4.0.

Edge effects also create several additional threats to the forest species already suffering from altered microclimates. Notably, disturbed edge conditions present a favourable environment for colonisation by fast growing and fast reproducing invasive species. (Threats posed by invasive species are discussed in more detail in Section 7.4). Those forest species that are not displaced by the altered microclimates and invasive species also face elevated predation risk. That is because trees that have died due to altered edge conditions provide suitable perches with clear views from which predatory birds can hunt (Sedláček et al., 2014). The degraded forest edge, sometimes resembling a savannah structure, also provides opportunities for woodland species such as snakes to enter the forests, pushing the remaining forest species even deeper into the forest (Freedman et al., 2009). For this reason, forest edge communities generally consist of widespread generalist species and invasive species, while specialist species that can hang on are, literally and figuratively, living on the edge.

The most devastating impact of edge effects is that edge effects beget further edge effects in a positive feedback loop leading to a rapidly disappearing ecosystem. First, expanding invasive (and generalist) species populations at habitat edges can easily overwhelm more sensitive habitat specialists. As habitat specialists are displaced at the contact zones, microclimatic conditions change, which allows for invasions even deeper into the fragmented habitat patch. In this way, invasions systematically penetrate deeper and deeper into the forest as microclimates are disturbed, habitat specialists are displaced, and new contact zones are created. The forest plants that die in the process also increase fuel loads, which, combined with drier and windier edge conditions, create an environment increasingly favourable for fire disturbance. Whether from lightning strikes or human activities, subsequent fires burn hotter and over a larger area (van Wilgen et al., 2007), disturbing and destroying more and more habitat each time. Through these mechanisms, edge effects can degrade entire ecosystems over time, harming both the native species and human livelihoods that depend on those areas.

> Edge effects beget further edge effects in a positive feedback loop leading to a rapidly disappearing ecosystem.

5.2 Drivers of Habitat Loss and Fragmentation

At present, Africa's biggest driver of habitat loss is agriculture (Potapov et al., 2017). African farmers have always cleared lands to meet their subsistence needs. Much of this clearing was traditionally and historically done in the form of **slash-and-burn agriculture** (also called shifting cultivation, Figure 5.4). To prepare land for crops, **smallholder farmers** would first cut down trees to clear the land and to obtain fuel wood. The remaining vegetation would then be burned away to release carbon and other nutrients, which increases land fertility. Farmers would grow crops on these cleared areas for two or three seasons. Then soil fertility would diminish, crop production would decline, and the farmers would abandon the area and clear new land, giving the natural ecosystem on the abandoned land time to regenerate.

Medical and technological advances, and the arrival of colonists, saw Africa's human population grow considerably since the 1800s. Feeding and accommodating the activities of this growing human population saw an increasing number of natural ecosystems replaced by agricultural land, and less area given the time to regenerate. An increasing number of people also started abandoning their rural

Figure 5.4 On a cloudless day, multiple fires raging in Mozambique's Zambezi River delta region can be seen from the International Space Station. Slash-and-burn techniques are often used to clear natural ecosystems for grazing and crops. Overly frequent fires, however, do not allow for ecosystem recovery, and are devastating to fire-sensitive ecosystems, such as tropical forests; instead of recovery, every fire creeps deeper and deeper into the forest until the entire ecosystem has been degraded. Image by NASA, https://commons.wikimedia.org/wiki/File:Zambezi_delta.jpg, CC0.

subsistence lifestyles for cities in search of jobs, financial freedom, and an easier life. As **urbanisation** increased (i.e. more people moved to cities) and competition for jobs intensified, an increasing number of city dwellers became dependent on collecting charcoal for cooking and cultivating cash crops, such as yams and cassava (Rudel, 2013). This saw even more natural ecosystems converted, particularly on the outskirts of cities. In the meantime, the remaining rural population became increasingly sedentary due to changing land tenure systems, which forced them into unsustainable farming practices as competition for land increased. These factors not only increased rates of habitat loss, but also changed the nutrient content in the soil which, in turn, reduce the land's ability to regenerate and to produce food (Drechsel et al., 2001; Wallenfang et al., 2015) which, in turn, leads to even more land clearing for agriculture.

While land clearing for smallholder agricultural needs continues to be an important driver of habitat loss (Tyukavina et al., 2018), its impact is increasingly dwarfed by the demands of commercial interests (Austin et al., 2017). The impact of **land grabbing** is of particular concern. Foreign companies from Asia and other parts of the world have acquired millions of hectares of land across Africa to stake a claim on the continent's rich natural resources, and to produce food and biofuels for their own people (von Braun and Meinzen-Dick, 2009). The foreign stakeholders, who often strike these land deals through loan agreements at the governmental level (i.e. with little to no local input), typically prioritise their own needs and profits over local interests with little care for the environment. These deals thus often end with a country saddled with debt they struggle to repay, and environmental damage that will take generations to reverse. Moreover, the foreign companies often employ migrant labourers with fewer protections and rights, compared to local peoples. In the process, while a modest number of local people may benefit from job creation, technology investment, and infrastructure development, a large number of local people become disenfranchised and displaced from the lands that previously supported their livelihoods. These foreign investments are a type of **neocolonialism** for their resemblance to Africa's earlier colonial era. They not only drive large-scale habitat loss, but in many instances also leave local people impoverished and desolate (Koohafkan et al., 2011).

To understand the impact of land grabbing on Africa's natural environment, one simply needs to consider their scale. For example, Chinese **bioenergy** producers recently secured over 48,000 km^2 of land in the DRC and Zambia (Smaller et al., 2012). Another deal, between the Ethiopian government and companies from India and Saudi Arabia, saw 5,000 km^2 of land (including sections of Gambella National Park) earmarked for commercial agriculture. At the time, this Ethiopian deal threatened both the second largest mammal migration on Earth (Ykhanbai et al., 2014) and the livelihoods of the local pastoralist Anuak community (Abbink, 2011). Fortunately, the Ethiopian government and developers were responsive to concerns raised by conservationists and human rights advocates, and agreed to set some areas aside for conservation, while also putting measures in place to maintain free movement of animals and pastoralists.

> The impacts of land clearing for smallholder farms are increasingly dwarfed by the outsized demands of commercial interests.

Infrastructure developments are also becoming an important driver of habitat loss. Offering access to previously unexploited areas, roads are perhaps the single biggest driver of habitat loss facing Africa's last remaining wildernesses (Figure 5.5). As prominent tropical biologist Bill Laurance eloquently noted, "Roads usually open a Pandora's Box of environmental problems—such as illegal fires, deforestation, overhunting and gold mining" (Laurance et al., 2014). A vast, growing body of literature from Africa supports these claims. For instance, research in the Congo Basin has shown how deforestation generally occurs within 2 km from roads (Mertens and Lambin, 1997)—more roads thus mean more deforestation. Roads also facilitate other drivers of forest loss, including the spread of invasive species, human settlements, fire, and pollution (Kalwij et al., 2008; Potapov et al., 2017). Providing access points for hunters, roads also facilitate unsustainable hunting; a recent review found that the wildlife reductions due to hunting could be detected as far as 40 km from the nearest road (Benítez-López et al., 2017).

Figure 5.5 New road developments, such as this one in the Congo Basin, represent one of the most immediate threats to biodiversity conservation. Road development provides access to previously unexploited areas, allowing more areas to be hunted, logged, farmed, and settled; increased human activity also exposes these areas to invasive species and pollution. Photograph by Charles Doumenge, https://www.flickr.com/photos/internetarchivebookimages/20689353531, CC0.

5.3 Habitat Loss' Impact on Africa's Ecosystems

5.3.1 Tropical forests

Occupying about 7% of all land surfaces, tropical forests are estimated to contain over 50% of the world's terrestrial species (Corlett and Primack, 2010). Due to these high levels of biodiversity, the complexity of biological interactions in tropical forests is unparalleled in other ecosystems, and consequently also their importance to humans. On a local scale, the timber and non-timber products from tropical forests sustain the traditions (Box 5.2), livelihoods, and financial well-being of millions of Africans. Tropical forests also have regional importance including protecting catchment areas (Section 4.2.4) and moderating climate (Section 4.2.3). Lastly, as reservoirs of carbon, tropical forests play a globally important role in mitigating the negative effects of anthropogenic climate change (Section 10.4), and with 17% of Earth's tropical forests, Africa plays a globally important role in tropical forest conservation efforts.

Despite the importance of tropical forests, their destruction has become synonymous with the rapid loss of biodiversity (Figure 5.6). Africa had already lost over 65% of its original tropical forests by 1990 (Sayer, 1992); human activities destroyed an additional 308,000 km^2 (an area larger than Italy) between 1990 and 2010 (Achard et

Box 5.2 The Conservation and Exploitation of East African Plants

John R. S. Tabuti

College of Agricultural and Environmental Sciences, Makerere University, Kampala, Uganda.

✉ *jtabuti@caes.mak.ac.ug*

Ethnobotany, as a scientific discipline, studies the relationships between people and plants: how people affect the survival and distribution of plants, and how plants influence human behaviour and cultures. For instance, local cuisines are shaped by available plant species, and people cultivate species that they consider useful. Conservation of plant diversity can be aided in many ways by recognising the importance of plants to people's livelihood and spiritual practices.

The people of East Africa identify and use a great many plant species that are essential for their well-being (Tabuti, 2006). Native plants are used for food, for construction, to treat the diseases of both people and livestock, and in numerous other ways. Some of the most important species include White's ginger (*Mondia whitei*) and red stinkwood (*Prunus africana*, VU) for medicine, African teak (*Milicia excels*, NT) for timber, shea tree (*Vitellaria paradoxa*, VU) for food and cosmetics, and African sandalwood (*Osyris lanceolate*, LC) as a source of fragrant oil.

Some plant species (and sometimes entire ecosystems, such as forests) are valued for religious or cultural reasons. The plants or forest areas themselves are considered sacred, the site of a deity or spirit, with certain rituals performed using those special plant species or the habitats they occupy. These sacred sites and species are protected by local taboos. For example, the powderbark gardenia (*Gardenia ternifolia*) is not harvested for firewood among the Balamogi people of Uganda because it is believed to bring bad luck. Among the Mijikenda people of Kenya, sacred forests known as Kaya are protected because people believe that the forests are inhabited by spirits and are places of prayer and held as a source of ritual power. Cutting down trees, grazing livestock, and farming are prohibited within the Kaya. One protective belief holds that cutting a tree in the Kaya with a machete can result in the machete rebounding and causing injury to the woodcutter. Another belief is that food cooked using wood from these sacred forests can cause sickness, and that a dwelling built with timber drawn from the forest will collapse. Consequently, more than 50 Kaya—ranging in size from 0.3 to 3 km^2 and home to 187 plants, 48 birds, and 45 butterfly species—have enjoyed unofficial protection due to religious and cultural beliefs.

Today, however, the plants and their natural communities on which people rely for their well-being are being threatened. By far the greatest threat is land use change and habitat conversion to agriculture to grow food for a growing population. Changing cultural and spiritual values in East Africa, as well as social and economic pressures, are threatening the existence of even sacred forests. For instance, the coronation site of the Paramount Chief of the Balamogi in Uganda was previously protected as a sacred forest by local lore, but it has now been cut down and converted into gardens by local people who no longer follow ancient traditions. Harvesting of plant species, such as the red stinkwood and East African sandalwood, for international markets is also a significant threat no longer held at bay by cultural norms.

Figure 5.B The edge of the Budongo Forest Reserve, Uganda, where researchers collaborate with local communities to refine methods for the sustainable utilisation of tropical plant products. Photograph by John Tabuti, CC BY 4.0.

Thankfully, several species continue to be actively protected by local communities and governments. According to Greger (2012), traditional healers aid conservation by replanting around 50% of the medicinal plant species that they consider to be important to their practice. For the relationship between people and plants to survive, scientific conservation and local tradition must work together. An example of such collaboration is on display in Uganda's Budongo Forest Reserve (Figure 5.B), where researchers at the Budongo Conservation Field Station are working with local communities to refine methods for sustainable management and utilisation of the region's local plants.

al., 2014). Losses were particularly severe in Burundi, Benin, and Mozambique, with each country holding less than 5% of its original forest cover (Sayer, 1992). Retaining about half of its original forest cover, the DRC is relatively better off, but current deforestation rates in this country are currently second highest globally (Weisse and

Goldman, 2019). Current deforestation rates are so severe in Equatorial Guinea that this country will lose all its forests within the next 20 years if current trends hold (Potapov et al., 2017). Despite these alarming trends—the destruction continues nonstop, particularly in Ghana and Côte d'Ivoire, which saw a 60% and 26% rise in forest loss (the highest rise globally), respectively, between 2017 and 2018 (Weisse and Goldman, 2019). Across Africa, logging is currently the dominant driver of tropical forest loss (causing 77% of total losses over the past decade), followed by agriculture (Potapov et al., 2017).

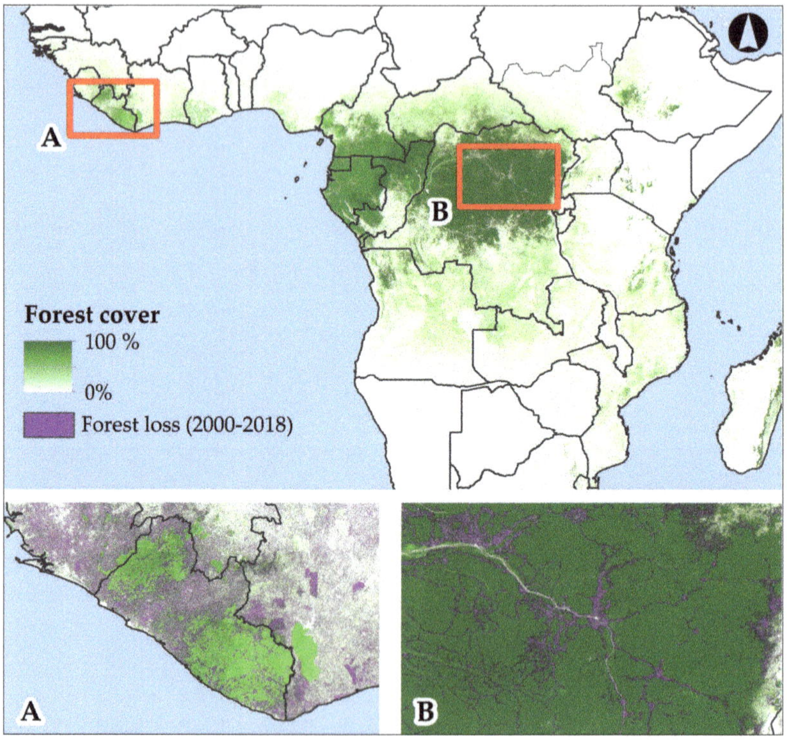

Figure 5.6 The extent of Sub-Saharan Africa's tropical forests in 2018, and the extent of tropical forests loss (A) around Liberia and (B) in the north-eastern part of the Congo Basin between 2000 and 2018. Note in (A) how deforestation follows country borders, and in (B) how deforestation follows road networks. Source: Hansen et al., 2013. Map by Johnny Wilson, CC BY 4.0.

5.3.2 Rivers and deltas

Due to our dependence on freshwater, humans have always preferred to live near rivers, streams, and lakes. Consequently, these aquatic environments have been destroyed at a scale at least equal to that of terrestrial environments. Rivers have taken a particularly hard hit from human activities, being polluted by industries and dammed to ensure a reliable, year-round supply of water for consumption and irrigation, and to generate hydroelectricity.

Dam construction holds several negative consequences for biodiversity and people. Aquatic organisms that cannot survive the altered river conditions downstream (reduced flow and dissolved oxygen, higher temperatures, and increased turbidity) are most vulnerable. For example, a study from South Africa found that native macroinvertebrate populations (often a good indicator of water quality) were reduced by 50%, and some insect orders virtually extirpated following dam construction (Bredenhand and Samways, 2009). Dams also displace aquatic organisms upsteam. In one well-studied example, back flooding of Mozambique's Massingir Dam facilitated river substrate changes and the spread of invasive species, which in turn forced sharptooth catfish (*Clarias gariepinus*, LC), tiger fish (*Hydrocynus vittatus*, LC), and Nile crocodiles (*Crocodylus niloticus*, LC) to change their diet. Increased stress levels due to these dietary and environmental changes leave the affected animals susceptible to pansteatitis (a condition where body fat becomes inflamed), leading to mass wildlife mortality events in South Africa's Kruger National Park (Woodborne et al., 2012). Lastly, dams reduce connectivity in freshwater ecosystems, preventing freshwater organisms from exchanging genetic material, migrating between upsteam and downstream areas, and adapting to changing conditions. For example, in West Africa, the damming of the Senegal River blocked the annual migration path for African river prawns (*Macrobrachium vollenhoveni*, LC), a major predator of snails which host schistosomiasis (bilharzia). Once the dam was completed, prawn populations collapsed, leading to a schistosomiasis epidemic in villages upstream from the dam (Sokolow et al., 2015).

Terrestrial ecosystems also suffer from dam construction. Of concern is the direct loss of riverine and **palustrine ecosystems** downstream from the dam due to reduced waterflow. For example, construction of Nigeria's Kainji Dam in the Niger River caused the drying of large wetlands and floodplains downstream, in the process displacing nearly 400,000 people who depended on the river's now-compromised seasonal flood cycles (Drijver and Marchand, 1985). Flooding of upland areas next to dammed rivers also displaces terrestrial wildlife and people. For example, construction of Mali's Manalati Dam flooded 430 km^2 of savannah and 120 km^2 of forest, which fractured the migration routes of the region's **nomadic** pastoralists, leading to overgrazing and soil erosion of the remaining grazing lands (deGeorges and Reilly, 2006), in addition to a 90% loss of fisheries downstream (Acreman, 1996).

> Damming rivers harms biodiversity and people both upsteam and downstream from these developments.

5.3.3 Wetlands

Throughout Africa, wetlands are being mined for valuable peat, or drained and/ or filled in for development and agriculture. Through these activities, the region has already lost approximately 43% of its wetlands, with current rates of loss among the highest in the world (Davidson, 2014). This is a major concern because wetlands serve as spawning grounds and nurseries for aquatic and amphibious wildlife and stop-over sites for migratory birds (Box 5.3). Wetlands also provide multiple important ecosystem services. For example, they prevent erosion and runoff by capturing large volumes of floodwater, which is then released slowly over time. This process also allows sediments and nutrients kicked up during flood events to settle out, creating fertile habitats for a wide diversity of animals and plants, as well as for agriculture. Water that leaves after this settling period is cleaner than when it entered, having been filtered by the soil, plants, and microbes of wetlands. This water purification and filtration service is generally cheaper and much more efficient than man-made filtrations systems. The loss of any wetlands, but especially at such large scales, is thus a grave concern not only because of the countless animals and plants threatened with extinction, but also the people that depend on all the valuable ecosystem services they offer.

Box 5.3 Migratory Birds of Africa: The Largest of the Last Great Migrations?

Abraham J. Miller-Rushing[1] *and John W. Wilson*

[1]*Acadia National Park, US National Park Service, Bar Harbor, ME, USA.*

How are Africa's bird migrations, the biggest in the world, faring in a rapidly changing world? Each year, about 2.1–5 billion birds (mostly songbirds, but also raptors, waterbirds, and many others) travel back and forth between their wintering grounds in Africa and breeding grounds in Europe and Asia (Figure 5.C). Of the 126 species involved in this migration, over 40% have continuously decreased in abundance since 1970 (Vickery et al., 2014). At first, populations that overwintered in open dry savannahs declined: examples include the Ortolan bunting (*Emberiza hortulana*, LC) and European turtle dove (*Streptopelia turtur*, VU) which decreased by 84% and 69% between 1980 and 2009, respectively. More recently, species overwintering in the humid Afrotropics also started declining: this includes songbirds, such as the common nightingale (*Luscinia megarhynchos*, LC) and river warbler (*Locustella fluviatilis*, LC)—populations of both declined by 63%—and waterbirds such as the black-tailed godwit (*Limosa limosa*, NT), which declined by 45%.

To survive their long journeys, migratory birds need favourable weather conditions, adequate food sources, and intact habitat not only at the end points where they breed or overwinter, but also along their routes where the migratory animals can rest and refuel (Runge et al., 2015). Disturbances in any of these places can lead to sharp population declines. For example, recent research showed that the habitat quality of a single stop-over site can determine whether a migration is successful or not (Gómez et al., 2017). Illustrating this point, a drought in the Sahel, an important migratory stop-over site, led to food shortages that killed 77% of the world's common whitethroats (*Sylvia communis*, LC); even today, this population has not yet fully recovered (Vickery et al., 2014).

Human activities have greatly contributed to the declines of Africa's migratory birds (Kirby et al., 2008; Vickery et al., 2014). For example, each year thousands of hectares of wetlands, forests, grasslands, and savannahs

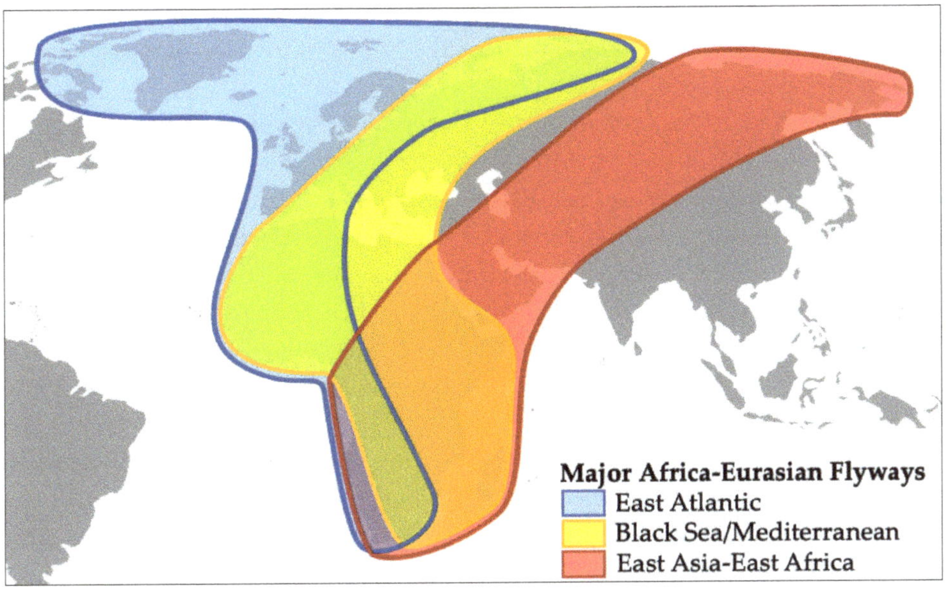

Figure 5.C The three major migratory flyways that African birds use to travel back and forth between their wintering grounds in Africa and breeding grounds in Europe and Asia each year. After BirdLife International, 2019, CC BY 4.0.

are being converted into farmlands and urban areas or polluted by rampant use of pesticides and herbicides. Migratory birds also need to deal with hunters and trappers, and an increasing number of human-made structures, such as high-rise buildings, wind turbines, and power lines that represent collision and electrocution hazards (e.g. Rushworth et al., 2014). Then there is the threat of inconsistent rainfall, which causes food shortages and direct mortality, and climate change, which causes temporal mismatches between migratory movements and abundance of key food resources (Both et al., 2006; Vickery et al., 2014).

Addressing these declines, governments, conservation organisations, and local communities all over Africa have started initiatives to protect migratory birds and their habitats. One such initiative is happening in Kenya's Tana River Delta, one of the most important stop-over sites along the Asian-East African Flyway. Every year, Basra reed warblers (*Acrocephalus griseldis*, EN) return from their Middle Eastern breeding grounds to overwinter in the Delta, which covers 1,300 km^2 and supports dozens of threatened species. The area, however, has been under serious threat from development for sugarcane and biofuel crops since 2008. These activities could reduce dry season water flow by up to one third. Local people and conservationists strongly oppose these developments because of its threat to local communities' ways of life and to wildlife populations. Their efforts gained international attention, and in 2012, Kenyan courts halted development until comprehensive management plans were developed that included environmental impact assessments and local stakeholder engagement (Neville, 2015). Today, local people gain benefit from more sustainable industries, including eco-charcoal audited by the Forest Stewardship Council (FSC), and solar-powered energy to reduce the need for wood.

Also, in West Africa, collaborative conservation initiatives are taking steps to protect the critical East Atlantic Flyway. For example, under the guidance of BirdLife International, Guinea-Bissau residents are now monitoring several wetlands in the Bijagós Archipelago to track how well migratory waterbirds are doing at this critically important stop-over site. Also, in Senegal, where two important stop-over sites (Saloum Delta and Djoudj wetlands) are located, the local non-profit NGO Nature Communautés Développement initiated an extensive conservation education programme aimed at safeguarding the region's birds.

Conserving migratory species that cover huge distances and rely on habitats in many areas is not easy. However, efforts like these in West Africa and Kenya (which combine the interests of local people and wildlife) provide excellent models for others to build from.

Mangrove losses around Africa have been extensive despite them providing an estimated US $57,000 worth of ecosystem services per hectare.

Mangrove swamps (sometimes called mangrove forests, though technically a wetland because their function and structure are primarily determined by hydrology, Lewis, 2005; Gopal, 2013) are one of Africa's most threatened wetland ecosystems. Characterised by woody plants that can tolerate saltwater, mangrove swamps occupy brackish waters in tropical coastal areas, typically where there are muddy bottoms. These areas are sparsely distributed; globally, mangrove swamps cover only 53,000 km^2 of land scattered across 118 countries (Dybas, 2015). Protecting Africa's mangrove swamps, comprising 21% of Earth's total, is important both biologically and economically. In addition to holding many unique species, mangrove swamps also protect coastal cities and villages from cyclone/hurricane and tsunami damage and provide important breeding and feeding grounds for marine shellfish and fish. One study estimated that mangrove swamps provide an estimated US $57,000 worth of ecosystem services per hectare (van Bochove et al., 2014). Yet, only 7% of Africa's mangrove swamps are protected. With so little protection, it comes as no surprise that a large percentage of Africa's mangrove swamps have been destroyed or damaged by agriculture, urban expansion, pollution, and commercial shellfish farming (Giri et al., 2011). In West Africa, the situation is particularly dire. Wood extraction for commercial fish smoking is one of the biggest drivers of mangrove losses, even within protected areas (Feka et al., 2009). With so much destruction, it should come as little surprise that about 40% of vertebrate species endemic to mangrove swamps are currently threatened with extinction (Luther and Greenberg, 2009).

5.3.4 Seasonal drylands

Africa is also rapidly losing its semi-arid savannahs, scrublands, and grasslands through conversion to agriculture (Box 5.4) and **desertification**—the systematic degradation of formerly complex and adaptive seasonal drylands into barren wastelands (Figure 5.7). When human populations were low, nomadic pastoralism and shifting cultivation enabled people to utilise seasonal drylands in a sustainable way. Today however, population growth, combined with restrictions placed on free movement by administrative borders and competition for land, forces people and animals living on drylands to be more sedentary. While these areas may initially support some agriculture and livestock, unsustainable techniques, such as overgrazing and excessive tilling, lead to soil erosion and the depletion of soil nutrients and natural **seed banks**. With the cover vegetation gone, the unprotected topsoil is easily lost to wind and flooding, leaving behind the deeper, infertile, and compact subsoil layers with little capacity tfo hold water. The result is something that closely resembles a man-made desert. However, rather than a functional ecosystem characterised by species adapted to life in the desert, these wastelands have lost their original productivity and biological communities, only to be revived through expensive and/or time-consuming **land reclamation** methods.

> Africa is rapidly losing semi-arid ecosystems due to desertification, the conversion of productive ecosystems into barren wastelands.

Box 5.4 Saving Critically Endangered Ground Nesting Birds from Habitat Loss

Bruktawit Abdu Mahamued[1,2]

[1]*Biology Department, Kotebe Metropolitan University, Addis Ababa, Ethiopia.*

[2]*Edge of Existence Fellow, Zoological Society of London, London, UK.*

✉ *brukabdu.m@gmail.com*

We are currently witnessing the start of the sixth mass extinction of species on our planet. From here onwards, biodiversity losses are expected to increase rapidly: a recent UN report estimated that about one million species are already threatened with extinction (IBPES, 2019). While the reasons behind these losses vary by region, in Africa, a major driver is habitat loss. With the current push for development, the impacts of habitat loss are increasing dramatically, affecting species both inside and outside of protected areas. Two Ethiopian birds (Figure 5.D), the Liben lark (*Heteromirafra archeri*, CR) and white-winged flufftail (*Sarothrura ayresi*, CR), exemplify many of the dilemmas associated with protecting biodiversity on unprotected lands where habitat loss is severe.

Figure 5.D (Left) A white-winged flufftail, one of Africa's most enigmatic birds, standing defensively in front of its nest (eggs can be seen in the background) in the flooded grassland on the Berga floodplain, Ethiopia. Photograph by Bruktawit Abdu Mahamued, CC BY 4.0. (Right) A Liben lark on its last remaining stronghold in the world, Ethiopia's Liben Plain. Photograph by Tommy P. Pedersen, CC BY 4.0.

The Liben Plain is part of the Borana rangelands, managed by Borana pastoralists under their traditional rangeland management system which is generally compatible with conservation ideals. The Borana's way of life was disrupted about 40 years ago due to pressure from a former Ethiopian government who wanted the Boranas to adopt a more sedentary lifestyle. For example, drilling of water wells in dry season grazing areas disrupted seasonal grazing systems, while fires that the Boranas used to maintain productive grazing lands and prevent shrub encroachment were prohibited. The Boranas also face pressure from changing land tenure systems. The Liben Plain grasslands are located on communal lands upon which nobody can claim ownership. However, if someone wants to farm here, they just pay a tax that in effect assures ownership of the land. The Boranas were initially slow to adopt this farming lifestyle, but when outside settlers started taking advantage of the government's farming incentives, the Boranas were pushed to do the same to prevent all their ancestral land from being turned over (Mahamued, 2016). The subsequent loss of fire management (and associated shrub encroachment) and cropland expansion, together with increased human and livestock populations, have led to a major loss of the Liben Plains' natural ecosystem.

The Liben lark is a ground-nesting bird that is near-endemic to Ethiopia (a second population in Somalia may already be extinct; Spottiswoode et al., 2013). Here, its main population is restricted to the open grasslands of the Liben Plain. Although it was previously common in this ecosystem, habitat loss and degradation have reduced the availability of suitable feeding and nesting sites. Further, the reduced population is also increasingly vulnerable to direct threats such as nest predation and trampling of nests by cattle (Spottiswoode et al., 2009). Due to these threats, the lark's numbers have decreased so dramatically in recent years that it was classified as *Critically Endangered* in 2009.

To prevent the extinction of the lark, the Ethiopian Wildlife and Natural History Society (EWNHS), BirdLife International, and other organizations collaborated with local authorities and community leaders in 2016 to establish enclosures for grassland regeneration. These enclosures are in effect communally-managed grassland reserves regulated under a subset of customary laws. These areas not only secure suitable habitats for the Liben lark, they also provide benefits to the Borana community like securing grazing lands for the dry season when the lark is not breeding. This initiative shows early promise—over 350 ha of grassland reserves have already been established, and over 1,000 ha of shrub have been cleared (Kariuki and Ndang'ang'a, 2018). But to truly secure the future of the Liben lark, more support is needed from the Ethiopian government, particularly in preventing further land conversion, supporting ecosystem restoration, and encouraging the Borana pastoralists' traditional way of life.

Another species facing imminent extinction due to habitat loss is the white-winged flufftail. One of Africa's most enigmatic birds, the flufftail is an intra-African migrant restricted to a few seasonal high-altitude wetlands in South Africa and Ethiopia. Like the lark, the flufftail is a ground-nester that struggles to find suitable nesting sites relatively free from disturbance. The Berga floodplain, the flufftail's Ethiopian stronghold, used to be covered by productive grasslands. This unspoiled landscape is now being replaced by settlements, crop farms, and eucalyptus plantations that generate quick profits. This, together with overgrazing, has led to extensive soil erosion, which in turn has altered the structure and grass composition of the floodplain. Today, the floodplain is encroached by invasive weeds and other less desirable vegetation (seen during EDGE project surveys in 2018) which, together with others forms of disturbance, have reduced the amount of suitable habitat available for the flufftail to such an extent that it is now considered *Critically Endangered*.

To prevent the extinction of the flufftail, the EWNHS along with the Middlepunt Trust and BirdLife South Africa have taken several steps to improve the outlook for the flufftail. Much of this work involved working with the people at Berga to improve their livelihoods, and to instil a sense of ownership of their local biodiversity. A prominent outcome of this collaboration was a primary school named after the flufftail; results from the project also contributed to a species action plan (Sande et al., 2008). But without continued maintenance, the progress made by this short-term initiative will have limited long-term value. The flufftail's future thus continues to be dire, as unsustainable land use practices continue to destroy the Berga floodplain. There is an urgent need for joint long-term efforts to reverse the fate of the species, including taking steps to establish protected areas, to initiate carefully-planned ecosystem restoration efforts, and to develop a new species management plan that will provide lasting benefits.

Figure 5.7 Desertification, the degradation of formerly complex and adaptive seasonal drylands into barren wastelands, is a growing threat to Africa's natural environment, its wildlife, and its people. It is a prominent problem in the Sahel region, such as the area pictured, in Burkina Faso. Photograph by Jose Navarro, https://www.flickr.com/photos/53871588@N05/5630241115, CC BY 4.0.

5.4 Population Growth and Consumption?

Until about 150 years ago, the rate of human population growth in Africa had been relatively slow, with the birth rate only slightly exceeding the death rate. Modern medical achievements and more reliable food supplies have changed this balance; they have reduced mortality rates while birth rates remain high. Consequently, Sub-Saharan Africa's human population has exploded to 1 billion people over the past decade (World Bank, 2019). Today, Sub-Saharan Africa is leading the world in human population growth, projected to increase by four-fold over the next century. Population growth rates for individual countries are similar, if not higher. For example, Ethiopia's human population has grown from 48 million in 1990—when the region experienced a famine crisis—to nearly 100 million in 2015; current projections forecast a population of 172 million by 2050. The human population of Tanzania's Dar es Salaam, a coastal city particularly vulnerable to sea level rise (Section 6.3.2), is expected to increase from

4 million to 21 million between 2015 and 2050, while Lagos in Nigeria is expected to grow from 21 million to 39 million people over the same time.

Simple math suggests that more people leads to less space for biodiversity (Figure 5.8), because humans and wildlife compete for the same resources, broadly speaking. With many countries in Africa already facing social, economic, and developmental challenges such as malnutrition, crime, and unemployment, one can almost understand why politicians prioritise socio-economic upliftment over biodiversity conservation. This is a grave mistake; as discussed in Chapter 4, biodiversity and human well-being are intricately linked. It is one of conservation biologists' most important tasks: to make the link between conservation and human welfare clear to policy scholars and politicians.

Figure 5.8 Night lights of Kinshasa, capital of the DRC and Africa's second largest city. To have more people leads to more competition for space, leaving less space to maintain biodiversity and ecosystem services. It also means more natural resources extracted, more pollution, and more greenhouse gas emissions. Photograph by MONUSCO/Abel Kavanagh, https://www.flickr.com/photos/monusco/23769991270, CC BY-SA 2.0.

In recent years, there has been an increasing tendency of economists, scientists, and politicians to shift the focus from population growth to consumption as the more important underlying driver of biodiversity loss. For many, the emphasis on consumption avoids politically charged topics, such as population control, which most people oppose on ethical or moral grounds, and because it is associated with divisive topics such as xenophobia, racism, and eugenics (Kolbert and Roberts, 2017). Others highlight that it is not the number of people per se, but how natural resources are consumed that is the main cause of environmental decline. Indeed, affluent people and affluent countries have a disproportionate impact on the natural environment because they consume a disproportionately large share of the world's natural resources. To use one example, the USA accommodates only 5% of the world's human population but uses 25% of the world's harvested natural resources each year (WRI, 2019). In fact, decorative Christmas lights in the USA alone use more energy than the annual energy

The major threats to biodiversity are all rooted in expanding human populations and unsustainable consumption patterns.

usage of the entirety of Ethiopia or Tanzania (Moss and Agyapong, 2015). And yet, the average USA citizen uses less than half of the energy (measured as carbon emissions) that an average citizen of Qatar uses (World Bank, 2019; see also Figure 5.9), Qatar being a small but wealthy Middle Eastern country.

Another important aspect to consider in the consumption argument is that, through increased **globalisation**, the impacts of consumption in industrialised countries are felt over much greater distances than before (Moran and Kanemoto, 2017). For instance, chocolate consumed in Europe was most likely made with cacao produced in West Africa (Gockowski and Sonwa, 2011); other crops, such as coffee and tea, produced in Africa are similarly enjoyed all over the world. In the best-case scenario, African farmers are satisfying a demand in a global market; at worst, foreign companies are establishing croplands with little benefit trickling down to local people. Supporters of the consumption argument rightfully point out that it would be very unjust to blame the local farmers for cleared forests when they simply produce commodity crops that the international market demands.

The global demand for natural resources such as coffee, cacao, palm oil, and timber is helping fuel habitat loss in Africa.

As with many other complex challenges, both sides of the population-consumption debate are correct. One method to link the impact (I) of a human population on the environment is through the formula $I = P \times A \times T$ (*IPAT* in short), where P is population size, A is Affluence (e.g. per capita GPD), and T is technology (e.g. per capita energy use) (Ehrlich and Goulder, 2007). The *IPAT* equation is similar in concept to the **ecological footprint** (Figure 5.9): both illustrate that human populations *and* consumption patterns interact to exacerbate human impacts on the environment. In other words, many poor Africans can have the same impact on the environment as just a few wealthy Americans, and vice versa.

Both the *IPAT* equation and ecological footprint concept are insightful as to the challenges facing Africa's ecosystems and people. Today, Africans are increasingly aspiring to attain the same levels of high consumption as industrial countries. These patterns generally lead to an inefficient, wasteful, and unsustainable use of natural resources (i.e. overconsumption). Population growth rates in many industrial countries are currently slowing; some countries are even experiencing long-term population declines, which allow conservation-minded individuals in those countries to focus their efforts on addressing consumption patterns. The situation is quite different in Africa, where we are faced with increasing per capita consumption *and* the fastest growing human population on Earth. In the face of the resulting increased competition for space, African conservation biologists must adopt a holistic approach to ensure that welfare standards are upheld or improved while our natural heritage is protected. One of the most important strategies involves championing sustainable economic development over unsustainable economic growth (Section 15.1). While conservation biologists differ in terms of how strongly they argue for addressing the population size issue, most also agree that conservation goals benefit from education, the empowerment of women, and wider access to family planning and reproductive health services.

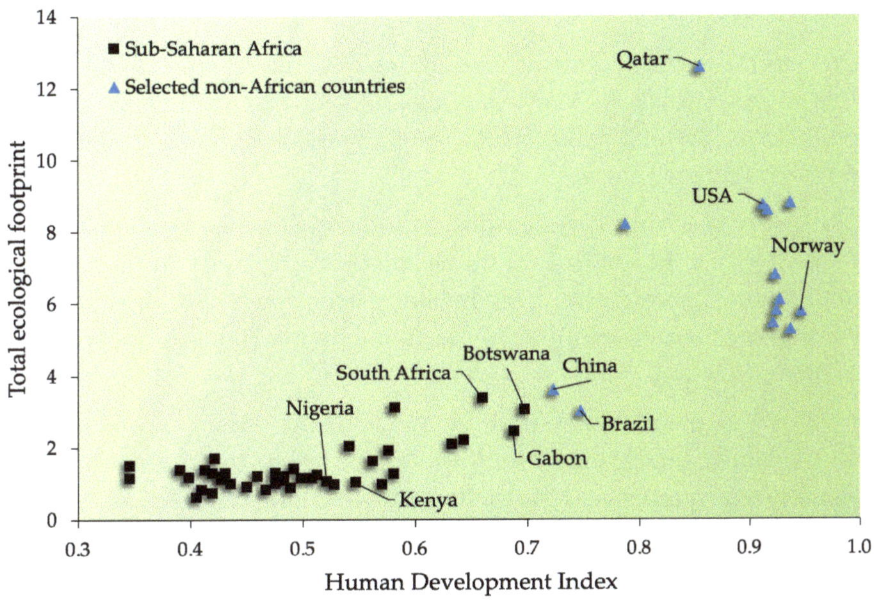

Figure 5.9 A nation's ecological footprint is calculated by estimating the amount of land needed to support the average resident of that nation. Although there is some disagreement as to the exact methods for these calculations, the overall message is clear: people in more developed nations use a disproportionately large amount of natural resources. However, the *overall* impacts of countries with huge populations, such as China, are also huge because of the cumulative impact from so many people. Source: GFN, 2017, CC BY 4.0.

5.5 Concluding Remarks

There is no doubt that agriculture, forestry, and infrastructure developments—the main drivers of habitat loss and fragmentation—play an important role in socio-economic development across Africa. Nevertheless, many (perhaps most) of these developments are set up to benefit a select few individuals and corporations primarily interested in short-term gains rather than a wide range of stakeholders over the long-term. To maintain biodiversity and improve our quality of life, governments across the region must ensure that the benefits of development are shared fairly across society and that industries are accountable for their fair share of the natural resources they use (Section 4.5.3). Also, the region's growing number of wealthy people who benefit most from development must re-evaluate their lifestyles (whether willingly or through government interventions, such as taxation) to avoid excessive consumption patterns. Some of the first steps may be relatively easy. For example, the water used to produce Sub-Saharan Africa's wasted food—a full third of all produced food (FAO, 2013)—equals the annual discharge of the mighty Zambezi River where it enters the Indian Ocean in Mozambique (Beilfuss and dos Santos, 2001). At the same time, we *must* all play our part in achieving **sustainable development**, by encouraging family planning activities and assisting industries to grow in a responsible way (Section 15.1). Neglecting that, we compromise our own futures, and that of our children.

5.6 Summary

1. One of the primary threats to biodiversity today are habitat loss and habitat fragmentation. Many species living in tropical forests, freshwater ecosystems, the marine environment, and seasonal drylands are at risk of extinction due to habitat loss.
2. The theory of island biogeography and the species-area relationship can be used to predict the numbers of species that will go extinct due to habitat loss. Both theories predict that large habitat patches are better able to maintain wildlife populations because they accommodate populations better buffered against extinction.
3. Habitat fragmentation describes the process when once large and widespread habitats (and hence wildlife populations) are divided into several increasingly smaller and isolated units. This process leads to extinctions because it impedes dispersal, colonisation, foraging, and reproduction.
4. Edge effects reduce the functional size of habitats because they alter microclimates and expose habitat specialists to displacement by invasive species, predators, and other disturbances.
5. Habitat loss and fragmentation are rooted in expanding human populations and excessive consumption of natural resources. The *IPAT* equation illustrates how population size, wealth, and technology together determine our impact on the environment.

5.7 Topics for Discussion

1. Why does oil palm (*Elaeis guineensis*, LC) cultivation represent a significant threat to biodiversity in Africa? (In addition to your own research, it might also be useful to read Box 6.1.)
2. Read Harris et al. (2009) on the decline of the world's mass wildlife migrations. Which lost African migration appeals to you most and why? What species were involved? What numbers of animals were involved? How do you think this migration can be revived?
3. Which ecosystem in your region would you consider the most damaged, and which would you consider the most pristine? Can you explain why these two ecosystems have such different fates?
4. Do you agree with the idea that human population growth is the primary driver of extinctions today? Why? How do we balance protecting biodiversity with providing for a growing human population, and the right of people to have children?

5.8 Suggested Readings

Arcilla, N., L.H. Holbech, and S. O'Donnell. 2015. Severe declines of understory birds follow illegal logging in Upper Guinea forests of Ghana, West Africa. *Biological Conservation* 188: 41–49. https://doi.org/10.1016/j.biocon.2015.02.010 Illegal logging is on the increase, and wildlife communities struggle to recover.

Caro, T., J. Darwin, T. Forrester, et al. 2012. Conservation in the Anthropocene. *Conservation Biology* 26: 185–88. https://doi.org/10.1111/j.1523-1739.2011.01752.x Even though human activities dominate large areas of the earth, it is important to remember and plan for the many places and ecosystems where human influence is still minimal.

Haddad, N.M., L.A. Brudvig, J. Clobert, et al. 2015. Habitat fragmentation and its lasting impact on Earth's ecosystems. *Science Advances* 1: e1500052. https://doi.org/10.1126/sciadv.1500052 There are many ways that fragmentation hurts biodiversity.

Harris, G., S. Thirgood, J.G.C. Hopcraft, et al. 2009. Global decline in aggregated migrations of large terrestrial mammals. *Endangered Species Research* 7: 55–76. https://doi.org/10.3354/esr00173 Habitat loss continues to threaten the world's remaining mass migrations.

Ibisch, P.L., M.T. Hoffmann, S. Kreft, et al. 2016. A global map of roadless areas and their conservation status. *Science* 354: 1423–27. https://doi.org/10.1126/science.aaf7166 Africa still holds large roadless areas; we need to keep it that way.

Laurance, W.F., J. Sayer, and K.G. Cassman. 2014. Agricultural expansion and its impacts on tropical nature. *Trends in Ecology and Evolution* 29: 107–16. https://doi.org/10.1016/j.tree.2013.12.001 Agriculture and roads will have severe impacts on Africa's ecosystems in the coming century.

Rudel, T.K. 2013. The national determinants of deforestation in sub-Saharan Africa. *Philosophical Transactions of the Royal Society B* 368: 20120405. https://doi.org/10.1098/rstb.2012.0405 One manuscript in a special issue on deforestation in Africa; other manuscripts in this issue are also worth scanning.

van der Hoeven, C.A., W.F. de Boer, and H.H. Prins. 2010. Roadside conditions as predictor for wildlife crossing probability in a Central African rainforest. *African Journal of Ecology* 48: 368–77. https://doi.org/10.1111/j.1365-2028.2009.01122.x Some species are highly reluctant to cross roads, even inside protected areas.

Woodborne, S., K.D.A. Huchzermeyer, D. Govender, et al. 2012. Ecosystem change and the Olifants River crocodile mass mortality events. *Ecosphere* 3: 1–17. https://doi.org/10.1890/ES12-00170.1 Damming rivers can lead to ecological disasters

Ykhanbai, H., R. Garg, A. Singh, et al. 2014. *Conservation and "Land Grabbing" in Rangelands: Part of the Problem or Part of the Solution*? (Rome: International Land Coalition). http://pubs.iied.org/pdfs/G03853.pdf Conservation biologists should work with local communities to prevent land grabs.

Bibliography

Abbink, J. 2011. 'Land to the foreigners': Economic, legal, and socio-cultural aspects of new land acquisition schemes in Ethiopia. *Journal of Contemporary African Studies* 29: 513–35. https://doi.org/10.1080/02589001.2011.603213

Achard, F., R. Beuchle, P. Mayaux, Het al. 2014. Determination of tropical deforestation rates and related carbon losses from 1990 to 2010. *Global Change Biology* 20: 2540–54. https://doi.org/10.1111/gcb.12605

Acreman, M.C. 1996. Environmental effects of hydro-electric power generation in Africa and the potential for artificial floods. *Water and Environment Journal* 10: 429–35. https://doi.org/10.1111/j.1747-6593.1996.tb00076.x

Arcilla, N., L.H. Holbech, and S. O'Donnell. 2015. Severe declines of understory birds follow illegal logging in Upper Guinea forests of Ghana, West Africa. *Biological Conservation* 188: 41–49. https://doi.org/10.1016/j.biocon.2015.02.010

Austin, K.G., M. González-Roglich, D. Schaffer-Smith, et al. 2017. Trends in size of tropical deforestation events signal increasing dominance of industrial-scale drivers. *Environmental Research Letters* 12: 054009. https://doi.org/10.1088/1748-9326/aa6a88

Beier, P., M. van Drielen, and B.O. Kankam. 2002. Avifaunal collapse in West African forest fragments. *Conservation Biology* 16: 1097–111. https://doi.org/10.1046/j.1523-1739.2002.01003.x

Beilfuss, R., and D. dos Santos. 2001. Patterns of hydrological change in the Zambezi Delta, Mozambique. *Working Paper* 2 (Baraboo: International Crane Foundation). https://doi.org/10.13140/RG.2.2.14255.12961

Benítez-López, A., R. Alkemade, A.M. Schipper, et al. 2017. The impact of hunting on tropical mammal and bird populations. *Science* 356: 180–83. https://doi.org/10.1126/science.aaj1891

BirdLife International. 2019. *Migratory birds and flyways*. https://www.birdlife.org/worldwide/programmes/migratory-birds

Blake, S., S.L. Deem, S. Strindberg, et al. 2008 Roadless wilderness area determines forest elephant movements in the Congo Basin. *PloS ONE* 3: e3546. https://doi.org/10.1371/journal.pone.0003546

Both, C., S. Bouwhuis, C.M. Lessells, et al. 2006. Climate change and population declines in a long-distance migratory bird. *Nature* 441: 81–83. https://doi.org/10.1038/nature04539

Bredenhand, E., and M.J. Samways. 2009. Impact of a dam on benthic macroinvertebrates in a small river in a biodiversity hotspot: Cape Floristic Region, South Africa. *Journal of Insect Conservation* 13: 297–307. http://doi.org/10.1007/s10841-008-9173-2

Cordeiro, N.J., H.J. Ndangalasi, J.P. McEntee, et al. 2009. Disperser limitation and recruitment of an endemic African tree in a fragmented landscape. *Ecology* 90: 1030–41. http://doi.org/10.1890/07-1208.1

Corlett, R., and R.B. Primack. 2010. *Tropical Rainforests: An Ecological and Biogeographical Comparison* (Malden: Wiley-Blackwell). https://doi.org/10.1002/9781444392296

Dale, S., K. Mork, R. Solvang, et al. 2000. Edge effects on the understorey bird community in a logged forest in Uganda. *Conservation Biology* 14: 265–76. https://doi.org/10.1046/j.1523-1739.2000.98340.x

Davidson, N.C. 2014. How much wetland has the world lost? Long-term and recent trends in global wetland area. *Marine and Freshwater Research* 65: 934–41. https://doi.org/10.1071/MF14173

deGeorges, A., and B.K. Reilly. 2006. Dams and large-scale irrigation on the Senegal River: impacts on man and the environment. *International Journal of Environmental Studies* 63: 633–44. https://doi.org/10.1080/00207230600963296

Drechsel, P., L. Gyiele, D. Kunze, et al. 2001. Population density, soil nutrient depletion, and economic growth in sub-Saharan Africa. *Ecological Economics* 38: 251–58. https://doi.org/10.1016/S0921-8009(01)00167-7

Drijver, C.A., and M. Marchand, 1985. *Taming the floods. Environmental aspects of the floodplain developments of Africa* (Leiden: Centre of Environmental Studies, University of Leiden).

Durant, S.M., M.S. Becker, S. Creel, et al. 2015. Developing fencing policies for dryland ecosystems. *Journal of Applied Ecology* 52: 544–51. https://doi.org/10.1111/1365-2664.12415

Dybas, C.L. 2015. Forests between the tides: Conserving Earth's vanishing mangrove ecosystems. *BioScience* 65: 1039–45. https://doi.org/10.1093/biosci/biv132

Ehrlich, P.R., and L.H. Goulder. 2007. Is current consumption excessive? A general framework and some indications for the United States. *Conservation Biology* 21: 1145–54. https://doi.org/10.1111/j.1523-1739.2007.00779.x

Estrada, A., P.A. Garber, A.B. Rylands, et al. 2017. Impending extinction crisis of the world's primates: Why primates matter. *Science Advances* 3: e1600946. https://doi.org/10.1126/sciadv.1600946

FAO (Food and Agriculture Organisation). 2013. *Food wastage footprint: Impacts on natural resources. Summary Report* (Rome: FAO). http://www.fao.org/docrep/018/i3347e/i3347e.pdf

Feka, N.Z., G.B. Chuyong, and G.N. Ajonina. 2009. Sustainable utilization of mangroves using improved fish-smoking systems: A management perspective from the Douala-Edea wildlife reserve, Cameroon. *Tropical Conservation Science* 2: 450–68. https://doi.org/10.1177/194008290900200406

FFI and FDA (Forestry Development Authority). 2013. *National action plan for the conservation of the pygmy hippopotamus in Liberia* (Cambridge: FFI; Monrovia: FDA).

Freedman, A.H., W. Buermann, M. Lebreton, et al. 2009. Modeling the effects of anthropogenic habitat change on savanna snake invasions into African rainforest. *Conservation Biology* 23: 81–92. https://doi.org/10.1111/j.1523-1739.2008.01039.x

GFN (Global Footprint Network). 2017. *National Footprint Accounts*. http://www.footprintnetwork.org

Giri, C., E. Ochieng, L.L. Tieszen, et al. 2011. Status and distribution of mangrove forests of the world using Earth observation satellite data. *Global Ecology and Biogeography* 20: 154–59. https://doi.org/10.1111/j.1466-8238.2010.00584.x

Gockowski, J., and D. Sonwa. 2011. Cocoa intensification scenarios and their predicted impact on CO_2 emissions, biodiversity conservation, and rural livelihoods in the Guinea Rain Forest of West Africa. *Environmental Management* 48: 307–21. https://doi.org/10.1007/s00267-010-9602-3

Gómez, C., N.J. Bayly, D.R. Norris, et al. 2017. Fuel loads acquired at a stopover site influence the pace of intercontinental migration in a boreal songbird. *Scientific Reports* 7: 3405. https://doi.org/10.1038/s41598-017-03503-4

Gopal, B. 2013. Mangroves are wetlands, not forests: Some implications for their management. In: *Mangrove Ecosystems of Asia*, ed. by I. Faridah-Hanum et al. (New York: Springer). https://doi.org/10.1007/978-1-4614-8582-7

Greger, M.M. 2012. *Traditional Healers as a Foundation Pillar of Medicinal Plant Conservation in Uganda*. M.Sc. Thesis (Ås: Norwegian University of Life Sciences).

Haddad, N.M., L.A. Brudvig, J. Clobert, et al. 2015. Habitat fragmentation and its lasting impact on Earth's ecosystems. *Science Advances* 1: e1500052. https://doi.org/10.1126/sciadv.1500052

Hansen, M.C., P.V. Potapov, R. Moore, et al. 2013. High-resolution global maps of 21st-century forest cover change. *Science* 342: 850–53. https://doi.org/10.1126/science.1244693 Data available on-line from: http://earthenginepartners.appspot.com/science-2013-global-forest

Harris, G., S. Thirgood, J.G.C. Hopcraft, et al. 2009. Global decline in aggregated migrations of large terrestrial mammals. *Endangered Species Research* 7: 55–76. https://doi.org/10.3354/esr00173

Hillers A., G.M. Buchanan, J.C. Garteh, et al. 2017. A mix of community-based conservation and protected forests is needed for the survival of the Endangered pygmy hippopotamus *Choeropsis liberiensis*. *Oryx* 51: 230–39. https://doi.org/10.1017/S003060531600020X

Hoffmann, M., J.W. Duckworth, K. Holmes, et al. 2015. The difference conservation makes to extinction risk of the world's ungulates. *Conservation Biology* 29: 1303–13. https://doi.org/10.1111/cobi.12519

Hopcraft, J.G.C., S.A.R. Mduma, M. Borner, et al. 2015. Conservation and economic benefits of a road around the Serengeti. *Conservation Biology* 29: 932–36. https://doi.org/10.1111/cobi.12470

Ibisch, P.L., M.T. Hoffmann, S. Kreft, et al. 2016. A global map of roadless areas and their conservation status. *Science* 354: 1423–27. https://doi.org/10.1126/science.aaf7166

IBPES. 2019. Nature's dangerous decline 'unprecedented': Species extinction rates accelerating. *IBPES media release*. https://www.ipbes.net/news/Media-Release-Global-Assessment

IUCN. 2019. *The IUCN Red List of Threatened Species*. http://www.iucnredlist.org

Kalwij, J.M., M.P. Robertson, and B.J. Rensburg. 2008. Human activity facilitates altitudinal expansion of exotic plants along a road in montane grassland, South Africa. *Applied Vegetation Science* 11: 491–98. https://doi.org/10.3170/2008-7-18555

Kariuki, M. and K. Ndang'ang'a, 2018. Can the Critically Endangered Liben lark be saved? Our latest update. *BirdLife News*. https://www.birdlife.org/worldwide/news/can-critically-endangered-liben-lark-be-saved-our-latest-update

Kirby, J.S., A.J. Stattersfield, S.H.M. Butchart, et al. 2008. Key conservation issues for migratory land-and waterbird species on the world's major flyways. *Bird Conservation International* 18: S49–S73. https://doi.org/10.1017/S0959270908000439

Kolbert, E., and D. Roberts. 2017. I'm an environmental journalist, but I never write about overpopulation. Here's why. *Vox*. https://www.vox.com/energy-and-environment/2017/9/26/16356524/the-population-question

Koohafkan, P., M. Salman, and C. Casarotto. 2011. Investments in land and water. In: *The State of the World's Land and Water Resources for Food and Agriculture (SOLAW)—Managing Systems at Risk* (Rome: FAO; London: Earthscan). http://www.fao.org/docrep/017/i1688e/i1688e.pdf

Laurance, W.F., J. Sayer, and K.G. Cassman. 2014. Agricultural expansion and its impacts on tropical nature. *Trends in Ecology and Evolution* 29: 107–16. https://doi.org/10.1016/j.tree.2013.12.001

Lehouck, V., T. Spanhove, L. Colson, et al. 2009. Habitat disturbance reduces seed dispersal of a forest interior tree in a fragmented African cloud forest. *Oikos* 118: 1023–34. https://doi.org/10.1111/j.1600-0706.2009.17300.x

Lewis, R.R. 2005. Ecological engineering for successful management and restoration of mangrove forests. *Ecological Engineering* 24: 403–18. https://doi.org/10.1016/j.ecoleng.2004.10.003

Luther, D.A., and R. Greenberg. 2009. Mangroves: A global perspective on the evolution and conservation of their terrestrial vertebrates. *BioScience* 59: 602–12. https://doi.org/10.1525/bio.2009.59.7.11

MacArthur, R.H., and E.O. Wilson. 2015. *Theory of Island Biogeography* (Princeton: Princeton University Press).

Mahamued, B.A. 2016. *Designing a Rangeland to Preserve Africa's Most Endangered Mainland Bird and a People's Way of Life.* Doctoral dissertation (Manchester: Manchester Metropolitan University).

Mertens, B., and E.F. Lambin. 1997. Spatial modelling of deforestation in southern Cameroon: Spatial disaggregation of diverse deforestation processes. *Applied Geography* 17: 143–62. https://doi.org/10.1016/S0143-6228(97)00032-5

Moran, D., and K Kanemoto. 2017. Identifying species threat hotspots from global supply chains. *Nature Ecology and Evolution* 1: 0023. https://doi.org/10.1038/s41559-016-0023

Morgan, B.J., E.E. Abwe, A.F. Dixson, et al. 2013. The distribution, status, and conservation outlook of the drill (*Mandrillus leucophaeus*) in Cameroon. *International Journal of Primatology* 34: 281–302. http://doi.org/10.1007/s10764-013-9661-4

Moss, T., and P. Agyapong. 2015. *US holiday lights use more electricity than El Salvador does in a year* (London: Center for Global Development). http://www.cgdev.org/blog/us-holiday-lights-use-more-electricity-el-salvador-does-year

Neville, K. 2015. The contentious political economy of biofuels. *Global Environmental Politics* 15: 21–40. https://doi.org/10.1162/GLEP_a_00270

Olson, D.M., and E. Dinerstein. 2002. The Global 200: Priority ecoregions for global conservation. *Annals of the Missouri Botanical Garden* 89: 199–224. https://doi.org/10.2307/3298564

Potapov, P., M.C. Hansen, L. Laestadius, et al. 2017. The last frontiers of wilderness: Tracking loss of intact forest landscapes from 2000 to 2013. *Science Advances* 3: e1600821. https://doi.org/10.1126/sciadv.1600821

Primack, R.B. 2012. *A Primer for Conservation Biology* (Sunderland: Sinauer).

Ransom, C, P.T. Robinson, and B. Collen. 2015. *Choeropsis liberiensis*. *The IUCN Red List of Threatened Species* 2015: e.T10032A18567171. http://doi.org/10.2305/IUCN.UK.2015-2.RLTS.T10032A18567171.en

Rudel, T.K. 2013. The national determinants of deforestation in sub-Saharan Africa. *Philosophical Transactions of the Royal Society B* 368: 20120405. https://doi.org/10.1098/rstb.2012.0405

Runge, C.A., J.E.M. Watson, S.H.M. Butchart, et al. 2015. Protected areas and global conservation of migratory birds. *Science* 350: 1255–58. https://doi.org/10.1126/science.aac9180

Rushworth, I., and S. Krüger. 2014. Wind farms threaten southern Africa's cliff-nesting vultures. *Ostrich* 8: 13–23. http://doi.org/10.2989/00306525.2014.913211

Sande, E., P.K. Ndang'ang'a, J. Wakelin, J., et al. 2008. International single species action plan for the conservation of the White-winged Flufftail *Sarothrura ayresi*. *CMS Technical Series* 19 (Bonn: CMS and AWEA). http://www.cms.int/sites/default/files/publication/Whitewinged_flufftail_3_0_0.pdf

Sayer, J. 1992. A future for Africa's tropical forests. In: *The Conservation Atlas of Tropical Forests: Africa*, ed. by J.A. Sayer et al. (London: Palgrave Macmillan). https://portals.iucn.org/library/sites/library/files/documents/1992-063.pdf

Sedláček, O., M. Mikeš, T. Albrecht, et al. 2014. Evidence for an edge effect on avian nest predation in fragmented afromontane forests in the Bamenda-Banso Highlands, NW Cameroon. *Tropical Conservation Science* 7: 720–32. https://doi.org/10.1177/194008291400700410

Smaller, C., Q. Wei, and L. Yalan. 2012. *Farmland and water: China invests abroad* (Winnipeg: IISD). https://www.iisd.org/pdf/2012/farmland_water_china_invests.pdf

Sokolow, S.H., E. Huttinger, N. Jouanard, et al. 2015. Reduced transmission of human schistosomiasis after restoration of a native river prawn that preys on the snail intermediate host. *Proceedings of the National Academy of Sciences* 112: 9650–55. https://doi.org/10.1073/pnas.1502651112

Spottiswoode, C.N., M. Wondafrash, M. Gabremicheal, et al. 2009. Rangeland degradation is poised to cause Africa's first recorded avian extinction. *Animal Conservation* 12: 249–57. https://doi.org/10.1111/j.1469-1795.2009.00246.x

Spottiswoode, C.N., U. Olsson, M.S. Mills, et al. 2013. Rediscovery of a long-lost lark reveals the conspecificity of endangered *Heteromirafra* populations in the Horn of Africa. *Journal of Ornithology* 154: 813–25. https://doi.org/10.1007/s10336-013-0948-1

Stabach, J.A., G. Wittemyer, R.B. Boone, et al. 2016. Variation in habitat selection by white-bearded wildebeest across different degrees of human disturbance. *Ecosphere* 7: e01428. https://doi.org/10.1002/ecs2.1428

Tabuti, J.R.S. 2006. Traditional knowledge in Bulamogi County - Uganda: Importance to sustainable livelihoods. In: *African Knowledge and Sciences: Understanding and Supporting the Ways of Knowing in Sub-Saharan Africa*, ed. by D. Millar et al. (Leusden: Compas).

Tucker, M.A., K. Böhning-Gaese, W.F. Fagan, et al. 2018. Moving in the Anthropocene: Global reductions in terrestrial mammalian movements. *Science* 359: 466–69. https://doi.org/10.1126/science.aam9712

Tyukavina, A., M.C. Hansen, P. Potapov, et al. 2018. Congo Basin forest loss dominated by increasing smallholder clearing. *Science Advances* 4: eaat2993. https://doi.org/10.1126/sciadv.aat2993

van Bochove, J., E. Sullivan, and T. Nakamura. 2014. *The Importance of Mangroves to People: A Call to Action* (Cambridge: UNEP). http://wedocs.unep.org/handle/20.500.11822/9300

van der Hoeven, C.A., W.F. de Boer, and H.H. Prins. 2010. Roadside conditions as predictor for wildlife crossing probability in a Central African rainforest. *African Journal of Ecology* 48: 368–77. https://doi.org/10.1111/j.1365-2028.2009.01122.

van Wilgen, B.W., J.L. Nel, and M. Rouget. 2007. Invasive alien plants and South African rivers: A proposed approach to the prioritization of control operations. *Freshwater Biology* 52: 711–23. https://doi.org/10.1111/j.1365-2427.2006.01711.x

Vickery, J.A., S.R. Ewing, K.W. Smith, et al. 2014. The decline of Afro-Palaearctic migrants and an assessment of potential causes. *Ibis* 156: 1–22. https://doi.org/10.1111/ibi.12118

Vogt, T. 2011. *Results of Sapo National Park bio-monitoring programme 2007–2009*. FFI-FDA Biomonitoring and Research Report (Cambridge: FFI; Monrovia: FDA).

von Braun, J., and R.S. Meinzen-Dick. 2009. *"Land grabbing" by foreign investors in developing countries: Risks and opportunities* (Washington: IFPRI). http://ebrary.ifpri.org/cdm/ref/collection/p15738coll2/id/14853

Wallenfang, J., M. Finckh, J. Oldeland, et al, 2015. Impact of shifting cultivation on dense tropical woodlands in southeast Angola. *Tropical Conservation Science* 8: 863–92. https://doi.org/10.1177/194008291500800402

Weisse, M., and E.D. Goldman. 2019. The world lost a Belgium-sized area of primary rainforest last year. *World Resources Institute Blog*. https://www.wri.org/blog/2019/04/world-lost-belgium-sized-area-primary-rainforests-last-year

Woodborne, S., K.D.A. Huchzermeyer, D. Govender, et al. 2012. Ecosystem change and the Olifants River crocodile mass mortality events. *Ecosphere* 3: 1–17. https://doi.org/10.1890/ES12-00170.1

World Bank. 2019. *World Bank Open Data*. http://data.worldbank.org/region/sub-saharan-africa

WRI (World Resources Institute). 2019. *Climate Analysis Indicators Tool: WRI's Climate Data Explorer*. http://cait2.wri.org

Ykhanbai, H., R. Garg, A. Singh, et al. 2014. *Conservation and "land grabbing" in rangelands: Part of the problem or part of the solution?* (Rome: International Land Coalition). http://www.landcoalition.org/sites/default/files/documents/resources/conservation%20and%20land%20grabbing%20in%20rangelands_web_en_0.pdf

6. Our Warming World

UN secretary general Ban Ki-moon introduces the Momentum for Change initiative at the UN Climate Change Conference, also known as COP17, held in Durban, South Africa in 2011. Photograph by UNFCCC, https://www.flickr.com/photos/unfccc/6470741719, CC BY 2.0.

 https://doi.org/10.11647/OBP.0177.06

Life-threatening heatwaves, drowning coastal towns, tens of thousands of displaced refugees… These words may very well describe a scene from the latest horror movie. But they also describe the nightmare scenario facing us humans in just a few decades if we continue to leave the threat of climate change under-addressed. This term, climate change (which is shorthand for **anthropogenic climate change**), refers to the complete set of climate characteristics—temperature; precipitation; pressure systems; wind patterns; and oceanic currents—that are changing both locally and regionally due to human influences. It is closely related to **global warming**, also called global heating, which describes the general trend of increasing global temperatures we see under climate change.

Climate change has the potential to render Earth unrecognisable from what any human has ever experienced. These changes will have an immense impact on ecosystem services, global economies, and our own quality of life. Yet, while there is much talk about these risks, there is too little action addressing its main causes. Some of the lack of action may be attributed to "climate change" and "future" often being used in the same sentence, giving politicians and industries a false impression that we can deal with climate change once we achieved sufficient economic growth. The reality could however not be further from the truth, as we already see signs of the changes to come here today (Table 6.1), including near-annual crop-failures, record-high temperatures, and record-strength coastal storms.

Thankfully, with the increase in understanding that our activities are creating a global crisis of epic proportions, the impacts of climate change are now being actively debated in the corridors of governments and major corporations. Politicians, the media, and others are also increasingly replacing "climate change" with more vivid language, like "climate crisis" and "climate emergency" (e.g. Carrington, 2019). This will hopefully encourage even more governments and industries to come to the table and cooperate like never before to address the fundamental drivers of climate change. Solving this global crisis requires an international multi-pronged approach that should include ecosystem protection and restoration (Chapter 10), direct species management (Chapter 11), and legislative action (Chapter 12). But before we consider the solutions, we will first investigate why climate change is happening, and how it will impact biodiversity over the coming decades.

6.1 Drivers of Climate Change

The climate change we are experiencing today is driven by human activities that increase greenhouse gas concentrations in Earth's atmosphere. Although we mainly hear about **greenhouse gases** in the context of their contribution to climate change, they are in fact essential for life on Earth. Consider for a moment carbon dioxide's (CO_2) critical role in photosynthesis, and water vapour's role in the formation of rain. Both of these gases are greenhouses gases. Greenhouse gases earn their name because they function much like the glass covering a greenhouse; they allow sunlight

Table 6.1 Some examples illustrating how climate change is already impacting Africa.

Impact	Evidence
Increased temperatures and incidence of heat waves	Global temperatures in 2016 were the warmest since modern recordkeeping began in 1880; the two previous records were set in 2015 and 2014 (Gillis, 2017). Heat waves are also hotter, longer, and over a larger area than before (Russo et al., 2016).
Widespread droughts	East Africa saw its worst drought in 60 years from mid-2011 to mid-2012. Over 250,000 people died; nearly 10 million more needed humanitarian assistance (Maxwell et al., 2014). The increased intensity of similar droughts in 2016 are directly attributable to climate change (Uhe et al., 2017).
Rising sea levels	Coastal floods disrupt lives and local economies in Ghana, Nigeria, and Benin almost every year. Coastal erosion has damaged commercial properties in The Gambia and Senegal, while the coastline retreated 35 m in some areas of Togo (Fagotto and Gattoni, 2016).
Earlier spring activity	Bloom dates for several plants, including commercially grown apple and pear trees in South Africa, are now between 1.6 and 4.2 days earlier per decade than 35 years ago (Grab and Craparo, 2011).
Shifts in species ranges	Malaria recently appeared in the highlands of Ethiopia, Kenya, Rwanda, and Burundi, in areas where it did not occur before (Siraj et al., 2014).
Wildlife population declines	Reporting rates for some bird species endemic to the Cape Floristic region declined by over 30% over the past 15 years (Milne et al., 2015).

to easily pass through the atmosphere but trap the reflected heat energy so that it stays close to Earth's surface. This **greenhouse effect** allows all the organisms on Earth, even us humans, to flourish. Without greenhouse gases, temperatures would drop, and our planet would be too cold to sustain life. However, high concentrations of greenhouse gases can also be harmful. Think for a moment of greenhouse gases as "blankets" covering the Earth's surface: more "blankets" will trap more heat, giving rise to higher temperatures. This is exactly what is happening today—human activities are currently increased greenhouse gas concentrations in the atmosphere so much, and at such a fast pace, that Earth is heating up too fast for biodiversity to adapt to the changes.

> Greenhouse gases are essential for life on Earth. But too much of them cause Earth to heat up too much too fast, leading to climate change.

At present, the single biggest cause of increased greenhouse gas concentrations is the burning of fossil fuels. Since the **Industrial Revolution** about 200–250 years ago, humans have become heavily dependent on the energy captured in these fuels—coal, oil, and natural gas—for activities such as transportation, heating, manufacturing, and electricity generation. Fossil fuels contain a high percentage of carbon, so when it is burned, that carbon is released into the atmosphere, generally as CO_2. Consequently, since human populations started exploding and have been using fossil fuels at increased rates, the greenhouse effect has been significantly amplified.

Africa's biggest contribution to climate change comes from the destruction of complex ecosystems, which leads to the loss of important carbon sinks.

While fossil fuel burning is currently the biggest overall driver of climate change, the greatest contribution from Africa is the destruction of **carbon sinks**, such as tropical forests (Box 6.1) and peatlands. Destroying these ecosystems contributes to rising atmospheric CO_2 concentrations directly through burning of vegetation that releases carbon, and indirectly through the loss of vegetation that would otherwise extract CO_2 from the atmosphere if they were still alive. The contribution of ecosystem loss to climate change is substantial: 13% of today's global carbon emissions can be accounted for by tropical deforestation (IPCC, 2014). This impact is much stronger in Africa where deforestation accounts for 35% of the region's overall climate change impacts (WRI, 2019). In comparison, Africa's energy and agricultural sectors contribute 30% and 24%, respectively.

Box 6.1 Does Oil Palm Agriculture Threaten Biological Diversity in Equatorial Africa?

Abraham J. Miller-Rushing

Acadia National Park, US National Park Service, Bar Harbor, ME, USA.

Oil palm (*Elaeis guineensis*, LC) is among the fastest expanding crops in the world. Native to West Africa, this species produces more oil per hectare than any other cultivated crop in the world. It should thus come as no surprise that it has become world's most popular source of vegetable oil. Tropical Africa is poised as a hotspot for new oil palm plantations (Linder, 2013; Vijay et al., 2016). Is this a good thing? Will the benefits from jobs and carbon sequestration outweigh the loss of native ecosystems?

To many people, oil palm cultivation presents a win-win situation. The industry provides jobs and economic stimulus (Figure 6.A) and claims that oil palms sequester carbon from the atmosphere (Burton et al., 2017). This could potentially help countries offset carbon emissions; they may even receive funding from carbon markets. Palm oil can also be used for cheap bioenergy production, and as an ingredient in food and household products (e.g. cooking oil, baked goods, salad dressings, shampoo, and soap). Consequently, demand is rapidly growing as sales of processed and packaged foods (today about 50% of packaged foods include palm oil as an ingredient) expand globally.

Figure 6.A A plantation worker getting ready to harvest oil palm fruit, locally known as red gold, in Côte d'Ivoire. Often associated with land grabbing, deforestation, biodiversity losses, and exploitation of local communities, there is a need for the palm oil industry to become more sustainable to provide lasting and meaningful benefits to local economies. Photograph by Donatien Kangah, https://commons.wikimedia.org/wiki/File:R%C3%A9colteur_de_r%C3%A9gimes_de_palme_1.jpg, CC BY-SA 4.0.

Oil palm plantations, however, are rarely developed in environmentally friendly ways that allow them to realise their potential value. Rather, it generally comes at a great ecological cost. For example, to ensure net positive carbon sequestration, oil palm plantations must be developed on degraded landscapes, rather than displacing intact ecosystems that are already very effective at sequestering carbon (Burton et al., 2017). In practice however, intact forests are more often logged to make space (and additional revenue) for oil palm plantations (Ordway et al., 2019), resulting in habitat loss and net positive carbon emissions. Oil palm plantations are also often associated with great societal costs, like land grabbing, exploitation of local people, and displacement of traditional activities (Linder and Palkovitz, 2016). The influx of migrant plantation workers puts further strain on the environment through unsustainable hunting of bushmeat. One study found that primate population sizes declined by 25–100% after palm plantation development in Côte d'Ivoire (Gonedelé et al., 2012).

Recently, Herakles Farms/SG Sustainable Oils, an American agribusiness company, attempted to develop a 730 km^2 oil palm plantation in Cameroon. This land grab would have been one of the largest palm oil projects in Africa, nestled deep within Cameroon's lowland tropical forests, one of the continent's most biologically diverse and threatened ecosystems. The forests threatened by this development is situated adjacent to four protected areas that include two national parks (Linder and Palkovitz, 2016), and host 14 species of threatened primates, including the Nigeria-Cameroon chimpanzee (*Pan troglodytes ellioti*, EN) (Linder, 2013). Residents and environmental groups opposed the plantation because of possibly illegal activities by the company, the ecological consequences of the project, and because the local people would have received little, if any, benefit from the project. After protracted debate and struggle, including intimidation and the arrest of local social and environmental activists, the company withdrew its Cameroonian plans in 2013.

It seems that there is potential for oil palm plantations to be good for economic development, job creation, and conservation. But in practice, companies establishing these plantations frequently exploit local people and degrade local ecosystems. They sometimes even do this under the auspices of sustainability, arguing that low-impact activities by traditional peoples indicate that the area is already degraded and thus suitable for development. Hopefully, one day we can live in a world where palm oil companies and robust legal systems truly consider the protection of biodiversity and the rights of local people in those operations.

The link between human-induced climate change and atmospheric CO_2 concentrations was first highlighted in the late 19th century (Arrhenius, 1896). However, it was not until the mid-1950s (e.g. Kaempffert, 1956) that scientists started to raise concerns about increasing CO_2 concentrations in the atmosphere. By the 1980s, as global annual mean temperatures started to rise, consensus about climate change linked to CO_2 began to spread among the broader public. Yet concrete steps to curb CO_2 emissions would only be initiated decades later (Section 12.2.1). In the meantime, CO_2 emissions continue to accelerate (Figure 6.1): more than 37 billion tonnes of carbon, a new record, were released into the atmosphere in 2018 (Jackson et al., 2018; Le Quéré et al., 2018). To put it in another way, during 2018, humans released on average over 100 million tonnes of CO_2 into the atmosphere *every day*.

The second-most important greenhouse gas that drives climate change is methane (CH_4). Methane is a natural by-product emitted from decaying organic matter, most notably from wetlands that inhibit the speed of decomposition. These important ecosystem processes release methane into the atmosphere, albeit in relatively low concentrations. However, human activities have boosted methane emissions significantly over the past few centuries, through wasted food decaying at landfills,

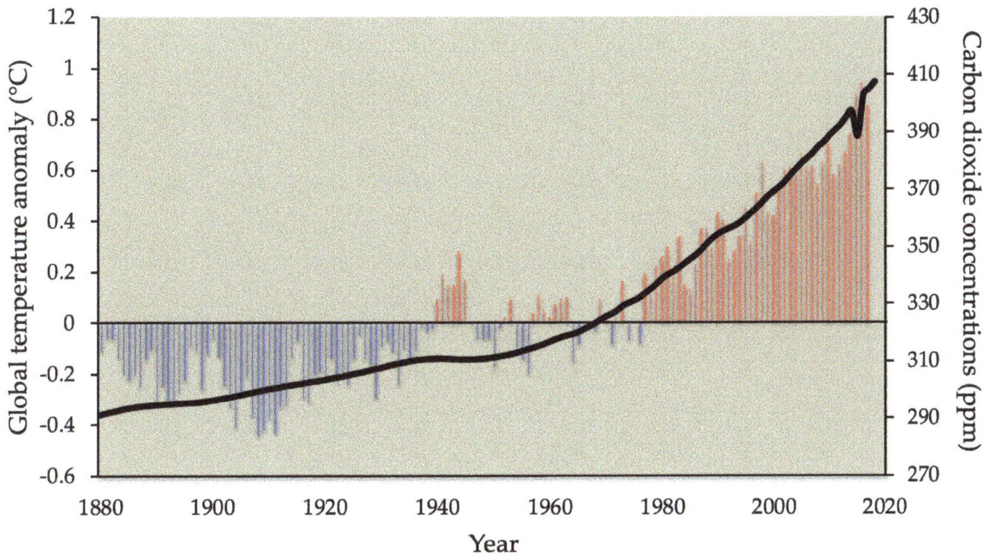

Figure 6.1 Human activities, notably the burning of fossil fuels and deforestation, have drastically increased atmospheric CO_2 concentrations over the past century. As a result, average annual global temperatures are now much higher than they have been in the past. Temperatures are reported in terms of difference (anomaly) from average annual temperature from 1910–2000. Source for climate data: NOAA, 2018a. Sources for CO_2 data: NASA, 2018 (before 2006); NOAA, 2018b (after 2005), CC BY 4.0.

leaks from natural gas wells, an increase of industrial-scale cattle and dairy farms, and large-scale destruction of swamps and peatlands. Warmer temperatures also result in the drying of wetlands and peatlands; this drying speed up decomposition of organic material, which increases the rate of methane release. Methane currently constitutes 16% of all global greenhouse gas emissions released by humans (IPCC, 2014). This may not seem to be a major contribution; however, methane is 72 times more effective than CO_2 in trapping radiation over a 20-year period (Forster et al., 2007), so even small increases in atmospheric methane can have dramatic effects.

The third important greenhouse gas that drives climate change is nitrous oxide (N_2O), also known as laughing gas. Nitrous oxide is a by-product of synthetic fertilisers used in agriculture, burning of fossil fuels, and several industrial processes, and accounts for 6% of all human-caused greenhouse emissions (IPCC, 2014). However, it is even more potent than methane, and stays in the atmosphere for about 114 years, so the impact of one tonne of N_2O is equivalent to 310 tonnes of CO_2 over 100 years (Forster et al., 2007).

6.2 Predicting Earth's Future Climate

Climate change forecasting is famously complex, with a great amount of uncertainty attached to the task. Most of us have been exposed to short-term (i.e. day-to-day) weather forecasts on television, radio, and newspapers. These daily weather forecasts are derived from current weather measurements while considering the historical record of past events that occurred during similar conditions. Some daily forecasts may also be created for as many as two weeks into the future, but these future outlooks are generally much less detailed. In contrast, forecasting climate change involves predicting novel weather conditions for several decades into the future. The **general circulation models (GCM)** used for climate change forecasting (Figure 6.2) also need to account for a great number of highly variable components, each affecting one another across the only planet we can adequately measure or examine (we have no other planet where we can test predictions). Among thousands of considerations, climatologists (scientists who study climate) need to account for how human activities might change over time, and how these activities will change the atmosphere's composition. They also need to account for how much CO_2 the world's oceans and plants will absorb, and how wind and fire might influence these processes. Combining all the component parts, climatologists then need to estimate how increased temperatures will affect the polar ice caps, how the melting ice will affect oceanic conditions and currents which, in turn, will affect terrestrial conditions and weather patterns. Uncertainty also exists over interactive effects of some drivers. For example, higher temperatures increase evaporation and cloud cover which, in turn, will have a cooling effect (a similar short-term cooling effect, caused by an albedo effect, is noted after an ecosystem is cleared due to the bare ground's ability to reflect more sunlight than it absorbs, Section 4.2.3). Because of the complexity of these and other variables going into climate models, a great number of research groups are encouraged to develop their own climate forecasts, each using a range of different scenarios on how human activity might change in the future.

To further improve upon climate change forecasting, in 1988, the UN appointed a group of leading scientists, collectively known as the **Intergovernmental Panel on Climate Change (IPCC)**, to study the implications of climate change. By regularly doing extensive reviews of all the evidence and climate science literature, the IPCC has found that, despite the complexity of climate models, results of all the models taken together exhibited significant agreement with changes already observed. Climate change models have also proven reliable in predicting responses of biodiversity to climate change (Fordham et al., 2018). Thus, while some fringe groups may continue to deny the validity of climate science, there is broad consensus among the world's scientists that increased atmospheric greenhouse gases—caused by human activities—are causing the world's climate to change, and it will continue to change in coming decades. While climatologists continue to improve on the finer details of their models, conservation biologists can and should confidently use the climate forecasts available for general conservation planning purposes.

Assuming human activities continue business as usual, and current greenhouse gas emission rates continue unabated, climatologists predict that average annual temperatures in Sub-Saharan Africa will increase by 0.5°C by 2050, compared to temperatures late in the 20th century (Serdeczny et al., 2017). The increase could be even greater, towards 4°C, if humans emit more greenhouse gases than predicted

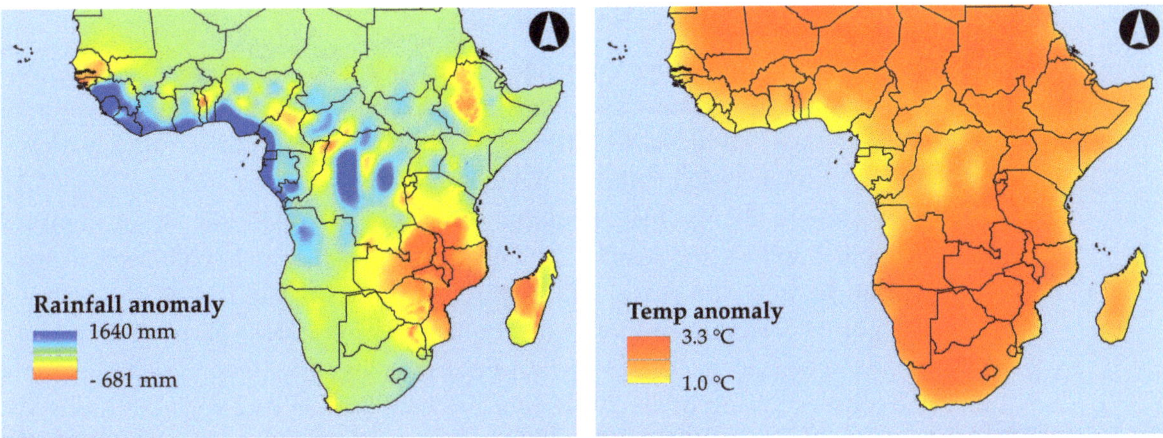

Figure 6.2 (Left) Annual precipitation (mm) and (Right) annual mean temperature (°C) shift predicted for Sub-Saharan Africa in 2070, assuming greenhouse gas emissions peak around 2080. Values presented as the amount of deviation from 1960-1990 averages. Some coastal areas of West and Central Africa are predicted of have more rain, but large areas of southeast Africa will get much drier. All of Africa is predicted to get hotter, with the greatest increases in southern Africa. Source: https://www.worldclim.org; model: GISS-E2-R. Map by Johnny Wilson, CC BY 4.0.

and Earth's carbon storage systems underperform. Conversely, temperatures could warm less or more slowly if we manage to slow greenhouse gases emissions and better protect natural carbon sinks. Unfortunately, current evidence suggests that the higher temperature estimates seem more likely. For example, 2016 was the hottest year (since modern record-keeping) globally for the third straight year (Gillis, 2017) with temperatures already 0.9°C above late 20th century averages. Another climate record was set in April 2018, which was Earth's 400th straight warmer-than-average month (NOAA, 2018c). Also, more locally, scientists observed that temperatures in some South African national parks reached temperature increases predicted for 2035 already in 2015 (van Wilgen et al., 2016).

6.3 The Impact of Climate Change

Climate change is not a new phenomenon. During the past 2 million years, there have been at least 10 cycles of global warming and cooling. When the polar ice caps melted during warming periods, sea levels rose to well above their earlier levels, and a larger portion of Earth experienced tropical climates. During cooling periods, the polar ice caps expanded, sea levels dropped, and tropical species' ranges contracted. Sometimes these changes occurred gradually, which enabled the affected species to adapt. But the onset of some climate change periods was abrupt, causing major ecosystem disruptions and global mass extinction events (Section 8.1). Yet, nature recovered every time; many of the species we see today are survivors of previous climate change events. It is thus fair to ask why today's climate change is of such concern to us.

6.3.1 Climate change's impact on people

History provides us with many lessons to illustrate the impact of climate change on human societies. These lessons start with the earliest well-documented example of a societal collapse—that of the Middle East's Natufian communities roughly 10,000 years ago—which has been attributed to climatic changes (Weiss and Bradley, 2001). Since then, climate change has regularly contributed to the collapse of complex human societies across the world. Notable examples of such collapses include the Akkadian Empire (the world's first empire) of the Middle East (Carolin et al., 2019), Egypt's Old Kingdom (who constructed the pyramids), Central America's Classic Mayan civilisation, the USA's first English colony (deMenocal, 2001), several Chinese dynasties (Wang et al., 2010), and the Late Bronze Age societies along the Mediterranean Sea (Kaniewski et al., 2013). Also, in Southern Africa, the fall of the Mapungubwe Kingdom has been attributed to crop failures and declining grazing lands due to regional droughts and warming cycles (O'Connor and Kiker, 2004).

Unlike the unavoidable natural climatic shifts that led to the historical societal collapses discussed above, we have brought today's climatic change impacts upon ourselves. Because of our general lack of response in addressing the drivers of climate change, thousands of people will suffer the consequences. Prominently, many parts of Africa are already seeing higher temperatures and longer droughts (Engelbrecht et al., 2009). These conditions are compromising our quality of life (Watts et al., 2017) by leading to more intense wildfires (Jolly et al., 2015; Strydom and Savage, 2016), increased incidences of malaria (Siraj et al., 2014), increased crop failures (Myers et al., 2014; Medek et al., 2017), and increased competition for water (Flörke et al., 2018). Many coastal areas are also seeing storms increasing in intensity and frequency, exposing people living near large rivers, deltas, and estuaries to more frequent flooding (Figure 6.3) and storm surges (Fitchett and Grab, 2014). **Sea level rise** is expected to leave many low-lying oceanic islands uninhabitable within a few decades (Storlazzi et al., 2018). With all these impacts expected to increase the competition for space under an increasing human population, it would be wise for the world's governments to start preparing for thousands of climate refugees that would need to be relocated in the near future (Merone and Tait, 2018).

To combat climate change, politicians of several countries have started to enact laws to reduce greenhouse gas emissions and habitat destruction (Section 12.2.1). Many industries are also hard at work developing "greener" technologies to enable us to live more sustainable lives. Conservation biologists also play a crucial role in mitigating the negative impacts of climate change. In addition to highlighting the plight of the natural world to society at large, we could work towards reducing the loss of ecosystem services and preventing species extinctions. To accomplish this task, we need to identify which species and ecosystems are most sensitive to climate change and develop strategies that will ensure the continued persistence of as many sensitive species and their habitats as possible. The rest of this chapter is dedicated to methods we can employ to understand which species are sensitive, and

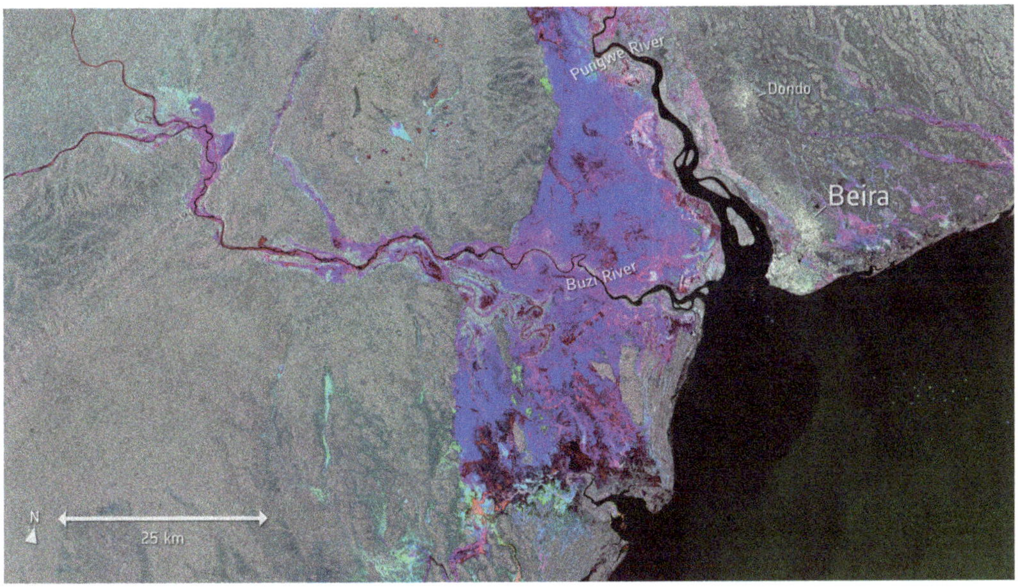

Figure 6.3 (Top) A Copernicus Sentinel-1 satellite image showing the extent of flooding (areas shown in blue) in central Mozambique after Cyclone Idai made landfall on 15 March, 2019. Photograph by European Space Agency, https://www.flickr.com/photos/europeanspaceagency/47477652401, CC BY-SA 2.0. (Bottom) People in Beira, Mozambique, taking refuge on rooftops to escape flooding brought by Cyclone Idai. Photograph by World Vision, https://www.flickr.com/photos/dfid/46570320385, CC BY 2.0. More than a thousand people died during this, one of the worst storms on record to have hit Africa. While no single flooding event can be attributed to climate change, it is undeniable that warmer oceans create conditions for hurricanes and cyclones to be stronger, bigger, and more frequent.

how they may respond to climate change, while Chapters 10–15 discuss methods we can employ to better address climate change.

6.3.2 Climate change's impact on terrestrial ecosystems

Aside from regional variations in temperature and precipitation, Earth's surface will be a few degrees warmer in future than the temperatures we experience today. In effect, that means that today's climatic zones will generally shift upslope in mountainous areas and towards the poles on lowlands, plains, and plateaus. To survive, climate-sensitive plants and animals will need to track these shifts so that they remain within their suitable **climatic envelopes** of temperature and precipitation.

Climate change on mountains

Species that live on mountain peaks are vulnerable to climate change because they may have nowhere else to go as the world heats up.

Species that live on mountains are at particular risk from climate change. Because temperatures decrease by roughly 0.65°C for every 100 m in elevation rise (known as temperature lapse rates), a 1°C increase suggests that climate-sensitive species living on a mountain would be displaced by *at least* 150 m (1.5 m/year) upslope between the years 2000 and 2100. Species that live on the lower slopes of mountains and are mobile enough to make such an adjustment may have opportunities to move to higher ground. However, species that live on or near peaks may have nowhere else to go as the world heats up, resulting in what biologists call **mountain-top extinctions**. While a mountain-top extinction has yet to be recorded in Africa, we have ample evidence to suggest that the region's wildlife is vulnerable to it. For example, due to climate change, populations of some bird species endemic to the Cape Floristic region's mountains have shrunk by 30% over the past two decades (Milne et al., 2015). Species inhabiting Tanzania's Eastern Arc Mountains (Dimitrov et al., 2012), Albertine Rift (Ponce-Reyes et al., 2017), and the Guinean Forests of West Africa (Carr et al., 2014) appear to have experienced similar declines. Given these observations, it is only a matter of time before one of Africa's mountain specialists follows the example of Costa Rica's once abundant Monteverde golden toad (*Bufo periglenes*, EX), the first known amphibian extinction attributed to climate change (Crump et al., 1992).

Climate change in the lowlands

The response of species living in lowlands and on plains tend be more variable and complex than those living on mountains. While some species may only need to make minor range adjustments, researchers estimate that some African taxa may need to move 500 km (Barbet-Massin et al., 2009)—maybe even 1,000 km (Hsiang and Sobel, 2016)—to keep up with climate shifts. For species, such as Tanzania's savannah birds that have already shifted their distributions by 200–300 km (Beale et al., 2013), adapting seems relatively easy thanks to their mobility and largely intact ecosystems. Unfortunately, the rate of climate change will likely outpace the ability for most species to adapt (Jezkova and Wiens, 2016; Wiens, 2016). For example, nearly 62%

of Sub-Saharan Africa's species are predicted to undergo range contractions (Hole et al., 2009), and 37% species are facing extinction if climate forecasts hold true (Thomas et al., 2004). Species living in Southern Africa's Miombo Woodlands are even more vulnerable, where as many as 90% of amphibians, 86% of birds, and 80% of mammals face extirpation (Warren et al., 2018).

Species of tropical lowland forests and deserts are also highly vulnerable to shifting climates. Many tropical species have narrow tolerances for temperature and rainfall variation, while desert specialists may be at the limits of their physiological heat and desiccation tolerances (Figure 6.4). Consequently, even small changes in the climate of these two ecosystems may have major effects on reproduction, species distributions, and hence ecosystem composition (Box 6.2). One species already impacted is the nocturnal aardvark (*Orycteropus afer*, LC): a study in Southern Africa's Kalahari Desert found over 80% mortality rates in this species during recent summers (Rey et al., 2017). The high levels of mortality in this species was attributed to above average temperatures, which subjected the animals to heat stress, leading to behavioural disruptions, declining body conditions, and eventually starvation. The impact of climate change on the aardvark is of concern because it is an ecosystem engineer: their burrows provide denning and refuge sites for multiple other species (Whittington-Jones et al., 2011).

Box 6.2 Desert Birds and Climate Change

Susan Cunningham[1] *and Andrew McKechnie*[2,3]

[1]*FitzPatrick Institute of African Ornithology, DST-NRF Centre of Excellence, University of Cape Town, South Africa.*

[2]*DST-NRF Centre of Excellence at the FitzPatrick Institute, Department of Zoology and Entomology, University of Pretoria, South Africa.*

[3]*South African Research Chair in Conservation Physiology, National Zoological Garden, South African National Biodiversity Institute, Pretoria, South Africa.*

✉ *susie.j.c@gmail.com*

Deserts, with their extreme temperatures and scarce and unpredictable rainfall, are among the most inhospitable environments on the planet. To survive and breed in arid regions, organisms must minimise their energy and water requirements, and avoid exposure to potentially lethal temperatures. Birds are generally small and diurnal; and are therefore among the groups of animals most vulnerable to even small increases in air temperatures associated with climate change. Studies of the effects of temperature on arid-zone birds can thus be highly informative in terms of identifying new conservation challenges posed by global warming, developing mitigation measures, and understanding the management interventions that may become necessary during the 21st Century.

Daytime temperatures in many deserts regularly exceed avian body temperature, creating conditions under which birds can avoid lethal heat stroke only by dissipating heat via evaporation. But rapid rates of evaporation increase the risk of birds becoming lethally dehydrated. Desert birds thus face life-or-death decisions between avoiding hyperthermia by evaporative cooling *versus* avoiding lethal dehydration by minimising water losses. Mass mortality events occasionally take place during extreme heat waves when air temperatures exceed birds' physiological tolerance limits. In Australia, for example, there are both historic and contemporary accounts of die-offs sometimes involving millions of birds. As Earth heats up under climate change, the risk of such die-offs in desert birds is expected to increase dramatically for the deserts of Australia and North America during the 21st Century (McKechnie and Wolf, 2010; Albright et al., 2017).

Africa's arid regions are also experiencing significant temperature increases which are predicted to continue over the next several decades (Conradie et al., 2019). Under these conditions, the impact of air temperature on avian physiology can be mediated by behaviour. Birds employ a trio of behavioural adjustments to manage heat load and keep their body temperatures within safe limits. These include shade-seeking, reducing activity to minimise metabolic heat production, and gaping the beak (panting, sometimes accompanied by gular flutter) to facilitate respiratory evaporative cooling (Figure 6.B). Although these behaviours can buffer birds against physiological costs of high temperature, they carry subtle but important costs of the own, notably via their impact on birds' ability to forage.

Figure 6.B A pair of southern pied babblers (*Turdoides bicolor*, LC), an endemic of Southern Africa's arid savannahs, gaping their beaks to facilitate respiratory evaporative cooling during a particularly hot summer afternoon. Photograph by Nicholas Pattinson, CC BY 4.0.

For desert birds, foraging is critically important for maintaining both energy and water balance, as most species obtain all their water from food. Reduced activity almost inevitably means reduced food intake via impacts on time available for foraging. Seeking shade also carries costs: for some species, returns on foraging effort in shaded locations are significantly lower than in the sun (e.g. Cunningham et al., 2013). Finally, respiratory evaporative cooling can severely restrict the ability of actively-foraging birds to acquire food due to mechanical constraints on simultaneously gaping the bill and using it for prey capture and handling (e.g. du Plessis et al., 2012).

Under climate change, the implications of these behavioural trade-offs between foraging and thermoregulation are non-trivial. Inability to balance water and energy budgets mean birds progressively lose body condition during heat waves (du Plessis et al., 2012). Compromised foraging also affects birds' capacity to provision offspring, resulting in reduced nest success and/or smaller, lighter fledglings which may struggle to survive and recruit into the breeding population (e.g. Cunningham et al., 2013, Wiley and Ridley, 2016).

Successfully balancing the trade-offs between foraging and thermoregulation, and between hyperthermia and dehydration, is the secret to success for birds in hot places. As the climate warms, achieving this balance will become ever more challenging. Sublethal behavioural costs of keeping cool kick in at temperatures cooler than those promoting mass mortalities. In some parts of the world, such as Southern Africa, the loss of birds from desert ecosystems may therefore occur through the insidious whittling away of fitness and weakening of populations (Conradie et al., 2019) before we even witness the dramatic die-off events for which Australia is already infamous.

Figure 6.4 The Namib sand gecko (*Pachydactylus rangei*), endemic to Namibia's Namib Desert, survives the scorching heat by being nocturnal and burrowing into loose sand with its webbed toes. Warmer conditions under climate change may make it much harder for the gecko and other desert species, operating at the limits of their physiological tolerances, to survive. Photograph by Marije Louwsma, https://www.inaturalist.org/observations/18594993, CC BY 4.0.

An additional concern for lowland ecosystems is that climate change will likely lead to the creation of novel (i.e. hotter) ecosystems unlike any others currently on Earth (Williams et al., 2007). These changes will lead to **biotic attrition**. The gradual impoverishment of biological communities of lowland ecosystems as species either go extinct or move away while tracking their climatic envelopes. What is not clear is how the niches left open by the net loss of species, and newly created niches in the novel ecosystems, will be filled. The most likely scenario is that more tolerant, generalist species will fill the empty niches. However, with the inevitable loss of some species, combined with the decoupling of important biological interactions (discussed below), some functions and services associated with lowland ecosystems are likely to eventually collapse. It is important to note that tropical lowland forests and deserts are by no means the only ecosystems vulnerable to biotic attrition. For example, researchers have found that even mild warming would expose the Ethiopian Highlands to biotic attrition (Kreyling et al., 2010).

Climate change and dispersal limitations

Across many diverse ecosystems, a great number of species are threatened by climate change because of their poor dispersal abilities. Because they lack appropriate dispersal mechanisms, species, such as slow maturing plants (Foden et al., 2007), mosses, and flightless insects may simply not be able to keep up with changing climatic conditions. The impacts of climate change on Africa's dispersal-limited species can already be seen. For example, the once abundant Aldabra banded snail (*Rhachistia aldabrae*, CR) is today so rare that this Lazarus species (Figure 6.5) was once believed to be extinct due to climate change (Battarbee, 2014). There are also fears that successive droughts in the Cape Floristic Region may have recently driven a rare sorrel species (*Oxalis hygrophila*, CR) to extinction (Zietsman et al., 2008). Next might be the cave katydid (*Cedarbergeniana imperfecta*, CR) and Marais' lace-winged katydid (*Pseudosaga maraisi*, CR); these highly threatened insects count among Africa's very few cave specialists, and yet, by living in highly restricted and restrictive ecosystems, they face major challenges in adapting to climate change (Bazelet and Naskrecki, 2014). Dispersal limitations will also greatly affect terrestrial species living on oceanic islands, which will find it near impossible to track their climatic niches as it moves over the ocean. One such species is Cabo Verde's Raso lark (*Alauda razae*, CR); with a population size that fluctuates in response to rainfall, climate change induced drought conditions have pushed this bird to the brink of extinction in recent years (BirdLife International, 2016).

Climate change and biological interactions

Species that are highly mobile are not entirely spared from the negative impacts of climate change. Consider migratory species for a moment. The same way the musicians of an orchestra rely on a conductor to remain synchronised, migratory species rely on environmental cues, such as daylength and temperature, to decide

Figure 6.5 The Seychelles' Aldabra banded snail was once thought to be one of the world's first species pushed to extinction by climate change. Luckily, a small isolated population has been discovered, offering conservation biologists a second chance to ensure this species' survival. Photograph by Catherina Onezia/Seychelles Islands Foundation, CC BY 4.0.

when they need to start moving from one area to the next. But because different species rely of different environmental cues to time their life cycles (e.g. breeding), not all species will adjust to climate change at the same rate. There is consequently a high likelihood that climate change will disrupt these synchronous movements that the animal kingdom has developed over thousands of years (Renner and Zohner, 2018). This disruption of timed aspects of species' life cycle, such as migration and breeding, is called phenological mismatch or trophic asynchrony. Researchers have already seen signs of **phenological mismatch**: some migratory birds that overwinter in Africa have started to migrate to their European breeding grounds at earlier dates than before (Both et al., 2006; Vickery et al., 2014). If these trends hold, they may soon start breeding before peak food availability, which could lead to lower fitness of offspring.

We can already see evidence of how climate change is disrupting migrations and mutualistic relationships that were developed over thousands of years.

Resident species are also vulnerable to phenological mismatch. While these species might not be known for large-scale movements around the globe, they may still have to adjust their ranges to keep track of their climatic niches. Considering the improbability of different species will adapt at the same pace, there is thus a danger that important mutualistic relationships might be pulled apart these during range adaptations. This is of concern for species with specialised feeding niches, as seen in some pollinators. For example, studies from South Africa have shown how necessary range adjustments under climate change threaten both sunbirds—which show low adaptability (Simmons et al., 2004)—and their host plants, if specialised pollinator niches are left vacant (Huntley and Barnard, 2012). Extinctions arising from this decoupling of mutualistic relationships are referred to as coextinction (Koh et al., 2004), while a series of linked coextinctions is called an extinction cascade (Section 4.2.1).

Climate change and reptiles

One may think that reptiles—often seen basking on sun-drenched rocks to obtain active body temperatures—may benefit from climate change. Yet, as a group, they are also expected to suffer under climate change. One reason is because many reptiles will also have to adapt their ranges to shifting climates (Houniet et al., 2009). Even more important, climate change will increase reptiles' vulnerability to **demographic stochasticity** (Section 8.7.2). Many reptiles—and some fish—have their sex determined by temperature during embryonic development, with warmer temperatures often leading to more females (Valenzuala and Lance, 2004). In general, females regulate their offspring's sex ratios by fine-scale breeding site selection. Under climate change, however, it might be harder for the females to find breeding sites with suitable microclimates. This situation is of concern at South Africa's iSimangaliso Wetland Park, where Nile crocodiles (*Crocodylus niloticus*, LC) are already struggling to find suitable breeding sites due to microclimate changes caused by invasive plant encroachment (Leslie and Spotila, 2001). Those species unable to adopt new mechanisms to control for offspring sex ratio bias may eventually go extinct, even under relatively small temperature shifts (Sinervo et al., 2010).

6.3.3 Climate change's impact on freshwater ecosystems

With Africa's freshwater ecosystems already strained by the demands of a growing human population, freshwater biodiversity will face several additional stressors associated with climate change. Climate change will impact water temperature, flow volume, and flow variability. Because these variables are three primary predictors of freshwater ecosystem composition (van Vliet et al., 2013; Knouft and Ficklin, 2017), it is expected that climate change will greatly affect freshwater ecosystem composition and functioning in the coming decades.

Warmer rivers and streams

Climatologists and hydrologists predict that freshwater ecosystems will generally experience temperature increases under climate change. These changes are already evident in Africa: for example, Lake Albert on the DRC-Uganda border, and Zambia's Lake Mweru Wantipa, have experienced surface temperature increases of 0.62°C and 0.56°C respectively over the past decade (O'Reilly et al., 2015). Like their terrestrial counterparts, many freshwater species are sensitive to temperature shifts (e.g. Reizenberg et al., 2019). Warmer water also holds less dissolved oxygen, and increased pollutant toxicity (Whitehead et al., 2009). In addition, longer growing seasons and higher water temperatures will lead to a general increase in **primary productivity** and decomposition rates, which in turn will lead to increased nutrient loads, algae blooms,

> Climate change will alter water temperature, flow volume, and flow variability, the three primary predictors of freshwater ecosystem composition.

and **eutrophication** (Whitehead et al., 2009). All these factors will force many freshwater species—even those not sensitive to temperature shifts—to adjust their ranges to keep track of suitable conditions. Many of these adjustments will be impeded by habitat fragmentation, notably by dams and other human constructs that block suitable dispersal pathways. As an additional complication, many aquatic organisms cannot travel overland, so are naturally limited to adjust their ranges along the rivers and streams in which they live. But the orientation of these rivers and streams may not follow suitable thermal isolines: consider a cold-water species that needs to disperse to higher elevation—and hence upstream—as its climate niche moves higher up a mountain. For some freshwater species, the impediments to adjusting their ranges as necessary may be insurmountable.

Changing flow regimes

Changing precipitation levels will have several impacts on freshwater ecosystems, particularly as it relates to changes in their flow regimes (Thieme et al., 2010; Knouft and Ficklin, 2017). For example, areas that are undergoing decreased precipitation will experience decreased runoff and increased drying of wetlands and small streams, while areas with increased precipitation will experience increased storm surges and flushing. These changes, together with the impacts of increased water extraction rates and evapotranspiration under a warmer world, will cause significant changes in water levels, flow rates, sediment loads, water turbidity, and the structure of the physical environment. With an estimated 80% of Africa's freshwater fishes predicted to experience significant flow regime changes (Thieme et al., 2010), the region will likely see substantial changes in the composition of freshwater communities in the coming decades.

Given these multiple stressors, there is a reasonable expectation that many freshwater species will go extinct or face significant population declines and range shifts over the next decades. These changes are of major concern in Africa, where so many people depend on fish and related natural resources for their livelihoods. Communities in Uganda, Malawi, Guinea, and Senegal are already finding it more difficult to meet their nutritional needs due to climate-induced freshwater fish declines (Allison et al., 2009). Also, at Lake Tanganyika—which supplies 20–40% of the surrounding countries' dietary protein—fish yields have decreased by 30% in recent years, also attributed to climate change (O'Reilly et al., 2004).

6.3.4 Climate change's impact on marine ecosystems

Like tropical forests, the world's oceans have historically provided a relatively stable environment in which marine organisms have evolved. While this stability promotes species diversity, it also leaves marine species more vulnerable to environmental changes. In fact, a recent study found that cold-blooded marine species are twice as vulnerable to the impacts of warmer oceans than their terrestrial counterparts (Pinsky et al., 2019). In addition to the impacts of storm surges (Figure 6.6) and **ocean warming** (which leads to rising sea levels and ocean deoxygenation), marine organisms also must deal with **ocean acidification**. These threats will likely have impacts like those faced in terrestrial and freshwater ecosystems, including range adjustments, biotic attrition, and decoupling or important interactions. Below we discuss the mechanisms that will lead to some of these changes in more detail.

Figure 6.6 Many species living in low-lying coastal regions may be pushed to extinction by more frequent cyclones/hurricanes, storm surges, and sea level rise resulting from climate change. One such example is the Knysna seahorse (*Hippocampus capensis*, EN); biologists have attributed the deaths of thousands of these unique animals to temperature fluctuations and flooding events that altered parts of their highly restricted range in South Africa (Pollom, 2017). Photograph by Brian Gratwicke, https://www.flickr.com/photos/briangratwicke/7108174613, CC BY 2.0.

Ocean acidification

As discussed earlier, human activities release massive amounts of CO_2 into the atmosphere each day. Although forests and other plant communities get considerable attention for CO_2 sequestration, the world's oceans also play a key role in keeping Earth's carbon balance in check. In fact, the world's oceans absorb an estimated 20–25% of our current CO_2 emissions (Khatiwala et al., 2009). Now, with more atmospheric CO_2 available, oceans absorb more carbon, which dissolves in seawater as carbonic acid. While this absorption may slow climate change, it also increases the acidity (i.e. lowing the pH levels) of the world's oceans. This process—known as **ocean acidification**—has several consequences that may directly and indirectly kill marine organisms. For example, it inhibits the ability of coral animals to deposit the calcium

> Climate change is causing sea level rise and increased seawater temperatures, with broad implications for marine ecology and people living in coastal areas.

used to build their reefs' structure (Mollica et al., 2018), and prevents shellfish from accumulating adequate amounts of calcium carbonate to develop shells strong enough for survival (Branch et al., 2013). Ocean acidification also disturbs predator-prey dynamics by impairing the senses of prey species (Leduc et al., 2013), and compromising the ability of marine creatures to communicate with conspecifics (Roggatz et al., 2016).

Sea level rise

Over the past 30–40 years, ocean surface temperatures have warmed by about 0.64°C (NOAA, 2016).Ocean warming has several implications, the most well-known being **sea level rise**, caused by the thermal expansion of ocean water combined with the released water from melting glaciers and polar ice caps. Current predictions suggest that sea levels in Sub-Saharan Africa will rise by 0.2–1.15 m over the next 100 years, compared to 2005 levels (Serdeczny et al., 2017). As the oceans creep further inland, the extent of low-lying coastal ecosystems such as rocky shores or sandy beaches will shrink, and so also the sizes of the wildlife populations living in those areas. The extinction of Australia's Bramble Cay melomys (*Melomys rubicola*, EX)—the world's first documented mammalian extinction caused by anthropogenic climate-change—has been attributed to sea level rise (Gynther et al., 2016).

Coral bleaching

The incredible diversity of corals reef ecosystems is attributable to the relative stability of tropical oceans. Because of this stability, individual coral species have adapted to very specialized niches. Many corals thus tolerate only narrow ranges in temperature, sunlight levels, water opacity, and nutrient loads. Climate change is disrupting this stability, by changing the temperature (ocean warming), depth (sea level rise), sediment and nutrient loads (increased erosion and runoff) of the environments where corals live. These changes are leading to a breakdown of critical mutualistic relationships between photosynthetic algae and corals. In the process, corals also lose their vibrant colours, revealing the corals' ghostly white skeletons, hence the name **coral bleaching** (Figure 6.7). This relationship breakdown deprives the corals of essential carbohydrates they obtain from the algae, causing the corals to starve to death if the stressful conditions continue for a prolonged time.

Africa's tropical oceans have experienced extensive coral bleaching events in recent years. For example, parts of Tanzania and Kenya have seen over 80% of their corals affected (McClanahan et al., 2007; Chauka, 2016). Coral bleaching also affects other species associated with coral reefs. For example, in the Seychelles, where coral bleaching at 70–99% of reefs was observed, butterfly fish exhibited a breakdown in territorial behaviour, making it hard for them to breed and feed (Samways, 2005). In Zanzibar, Tanzania, eroded fish communities showed little signs of recovery multiple years after a bleaching event (Garpe et al., 2006).

Figure 6.7 A marine biologist surveys bleached corals in Curieuse Marine National Park, Seychelles. Bleaching events occur when heat stress kills corals, leaving only white skeletons where a once-vibrant coral reef community existed. Photograph by Emma Camp, CC BY 4.0.

Ocean deoxygenation

Marine fish and invertebrates rely on dissolved oxygen that enters the water either through the atmosphere, or by photosynthetic plankton. But because warmer water absorbs less oxygen, scientists predict that some areas of the ocean will see a 3–6% drop in dissolved oxygen concentrations under climate change (IPCC, 2014). This process, known as **ocean suffocation** or ocean deoxygenation (Ito et al., 2017), will leave parts of the ocean unsuitable for marine fishes and invertebrates. The impact of ocean deoxygenation will also be felt by economically important fisheries, notably along West Africa (Long et al., 2016) where climate change is predicted to lead to fisheries-related economic losses upwards of US $311 million each year (Lam et al., 2012).

6.3.5 Climate change interacts with habitat loss

Habitat loss and climate change each cause negative impacts on biodiversity; however, these threats also interact to have an overall larger negative impact than the sum of these threats independently. Prominently, because of habitat loss, many species will be unable to adequately adjust their ranges to keep track of their shifting climatic niches. For example, some species might not be able to adapt their ranges because suitable habitat in their future ranges will be destroyed by human activity.

> Climate change interacts with habitat loss, by impeding species' ability to adapt, and by bringing dispersing wildlife into conflict with humans.

Range-shift gaps describes a habitat gap that prevents a species from dispersing from its current to future ranges (Figure 6.8). These gaps, which may occur naturally or because of habitat fragmentation, may also impede range adjustments under climate change. While the impact of range-shift gaps is an active area of research, it is expected that mountain-top species may be inherently vulnerable to range-shift gaps, particularly if they are unable to first disperse downslope before they can reach climatically suitable locations at higher elevation elsewhere. For example, over 60% of herbaceous plants living on Ethiopia's Arsi Mountains might face range-shift gaps soon (Mekasha et al., 2013). But even highly mobile species might be vulnerable, with many African birds expected to face range-shift gaps as they adjust their ranges (La Sorte et al., 2014).

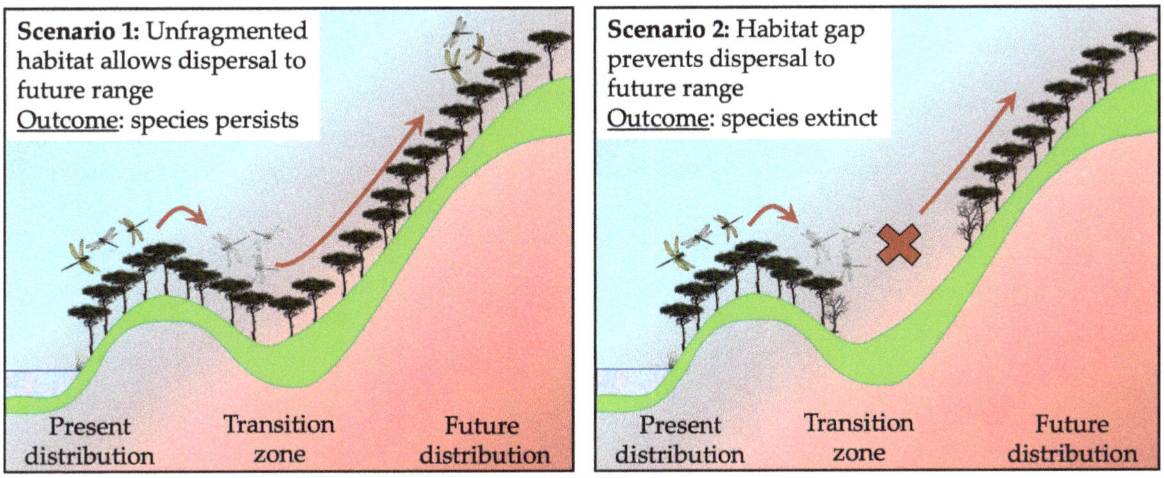

Figure 6.8 Hypothetical example of a species adjusting its range to climate change along unfragmented habitat (Scenario 1) and a species unable to adjust its range due to a range-shift gap (Scenario 2). In Scenario 1, the species persists; in Scenario 2 the species goes extinct because a gap in available habitat prevents dispersal into suitable areas. CC BY 4.0.

Habitat loss and climate change are also expected to exacerbate human-wildlife conflicts (Section 14.4). Sub-Saharan Africa will face losses of up to 2.5 million km^2 in arable land between 2010 and 2100 (Zabel et al., 2014). These losses will see even more natural ecosystems converted for agriculture which, in turn, will further increase competition among and between humans and wildlife for resources such as food, water, and suitable habitat (Serdeczny et al., 2017). As the human footprint expands across Earth, agriculture and infrastructure will impede the ability of specialist species to find food and adapt to changing conditions, while generalist species will be forced into agricultural lands and nearby human habitation as they search for resources and/or disperse across the landscape. Such a scenario will likely exacerbate human-wildlife conflict in areas like Kenya's Amboseli region, where lions living in fragmented ecosystems with diminishing natural prey populations are increasingly prone to wandering beyond protected area boundaries into ranching areas in search of food (Tuqa et al., 2014).

6.4 Beneficiaries of Climate Change

To be clear, not all species will suffer equally from climate change. In fact, there are some species that will be **resilient**, and others that will even benefit from a warming world. Primary among the beneficiaries are plants in the northern areas of Europe, Asia, and North America (Zabel et al., 2014), and to a lesser extent in southern South America and New Zealand. In these areas, plants will benefit from longer growing seasons (earlier springs and shorter winters) and increased CO_2 concentrations (which will increase photosynthesis rates).

Closer to home, a variety of African species are also expected to benefit from climate change. These include generalist species currently limited by interactions with localised specialists that are—at least at present—better competitors for limiting resources. Some tropical species may thrive as their habitats become hotter and wetter. Species with high genetic diversity that reproduce quickly (allowing for rapid adaption to environmental changes) are also likely to benefit. Unfortunately, many species that exhibit these traits carry diseases (Box 6.3) and are agricultural pests (Serdeczny et al., 2017). For example, populations of the coffee berry borer (*Hypothenemus hampei*)—Africa's most notorious coffee pest—are expected to greatly increase in a warmer world (Jaramillo et al., 2011). This growing threat is particularly worrying given that higher temperatures have already reduced coffee harvests in countries such as Tanzania by as much as 50% (Craparo et al., 2015).

Generalist species with high genetic diversity and that reproduce quickly are likely to benefit from climate change. Many species that exhibit these traits carry diseases and are agricultural pests.

Box 6.3 Habitat Alteration, Climate Change, and Mosquito-Borne Diseases

Kevin Njabo

Center for Tropical Research,
UCLA Institute of the Environment and Sustainability,
Los Angeles, CA, USA.

kynjabo@hotmail.com

With unprecedented climate change looming, mosquito-borne diseases, including malaria and dengue fever, will impact humans and wildlife in novel and unpredictable ways. While climate change is global in nature, changes due to habitat alteration are occurring more rapidly on a local scale, and are having significant effects on mosquito-borne diseases (Figure 6.C). For example, destruction of Peruvian rainforests unleashed more than 120,000 cases of malaria in the late 1990s, compared to fewer than 150 nine years earlier (Vitor et al., 2006).

The rainforests of the Congo Basin harbour roughly 20% of all known plant and animal species on Earth. Yet, habitat alteration continues at an alarming rate (Harris et al., 2012). Exacerbating these threats is the fact that Africa (Boko et al.,

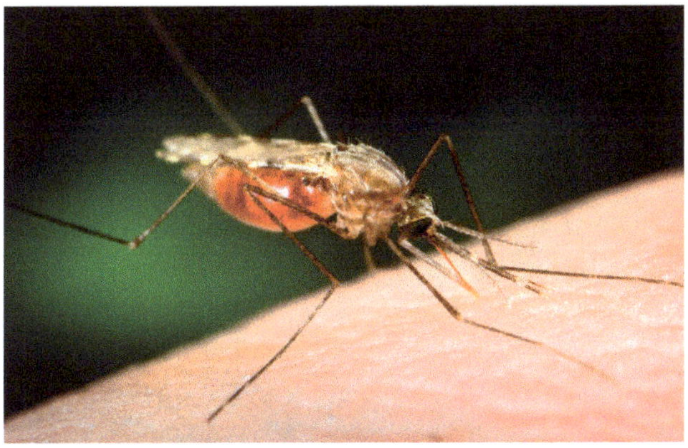

Figure 6.C (Top) Trucks transporting recently logged trees in Gabon. Photograph by David Stanley, https://www.flickr.com/photos/davidstanleytravel/46170117302, CC BY 2.0. (Bottom) *Anopheles funestus*, one of the most important vectors of malaria in Africa. Photograph by USCDCP, CC0.

2007), and Central Africa in particular (McClean et al., 2006), are predicted to be some of the most severely affected by climate change. Predicted temperature increases would lead to longer seasons of malaria transmission and a 5–7% extension of the disease into higher latitudes (Craig et al., 1999, Boko et al., 2007). Coupled with projected population growth, climate change would nearly double the number of people at risk of infection from dengue fever by 2080. This is concerning because Africa is particularly vulnerable to environmental changes due to its limited adaptive capacity, widespread poverty, and low levels of development.

How then, will habitat alteration and climate change affect mosquito-borne diseases such as malaria? The relationship between disease transmission, habitat alteration, and climate change is complex. Though deforestation increases the risk of disease transmission (Vitor et al., 2006), different malaria-carrying mosquitoes (*Anopheles* spp.) are adapted to different microclimates. Ironically, our multi-faceted ecosystems both play the role of maintaining transmission cycles with cross-infections to humans and regulating those cycles while controlling spill-over into human populations. The balance between these factors is influenced by the availability of suitable habitat for mosquitoes and of reservoir hosts of infection. In an ideal world, transmission cycles are regulated by density-dependent processes such as acquired immunity to infectious diseases, and by limits on the carrying capacity of the environment to support insects and hosts.

Altered natural habitats and possible increases in disease transmission from animals to people also increase potential risks of new pathogens adapting to human hosts. Only about 2,000 of an estimated 1 million unique viruses carried by wild vertebrate species with potential zoonotic threats have been described. For example, when a lentivirus of chimpanzees first jumped into humans in the 1930s, not many people died. But the disease carved a foothold in the rapidly growing African city of Kinshasa in DRC and evolved into a form that efficiently preyed upon humans. More than 78 million people were infected between 1981 and 2015. To date, the disease it causes, AIDS, has killed more than 39 million people, while another estimated 37 million people are living with HIV.

Today, habitat alteration, such as deforestation, is not only driving species extinct and emitting lots of climate-changing carbon dioxide, it is also increasing opportunities for mosquito-borne diseases, such as malaria and dengue, to infect more humans in new places. Technological advances, including mathematical and computer modelling, genomics, and satellite tracking, will hopefully allow us to predict future disease outbreaks better. But we can also reduce outbreak opportunities by taking better care of our environment.

One group of species currently threatened with extinction that may benefit from a warmer world is marine turtles. Researchers working on Cabo Verde speculate that the island nation's loggerhead turtle (*Caretta caretta* VU) populations will benefit from an increasing female-biased sex ratio (as expected under warmer conditions) given that a single male can breed with several females (Laloë et al., 2014). However, the researchers note that this population requires continued monitoring as insurance against demographic stochasticity (Section 8.7.2) that may become a larger threat under climate change.

Climate change has the potential to greatly restructure the world's ecosystems, ecosystem services, and national economies.

6.5 The Overall Impact of Climate Change

It should be clear to anyone that climate change has the potential to greatly restructure the world's ecosystems, ecosystem services, and national economies. Many coastal areas will experience rising sea levels and increased flooding, while inland areas may experience desertification and less favourable crop growing conditions. Poor Africans will suffer the consequences disproportionately because of their limited mobility, high dependence on ecosystem services, and general lack of disaster management infrastructure (Serdeczny et al., 2017).

If we are to mitigate the far-reaching impacts of climate change, we must carefully monitor and study changes in biological communities and ecosystem functioning, and how they relate to changes in climate and other stressors. While we may lose some species in a warmer world, we can also prevent many extinctions with pro-active wildlife management (Section 11.4). It is likely that many existing protected areas will no longer preserve some of the rare and threatened species that currently live in them (Hole et al., 2009; Smith et al., 2016, but see Beale et al., 2013), necessitating careful planning when establishing new protected areas (Section 13.7.2). Even if climate change is not as severe as predicted, the steps we take now to safeguard biodiversity can only help in future.

In 2007, the world economy was close to collapse because of the misdeeds of the financial services industry. Considering climate change's record in causing societal disruption and suffering, and our increasingly globalised world (in which regional disruptions are felt much wider than before), politicians are rightfully concerned about our ability to adapt to a widespread restructuring of the world's natural resources (Dietz et al., 2016). While the consequences of climate change are closely associated with the environmental sciences, it is truly, at its core, a human rights concern.

The widespread and dramatic impacts of climate change rightfully deserve much attention. But it is also important to remember that we continue to destroy habitat at a massive scale and increasing pace, and this loss of habitat is currently the main cause of species extinctions. The highest priorities for conservation must continue to be the preservation of healthy, intact, and connected ecosystems, and the restoration of degraded ecosystems. These actions will simultaneously reduce the impacts of climate change, by reducing carbon emissions, increasing carbon sequestration, and giving wildlife more opportunities to adjust their ranges, in their own time, as the world's climate changes.

6.6 Summary

1. While climate change is often thought of as a future challenge, we can already see its impacts today, as shown by record-high temperatures and changing rainfall patterns. These changes are happening because human activities release large amounts of greenhouse gases into the atmosphere on a daily basis.

2. Habitat loss contributes to climate change directly through the destruction of complex ecosystems (i.e. carbon sinks) which releases stored CO_2, and indirectly through the loss of vegetation that would otherwise sequester CO_2 from the atmosphere.

3. Some climatic shifts are predicted to be so rapid in coming decades that many species will be unable to adjust their ranges to keep up with environmental changes. Species with dispersal limitations, special habitat requirements, and important mutualistic relationships are at especially high risk of extinction.
4. Mitigating the negative impacts of climate change will require an international multi-pronged approach that includes ecosystem protection and restoration, direct species management, and legislative action.
5. Species are seldom exposed to only one threat; rather, different threats interact with climate change so that their combined impact is greater than their individual effects. A successful conservation strategy needs to deal with these threats collectively.

6.7 Topics for Discussion

1. Think of any particular ecosystem in your region. How do you think climate change will impact that ecosystem? What single measure do you think can be implemented to reduce the impact of climate change on that ecosystem? Can you think of the resources you will need to implement that measure?
2. Which groups of people and wildlife in Africa do you think will benefit the most from climate change and why? Who do you think will suffer the most and why?
3. How should we deal with species that have nowhere to go under climate change? Should we let them go extinct? What if it is an economically important species, like one that supports an important fishery or ecotourism industry? What do you think are our best options to save such species?

6.8 Suggested Readings

Allison, E.H., A.L. Perry, M.-C. Badjeck, et al. 2009. Vulnerability of national economies to the impacts of climate change on fisheries. *Fish and Fisheries* 10: 173–96. https://doi.org/10.1111/j.1467-2979.2008.00310.x African fisheries are highly vulnerable to climate change.

Dietz, S., A. Bowen, C. Dixon, et al. 2016. 'Climate value at risk' of global financial assets. *Nature Climate Change* 6: 676–79. https://doi.org/10.1038/nclimate2972 Climate change will cause financial losses of up to US $24 trillion.

Hole, D.G., S.G. Willis, D J. Pain, et al. 2009. Projected impacts of climate change on a continent-wide protected area network. *Ecology Letters* 12: 420–43. https://doi.org/10.1111/j.1461-0248.2009.01297.x In coming decades, many species will not be able to survive in their present locations because of climate change.

IPCC. 2014. *Climate Change 2014: AR5 Synthesis Report* (Geneva: IPCC). https://www.ipcc.ch/report/ar5/syr Comprehensive presentation of the evidence for global climate change, along with predictions for the coming decades.

Jaramillo J., E. Muchugu, F.E. Vega, et al. 2011. Some like it hot: The influence and implications of climate change on coffee berry borer (Hypo*thenemus hampei*) and coffee production in East Africa. *PLoS ONE* 6: e24528. https://doi.org/10.1371/journal.pone.0024528 Some important crop pests will benefit from a warming world.

La Sorte, F.A., S.H.M. Butchart, W. Jetz, et al. 2014. Range-wide latitudinal and elevational temperature gradients for the world's terrestrial birds: Implications under global climate change. *PLoS ONE* 9: e98361. https://doi.org/10.1371/journal.pone.0098361 African birds are particularly vulnerable to climate change.

Niang, I., O.C. Ruppel, M.A. Abdrabo, et al. 2014. Africa. In: *Climate Change 2014: Impacts, Adaptation, and Vulnerability*, ed. by V.R. Barros, et al. (Cambridge: Cambridge University Press). https://www.ipcc.ch/site/assets/uploads/2018/02/WGIIAR5-Chap22_FINAL.pdf A review of climate change impacts across Africa.

Thieme, M.L., B. Lehner, R. Abell, et al. 2010. Exposure of Africa's freshwater biodiversity to a changing climate. *Conservation Letters* 3:324–31. https://doi.org/10.1111/j.1755-263X.2010.00120.x Climate change will have wide-ranging impacts on Africa's freshwater ecosystems.

Wiens, J.J. 2016. Climate-related local extinctions are already widespread among plant and animal species. *PLoS Biology* 14: e2001104. https://doi.org/10.1371/journal.pbio.2001104 Hundreds of species have already been subjected to local extinctions due to climate change.

Bibliography

Albright, T.P., D. Mutiibwa, A.R. Gerson, et al. 2017. Mapping evaporative water loss in desert passerines reveals an expanding threat of lethal dehydration. *Proceedings of the National Academy of Sciences* 114: 2283–88. https://doi.org/10.1073/pnas.1613625114

Allison, E.H., A.L. Perry, M.-C. Badjeck, et al. 2009. Vulnerability of national economies to the impacts of climate change on fisheries. *Fish and Fisheries* 10: 173–96. https://doi.org/10.1111/j.1467-2979.2008.00310.x

Arrhenius, S. 1896. On the influence of carbonic acid in the air upon the temperature of the Earth. *Philosophical Magazine and Journal of Science* 41: 237–76. https://doi.org/10.1080/14786449608620846

Barbet-Massin, M., B.A. Walther, W. Thuiller, et al. 2009. Potential impacts of climate change on the winter distribution of Afro-Palaearctic migrant passerines. *Biology Letters* 5: 248–51. https://doi.org/10.1098/rsbl.2008.0715

Battarbee, R.W. 2014. The rediscovery of the Aldabra banded snail, *Rhachistia aldabrae*. *Biology Letters* 10: 20140771. https://doi.org/10.1098/rsbl.2014.0771

Bazelet, C., and P. Naskrecki. 2014. *Pseudosaga maraisi*. *The IUCN Red List of Threatened Species* 2014: e.T62452865A62452868. http://doi.org/10.2305/IUCN.UK.2014-3.RLTS.T62452865A62452868.en

Beale, C.M., N.E. Baker, M.J. Brewer, et al. 2013. Protected area networks and savannah bird biodiversity in the face of climate change and land degradation. *Ecology letters* 16: 1061–68. https://doi.org/10.1111/ele.12139

BirdLife International. 2016. *Alauda razae*. *The IUCN Red List of Threatened Species* 2016: e.T22717428A94531580. http://doi.org/10.2305/IUCN.UK.2018-2.RLTS.T22717428A131103086.en

Boko, M., I. Niang, A. Nyong, et al. 2007. Africa. In: *Climate Change 2007: Impacts, Adaptation and Vulnerability*, ed. by S. Solomon et al. (Cambridge: Cambridge University Press). https://www.ipcc.ch/site/assets/uploads/2018/02/ar4-wg2-chapter9-1.pdf

Both, C., S. Bouwhuis, C.M. Lessells, et al. 2006. Climate change and population declines in a long-distance migratory bird. *Nature* 441: 81–83. https://doi.org/10.1038/nature04539

Branch, T.A., B.M. DeJoseph, L.J. Ray, et al. 2013. Impacts of ocean acidification on marine seafood. *Trends in Ecology and Evolution* 28: 178–86. https://doi.org/10.1016/j.tree.2012.10.001

Burton, M.E.H., J.R. Poulsen, M.E. Lee, et al. 2017. Reducing carbon emissions from forest conversion for oil palm agriculture in Gabon. *Conservation Letters* 10: 297–307. https://doi.org/10.1111/conl.12265

Carolin, S.A., R.T. Walker, C.C. Day, et al. 2019. Precise timing of abrupt increase in dust activity in the Middle East coincident with 4.2 ka social change. *Proceedings of the National Academy of Sciences* 116: 67–72. https://doi.org/10.1073/pnas.1808103115

Carrington, D. Why the Guardian is changing the language it uses about the environment. *Guardian. https://gu.com/p/bfgxm*

Carr, J.A., A.F. Hughes, and W.B. Foden. 2014. *A climate change vulnerability assessment of West African species*. Technical Report (Cambridge: UNEP-WCMC). http://parcc.protectedplanet.net/assets/IUCN_species_vulnerability-181b4593dd469dcba033b1f06aaa3cd7c7678424c3a2b056578c9582bd5bf7fb.pdf

Chauka, L.J. 2016. Tanzanian reef building corals may succumb to bleaching events: Evidences from coral-symbiodinium symbioses. In: *Estuaries: A Lifeline of Ecosystem Services in the Western Indian Ocean*, ed. by S. Diop et al. (Cham: Springer). https://doi.org/10.1007/978-3-319-25370-1

Conradie, S.R., S.M. Woodbourne, S.J. Cunningham, et al. 2019. Chronic, sublethal effects of high temperatures will cause severe declines in southern African arid-zone birds during the 21st Century. *Proceedings of the National Academy of Sciences* 116: *in press*.

Craig, M.H., R.W. Snow, and D. le Sueur. 1999. A climate-based distribution model of malaria transmission in sub-Saharan Africa. *Parasitology Today* 15: 105–11. https://doi.org/10.1016/S0169-4758(99)01396-4

Craparo, A.C.W., P.J.A. van Asten, P. Läderach, et al. 2015. *Coffea arabica* yields decline in Tanzania due to climate change: Global implications. *Agricultural and Forest Meteorology* 207: 1–10. https://doi.org/10.1016/j.agrformet.2015.03.005

Crump, M.L., F.R. Hensley, and K.L. Clark, 1992. Apparent decline of the golden toad: Underground or extinct? *Copia* 1992: 413–20.

Cunningham, S.J., R.O. Martin, C.L. Hojem, et al. 2013. Temperatures in excess of critical thresholds threaten nestling growth and survival in a rapidly-warming arid savanna: A study of common fiscals. *PLoS ONE* 8: e74613. https://doi.org/10.1371/journal.pone.0074613

DeMenocal, P.B. 2001. Cultural responses to climate change during the late Holocene. *Science* 292: 667–73. https://doi.org/10.1126/science.1059287

Dietz, S., A. Bowen, C. Dixon, et al. 2016. 'Climate value at risk' of global financial assets. *Nature Climate Change* 6: 676–79. https://doi.org/10.1038/nclimate2972

Dimitrov, D., D. Nogués-Bravo, and N. Scharff. 2012. Why do tropical mountains support exceptionally high biodiversity? The Eastern Arc Mountains and the drivers of *Saintpaulia* diversity. *PloS ONE* 7: e48908. https://doi.org/10.1371/journal.pone.0048908

du Plessis, K.L., R.O. Martin, P.A.R. Hockey, et al. 2012. The costs of keeping cool in a warming world: Implications of high temperatures for foraging, thermoregulation and body condition of an arid-zone bird. *Global Change Biology* 18: 2063–3070. https://doi.org/10.1111/j.1365-2486.2012.02778.x

Engelbrecht, F.A., J.L. McGregor, and C.J. Engelbrecht. 2009. Dynamics of the Conformal-Cubic Atmospheric Model projected climate-change signal over southern Africa. *International Journal of Climatology* 29: 1013–33. https://doi.org/10.1002/joc.1742

Fagotto, M., and M. Gattoni. 2016. West Africa is being swallowed by the sea. *Foreign Policy.* http://atfp.co/2tUZCaM

Fitchett, J.M., and S.W. Grab. 2014. A 66-year tropical cyclone record for south-east Africa: Temporal trends in a global context. *International Journal of Climatology* 34: 3604–15. https://doi.org/10.1002/joc.3932

Flörke, M., C. Schneider, and R.I. McDonald. 2018. Water competition between cities and agriculture driven by climate change and urban growth. *Nature Sustainability* 1: 51–58. https://doi.org/10.1038/s41893-017-0006-8

Foden, W., G.F. Midgley, G. Hughes, et al. 2007. A changing climate is eroding the geographical range of the Namib Desert tree Aloe through population declines and dispersal lags. *Diversity and Distributions* 13: 645–53. https://doi.org/10.1111/j.1472-4642.2007.00391.x

Fordham, D.A., C. Bertelsmeier, B.W. Brook, et al. 2018. How complex should models be? Comparing correlative and mechanistic range dynamics models. *Global Change Biology* 24: 1357–70. https://doi.org/10.1111/gcb.13935

Forster, P., V. Ramaswamy, P. Artaxo, et al. 2007. Changes in atmospheric constituents and in radiative forcing. In: *Climate Change 2007: The Physical Science Basis*, ed. by S. Solomon et al. (Cambridge: Cambridge University Press). https://www.ipcc.ch/site/assets/uploads/2018/02/ar4-wg1-chapter2-1.pdf

Garpe, K.C., S.A.S. Yahya, U. Lindahl, et al. 2006. Long-term effects of the 1998 coral bleaching event on reef fish assemblages. *Marine Ecology Progress Series* 315: 237–47. https://doi.org/10.3354/meps315237

Gillis, J. 2017. Earth sets a temperature record for the third straight year. *New York Times.* https://nyti.ms/2jAdWlA

Gonedelé B.S., I. Koné, A.E., Bitty, et al. 2012. Distribution and conservation status of catarrhine primates in Côte d'Ivoire (West Africa). *Folia Primatologica* 83: 11–23. https://doi.org/10.1159/000338752

Grab, S., and A. Craparo. 2011. Advance of apple and pear tree full bloom dates in response to climate change in the southwestern Cape, South Africa: 1973–2009. *Agricultural and Forest Meteorology* 151: 406–13. http://doi.org/10.1016/j.agrformet.2010.11.001

Gynther, I., N. Waller, and L.K.-P. Leung. 2016. *Confirmation of the extinction of the Bramble Cay melomys* Melomys rubicola *on Bramble Cay, Torres Strait* (Brisbane: EHP). https://environment.des.qld.gov.au/wildlife/threatened-species/documents/bramble-cay-melomys-survey-report.pdf

Harris, N.L., S. Brown, S.C. Hagen, et al. 2012. Baseline map of carbon emissions from deforestation in tropical regions. *Science* 336: 1573–76. https://doi.org/10.1126/science.1217962

Hole, D.G., S.G. Willis, D.J. Pain, et al. 2009. Projected impacts of climate change on a continent-wide protected area network. *Ecology Letters* 12: 420–31. https://doi.org/10.1111/j.1461-0248.2009.01297.x

Houniet, D.T., W. Thuiller, and K.A. Tolley. 2009. Potential effects of predicted climate change on the endemic South African Dwarf Chameleons, *Bradypodion*. *African Journal of Herpetology* 58: 28–35. https://doi.org/10.1080/21564574.2009.9635577

Hsiang, S.M., and A.H. Sobel. 2016. Potentially extreme population displacement and concentration in the tropics under non-extreme warming. *Scientific Reports* 6: 25697. https://doi.org/10.1038/srep25697

Huntley, B., and P. Barnard. 2012. Potential impacts of climatic change on southern African birds of fynbos and grassland biodiversity hotspots. *Diversity and Distributions* 18: 1–13. https://doi.org/10.1111/j.1472-4642.2012.00890.x

IPCC. 2014: *Climate Change 2014: Synthesis Report* (Geneva: IPCC). https://www.ipcc.ch/report/ar5/syr

Ito, T., S. Minobe, M.C. Long, et al. 2017. Upper ocean O_2 trends: 1958–2015. *Geophysical Research Letters* 44: 4214–23. https://doi.org/10.1002/2017GL073613

Jackson, R.B., C. Le Quéré, R.M. Andrew, et al. 2018. Global energy growth is outpacing decarbonization. *Environmental Research Letters* 13: 120401. https://doi.org/10.1088/1748-9326/aaf303

Jaramillo J., E. Muchugu, F.E. Vega, et al. 2011. Some like it hot: The influence and implications of climate change on coffee berry borer (*Hypothenemus hampei*) and coffee production in East Africa. *PLoS ONE* 6: e24528. https://doi.org/10.1371/journal.pone.0024528

Jezkova, T., and J.J. Wiens. 2016. Rates of change in climatic niches in plant and animal populations are much slower than projected climate change. *Proceedings of the Royal Society B* 283: 20162104. https://doi.org/10.1098/rspb.2016.2104

Jolly, W.M., M.A. Cochrane, P.H. Freeborn, et al. 2015. Climate-induced variations in global wildfire danger from 1979 to 2013. *Nature Communications* 6: 8537. https://doi.org/10.1038/ncomms8537

Kaempffert, W. 1956. Warmer climate on the Earth may be due to more carbon dioxide in the air. *New York Times*. https://nyti.ms/2zYC2Ot

Kaniewski, D., E. van Campo, J. Guiot, et al. 2013. Environmental roots of the Late Bronze Age crisis. *PLoS ONE* 8: e71004. https://doi.org/10.1371/journal.pone.0071004

Khatiwala, S., F. Primeau, and T. Hall. 2009. Reconstruction of the history of anthropogenic CO_2 concentrations in the ocean. *Nature* 462: 346–49. https://doi.org/10.1038/nature08526

Knouft, J.H., and D.L. Ficklin. 2017. The potential impacts of climate change on biodiversity in flowing freshwater systems. *Annual Review of Ecology, Evolution, and Systematics* 48: 111–33. https://doi.org/10.1146/annurev-ecolsys-110316-022803

Koh, L.P., R.R. Dunn, N.S. Sodhi, et al. 2004. Species coextinctions and the biodiversity crisis. *Science* 305: 1632–34. https://doi.org/10.1126/science.1101101

Kreyling, J., D. Wana, and C. Beierkuhnlein. 2010. Potential consequences of climate warming for tropical plant species in high mountains of southern Ethiopia. *Diversity and Distributions* 16: 593–605. https://doi.org/10.1111/j.1472-4642.2010.00675.x

La Sorte, F.A., S.H.M. Butchart, W. Jetz, et al. 2014. Range-wide latitudinal and elevational temperature gradients for the world's terrestrial birds: Implications under global climate change. *PLoS ONE* 9: e98361. https://doi.org/10.1371/journal.pone.0098361

Laloë, J.-O., J. Cozens, B. Renom, et al. 2014. Effects of rising temperature on the viability of an important sea turtle rookery. *Nature Climate Change* 4: 513–18. https://doi.org/10.1038/nclimate2236

Lam, V.W.Y., W.W.L. Cheung, W. Swartz, et al. 2012. Climate change impacts on fisheries in West Africa: Implications for economic, food and nutritional security. *African Journal of Marine Science* 34: 103–17. http://doi.org/10.2989/1814232X.2012.673294

Le Quéré, C.L., R.M. Andrew, P. Friedlingstein, et al. 2018. Global carbon budget 2018. *Earth System Science Data* 10: 2141–94. https://doi.org/10.5194/essd-10-2141-2018

Leduc, A.O.H.C., P.L. Munday, G.E. Brown, et al. 2013. Effects of acidification on olfactory-mediated behaviour in freshwater and marine ecosystems: A synthesis. *Philosophical Transactions of the Royal Society* B 368: 20120447. http://doi.org/10.1098/rstb.2012.0447

Leslie, A.J., and J.R. Spotila. 2001. Alien plant threatens Nile crocodile (*Crocodylus niloticus*) breeding in Lake St. Lucia, South Africa. *Biological Conservation* 98: 347–55. https://doi.org/10.1016/S0006-3207(00)00177-4

Linder, J.M. 2013. African primate biodiversity threatened by "new wave" of industrial oil palm expansion. *African Primates* 8: 25–38.

Linder, J.M., and R.E. Palkovitz. 2016. The threat of industrial oil palm expansion to primates and their habitats. In: *Ethnoprimatology*, ed. by M. Waller (Cham: Springer). https://doi.org/10.1007/978-3-319-30469-4

Long, M.C., C. Deutsch, and T. Ito. 2016. Finding forced trends in oceanic oxygen. *Global Biogeochemical Cycles* 30: 381–97. https://doi.org/10.1002/2015GB005310

Maxwell, D., N. Majid, H. Stobaugh, et al. 2014. *Lessons learned from the Somalia famine and the greater Horn of Africa crisis 2011–2012* (Medford: Feinstein International Center, Tufts University). http://fic.tufts.edu/publication-item/famine-somalia-crisis-2011-2012

McClanahan, T.R., M. Ateweberhan, C.A. Muhando, et al. 2007. Effects of climate and seawater temperature variation on coral bleaching and mortality. *Ecological Monographs* 77: 503–25. https://doi.org/10.1890/06-1182.1

McClean, C.J., N. Doswald, W. Küper, et al. 2006. Potential impacts of climate change on Sub-Saharan African plant priority area selection. *Diversity and Distributions* 12: 645–55. https://doi.org/10.1111/j.1472-4642.2006.00290.x

McKechnie, A.E., and B.O. Wolf. 2010. Climate change increases the likelihood of catastrophic avian mortality events during extreme heat waves. *Biology Letters* 6: 253–56. https://doi.org/10.1098/rsbl.2009.0702

Medek, D.E., J. Schwartz, and S.S. Myers. 2017. Estimated effects of future atmospheric CO_2 concentrations on protein intake and the risk of protein deficiency by country and region. *Environmental Health Perspectives* 125: 087002. https://doi.org/10.1289/EHP41

Mekasha, A., L. Nigatu, K. Tesfaye, et al. 2013. Modeling the response of tropical highland herbaceous grassland species to climate change: The case of the Arsi Mountains of Ethiopia. *Biological Conservation* 168: 169–75. https://doi.org/10.1016/j.biocon.2013.09.020

Merone, L, and P. Tait. 2018. 'Climate refugees': Is it time to legally acknowledge those displaced by climate disruption? *Australian and New Zealand Journal of Public Health* 6: 508–09. https://doi.org/10.1111/1753-6405.12849

Milne, R., S.J. Cunningham, A.T. Lee, et al. 2015. The role of thermal physiology in recent declines of birds in a biodiversity hotspot. *Conservation Physiology* 3: p.cov048. https://doi.org/10.1093/conphys/cov048

Mollica, N.R., W. Guo, A.L. Cohen, et al. 2018. Ocean acidification affects coral growth by reducing skeletal density. *Proceedings of the National Academy of Sciences* 115: 1754–59. https://doi.org/10.1073/pnas.1712806115

Myers, S.S., A. Zanobetti, I. Kloog, et al. 2014. Increasing CO_2 threatens human nutrition. *Nature* 510: 139–42. https://doi.org/10.1038/nature13179

NASA. 2018. *Forcings in GISS Climate Model: Historical Data*. https://data.giss.nasa.gov/modelforce/ghgases

NOAA. 2016. *Extended Reconstructed Sea Surface Temperature (ERSST)*, v. 4. http://doi.org/10.7289/V5KD1VVF

NOAA. 2018a. *Climate at a Glance: Global Time Series*, December 2018. https://www.ncdc.noaa.gov/cag

NOAA. 2018b. *NOAA Earth System Research Laboratory: Global Monitoring Division*, December 2018. https://www.esrl.noaa.gov/gmd/ccgg/trends

NOAA. 2018c. *State of the Climate: Global Climate Report for April 2018*. https://www.ncdc.noaa.gov/sotc/global/201804

O'Connor, T.G., and G.A. Kiker. 2004. Collapse of the Mapungubwe society: Vulnerability of pastoralism to increasing aridity. *Climatic Change* 66: 49–66. https://doi.org/10.1023/B:CLIM.0000043192.19088.9d

O'Reilly, C.M., S. Sharma, D.K. Gray, et al. 2015. Rapid and highly variable warming of lake surface waters around the globe. *Geophysical Research Letters* 42: 10773–81. https://doi.org/10.1002/2015GL066235

O'Reilly, C.M., S.R. Alin, P.-D. Plisnier, et al. 2003. Climate change decreases aquatic ecosystem productivity of Lake Tanganyika, Africa. *Nature* 424: 766–68. https://doi.org/10.1038/nature01833

Ordway, E.M., R.L. Naylor, R.N. Nkongho, et al. 2019. Oil palm expansion and deforestation in Southwest Cameroon associated with proliferation of informal mills. *Nature Communications* 10: 114. https://doi.org/10.1038/s41467-018-07915-2

Pinsky, M.L., A.M. Eikeset, D.J. McCauley, et al., 2019. Greater vulnerability to warming of marine versus terrestrial ectotherms. *Nature* 569: 108–11. https://doi.org/10.1038/s41586-019-1132-4

Pollom, R. 2017. *Hippocampus capensis*. *The IUCN Red List of Threatened Species* 2017: e.T10056A54903534. http://doi.org/10.2305/IUCN.UK.2017-3.RLTS.T10056A54903534.en

Ponce-Reyes, R., A.J. Plumptre, D. Segan, et al. 2017. Forecasting ecosystem responses to climate change across Africa's Albertine Rift. *Biological Conservation* 209: 464–72. https://doi.org/10.1016/j.biocon.2017.03.015

Reizenberg, J.-L., L.E. Bloy, O.L.F. Weyl, et al. 2019. Variation in thermal tolerances of native freshwater fishes in South Africa's Cape Fold Ecoregion: Examining the east-west gradient in species' sensitivity to climate warming. *Journal of Fish Biology* 94: 103–12. https://doi.org/10.1111/jfb.13866

Renner, S.S., and C.M. Zohner. 2018. Climate change and phenological mismatch in trophic interactions among plants, insects, and vertebrates. *Annual Review of Ecology, Evolution, and Systematics* 49: 165–82. https://doi.org/10.1146/annurev-ecolsys-110617-062535

Rey, B., A. Fuller, D. Mitchell, et al. 2017. Drought-induced starvation of aardvarks in the Kalahari: An indirect effect of climate change. *Biology Letters* 13: 20170301. https://doi.org/10.1098/rsbl.2017.0301

Roggatz, C.C., M. Lorch, J.D. Hardege, et al. 2016. Ocean acidification affects marine chemical communication by changing structure and function of peptide signalling molecules. *Global Change Biology* 22: 3914–26. https://doi.org/10.1111/gcb.13354

Russo, S., A.F. Marchese, J. Sillmann, et al. 2016. When will unusual heat waves become normal in a warming Africa? *Environmental Research Letters* 11: 054016. https://doi.org/10.1088/1748-9326/11/5/054016

Samways, M.J. 2005. Breakdown of butterflyfish (Chaetodontidae) territories associated with the onset of a mass coral bleaching event. *Aquatic Conservation* 15: S101–S107. https://doi.org/10.1002/aqc.694

Serdeczny, O., S. Adams, F. Baarsch, et al. 2017. Climate change impacts in Sub-Saharan Africa: From physical changes to their social repercussions. *Regional Environmental Change* 17: 1585–600. https://doi.org/10.1007/s10113-015-0910-2

Simmons, R.E., P. Barnard, W.R.J. Dean, et al. 2004. Climate change and birds: Perspectives and prospects from southern Africa. *Ostrich* 75: 295–308. https://doi.org/10.2989/00306520409485458

Sinervo, B., F. Mendez-De-La-Cruz, D.B. Miles, et al. 2010. Erosion of lizard diversity by climate change and altered thermal niches. *Science* 328: 894–99. https://doi.org/10.1126/science.1184695

Siraj, A.S., M. Santos-Vega, M.J. Bouma, et al. 2014. Altitudinal changes in malaria incidence in highlands of Ethiopia and Colombia. *Science* 343: 1154–58. https://doi.org/10.1126/science.1244325

Smith, A., M.C. Schoeman, M. Keith, et al. 2016. Synergistic effects of climate and land-use change on representation of African bats in priority conservation areas. *Ecological Indicators* 69: 276–83. http://doi.org/10.1016/j.ecolind.2016.04.039

Storlazzi, C.D., S.B. Gingerich, A. van Dongeren, et al. 2018. Most atolls will be uninhabitable by the mid-21st century because of sea-level rise exacerbating wave-driven flooding. *Science Advances* 4: eaap9741. https://doi.org/10.1126/sciadv.aap9741

Strydom, S., and M.J. Savage. 2016. A spatio-temporal analysis of fires in South Africa. *South African Journal of Science* 112: 1–8. https://doi.org/10.17159/sajs.2016/20150489

Thieme, M.L., B. Lehner, R. Abell, et al. 2010. Exposure of Africa's freshwater biodiversity to a changing climate. *Conservation Letters* 3:324–31. https://doi.org/10.1111/j.1755-263X.2010.00120.x

Thomas, C.D., A. Cameron, R.E. Green, et al. 2004. Extinction risk from climate change. *Nature* 427: 145–48. https://doi.org/10.1038/nature02121

Tuqa, J.H., P. Funston, C. Musyoki, et al. 2014. Impact of severe climate variability on lion home range and movement patterns in the Amboseli ecosystem, Kenya. *Global Ecology and Conservation* 2: 1–10. https://doi.org/10.1016/j.gecco.2014.07.006

Uhe, P., S. Philip, S. Kew, et al. 2017. *Kenya drought, 2016*. https://wwa.climatecentral.org/analyses/kenya-drought-2016

Valenzuela, N., and V. Lance. 2004. *Temperature-Dependent Sex Determination in Vertebrates* (Washington: Smithsonian Books). https://doi.org/10.5479/si.9781944466213

van Vliet, M.T., D. Ludwig, and P. Kabat. 2013. Global streamflow and thermal habitats of freshwater fishes under climate change. *Climate Change* 121: 739–54. https://doi.org/10.1007/s10584-013-0976-0

van Wilgen, N.J., V. Goodall, S. Holness, et al. 2016. Rising temperatures and changing rainfall patterns in South Africa's national parks. *International Journal of Climatology* 36: 706–21. https://doi.org/10.1002/joc.4377

Vickery, J.A., S.R. Ewing, K.W. Smith, et al. 2014. The decline of Afro-Palaearctic migrants and an assessment of potential causes. *Ibis* 156: 1–22. https://doi.org/10.1111/ibi.12118

Vijay, V., S.L. Pimm, C.N. Jenkins, et al. 2016. The impacts of oil palm on recent deforestation and biodiversity loss. *PloS ONE* 11: e0159668. https://doi.org/10.1371/journal.pone.0159668

Vittor, A.Y., R.H. Gilman, J. Tielsch, et al. 2006. The effect of deforestation on the human-biting rate of *Anopheles darlingi*, the primary vector of falciparum malaria in the Peruvian Amazon. *American Journal of Tropical Medicine and Hygiene* 74: 3–11. https://doi.org/10.4269/ajtmh.2006.74.3

Wang, X., F. Chen, J. Zhang, et al. 2010. Climate, desertification, and the rise and collapse of China's historical dynasties. *Human Ecology* 38: 157–72. https://doi.org/10.1007/s10745-009-9298-2

Warren, R., J. Price, J. VanDerWal, et al. 2018. The implications of the United Nations Paris Agreement on climate change for globally significant biodiversity areas. *Climatic Change* 147: 395–409. https://doi.org/10.1007/s10584-018-2158-6

Watts, N., W.N. Adger, S. Ayeb-Karlsson, et al. 2017. The Lancet Countdown: Tracking progress on health and climate change. *Lancet* 389: 1151–64. https://doi.org/10.1016/S0140-6736(16)32124-9

Weiss, H., and R.S. Bradley. 2001. What drives societal collapse? *Science* 291: 609–10. https://doi.org/10.1126/science.1058775

Whitehead, P., R. Wilby, R. Battarbee, et al. 2009. A review of the potential impacts of climate change on surface water quality. *Hydrological Sciences Journal* 54:101–23. https://doi.org/10.1623/hysj.54.1.101

Whittington-Jones, G.M., R.T.F. Bernard, and D.M. Parker. 2011. Aardvark burrows: A potential resource for animals in arid and semi-arid environments. *African Zoology* 46: 362–70. https://doi.org/10.3377/004.046.0215

Wiens, J.J. 2016. Climate-related local extinctions are already widespread among plant and animal species. *PLoS Biology* 14: e2001104. https://doi.org/10.1371/journal.pbio.2001104

Wiley, E.M., and A.R. Ridley. 2016. The effects of temperature on offspring provisioning in a cooperative breeder. *Animal Behaviour* 117: 187–95. https://doi.org/10.1016/j.anbehav.2016.05.009

Williams, J.W., S.T. Jackson, and J.E. Kutzbach. 2007. Projected distributions of novel and disappearing climates by 2100 AD. *Proceedings of the National Academy of Sciences* 104: 5738–42. https://doi.org/10.1073/pnas.0606292104

WRI (World Resources Institute). *2019 Climate analysis indicators tool: WRI's climate data explorer.* http://cait2.wri.org

Zabel, F., B. Putzenlechner, and W. Mauser. 2014. Global agricultural land resources—a high resolution suitability evaluation and its perspectives until 2100 under climate change conditions. *PLoS ONE* 9: e107522. https://doi.org/10.1371/journal.pone.0107522

Zietsman, J., L.L. Dreyer, and K.J. Esler. 2008. Reproductive biology and ecology of selected rare and endangered *Oxalis* L. (Oxalidaceae) plant species. *Biological Conservation* 141: 1475–83. http://doi.org/10.1016/j.biocon.2008.03.017

7. Pollution, Overharvesting, Invasive Species, and Disease

A pair of southern white rhinoceros (*Ceratotherium simum simum*, NT) in Tshukudu Private Game Reserve, adjacent to South Africa's Kruger National Park. Once nearly extinct due to uncontrolled hunting, this species rebounded to over 20,000 individuals. Now, poaching is taking a toll again, even in premier protected areas such Kruger, where at least 662 rhinos were lost in 2016. Rangers have cut off the horn of one of the rhinos in the photo in a desperate attempt to prevent poaching. Photograph by Jan Fleishmann, https://commons.wikimedia.org/wiki/File:Nw_9302_white_rhinos_Tshukudu_JF.jpg, CC BY-SA 4.0.

https://doi.org/10.11647/OBP.0177.07

Conservation biologists aim to preserve all the components, interactions, and processes within and between ecosystems, natural communities, species, and populations. The main obstacle to accomplishing this goal is habitat loss, while climate change will also play an increasingly important role. But let us for a moment consider widespread species and migratory populations. These species and populations typically live in different habitats and encounter different climates as they move across the landscape. We might think that tolerance for variety would make these groups robust against habitat loss and climate change. And yet, they are also declining, even in seemingly intact ecosystems and protected areas (Craigie et al., 2010; Lindsey et al., 2014). How can it be that populations apparently buffered from the two main extinction drivers are also subjected to population declines and extirpations?

While habitat loss and climate change are the most prominent threats facing biodiversity at present, they are not alone. Nearly all human activities place additional pressures on populations, even those that already suffer from habitat loss and climate change. These additional pressures are primarily from pollution, overharvesting, **persecution**, invasive species, and disease (Maxwell et al., 2016). Because these threats are associated with and/or exacerbated by human activities, they can be dynamic in their nature, develop rapidly, and persist at such large scales that wildlife populations have little opportunity to adapt or move to safer areas. Moreover, these threats may interact with each other, as well as with climate change and habitat loss, so that their combined impact is greater than their individual effects. In this chapter, we explore how each of these threats impact wildlife and natural communities, and how they could push populations and species to extinction. Methods for lessening the impact of these threats are integrated into Chapters 9–15.

> Comprehensive conservation efforts must recognise that biodiversity faces multiple threats that need to be dealt with at different scales.

7.1 Pollution in Its Many Forms

Rachel Carson's 1962 book, *Silent Spring*, described the dangers of pollution—pesticide pollution in particular—with a clarity that captured the public's attention for many years afterwards. Carson, an American biologist, was particularly successful in drawing attention to **biomagnification** (also called bioaccumulation), a process through which pesticides and other toxins accumulate and become more concentrated in animals at higher levels of the **food chain** (Figure 7.1). Her work drew on research that found that dichlorodiphenyltrichloroethane (DDT), sprayed on crops to kill pest insects and on water bodies to kill malaria mosquito (*Anopheles* spp.) larvae, was also harming non-target organisms that consumed insects and fish exposed to DDT. Of note is that non-target organisms high on food chains, particularly fish-eating birds, such as eagles, pelicans, and egrets, often had high levels of DDT concentrated in their tissues. The affected birds were generally weakened, and the shells of their eggs were thin and prone to cracking during incubation. Consequently, bird populations declined dramatically in areas where DDT was used, as adults died and failed to raise young.

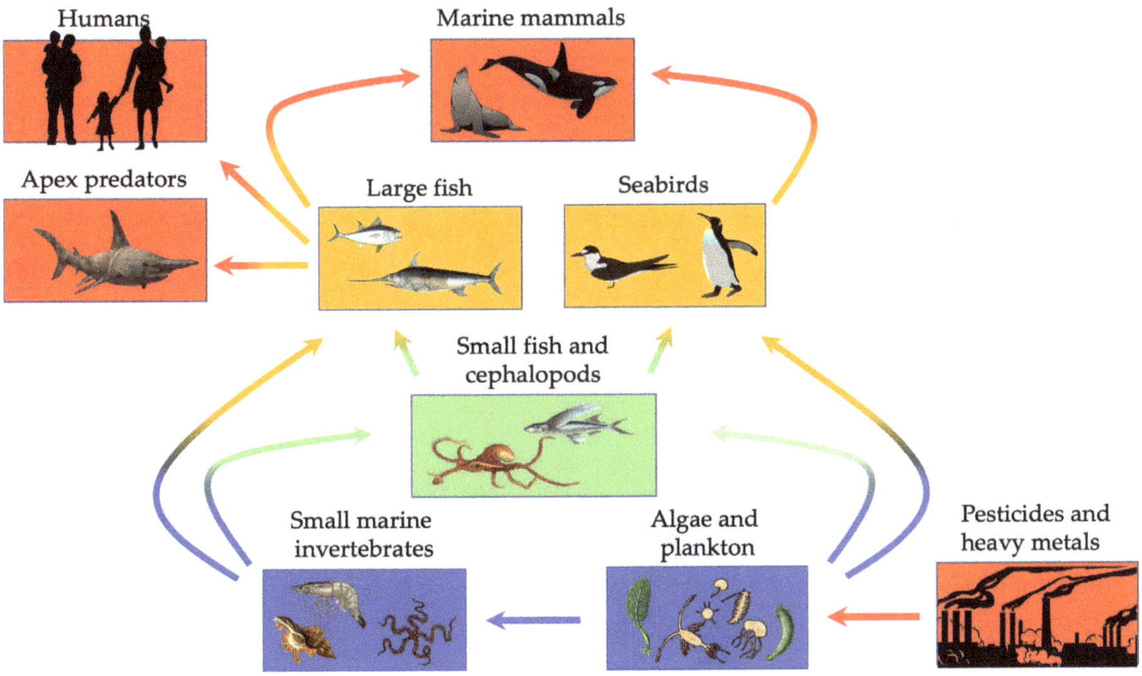

Figure 7.1 A simplified marine food web showing how sharks, marine mammals, seabirds, and even humans are all vulnerable to health problems associated with bioaccumulation where pesticides, heavy metals, and other harmful chemicals become concentrated at higher trophic levels. After Ross and Birnbaum, 2003, CC BY 4.0.

In the 1970s, many industrialised countries recognised the dire situation and banned the use of DDT, which eventually allowed for the partial recovery of the affected bird populations. Unfortunately, while some countries have switched to safer alternatives (e.g. Hargrove, 2003), DDT continues to be widely used in Africa to control malaria mosquito, tsetse fly (*Glossina* spp.), and other disease vectors. Researchers recently observed complete absences of breeding fish-eating birds in some African wetlands, and some of the highest-ever recorded DDT levels in seed-eating birds (Bouwman et al., 2013). This is cause for concern, not only for wildlife, but also for the long-term effects on people, particularly the consumers of the food products exposed to these chemicals (e.g. Manaca et al., 2011) and the workers who handle these chemicals in the field.

> Pollution does not always lead to immediate mortality, but instead can have sublethal impacts that compromise organisms' fitness over time, with population declines as the end result.

DDT is however not the only form of pollution we battle today. With the impacts of a growing human population becoming gradually more pervasive, pollution is compromising water, soil, and air quality at rates faster than ever before. Some forms of pollution can be highly visible, and with dramatic consequences (Figure 7.2). But importantly, there are many less detectable forms of pollution. While it may not always lead to immediate mortality, these insidious forms of pollution have sublethal impacts that compromise organisms' fitness over time, with early death and population declines still being the end result. Responding to the silent threats of subtle and easily-overlooked pollution is often delayed, especially when the negative effects are felt only years after exposure. In their totality, pesticides and other pollutants claim 1.4–2.2 million human lives in Africa each year; globally, they claim 9 million lives, which is over three times more than the total impact of AIDS, malaria, and tuberculosis, together (Landrigan et al., 2018). Yet we continue to tolerate these threats, in part because the impact of pollution on our health is not always that apparent, especially when pollution deaths are expressed as a stroke, heart disease, respiratory infections, diarrhoea, or cancer, among other health issues.

Figure 7.2 A young boy next to an open sewer in Nairobi, Kenya. In addition to the dangers to human welfare and livelihoods, polluted waterways kill millions of native animals and plants each year, and harm countless number of ecosystems. Photograph by Eoghan Rice/Trócaire, https://commons.wikimedia.org/wiki/File:A_young_boy_sits_over_an_open_sewer_in_the_Kibera_slum,_Nairobi.jpg, CC BY 2.0.

One of the most challenging aspects when trying to prevent pollution is identifying the source. Many forms of pollution can easily be transported away from their source through the air, via rivers, even in groundwater. This transport of pollutants (called **pesticide drift** in the case of pesticides) means that a substantial burden (perhaps as much as 95%, Miller, 2004) of impacts are being felt by non-target species, including economically important non-target organisms. For example, pesticide drift from cotton fields in Benin has caused extirpations of freshwater fish (Agbohessi et al., 2015), while beneficial pollinating insects are also often negatively impacted (Pettis et al., 2013). Studies on fish in Nigeria (Adeogun et al., 2016), large mammals in South Africa (Bornman et al., 2010), and frogs in Kenya (Hayes and Menendez, 1999) have shown that beneficial organisms that survive this secondary pesticide exposure have disrupted reproductive and endocrine systems, and hence reduced fitness. Even humans may be exposed to secondary poisoning from pesticides, as toxic pesticide levels have been found in edible oysters and mussels in Ghana (Dodoo et al., 2013), prawn in Côte d'Ivoire (Roche and Tidou, 2009), and even chickens in South Africa (Barnhoorn et al., 2009).

To make matters worse, many pollutants take many years to **biodegrade** (i.e. break down in nature), and thus continue to pose a threat to wildlife and humans long after entering the environment. One important class of such long-lived pollutants is **persistent organic pollutants (POP)**. Several types of pesticides qualify as POPs, which are prone to bioaccumulation and drift. The most famous POP is DDT; in the USA, biologist continue to see eggshell thinning and bird deaths, nearly 50 years after DDT was banned in that country (Burnett et al., 2013). This is a concern in places like Ethiopia's Lake Koka, where recent studies have found DDT residues in every sample of fish tissue (from several different species) tested (Deribe et al., 2011). More information on POPs, many which are banned from use by signatories of the *Stockholm Convention on Persistent Organic Pollutants*, can be found on the *Stockholm Convention* website (http://pops.int).

There are also many types of persistent inorganic pollutants that find their way into the environment on a daily basis. One important class of persistent inorganic pollutants that also bioaccumulate is heavy metals; these include mercury, cobalt, copper, lead, and arsenic. A study from Zambia traced cobalt contamination in living trees to soil pollution from mining activities that occurred the mid-1970s (Mihaljevič et al., 2011). Some everyday products can also persist in the environment. For example, an aluminium can takes about 200 years to break down, while a plastic bag takes between 100–1,000 years to break down. The continued use of these products should thus raise alarm to anyone concerned about the environment and human health. But it also provides opportunities for any person to contribute to conservation by reducing use of these products and reusing/recycling those products that find their way into the supply chain.

Many pollutants take many years to biodegrade, and thus continue to pose a threat to wildlife and humans long after entering the environment.

7.1.1 Water pollution

Water pollution, the accidental or intentional dumping of pesticides; herbicides; oil products; fertilisers; sewage; industrial waste; detergents; and other foreign chemicals and objects into aquatic environments, is arguably the biggest current pollution concern in Africa (Prüss-Ustün et al., 2016; Landrigan et al., 2018).

The dumping of products containing heavy metals into aquatic environments is particularly concerning because heavy metals are toxic even in small concentrations, and likely to biomagnify. When aquatic organisms process contaminated water, they absorb or ingest the heavy metals along with other essential nutrients. With each additional step along the food chain, organisms ingest and accumulate increasingly higher concentrations of these toxic elements (see Figure 7.1). In this way, even small amounts of heavy metals can become lethal across several levels of the **food web** over time. Biomagnification is especially a concern with long-lived predatory marine fishes that people consumed as food, such as swordfish (*Xiphias gladius*, LC), marlins, sharks, and some tunas and sea basses. For example, mercury (emitted mainly during fossil fuel use), lead, and arsenic have bioaccumulated so much in sharks off South Africa that many species are now considered unsafe for human consumption (McKinney et al., 2016; Bosch et al., 2016; Merly et al., 2019). Recent studies also found unsafe levels of mercury in freshwater fish from regions as wide as Central Africa's Great Lakes (Campbell et al., 2008), Ethiopia's Lake Awassa (Desta et al., 2006), and several reservoirs in West Africa (Quédraogo and Amyot, 2013).

> Because of biomagnification, many long-lived predatory marine fishes are now considered unsafe for human consumption.

Oil pollution involves the release of petroleum products into the environment, which can originate from damaged ships, failed drilling rigs, leaking offshore platforms, or other unexpected events. The released oil causes mammals and birds to lose the insulating abilities of their fur and feathers, leaving those animals vulnerable to hypothermia and drowning. Other aquatic animals, including fish and shellfish, may ingest oil products, causing them to sicken and die. Because of the way oil is extracted and transported, marine ecosystems are particularly at risk. Furthermore, because of the massive amount of oil that are involved in oil extraction and transport, an oil pollution event often represents a serious ecological disaster (Figure 7.3). Africa has been hit hard by oil spills in recent years, particularly around oil-producing countries like Angola and Nigeria, and along shipping lanes passing along the coasts of Namibia, South Africa, and Mozambique. Nigeria is perhaps the biggest victim of oil spills; between 1976 and 2001, there were an estimated 6,817 oil spills around Africa's largest wetland, the Niger Delta (UNDP, 2006)! These oil spills have destroyed thousands of hectares of mangrove swamps, estuarine wetlands, and other coastal ecosystems, causing severe hardship to marginalised local communities who depended on those areas for subsistence fishing and farming (Fentiman and Zabbey, 2015).

Plastic pollution is fast becoming a ubiquitous threat to Africa's environment, its wildlife, and its people. To visualise the magnitude of the problem, consider that there are more than 1.6 trillion pieces of plastic, collectively weighing over 70,000 tonnes, currently floating in the Atlantic and Indian Oceans surrounding Africa (Eriksen et al., 2014). While many of these plastic items were dumped directly in the ocean, many also have a terrestrial origin. For example, if someone throws a plastic wrapper on a sidewalk, there is a good chance that the wrapper will find its way into a nearby stream at some point, carried by wind or rain runoff. From here, the wrapper will float along various

> There are more than 1.6 trillion pieces of plastic, collectively weighing over 70,000 tonnes, currently floating in the Atlantic and Indian Oceans surrounding Africa.

Figure 7.3 (Top) Staff and volunteers from the seabird rescue centre SANCCOB are caring for some of the 19,000 African penguins (*Spheniscus demersus*, EN) that were rescued after a stricken iron ore carrier spilled 1,400 tonnes of oil off South Africa in June 2000 (Wolfaardt et al., 2009). (Bottom) SANCCOB volunteers releasing a group of African penguins that were rescued from the oil spill. Photographs by SANCCOB, CC BY 4.0.

streams and rivers until it reaches the ocean. A recent review found that 88–95% of plastics floating into the world's oceans originated from just 10 rivers, which include West Africa's Niger River and East Africa's Nile River (Schmidt et al., 2017; Lebreton et al., 2017). In the process, thousands of seabirds, dolphins, whales, turtles, seals and fish die each year from suffocation or starvation after ingesting plastics and other pieces of trash that they confused with food (Wilcox et al., 2015). This plastic pollution also impacts humans: researchers recently found microfibers (many of which are plastic) in over 80% of tap water samples from Uganda (Kosuth et al., 2017), as well as food-grade commercial sea salt originating from South Africa (Karami et al., 2017).

Some of the biggest impacts from plastic pollution are caused not by visible scraps of plastic, but by **microplastics**, the collective name for plastic particles smaller than 1 mm (some are microscopic). Microplastics may originate from the breakdown of larger pieces of plastic and polystyrene products, or they may be manufactured intentionally small, such as beads added to cosmetics and other personal care products that are flushed down drains after use. Because microplastics are so small, they easily pass through the standard filters used at sewage treatment plants. Consequently, microplastics generally end up in the aquatic environment, where they are unintentionally consumed by crustaceans (crabs, lobsters, and krill), molluscs (mussels, oysters, and clams), echinoderms (sea stars, sea urchins, sea cucumbers), and baby fish. This consumption can block or damage the victim's digestive and respiratory systems, cause reduced food uptake by creating a false sense of satiation, or even poison animals through leeching of synthetic chemicals. Each of these threats increases death rates and lowers reproductive rates (Sussarellu et al., 2016). Just as with the biomagnification we discussed earlier, the consumption of microplastics also affects other consumers (including humans), because the small organisms that ingest the microplastics are often food for other animals, allowing plastic pollution to move through an entire food chain. For example, a recent study from Lake Victoria found microplastics imbedded in the digestive tracts of perch and tilapia bought at a local market and meant for human consumption (Biginagwa et al., 2016). Because microplastics are so hard to remove once in an ecosystem, the best method for their containment may be to reduce plastic use, to ban products containing microplastics, or to develop microplastics that are biodegradable within a reasonable timeframe. But for this to happen, there is a need to educate the public and lawmakers (Galloway and Lewis, 2016) about the dangers posed by this threat to the environment and local economies.

Nutrient pollution represents another growing threat to Africa's aquatic environments. Many lakes, streams, and other freshwater and marine environments naturally contain low concentrations of essential nutrients, such as nitrates and phosphates. In order to survive, the species living in these nutrient-poor waters must then be adapted to this natural nutrient scarcity. However, raw sewage, agricultural fertilisers, concentrated animal feeding operations, and industrial processes release large amounts of additional nitrates and phosphates into the environment, which are washed into the aquatic environment. Minor additions of essential nutrients stimulate plant growth, providing more food for organisms at higher trophic levels. However, at high concentrations, the system become subjected to nutrient pollution.

> Nutrient pollution, caused in part by excessive fertiliser use, can led to eutrophication, famous for causing algae blooms, aquatic dead zones, and fish kills.

One of the worst outcomes of nutrient pollution is **eutrophication**. During eutrophication, surface algae grow so rapidly (known as an algae bloom) that it starts blocking sunlight from reaching aquatic organisms below the surface. Because each individual alga is short-lived, their rapid growth also adds large amounts of decaying matter to the environment. In response, decomposers that feed on the dead algae can become so abundant that they consume most of the water's dissolved oxygen. Without oxygen and sunlight, aquatic plant and animal life may die off in large numbers. The resultant dead zones are sometimes visibly in the form of fish kills, with large numbers of dead fish floating on the surface of the affected water body. The organisms that die during this process is generally also toxic to humans because of bacteria build-up and other imbalances. Eutrophication is an increasingly common problem in Africa; for example, a recent review found that 41–76% of South Africa's lakes may be eutrophic (Harding, 2015). Eutrophication has already negatively impacted Africa's tourism and fisheries sectors (Nyenje et al., 2010), and even led a temporary shutdown of water supplies on the Kenyan side of Lake Victoria (Sitoki et al., 2012). Preventing further eutrophication should thus be a high priority—not only will it prevent harmful algae blooms but may even play an important role in controlling invasive aquatic plants such as the water hyacinth (*Eichhornia crassipes*) (Coetzee and Hill, 2012; Bownes et al., 2013).

Groundwater pollution—the release of pollutants into aquifers and other sources of groundwater—is also becoming a serious issue across Africa. This type of pollution generally originates from landfills, on-site sanitation systems, leaking sewage systems, mining leachate, **agriculture runoff** (fertiliser, pesticides, animal waste, etc.), and other types of waste dumping. The pollutants may sometimes be released directly into aquifers; however, more often the contaminants and pathogens leak into the soil, from where it seeps into groundwater.

One of the most important emerging threats to groundwater in Africa is **hydrological fracturing** or fracking, in short. During this process, pressurised liquids that contain suspended particles and thickening agents are blasted into rock formations deep underground to break them open. When the pressure and liquids are removed, the suspended particles keep the fractures open, which enables extraction of natural gas and petroleum. While fracking was initially hailed as a method to access previously inaccessible fossil fuels, scientists subsequently found that it poses a wide variety of very serious environmental and health risks. Most importantly, the liquids used in fracking contain toxic chemicals which pose a high risk for groundwater pollution (Osborne et al., 2011), which in turn lead to miscarriages and birth defects (McKenzie et al., 2014), cancer (McKenzie et al., 2012), as well as skin and respiratory diseases (Rabinowitz et al. 2015). In addition, fracking increases greenhouse gas emissions (Howarth, 2014) and induces infrastructure-damaging earthquakes (Ellsworth, 2013). Because of these myriad serious risks, several national governments in Europe, and several local governments in the USA, UK, Canada, and Australia have banned the practice from their lands (https://keeptapwatersafe.org/global-bans-on-fracking). In contrast, and despite opposition from civil society, several countries in Africa (e.g. South Africa: Roelf, 2016; Botswana: Barbee, 2015) recently approved this harmful practice.

Because fracking poses many serious risks, governments across the world have banned the practice from their lands.

7.1.2 Air pollution

In the past, people and industries thought that the atmosphere was so vast that any gases or particles released into the air would disperse and dilute to the point that they would post no ill effects. But as air quality has diminished over time, scientists have documented that air pollution can cause irreparable harm to ecosystems and human health, often far from the original sources. A striking example comes from West Africa's Lake Chad, which shrank by 95% between 1963 and 1998 (Figure 7.4). Experts generally thought that the shrinkage was caused by unsustainable water use in the region, but recent evidence suggests that air pollution from Europe which reduced rainfall in the Lake's catchment area may also have contributed to this ecological disaster (Hwang et al., 2013). The Lake's water level has risen since 2007, likely due, in part, to clean air regulations implemented by the European Union. Despite this positive turn around, air pollution continues to be a serious problem (Amegah and Agyei-Mensah, 2017) that threatens humans and wildlife throughout Africa.

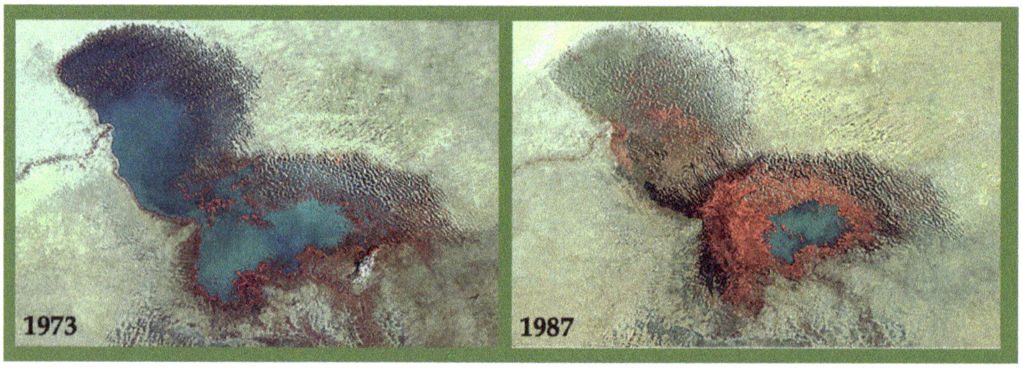

Figure 7.4 Changing rainfall patterns attributed to air pollution may have contributed to West Africa's Lake Chad shrinking by 95% between 1963 and 1998. An ecological disaster ensued, as the 68 million people whose livelihoods were at risk put additional strains on the environment while trying to replace the natural resources the lake previously provided. Images by NASA/GSFC, https://svs.gsfc.nasa.gov/2065, CC BY 4.0.

An important form of air pollution is **hydrocarbons**, which are released during fossil fuel burning, particularly during transport, power generation, and other industrial activities (Karagulian et al., 2015). Pollution from airborne hydrocarbon compounds can sometimes be sensed without scientific equipment, by the bad smells, high air turbidity, and eye and lung irritation a person may experience in large cities with highly polluted air. When exposed to sunlight, these chemicals can react with other gases and particles in the atmosphere to produce **photochemical smog**, which is made up of ozone and other secondary compounds. In the upper atmosphere, ozone filters harmful ultraviolet radiation, which benefits most living things; but at ground level, high concentrations of ozone pose several dangers. For example, it damages plant tissues which make them brittle; high surface ozone levels have found to cause crop damage in Botswana and South Africa (Zunckel et al., 2004). Hydrocarbon exposure also poses several threats to humans: it altered some people's DNA—often a cancer precursor—in Benin (Fanou et al., 2006), caused lung damage in Côte d'Ivoire (Kouassi et al., 2010), and subjected people to **carcinogenic compounds** in the DRC and Ghana (Tuakuila, 2013; Bortey-Sam et al., 2017). The lack of air monitoring and standards over much of Sub-Saharan (Petkova et al., 2013), and lack of awareness—people often confuse photochemical smog with natural mist and early-morning fog—should thus be of serious concern both to conservation biologists and society at large.

Air pollution from hydrocarbons often manifests itself as photochemical smog. Hanging like a thick cloud over industrial areas, people sometimes confuse it with natural mist and early-morning fog.

Burning fossil fuels also releases sulphur oxides (SO_x) and nitrogen oxides (NO_x) into the atmosphere, where they combine with water vapor to produce nitric and sulphuric acids. These acids later return to the ground as **acid rain**, with dramatically low pH relative to normal rainwater. Prevailing winds can transport acid rain clouds over long distances, so the effects of acid rain may occur hundreds of kilometres from its sources. Because the acid rain is closely tied to the water cycle, aquatic and soil organisms are particularly vulnerable to the negative effects of acid rain. Plants exposed to acid rain, either directly or after absorbing contaminated water from the ground, are often left severely weakened or even killed: it has even caused plant extirpations in Zambia (UNEP, 2006).

Another important contributor to air pollution is domestic fuel burning (Karagulian et al., 2015). During these activities, very small pollutant particles are released into the air. Because these particles are so small, they are difficult to filter from the air, and can easily be inhaled. Once inhaled, the particles can pass into the victim's bloodstream, from where they negatively impact cardiovascular health, neurodevelopment, and cognitive function (WHO, 2013). Despite the harmful impact of these particles in the environment, their monitoring is virtually non-existent in Africa, making it very hard to guide air quality policy decisions and legislations. In contrast, measures that mitigate pollution from domestic fuel burning may even help slow the rate of habitat loss (Chapter 5), as this type of pollution is associated with inefficient wood stoves, slash-and-burn agriculture, and the artisanal charcoal industry.

7.1.3 Soil pollution

Soil pollution occurs when soil meets foreign chemicals and other pollutants. This type of pollution is often associated with industrial activities that extract resources from the earth, agricultural runoff, pesticide use, oil spills, acid rain, improper treatment of sewage, and improper disposal of waste. People and wildlife can then become sick through direct contact with contaminated soils, or through secondary contamination via polluted groundwater or eating food grown in contaminated soil. For example, a recent review reported how soil pollution has left medicinal plants toxic to humans in countries such as Botswana; Ghana; and Mali, at times with fatal consequences (Street, 2012).

The improper disposal of electronic waste (or e-waste in short) is a particularly serious form of soil pollution. Because electronic products contain toxic heavy metal contaminants that are expensive to recycle, discarded electronic products usually end up in dump yards (Figure 7.5). Here, open burning of electronic and other waste materials releases the toxic compounds into the soil, as well as the air and water (Robinson, 2009), from where it also accumulates in the environment.

Figure 7.5 Nearly all electronic goods contain parts with toxic chemicals that are expensive to recycle. Instead, such components end up in dump yards such as this one in Ghana, from where the toxic compounds pollute the air, water, and soil, posing many human and environmental risks. Photograph by Agbogbloshie Makerspace Platform, https://www.flickr.com/photos/qampnet/14937188796, CC BY-SA 2.0.

7.1.4 Light pollution

Light pollution describes the addition of excessive, ill-timed, or poorly designed artificial light to the natural world. A consequence of an increasingly industrialised world (Falchi et al., 2016), light pollution has increased dramatically over the past decades as more people have gained greater access to electricity (Figure 7.6). Behavioural disruption is perhaps the most well-known consequence of increased light pollution—consider all the moths and other nocturnal insects (and insect predators, such as bats and geckos) attracted to artificial night lights. Light pollution also interferes with the navigation abilities of nocturnal species, which often use the stars, moon, and light reflectance from water surfaces to orientate themselves. For example, work in Gabon has shown how artificial lights disorientate sea turtle hatchlings trying to reach the sea (Bourgeois et al., 2009), while others have highlighted the significance of light-induced seabird mortality (Black, 2005). These and other behavioural disruptions—which include attraction to and repelling away from artificial light—may seem to only affect a small number of individuals around a few lights in your home. But the systemic impact of thousands of lights every night has wide-ranging ecosystem impacts when considering the cumulative impact of reduced reproductive performance (Firebaugh and Haynes, 2016), disrupted predator-prey dynamics (Minnaar et al., 2015) and disturbed night-time pollination services (Knop et al., 2017) on the many thousands of organisms impacted every night.

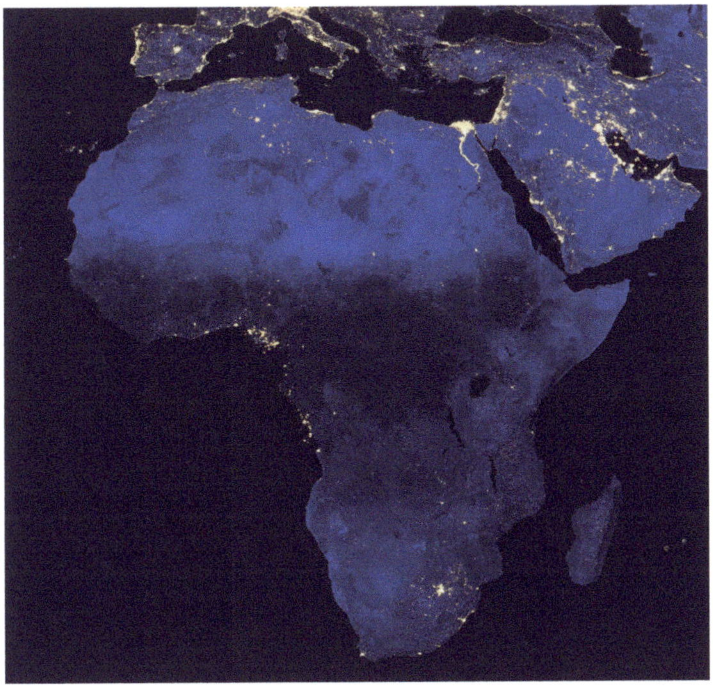

Figure 7.6 Night-time composite of Africa and parts of Europe and Asia, taken by the Suomi NPP satellite in 2012. City lights of every major city in Africa can be seen; lights are particularly concentrated in South Africa and West Africa's Gulf of Guinea. Image by NASA/GSFC, https://www.flickr.com/photos/gsfc/8246931247, CC BY 2.0.

Light pollution also disrupts the natural day-night cycles with which most species evolved. These disruptions interfere with **circadian rhythms**, which negatively affect living organisms' physiology. For example, one study showed that night-time light pollution disrupted natural sleep patterns in birds, leaving the affected individuals more susceptible to malaria infections (Ouyang et al., 2017). Circadian rhythm disruptions from light pollution (especially from high frequency "blue" light) also impact humans by increasing stress, fatigue, and anxiety, and susceptibility to obesity (Rybnikova et al., 2016) and cancer (Haim and Portnov, 2013). It is important to note that light pollution does not mean that the use of light is inherently bad—light has and will continue to play an important role in our daily lives. However, it does mean that we need to be more thoughtful about the consequences of light pollution and put measures in place to mitigate its impacts on the natural world and our own lives.

7.1.5 Noise pollution

Many people find a sense of freedom when they are in natural surroundings, with peace and quiet facilitating a much-needed connection to nature. These experiences are increasingly being threatened by **noise pollution.** However, noise pollution (also called acoustic pollution)—caused by human activities, such as industrial, military, and transportation systems—affects more than just the appealing tranquillity of nature. It also prevents animals from hearing each other, predators, and prey, all which could interfere with feeding, reproduction, navigation, and predator-avoidance behaviours. While African studies on the impact of noise pollution on wildlife are near-absent (Shannon et al., 2015), one study that did investigate the topic found that traffic noise increased dwarf mongooses' (*Helogale parvula,* LC) alertness but also reduced responsiveness to alarm calls (Kern and Radford, 2016). Such responses could leave the affected individuals less fit and more vulnerable to predators.

One would think that marine organisms living in the vast oceans may be spared from noise pollution, but this is not the case (Koper and Plön, 2012; Kunc et al., 2017). Sound carries much further in salt water than air, so noises from ship propellers; military sonar; seismic activities, and construction have significantly increased the level of ambient noise levels marine organisms experience. This increased level of ambient noise not only disrupts communication in sea animals (e.g. Cerchio et al., 2014), but can even lead to death (some mass whale strandings have been attributed to noise pollution: Morell et al., 2017; Williams et al., 2017). As with light pollution, there is a general need to be more thoughtful about the consequences of sound pollution on the natural world and to put measures in place (see e.g. Koper and Plön, 2012) to mitigate its impacts.

> Noise pollution prevents interferes with communication, feeding, reproduction, navigation, and predator-avoidance behaviours; it may even contribute to mass strandings of whales.

7.1.6 Thermal pollution

Thermal pollution describes localised human-induced temperature changes to the natural world. Aquatic ecosystems represent one of the ecosystems most vulnerable to thermal pollution. For example, when water is released from big dams, it comes from the colder middle and lower strata of the reservoir, leading to rapid cooling of aquatic ecosystems further downstream. The opposite is true at power plants that use river water as a coolant; turbines release their heat to the circulating water and then the warmed water is released back into the environment. These abrupt releases of thermally discordant water often lead to **thermal shock** which can be lethal to fish and other aquatic organisms. For example, studies from South Africa have shown that thermal shock can kill fish embryos and larvae and caused deformities in the young of Clanwilliam yellowfish (*Barbus capensis,* VU) (King et al., 1998).

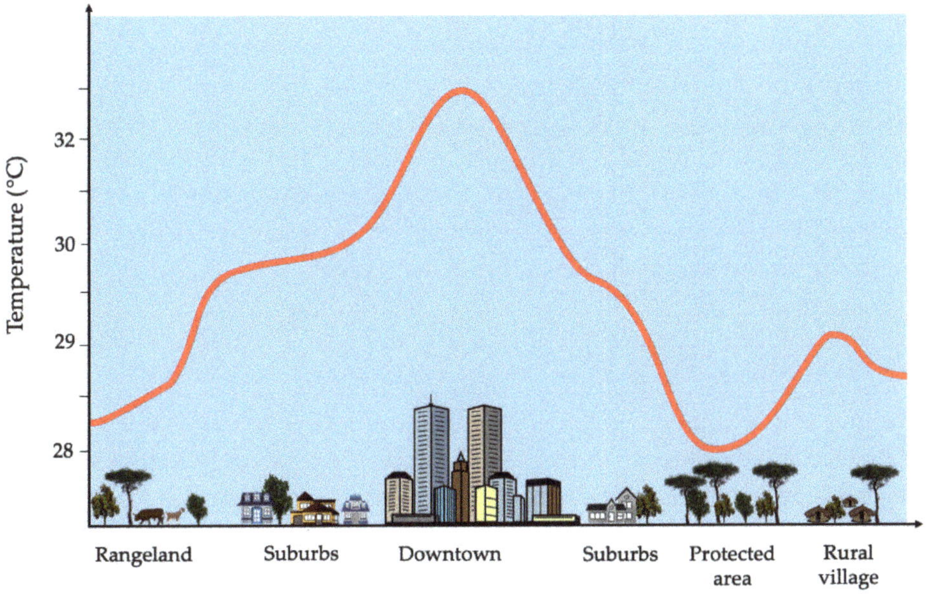

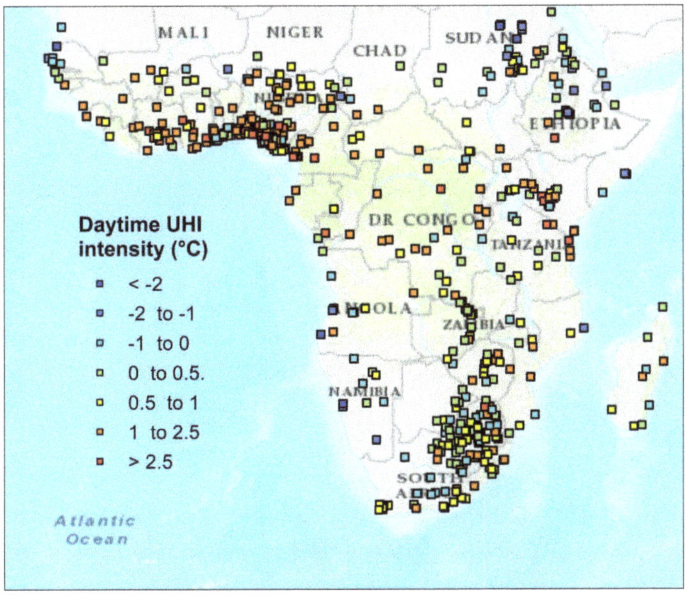

Figure 7.7 (Top) Man-made surfaces such as roofs, roads, and pavements do not reflect, but rather absorb solar energy as heat, causing built-up areas to be warmer than the surrounding rural areas, CC BY 4.0. (Bottom) Data derived from 16-year mean daytime temperatures obtained by TERRA and AQUA satellites show how the urban heat island effect increases daytime temperatures of several African cities (represented as squares). Note how areas with high levels of deforestation (e.g. West Africa and Albertine Rift) also show the highest temperature increases (> 2.5°C). Map by T.C. Chakrabotry, after Chakrabotry and Lee, 2018, CC BY 4.0.

The **urban heat island effect** represents a terrestrial form of thermal pollution. Urban and other developed areas are generally covered with large swaths of man-made surfaces (e.g. asphalt roads, pavement surfaces, and building roofs), which absorb solar energy rather than reflect it. This absorbed heat, in combination with heat outputs from industrial activities, cause urban areas to function like "islands of heat" that are several degrees warmer (Figure 7.7) than surrounding rural areas (Feyisa et al., 2014; Chakraborty and Lee, 2018). The urban heat island effect reduces the quality of life for people and wildlife by reducing comfort and water availability (due to increasing evaporation). It also increases energy consumption to offset the heat increases which, in turn, contributes to air pollution and climate change.

7.2 Overharvesting

People have always hunted, collected, trapped, or otherwise harvested the food and other natural resources they need to survive. When human populations were small, at least relative to the abundance of their resources, and collection methods were relatively unsophisticated, people could sustainably harvest and hunt wildlife in their local environments. However, as human populations have increased, and roads have provided access to previously remote areas, our impact on the environment has escalated. At the same time, our methods of harvesting have become dramatically more efficient. Guns are now used instead of blowpipes, spears, or arrows, while networks of wire snares indiscriminately catch animals of all types, even young and pregnant females. Populations of species that mature and reproduce rapidly can often recover quickly after harvests and can thus be exploited sustainably; however, species that are slow-maturing and slow-reproducing cannot sustain current harvest levels. Consequently, many species are threatened due to **overharvesting**, the unsustainable collection of natural resources (Maxwell et al., 2016). Overharvesting may take on many forms, including hunting, fishing, logging, and gathering of plants and animals for medicine, captive collections, subsistence, commerce, or recreation purposes (Figure 7.8).

7.2.1 The Bushmeat Crisis

Bushmeat harvesting is one of Africa's most prominent overharvesting concerns (see Box 4.1). **Bushmeat**—wild sources of protein obtained on land by hunting and collecting birds, mammals, snails, and caterpillars—provides much of the protein in people's diets in large parts of Africa. For example, in Nigeria and Cameroon, 12,000 tonnes of bushmeat—two tonnes obtained from bay duiker (*Cephalophus dorsalis*, NT) alone—are sold at markets in the Cross-Sanaga rivers region each year (Fa et al., 2006). Similarly, more than 9,000 primates are killed annually for a single market in Côte d'Ivoire (Covey and McGraw, 2014); people from Central Africa harvest an astonishing 5.3 million tonnes of mammalian bushmeat annually (Fa et al., 2002). Usually seen as a conservation challenge in Africa's tropical forests, the **bushmeat crisis** also impacts savannah regions (reviewed in Lindsey et al., 2013). For example, bushmeat hunters, numbering between 1.500 and 2,000, remove over 600,000 kg of herbivore biomass

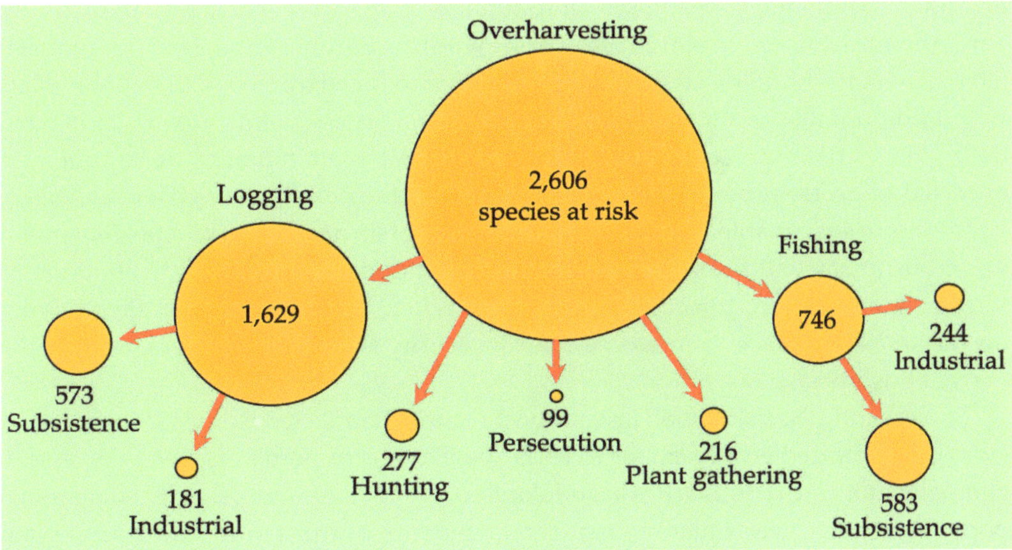

Figure 7.8 Overharvesting in Sub-Saharan Africa at scale: Over 60% of species that are threatened by overharvesting are also threatened (directly and indirectly) by logging. Nearly 30% of all species are threatened by fishing, and 10% by hunting and trapping. Source: IUCN, 2019, CC BY 4.0.

from Botswana's Okavango Delta each year, despite the region's protected status and importance for ecotourism sectors (Rogan et al., 2017).

Outside influences play a prominent role in the harvesting pressure associated with the bushmeat crisis. In Section 5.2, we discussed neocolonialism, where jobs associated with land-grabbing industries are frequently reserved for migrant labourers. Poorly paid and with limited rights, migrant labourers are often forced to turn to local natural resources to fulfil their basic needs (Thibault and Blaney, 2003). The impact of these migrant labourers on the local environment is massive compared to traditional (and many other local) peoples that prioritise sustainability. For example, immigrants working at logging concessions in the northern parts of the Republic of the Congo hunt 72% of all bushmeat harvested in the region (Poulsen et al., 2009). The increased commercialisation of bushmeat also poses challenges (Lindsey et al., 2013); for example, in the broader Congo Basin, commercial hunters are exploiting bushmeat at scales 27 times that of the area's traditional peoples (Fa et al., 2016). In addition to hunting for local markets, illegal exports also play an important role. For example, more than 50 tonnes of wild fish and bushmeat enters France from Africa each week (Chaber et al., 2010); similar amounts were also estimated for airports in Switzerland (Wood et al., 2014).

> The massive wildlife declines caused by the bushmeat crisis are also threatening ecosystem services, food security and people's livelihoods.

Very few animal populations can withstand such high extraction rates. Consequently, regions dependent on bushmeat have already seen substantially wildlife declines (Lindsey et al., 2013). West Africa, where forest mammal populations are down an estimated 80%, have been hot particularly hard (Benítez-López et al., 2019). These wildlife declines also lead to reduced harvests—some hunters have seen their harvests reduced by over 80%, with impacts to wildlife notable as far as 40 km from hunters' access points along roads from their villages (Benítez-López et al., 2017). At current exploitation rates, supplies are expected to decrease by an additional 80% within the next 50 years (Fa et al., 2003). Unless more sustainable, alternative sources of protein are found, people dependent upon bushmeat will see increased malnutrition and compromised livelihoods as bushmeat species are pushed to extinction. When that happens, families relying on bushmeat will face even worse food insecurity than that which is driving the current bushmeat crisis.

Exacerbating the risk of food insecurity, people in the affected regions will also suffer from compromised ecosystem services as populations of predators, seed dispersers, and pollinators are reduced (Rosin and Poulsen, 2016). For example, reduced mammal populations have been linked to reduced abundance of fruits and other useful plant products available for human consumption (Vanthomme et al., 2010). Some areas are already suffering from "empty forest syndrome"—a condition where a forest appears to be green and healthy, but is practically devoid of animals, and in which ecological processes have been irreversibly altered such that the forest's species composition will change over subsequent decades (Nasi et al., 2011; Benítez-López et al., 2019). The bushmeat crisis is thus a major concern to people concerned about biodiversity and/or human well-being.

7.2.2 Overfishing

Pressure on biodiversity in aquatic environments is also increasing as people continue to harvest fish, sea turtles, dolphins, shellfish, and manatees for meat at increasing rates. Modernised fishing methods play a major role. For example, a motorised fishing fleet that faces few restrictions has caused a 75% decline in fish populations at Ethiopia's Lake Tana in recent years (de Graaf et al., 2004). Also, in the marine environment, motorised fleets and enormous factory ships can now spend months at sea where they catch fish to sell at local and global markets (Ramos and Grémillet, 2013; Pauly et al., 2014). Some estimates suggest that wild-caught seafood could be virtually absent by 2050 if current exploitation levels persist (Worm et al., 2006).

For many aquatic organisms, the indirect impacts of modern commercial fishing methods outweigh direct exploitation (Figure 7.9). One example is **ghost fishing,** which causes thousands of animals to die each year after becoming entangled in dumped, abandoned, and lost fishing gear. Similarly, approximately 25% of fish harvests are considered **bycatch**—animals that are accidentally caught, injured, or killed during fishing operations. Recent declines in skates, rays, turtles, sharks, dolphins, and seabirds have all been linked to incidental deaths as bycatch (Cox et al., 2007; Carruthers et al., 2009). Seabird biologists from South Africa have been at the forefront of solving bycatch problems in recent years (Box 7.1).

Box 7.1 Solving Seabird Bycatch Problems: From Theory to Practice

Ross Wanless

DST-NRF Centre of Excellence at the FitzPatrick Institute of African Ornithology, University of Cape Town, South Africa.

rosswanless@gmail.com

The global problem of seabird bycatch in fisheries—the accidental death of seabirds during fishing—is one of the biggest threats to pelagic seabirds (Croxall et al., 2012). Ironically, it is both one of the easiest and one of the most challenging problems to solve. How so? Simple technical fixes to stop birds from getting snagged on fishing gear and drowning can work amazingly well, but fishermen must be convinced to use them.

Techniques for preventing bycatch break down into two basic approaches. The first approach is to prevent access to the danger point (the baited hook or the cables that birds strike). Fishing only at night eliminates up to 80% of the problem, but still jeopardises nocturnal foragers and diurnal species during a full moon. Another option is bird-scaring lines (Figure 7.A) consisting of a mainline flown from the stern of a boat with hanging streamers that scare birds away from danger areas behind the vessel. The second approach, primarily used in longline fishing, is to remove the risky gear (baited hooks) as quickly as possible; essentially this involves adding weights to lines to sink them faster. It has reduced seabird bycatch on some fisheries by 90–95%.

Despite clear harm to seabirds caused by fisheries and the simple, effective fixes at hand, implementing these mitigation measures has been patchy at best in most fisheries where seabird bycatch occurs. There are some exceptions, and it is useful to examine what elements led certain fisheries to fix the problem. A good case study is in South Africa, where BirdLife South Africa's Albatross Task Force (ATF) demonstrated in 2006 that trawl fishing for hake (*Merluccius* spp.) was killing around 18,000 seabirds each year (Watkins et al., 2008). The fishery involved had Marine Stewardship Council (MSC) certification, which gives a fishery access to premium European markets on the condition that it meets environmentally friendly and sustainable metrics, including no significant bycatch impacts. This provided a powerful incentive for fleet-wide implementation of a bycatch mitigation measure; failure to do so would have resulted in a loss of certification, with catastrophic financial implications.

Despite this strong incentive, it required another five years of work from the ATF to assess bird scaring lines and refine the design, overcome resistance to their use, and close loopholes in regulations. In 2014, the ATF published an assessment of the effectiveness of their bird scaring lines—a single measure to

Figure 7.A Bird-scaring lines (in orange) keep a variety of albatross, petrels, and gannets at a safe distance from the baited hooks deployed off a trawler off South Africa. Photograph by BirdLife South Africa Albatross Task Force Programme, CC BY 4.0.

prevent the accidental and avoidable deaths of around 10,000 albatrosses and large numbers of other species. When used correctly, the system eliminated 90–95% of seabird bycatch (Maree et al., 2014). Why did it take so long for the fleet to adopt this measure, despite it costing almost nothing, requiring no skill or time to use, and posing no meaningful operational problems? And why have identical fishing industries in many other countries failed to follow suite?

The answer is complex. ATF teams are present in South Africa (and now also in Namibia), providing sustained pressure and constant presence. South Africa had standing legislation, yet compliance from the South African fleet was initially minimal (and remains less than perfect today). MSC certification certainly created an enabling environment (Wanless and Maree, 2014) and incentive to drive change, yet it took more than that to change the entire fleet. Constant lobbying from BirdLife and regular dialogue from deck to boardroom were also critical ingredients. A legislative framework that provides some hope of censure against non-compliant vessels meant that there was internal pressure within the industry to "*tow* the line"—pun intended. Ultimately, widespread change became possible when there was a credible, independent observer programme to verify deck practice and give teeth to agencies when addressing non-compliance.

Figure 7.9 (Top) Discarded fishing gear, such as this ghost net, poses an entanglement hazard to marine wildlife. Photograph by Tim Sheerman-Chase, https://www.flickr.com/photos/tim_uk/2692835363, CC BY 2.0. (Bottom) A wandering albatross (*Diomedea exulans*, VU) that was a victim of bycatch, the accidental catching of non-target species during fishing operations. Photograph by Graham Robertson, CC BY 4.0.

7.2.3 The impact of traditional medicine

Africa has a long history of sustainable use of traditional medicines. Unfortunately, as the number of people living in Africa has increased, so has the demand for traditional medicine. Today, harvesting for traditional medicine is putting unsustainable pressure on species exploited for this purpose (Williams et al., 2014). One prominent example is vultures: the demand of vulture body parts, believed to bestow clairvoyant abilities, is driving massive vulture population declines across Africa (see Box 4.4). The growth in traditional medicine markets in East Asian countries such as China, Thailand, Cambodia, and Vietnam exacerbates these problems. For example, as tigers (*Panthera tigris*, EN) and rhinoceros have become scarce in Asia, Asian traditional healers are increasingly targeting African predators and rhinoceros to satisfy their market demands. Another group of species threatened by the Asian traditional medicine trade is sea horses (*Hippocampus* spp.). Due to population declines from overharvesting, sea horse exports from Kenya and Tanzania to East Asia have halved over recent years; yet, more than 600 kg of dried sea horses (over 254,000 individuals) continue to be exported annually (McPherson and Vincent, 2004). Exploitation for Asian traditional medicine markets has already pushed the western black rhinoceros (*Diceros bicornis longipes*, EX) to extinction. In a similarly perilous position is the northern white rhinoceros (*Ceratotherium simum cottoni*, CR); with only two non-reproductive females left in the world, this species is now considered **committed to extinction** (see Section 8.3; Box 11.4). A group of species sought after by both African and Asian traditional medicine markets is pangolins, thought to be the most heavily poached animals on Earth. For example, between 2012 and 2016, more than 20 tonnes of African pangolin scales (involving up to 30,000 animals) were seized during law enforcement operations across the region (Andersen, 2016). The problem is also getting worse: authorities intercepted 13 tonnes of scales in Singapore in 2019, all from a single shipment believed to travel from Nigeria to Vietnam (Geddie, 2019). With such a large active operational scale, it comes as no surprise that all four African pangolin species are now threatened with extinction (IUCN, 2019).

7.2.4 The impact of live animal trade

Millions of non-domesticated animals are sold as pets around the world each year (Table 7.1). Given that many of these pets were originally collected in the wild, it is no surprise that the most popular species tend to be at a high risk of extinction (Bush et al., 2014). These huge numbers are magnified by the extra millions of animals needed to compensate for deaths during collection and shipping. Collection of wild animals for pets and other purposes has a massive impact of biodiversity in Africa, the world's largest pet trade exporter (Bush et al. 2014).

Among the most popular groups of wildlife traded are Africa's parrots (Figure 7.10). For example, 32,000 wild-sourced African grey parrots (*Psittacus erithacus*, EN) were imported into the European Union in 2005 (UNEP-WCMC, 2007). Combined with habitat loss, the wild bird trade has already caused extirpations of this species in some areas of West Africa (Annorbah et al., 2015). Similarly, 82 of the 291 species of African freshwater fish known to occur in the pet trade are considered threatened with extinction (UNEP-WCMC, 2008). While it is true that collecting wild animals for the pet trade sustains many people's livelihoods, research on harvesting of ornamental fish in Cameroon has shown that this practice is not sustainable in the long term (Brummet et al., 2010). It is therefore critical to find ways to make these practices more sustainable, for the sake of the pet collectors and biodiversity.

Table 7.1 Examples of groups targeted in global wildlife trade, and their levels of exploitation.

Group	Number traded each year	Notes
Orchids	250 million	Mainly cultivated, but about 10% sourced from the wild. Illegal trade—and mislabelling to avoid regulation—a major problem.
Succulent plants	35 million	Mainly cultivated, but about 15% sourced from the wild. Illegal trade remains a major problem.
Corals	13 million	Collected using destructive methods; used for aquarium decor and jewellery.
Reptiles	7.2 million	Mainly sourced from the wild for zoos and pet trade, but increasingly from farms. Does not include large skin trade.
Birds	2.3 million	Mostly perching birds destined for zoos and pet trade. Also includes legal and illegal trade of parrots.
Ornamental fish	2 million	Most originate from wild reefs, caught by illegal methods that damage the surrounding coral reef and other wildlife.
Primates	148,000	Used for biomedical research, while many also destined for pets, circuses, zoos, and private collections.

Sources: http://cites-dashboards.unep-wcmc.org, data presented as live specimens exported from 2011–2015. Data generally do not include illegal traded specimens, which are usually not reported to CITES.

Figure 7.10 Wild-caught African grey parrots crammed into a travel crate before export to Asia. Researchers estimate that over 60% of smuggled parrots die from stress, dehydration, and smothering during transit (Mcgowan, 2008). Because of trade-driven population declines, CITES banned all international trade in this species in October 2016. Photograph by Lwiro Primates, CC BY 4.0.

7.2.5 Overharvesting of plant products

Overharvesting is not restricted to animals and animal products. While legal and illegal timber and firewood extraction is a major source of deforestation throughout Africa, it is also an important extinction driver. In fact, logging and other forms of wood harvesting have already contributed to the extinction of at least six plant species in Sub-Saharan Africa, with an additional 116 species considered *Critically Endangered* in part due to these threats (IUCN, 2019). Other plant species face extinction due to exploitation for medicines, spices, fragrances, and ornaments. For example, White's ginger (*Mondia whitei*)—reputed to have aphrodisiac and antidepressant properties—has been harvested to extirpation in parts of central Kenya and South Africa (Aremu et al., 2011). Similarly, harvesting rates of African blackwood (*Dalbergia melanoxylon*, NT)—popular for making musical instruments and fine furniture—are currently unsustainable because the tree is slow-growing, has low germination rates, and extractions are seldom offset with planting of new seeds or seedlings (Amri et al., 2009).

7.2.6 Challenges in managing overharvesting

One of the biggest challenges in combatting overharvesting is the non-enforcement and/or outright absence of legal controls to protect exploited species. But even where strong regulatory frameworks exist, the sheer scale of the problem poses practical challenges for effective enforcement (discussed in Chapter 12), as billions of dollars flow among participants in illegal wildlife trade, which include local people trying to make a living, professional poachers, corrupt government officials, unethical dealers, and wealthy buyers who are not concerned about how the wildlife products they use were obtained. The illegal wildlife trade has hit Africa's megafauna particularly hard. For example, even though there has been an international ban on the ivory trade since 1989, thousands of African elephants continue to be illegally killed on an annual basis (Box 7.2). Similarly, despite a ban on rhinoceros horn trade since 1977, an increasing number of rhinoceros succumb to poaching every year (Figure 7.11). Worse yet, the illegal wildlife trade shares many characteristics and practices with the illegal trade in drugs and weapons; in some cases, the same syndicates run these various criminal enterprises (Christy and Stirton, 2015). Apprehending these criminal networks is generally very dangerous, requiring vast resources.

Now, consider a hypothetical conservationist concerned about the bushmeat crisis' impact on biodiversity. This person may very well think that effective enforcement of a hunting ban would be the best solution to prevent further overharvesting. Unfortunately, solving complex challenges with simplistic steps runs a high risk of being counterproductive. For example, local bushmeat markets provide important contributions to food and financial security in many rural parts of Africa (van der Merwe et al., 2015). Replacing bushmeat with livestock and crops production—two primary drivers of habitat loss (Chapter 5)—also carries risks. For example, an estimated 250,000 km^2 of forest will need to be converted to pastureland to replace the bushmeat trade just

Box 7.2 Conserving Elephants in the Anthropocene

David H.M. Cumming[1,2]

[1]*FitzPatrick Institute of African Ornithology, University of Cape Town, Cape Town, South Africa.*

[2]*Tropical Resource Ecology Programme, University of Zimbabwe, Harare, Zimbabwe.*

cummingdhm@gmail.com

As our increasingly human-dominated planet enters a new geological era, will there still be room for Earth's largest land mammals? Or will there be, as happened to mammoths, sabre-toothed cats, and giant sloths during the Pleistocene (see Box 8.1), another hominid-induced extinction of large mammals? Our new human-dominated era has become known as the Anthropocene (Waters et al., 2015), and the animals are, of course, Africa's elephants.

Elephants encapsulate the dilemmas of conserving large charismatic mammals. They are dominant ecosystem engineers that, depending on their densities, can facilitate or adversely impact species diversity and ecosystem processes (Section 4.2.1). They are also economically important to ecotourism industries and revered by many; ivory ornaments and carvings have been valued highly by many cultures past and present. But elephants are also regarded as dangerous pests by expanding small-scale farming communities, responsible for destroying crops and killing people. While retaliatory killings and habitat loss (primarily through agricultural expansion) certainly contribute to the endangerment of elephants, poaching to supply Asian markets (see Figure 12.1) is the primary cause behind massive population reductions we are currently witnessing (Wittemyer et al., 2014).

Africa's elephants have, in the past, been greatly exploited, first for their meat and later also for their ivory. In 1887, about 1,000 tonnes of ivory were being exported from Africa (Spinage, 1973) and, by 1900, elephant populations in many African countries had all but collapsed. In Southern Africa, for example, it was feared that they might soon go extinct. However, with effective protection, elephant populations increased twentyfold, to more than 200,000, south of the Zambezi River by the 1970s. Elsewhere, elephant numbers also recovered, and, in the mid-1970s, the continental elephant population was estimated to be more than 1 million (Table 7.B). But a rapid escalation of the illegal killing of elephants for ivory and meat soon followed, accompanied by a steep rise in the price of ivory. In response, African elephants were placed on CITES Appendix II in 1976 to control the international trade in ivory. Elephants in some Southern African countries were well protected, so numbers continued to grow. However, elsewhere poaching and illegal trade in ivory continued and, in 1989, the African elephant was placed on CITES Appendix I, which banned all international trade in elephants and elephant products. The result was a decline in the price of ivory, and recovery of many populations.

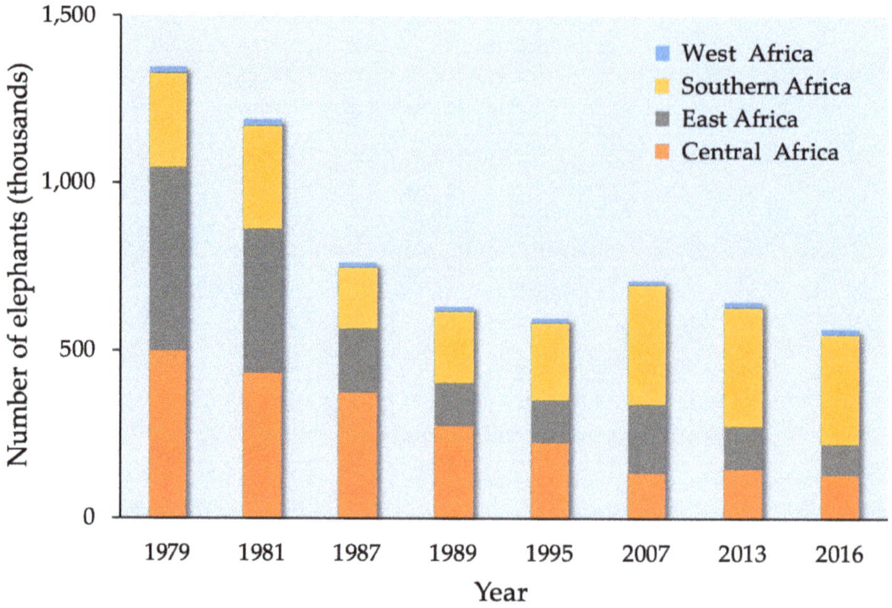

Figure 7.B Estimated number of elephants in West, Central, East and Southern Africa between 1979 and 2016. Source: http://africanelephantdatabase.org, CC BY 4.0.

Conservationists, and the world at large, traditionally regarded the African elephant as a single species. Recent morphometric and genetic evidence has revealed that forest and savannah elephants represent two distinct species, with forest elephants (*Loxodonta cyclotis*) occupying the West and Central African forests, and the larger savannah elephants (*Loxodonta africana*) being widespread in non-forested regions of Sub-Saharan Africa (Roca et al., 2015). The distinction has important implications for their conservation, as each of these elephant species now viewed on its own is even more sensitive to population declines (CBD, 2015).

Since about 2006, poaching of elephants again began to escalate, in part a response to an increase in the price of ivory, poorly funded wildlife agencies, and corruption (Hauenstein et al., 2019). The scale of these killings is extraordinary. For example, an estimated 30,000 elephants were killed in 2013 alone. Forest elephants declined by about 60% (Maisel et al., 2013). While population trends for savannah elephants vary across the region, they too face increased poaching pressure (Chase et al., 2016). Due to these large-scale killings, combined with the impact of habitat loss from agricultural expansion, West Africa's elephants are today confined to small isolated protected areas with a total population of about 17,000 (Maisel et al., 2013). Elephant population trends in East Africa vary: numbers are increasing in Uganda and Kenya, but Tanzania has lost some 60,000 elephants in the last few years. In Southern Africa, Botswana has the largest elephant population, estimated in 2014 to number at least 130,000. Neighbouring Zimbabwe has a population of 83,000 elephants, much the same number as it had in 2001. However, two of Zimbabwe's four regional populations declined significantly between 2006 and 2014 with a loss of at least 20,000 elephants (Figure 7.C).

Figure 7.C Savannah elephants at a waterhole in Hwange National Park. Hwange, Zimbabwe's flagship protected area, has been a hotspot of poacher activity in recent years. By lacing waterholes and salt licks with cyanide during the dry season, poachers killed over 100 elephants here in 2013 (Cruise, 2017); several elephant poisoning events have occurred since then. Photograph by D.H.M. Cumming, CC BY 4.0.

Global and national efforts to curb elephant poaching are currently focused on improving law enforcement on the ground, intercepting ivory shipments to Asia, closing ivory markets in Africa and Asia, and leading campaigns to reduce demand for ivory in major consuming countries in Asia, particularly in China. Importantly, while these initiatives are relieving poaching pressure on African elephant populations, they fail to address core issues relating to the interactions between people and elephants in the rural areas of Africa. A high proportion of elephant ranges lie outside protected areas where they overlap with people. Relieving continued pressures on elephants, outside as well as inside protected areas, would only happen if people who are harmed by elephants derive enough benefits from elephants and other wildlife to outweigh the direct and indirect costs of sharing land with them. Community-based natural resource management (CBNRM) projects, such as those in Namibia (Section 14.3) show that this can be achieved. Establishing secure and sustainable funding streams through payments for ecosystem services (Section 15.3), or payments for co-existing with large dangerous mammals such as elephants (Section 14.4), could extend these promising initiatives even further.

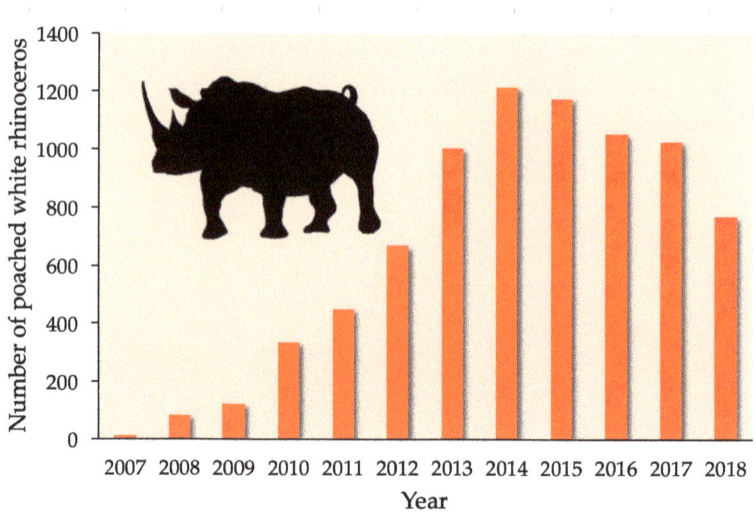

Figure 7.11 More than 7,200 rhinoceros were illegally killed in South Africa between 2007 and 2017. Fortunately, illegal killings have declined in recent years, thanks to massive anti-poaching campaigns and increased law enforcement efforts. Source: SADEA, CC BY 4.0.

for the Congo Basin (Nasi et al., 2011). Clearly, there is a need for the bushmeat trade, albeit in a sustainable manner to ensure long-term viability of the local biodiversity, as well as the prosperity of the people who inhabit these areas. We delve deeper into solutions for these kinds of complex challenges from Chapter 9 onwards.

7.3 Persecution

Persecution involves the indiscriminate abuse or killing of a group of animals, generally used as a strategy to prevent property damage (e.g. crop-raiding elephants) and livestock depredation (e.g. lions endangering cattle). Some animals are also persecuted because of the real or perceived dangers they pose to humans; this includes the indiscriminate killing of sharks, snakes, and spiders to avoid bites, and culling of bats to prevent spread of **zoonotic diseases** (Schneeberger and Voigt, 2016). Lastly, local folklore also contributes to persecution: animals, such as moles, chameleons, and owls are sometimes indiscriminately killed because of cultural beliefs that they bring bad luck.

While killing a (potential or perceived) problem animal may bring a certain instant gratification, it is a short-term solution that often causes more harm than good. Culling bats, for example, may in fact increase the prevalence of the same zoonotic diseases people are trying to control (Schneeberger and Voigt, 2016). Persecution also leads to the loss of ecosystem services, as it often targets ecosystem engineers and keystone species (Section 4.2.1). Retaliatory poisoning is particularly harmful for all the other useful organisms that may be killed in the process. One study in Namibia estimated that about 100 non-target animals are killed for every target animal, putting harmless species such

While killing a (potential or perceived) problem animal may bring a certain instant gratification, it is a short-term solution that often causes more harm than good.

as aardwolf (*Proteles cristatus*, LC), bat-eared fox (*Otocyon megalotis*, LC), and Cape fox (*Vulpes chama*, LC) also at risk (Brown, 2006). Among the most vulnerable to such poisoning are vultures, which may be accidentally poisoned by bait set out for problem predators, or directly targeted for traditional medicine and to hide poaching activities. In June 2019, 537 vultures (comprising five different *Endangered* and *Critically Endangered* species) were killed in Botswana after scavenging from three poisoned elephant carcass (de Greef, 2019). This mass poisoning event was particularly devastating because it was during the vulture breeding season, so many vulture chicks likely also died from starvation if not from eating tainted meat brought back to the nest by their parents.

7.4 Invasive Species

An **exotic species** is a species that occurs beyond its native range, most often because humans have moved it, whether intentionally or not. Most exotic species do not establish viable populations in the new areas to which they have been moved because the new environments may not meet their needs, because native species may outcompete them or otherwise displace them, or because there may not be a sufficient number of individuals to become established. However, a small number of exotic species go on to become **invasive species**—exotic species that rapidly spread and increase in abundance at the expense of native species and ecosystems. While there is no definitive list of qualities that predict which exotic species can become invasive, many invasive species have the following in common: (1) they begin to reproduce at an early age; (2) they can reproduce rapidly; (3) they lack sufficient predators in their introduced range; (4) they disperse easily; (5) and they are generalist species, able to survive in a variety of ecosystems.

> Invasive species displace native species through competition, predation, and habitat alterations. They often thrive in environments disturbed by human activities.

7.4.1 Spread of invasive species

As stated above, invasive species spread and invade new areas because human activities move them there. Some of the most prominent means by which human activities facilitate the spread of invasive species include:

- *Agriculture*: Large industries exist to grow agricultural plants for crop production, ornamental plants for gardens, grasses for pastures, and livestock for food. Many of these organisms later escape from cultivation and captivity and go on to invade and harm local ecosystems. Other species spread when industry workers accidentally harvest the seeds of weedy plants along with commercial seeds, and then sow those seeds in new localities, while microbes, parasitic organisms, and insects may be transported with plant leaves and roots, in potting soil, or even attached to transported animals. One of Africa's worst plant invaders, the triffid weed (*Chromolaena odorata*), continues to spread because of its sustained use to boost soil fertility on agricultural lands (Uyi et al., 2014).

- *Accidental transport*: Many invasive species spread to new areas because people transported them unintentionally. Domestic rats (*Rattus* spp.) and mice (*Mus* spp.) are classic examples: they have spread around the world as stowaways aboard ships (Box 7.3). A great number of invasive plants that currently occur in South Africa's Cape Floristic Province arrived by accident by clinging onto the luggage and hiking gear of tourists (Anderson et al., 2015). The ballast tanks of ships are also common hiding places for invasive aquatic species: DNA analyses have shown how a German ship carried tiny snails in this way along the entire length of Africa's Atlantic coast, providing invasion opportunities across this entire shipping route (Ardura et al., 2015).

- *Biological control*: Environmental and agricultural organisations sometimes use biological control (Section 4.2.7) to manage the spread of, and harm caused by, invasive species. While this approach can be very effective, in rare cases the biocontrol agent can become invasive and harm native species, rather than its intended target. For example, domestic cats that were introduced to Marion Island off Africa's south coast feasted on native seabirds—in some cases, even causing seabird extirpations—instead of the rats and mice they were meant to control (Bloomer and Bester, 1992). For this reason, very careful research is necessary to test the appropriateness of a biocontrol agent before being released.

- *Deliberate introductions*: Soon after their arrival, colonists released hundreds of European birds and mammals into countries like South Africa and Kenya to make the African countryside feel more familiar. Other species, especially fish (e.g. trout, bass, and carp), were released to provide food and recreational opportunities. Many of these species have subsequently become so successful that they harm native species. For example, mesquite (*Prosopis juliflora*), introduced to Ethiopia from Mexico to reduce soil salinity, proved such a successful invader that it completely displaced local plants; the subsequent encroachment even threatens the viability of Ethiopia's few remaining Grevy's zebras (*Equus grevyi*, EN) (Kebede and Coppock, 2015). The introduction of the Nile perch (*Lates niloticus*, LC) to the Rift Valley to boost local fisheries likely led to the extinction of hundreds of fish species endemic in Lake Victoria (Pringle, 2005).

- *Captive escapees*: Many invasive species were originally kept as pets or ornamental plants but have escaped from captivity to establish feral populations that harm local wildlife. Some of the most problematic aquatic invasive species are common pets that people dumped in streams, lakes, or storm drains because they could not care for them anymore. Finding these escapees before they establish should thus be a priority. For example, a recent survey found that 258 alien ornamental plant species growing in South Africa's Kruger National Park are at risk of becoming invasive—most of these plants were subsequently removed from the park (Foxcroft et al., 2008).

Box 7.3 Aliens on Islands: Damage and Control

Peter Ryan

FitzPatrick Institute of African Ornithology, DST-NRF Centre of Excellence, University of Cape Town, South Africa.

peter.ryan@uct.ac.za

Invasive species are one of the main threats to biodiversity. Island ecosystems are particularly vulnerable; of the 156 bird species that have gone extinct in the last 500 years, more than 90% lived on oceanic islands (IUCN, 2019). This vulnerability is mainly due to species on oceanic islands evolving in the absence of competing species or predators, and thus lacking adequate defences against introduced species (including humans).

Invasive species pose many threats to island biodiversity. Newly arrived mammalian predators have exacted the greatest toll. Seemingly unable to appreciate the danger posed by these strange new arrivals, the ecologically-naïve adult birds simply remain on their nests to be eaten rather than fleeing. Introduced herbivores can have devastating impacts, because many island plants lack defences like tough leaves or thorns. Most devastating are domestic goats and rabbits, introduced by early island explorers to provide a source of food in the case of shipwreck, that have grazed many once-lush islands down to the ground. Introduced plants can also outcompete native plants. For example, the Mexican thorn (also called mesquite, *Prosopis juliflora*) has formed dense thickets on the once sparsely vegetated lowlands of Ascensión Island, making those areas unsuitable for both nesting seabirds and sea turtles.

Some of Africa's least-transformed islands are the sub-Antarctic Prince Edward Islands, 2,000 km southeast of Cape Agulhas, and Gough Island, 2,800 km west of Cape Town. Their small size, isolation, and lack of sheltered harbours prevented human settlement. Nevertheless, their large seal and seabird populations were frequently exploited for oil and skins in the 19th and early 20th centuries. In the 19th century, sealing parties accidentally introduced house mice (*Mus musculus*) to Marion Island, the larger of the two Prince Edward Islands, and Gough Island. The mice flourished by eating native invertebrates and plants, probably causing the local extinction of one flightless moth on Marion Island. A few domestic cats were brought in to control the mice at Marion Island's weather station, established in 1948. Instead, the cats targeted the island's birds, which were easier prey. By the 1970s, some 2,000 cats were killing an estimated 450,000 seabirds each year, greatly reducing the island's burrow-nesting petrels and even driving some species to local extinction (Bloomer and Bester, 1992). The events on Marion contrasted with nearby predator-free Prince Edward Island that continued to support vast breeding populations of burrowing petrels.

A pioneering initiative eradicated Marion Island's cats in 1991, using a combination of introduced cat influenza, hunting, trapping, and poisoning (Bloomer and Bester, 1992). Researchers hoped Marion's seabird populations would recover within a decade but had not considered the impact of mice once the cats were removed. The precedent was set on Gough Island; in 2001, introduced mice were discovered to predate on large numbers of seabird chicks, including Tristan albatross (*Diomedea dabbenena*, CR) chicks more than 100 times larger than themselves (Davies et al., 2015; Dilley et al., 2015a). It was hypothesised that mice are more likely to attack seabirds when they are the sole introduced predators on an island. Sure enough, the first attacks on Marion's albatross chicks were recorded in 2003 (Figure 7.D); by 2015, the attacks had increased dramatically (Dilley et al., 2015b).

Figure 7.D Two grey-headed albatross fledglings (*Thalassarche chrysostoma*, EN), still on the nest, suffering from injuries after predatory mouse attacks on Marion Island, southern Indian Ocean. The bird in the foreground already has its scalp exposed. Attacks usually occur at night; mice will come back to the attack site over successive nights until the victims succumb to their injuries. Photograph by P.G. Ryan,

Fortunately, it is possible to eradicate invasive species from islands. In 2014, Australia removed mice, rats, and rabbits from sub-Antarctic Macquarie Island (Parks and Wildlife Service, 2014), which is almost twice the size of Gough Island. Plans are now also in place to eradicate Gough's mice in 2019. The island's isolation facilitates this effort—damage to other species can be minimized, and possible spread of toxins or diseases will be confined. To access areas inaccessible on foot, helicopters will be used to spread poison bait from specially designed hoppers slung under the aircraft. Some poisoning of non-target native individuals is inevitable, but this is a small price to pay compared to extinctions of those species. If adequate measures are put in place to prevent subsequent reintroductions, there is hope of restoring at least part of the island's natural balance.

7.4.2 Impact of invasive species

Invasive species have many negative consequences for native biodiversity: they displace native species through competition, alter the structure and composition of natural communities, and sometimes also hybridise with native species. These impacts may also translate to financial losses, as invasive species compromise ecosystem services (Figure 7.12), damage infrastructure, and spread infectious diseases.

Figure 7.12 The tickberry (*Lantana camara*) may be pretty, but it is a serious invasive species across Africa. Where it invades, it reduces the productivity of natural ecosystems and agricultural lands by forming dense thickets that outcompete native plants. This plant is also toxic to animals, including livestock and pets. Photograph by Alves Gaspar, https://en.wikipedia.org/wiki/File:LantanaFlowerLeaves-3.jpg, CC BY-SA 3.0.

Invasive species often become pervasive because they outcompete and displace native species. One such example is the Mediterranean mussel (*Mytilus galloprovincialis*), which was accidentally introduced to South Africa in the mid-1970s via European ships. A superior competitor, the exotic mussel soon started displacing native mussels and limpets, especially in the inter-tidal zone of South Africa's west coast (Branch and Steffani, 2004). Considered South Africa's most successful marine invasive, recent evidence suggests that the Mediterranean mussel is continuing to spread north into Namibia and along South Africa's east coast towards Mozambique.

Another superior competitor is the water hyacinth. A native to South America's Amazon forest, this species was intentionally introduced as a showy ornamental plant to dams, ponds, and lakes across Africa in the early 20th century. The plant established well, but then started reproducing and spreading at such rapid rates that water bodies across the region were soon covered by a dense mat of leaves. With little surface exposure and water movement, eutrophication and suffocation followed, leading to the deaths of countless fish and other aquatic organisms (Villamagna and Murphy, 2010). A biological control program targeting hyacinth showed promise during the 1990s; however, eutrophication from fertiliser overuse (which stimulate growth of hyacinth and other invasive aquatic plants) may be contributing to this species' recent resurgence (Coetzee and Hill, 2012; Bownes et al., 2013).

Natural communities are at particular risk in cases where invasive species change ecosystem structure and functioning so much that native species can no longer survive. Such is the case across many parts of Africa, where invasive Australian gum (*Eucalyptus* spp.) and pine (*Pinus* spp.) trees (both widely planted for timber) transpire so much water through their leaves (as much as 50,000 litres of water per tree/year; Dzikiti et al., 2016) that they can reduce the availability of surface- and groundwater in an area by as much as 70% (le Maitre et al., 2016). In addition to creating drought conditions, the closed canopies created by these invasive trees reduce the amount of solar radiation reaching the ground, greatly limiting thermoregulatory opportunities for taxa such as reptiles (Schreuder and Clusella-Trullas, 2016). Habitat degradation caused by invasions of the cinnamon tree (*Cinnamomum verum*), originally from Sri Lanka, has already caused at least ten invertebrate extinctions in the Seychelles (IUCN, 2019).

> A single gum or pine tree can transpire as much as 50,000 litres of water per year, while plantations of these trees can reduce water resources in an area by as much as 70%.

While many of the invasive species mentioned earlier originated from outside Africa, it is important to note that African species can also become invasive in other parts of Africa when they are moved outside of their native ranges. When invading nearby areas, non-native species can come in close contact with closely-related species, creating a high risk of genetic mixing—also called **genetic pollution** or genetic swamping—which describes the hybridisation of invasive species with native species. For example, hybridisation with the widespread banded tilapia (*Tilapia sparmanii*, LC) threatens the survival of Namibia's Otjikoto tilapia (*Tilapia guinasana*, CR), globally restricted to the $< 1\ km^2$ Lake Guineas (Bills 2007). The Cape platanna (*Xenopus gilli*, EN), an endemic to South Africa's Cape Floristic Region, is similarly threatened by hybridisation with the widespread African clawed frog (*Xenopus laevis*, LC) (Fogell et al., 2013).

7.4.3 Genetically modified organisms

A topic of conflict among conservation biologists is the increased popularity of **genetically modified organisms (GMO)**. A GMO is an organism whose genetic material has been altered to provide useful or improved products and services. To do this, scientists typically use genome editing technologies to transfer genes from a "source" organism into the DNA of the target organism. For example, scientists can transfer a bacterial gene that produces an insect toxin into a crop, such as maize, to obtain a GMO that can resist insect herbivory. Farmers using this GMO maize would then be able to increase production *and* reduce pesticide use (Gewin, 2003). While GMOs are usually associated with the development of pest-resistant and drought-resistant crops, uses are highly varied. For example, in Senegal, GMO technologies are used to produce tilapia that are better adapted to local ecosystems (Eknath et al., 2007). GMO technologies are being used to develop new and cheaper medicines (Concha et al., 2017), and to combat important diseases: trials in Burkina Faso shows that fungi genetically engineered to produce spider toxins caused a 99% collapse in malaria-carrying *Anopheles* mosquito populations within 45 days (Lovett et al., 2019). GMOs can even be used for conservation purposes, like developing new methods to combat invasive species (Esvelt et al., 2014), creating more effective bioenergy sources (Beer et al., 2009), and making vulnerable species more **resistant** to climate change (Piaggio et al., 2017). Some scientists even hope to combat plastic pollution by creating a genetically modified bacterium able to consume plastic waste (Austin et al., 2018).

The use of GMOs is not a new phenomenon. Selective breeding, hybridisation, and other forms of artificial selection—techniques that have been used for much of human history—all result in different forms of genetically modified crops and animal species. However, technological advances in genetic engineering have enabled scientists to transfer genes from and between different taxa that have not previously been used in selective breeding programmes (i.e. viruses, bacteria, insects, fungi, and shellfish). GMO technologies that transfer genetic material between wholly disparate taxa has led to concern about the unknown and unintended consequences of such "crossovers". Some people are also concerned that GMOs that escape from captivity or cultivation (e.g. Gilbert, 2010) could hybridise with closely-related wild species, endangering native wildlife while resulting in new, aggressive weeds and virulent diseases. Additionally, the use of GMO crops could potentially harm non-crop species (e.g. insects, birds, and soil organisms) that live in, on, or near the GMO crops. Concerns have also been raised about the potential effects to people eating GMO foods, leading some governments to regulate GMO research and commercial applications differently than traditional agriculture. However, after decades of research, it appears that GMO food is safe to eat (Blancke, 2015).

> GMOs offer benefits such as increased production and reduced pesticide use, but there are concerns about unknown and unintended consequences like hybridisation and invasiveness.

Clearly, GMOs offer a wide range of opportunities which could directly and indirectly benefit biodiversity conservation. However, the benefits of GMOs must be examined and weighed against the potential risks on a case by case basis. It is probably wise to proceed cautiously, to study GMOs thoroughly, and to monitor their impacts on ecosystems and human health in the areas where they are used. These investigations could involve workshops where experts come together to perform **environmental impact assessments (EIA)** (Section 12.2.2). The potential but unknown impacts of GMOs can also be mitigated by limiting the ability of these organisms to spread or reproduce (Muir and Howard, 2004).

7.6 Parasites and Diseases

Parasites and diseases have always been an important natural factor in regulating the ecology of wildlife, especially in wild populations that have become unsustainably large. Today, however, human activities are facilitating increased spread and transmission of parasites (Box 7.4) and other pathogens, sometimes even creating conditions for epidemics to develop (Figure 7.13). Consequently, parasites and diseases have become a major threat to wildlife, including those already suffering under low population sizes and densities.

Box 7.4 Promoting African and Global Honeybee Health

Vincent Dietemann

Agroscope, Swiss Bee Research Center, Schwarzenburgstrasse, Switzerland.

✉ *vincent.dietemann@alp.admin.ch*

Modern society imposes increased pressure on animals and plants to secure the food needed for a growing human population. Honeybees especially contribute to crop productivity in a crucial way thanks to their pollination services. Unfortunately, the number of bee colony losses has surged in recent years in several regions of the world (Goulson et al., 2015), worrying scientists, politicians, and the public. Some regions in Africa, however, have maintained healthy domesticated and wild honeybee colonies (Pirk et al., 2016), which continue to pollinate flowers and enhance the production of many fruit and vegetable crops.

The power of comparisons

Comparative studies have always been important to biological research. Studying a model organism or system under different conditions allows scientists to identify how these organisms or systems react and adapt to their environment. This can even be done at the continental scale and could help us understand the effect of human pressure exerted on honeybees. In general, beekeeping in North America and Europe has been widely industrialised, involving large-scale operations and modern technology, whereas in Africa beekeeping has remained small-scale and mostly low-tech. This gives us an opportunity to determine the effects of beekeeping management and trade on honeybee health.

Different contexts

The varroa mite (*Varroa destructor*) originally parasitised the eastern honeybee (*Apis cerana*). In the wake of global honeybee trade, this parasite has invaded most regions of the world that are home to the western honeybee (*Apis mellifera*) and resulted in major colony losses (Figure 7.E). Although eastern and western honeybees are closely-related species, the western honeybee did not coevolve with this parasite and, thus, has few natural defences against it, with colonies dying within a few years after infestation. Consequently, only those colonies treated against the parasite by beekeepers can survive, and most wild honeybee populations have been decimated. However, there have been exceptions: colonies of the western honeybee in the southern parts of Africa are resistant to the parasite, and large wild populations remain. Several international teams have now turned their attention to resistant African honeybee populations to understand the basis of their survival (Strauss et al., 2016). Researchers hope to use this knowledge to promote the breeding of surviving colonies in currently susceptible populations both within and outside of Africa.

Figure 7.E During its development, this young honey bee worker has been damaged by the varroa mite it carries on the front end of its abdomen. Its deformed wings are a consequence of this parasitism. Photograph by Vincent Dietemann, CC BY 4.0.

Africa also differs strongly in land use and crop management techniques that are likely to influence honeybees' nutrition and health. In many areas, small-scale farming prevails, with a lower use of pesticides than in other areas of the world. Understanding how pesticide and other chemical use, as well as how nectar and pollen variety and quality, impact these pollinators will be key to their survival. Therefore, the rest of the world may learn how to maintain healthy honeybees from Africa. Africa, in return, might benefit from global efforts to maintain sustainable pollination services and promote food security.

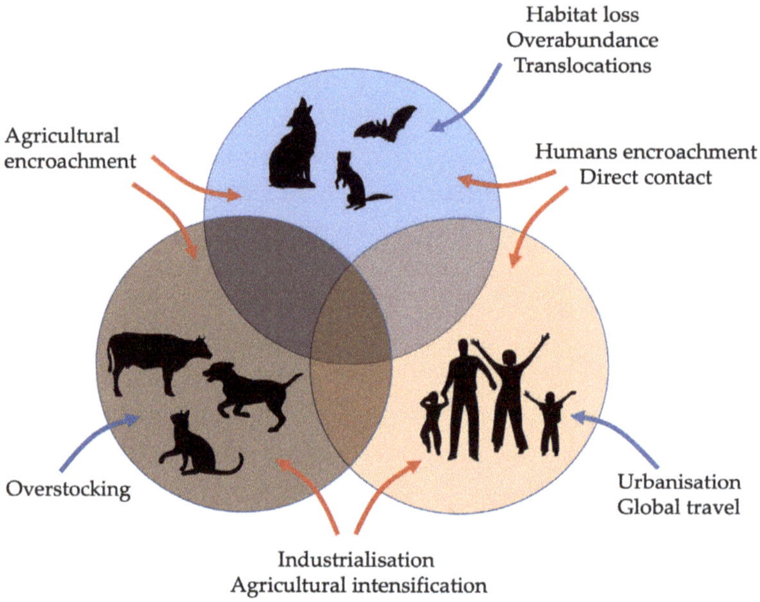

Figure 7.13 Habitat loss and encroachment of human activities into natural areas increase the rates of transmission of infectious diseases such as influenza, rabies, canine distemper virus, and Ebola between wildlife, domestic animals, and human. This figure illustrates the infection and transmission routes of rabies. Blue arrows represent factors that leads to higher infection rates, while red allows represent factors that contribute to the spread of the disease among the three vulnerable groups. After Daszak et al., 2000, CC BY 4.0.

One way in which humans elevate the impact of parasites and diseases on wildlife is by exposing native species to harmful organisms that they have never previously encountered, and thus have no evolved coping mechanisms. For example, population declines and extirpations of about 200 frog species across the world, including in Africa (Tarrant et al., 2013; Hirschfeld et al., 2016), is due, in part, to a disease caused by the chytrid fungus (*Batrachochytrium dendrobatidis*). This disease, known as chytridiomycosis (Figure 7.14), affects a frog's ability to absorb water and electrolytes through the skin (Alroy, 2015). It likely originated in the Korean Peninsula (O'Hanlon et al., 2018), and spread across the world through trade with African clawed frogs (*Xenopus laevis*, LC) (Weldon et al., 2004). As of yet, there is no cure for this disease, and it continues to be seen as one of the biggest threats currently facing the world's amphibians.

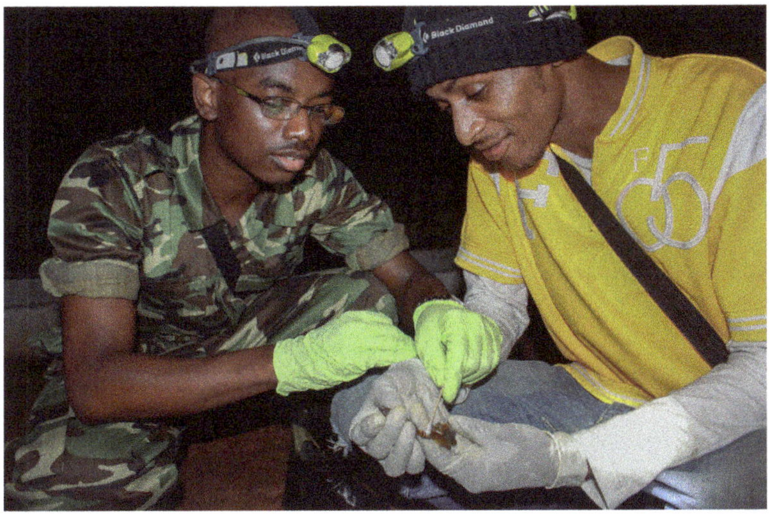

Figure 7.14 Biologists swabbing an olive striped frog (*Phlyctimantis leonardi,* LC) in Gabon to screen for the chytrid fungus. Frogs from Cameroon has tested positive (Baláž et al., 2012) which suggests the species might be at risk; however, thus far no ill effects have been observed. Photograph by Brian Gratwicke, https://www.flickr.com/photos/briangratwicke/4395505435, CC BY 2.0.

Disease transmissions can also occur when humans and their pets or livestock interact with wildlife (Cumming and Cumming, 2015). For example, during the early 1990s about 25% of lions in Tanzania's Serengeti National Park were killed by canine distemper virus which they contracted from domestic dogs living near the park (Kissui and Packer, 2004). Because of the many biological similarities between apes and humans, gorillas (*Gorilla* spp.), chimpanzees (*Pan troglodytes*, EN) and bonobos (*P. paniscus*, EN) are particularly vulnerable to **anthroponotic diseases,** such as measles, influenza, and pneumonia which can be transferred from humans to animals. But even chytridiomycosis (discussed above) can become an anthroponotic disease, transferred from frog to frog by a careless biologist that handles a healthy frog after a sick one without taking precautions against transmission. Some diseases (e.g. Ebola; flu; and tuberculosis) can be anthroponotic and **zoonotic** (transferred from animals to humans). While the impact of Ebola on humans in Africa is well-known, it is worth noting that gorillas suffer > 90% mortality when exposed to Ebola, compared to 50% mortality in humans. In fact, it was an Ebola outbreak in 2004 that caused the western lowland gorilla (*Gorilla gorilla gorilla*, CR) to be classified as highly threatened by the IUCN (Genton et al., 2012).

Humans also indirectly facilitate the transmission and spread of parasites and pathogens. While there are some exceptions (notably social insects), transmission and infection rates are typically low for wildlife living in large, complex ecosystems because they have space to move away from disease-carrying droppings, saliva, old skin, and other sources of infection. However, these natural buffers against pathogens and parasites are removed when humans confine those organisms to small areas (such as small fenced reserves) or keep them in crowded conditions. In addition to forcing those organisms to remain in close

> Human activities often facilitate the emergence and spread of infectious diseases, which threaten wildlife, domestic species, and humans alike.

contact with potential sources of infection, crowded conditions lead to deterioration of habitat quality and food availability. Both these factors increase the organisms' stress levels and reduce their body conditions which, in turn, lowers their resistance to parasites and diseases (reviewed in Gottdenker et al., 2014).

Human-induced extirpations indirectly facilitate the transmission and spread of parasites and pathogens, even to humans. Such is the case with schistosomiasis (also known as bilharzia), a zoonotic disease carried by a few freshwater snail species. In the 1980s, health care professionals observed an increased incidence of human schistosomiasis around Lake Malawi after overfishing depleted snail-eating fish populations, followed by decreased incidence of schistosomiasis as fish populations recovered in the 1990s (Stauffer et al., 2006). A similar situation occurred in East Africa, where the elimination of apex predators resulted in increased olive baboon (*Papio anubis*, LC) populations, which not only worsened crop raiding, but also increased parasite infection rates among local peoples (Brashares et al., 2010).

Parasites and diseases also threaten captive wildlife populations, including those kept at zoos and other **ex situ conservation** facilities (Section 11.5). Because of the proximity in which different species are kept, captive conditions may allow for easier spread of diseases. An added complication with captive populations is that some individuals may function as disease reservoirs. These individuals generally appear healthy because they are fairly resistant to the disease they carry, yet they are able to infect other susceptible individuals. Disease reservoirs frequently limit opportunities for translocation of captive populations (Section 11.2), even when dealing with threatened species. For example, well-meaning people often bring raggedy-looking yet healthy penguins in moult to rehabilitation centres, hoping the penguins will be released once "better". Yet, those animals might never be released back in the wild to avoid the risk of transmitting diseases to wild penguin populations (Brossy et al., 1999).

The impacts of diseases are bound to become more important in the future of conservation biology, especially as growing human populations and increased competition for space increase the need for single-species management and ex situ conservation (Chapter 11). Disease management should therefore always be taken very seriously, and appropriate steps taken to avoid disease transmissions.

7.7 Summary

1. Many threats to biodiversity do not lead to immediate and/or direct mortality, but instead have sublethal impacts that compromise organisms' fitness over time. Responses to these silent, insidious, and easily-overlooked threats are often delayed, especially when the negative effects are felt only years after exposure.
2. Environmental pollution leaves ecosystems uninhabitable for native wildlife, and cause sickness and death in wildlife and people. Common causes of

pollution include pesticides, heavy metals, plastic, fossil fuels, fertilisers, light, heat, and noise, leading to pollution of water, groundwater, air, and soil.

3. Overharvesting is becoming an increasingly damaging threat to biodiversity because people have better access to previously unexploited areas and are adopting increasingly efficient methods for harvesting wildlife products. Persecution, which has its roots in human-wildlife conflict, is becoming an important threat because a growing human population is increasingly encroaching on the shrinking remaining natural habitats.
4. Invasive species outcompete local species and change the structure and composition of their native ecosystems. Human activity is responsible for these invasions, by accidentally or deliberately moving wildlife to new regions of the world. Some invasive species require a great amount of effort and resources to manage.
5. Disease transmission and spread increase when wildlife is confined to small areas and/or crowded conditions. Diseases may also be transmitted between wildlife, domesticated species, and even humans. Managing for diseases is also important in zoos and other ex situ facilities, because diseases spreading from one individual to another can prevent those individuals from being released into the wild.

7.8 Topics for Discussion

1. Which forms of environmental pollution are most prominent in the region where you live? Which natural ecosystems are impacted most severely by this pollution? How are humans affected? What do you think can be done to reduce or even eliminate this pollution?
2. Consider all the fishing, hunting, trapping, collecting, logging, and other wildlife harvesting activities in your region. Which activities are well managed, and which are not? Why is it so difficult to regulate these activities, when so many people know that overharvesting would eventually harm local economies, their families, and their livelihoods? What measures do you think can be put in place to control overharvesting in your region, or at least reduce its impact?
3. Briefly describe what biological control is, and what its benefits are. What are the risks involved in biological control? How can these risks be predicted and avoided?
4. Can you name a few diseases that can be transmitted from people to wildlife, and from wildlife to people? Which species are involved in each of these diseases? How does the transmission of each of these diseases impact the conservation management of species involved?

7.9 Suggested Readings

Benítez-López, A., L. Santini, A.M. Schipper, et al. 2019. Intact but empty forests? Patterns of hunting-induced mammal defaunation in the tropics. *PLoS ONE* 17: e3000247. https://doi.org/10.1371/journal.pbio.3000247 African mammals are undergoing catastrophic declines due to unregulated hunting.

Bouwman, H., I.M. Viljoen, L.P. Quinn, et al. 2008. Halogenated pollutants in terrestrial and aquatic bird eggs: Converging patterns of pollutant profiles and impacts and risks from high level. *Environmental Research* 126: 240–52. https://doi.org/10.1016/j.envres.2013.06.003 DDT threaten African biodiversity and people even today

Campbell, L., P. Verburg, D.G. Dixon, et al. 2008. Mercury biomagnification in the food web of Lake Tanganyika (Tanzania, East Africa). *Science of The Total Environment* 402: 184–91. https://doi.org/10.1016/j.scitotenv.2008.04.017 Biomagnification of heavy metals are making some fish unsafe to eat.

Galloway, T.S., and C.N. Lewis. 2016. Marine microplastics spell big problems for future generations. *Proceedings of the National Academy of Sciences* 113: 2331–33. https://doi.org/10.1073/pnas.1600715113 The threat of plastic pollution to ocean life.

Gottdenker, N.L., D.G. Streicker, C.L. Faust, et al. 2014. Anthropogenic land use change and infectious diseases: A review of the evidence. *EcoHealth* 11: 619632. https://doi.org/10.1007/s10393-014-0941-z Linking environmental degradation to disease risk in humans and wildlife.

le Maitre, D.C., G.G. Forsyth, S. Dzikiti, et al. 2016. Estimates of the impacts of invasive alien plants on water flows in South Africa. Water SA 42: 659–72. http://dx.doi.org/10.4314/wsa.v42i4.17 The impact of invasive plants on water cycles

Maxwell, S.L., R.A. Fuller, T.M. Brooks, et al. 2016. The ravages of guns, nets and bulldozers. *Nature* 536: 143–45. https://doi.org/10.1038/536143a Biodiversity faces many anthropogenic threats.

Pringle, R.M. 2005. The origins of the Nile perch in Lake Victoria. *BioScience* 55: 780–87. https://doi.org/10.1641/0006-3568(2005)055[0780:TOOTNP]2.0.CO;2 The devastating impact of invasive fish in Lake Victoria.

Stauffer, J.R., H. Madsen, K. McKaye, et al. 2006. Schistosomiasis in Lake Malawi: Relationship of fish and intermediate host density to prevalence of human infection. *EcoHealth* 3: 22-27. https://doi.org/10.1007/s10393-005-0007-3 Trophic cascades can even affect human health.

Bibliography

Adeogun, A.O., O.R. Ibor, S.D. Adeduntan, et al. 2016. Intersex and alterations in reproductive development of a cichlid, *Tilapia guineensis*, from a municipal domestic water supply lake (Eleyele) in southwestern Nigeria. *Science of the Total Environment* 541: 372–82. https://doi.org/10.1016/j.scitotenv.2015.09.061

Agbohessi, P.T., I.I. Toko, A. Ouédraogo, et al. 2015. Assessment of the health status of wild fish inhabiting a cotton basin heavily impacted by pesticides in Benin (West Africa). *Science of the Total Environment* 506: 567–84. https://doi.org/10.1016/j.scitotenv.2014.11.047

Alroy, J. 2015. Current extinction rates of reptiles and amphibians. *Proceedings of the National Academy of Sciences* 112: 13003–08. https://doi.org/10.1073/pnas.1508681112

Amegah, A.K., and S. Agyei-Mensah. 2017. Urban air pollution in Sub-Saharan Africa: Time for action. *Environmental Pollution* 220: 738–43. https://doi.org/10.1016/j.envpol.2016.09.042

Amri, E., Z.L. Kanyeka, H.V.M. Lyaruu, et al. 2009. Evaluation of genetic diversity in *Dalbergia elanoxylon* populations using random amplified polymorphic DNA markers. *Research Journal of Cell and Molecular Biology* 3: 71–79. https://doi.org/10.15406/mojbm.2017.01.00015

Andersen, I. 2016. *Seizure of huge African pangolin scale shipment points to worrying increase in trafficking* (Gland: IUCN). https://www.iucn.org/news/secretariat/201606/seizure-huge-african-pangolin-scale-shipment-points-worrying-increase-trafficking

Anderson L.G., S. Rocliffe, N.R. Haddaway, et al. 2015. The role of tourism and recreation in the spread of non-native species: A systematic review and meta-analysis. *PLoS ONE* 10: e0140833. https://doi.org/10.1371/journal.pone.0140833

Annorbah, N.N.D., N.J. Collar, and S.J. Marsden. 2016. Trade and habitat change virtually eliminate the Grey Parrot *Psittacus erithacus* from Ghana. *Ibis* 158: 82–91. https://doi.org/10.1111/ibi.12332

Ardura, A., A. Zaiko, J.L. Martinez, et al. 2015. Environmental DNA evidence of transfer of North Sea molluscs across tropical waters through ballast water. *Journal of Molluscan Studies* 81: 495–501. https://doi.org/10.1093/mollus/eyv022

Aremu, A.O., L. Cheesman, J.F. Finnie, et al. 2011. *Mondia whitei* (Apocynaceae): A review of its biological activities, conservation strategies and economic potential. *South African Journal of Botany* 77: 960–71. https://doi.org/10.1016/j.sajb.2011.06.010

Austin, H.P., M.D. Allen, B.S. Donohoe, et al. 2018. Characterization and engineering of a plastic-degrading aromatic polyesterase. *Proceedings of the National Academy of Sciences* 2018: 201718804. https://doi.org/10.1073/pnas.1718804115

Baláž, V., O. Kopecký, and V. Gvoždík. 2012. Presence of the amphibian chytrid pathogen confirmed in Cameroon. *Herpetological Journal* 22: 191–94.

Barbee, J. 2015. Botswana sells fracking rights in national park. *Guardian*. https://gu.com/p/4eck4

Barnhoorn, I.E.J., M.S. Bornman, C. Jansen van Rensburg, et al. 2009. DDT residues in water, sediment, domestic and indigenous biota from a currently DDT-sprayed area. *Chemosphere* 77: 1236–41. https://doi.org/10.1016/j.chemosphere.2009.08.045

Beer, L.L., E.S. Boyd, J.W. Peters, et al. 2009. Engineering algae for biohydrogen and biofuel production. *Current Opinion in Biotechnology* 20: 264–71. https://doi.org/10.1016/j.copbio.2009.06.002

Benítez-López, A., L. Santini, A.M. Schipper, et al. 2019. Intact but empty forests? Patterns of hunting-induced mammal defaunation in the tropics. *PLoS ONE* 17: e3000247. https://doi.org/10.1371/journal.pbio.3000247

Benítez-López, A., R. Alkemade, A.M. Schipper, et al. 2017. The impact of hunting on tropical mammal and bird populations. *Science* 356: 180–83. https://doi.org/10.1126/science.aaj1891

Biginagwa, F.J., B.S. Mayoma, Y. Shashoua, et al. 2016. First evidence of microplastics in the African Great Lakes: Recovery from Lake Victoria Nile perch and Nile tilapia. *Journal of Great Lakes Research* 42: 146–49. https://doi.org/10.1016/j.jglr.2015.10.012

Bills, R. 2007. *Tilapia guinasana. The IUCN Red List of Threatened Species* 2007: e.T63354A12662434. http://doi.org/10.2305/IUCN.UK.2007.RLTS.T63354A12662434.en

Black, A. 2005. Light induced seabird mortality on vessels operating in the Southern Ocean: Incidents and mitigation measures. *Antarctic Science* 17: 67–68. https://doi.org/10.1017/S0954102005002439

Blancke, S. 2015. Why people oppose GMOs even though science says they are safe. *Scientific American.* https://www.scientificamerican.com/article/why-people-oppose-gmos-even-though-science-says-they-are-safe

Bloomer, J.P., and M.N. Bester. 1992. Control of feral cats on sub-Antarctic Marion Island, Indian Ocean. *Biological Conservation* 60: 211–19. https://doi.org/10.1016/0006-3207(92)91253-O

Bornman, M.S., I.E.J. Barnhoorn, C. de Jager, et al. 2010. Testicular microlithiasis and neoplastic lesions in wild eland (*Tragelaphus oryx*): possible effects of exposure to environmental pollutants? *Environmental Research* 110: 327–33. https://doi.org/10.1016/j.envres.2010.02.003

Bortey-Sam, N., Y. Ikenaka, O. Akoto, et al. 2017. Oxidative stress and respiratory symptoms due to human exposure to polycyclic aromatic hydrocarbons (PAHs) in Kumasi, Ghana. *Environmental Pollution* 228: 311–20. https://doi.org/10.1016/j.envpol.2017.05.036

Bosch, A.C., B. O'Neill, G.O. Sigge, et al. 2016. Heavy metal accumulation and toxicity in smoothhound (*Mustelus mustelus*) shark from Langebaan Lagoon, South Africa. *Food Chemistry* 190: 871–78. https://doi.org/10.1016/j.foodchem.2015.06.034

Bourgeois, S., E. Gilot-Fromont, A. Viallefont, et al. 2009. Influence of artificial lights, logs and erosion on leatherback sea turtle hatchling orientation at Pongara National Park, Gabon. *Biological Conservation* 142: 85–93. https://doi.org/10.1016/j.biocon.2008.09.028

Bouwman, H., I.M. Viljoen, L.P. Quinn, et al. 2008. Halogenated pollutants in terrestrial and aquatic bird eggs: Converging patterns of pollutant profiles and impacts and risks from high level. *Environmental Research* 126: 240–52. https://doi.org/10.1016/j.envres.2013.06.003

Bownes, A., M.P. Hill, and M.J. Byrne. 2013. The role of nutrients in the responses of water hyacinth, *Eichhornia crassipes* (Pontederiaceae) to herbivory by a grasshopper *Cornops aquaticum* Brüner (Orthoptera: Acrididae). *Biological Control* 67: 555–62. https://doi.org/10.1016/j.biocontrol.2013.07.022

Branch, G.M., and C.N. Steffani. 2004. Can we predict the effects of alien species? A case-history of the invasion of South Africa by *Mytilus galloprovincialis* (Lamarck). *Journal of Experimental Marine Biology and Ecology* 300: 189–215. https://doi.org/10.1016/j.jembe.2003.12.007

Brashares J.S., C.W. Epps, and C.J. Stoner. 2010. Ecological and conservation implications of mesopredator release. In: *Trophic Cascades*, ed. by J. Terborgh and J. Estes (Washington: Island Press).

Brossy, J.J., A.L. Plös, J.M. Blackbeard, et al. 1999. Diseases acquired by captive penguins: What happens when they are released into the wild? *Marine Ornithology* 27: 185–86.

Brown, C.J. 2006. *Historic distribution of large mammals in the Greater Fish River Canyon Complex, southern Namibia, and recommendations for re-introductions* (Windhoek: Namibia Nature Foundation). http://www.the-eis.com/data/literature/Greater%20Fish%20River%20Canyon%20Complex%20Historic%20distribution%20of%20mammals.pdf

Brummet, R.E., C. Cargill, L.M. Lekunze, et al. 2010. Stream degradation, fish abundance and the potential viability of ornamental fisheries in south-western Cameroon. *African Journal of Aquatic Science* 35: 155–64. https://doi.org/10.2989/16085914.2010.497650

Burnett, L.J., K.J. Sorenson, J. Brandt, et al. 2013. Eggshell thinning and depressed hatching success of California Condors reintroduced to central California. *Condor* 115: 477–91. https://doi.org/10.1525/cond.2013.110150

Bush, E.R., S.E. Baker, and D.W. Macdonald. 2014. Global trade in exotic pets 2006–2012. *Conservation Biology* 28: 663–76. https://doi.org/10.1111/cobi.12240

Campbell, L., P. Verburg, D.G. Dixon, et al. 2008. Mercury biomagnification in the food web of Lake Tanganyika (Tanzania, East Africa). *Science of The Total Environment* 402: 184–91. https://doi.org/10.1016/j.scitotenv.2008.04.017

Cannon, J.C. 2018. Half a ton of pangolin scales seized on the way to Asia from Benin. *Mongabay*. https://news.mongabay.com/2018/04/half-a-ton-of-pangolin-scales-seized-on-the-way-to-asia-from-benin

Carruthers, E.H., D.C. Schneider, and J.D. Neilson. 2009. Estimating the odds of survival and identifying mitigation opportunities for common bycatch in pelagic longline fisheries. *Biological Conservation* 142: 2620–30. https://doi.org/10.1016/j.biocon.2009.06.010

Carson, R. 1962. *Silent Spring* (Boston: Houghton Mifflin).

CBD (Center for Biological Diversity). 2015. *Petition to reclassify and uplist African elephants from Threatened to Endangered under the Endangered Species Act as two separate species: Forest elephants* (Loxodonta cyclotis) *and savannah elephants* (Loxodonta africana) (Tucson: Center for Biological Diversity). https://www.biologicaldiversity.org/species/mammals/pdfs/African_Elephant_Uplisting_Petition.pdf

Cerchio, S., S. Strindberg, T. Collins, et al. 2014. Seismic surveys negatively affect humpback whale singing activity off northern Angola. *PloS ONE* 9: e86464. https://doi.org/10.1371/journal.pone.0086464

Chaber, A.-L., A. Allebone-Webb, Y. Lignereux, et al. 2010. The scale of illegal meat importation from Africa to Europe via Paris. *Conservation Letters* 3: 317–21. https://doi.org/10.1111/j.1755-263X.2010.00121.x

Chakraborty, T., and X. Lee. 2018. A simplified urban-extent algorithm to characterise surface urban heat islands on a global scale and examine vegetation control on their spatiotemporal variability. *International Journal of Applied Earth Observation and Geoinformation* 74: 269–80. https://doi.org/10.1016/j.jag.2018.09.015

Chase, M.J., S. Schlossberg, C.R. Griffin, et al. 2016. Continent-wide survey reveals massive decline in African savannah elephants. *PeerJ* 4:e2354. https://doi.org/10.7717/peerj.2354

Christy, B., and B. Stirton. 2015. How killing elephants finances terror in Africa. *National Geographic*. http://on.natgeo.com/1I5N2aO

Coetzee, J.A., and M.P. Hill. 2012. The role of eutrophication in the biological control of water hyacinth, *Eichhornia crassipes*, in South Africa. *BioControl* 57: 247–61. https://doi.org/10.1007/s10526-011-9426-y

Concha, C., R. Cañas, J. Macuer, et al. 2017. Disease prevention: An opportunity to expand edible plant-based vaccines? *Vaccines* 5: 14. https://doi.org/10.3390/vaccines5020014

Covey, R., and W.S. McGraw. 2014. Monkeys in a West African bushmeat market: Implications for Cercopithecid conservation in eastern Liberia. *Tropical Conservation Science* 7: 115–25. https://doi.org/10.1177/194008291400700103

Cox, T.M., R.L. Lewison, R. Zydelis, et al. 2007. Comparing effectiveness of experimental and implemented bycatch reduction measures: The ideal and the real. *Conservation Biology* 21: 1155–64. https://doi.org/10.1111/j.1523-1739.2007.00772.x

Craigie, I.D., J.E.M. Baillie, A. Balmford, et al. 2010. Large mammal population declines in Africa's protected areas. *Biological Conservation* 143: 2221–28. https://doi.org/10.1016/j.biocon.2010.06.007

Croxall, J.P., S.H. Butchart, B. Lascelles, et al. 2012. Seabird conservation status, threats and priority actions: A global assessment. *Bird Conservation International* 22: 1–34. https://doi.org/10.1017/S0959270912000020

Cruise, A. 2017. Ten more elephants poisoned by poachers in Zim. *Guardian*. https://gu.com/p/6k6tx

Cumming, D.H.M., and G.S. Cumming. 2015. One Health: An ecological and conservation perspective. In: *One Health: The Theory and Practice of Integrated Health Approaches*, ed. by J. Zinsstag, et al. (Wallingford: CAB International).

Daszak, P., A.A. Cunningham, and A.D. Hyatt. 2000. Emerging infectious diseases of wildlife—threats to biodiversity and human health. *Science* 287: 443–49. https://doi.org/10.1126/science.287.5452.443

Davies, D., B.J. Dilley, A.L. Bond, et al. 2015. Trends and tactics of mouse predation on Tristan Albatross *Diomedea dabbenena* chicks at Gough Island, South Atlantic Ocean. *Avian Conservation and Ecology* 10: 5. http://doi.org/10.5751/ACE-00738-100105

de Graaf, M., M.A.M. Machiels, T. Wudneh, et al. 2004. Declining stocks of Lake Tana's endemic *Barbus* species flock (Pisces, Cyprinidae): Natural variation or human impact? *Biological Conservation* 116: 277–87. http://doi.org/10.1016/S0006-3207(03)00198-8

de Greef, K. 2019. 500 vultures killed in Botswana by poachers' poison, government says. *New York Times*. https://nyti.ms/2FpkfTf

Deribe, E., B.O. Rosseland, R. Borgstrøm, et al. 2011. Bioaccumulation of persistent organic pollutants (POPs) in fish species from Lake Koka, Ethiopia: The influence of lipid content and trophic position. *Science of The Total Environment* 410–11: 135–45. http://doi.org/10.1016/j.scitotenv.2011.09.008

Desta, Z., R. Borgstrøm, B.O. Rosseland, et al. 2006. Major difference in mercury concentrations of the African big barb, *Barbus intermedius* (R.) due to shifts in trophic position. *Ecology of Freshwater Fish* 15: 532–43. https://doi.org/10.1111/j.1600-0633.2006.00193.x

Dilley, B.J., D. Davies, A.L. Bond, et al. 2015a. Effects of mouse predation on burrowing petrel chicks at Gough Island. *Antarctic Science* 27: 543–53. https://doi.org/10.1017/S0954102015000279

Dilley, B.J., S. Schoombie, J. Schoombie, et al. 2015b. 'Scalping' of albatross fledglings by introduced mice spreads rapidly at Marion Island. *Antarctic Science* 28: 73–80. https://doi.org/10.1017/S0954102015000486

Dodoo, D.K., D.K. Essumang, and J.W.A. Jonathan. 2013. Accumulation profile and seasonal variations of polychlorinated biphenyls (PCBs) in bivalves *Crassostrea tulipa* (oysters) and *Anadara senilis* (mussels) at three different aquatic habitats in two seasons in Ghana. *Ecotoxicology and Environmental Safety* 88: 26–34. http://doi.org/10.1016/j.ecoenv.2012.10.013

Dzikiti, S., M.B. Gush, D.C. le Maitre, et al. 2016. Quantifying potential water savings from clearing invasive alien *Eucalyptus camaldulensis* using in situ and high resolution remote sensing data in the Berg River Catchment, Western Cape, South Africa. *Forest Ecology and Management* 361: 69–80. https://doi.org/10.1016/j.foreco.2015.11.009

Eknath, A.E., H.B. Bentsen, P.W. Ponzoni, et al. 2007. Genetic improvement of farmed tilapias: Composition and genetic parameters of a synthetic base population of *Oreochromis niloticus* for selective breeding. *Aquaculture* 273: 1–14. https://doi.org/10.1016/j.aquaculture.2007.09.015

Ellsworth, W.L. 2013. Injection-induced earthquakes. *Science* 341: 1225942. https://doi.org/10.1126/science.1225942

Eriksen M., L.C.M. Lebreton, H.S. Carson, et al. 2014. Plastic pollution in the world's oceans: More than 5 trillion plastic pieces weighing over 250,000 tons afloat at sea. *PLoS ONE* 9: e111913. https://doi.org/10.1371/journal.pone.0111913

Esvelt, K.M., A.L. Smidler, F. Catteruccia, et al. 2014. Emerging technology: Concerning RNA-guided gene drives for the alteration of wild populations. *eLife* 3: e03401. https://doi.org/10.7554/eLife.03401.001

Fa, J.E., C.A. Peres, and J. Meeuwig. 2002. Bushmeat exploitation in tropical forests: An intercontinental comparison. *Conservation Biology* 16: 232–37. https://doi.org/10.1046/j.1523-1739.2002.00275.x

Fa, J.E., D. Currie, and J. Meeuwig. 2003. Bushmeat and food security in the Congo Basin: Linkages between wildlife and people's future. *Environmental Conservation* 30: 71–78. https://doi.org/10.1017/S0376892903000067

Fa, J.E., J. Olivero, M.A. Farfán, et al. 2016. Differences between Pygmy and non-Pygmy hunting in Congo Basin forests. *PLoS ONE* 11: e0161703. https://doi.org/10.1371/journal.pone.0161703

Fa, J.E., S. Seymour, J.E.F. Dupain, et al. 2006. Getting to grips with the magnitude of exploitation: Bushmeat in the Cross-Sanaga rivers region, Nigeria and Cameroon. *Biological Conservation* 129: 497–510. https://doi.org/10.1016/j.biocon.2005.11.031

Falchi, F., P. Cinzano, D. Duriscoe, et al. 2016. The new world atlas of artificial night sky brightness. *Science Advances* 2: e1600377. https://doi.org/10.1126/sciadv.1600377

Fanou, L.A., T.A. Mobio, E.E. Creppy, et al. 2006. Survey of air pollution in Cotonou, Benin—air monitoring and biomarkers. *Science of The Total Environment* 358: 85–96. https://doi.org/10.1016/j.scitotenv.2005.03.025

Fentiman, A., and N. Zabbey. 2015. Environmental degradation and cultural erosion in Ogoniland: A case study of the oil spills in Bodo. *Extractive Industries and Society* 2: 615–24. https://doi.org/10.1016/j.exis.2015.05.008

Feyisa, G.L., K. Dons, and H. Meilby. 2014. Efficiency of parks in mitigating urban heat island effect: An example from Addis Ababa. *Landscape and Urban Planning* 123: 87–95. https://doi.org/10.1016/j.landurbplan.2013.12.008

Firebaugh, A., and K.J. Haynes. 2016. Experimental tests of light-pollution impacts on nocturnal insect courtship and dispersal. *Oecologia* 182: 1203–11. https://doi.org/10.1007/s00442-016-3723-1

Fogell, D.J., K.A. Tolley, and G.J. Measey. 2013. Mind the gaps: Investigating the cause of the current range disjunction in the Cape Platanna, *Xenopus gilli* (Anura: Pipidae). *PeerJ* 1: e166. https://doi.org/10.7717/peerj.166

Foxcroft, L.C., D.M. Richardson, and J.R.U. Wilson. 2008. Ornamental plants as invasive aliens: Problems and solutions in Kruger National Park, South Africa. *Environmental Management* 41: 32–51. https://doi.org/10.1007/s00267-007-9027-9

Galloway, T.S., and C.N. Lewis. 2016. Marine microplastics spell big problems for future generations. *Proceedings of the National Academy of Sciences* 113: 2331–33. https://doi.org/10.1073/pnas.1600715113

Geddie, J. 2019. Singapore seizes record haul of pangolin scales enroute to Vietnam. *Reuters*. https://reut.rs/2WMDtsh

Genton, C., R. Cristescu, S. Gatti, et al. 2012. Recovery potential of a western lowland gorilla population following a major Ebola outbreak: Results from a ten-year study. *PLoS ONE* 7: e37106. https://doi.org/10.1371/journal.pone.0037106

Gewin, V. 2003. Genetically modified corn—Environmental benefits and risks. *PLoS Biology* 1: e8. https://doi.org/10.1371/journal.pbio.0000008

Gilbert, N. 2010. GM crop escapes into the American wild. *Nature News* https://doi.org/10.1038/news.2010.393

Gottdenker, N.L., D.G. Streicker, C.L. Faust, et al. 2014. Anthropogenic land use change and infectious diseases: A review of the evidence. *EcoHealth* 11: 619632. https://doi.org/10.1007/s10393-014-0941-z

Goulson, D., E. Nicholls, C. Botías, et al. 2015. Bee declines driven by combined stress from parasites, pesticides, and lack of flowers. *Science* 347: 1255957. https://doi.org/10.1126/science.1255957

Haim, A., and B.A. Portnov. 2013. *Light Pollution as a New Risk Factor for Human Breast and Prostate Cancers* (Dordrecht: Springer). https://doi.org/10.1007/978-94-007-6220-6

Harding, W.R. 2015. Living with eutrophication in South Africa: A review of realities and challenges. *Transactions of the Royal Society of South Africa* 70: 155–71. https://doi.org/10.1080/0035919X.2015.1014878

Hargrove, J.W. 2003. *Tsetse eradication: Sufficiency, necessity and desirability* (Edinburgh: DFID Animal Health Programme, Centre for Tropical Veterinary Medicine, University of Edinburgh). https://assets.publishing.service.gov.uk/media/57a08d1be5274a31e0001656/RLAHtsetse_Erad.pdf

Hauenstein, S., M. Kshatriya, J. Blanc, et al. 2019. African elephant poaching rates correlate with local poverty, national corruption and global ivory price. *Nature Communications* 10: 2242. https://doi.org/10.1038/s41467-019-09993-2

Hayes, T.B., and K.P. Menendez. 1999. The effect of sex steroids on primary and secondary sex differentiation in the sexually dichromatic reedfrog (*Hyperolius argus*: Hyperolidae) from the Arabuko Sokoke Forest of Kenya. *General and Comparative Endocrinology* 115: 188–99. https://doi.org/10.1006/gcen.1999.7321

Hirschfeld M., D.C. Blackburn, T.M. Doherty-Bone, et al. 2016. Dramatic declines of montane frogs in a central African biodiversity hotspot. *PLoS ONE* 11: e0155129. https://doi.org/10.1371/journal.pone.0155129

Howarth, R.W. 2014. A bridge to nowhere: Methane emissions and the greenhouse gas footprint of natural gas. *Energy Science and Engineering* 2: 47–60. https://doi.org/10.1002/ese3.35

Hwang, Y.-T., D.M.W. Frierson, and S.M. Kang. 2013. Anthropogenic sulfate aerosol and the southward shift of tropical precipitation in the late 20th century. *Geophysical Research Letters* 40: 2845–50. https://doi.org/10.1002/grl.50502

IUCN. 2019. *The IUCN Red List of Threatened Species.* http://www.iucnredlist.org

Karagulian, F., C.A. Belis, C.C.C. Dora, et al. 2015. Contributions to cities' ambient particulate matter (PM): A systematic review of local source contributions at global level. *Atmospheric Environment* 120: 475–83. https://doi.org/10.1016/j.atmosenv.2015.08.087

Karami, A., A. Golieskardi, C.K. Choo, et al. 2017. The presence of microplastics in commercial salts from different countries. *Scientific Reports* 7: 46173. https://doi.org/10.1038/srep46173

Kebede, A.T., and D.L. Coppock. 2015. Livestock-mediated dispersal of *Prosopis juliflora* imperils grasslands and the endangered Grevy's zebra in northeastern Ethiopia. *Rangeland Ecology and Management* 68: 402–07. https://doi.org/10.1016/j.rama.2015.07.002

Kern, J.M., and A.N. Radford. 2016. Anthropogenic noise disrupts use of vocal information about predation risk. *Environmental Pollution* 218: 988–95. http://doi.org/10.1016/j.envpol.2016.08.049

King J., J.A. Cambray, and N.D. Impson. 1998. Linked effects of dam-released floods and water temperature on spawning of the Clanwilliam yellowfish *Barbus capensis. Hydrobiologia* 384: 245–65. https://doi.org/10.1023/A:1003481524320

Kissui, B.M., and C. Packer. 2004. Top-down population regulation of a top predator: Lions in the Ngorogonro Crater. *Proceedings of the Royal Society B* 271: 1867–74. https://doi.org/10.1098/rspb.2004.2797

Knop, E., L. Zoller, R. Ryser, et al. 2017. Artificial light at night as a new threat to pollination. *Nature* 548: 206–09. https://doi.org/10.1038/nature23288

Koper, R.P., and S. Plön. 2012. The potential impacts of anthropogenic noise on marine animals and recommendations for research in South Africa. *EWT Research and Technical Paper* 1 (Johannesburg: EWT).

Kosuth, M., S.A. Mason, and C. Tyree. 2017. *Synthetic Polymer Contamination in Global Drinking Water* (Washington: Orb Media). https://orbmedia.org/stories/Invisibles_plastics

Kouassi, K.S., S. Billet, G. Garçon, et al. 2010. Oxidative damage induced in A549 cells by physically and chemically characterized air particulate matter (PM2. 5) collected in Abidjan, Côte d'Ivoire. *Journal of Applied Toxicology* 30: 310–20. https://doi.org/10.1002/jat.1496

Kunc, H.P., K.E. McLaughlin, and R. Schmidt. 2016. Aquatic noise pollution: Implications for individuals, populations, and ecosystems. *Proceedings of the Royal Society B* 283: 20160839. https://doi.org/10.1098/rspb.2016.0839

Landrigan, P.J., R. Fuller, N.J.R. Acosta, et al. 2018. The Lancet Commission on pollution and health. *Lancet* 391: P462–P512. https://doi.org/10.1016/S0140-6736(17)32345-0

le Maitre, D.C., G.G. Forsyth, S. Dzikiti, et al. 2016. Estimates of the impacts of invasive alien plants on water flows in South Africa. *Water SA* 42: 659–72. http://dx.doi.org/10.4314/wsa.v42i4.17

Lebreton, L.C.M., J. van der Zwet, J.-W. Damsteeg, et al. 2017. River plastic emissions to the world's oceans. *Nature Communications* 8: 15611. https://doi.org/10.1038/ncomms15611

Lindsey P.A., V.R. Nyirenda, J.L. Barnes, et al. 2014. Underperformance of African protected area networks and the case for new conservation models: Insights from Zambia. *PLoS ONE* 9: e94109. https://doi.org/10.1371/journal.pone.0094109

Lindsey, P.A., G. Balme, M. Becker, et al. 2013. The bushmeat trade in African savannas: Impacts, drivers, and possible solutions. *Biological Conservation* 160: 80–96. https://doi.org/10.1016/j.biocon.2012.12.020

Lovett, B., E. Bilgo, S.A. Millogo, et al., 2019. Transgenic *Metarhizium* rapidly kills mosquitoes in a malaria-endemic region of Burkina Faso. *Science* 364: 894897. https://doi.org/10.1126/science.aaw8737

Maisels, F., S. Strindberg, S. Blake, et al. 2013. Devastating decline of forest elephants in Central Africa. *PLoS ONE* 8: e59469. https://doi.org/10.1371/journal.pone.0059469

Manaca, M.N., J.O. Grimalt, J. Sunyer, et al. 2011. Concentration of DDT compounds in breast milk from African women (Manhiça, Mozambique) at the early stages of domestic indoor spraying with this insecticide. *Chemosphere* 85: 307–14. https://doi.org/10.1016/j.chemosphere.2011.06.015

Maree, B.A., R.M. Wanless, T.P. Fairweather, et al. 2014. Significant reductions in mortality of threatened seabirds in a South African trawl fishery. *Animal Conservation* 17: 520–29. https://doi.org/10.1111/acv.12126

Maxwell, S.L., R.A. Fuller, T.M. Brooks, et al. 2016. The ravages of guns, nets and bulldozers. *Nature* 536: 143–45. https://doi.org/10.1038/536143a

Mcgowan, P. 2008. *WG6 CS1: African grey parrot* Psittacus erithacus *case study* (Cancun: NDF Workshop). https://www.cites.org/sites/default/files/ndf_material/WG6-CS1-S.pdf

McKenzie, L.M., R. Guo, R.Z. Witter, et al. 2014. Birth outcomes and maternal residential proximity to natural gas development in rural Colorado. *Environmental Health Perspectives* 122: 412–17. https://doi.org/10.1289/ehp.1306722

McKenzie, L.M., R.Z. Witter, L.S. Newman, et al. 2012. Human health risk assessment of air emissions from development of unconventional natural gas resources. *Science of the Total Environment* 424: 79–87. https://doi.org/10.1016/j.scitotenv.2012.02.018

McKinney, M.A., K. Dean, N.E. Hussey, et al. 2016. Global versus local causes and health implications of high mercury concentrations in sharks from the east coast of South Africa. *Science of the Total Environment* 541: 176–83. https://doi.org/10.1016/j.scitotenv.2015.09.074

McPherson, J.M., and A.C.J. Vincent. 2004. Assessing East African trade in seahorse species as a basis for conservation under international controls. *Aquatic Conservation* 14: 521–38. https://doi.org/10.1002/aqc.629

Merly, L., L. Lange, M. Meÿer, et al. 2019. Blood plasma levels of heavy metals and trace elements in white sharks (*Carcharodon carcharias*) and potential health consequences. *Marine Pollution Bulletin* 142: 85–92. https://doi.org/10.1016/j.marpolbul.2019.03.018

Mihaljevič, M., V. Ettler, O. Šebek, et al. 2011. Lead isotopic and metallic pollution record in tree rings from the Copperbelt mining-smelting area, Zambia. *Water, Air, and Soil Pollution* 216: 657–68. http://doi.org/10.1007/s11270-010-0560-4

Miller, G.T., and S. Spoolman 2011. *Sustaining the Earth* (Pacific Grove: Thompson Learning).

Minnaar, C., J.G. Boyles, I.A. Minnaar, et al. 2015. Stacking the odds: Light pollution may shift the balance in an ancient predator-prey arms race. *Journal of Applied Ecology* 52: 522–31. https://doi.org/10.1111/1365-2664.12381

Morell, M., A. Brownlow, B. McGovern, et al., 2017. Implementation of a method to visualize noise-induced hearing loss in mass stranded cetaceans. *Scientific Reports* 7: 41848. https://doi.org/10.1038/srep41848

Muir, W.M., and R.D. Howard. 2004. Characterization of environmental risk of genetically engineered (GE) organisms and their potential to control exotic invasive species. *Aquatic Sciences* 66: 414–20. https://doi.org/10.1007/s00027-004-0721-x

Nasi, R., A. Taber, and N. van Vliet. 2011. Empty forests, empty stomachs? Bushmeat and livelihoods in the Congo and Amazon Basins. *International Forestry Review* 13: 355–68. http://doi.org/10.1505/146554811798293872

Nyenje, P.M., J.W. Foppen, S. Uhlenbrook, et al. 2010. Eutrophication and nutrient release in urban areas of sub-Saharan Africa—a review. *Science of The Total Environment* 408: 447–55. https://doi.org/10.1016/j.scitotenv.2009.10.020

O'Hanlon, S.J., A. Rieux, R.A. Farrer, et al. 2018. Recent Asian origin of chytrid fungi causing global amphibian declines. *Science* 360: 621–27. https://doi.org/10.1126/science.aar1965

Osborn, S.G., A. Vengosh, N.R. Warner, et al. 2011. Methane contamination of drinking water accompanying gas-well drilling and hydraulic fracturing. *Proceedings of the National Academy of Sciences* 108: 8172–76. https://doi.org/10.1073/pnas.1100682108

Ouyang, J.Q., M. de Jong, R.H.A. van Grunsven, et al. 2017. Restless roosts: Light pollution affects behavior, sleep, and physiology in a free-living songbird. *Global Change Biology* 23: 4987–94. https://doi.org/10.1111/gcb.13756

Parks and Wildlife Service 2014. *Evaluation report August 2014: Macquarie Island pest eradication project* (Hobart: DPIPWE). https://www.parks.tas.gov.au/file.aspx?id=31160

Pauly, D., D. Belhabib, R., Blomeyer, et al., 2014. China's distant-water fisheries in the 21st century. *Fish and Fisheries* 15: 474–88. https://doi.org/10.1111/faf.12032

Petkova, E.P., D.W. Jack, N.H. Volavka-Close, et al. 2013. Particulate matter pollution in African cities. *Air Quality, Atmosphere and Health* 6: 603–14. https://doi.org/10.1007/s11869-013-0199-6

Pettis, J.S., E.M. Lichtenberg, M. Andree, et al. 2013. Crop pollination exposes honey bees to pesticides which alters their susceptibility to the gut pathogen *Nosema ceranae*. *PloS ONE* 8.7: e70182. https://doi.org/10.1371/journal.pone.0070182

Piaggio, A.J., G. Segelbacher, P.J. Seddon, et al. 2017. Is it time for synthetic biodiversity conservation? *Trends in Ecology and Evolution* 32: 97–107. https://doi.org/10.1016/j.tree.2016.10.016

Pirk, C.W.W., U. Strauss, A.A. Yusuf, et al. 2015. Honeybee health in Africa—a review. *Apidologie* 1–25. https://doi.org/10.1007/s13592-015-0406-6

Poulsen, J.R., C.J. Clark, G. Mavah, et al. 2009. Bushmeat supply and consumption in a tropical logging concession in northern Congo. *Conservation Biology* 23: 1597–608. https://doi.org/10.1111/j.1523-1739.2009.01251.x

Pringle, R.M. 2005. The origins of the Nile perch in Lake Victoria. *BioScience* 55: 780–87. https://doi.org/10.1641/0006-3568(2005)055[0780:TOOTNP]2.0.CO;2

Prüss-Ustün, A., J. Wolf, C. Corvalán, et al. 2016. *Preventing disease through healthy environments: A global assessment of the burden of disease from environmental risks* (Geneva: WHO). http://apps.who.int/iris/bitstream/10665/204585/1/9789241565196_eng.pdf

Quédraogo, O., and M. Amyot. 2013. Mercury, arsenic and selenium concentrations in water and fish from sub-Saharan semi-arid freshwater reservoirs (Burkina Faso). *Science of The Total Environment* 444: 243–54. https://doi.org/10.1016/j.scitotenv.2012.11.095

Rabinowitz, P.M., I.B. Slizovskiy, V. Lamers, et al. 2015. Proximity to natural gas wells and reported health status: Results of a household survey in Washington County, Pennsylvania. *Environmental Health Perspectives* 123: 21–26. https://doi.org/10.1289/ehp.1307732

Ramos, R., and D. Grémillet. 2013. Overfishing in West Africa by EU vessels. *Nature* 496: 300. https://doi.org/10.1038/496300a

Robinson, B.H. 2009. E-waste: An assessment of global production and environmental impacts. *Science of The Total Environment* 408: 183–91. https://doi.org/10.1016/j.scitotenv.2009.09.044

Roca, A.L., Y. Ishida, A.L. Brandt, et al. 2015. Elephant natural history; a genomic perspective. *Annual Review of Animal BioScience* 3: 139–67. https://doi.org/10.1146/annurev-animal-022114-110838

Roche, H., and A. Tidou. 2009. First ecotoxicological assessment assay in a hydroelectric reservoir: The Lake Taabo (Côte d'Ivoire). *Bulletin of Environmental Contamination and Toxicology* 82: 322–26. https://doi.org/10.1007/s00128-008-9572-9

Roelf, W. 2016. South Africa to start shale gas exploration in next year. *Reuters*. http://reut.rs/1U1RNZQ

Rogan, M.S., P.A. Lindsey, C.J. Tambling, et al. 2017. Illegal bushmeat hunters compete with predators and threaten wild herbivore populations in a global tourism hotspot. *Biological Conservation* 210: 233–42. https://doi.org/10.1016/j.biocon.2017.04.020

Rosin, C., and J.R. Poulsen. 2016. Hunting-induced defaunation drives increased seed predation and decreased seedling establishment of commercially important tree species in an Afrotropical forest. *Forest Ecology and Management* 382: 206–13. https://doi.org/10.1016/j.foreco.2016.10.016

Ross, P.S., and L.S. Birnbaum. 2003. Integrated human and ecological risk assessment: A case study of persistent organic pollutants (POPs) in humans and wildlife. *Human and Ecological Risk Assessment* 9: 303–24. https://doi.org/10.1080/727073292

Rybnikova, N.A., A. Haim, and B.A. Portnov. 2016. Does artificial light-at-night exposure contribute to the worldwide obesity pandemic? *International Journal of Obesity* 40: 815–23. https://doi.org/10.1038/ijo.2015.255

Schmidt, C., T. Krauth, and S. Wagner. 2017. Export of plastic debris by rivers into the sea. *Environmental Science and Technology* 51: 12246–53. https://doi.org/10.1021/acs.est.7b02368

Schneeberger, K., and C.C. Voigt. 2016. Zoonotic viruses and conservation of bats. In: *Bats in the Anthropocene: Conservation of Bats in a Changing World*, ed. by C.C. Voigt and T. Kingston (Cham: Springer). https://doi.org/10.1007/978-3-319-25220-9

Schreuder, E., and S. Clusella-Trullas. 2016. Exotic trees modify the thermal landscape and food resources for lizard communities. *Oecologia* 182: 1213–25. https://doi.org/10.1007/s00442-016-3726-y

Shannon, G., M.F. McKenna, L.M. Angeloni, et al. 2015. A synthesis of two decades of research documenting the effects of noise on wildlife. *Biological Reviews* 91: 982–1005. https://doi.org/10.1111/brv.12207

Sitoki, L., R. Kurmayer, and E. Rott. 2012. Spatial variation of phytoplankton composition, biovolume, and resulting microcystin concentrations in the Nyanza Gulf (Lake Victoria, Kenya). *Hydrobiologia* 691: 109–22. https://doi.org/10.1007/s10750-012-1062-8

Spinage, C.A. 1973. A review of ivory exploitation trends in Africa. *East African Wildlife Journal* 11: 281–89. https://doi.org/10.1111/j.1365-2028.1973.tb00093.x

Stauffer, J.R., H. Madsen, K. McKaye, et al. 2006. Schistosomiasis in Lake Malawi: Relationship of fish and intermediate host density to prevalence of human infection. *EcoHealth* 3: 22–27. https://doi.org/10.1007/s10393-005-0007-3

Strauss, U, V. Dietemann, H. Human, et al. 2016. Resistance rather than tolerance explains survival of savannah honeybees (*Apis mellifera scutellata*) to infestation by the parasitic mite, *Varroa destructor*. *Parasitology* 143: 374–87. https://doi.org/10.1017/S0031182015001754

Street, R.A. 2012. Heavy metals in medicinal plant products—an African perspective. *South African Journal of Botany* 82: 67–74. https://doi.org/10.1016/j.sajb.2012.07.013

Sussarellu, R., M. Suquet, Y. Thomas, et al. 2016. Oyster reproduction is affected by exposure to polystyrene microplastics. *Proceedings of the National Academy of Sciences* 113: 2430–35. https://doi.org/10.1073/pnas.1519019113

Tarrant, J, D. Cilliers, L.H. du Preez, et al. 2013. Spatial assessment of amphibian chytrid fungus (*Batrachochytrium dendrobatidis*) in South Africa confirms endemic and widespread Infection. *PLoS ONE* 8: e69591. https://doi.org/10.1371/journal.pone.0069591

Thibault, M., and S. Blaney. 2003. The oil industry as an underlying factor in the bushmeat crisis in Central Africa. *Conservation Biology* 17: 1807–13. https://doi.org/10.1111/j.1523-1739.2003.00159.x

Tuakuila, J. 2013. S-phenylmercapturic acid (S-PMA) levels in urine as an indicator of exposure to benzene in the Kinshasa population. *International Journal of Hygiene and Environmental Health* 216: 494–98. https://doi.org/10.1016/j.ijheh.2013.03.012

UNDP. 2006. *Niger Delta human development repor*t (Abuja: UNDP). http://hdr.undp.org/sites/default/files/nigeria_hdr_report.pdf

UNEP-WCMC. 2007. *Wildlife Trade 2005: An analysis of the European community and candidate countries' annual reports to CITES* (Cambridge: UNEP-WCMC). http://ec.europa.eu/environment/cites/pdf/2007_yearbook.pdf

UNEP-WCMC. 2008. *Monitoring of international trade in ornamental fish* (Cambridge: UNEP-WCMC). http://ec.europa.eu/environment/cites/pdf/reports/ornamental_fish.pdf

UNEP. 2006. Report on atmosphere and air pollution. In: *African Regional Implementation Review for the 14th Session of the Commission on Sustainable Developmen*t (New York: UNEP). https://sustainabledevelopment.un.org/content/documents/ecaRIM_bp2.pdf

Uyi, O.O., F. Ekhator, C.E. Ikuenobe, et al. 2014. *Chromolaena odorata* invasion in Nigeria: A case for coordinated biological control. *Management of Biological Invasions* 5: 377–93. http://doi.org/10.3391/mbi.2014.5.4.09

van der Merwe, P., Saayman, M., and Rossouw, R. 2015. The economic impact of hunting in the Limpopo Province. *Journal of Economic and Financial Sciences* 8: 223–42. https://hdl.handle.net/10520/EJC170564

Vanthomme, H., B. Bellé, and P.M. Forget. 2010. Bushmeat hunting alters recruitment of large-seeded plant species in Central Africa. *Biotropica* 42: 672–79. https://doi.org/10.1111/j.1744-7429.2010.00630.x

Villamagna, A.M., and B.R. Murphy. 2010. Ecological and socio-economic impacts of invasive water hyacinth (*Eichhornia crassipes*): A review. *Freshwater Biology* 55: 282–98. https://doi.org/10.1111/j.1365-2427.2009.02294.x

Wanless, R.M., and B.A. Maree. 2014. Problems and solutions for seabird bycatch in trawl fisheries. *Animal Conservation* 17: 534. https://doi.org/10.1111/acv.12183

Waters, C.N., J. Zalasiewicz, C. Summerhayes, et al. 2016. The Anthropocene is functionally and stratigraphically distinct from the Holocene. *Science* 351: aad2622. https://doi.org/10.1126/science.aad2622

Watkins, B.P., S.L. Petersen, and P.G. Ryan. 2008. Interaction between seabirds and deep-water hake trawl gear: An assessment of impacts in South African waters. *Animal Conservation* 11: 247–54. http://doi.org/10.1111/j.1469-1795.2008.00192.x

Weldon, C, L.H. Du Preez, A.D. Hyatt, et al. 2004. Origin of the amphibian chytrid fungus. *Emerging Infectious Diseases* 10: 2100–05. https://doi.org/10.3201/eid1012.030804

WHO (World Health Organisation). 2013. *Review of evidence on health aspects of air pollution—REVIHAAP project* (Bonn: WHO). http://www.euro.who.int/__data/assets/pdf_file/0004/193108/REVIHAAP-Final-technical-report.pdf

Wilcox, C., E. van Sebille, and B.D. Hardesty. 2015. Threat of plastic pollution to seabirds is global, pervasive, and increasing. *Proceedings of the National Academy of Sciences* 112: 11899–904. https://doi.org/10.1073/pnas.1502108112

Williams V.L., A.B. Cunningham, A.C. Kemp, et al. 2014. Risks to birds traded for African traditional medicine: A quantitative assessment. *PLoS ONE* 9: e105397. https://doi.org/10.1371/journal.pone.0105397

Williams, T.M., T.L. Kendall, B.P. Richter, et al. 2017. Swimming and diving energetics in dolphins: A stroke-by-stroke analysis for predicting the cost of flight responses in wild odontocetes. *Journal of Experimental Biology* 220: 1135–45. https://doi.org/10.1242/jeb.154245

Wittemyer, G., J.M. Northrup, J. Blanc, et al. 2014. Illegal killing for ivory drives global decline in African elephants. *Proceedings of the National Academy of Sciences* 111: 13117–21. https://doi.org/10.1073/pnas.1403984111

Wolfaardt, A.C., A.J. Williams, L.G. Underhill, et al. 2009. Review of the rescue, rehabilitation and restoration of oiled seabirds in South Africa, especially African penguins *Spheniscus demersus* and Cape gannets *Morus capensis*, 1983–2005. *African Journal of Marine Science* 31: 31–54. https://doi.org/10.2989/AJMS.2009.31.1.3.774

Wood, K.L., B. Tenger, N.V. Morf, et al. 2014. *Report to CITES: CITES-listed species at risk from the illegal trafficking of bushmeat; Results of a 2012 study in Switzerland's international airports* (Bonn CITES). https://doi.org/10.5167/uzh-111850

Worm, B., E.B. Barbier, N. Beaumont, et al. 2006. Impacts of biodiversity loss on ocean ecosystem services. *Science* 314: 787–90. https://doi.org/10.1126/science.1132294

Zunckel, M., K. Venjonoka, J.J. Pienaar, et al. 2004. Surface ozone over southern Africa: Synthesis of monitoring results during the Cross-border Air Pollution Impact Assessment Project. *Atmospheric Environment* 38: 6139–47. http://doi.org/10.1016/j.atmosenv.2004.07.029

8. Extinction Is Forever

Extinctions are seldom attributable to only one threat; rather, multiple stressors may act synergistically to drive the demise of a species. Pictured here is a Hewitt's ghost frog (*Heleophryne hewetti*, EN), which is globally restricted to an area of 140 km^2 in the Cape Floristic Region. It is threatened by alien vegetation, overly frequent fires, erosion, siltation, and construction of roads and reservoirs, all of which deteriorate its clear, fast-flowing stream habitat. Photograph by Werner Conradie, CC BY 4.0.

https://doi.org/10.11647/OBP.0177.08

Species have evolved and disappeared since the very first species (thought to be microorganisms living in hydrothermal vents) made an appearance on Earth. Some species outcompeted others for access to limiting resources; some were driven to extinction by dangerous pathogens; some just found it hard to survive in constantly evolving ecosystems. While many extinction events have been rather limited in scope and, hence, caused only one or a few extinctions at a time, there have been instances where perturbations were so impactful that they drove very large numbers of species to extinction over a short period of time. There have been five such past **mass extinction events**—periods marked by the sudden and dramatic loss of a large percentage of species (Figure 8.1). But these mass extinctions have also been followed by periods that favoured increased rates of speciation, during which new species evolved to fill the niches left empty by the extinctions.

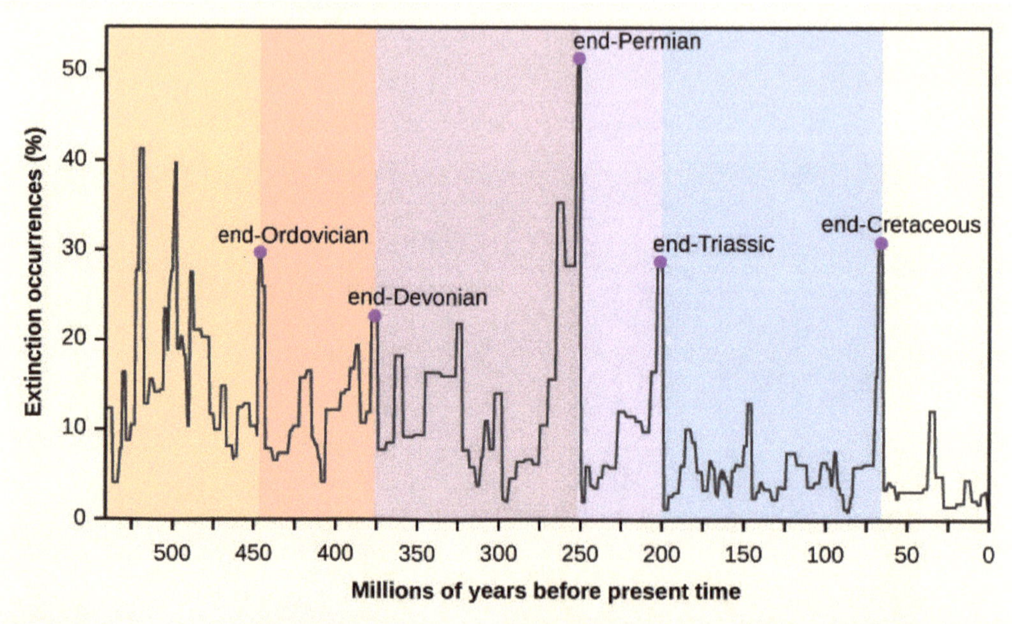

Figure 8.1 There have been five past mass extinction events—periods when natural events changed Earth's environment so dramatically that between 60–95% of species were wiped away forever—over Earth's geological history. So far, the most dramatic extinction event occurred at the end of the Permian period, about 250 million years ago, and thought to be the result of widespread volcanic activity and climate change. The most recent mass extinction, at the end of the Cretaceous period about 65 million years ago and thought to be the result of a massive asteroid impact, saw the disappearance of non-avian dinosaurs. Source: OpenStax, 2019, CC BY 4.0.

Nature's ability to balance extinctions with speciation was greatly disturbed around 300,000 years ago, when *Homo sapiens* made their appearance on Earth. Since then, humans have gradually increased their dominance on the natural world, leading to large-scale restructuring and destruction of biological communities. Human modifications of Earth's climatic, biological, and geochemical environments accelerated greatly during the rise of agriculture (12,000–15,000 years ago) and again during the **Industrial Revolution** (1760–1840), when fossil fuel usage and urbanisation became the norm. Now, many scientists recognise today's new and distinct human-dominated **geological epoch**, the **Anthropocene** (Waters et al., 2015). One notable feature of the Anthropocene is that species extinctions are increasing at such rapid rates that many conservation biologists now recognise that we are also witnessing the beginnings of Earth's **sixth extinction episode** (Barnosky et al., 2011; Ceballos et al., 2017). However, unlike previously, this extinction episode is caused by human activities rather than natural events.

8.1 What is Extinction?

The term "extinct" has several nuances in conservation biology, and its meaning can vary somewhat depending on the context:

- A species is **globally extinct** when no individuals of that species remains alive anywhere in the world. The bluebuck (*Hippotragus leucophaeus*, EX) has been globally *Extinct* since the last individual was shot around 1800 (Kerley et al., 2009).

> Over ninety-nine percent of recent extinctions have been caused by human activities.

- Four (possibly seven) species of cycad (*Encephalartos* spp.)—ancient seed plants that were dominant in the age of the dinosaurs—are currently considered **extinct in the wild**; in other words, they exist only in cultivation; in captivity; or another human-managed situation (IUCN, 2019).
- A species is **locally extinct**, also called **extirpated**, when it is extinct in a part of its historic range but can still be found elsewhere in the world. Cheetahs (*Acinonyx jubatus*, VU) once roamed throughout much of Africa, but are now extirpated in over 90% of their historical range (Durant et al., 2017).
- A species is **ecologically extinct** (also called **functionally extinct**) if it persists at such low numbers that its role in an ecosystem is negligible. Africa's vultures are ecologically extinct over much of their range and thus unable to remove diseased carcasses from the environment, posing both an ecological and socio-economic hazard (see Box 4.4).

8.2 Rates of Extinction

If extinction and speciation are natural processes, an obvious question follows: "Why should we care about the loss of biodiversity?" The answer concerns not individual species extinctions as much as the *increasing rate* of these extinctions (Figure 8.2). While a species can be wiped off Earth over a relatively short period of time, speciation typically occurs slowly as the genetic makeup of a population shifts over thousands of years. Unfortunately, we are currently losing species 1,000 times faster than natural **background extinction rates** (for mammals estimated to be 1.8 extinctions per 10,000 species per 100 years, Barnosky et al., 2011), and future rates may be 10,000 times higher that background rates (de Vos et al., 2015). Because over 99% of current species extinctions have been linked to human activity rather than natural processes (Pimm et al., 2014), observations on past extinctions and subsequent speciation may not apply to the present. Moreover, unlike before, humans now share the planet with the species we are wiping out. These losses mean that we are also losing the benefits we gain from nature (Chapter 4) at unprecedented rates.

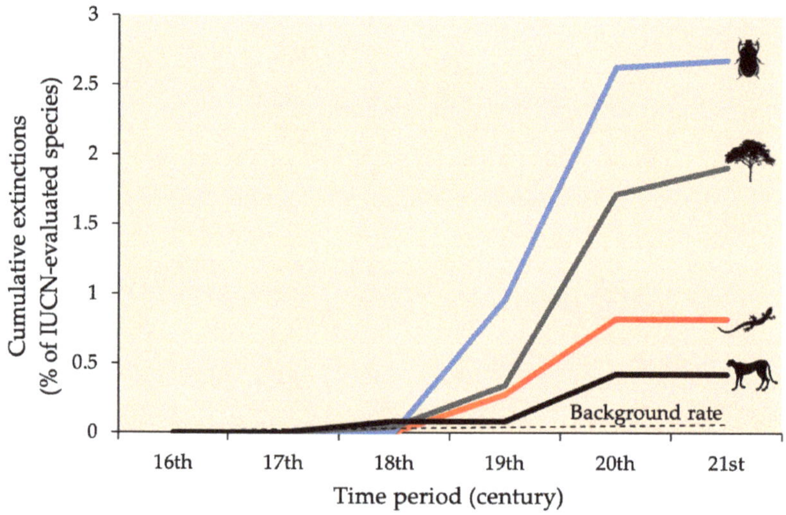

Figure 8.2 Percentage of Sub-Saharan Africa invertebrates, plants, reptiles, and mammals that have gone *Extinct*, *Extinct in the Wild*, and likely *Extinct* since the year 1500. Dashed line represents the natural rate of extinctions expected without human influences. After Ceballos et al., 2015, CC BY 4.0.

8.3 When is a Species Extinct?

While the term "extinction" is relatively easy to define (Section 8.1), determining whether a species is indeed extinct is a more difficult task. One of the most important questions conservation biologists grapple with is deciding how long to wait after the *last* observation before declaring a species extinct. Answering this question is particularly complicated when considering cryptic and shy species that are difficult to survey, sparsely distributed animals that are hard to find, or plants that are difficult to identify when not in flower.

To complicate matters, over the last few decades, biologists and their colleagues have rediscovered several species that were once thought to be extinct. These rediscovered species are often called **Lazarus species**, in reference to their apparent return to life. Recent examples include Burundi's Bururi long-fingered frog (*Cardioglossa cyaneospila*. NT) rediscovered after a 60-year absence (Blackburn et al., 2016), a Tanzanian coral tree, (*Erythrina schliebenii*, CR) originally known from only one specimen collected from a deforested region in the 1930s (Clarke et al., 2011), and the coelacanth (*Latimeria chalumnae*, CR), a fish that was once thought to be extinct for millions of years (Balon et al., 1988). To avoid declaring more extant species as extinct, there is currently a practice of only declaring a species extinct after several decades of intensive searching and "there is no reasonable doubt that the last individual has died" (IUCN, 2012). Consequently, species, such as the black-spotted false shieldback katydid (*Aroegas nigroornatus*, CR) and the Ethiopian sedge, *Cyperus chionocephalus*, *Critically Endangered*, last seen in 1916 (Bazelet and Naskrecki, 2014) and 1836 (Contu, 2013) respectively, have not yet been declared extinct, even though the last individual may have died a long time ago. Similarly, as many as 15 African orchid species—a group that includes some of the most beautiful and specialised plants on Earth, some of which have not seen since 1890—are currently considered *Critically Endangered* but may actually be extinct (IUCN, 2019).

The rediscovery of species once thought to be extinct should not necessarily be considered a sign of conservation progress. In many cases, Lazarus species were simply overlooked because they were extremely rare and restricted to isolated locations. Such is the case for two forest birds from the island nation of São Tomé and Príncipe, namely the São Tomé grosbeak (*Crithagra concolor*, CR) and Newton's fiscal (*Lanius newtoni*, CR). The grosbeak, the world's largest canary (50% larger than the second largest canary), was for a long time known only from three specimens collected in 1888–1890; it was thus considered extinct, until its rediscovery over 100 years later, in 1991 (BirdLife International, 2018a). The fiscal shares a remarkably similar history: it was previously known only from records in 1888 and 1928, until its rediscovery in 1990 (BirdLife International, 2018b). Despite these rediscoveries, both species persist as very small (< 250 individuals) populations that are at risk from extinction due to ongoing habitat loss and the impact of invasive predators.

Because extinctions may not always happen immediately after a disturbance, conservation biologists must also consider the lag time between destructive human activities and eventual extinctions. This is illustrated in a study from Kenya's Kakamega Forest, which found that only half of the species that will eventually go extinct due to habitat loss do so in the first 50 years following habitat fragmentation (Figure 8.3). Long-lived plants can have particularly long extinction lag times, sometimes of several centuries. For example, populations of the Saint Helena olive (*Nesiota elliptica*, EX) fell below viable levels in the mid-1800s, but the last individual died only in 2003, when the species was officially declared extinct (Cronk, 2016). Species that are doomed to eventual extinction are considered **committed to extinction** (also called functionally extinct), while the total number of species committed to extinction is referred to as an area's **extinction debt**. In one study, researchers used the island biogeography theory to estimate that the average extinction debt for African forest primates was over 30%—that is, more than 30% of forest primates are predicted to go extinct because of habitat destruction and other human activities that have already happened (Cowlishaw, 1999).

On a more positive note, extinction debts may also provide hope for conservation biologists, as the lingering presence of seriously imperilled species affords

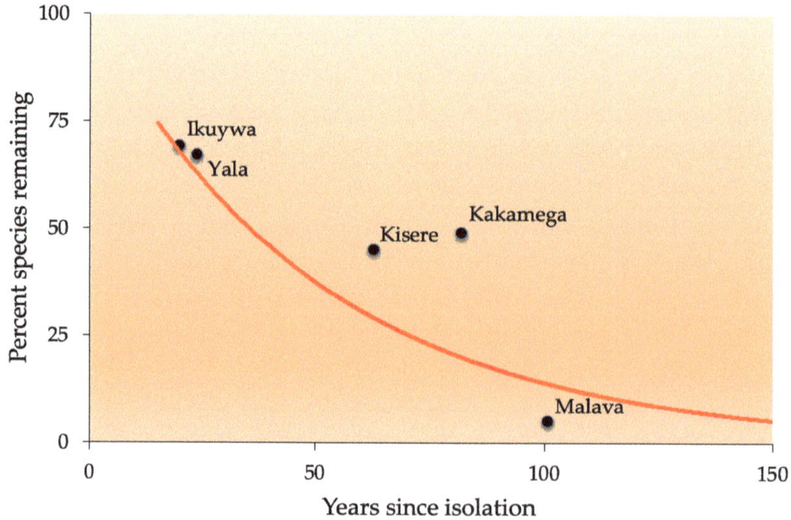

Figure 8.3 Percentage of bird species expected to persist over time in isolated forest patches in western Kenya. Because of extinction debt, not all species are expected to be extirpated immediately after fragmentation; instead there is a time lag between habitat loss and species losses. The image also illustrates how forest size and degree of isolation influences the speed of losses: Kakamega (the largest forest) loses species much slower than Malava, the smallest and most isolated forest. After Brooks et al., 1999, CC BY 4.0.

opportunities to prevent impending extinctions. Conservation biologists are currently illustrating how this can be done by preventing the extinction of three species of pale-coated, desert-adapted ungulates that were formerly common and widespread across the Sahel-Sahara region, namely the scimitar-horned oryx (*Oryx dammah*, EW), dama gazelle (*Nanger dama*, CR), and addax (*Addax nasomaculatus*, CR) (Durant et al., 2014; Brito et al., 2018; IUCN, 2019). The oryx once numbered around one million individuals, with herds of 10,000 seen as recently as 1936. But a population collapse soon followed: by 1985 only 500 oryx survived, and by 2000 it was declared *Extinct in the Wild*. The addax, relatively common as recently as in the 1970s, also experienced precipitous declines; today fewer than 30 individuals remain in the wild. Similarly, the once-common dama gazelle's current global population numbers fewer than 250 individuals, fragmented among five subpopulations in Chad, Mali, and Niger. Conservationists noted initial declines already in the 1960s and 1970s, when wild individuals of all three species were caught to initiate captive breeding programmes. Luckily, all three species responded well to these programmes, and captive populations have grown so strong that reintroduction programmes (Section 11.2) have been initiated for the addax (in 1985, in Tunisia), dama gazelle (in 2015, in Morocco), and oryx (in 2016, in Chad). With several reintroductions seemingly successful, there is hope that viable populations of these iconic species may one day again roam free in their previous strongholds. This will only happen if we can reverse or mitigate the threats that causes their population collapses in the first place, namely uncontrolled and illegal hunting, as well as disturbances associated with agriculture, oil exploration, and inconsiderate drilling of wells for groundwater extraction.

8.4 History of Extinctions in Sub-Saharan Africa

Many people hold on to the romanticised belief that historical human societies lived in harmony with nature. Accumulated evidence however indicates that this is not true; early humans have caused extensive ecosystem changes and species extinctions since *Homo sapiens* appeared on Earth about 300,000 years ago. In fact, even before the arrival of humans, our ancestors had made a mark, by driving species to extinction as early as during the Pleistocene period, which started about 2.5 million years ago (Box 8.1). The impact of early humans was particularly devastating to the wildlife of North America, South America, and Australia, which saw the demise of nearly all their large (> 100 kg) mammals, most famously megaherbivores such as the mammoths (*Mammuthus* spp.). The Pleistocene extinctions were somewhat less devastating to wildlife in Africa, Europe, and Asia, possibly because large mammals on these continents evolved with human predators, allowing them to develop appropriate defence/escape mechanisms. Nevertheless, Africa's wildlife did not completely escape the Pleistocene extinctions, as increasingly-sophisticated human activities during that time ensured the demise of as many as 28 large mammal groups, which included Africa's sabre-toothed cats (Barbourofelidae), nearly all the elephant relatives (Proboscidae), as well as giant hartebeests (*Megalotragus* spp.), giant buffaloes (*Pelorovis* spp.), giant hyenas (*Pachycrocuta* spp.), and giant giraffes (*Sivatherium* spp.).

> Early humans have caused extensive ecosystem changes and species extinctions even before *Homo sapiens* appeared on Earth about 300,000 years ago.

Box 8.1 Pleistocene Extinctions: Climate Change, Hominin Predation, or Both?

David H.M. Cumming[1,2]

[1]*FitzPatrick Institute of African Ornithology, University of Cape Town, South Africa.*

[2]*Tropical Resource Ecology Programme, University of Zimbabwe, Harare, Zimbabwe.*

✉ *cummingdhm@gmail.com*

Many scientists consider the present rapid loss of biodiversity to be the start of the 6th mass extinction following five previous extinction episodes (see Figure 8.1), each of which led to large-scale restructuring of Earth's biodiversity. The 5th global mass extinction took place 65 million years ago (Ma) when a massive meteorite collided with the Earth, and resulted in the extinction of all non-avian dinosaurs, and much else besides. This 5th mass extinction event also marked the transition from the Cretaceous to the Tertiary epoch (65 to 2.5 Ma). The Tertiary epoch was followed by the Quaternary, which includes the Pleistocene period (2.5 million to ~ 12 thousand years ago) and more recently the Holocene period—marked by the development of agriculture, and the subsequent domination of the Earth's resources by *Homo sapiens*. The Pleistocene is known for a mini mass-extinction of sorts, which saw the demise of species such as mammoths and sabre-tooth cats. However, unlike previous comprehensive mass extinctions, the Pleistocene was characterised by the extinction of mostly large mammals and very large island birds.

Attempts by scientists to explain these Pleistocene extinctions have been characterised by two centuries of controversy over whether they were caused by climate change or by predatory hominins—the evolutionary line of primates that gave rise to modern humans. The four main hypotheses advanced to account for the loss of Pleistocene fauna are: (i) climate change with minimal if any hominin influence (e.g. Faith et al., 2018); (ii) climate change together with some hominin influence (e.g. Barnosky et al., 2004); (iii) selective hominin predation aided by climate change (e.g. Bartlett et al., 2015), and (iv) hominin predation helped by other large predators without the influence of climate change (e.g. Janzen, 1983, Ripple and Van Valkenburgh, 2010).

The very close relationship between the dispersal of hominins out of Africa, the timing of their arrival elsewhere in the world, and the subsequent extinction of large mammals and birds, provided the primary (if challenged) evidence for human agency in non-African Late- Pleistocene extinctions (e.g. Surovell et al., 2005; Johnson, 2009; Ripple and Van Valkenburgh, 2010). As Haynes (2018) has remarked, "…the proponents of climate change as the only cause of the Late Pleistocene extinctions have not clearly explained how or why so many of the extinct megafaunal genera had survived numerous earlier climate changes."

Similarly, Faith et al. (2018) have stated that the failure of Pleistocene megaherbivores to adapt to the emergence of C4 grasses was the primary driver of their extinction. However, this claim ignores evidence that many extinct herbivore genera and species previously survived changes in diet over time (Ripple and Van Valkenburgh, 2010), that the diets of particular species were known to vary with location (Ferranec, 2004), and that many large species, which are typically highly mobile generalists, would have had little trouble adapting their ranges and diets to changing climates.

Research and debate on Pleistocene extinctions have tended to focus on the demise of non-African large mammals in the Late Pleistocene, which coincided with the period when hominins (*Homo erectus* and later also *H. sapiens*) dispersed across the globe starting at about 2 Ma. In Africa, however, earliest hominins appeared some 7 Ma. It didn't take long for these early African hominins to develop the skills necessary to manipulate the environment to their advantage. Setting the pace were the Australopithecines, who used stone tools to butcher mammalian carcasses between 4–3 Ma. The Australopithecines and the rest of a diverse group of large predators were joined by *H. erectus* at the beginning of the Pleistocene, about 2 Ma. This new, qualitatively unique, hunter was able to hunt collaboratively in bands, and was anatomically adapted to throw projectiles forcefully and accurately at large prey (Lombardo and Deaner, 2018). A large brain also placed high nutritional and energetic demands on *H. erectus*, that could best be met by obtaining meat and bone marrow from proboscidians, the elephant relatives (Surovell et al., 2005; Boschian et al., 2019). Early African hominins were thus well adapted to hunt large prey and contribute to the demise of wildlife, particularly megaherbivores (those over 1,000 kg), through the Early and Middle Pleistocene (Figure 8.A). Other large carnivores at the time may very well have helped hominins drive many Pleistocene herbivores to extinction (Janzen, 1983, Ripple and Van Valkenburgh, 2010; Van Valkenburgh et al., 2016). But this combination may ironically also have led to demise of many of the large Pleistocene predators, through co-extinctions, after their main prey base disappeared (Werdelin and Lewis, 2013).

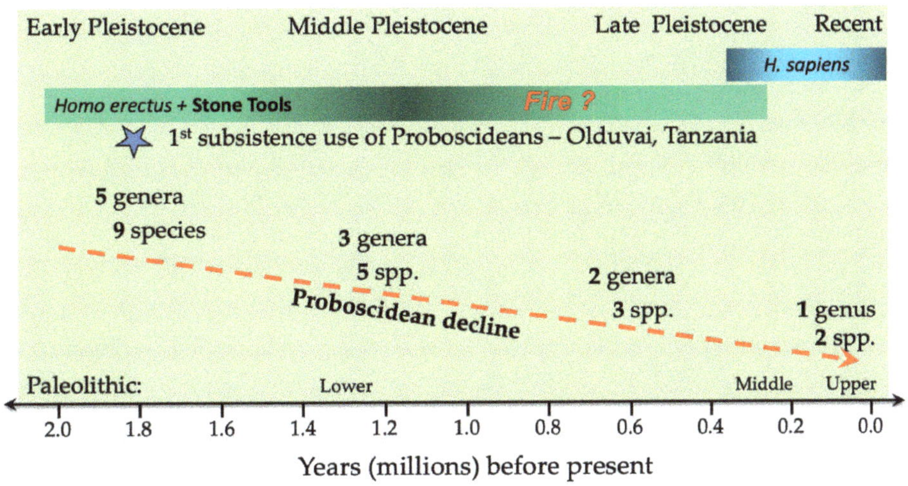

Figure 8.A Decline in African proboscidian (i.e. elephant relatives) diversity through the Early, Middle and Late Pleistocene in relation to the emergence of *Homo erectus* and *H. sapiens*. Similar patterns also occurred in the extinction of large carnivores and giant pigs/hogs. Source: Cumming, 2007, CC BY 4.0.

It thus seems likely that the emergence of a novel and increasingly effective predator during the Early Pleistocene, rather than climate change, was the ultimate factor that tipped the balance against the iconic species that disappeared soon after hominins appeared on Earth. It is worth noting that there is a clear relationship between body size and extinction risk (Figure 8.B), the result of large animals' relatively long generation lengths, long gestation periods, long periods of caring for young, and an abundance of meat presented to eager hunters. Consequently, even a small increase in mortality may very well result in a large animal's annual mortality exceeding its generational mortality, the end result being extinction. This relationship also partly explains why present-day elephant populations are unable to withstand poaching in many parts of Africa (Box 7.2).

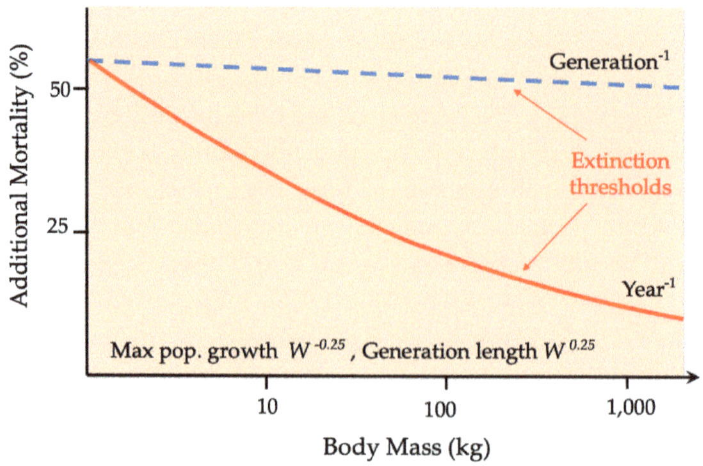

Figure 8.B The impact of additional mortality per year, as it relates to mammalian body mass. The graph shows how even slight increases in annual mortality can rapidly drive megaherbivore species to extinction. After Brook and Bowman, 2005, CC BY 4.0.

The lessons from the Pleistocene extinctions are relevant also today. As explained above, accumulated evidence indicates that early humans have caused extensive ecosystem changes and species extinctions extending over more than a million years. *Homo sapiens* had emerged in Africa about 300,000 years ago (Callaway, 2017). As early humans mastered the use of fire, poison-tipped spears and arrows, pitfall traps, and a host of additional hunting techniques, this unique apex predator proceeded to influence the structure and composition of African (and global) landscapes, and the plant and vertebrate assemblages of the continent (Smith et al., 2019). For the last two million years our ancestors have set in motion a series of trophic cascades that continue to this day and are resulting in increasing loss of diversity of the flora and fauna of the African continent and the rest of the world.

While early extinctions were generally isolated and selective, extinction rates increased rapidly after the rise of agriculture, and especially after European settlers started colonising Africa from the 17th century onward. By no coincidence, the area where most of the extinctions and extirpations during colonialism occurred was on the southwestern tip of Africa, the location of the earliest intensive European settlements on the continent. For example, by 1700, hunting caused the extirpation of every single land animal over 50 kg within 200 km of Cape Town (Rebelo, 1992). As hunters moved further afield in search of targets, Africa saw its first post-colonial large mammal extinctions, namely the bluebuck (Figure 8.4), quagga (*Equus quagga quagga*, EX), and Cape warthog (*Phacochoerus aethiopicus aethiopicus*, EX).

Figure 8.4 One of only four remaining skins of the bluebuck at the Vienna Museum of Natural History, Austria. Once a prized hunting target, it was the first known African antelope hunted to extinction. In the background is a quagga (*Equus quagga quagga*, EX), another African animal hunted to extinction. Photograph by Sandstein, https://en.wikipedia.org/wiki/Bluebuck#/media/File:Hippotragus_leucophaeus,_Naturhistorisches_Museum_Wien.jpg, CC BY 3.0

Following the demise of many of the Cape Floristic Province's large animals, humans have driven African species to extinction at an increasing pace. Today, at least 84 Sub-Saharan African species have been confirmed *Extinct* (Figure 8.5), nine species are *Extinct in the Wild*, and as many as 202 species are considered possibly *Extinct* (IUCN, 2019). Among the extinct species are two wildflowers (*Acalypha dikuluwensis*, EX; *Basananthe cupricola*, EX) wiped out by mining activities in the DRC; and from Seychelles, an endemic parakeet (*Psittacula wardi*, EX) that was hunted to extinction. Among the species that persist only in captivity is the Kihansi spray toad (*Nectophrynoides asperginis*, EW), whose population crashed from more than 20,000 individuals in June 2003 to only five individuals in January 2004 after the establishment of a hydropower plant in eastern Tanzania (Channing et al., 2006). While some species that are *Extinct in the Wild* may be released back into the wild at some point in future, the four (or possibly seven) cycad species that persist only in captivity will probably not be reintroduced due to ongoing concerns about poaching by plant collectors (Okubamichael et al., 2016).

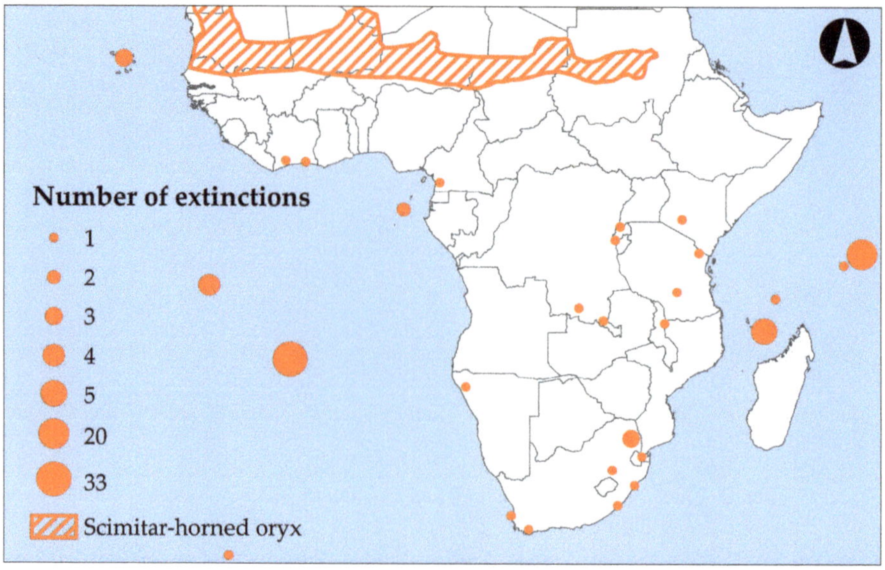

Figure 8.5 The locations of Sub-Saharan Africa's wildlife extinctions (including *Extinct* and *Extinct in the Wild* species) since 1500. Note how the largest number of extinctions involve species with restricted distributions, particularly those that occurred on islands. The scimitar-horned oryx is an example of mainland species that went extinct despite its large original range. Source: IUCN, 2019. Map by Johnny Wilson, CC BY 4.0.

While most of Africa's extinctions—at least until now—were isolated events involving one or two species at a time, the region also provides one of the best-studied examples of a recent man-made mass-extinction event. In the mid-1950s, the Uganda Game and Fisheries Department introduced the predatory Nile perch (*Lates niloticus*) to Lake Victoria to bolster the local fishery industry (Pringle, 2005). An ecological and economic disaster followed, pushing the entire ecosystem to the brink of collapse. First, the local people continued to prefer smaller endemic cichlids—which they could preserve by drying in the sun—to the perch with its oily flesh. This allowed the predatory perch's population to grow unchecked which, in turn, reduced over 500 endemic cichlid species' populations by 80% in just a few years (Witte et al., 1992). As the cichlid populations crashed, some local people started consuming perch for protein; however, they preferred smoking the perch over wood fires. To obtain firewood and charcoal, trees were logged around the lake, which in turn increased eutrophication, as well as erosion and **siltation**. Despite this array of emerging ecological threats, the local fishery continued to harvest the rapidly declining cichlid population. Consequently, as many as 200 cichlid species may have been driven to extinction in the decade following the perch introduction (Goldschmidt et al., 1993).

Judging by the number of extirpations over the last few decades, Africa will undoubtedly see more species pushed to extinction in the coming decades. Of particular concern is West and Southern Africa, which have lost over 75% of its large mammal populations over the past few decades; losses across Sub-Saharan Africa as a whole generally amount to over 50% (Ceballos et al., 2017). Some species will hopefully be spared this fate with the help of people and organisations fighting for their continued survival (Box 8.2). Many species will not be so lucky. The world's last western black rhinoceros (*Diceros bicornis longipes*, EX) died in Cameroon in 2011; the northern white rhinoceros (*Ceratotherium simum cottoni*, CR) may follow this fate within the next few years (see Box 11.4). Lions (*Panthera leo*, EN) have been extirpated from as many as 16 African nations (Bauer et al., 2015), while cheetahs (Figure 8.6) occur in less than 9% of their historic range (Durant et al., 2017).

Box 8.2 Swimming Dangerously Close to Extinction: Population Crash in Lesotho's Endemic Maloti Minnow

Jeremy Shelton

Freshwater Research Centre (FRCSA), Kommetji, South Africa.

✉ *jembejem@gmail.com*

Lesotho's iconic Maloti minnow (*Pseudobarbus quathlambae*, EN) (Figure 8.C) is a small, stream-dwelling cyprinid, and is the only freshwater fish species endemic to the country. Historically, the species was widespread, but its distribution has become increasingly restricted and fragmented in recent times due to interactions with non-native fishes and habitat degradation (Skelton et al., 2001), leading to it being classified as *Endangered* by the IUCN (Chakona and Kubheka, 2018). Genetic research has revealed that what was previously considered a single widespread species comprises two genetically distinct lineages: a "Mohale lineage" found in the Mohale catchment, and an "Eastern lineage" which includes populations in five catchments east of the Mohale catchment (Skelton et al., 2001). The genetic divergence between the two Maloti minnow lines is a result of a long period of geographic isolation and warrants that they be conserved as separate evolutionary significant units (ESU). Furthermore, the Mohale lineage, which comprises 77% of the species' known distribution, is of critical importance for continued survival of the species (Skelton et al., 2001).

Past surveys (e.g. Steyn et al., 1996) have revealed that the Maloti minnow was the only fish species inhabiting the rivers flowing into the Mohale Reservoir. Situated 4 km below the Reservoir, the 20 m high Semongkoaneng waterfall has historically prevented larger fish species from moving upstream into the upper catchment. Following the filling of the Mohale Reservoir in 2003, an inter-basin transfer (IBT) tunnel linking it to Katse Reservoir was opened. Biologists working in the catchment subsequently expressed concern that non-native fishes might colonise the Mohale Reservoir via the IBT tunnel, and from there invade the influent rivers (Rall and Sephaka, 2008). Because the Maloti minnow evolved in the absence of large-bodied fishes, it would not have had an opportunity to evolve

Figure 8.C The Maloti minnow, the only freshwater fish species endemic to the highlands of Lesotho, faces extinction due to habitat degradation and invasive species. Photograph by Craig Garrow, CC BY 4.0.

adaptations to cope with competition from and predation by larger species and may, therefore, be particularly sensitive to the arrival of other fish.

In 2006, the smallmouth yellowfish (*Labeobarbus aeneus*, LC), a larger, more aggressive cyprinid, was recorded in Mohale Reservoir (Rall and Sephaka, 2008), suggesting that it had dispersed from Katse Reservoir through the IBT tunnel. By 2013, it had spread into the major influent rivers in that system and coinciding with this was a virtual disappearance of the Maloti minnow from this former stronghold for the Mohale lineage. To illustrate this, surveys in previous decades described healthy populations of several thousand fish (e.g. Steyn et al., 1996), while only five individuals were recorded from the same sites in 2013 (Shelton et al., 2017).

Interestingly, the saving grace for this lineage may have originated from the same source that landed them in this predicament in the first place: human intervention. Prompted by the opening of the Kaste-Mohale IBT, a small team of passionate conservation scientists translocated several Maloti minnows to sections of stream above tall waterfalls, upstream of their natural distribution range (Rall and Sephaka, 2008). These sections, they knew, would be unreachable by larger species swimming upstream from Mohale Reservoir. This assisted colonisation approach has been viewed as controversial, but it may also have saved a tiny minnow from almost certain extinction in the wild. The prospect of losing a charismatic species like the Maloti minnow showcases how projects like the Lesotho Highlands Water Project can easily damage sensitive ecosystems that were not consider in development plans. In order to save the Maloti minnow from extinction, the next step will be to assess the success of translocation efforts and develop a rescue plan for the species.

Figure 8.6 A cheetah mother and her cubs in Tanzania's Tarangire National Park. Once found across much of Africa, cheetahs are now extirpated in 90% of their historic range. Photograph by Markus Lilje, CC BY 4.0.

8.5 Which Species are at Risk of Extinction?

An important task for conservation biologists is to identify and prioritise those species in greatest danger of extinction. Accomplishing this task requires biologists to collect and review all the information we have on each species. To facilitate this major undertaking, the IUCN has formalised the evaluation and reporting of threatened species assessments using an internationally accepted standard of conservation categories to reflect a taxon's risk of extinction. These nine categories (Figure 8.7), known as **Red List** Assessments (IUCN, 2017), are:

> The IUCN has formalised the evaluation of threatened species using an internationally accepted standard of conservation categories describing a taxon's risk of extinction.

- *Extinct* (*EX*). These species are no longer known to exist. As of mid-2019, the IUCN has listed 84 Sub-Saharan African species as *Extinct*.
- *Extinct in the Wild* (*EW*). These species exist only in cultivation, in captivity, or other human-managed situations. As of mid-2019, the IUCN has listed nine Sub-Saharan African species as *Extinct in the Wild*.
- *Critically Endangered* (*CR*). These species have an extremely high risk of going extinct in the wild. As of mid-2019, the IUCN has listed 880 Sub-Saharan African species as *Critically Endangered*. Also included in this category are the 202 Sub-Saharan African species that the IUCN considered *possibly Extinct*.
- *Endangered* (*EN*). These species have a very high risk of extinction in the wild. As of mid-2019, the IUCN has listed 1,600 Sub-Saharan African species as *Endangered*.

- *Vulnerable* (*VU*). These species have a moderately high risk of extinction in the wild. As of mid-2019, the IUCN has listed 2,153 Sub-Saharan African species as *Vulnerable*.
- *Near Threatened* (*NT*). These species are close to qualifying for a threatened category but are not currently considered threatened. As of mid-2019, the IUCN has listed 1,034 Sub-Saharan African species as *Near Threatened*.
- *Data Deficient* (*DD*). Inadequate information exists to determine the risk of extinction for these species. As of mid-2019, the IUCN has listed 2,441 Sub-Saharan African species as *Data Deficient*.
- *Least Concern* (*LC*). These species are not considered *Near Threatened* or threatened. (Widespread and abundant species are included in this category.) As of mid-2019, the IUCN has listed 11,776 Sub-Saharan African species as *Least Concern*.
- *Not Evaluated* (*NE*). Species that have not yet been evaluated. Most species fall in this category.

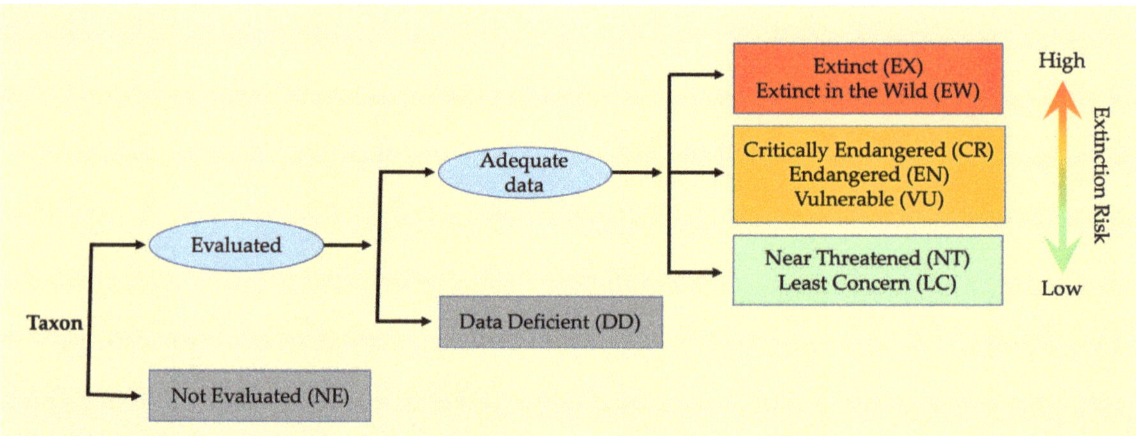

Figure 8.7 Flow diagram illustrating the structure of the IUCN categories of conservation status. An evaluated species can be considered at lower risk of extinction, at high risk of extinction (i.e. threatened), or extinct. A species for which not enough data are available for evaluation is considered *Data Deficient* (DD). After IUCN, 2017, CC BY 4.0.

These categories, and the **Red List Criteria** (Table 8.1) used to classify each species, are broadly based on population viability analysis (Section 9.2), and consider population size, population trends, and habitat availability. Species that are *Extinct in the Wild, Critically Endangered, Endangered,* and *Vulnerable* categories are officially considered "threatened with extinction". The advantage of this system is that it provides a standard protocol by which decisions can be reviewed and evaluated according to widely accepted yet flexible criteria. Consequently, species, subspecies, varieties, populations, and subpopulations can be assessed on a global or regional level, all under a unified set of standards. The resultant threat status assessment forms the basis of Red Data Books and Red Lists: detailed lists of threatened wildlife by group and/or by region compiled by the IUCN and its affiliate organisations. All global (and many regional) Red List assessments are freely available at http://www.iucnredlist.org, with feedback links provided from which anyone can alert the IUCN if they find errors or have suggestions for improvements.

Table 8.1 The IUCN's Red List criteria for evaluating a taxon's threat status. A species that meet any one of criteria A–E could be classified as *Critically Endangered.*

Red List criteria A–E	Summary criteria used to evaluate a taxon as *Critically Endangered*[a]
A. Population size declining	The population size has declined by 90% (or more) over last 10 years or 3 generations (whichever is longer).
B. Geographical range declining	The species is restricted to < 100 km^2 and it occurs at a single location *and* its distribution range is observed/expected to decline.
C. Small and declining populations	There are less than 250 mature individuals left *and* population has declined by 25% (or more) over last 3 years or 1 generation (whichever is longer).
D. Small populations	There are less than 50 mature individuals left.
E. Population viability analysis	There is a 50% (or greater) chance of extinction within 10 years or 3 generations.

[a] *Additional criteria for Critically Endangered, as well as criteria for Endangered and Vulnerable listings can be found at http://www.iucnredlist.org.*

While nearly 20,000 Sub-Saharan African species have been evaluated as of mid-2019 (IUCN, 2019), these assessments only cover a small proportion of the region's overall biodiversity. Consider for example that as of mid-2019, just over 4,900 Sub-Saharan Africa's plants have been listed on the IUCN Red List website. Yet, the Cape Floristic Region alone hosts over 6,200 endemic plant species. The assessment gaps are even more conspicuous for lesser-known taxa; for example, only seven species of bryophytes (a group of non-vascular plants that includes mosses) have been assessed as of mid-2019; some readers of this textbook will have more bryophyte species in their gardens. The reasons for such assessment gaps are many, but most boil down to manpower and funding limitations, which restrict our ability to obtain the data needed for comprehensive assessments. It is thus important to understand that the lack of information on these and other poorly known groups does not mean there is no threat. For example, as of mid-2019, no African abalone (*Haliotis* spp.) have been assessed, even though these highly valued molluscs are some of Africa's most heavily exploited (and heavily poached) marine organisms (Minnaar et al., 2018). A lack of information about a species' threats and populations trends is thus a good argument that more studies are needed, sometimes urgently. Similarly, continued monitoring of species thought to be common is also important, as it can shed light on how new threats may emerge or escalate over time.

8.5.1 Course-filter assessments

To fill Red List species assessment gaps, conservation biologists are increasingly relying on broader metrics, or **coarse-filter assessments**, to identify groups of species that are threatened with extinction. One such method, which reduces the need to evaluate every individual species, is to identify ecosystems that are threatened. This premise rests on the assumption that any threatened ecosystem will contain many threatened species. Hence, protecting and restoring threatened ecosystems will simultaneously allow many populations living in those ecosystems to recover. To facilitate this type of coarse-filter assessment, the IUCN recently established a Red List of Ecosystems (RLE, http://iucnrle.org). The RLE assesses ecosystem status against five criteria: (1) distribution declines, (2) distribution restrictions, (3) environmental degradation, (4) disruption of ecological processes and interactions, and (5) quantitative estimates for risk of ecosystem collapse (Keith et al., 2013). While the ecosystem assessment protocol was only recently developed—only three African ecosystems have been assessed as of mid-2019—its holistic strategy promises a more comprehensive accounting of local biodiversity which could be more informative than an accumulation of single species assessments.

8.6 Characteristics of Threatened Species

While a great number of factors may make a species vulnerable to extinction, conservation biologists have observed that species most vulnerable to extinction generally fall under one of six main groups:

- *Species with small populations*: Some species have very small populations, consisting of just a few individuals. Such small populations are highly vulnerable to random variations in demography or environmental conditions, and to the loss of genetic diversity—all factors that increase the risk of extinction (Section 8.7). Species whose population sizes naturally fluctuate between large and small populations also fall in this category, as they are at an increased risk of extinction during the small population phases of those fluctuations.

- *Species with declining populations*: Trends in population sizes tend to persist, so populations that are declining in abundance face a high risk of extinction (Caughley, 1994) unless conservation managers identify and address the causes of decline. Species impacted by the threats discussed in Chapters 5–7 generally also have declining populations.

- *Species with restricted distribution ranges*: Some species, such as those that are restricted to oceanic islands; mountains peaks; or isolated lakes, can be found only in a limited geographic range. A major disturbance, such as a cyclone/hurricane or drought, could easily affect that entire species' range, potentially driving the species to extinction.

- *Species with only one or a few populations*: A sufficiently large disturbance—such as a wildfire, storm, or disease outbreak—can wipe out a single population of a species. For a species with only one population, that means its extinction, while the loss of even a single population leaves species with only a few populations more vulnerable to the next disturbance. Species in this category (few populations) overlap with those in the previous category (restricted distribution ranges) because species with few populations tend to have restricted ranges.

- *Species that are exploited by people*: Overharvesting can easily reduce a population to the point of extinction (Section 7.2). Even if overharvesting is stopped just before the point of extinction, it may still have reduced a population to a size where it becomes susceptible to one or more of the three additional pressures faced by small populations (Section 8.7).

- *Species with critical symbiotic relationships*: Species that are members of obligate **symbiotic relationships** (where one species cannot survive without another) will go extinct if its host disappears. For instance, larvae of the rhinoceros stomach botfly (*Gyrostigma rhinocerontis*) mature in the stomach lining of African rhinoceros, and no other species (Barraclough, 2006). Thus, if the host species (the rhinoceros) were to go extinct, so would the botfly, Africa's largest fly species. This phenomenon in which one species' extinction leads to the extinction of other is called a coextinction (Koh et al., 2004), while a series of linked coextinctions is called an extinction cascade (Section 4.2.1).

The following characteristics are also linked with extinction, although the links are not as strong as is the case with the previous six categories:

- *Animal species with large body sizes*: Large animals generally require large ranges and more food, have lower rates of reproduction, and have smaller population sizes relative to smaller animals. Often, they are harvested by humans for material benefits (see Box 8.1). Consequently, within groups of related species, the largest are generally also the most vulnerable to extinction—that is, a larger species of carnivore, ungulate, or whale is more likely to go extinct than a smaller carnivore, ungulate, or whale.

- *Species that require a large home range*: Individuals or social groups of some species must forage over wide areas to fulfil their needs. When portions of their range are being degraded or fragmented, the remaining area will eventually be too small to support a viable population.

- *Species that are poor dispersers*: Moving to more suitable habitat is a common survival response following altered environmental conditions. But species with poor dispersal abilities may be doomed to extinction if they are unable to move to more suitable areas elsewhere (see e.g. discussion on range-shift gaps, Section 6.3.5).

- *Seasonal migrants*: A migratory species depends on intact ecosystems at two or more locations to complete its life cycle (see Box 5.3). If those ecosystems, either at stop-over sites along migration routes and/or at migratory endpoints, are damaged, the species may be at risk of extinction.
- *Species with low genetic diversity*: Because genetic diversity (Section 3.2) enables species to adapt to changing environmental conditions, species with low genetic diversity are more vulnerable to extinction because they have less ability to adapt to new diseases, new predators, or recent changes in their ecosystems.
- *Species that evolved in stable ecosystems*: Species that evolved in relatively stable environments (e.g. tropical ecosystems) are often threatened with extinction because under stabile conditions, a species is unlikely to retain the ability to adapt to environmental changes such as altered microclimates.
- *Species with specialised requirements*: Specialist species are often threatened with extinction because they are unable to adapt to altered ecosystems.
- *Group-living species*: A range of factors leaves group-living species at risk of extinction. For example, a herd of ungulates, a flock of birds at their night-time roost, or a school of fish can be harvested in its entirely by people using highly effective techniques. Even if some individuals remain, the harvesting may still leave the population below a critical threshold needed for effective foraging, mating, or territorial defence. This link between population size/density and individual fitness is termed the **Allee effect** (Section 8.7.2).
- *Species that have had no prior contact with people*: Species that encounter people for the first time are **ecologically naïve**—they lack avoidance strategies that promote survival during these encounters. Ecologically naïve species thus have a higher chance of extinction than species that have already survived human contact.
- *Species closely related to species that recently went extinct*: Groups of closely-related taxa, where some members are threatened or already extinct, often share characteristics that elevate their threat of extinction. Groups of related taxa that include many threatened species include apes, cranes, sea turtles, and cycads.
- *Species that live on islands*: Island species generally exhibit many of the characteristics mentioned above. In addition, the mere fact that an island is surrounded by ocean means that species that are unable to swim or fly have nowhere to go when they need to escape a threat.

8.7 Problems of Small Populations

While some small populations have persisted against the odds, sufficiently large populations are generally needed to prevent eventual extinction (Halley et al., 2016, see also Section 9.2). Small populations—which include species that have always had small populations and previously large populations that have been reduced to a few individuals—face three additional inherent and unavoidable pressures beyond the threats discussed in Chapters 5–7. These three additional pressures are: (1) loss of genetic diversity; (2) demographic stochasticity; and (3) environmental stochasticity and natural catastrophes. We will now examine how each of these pressures can lead a small population to eventual extinction. Much of this discussion is based on a groundbreaking manuscript by New Zealand ecologist Graeme Caughley, which discusses at length the threats faced by small and declining wildlife populations (Caughley, 1994).

8.7.1 Loss of genetic diversity

> Small populations are at risk of losing genetic variation much faster than large populations.

Species with high genetic diversity are generally more able to adapt to and reproduce under new conditions such as those brought by environmental changes (Section 3.2). These adaptations can occur at both individual and population levels. For example, under climate change, some genes may allow some populations to adapt their ranges faster or better tolerate warmer and wetter environments, while **phenotypic plasticity**—the ability of one gene to express itself differently under different conditions—may allow certain individuals to better adapt to a changing environment. One species that displays remarkable phenotypic plasticity is the crystalline iceplant (*Mesembryanthemum crystallinum*); by regulating its photosynthetic pathways, an individual plant can adjust its water needs based on the amount of salt and moisture available in the environment (Tallman et al., 1997). Such flexibility may explain why this species, native to southwestern Africa, North Africa, and Europe, has been a successful invader in environments as diverse as those in South America, North America, and Australia.

While populations with many individuals usually also have high levels of genetic diversity, small populations regularly suffer from low levels of genetic diversity. This low genetic diversity not only leaves those populations unable to adapt to changing conditions, but also makes them more susceptible to a variety of deleterious genetic effects (Caughley, 1994). Each of these effects leads to even greater loss of fitness and genetic diversity, hence even larger population declines, and eventually extinction. In the next sections, we discuss further why these deleterious genetic effects are so harmful to small populations.

Genetic drift

In wildlife populations, there are always some alleles that are relatively common, and others that are relatively rare. The relative abundance of any of these alleles may however change from one generation to another purely by chance. While common alleles generally tend to stay common, rare alleles have a high chance of being randomly lost in subsequent generations. Consider how each parent only passes on half of their genetic code to each offspring; this means that the ability of a rare allele to persist is dependent on how many individuals carry it, which individuals produce offspring, and how many offspring those individuals produce. Another important factor is population size (Figure 8.8): in any small population, only a limited number of individuals can carry any single allele, so the smaller the population, the higher the likelihood that alleles are lost to the next generation. This loss of alleles is called **genetic drift**.

While genetic drift equates to a loss of genetic diversity, there are some cases where populations show no obvious ill effects. Such may have been the case for female elephants in South Africa's Addo Elephant National Park. Hunting once nearly killed off this entire population; by the time they were adequately protected in 1931, only 11 animals remained, eight of which were female. Of those eight females, at least four were tuskless, while only two, maybe three, females carried both tusks. Over the next decades, Addo's female elephants have shown increasing degrees of tusklessness; by 2002, only 2% of females had tusks (by comparison, 96–98% of elephant females are normally expected to develop tusks, Maron, 2018). One can therefore postulate that the allele responsible for the tusk development in female elephants became rare, and that the progressive loss of tusked females is a sign of genetic drift (Whitehouse, 2002). While Addo's female elephants do not show any known limitations from being tuskless, the loss of alleles can also be devastating to the population suffering from genetic drift if, for example, the lost allele(s) coded for traits that would have allowed a species to adapt to a changing environmental condition.

It is important to note that genetic drift is distinct from natural selection. That is, genetic drift involves random changes in the frequency of alleles, whereas natural selection involves changes in traits in response to sexual selection or specific environmental conditions. For example, reduced tusk size in some heavily-hunted elephants in Africa (e.g. Chiyo et al., 2015) is a selective pressure in response to hunting that favour large tusks—this is distinct from Addo's female elephants that have lost their tusks even in the absence of selective hunting pressure.

> Mating among closely related individuals, which occurs in small populations, often results in lower reproductive success and weaker offspring.

Inbreeding depression

In large populations, a variety of instinctive mechanisms are in place to promote **heterosis**, which occur when offspring have a level of genetic variation that improves their individual evolutionary fitness. Some species are predisposed to disperse from their place of birth to prevent sibling–sibling or parent–offspring mating, while others are restrained from mating with close relatives through

Genetic Variation Worksheet

Small and isolated populations are expected to experience significant losses of genetic variation over time. To illustrate this point, consider an isolated heterozygous population that is not subjected to mutation. The proportion of original heterozygosity (*H*) that remains after one generation can be estimated using the formula (Wright 1931):

$$H = 1 - 1/[2N_e]$$

where (N_e) is effective population size. A population with $N_e = 2$ (two breeding individuals) would thus retain 75% of its original heterozygosity after one generation:

$$H = 1 - 1/4 = 1.00 - 0.25 = 0.75$$

But without a source for new genetic material, this population's heterozygosity will continue to decrease. After *t* generations (H_t):

$$H_t = H^t$$

The heterozygosity remaining in our population with two breeding individuals would thus be 56% after two generations (0.75 x 0.75), 24% after five generations, and 6% after ten generations. In contrast, a population with one hundred breeding individuals would retain 99.5% of its original heterozygosity after one generation, 98% after five generations, 95% after ten generations.

The migration of only a few individuals each generation or so can be enough to counter genetic drift in small populations. Mutation can theoretically also help, but natural mutation rates—estimated to be between 1 in 10,000 and 1 in 1,000,000 per gene per generation—are too rare to have an impact.

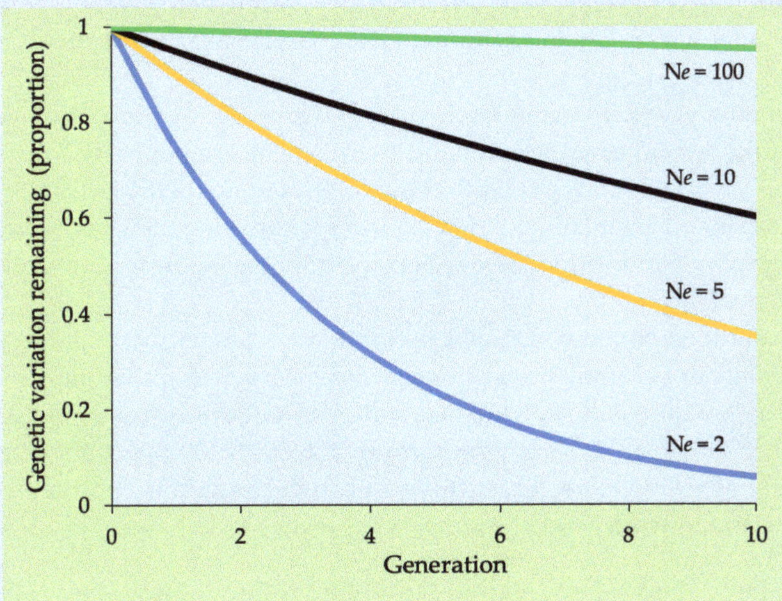

Figure 8.8 The amount of genetic diversity that is randomly lost over time due to genetic drift is highly dependent on a population's effective population size (*Ne*). A theoretical population with *Ne* = 2 may lose approximately 95% of its genetic diversity over 10 generations, while a population with *Ne* = 100 may lose only 5%. After Meffe and Carroll, 1997, CC BY 4.0.

sensory cues such as individual odours. Many plants have morphological and physiological traits that facilitate cross-pollination and reduce self-pollination.

However, in small populations with few unrelated mates, the urge to breed might be stronger than the mechanisms that promote heterosis. Under these conditions, rather than forgoing reproduction, breeding among closely-related individuals (or **inbreeding**) can occur. This breeding among close relatives might result in **inbreeding depression**, which can occur when closely-related parents give their offspring two copies of a deleterious allele. Individuals suffering from inbreeding depression typically have fewer offspring or have offspring that are weak or fail to reproduce. Such is the case for some mountain gorillas (*Gorilla beringei beringei*, EN): genetic studies have shown how birth defects in several small populations can be attributed to inbreeding depression (Xue et al., 2015). Inbreeding depression has also been identified as the reason why some small lion populations are more susceptible to diseases (Trinkel et al., 2011). Inbreeding depression can result in a vicious cycle for declining population sizes, where such declines can lead to even more inbreeding depression, and eventually extinction (see Section 8.7.4).

Outbreeding depression

Large populations have many ecological, behavioural, and physiological mechanisms that prevent hybridisation, the production of offspring among genetically distant taxa, whether they be individuals of different species, or individuals of the same species but with different adaptations (the latter being intraspecific hybridisation). As with inbreeding depression, these mechanisms may fail in small populations, leading to **outbreeding depression** (Frankham et al., 2011). Because offspring that result from outbreeding depression have traits that are intermediate to their parents, they may not be adapted to either of the parents' ecosystems. For example, one study found that plants suffering from outbreeding depression have weakened defences against herbivory (Leimu and Fischer, 2010). Outbreeding depression may also lead to a breakdown in physiological and biochemical compatibility between would-be parents—hybrid sterility is a well-known consequence of this breakdown. Consequently, species and populations suffering from outbreeding depression often show similar symptoms to inbreeding depression, including lower fitness, weakness, and high rates of mortality.

The opposite of outbreeding depression is hybrid vigour. Under these conditions, the hybrid offspring can be quite strong in an evolutionary sense; they may even outcompete their parent species. Such is the case with the South African endemic black wildebeest (*Connochaetes gnou*, LC); having recovered from near-extinction, poorly planned translocations are now threatening this species, which readily hybridises with the widespread common wildebeest (*Connochaetes taurinus*, LC) in areas of contact (Grobler et al., 2011).

Population bottlenecks

In some taxa, such as butterflies, annual plants, and amphibians, population size varies dramatically from generation to generation. During some years, populations can be so large that they appear to face little risk of extinction. However, abundant years can be misleading when followed by successive years of low abundance. Generally, in a population that undergoes extreme size fluctuations, the population size required to ensure continued persistence (i.e., the **minimum viable population** (MVP), Section 9.2) is in effect much nearer the lowest than the highest number of individuals in any given year. However, during years with low abundance, a phenomenon known as a **population bottleneck** may occur—that is, the small population size may lead to the loss of rare alleles from one generation to the next. Population bottlenecks may lead to more inbreeding depression which, in turn, reduces reproductive success (Heber and Briskie, 2010) and increases vulnerability to diseases (Dalton et al., 2016). Low genetic diversity in great white sharks (*Carcharodon carcharias*, VU) living in South Africa's Indian Ocean is thought to be the result of a population bottleneck (Andreotti et al., 2015).

New populations founded by only a few individuals are vulnerable to a special type of population bottleneck, the **founder effect**. The founding individuals of a new population by definition start off with low genetic diversity, much less than the original population that the founders left behind. This low genetic diversity puts the new population at risk of further genetic diversity declines, which have lasting effects through time. This situation can occur naturally when only a small number of individuals disperse to establish a new population or when founder individuals come from a small population that already suffered from low genetic diversity. Being mindful of these concerns is especially important for translocation (Section 11.2) or captive breeding (Section 11.5) projects. For example, to prevent extinction of the world's smallest gazelle, the Speke's gazelle (*Gazella spekei*, EN), a captive population of this species, almost entirely restricted to Somalia, was established in the USA. The founder population for this captive breeding project consisted of only one male and three females, leading to severe levels of inbreeding depression and high mortality rates in offspring (Kalinowski et al., 2000). Understanding the importance of managing for genetic diversity can help avoid these and other challenges that can threaten the success of translocation projects.

> Populations founded by only a few individuals by definition start off with low genetic diversity, having lasting effects in the population through time.

8.7.2 Demographic stochasticity

Demographic stochasticity (also known as demographic variation) refers to random variations in a population's demographic traits (e.g. sex ratios, birth rates, death rates), the cumulative effect of variation in individual organisms' fitness. In any natural population, some individuals will produce fewer offspring than average, while others will produce more than average; some individuals will produce no offspring at all. Similarly, some individuals die younger than average, while others live longer than average. For populations that are sufficiently large, average birth and death rates provide relatively stable descriptions of key aspects of that population's demography. However, when a population's size decreases to below a certain threshold, variations in fitness of a small number of individuals can have a large impact on the overall populations' demographic parameters, causing population size and other characters to fluctuate up or down unpredictably (Schleuning and Matthies, 2009). Consider, for example, an isolated population of crocodiles with only a few females. As with many other reptiles, offspring sex ratios of crocodiles are determined by the environmental temperature during incubation (Hutton 1987). If, by chance, the population experiences two years of high temperatures, which favour male offspring, and the few females die by chance, the all-male population may be doomed for extinction unless some female crocodiles immigrate from elsewhere.

Small population sizes or low densities can also disrupt social interactions among individuals—especially interactions that affect reproduction—which can cause populations to become demographically unstable. This situation, referred to as the Allee effect, can result in further declines in population size, population density, and population growth rate. Obligate **cooperative breeders,** such as African wild dogs (*Lycaon pictus*, EN), are especially vulnerable to the Allee effect (Courchamp et al., 2000) since they need a certain number of individuals to protect their territories and obtain enough food for their offspring (Figure 8.9). Allee effects might also prevent impact group-living species that are not cooperative breeders—recalling the "safety in numbers" mantra, Allee effects seem to prevent the recovery of locally-rare sable antelope (*Hippotragus niger*, LC) populations in South Africa's Kruger National Park, as reduced herd sizes increases their exposure to predation (Owen-Smith et al., 2012). But even solitary species that live at low densities are susceptible to Allee effects, since they may find it hard to locate mates once the population density drops below a certain level.

> The social systems of group-living animals can easily be disrupted when their population size or density falls below a critical level.

8.7.3 Environmental stochasticity and catastrophes

Environmental stochasticity, the unpredictable variation in environmental conditions, can cause dramatic population size fluctuations over time, and hence, substantially increase the risk of extinction. Consider, for example, how the development rate of many insects is strongly temperature-dependent (e.g. Rebaudo and Rabhi, 2018). In an average or warm year, young insects that hatch on time and feed well may result in ecologically fit adults that produce many young, whereas unusually cold years might reduce hatching success and larval activity, which could also reduce adult fitness

Figure 8.9 A pack of African wild dogs on a hunt in Madikwe Game Reserve, South Africa. Due to their vulnerability to Allee effects, African wild dog populations can only be sustained if packs are above a certain threshold that allows them to hunt, feed their young, and protect themselves effectively. Photograph by flowcomm, https://www.flickr.com/photos/flowcomm/13945572529, CC BY 2.0.

(Gibert et al., 2001). So, highly unfavourable conditions in any one year can cause dramatic population declines, or even push a species to extinction if conditions persist over successive years across its range.

> Even though a small population may appear to be stable or increasing, an environmental catastrophe can severely reduce population size or even cause extirpation or extinction.

The increased risk of extinction from environmental stochasticity also applies to natural catastrophes that can occur at unpredictable intervals (e.g. droughts, storms, earthquakes, and fires). Range-restricted species are particularly vulnerable to this kind of threat. For example, the biodiversity living in and around several African crater lakes are vulnerable to a rather unique natural phenomenon called "lake burping". Volcanic chambers underneath some of these lakes are rich in CO_2. Small amounts of CO_2 may sometimes (or constantly, in some cases) seep up through the lake bed into the surrounding water. Because these lakes are thermally stratified—layers of cold, dense water settle near the bottom while warm, less dense water floats near the top—the CO_2-saturated water remains near the bottom of the lake. However, when there is a geologic disturbance, such as a landslide or earthquake, massive amounts of CO_2 may suddenly be released, first saturating the warmer water at higher levels with CO_2 (killing fish and other oxygen-dependent species in the process), before displacing the breathable surface air in and around the lake. In 1986, one such CO_2 eruption killed 1,800 people and 3,500 heads of livestock near Cameroon's Lake Nyos (Krajick, 2003). Some scientists fear that increased deforestation (which may trigger erosion and landslides) and hydraulic fracturing (which may trigger earthquakes, Section 7.1.1) could trigger similar events at other crater lakes in the region.

Environmental stochasticity tends to increase the probability of extinction more than does demographic stochasticity. As discussed, this is especially true for small populations and range-restricted species.

8.7.4 The extinction vortex

As populations decline in size, they become increasingly vulnerable to the combined impacts from the loss of genetic diversity, inbreeding depression, Allee effects, environmental stochasticity, and demographic stochasticity. All these factors tend to lower reproduction, increase mortality rates, and reduce population size even more, in turn driving populations to extinction at increasingly faster rates over time (Fagan and Holmes, 2006). Conservationists sometimes compare this phenomenon to a vortex, spiralling inward, moving faster (or declining faster in the case of a population) as it gets closer to the centre. At the centre of this **extinction vortex** (Gilpin and Soulé, 1986) is oblivion—the extinction of the species (Figure 8.10).

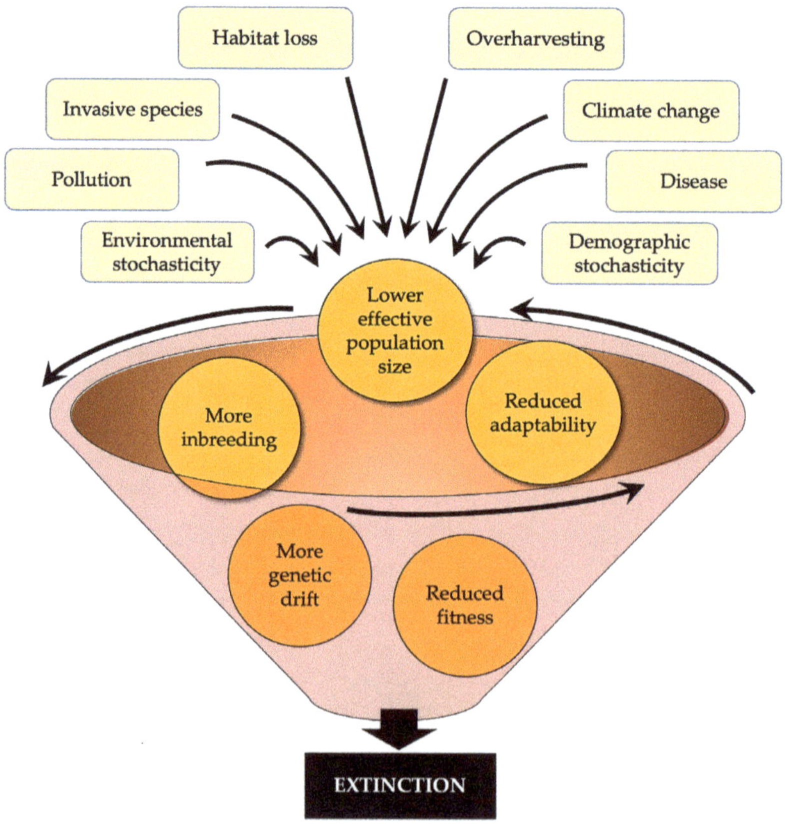

Figure 8.10 The extinction vortex describes a process whereby the factors that affect small populations can drive its size progressively downward towards extinction. CC BY 4.0.

The demise of the bluebuck—the first large mammal of Africa to face this fate after European colonisation—may have been the result of an extinction vortex. When European colonists first arrived in South Africa, this ungulate already persisted as a single, small population of an estimated 370 individuals (**effective population size** at 100 individuals) and a highly restricted (4,300km^2) distribution. Considering this small and restricted population's vulnerable to deleterious genetic factors and demographic stochasticity, a recent study showed that this species was probably caught in an extinction vortex by the time the first colonist shot the first bluebuck (Kerley et al., 2009). This species would thus likely have gone extinct even in the absence of hunting and habitat loss, which only hastened its departure.

8.7.5 Is there any hope for small populations?

Despite the odds and the many threats facing Africa's wildlife, many species that were once on the brink of extinction have clawed their way back from the abyss towards stable, and sometimes even growing populations. Prime examples include the Pemba flying fox (*Pteropus voeltzkowi*, VU); considered *Critically Endangered* in 1996, conservation education programmes raised awareness of this unique bat, which now has considered *Vulnerable*, having recovered to more than 28,000 individuals (Entwistle and Juma, 2016). Similarly, because of habitat destruction and introduced predators, the future of the Seychelles magpie-robin (*Copsychus sechellarum*, EN) looked rather bleak in 1970, when only 16 individuals remained, all on one island. Today, thanks to habitat restoration efforts, supplemental feeding, invasive species eradication, provisioning of nest boxes, and a translocation programme, there are more than 280 Seychelles magpie-robins scattered across five islands (Burt et al., 2016). Another remarkable conservation success story involves the rescue of the southern white rhinoceros (*Ceratotherium simum simum*, NT), which was reduced to about 20 individuals in a single protected area in the late 1880s. Dedicated conservation efforts since then have seen this iconic species recover to more than 20,000 individuals, with individuals introduced and reintroduced all over Africa and zoos throughout the world. None of these species would have been alive today if it wasn't for intensive multi-year efforts by dedicated conservation biologists to pull them out of their individual extinction vortices.

Bringing species with small populations back from the edge of extinction requires dedication, careful planning, and significant amounts of resources. It also requires careful population management to mitigate the negative impacts of founder effects and both demographic and environmental stochasticity (Box 8.4; see also Chapter 11). As these examples show, it can be done. But, given the challenges, it should always be a priority to prevent a species from declining to very low numbers in the first place.

> Bringing species with small populations back from the edge of extinction requires dedication, careful planning, and significant amounts of resources.

8.8 Is De-extinction a Solution?

One of the more interesting conservation debates to have emerged in recent years involve efforts to reverse extinction. This field, known as **de-extinction** or resurrection biology aims to revive extinct species, and eventually to reintroduce viable populations to their original locations (Seddon, 2017). One possible method, called "breeding back", aims to produce individuals genetically similar to an extinct species by selective

Box 8.3 Fenced Reserves Conserving Cheetahs and African Wild Dogs in South Africa

Kelly Marnewick[1,2]

[1]*Carnivore Conservation Programme, Endangered Wildlife Trust, Johannesburg, South Africa.*

[2]*Current address: Department of Nature Conservation, Tshwane University of Technology, Pretoria, South Africa.*

✉ *marnewickKA@tut.ac.za*

South Africa is one of the few countries in Africa where numbers of many large carnivore species are stable and, in some cases, increasing. Much of this success can be attributed to the managed metapopulation approach, which involves the reintroduction and subsequent translocation and management of populations in geographically isolated fenced reserves, between which natural dispersal is highly unlikely. As of 2016, more than 300 cheetahs are being managed in 51 reserves encompassing 10,995 km^2 (mean: 195 km^2 range: 20–1,000 km^2) and nearly 250 African wild dogs in 11 reserves encompassing 5,086 km^2 (mean: 216 km^2 range: 19–1,000 km^2). The reserves are situated across the country within a variety of land tenure systems including state and provincial protected areas and privately owned and community-run game reserves. Most reserves derive income primarily from ecotourism.

Each reserve forms part of the national network. Animals are moved between reserves to maintain the genetic integrity and demographic balance of individual subpopulations, but also to minimise direct management in the long term. Translocations are planned to mimic natural processes as far as possible but, due to the intricacies involved in managing animals between several reserves, this is not always possible. For wild dogs, small groups of unrelated adult males and females are artificially bonded to form packs, which mimics natural pack formation in the wild. For cheetahs, sub-adults are removed once they disperse from their maternal range. The animals are generally immobilised in the field and transported awake in crates on vehicles to their new reserves. Soft releases (Section 11.2.1) are preferred: these involve the animals being kept in temporary holding bomas of approximately 1 ha in size for about three months. The formation of artificial social groups is also done during this period. Intensive post-release monitoring is done at intervals reliant on reserve resources, but daily monitoring is recommended. The success rate of reintroductions has been high and, for wild dogs, has been strongly linked to the social cohesion of released groups (Marneweck et al., 2019), and the integrity of perimeter fences (Gusset et al., 2008).

This highly collaborative process involves multiple stakeholders, including conservation NGOs, provincial government conservation departments, private reserve owners and managers, researchers, local communities, and tourists. Effective and responsible population management tools help to prevent local populations growing too large or too small, and best practice guidelines ensure the ethical handling and management of animals. Individual reserves are responsible for providing infrastructure and other requirements including managing sustainable prey populations, perimeter fences, bomas and post release monitoring, as well as ensuring that a management plan is in place and adhered to. In many cases, students or volunteer organisations conduct post-release monitoring. National, high-level management is coordinated by the Endangered Wildlife Trust (EWT) and is funded through donations from corporations, individual philanthropists, conservation trusts, and foundations.

The managed metapopulation approach to carnivore conservation has increased the number and distribution of both cheetahs and African wild dogs in South Africa and built technical capacity in the country for metapopulation management (Davies-Mostert and Gusset, 2013), which has also been applied to species, such as lions, elephants, and black rhinoceros (*Diceros bicornis*, CR). Opportunities abound in other countries to use lessons learned in South Africa for the recolonisation of other areas where large mammals have been locally or regionally extirpated. Additionally, projected human population expansion, and the habitat fragmentation that comes with it, means that this approach is likely to become an indispensable tool in maintaining the viability of populations in disconnected landscapes.

breeding of extant species that carry genetic material of their extinct relatives. This is the main method currently being used to revive the aurochs (*Bos primigenius*, EX), the ancestor of today's domestic cattle (Stokstad, 2015). Other "breeding back" projects place less emphasis on genetics and more on morphology, by selectively breeding individuals with certain traits to produce individuals that visually appear similar to the extinct species. Such is the case at The Quagga Project, where selectively breeding of plains zebras (*Equus quagga*, NT) with quagga-like characteristics (reduced striping and brown hues) are resulting in animals (Figure 8.11) that look increasingly like extinct quaggas (Harley et al., 2009).

The second popular method used for de-extinction is **cloning**. This involves the transfer of viable genetic material from an extinct species to the eggs (or embryo) of a closely related surrogate mother, who will hopefully give birth to an individual of the extinct species. Cloning has been used in selective breeding of livestock for many years, and plans are also currently underway to use cloning to prevent the extinction of highly threatened species such as the northern white rhinoceros (see Box 11.4). Despite the promise that cloning offers for reviving extant and recently extinct species,

Figure 8.11 Not extinct anymore? A selective breeding programme in South Africa has raised several plain's zebras that closely resemble the extinct quagga. The progress from one generation to the next can be seen in this photo, with the adults showing reduced striping, and the foal showing brown hues. Photograph by The Quagga Project, CC BY 4.0.

cloning species that went extinct many years ago has been more challenging. So far, attempts to clone Spain's Pyrenean ibex (*Capra pyrenaica pyrenaica*, EX) and Australia's gastric-brooding frog (*Rheobatrachus silus*, EX) have produced individuals that lived for only a few minutes (Ogden, 2014).

Despite the progress made, de-extinction is one of the most controversial and polarising debates to emerge among conservation biologists in recent years. Proponents of de-extinction hope that the early work described above paves the way for the resurrection of extinct species once the threats that drove them to extinction have been managed. Many resurrection biologists have even started establishing banks where genetic material of threatened species is **cryopreserved** for future use. They also hope that their work will inspire more people to be interested in science in general and especially in conservation. Protected areas with extinct species may even draw tourists that can fund conservation projects, while reintroductions of once-extinct species can revive lost ecosystem services.

Bringing a species back from extinction is, however, highly controversial for several reasons. First, there is the argument that the limited funds available for conservation are better spent on species currently facing extinction rather than on projects with possibly insurmountable technical challenges. Others argue that there is no point in spending millions of dollars to bring back an extinct species if we cannot even solve the extinctions drivers that caused the demise of the extinct species in the first place. Conservationists also wonder how the issues facing small populations will be managed, especially early in the process. Many believe that these small compromised populations will simply occupy valuable space in zoos and protected areas that can be better used for protecting extant species. There are also major misgivings about whether the resurrected species will fill the same ecosystem function as before since they may behave differently; in fact, some worry that unpredictable behaviours may introduce new harmful threats to ecosystems. Many conservationists are also worried that the public's concern for species currently threatened may fade if there is a perception that we can simply revive the species after the last individual died.

Lastly, ethical questions are frequently raised about humans essentially trying to "play God" with these "vanity projects", and the possibility that this entire field will undermine one of the most important foundations of conservation biology—that we need to act now because extinction is forever. Clearly there are some advantages, but also disadvantages, to de-extinction. Most importantly, much research still needs to happen for this to be a viable idea.

8.9 Summary

1. The rates of species extinction are currently 1,000 times greater than natural background levels; this may soon increase to 10,000 times. Over 99% of modern extinctions can be attributed to human activity. These extinctions are leading to the rapid loss of ecosystem services.
2. The IUCN has developed quantitative criteria that assign species to nine conservation categories based on their risk of extinction. Species that are *Extinct in the Wild* (EW), *Critically Endangered* (CR), *Endangered* (EN), and *Vulnerable* (VU) are officially considered "threatened with extinction". The five other categories are *Extinct* (EX), *Near Threatened* (NT), *Least Concern* (LC), *Data Deficient* (DD), and *Not Evaluated* (NE).
3. Species with the following characteristics are particularly vulnerable to extinction: species with small populations, species with declining populations, species with restricted distribution ranges, species with one or only a few small populations, species that are exploited by people, and species with critical symbiotic relationships.
4. Small populations are at high risk of extinction because they are vulnerable to several deleterious genetic factors, as well as demographic and environmental stochasticity. Such populations often require intensive management to prevent them from becoming a victim of an extinction vortex.
5. De-extinction as a scientific field aims to revive extinct species using methods such as selective breeding and cloning. But this practice is controversial, and not practical with current technology.

8.10 Topics for Discussion

1. Why should you, or anyone else, be concerned if a species becomes extirpated (also known as locally extinct)? How does this concern compare to when a species becomes globally extinct?
2. Use the IUCN Red List (http://www.iucnredlist.org) to identify one species in your country that is currently threatened with extinction. How might this species be affected by the various challenges facing small populations?

Consider genetic, physiological, behavioural, and ecological factors, as appropriate.

3. Think of an imaginary animal that was recently discovered. Imagine this animal is also threatened with extinction. Name three characteristics that make your imaginary animal vulnerable to its threats. Now discuss some steps that can be implemented to ensure that your animal will continue to survive.

4. A herd of 80 rhinoceros have been moved from South Africa to Australia to "save the species" because "there is no safe place in Africa for rhinos today" (see Hayward et al., 2017). What do you think of this plan? Do you think the project will be successful? What are the main opportunities and challenges? Once you're done answering the question read Lundgren et al. (2017) and decide if you still feel the same.

8.11 Suggested Readings

Bonebrake, T.C., F. Guo, C. Dingle, et al. 2019. Integrating proximal and horizon threats to biodiversity for conservation. *Trends in Ecology and Evolution* 34: *in press*. https://doi.org/10.1016/j.tree.2019.04.001 Coordination is critical when combatting multiple threats.

Caughley, G. 1994. Directions in conservation biology. *Journal of Animal Ecology* 63: 215–24. https://doi.org/10.2307/5542 The classic overview of the challenges faced by small and declining populations.

Ceballos, G., P.R. Ehrlich, and R. Dirzo. 2017. Biological annihilation via the ongoing sixth mass extinction signaled by vertebrate population losses and declines. *Proceedings of the National Academy of Sciences* 114: E6089–E6096. https://doi.org/10.1073/pnas.1704949114 Extinctions by the numbers.

IBPES. 2019. Nature's dangerous decline 'unprecedented': Species extinction rates accelerating. *IBPES media release*. https://www.ipbes.net/news/Media-Release-Global-Assessment We are in the midst of a biodiversity crisis.

IUCN. 2012. *IUCN Red List Categories and Criteria*: v. 3.1 (Gland: IUCN). https://portals.iucn.org/library/sites/library/files/documents/RL-2001-001-2nd.pdf A summary of the IUCN Red List classifications.

Jones, H.P. 2010. Seabird islands take mere decades to recover following rat eradication. *Ecological Applications* 20: 2075–80. https://doi.org/10.1890/10-0118.1 Control of invasive species can be an effective method to achieve recovery of threatened species.

Koh, L.P., R.R. Dunn, N.S. Sodhi, et al. 2004. Species coextinctions and the biodiversity crisis. *Science* 305: 1632–34. https://doi.org/10.1126/science.1101101 Thousands of species face extinction due to the decoupling of important symbiotic relationships.

McClenachan, L., A.B. Cooper, K.E. Carpenter, et al. 2012. Extinction risk and bottlenecks in the conservation of charismatic marine species. *Conservation Letters* 5: 73–80. https://doi.org/10.1111/j.1755-263X.2011.00206.x Most marine species have not been evaluated for their risk of extinction.

Stearns, B.P., and S.C. Stearns. 2010. Still watching, from the edge of extinction. *BioScience* 60: 141–46. https://doi.org/10.1525/bio.2010.60.2.8 Many threatened species increasingly rely on human actions to avoid extinction.

Tranquilli, S., M. Abedi-Lartey, F. Amsini, et al. 2012. Lack of conservation effort rapidly increases African great ape extinction risk. *Conservation Letters* 5: 48–55. https://doi.org/10.1111/j.1755-263X.2011.00211.x Neglecting species in conservation need increases extinction risk.

Bibliography

Andreotti, S., S. Heyden, R. Henriques, et al. 2016. New insights into the evolutionary history of white sharks, *Carcharodon carcharias*. *Journal of Biogeography* 43: 328–39. https://doi.org/10.1111/jbi.12641

Balon, E.K., M.N. Bruton, and H. Fricke. 1988. A fiftieth anniversary reflection on the living coelacanth, *Latimeria chalumnae*: Some new interpretations of its natural history and conservation status. *Environmental Biology of Fishes* 23: 241–80. https://doi.org/10.1007/BF00005238

Barnosky, A.D., N. Matzke, S. Tomiya, et al. 2011. Has the Earth's sixth mass extinction already arrived? *Nature* 471: 51–57. https://doi.org/10.1038/nature09678

Barnosky, A.D., P.L. Koch, R.S. Feranec, et al. 2004. Assessing the causes of late Pleistocene extinctions on the continents. *Science* 306: 70–75. https://doi.org/10.1126/science.1101476

Barraclough, D.A. 2006. Bushels of bots. *Natural History* 115: 18–21.

Bartlett, L.J., D.R. Williams, G.W. Prescott, et al. 2015. Robustness despite uncertainty: Regional climate data reveal the dominant role of humans in explaining global extinctions of Late Quaternary megafauna. *Ecography* 39: 152–61. https://doi.org/10.1111/ecog.01566

Bauer, H., C. Packer, P.F. Funston, et al. 2015. *Panthera leo. The IUCN Red List of Threatened Species* 2015: e.T15951A79929984. https://doi.org/10.2305/IUCN.UK.2016-3.RLTS.T15951A107265605.en

Bazelet, C., and P. Naskrecki. 2014. *Aroegas nigroornatus*. *The IUCN Red List of Threatened Species 2014*: e.T20639917A56180093. https://doi.org/10.2305/IUCN.UK.2014-1.RLTS.T20639917A56180093.en

BirdLife International 2018a. *Crithagra concolor*. *The IUCN Red List of Threatened Species* 2018: e.T22720310A128249895. https://doi.org/10.2305/IUCN.UK.2018-2.RLTS.T22720310A128249895.en

BirdLife International 2018b. *Lanius newtoni*. *The IUCN Red List of Threatened Species* 2018: e.T22705080A131390093. https://doi.org/10.2305/IUCN.UK.2018-2.RLTS.T22705080A131390093.en

Blackburn, D.C., C. Boix, E. Greenbaum, et al. 2016. The distribution of the Bururi Long-fingered Frog (*Cardioglossa cyaneospila*, family Arthroleptidae), a poorly known Albertine Rift endemic. *Zootaxa* 4170: 355–64. http://doi.org/10.11646/zootaxa.4170.2.8

Boschian, G., D. Caramella, D. Saccà, et al. 2019. Are there marrow cavities in Pleistocene elephant limb bones, and was marrow available to early humans? New CT scan results from the site of Castel di Guido (Italy). *Quaternary Science Reviews 215* 86–97. https://doi.org/10.1016/j.quascirev.2019.05.010

Brito, J.C., S.M. Durant, N. Pettorelli, et al. 2018. Armed conflicts and wildlife decline: Challenges and recommendations for effective conservation policy in the Sahara-Sahel. *Conservation Letters* 11: e12446. http://doi.org/10.1111/conl.12446

Brook, B.W., and M.J.S. Bowman. 2005. One equation fits overkill: Why allometry underpins both prehistoric and modern body-sized extinctions. *Population Ecology* 47: 137–41. https://doi.org/10.1007/s10144-005-0213-4

Brooks, T.M., S.L. Pimm, and J.O. Oyugi. 1999. Time lag between deforestation and bird extinction in tropical forest fragments. *Conservation Biology* 13: 1140–50. https://doi.org/10.1046/j.1523-1739.1999.98341.x

Burt, A.J., J. Gane, I. Olivier, et al. 2016. The history, status and trends of the Endangered Seychelles Magpie-robin *Copsychus sechellarum*. *Bird Conservation International* 26: 505–23. https://doi.org/10.1017/S0959270915000404

Callaway, E. 2017. Oldest *Homo sapiens* fossil claim rewrites our species' history. *Nature News*. https://doi.org/10.1038/nature.2017.22114

Caughley, G. 1994. Directions in conservation biology. *Journal of Animal Ecology* 63: 215–24. https://doi.org/10.2307/5542

Ceballos, G., P.R. Ehrlich, A.D. Barnosky, et al. 2015. Accelerated modern human-induced species losses: Entering the sixth mass extinction. *Science Advances* 1: e1400253. https://doi.org/10.1126/sciadv.1400253

Ceballos, G., P.R. Ehrlich, and R. Dirzo. 2017. Biological annihilation via the ongoing sixth mass extinction signaled by vertebrate population losses and declines. *Proceedings of the National Academy of Sciences* 114: E6089–E6096. https://doi.org/10.1073/pnas.1704949114

Chakona, A, and S. Kubheka. 2018. *Pseudobarbus quathlambae*. *The IUCN Red List of Threatened Species* 2018: e.T18475A100171498. http://doi.org/10.2305/IUCN.UK.2018-2.RLTS.T18475A100171498.en

Channing, A., K.S. Finlow-Bates, S.E. Haarklau, et al. 2006. The biology and recent history of the Critically Endangered Kihansi Spray Toad *Nectophrynoides asperginis* in Tanzania. *Journal of East African Natural History* 95: 117–38. https://doi.org/10.2982/0012-8317(2006)95[117:TBARHO]2.0.CO;2

Chiyo, P.I., V. Obanda, and D.K. Korir. 2015. Illegal tusk harvest and the decline of tusk size in the African elephant. *Ecology and Evolution* 5: 5216–29. https://doi.org/10.1002/ece3.1769

Clarke, G.P., N.D. Burgess, F.M. Mbago, et al. 2011. Two 'extinct' trees rediscovered near Kilwa, Tanzania. *Journal of East African Natural History* 100: 133–40. https://doi.org/10.2982/028.100.0109

Contu, S. 2013. *Cyperus chionocephalus*. *The IUCN Red List of Threatened Species* 2013: e.T44393328A44490119. http://doi.org/10.2305/IUCN.UK.2013-1.RLTS.T44393328A44490119.en

Courchamp, F., T. Clutton-Brock, and B. Grenfell. 2000. Multipack dynamics and the Allee effect in the African wild dog, *Lycaon pictus*. *Animal Conservation* 3: 277–85. https://doi.org/10.1111/j.1469-1795.2000.tb00113.x

Cowlishaw, G. 1999. Predicting the pattern of decline of African primate diversity: An extinction debt from historical deforestation. *Conservation Biology* 13: 1183–93. https://doi.org/10.1046/j.1523-1739.1999.98433.x

Cronk, Q. 2016. Plant extinctions take time. *Science* 353: 446–47. https://doi.org/10.1126/science.aag1794

Cumming, D.H.M. 2007. Of elephants, predators, and plants in protected areas: A case of classic trophic cascades. Paper presented in Symposium: *Sharing the range: Elephants, people and biological conservation in Africa*. SCB Annual Meeting, 1–5 July, Port Elizabeth, South Africa.

Dalton, D.L., E. Vermaak, H.A. Smit-Robinson, et al. 2016. Lack of diversity at innate immunity Toll-like receptor genes in the Critically Endangered White-winged Flufftail (*Sarothrura ayresi*). *Scientific Reports* 6: 36757. https://doi.org/10.1038/srep36757

Davies-Mostert H.T., and M. Gusset. 2013. Restoring African wild dogs in South Africa: A managed metapopulation approach. *WAZA Magazine* 14: 41–44

de Vos, J.M., L.N. Joppa, J.L. Gittleman, P.R. Stephens, et al. 2015. Estimating the normal background rate of species extinction. *Conservation Biology* 29: 452–62. https://doi.org/10.1111/cobi.12380

Durant, S.M., T. Wacher, S. Bashir, et al. 2013. Fiddling in biodiversity hotspots while deserts burn? Collapse of the Sahara's megafauna. *Diversity and Distributions* 20: 114–22. https://doi.org/10.1111/ddi.12157

Durant, S., N. Mitchell, A. Ipavec, et al. 2017. The global decline of cheetah *Acinonyx jubatus* and what it means for conservation. *Proceedings of the National Academy of Sciences* 114: 528–33. https://doi.org/10.1073/pnas.1611122114

Entwistle, A.C., and J. Juma. 2016. *Pteropus voeltzkowi. The IUCN Red List of Threatened Species* 2016: e.T18768A22089205. http://doi.org/10.2305/IUCN.UK.2016-1.RLTS.T18768A22089205.en

Fagan, W.F., and E.E. Holmes. 2006. Quantifying the extinction vortex. *Ecology Letters* 9: 51–60. https://doi.org/10.1111/j.1461-0248.2005.00845.x

Faith, J.T., J. Rowan, A. Du, et al. 2018. Plio-Pleistocene decline of Africa's megaherbivores: No evidence for ancient hominin impacts. *Science* 362: 938–41. https://doi.org/10.1126/science.aau2728

Feranec, R.S. 2004. Geographic variation in the diet of hypsodont herbivores from the Rancholabrean of Florida. *Palaeogeography, Palaeoclimatology, Palaeoecology* 207: 359369. https://doi.org/10.1016/j.palaeo.2003.09.031

Frankham, R., J.D. Ballou, M.D.B. Eldridge, et al. 2011. Predicting the probability of outbreeding depression. *Conservation Biology* 25: 465–75. https://doi.org/10.1111/j.1523-1739.2011.01662.x

Gibert, P., R.B. Huey, and G.W. Gilchrist. 2001. Locomotor performance of *Drosophila melanogaster*: interactions among developmental and adult temperatures, age, and geography. *Evolution* 55: 205–09. https://doi.org/10.1111/j.0014-3820.2001.tb01286.x

Gilpin, M.E., and M.E. Soulé. 1986. Minimum viable populations: Processes of species extinction. In: *The Science of Scarcity and Diversity*, ed. by M.E. Soulé (Sunderland: Sinauer).

Goldschmidt, T., F. Witte, and J. Wanink. 1993. Cascading effects of the introduced Nile perch on the detritivorous/phytoplanktivorous species in the sublittoral areas of Lake Victoria. *Conservation Biology* 7: 686–700. https://doi.org/10.1046/j.1523-1739.1993.07030686.x

Grobler, J.P., I. Rushworth, J.S. Brink, et al. 2011. Management of hybridization in an endemic species: Decision making in the face of imperfect information in the case of the black wildebeest—*Connochaetes gnou*. *European Journal of Wildlife Research* 57: 997–1006. https://doi.org/10.1007/s10344-011-0567-1

Guerrant, E.O. 1992. Genetic and demographic considerations in the sampling and reintroduction of rare plants. In: *Conservation Biology: The Theory and Practice of Nature Conservation, Preservation and Management*, ed. by P.L. Fiedler and S.K. Jain (New York: Springer). https://doi.org/10.1007/978-1-4684-6426-9

Gusset, M., S.J. Ryan, M. Hofmeyr, et al. 2008. Efforts going to the dogs? Evaluating attempts to re-introduce endangered wild dogs in South Africa. *Journal of Applied Ecology* 45: 100–08. https://doi.org/10.1111/j.1365-2664.2007.01357.x

Halley, J.M., N. Monokrousos, A.D. Mazaris, et al. 2016. Dynamics of extinction debt across five taxonomic groups. *Nature Communications* 7: 12283. https://doi.org/10.1038/ncomms12283

Harley, E.H., M.H. Knight, C. Lardner, et al. 2009. The Quagga project: Progress over 20 years of selective breeding. *South African Journal of Wildlife Research* 39: 155–63. https://doi.org/10.3957/056.039.0206

Haynes G. 2018. The evidence for human agency in the Late Pleistocene megafaunal extinctions. In *Encyclopedia of the Anthropocene v. 1*, ed. by D.A. DellaSala, and M.I. Goldstein (Oxford: Elsevier).

Hayward, M.W., W.J. Ripple, G.I.H. Kerley, et al. 2017. Neocolonial conservation: Is moving rhinos to Australia conservation or intellectual property loss? *Conservation Letters* 11: e12354. https://doi.org/10.1111/conl.12354

Heber, S., and J.V. Briskie. 2010. Population bottlenecks and increased hatching failure in endangered birds. *Conservation Biology* 24: 1674–78. https://doi.org/10.1111/j.1523-1739.2010.01553.x

Hutton, J.M. 1987. Incubation temperatures, sex ratios and sex determination in a population of Nile crocodiles (*Crocodylus niloticus*). *Journal of Zoology* 211: 143–55. http://doi.org/10.1111/j.1469-7998.1987.tb07458.x

IUCN. 2012. *IUCN Red List categories and criteria*: v. 3.1 (Gland: IUCN). https://portals.iucn.org/library/sites/library/files/documents/RL-2001-001-2nd.pdf

IUCN. 2017. *Guidelines for using the IUCN Red List categories and criteria*, v. 13 (Gland: IUCN). http://nc.iucnredlist.org/redlist/content/attachment_files/RedListGuidelines.pdf

IUCN. 2019. *The IUCN Red List of Threatened Species.* http://www.iucnredlist.org

Janzen, D.H. 1983. The Pleistocene hunters had help. *American Naturalist* 121: 598–99. https://doi.org/10.1086/284088

Johnson, C.N. 2009. Ecological consequences of Late Quaternary extinctions of megafauna. *Proceedings of the Royal Society of London B* 276: 2509–19. https://doi.org/10.1098/rspb.2008.1921

Kalinowski, S.T., P.W. Hedrick, and P.S. Miller. 2000. Inbreeding depression in the Speke's gazelle captive breeding program. *Conservation Biology* 14: 1375–84. https://doi.org/10.1046/j.1523-1739.2000.98209.x

Keith, D.A., J.P. Rodríguez, K.M. Rodríguez-Clark, et al. 2013. Scientific foundations for an IUCN Red List of Ecosystems. *PLoS ONE* 8: e62111. https://doi.org/10.1371/journal.pone.0062111

Kerley, H.I.H., R. Sims-Castley, A.F. Boshoff, et al. 2009. Extinction of the blue antelope *Hippotragus leucophaeus*: Modelling predicts non-viable global population size as the primary driver. *Biodiversity and Conservation* 18: 3235. https://doi.org/10.1007/s10531-009-9639-x

Koh, L.P., R.R. Dunn, N.S. Sodhi, et al. 2004. Species coextinctions and the biodiversity crisis. *Science* 305: 1632–34. https://doi.org/10.1126/science.1101101

Krajick, K. 2003. Defusing Africa's killer lakes. *Smithsonian Magazine* 34: 46–55. https://www.smithsonianmag.com/science-nature/defusing-africas-killer-lakes-88765263

Leimu, R., and M. Fischer. 2010. Between-population outbreeding affects plant defence. *PLoS ONE* 5: e12614. https://doi.org/10.1371%2Fjournal.pone.0012614

Lombardo, M.P., and R.O Deaner. 2018. Born to throw: The ecological causes that shaped the evolution of throwing in humans. *Quarterly Review of Biology* 93: 1–16. https://doi.org/10.1086/696721

Lundgren, E.J., D. Ramp, W.J. Ripple, et al. 2017. Introduced megafauna are rewilding the Anthropocene. *Ecography* 41: 857–66. https://doi.org/10.1111/ecog.03430

Marneweck, C., P.A. Becker, G. Beverley, et al. 2019. Factors affecting the success of artificial pack formation in an endangered, social carnivore: The African wild dog. *Animal Conservation* 22: *in press*. https://doi.org/10.1111/acv.12490

Maron, D.F. 2018. Under poaching pressure, elephants are evolving to lose their tusks. *National Geographic*. https://on.natgeo.com/2z9mE1x

Meffe, G.C., and C.R. Carroll. 1997. *Principles of Conservation Biology* (Sunderland: Sinauer).

Minnaar, A., L. van Schalkwyk, and S. Kader. 2018. The difficulties in policing and combatting of a maritime crime: The case of Abalone poaching along South Africa's coastline. *Journal of the Indian Ocean Region* 14: 71–87. https://doi.org/10.1080/19480881.2018.1421448

Ogden, L.E. 2014. Extinction is forever… or is it? *BioScience* 64: 469–75. https://doi.org/10.1093/biosci/biu063

Okubamichael, D.Y., S. Jack, J. de Wet Bösenberg, et al. 2016. Repeat photography confirms alarming decline in South African cycads. *Biodiversity and Conservation* 25: 2153–70. http://doi.org/10.1007/s10531-016-1183-x

OpenStax. 2019. *Biology 2nd ed*. OpenStax CNX. https://openstax.org/details/books/biology-2e

Owen-Smith, N., G.J. Chirima, V. Macandza, et al. 2012. Shrinking sable antelope numbers in Kruger National Park: What is suppressing population recovery? *Animal Conservation* 15: 195–204. https://doi.org/10.1111/j.1469-1795.2011.00504.x

Pimm, S.L., C.N. Jenkins, R. Abell, et al. 2014. The biodiversity of species and their rates of extinction, distribution, and protection. *Science* 344: 1246752. https://doi.org/10.1126/science.1246752

Pringle, R.M. 2005. The origins of the Nile perch in Lake Victoria. *BioScience* 55: 780–87. https://doi.org/10.1641/0006-3568(2005)055[0780:TOOTNP]2.0.CO;2

Rall, J.L., and T. Sephaka. 2008. *Re-evaluation of the relevance to construct barriers as in-situ conservation measures for the protection of the Maloti minnow in the Senqunyane Catchment* (Florida: ECOSUN Environmental Consultants).

Rebaudo, F., and V.-B. Rabhi. 2018. Modeling temperature-dependent development rate and phenology in insects: Review of major developments, challenges, and future directions. *Entomologia Experimentalis et Applicata* 166: 607–17. https://doi.org/10.1111/eea.12693

Rebelo, A.G. 1992. Red Data Book species in the Cape Floristic Region: Threats, priorities and target species. *Transactions of the Royal Society of South Africa* 48: 55–86. https://doi.org/10.1080/00359199209520256

Ripple, W.J., and B. Van Valkenburgh. 2010. Linking top-down forces to the Pleistocene megafaunal extinctions. *BioScience* 60: 516–26. https://doi.org/10.1525/bio.2010.60.7.7

Schleuning, M., and D. Matthies. 2009. Habitat change and plant demography: Assessing the extinction risk of a formerly common grassland perennial. *Conservation Biology* 23: 174–83. https://doi.org/10.1111/j.1523-1739.2008.01054.x

Seddon, P.J. 2017. The ecology of de-extinction. *Functional Ecology* 31: 992–95. https://doi.org/10.1111/1365-2435.12856

Shelton, J.M., B.M. Clark, T. Sephaka, et al. 2017. Population crash in Lesotho's endemic Maloti minnow *Pseudobarbus quathlambae* following invasion by translocated smallmouth yellowfish *Labeobarbus aeneus*. *Aquatic Conservation* 65–77. https://doi.org/10.1002/aqc.2633

Skelton, P.H., J.L. Rall, E.R Swartz, et al. 2001. Maloti minnow conservation project, *LHDA Report*. 1041 (Grahamstown: JLB Smith Institute of Ichthyology).

Smith, F.A., R.E. Elliott Smith, S.K. Lyons, et al. 2019. The accelerating influence of humans on mammalian macroecological patterns over the late Quaternary. *Quaternary Science Reviews* 211: 1–16. https://doi.org/10.1016/j.quascirev.2019.02.031

Steyn, G., J.L. Rall, V. Rall, et al. 1996. Fish. In: Baseline biology survey and reserve development: Phase 1B. v. 3-Fauna Lesotho, *LHDA Report* 1008, ed. by AfriDev Consultants (Maseru: Lesotho Highlands Development Authority).

Stokstad, E. 2015. Bringing back the aurochs. *Science* 350: 1144–47. https://doi.org/10.1126/science.350.6265.1144

Surovell, T., N. Waguespack, and P.J. Brintingham. 2005. Global archaeological evidence for proboscidians overkill. *Proceedings of the National Academy of Science* 102: 6321–26. https://doi.org/10.1073/pnas.0501947102

Tallman, G., J. Zhu, B.T. Mawson, et al. 1997. Induction of CAM in *Mesembryanthemum crystallinum* abolishes the stomatal response to blue light and light-dependent zeaxanthin formation in guard cell chloroplasts. *Plant and Cell Physiology* 38: 236–42. https://doi.org/10.1093/oxfordjournals.pcp.a029158

Trinkel, M., D. Cooper, C. Packer, et al. 2011. Inbreeding depression increases susceptibility to bovine tuberculosis in lions: An experimental test using an inbred-outbred contrast through translocation. *Journal of Wildlife Diseases* 47: 494–500. https://doi.org/10.7589/0090-3558-47.3.494

Van Valkenburgh, B., M.W. Hayward, W.J. Ripple, et al. 2016. The impact of large terrestrial carnivores on Pleistocene ecosystems. *Proceedings of the National Academy of Sciences* 113: 862–67. https://doi.org/10.1073/pnas.1502554112

Waters, C.N., J. Zalasiewicz, C. Summerhayes, et al. 2016. The Anthropocene is functionally and stratigraphically distinct from the Holocene. *Science* 351: aad2622. https://doi.org/10.1126/science.aad2622

Werdelin, L., and M.E. Lewis. 2013. Temporal change in functional richness and evenness in the eastern African Plio-Pleistocene carnivoran guild. *PLoS ONE* 8: e57944. https://doi.org/10.1371/journal.pone.0057944

Whitehouse, A.M. 2002. Tusklessness in the elephant population of the Addo Elephant National Park, South Africa. *Journal of Zoology* 257: 249–54. https://doi.org/10.1017/S0952836902000845

Witte, F., T. Goldschmidt, P.C. Goudswaard, et al. 1991. Species extinction and concomitant ecological changes in Lake Victoria. *Netherlands Journal of Zoology* 42: 214–32. https://doi.org/10.1163/156854291X00298

Wright, S. 1931. Evolution in Mendelian populations. *Genetics* 16: 97–159. http://www.ncbi.nlm.nih.gov/pubmed/17246615

Xue, Y., J. Prado-Martinez, P.H. Sudmant, et al. 2015. Mountain gorilla genomes reveal the impact of long-term population decline and inbreeding. *Science* 348: 242–45. https://doi.org/10.1126/science.aaa3952

9. Applied Population Biology

Staff from the Frankfurt Zoological Society (FZS) conducting an aerial survey over Selous Game Reserve, Tanzania. Photograph by Daniel Rosengren, https://commons.wikimedia.org/wiki/File:FZS_plane_conducting_an_aerial_survey_in_Selous_Game_Reserve,_Tanzania.jpg, CC BY 4.0.

https://doi.org/10.11647/OBP.0177.09

Even without human influences, the size of any wildlife population may be stable, increasing, decreasing, or even fluctuating. These population changes, combined with occasional natural perturbations, can and have driven some species and populations to extinction. Such natural extinction events generally occur at local scales, and are interspersed by long periods of little change, so that overall ecosystem stability is not compromised. Moreover, as explained by the intermediate disturbance hypothesis (e.g. Bongers et al., 2009), localised disturbances and subsequent local extinctions play an important role in maintaining regional biodiversity, as they increase opportunities for a greater variety of species to live in an area (Figure 9.1), at least until succession drives them out again. Some species that colonise the empty niches left by extinctions or extirpations may even evolve to become new species over time.

Figure 9.1 A treefall gap allowing sunshine to penetrate the canopy in the sacred Bubi Forest on Bioko Island, Equatorial Guinea. Treefall gaps and other localised natural disturbances benefit regional biodiversity because they provide opportunities for a greater variety of species to eke out an existence. Responses do vary, however, from ecosystem to ecosystems: while fire disturbance maintains most grassland and savannah ecosystems, it has an overall negative impact on tropical forests. Photograph by Luke L. Powell/ Biodiversity Initiative, CC BY 4.0.

Human-driven disturbances often occur at larger scales and more frequently than natural perturbations. Consider, for example, the large amount of natural forests that are converted to agricultural land every year, or climate change impacts that are affecting every ecosystem on Earth. Because these disturbances are so widespread and occur with such regularity, they are causing a wholesale destabilisation of the natural environment. Many species and populations are unable to cope with these fast and vast changes and are consequently undergoing sharp declines. The human-driven extinctions that follow are leaving compromised ecosystems more vulnerable to invasions by widespread generalist species and exotic species. What remains is an environment dominated by only a few species unable to offer many of the ecosystem services we depend upon. To prevent further harm, we need to identify the most vulnerable species and ecosystems and develop strategies that can slow or even reverse current extinction rates. But how can we identify the species most likely to go extinctions soon, and how can we determine which actions should be taken to save them? The field of population biology, defined as the study of population dynamics over time and space, provides us the tools to answer many of these questions.

9.1 Monitoring Population Size

The primary aim of population monitoring is to detect changes in the environment, population size, and species distribution over time. Such monitoring efforts frequently focus on a particular area or a population of concern, but it can also target more common but sensitive species, such as butterflies and macroinvertebrates, which can function as indicator species to assess ecosystem condition (Section 4.2.6). The great number of methods (which are all types of surveys) used to monitor populations usually fall into one of three different categories: biodiversity inventories, population censuses, and demographic studies.

9.1.1 Biodiversity inventories

A biodiversity inventory is an attempt to document which species are present in some defined locality. Such an effort can focus on one specific taxa (e.g. a bird survey) or several taxa, on a small area (e.g. a city park) or large area (e.g. a large national park), over a short period of time (e.g. a few hours) or long period of time (e.g. several years, Box 9.1). There are many methods to compile a biodiversity inventory, ranging from uncomplicated to highly organised, performed by a single person or a large team of experts. Some of the most popular methods for biodiversity inventories include site visits by professional naturalists and questionnaires distributed among local people. To tap into the knowledge and eagerness of amateur naturalists, conservation biologists are also increasingly compiling biodiversity inventories using citizen science surveys (see Box 15.3). Rapid biodiversity assessments (RAP) are sometimes used to compile an inventory under tight deadlines to answer urgent questions and inform urgent decisions. A bioblitz is a special type of biodiversity inventory during which experts on a range of taxa come together to record all the living species within a designated area over a brief period (usually over 24 hours).

> The primary aim of population monitoring is to detect changes in the environment, population size, and species' distributions over time.

Box 9.1 The Role of Biodiversity Inventories in the Management of Gorongosa National Park

Marc Stalmans[1] *and Piotr Naskrecki*[2]

[1] *Scientific Services, Gorongosa National Park, Mozambique.*

[2] *E.O. Wilson Biodiversity Laboratory, Gorongosa National Park, Mozambique.*

✉ *stalmans@gorongosa.net and pnaskrec@oeb.harvard.edu*

The 4,000 km^2 Gorongosa National Park in central Mozambique was proclaimed in 1960 to protect one of the highest densities of large herbivores at the southern end of Africa's Great Rift Valley (Tinley, 1977). National Parks are often victims of war and political instability and Gorongosa National Park is no exception. It suffered grievously during the protracted period of civil war from the early 1970s to early 1990s. During this time the park lost 90–99% of its elephants, common hippopotamuses (*Hippopotamus amphibious*, VU), African buffalo (*Syncerus caffer*, NT), plains zebras, and common wildebeest (*Connochaetes taurinus*, LC) through poaching by warring parties and hunters from nearby cities and rural communities.

Since then, restoration efforts that started in earnest in 2004 have brought about a spectacular recovery of several affected large mammal populations (Bouley et al., 2018; Stalmans et al., 2019). But conservation management cannot only focus on these flagship species. Considering, amongst others, the impact of climate change and the importance of agriculture (with its associated pests and pollinators) to rural communities, it is vital that conservationists understand the breadth of biodiversity and its relationship to ecosystem functioning. To accomplish this, a programme of systematic biodiversity surveys is currently being undertaken in Gorongosa. Each year, a group of international and national specialists team up with park technicians and rangers to conduct a three-week long bioblitz in a subsection of the park. These surveys also serve as training opportunities to prepare young Mozambican scientists to apply modern biological survey methods and technologies.

By the beginning of 2019, a total of nearly 5,900 species represented by some 44,000 observations have been entered into the park's biodiversity database. Based on these data, initial projections suggest that Gorongosa protects 37,500–76,500 different species. Vertebrates are likely to number 850–1,000 species, while plants are estimated to number 2,000–3,000 species. Single orders of insects far exceed those numbers; for example, there may be 3,000–5,000 species of wasps, and 4,000–6,000 species of moths of which 15–25% may be new to science. Local ecosystems are also particularly rich. For the surveys around the Bunga inselbergs (Figure 9.A) in 2015, at least 580 species of butterflies and moths were collected, most of them never before recorded from the Park. Forty species of katydids were recorded, with two species of significance. A large population of *Debrona cervina,* a large arboreal katydid was discovered, until now known only from two type specimens collected in 1890. Also collected was the predaceous katydid *Peringueyella macrocephala,* Mozambique's largest katydid, previously known only from a handful of specimens collected between 1850 and 1965. About 100 species of grasshoppers were recorded, including two species new to science. Additionally, about 30 species of mantids were recorded, including *Rhomboderella thorectes,* a species previously known from the single holotype collected in the early 1900s. It is expected that the full inventory of the park's biodiversity will span a period of 20 years.

Figure 9.A Surveys in little-explored corners of Africa often yield biological discoveries. (Top) A recently described gecko species new to science, *Afroedura gorongosa,* discovered in 2015. (Bottom) The Mozambique girdled lizard (*Smaug mossambicus*) was previously known only from a small population on Mount Gorongosa and a single record in Manica province. Surveys in 2015 found a new population on Bunga inselberg. Photographs by Piotr Naskrecki, CC BY 4.0.

Biodiversity inventories play an important role in the management of Gorongosa and other national parks, especially in long-term conservation planning. Amongst others, baseline data obtained from our surveys will be used in future to measure overall biodiversity responses to large mammal population changes, evolving patterns of land use around the park, and the impact of climate change. Future biodiversity surveys will also target little known areas adjacent to the park to provide information required for corridor planning.

While biodiversity inventories seldom offer the kinds of detailed data required to predict likelihood of a species' persistence, they have several uses in conservation. First, a biodiversity inventory can be a comparatively inexpensive and straightforward method to broadly monitor an area's species and populations. Biodiversity inventories conducted over a wide area can also help determine the distribution of a species, while a comparison with follow-up inventories can highlight distribution changes (which often correspond to population changes). This was well illustrated in a study that used repeated citizen scientist surveys to investigate how songbird distributions have changed across South Africa, Lesotho, and eSwatini between 1987 and 2013 (Péron and Altwegg, 2015).

9.1.2 Population censuses

A population **census** (also called a count) uses a repeatable sampling protocol to estimate the abundance or density of a population or species which, in turn, can tell us whether a population is doing well or not. When a species is easy to detect, relatively sedentary, and the sampling area is small, a comprehensive census of all individuals may be possible. However, comprehensive censuses are generally very difficult, if not impossible, to conduct when implemented on large or highly mobile populations, or over large areas. In these cases, it may be better to restrict the census to a more manageable subsection of the population, by dividing the area of interest into sampling units, and randomly censusing only some of the units. Population estimates that capture only a fraction of the overall population can then serve as an index for broader trends, or it can be used to estimate the total population size through **extrapolation**, if the researcher knows which fraction of the population or area was counted.

Some popular methods for censusing subsections of wildlife populations are, sampling plots, distance sampling, and mark-recapture surveys. Sampling plots are popular in studies focussing on plants and invertebrates, allowing biologists to systematically count each individual observed in a small area (Figure 9.2). Birds and mammals are often censused using distance sampling, during which all observed individuals on predetermined transects or from points are tallied. The number of individuals observed in the count area can then be extrapolated to obtain population size (or density) estimates for individual (or multiple) species observed across the entire area of interest. Aerial censuses are often used to conduct distance sampling transects over large and open areas, while point counts and walked line-transects are more popular for small areas or closed-canopy ecosystems (White and Edward, 2000). Mark-recapture surveys, mark-resight surveys, and sight-resight surveys are popular for species that are easy to catch, trap, or individually recognised. In this case, captured (and thus counted) individuals would be marked for future identification, after which the total population in an area is estimated by accounting for the proportion of marked and unmarked individuals seen on subsequent visits. The marking of animals can be done with a variety of procedures, including using highly visible tags, paint approved for animal use, or unique marks on the animal itself. One creative study used tourists' photographs to generate a mark-recapture dataset, which was used to estimate the size of cheetah and African wild dog (*Lycaon pictus*, EN) populations in South Africa's Kruger National Park (Marnewick et al., 2014). Like inventories, population censuses can sometimes also lead to unexpected yet important findings: the first comprehensive population survey of sea turtles breeding on Africa's Atlantic coast recently alerted marine biologists to the fact that Gabon hosts several globally important rookeries (Box 9.2).

Box 9.2 Sea Turtle Conservation along Africa's Atlantic Coast

Angela Formia

Wildlife Conservation Society,
Global Conservation Program,
New York, NY, USA.

✉ *aformia@wcs.org*

Virtually all the characteristics of sea turtles' life histories make them difficult to study and conserve. They are long-lived, slow growing, migratory, and almost entirely ocean-dwelling. Although they return to their natal beaches to reproduce, these are usually thousands of kilometres from their developmental and adult foraging grounds. In addition, sea turtle habitat often overlaps with areas of high human use such as developed coastlines and intensive fisheries. Describing population ranges and assessing interaction with human threats is thus critical to their survival.

Over recent decades, we have learnt much about sea turtles along the coastline of Africa (Figure 9.B) thanks to extensive research efforts. For instance, we know that these coasts host globally important populations of green turtles (*Chelonia mydas*, EN) in Mauritania, Guinea Bissau, Equatorial Guinea and Republic of the Congo; loggerheads (*Caretta caretta*, VU) on Cabo Verde;

hawksbills (*Eretmochelys imbricata*, CR), on São Tomé and Principe; leatherbacks (*Dermochelys coriacea*, VU) in Equatorial Guinea and Gabon; and olive ridleys (*Lepidochelys olivacea*, VU) in Gabon and Angola.

Figure 9.B (Top) One of thousands of leatherback sea turtle females nesting in Gabon every year. Photograph by M.J. Witt, CC BY 4.0. (Bottom) An olive ridley turtle hatchling makes its way to sea on a northern Angolan beach where the local community ensures its protection. Photograph by A. Formia, CC BY 4.0.

One of Africa's most remarkable sea turtle populations is Gabon's leatherback rookery, the biggest in the world with as many as 15,000 to 41,000 nesting females (Witt et al., 2009). Gabon also hosts the largest olive ridley rookery in the Atlantic (Metcalfe et al., 2015), and foraging grounds for green and hawksbill turtles. Until the late 1990s, virtually nothing was known about these populations, other than the fact that eggs and adults were frequently collected for human consumption. Since then, a multi-pronged approach has been adopted to describe and protect Gabon's sea turtles. Intensive coastal monitoring has allowed scientists to assess spatio-temporal trends in nesting frequency and abundance, and levels of nest-site fidelity and reproductive success. Using techniques, such as satellite telemetry, flipper tagging, oceanic modelling, and dispersal simulations, and genetic and isotopic analyses, researchers have been able to map sea turtle behaviour at sea, in Gabon's coastal waters, and during post-nesting migrations to foraging grounds off South America and South Africa (i.e. Formia et al., 2006, Maxwell et al., 2011, Witt et al., 2011, Pikesley et al., 2018).

Building upon this knowledge, measures have been established to quantify and reduce the impact of threats to Gabon's sea turtles. In 2002, the Gabonese government created a system of national parks and protected areas encompassing approximately 80% of Gabon's sea turtle nests; in 2017, a new network of 20 marine protected areas (MPA) was officially created, covering 26% of Gabon's territorial waters (Parker, 2017). Laws enacted in 2011 prohibit all hunting, capture, and commercialisation of sea turtles. Trained observers on-board industrial fishing vessels quantify sea turtle bycatch from bottom trawling and tuna seiners and reduce mortality by treating and releasing captured turtles. In addition, the Gabonese government now requires that all shrimp trawlers use turtle excluder devices (TED), aluminium grids sewn into the nets allowing sea turtles and other large bycatch to escape, while conserving shrimp catch; similar devices are being developed for fish trawlers. Ongoing efforts are shifting traditional turtle hunting and other destructive practices toward more sustainable fisheries. Turtle-watching ecotourism also represents a growing potential to increase awareness and incentivize conservation efforts.

Nevertheless, African sea turtle conservation remains a formidable challenge. Although the economic context is changing rapidly, impoverished coastal villagers in many countries continue to collect turtles and eggs for local consumption or market sale, and many wealthier urbanites continue to consider them delicacies. These problems are often compounded by corruption, political instability, inadequate law enforcement, and development priorities focused on destructive exploitation. With funding deficits, combating these challenges sometimes seems like a losing battle, but public attitudes are slowly shifting. Even in remote beach villages, the idea that a turtle alive is worth more than dead is no longer such a bizarre concept.

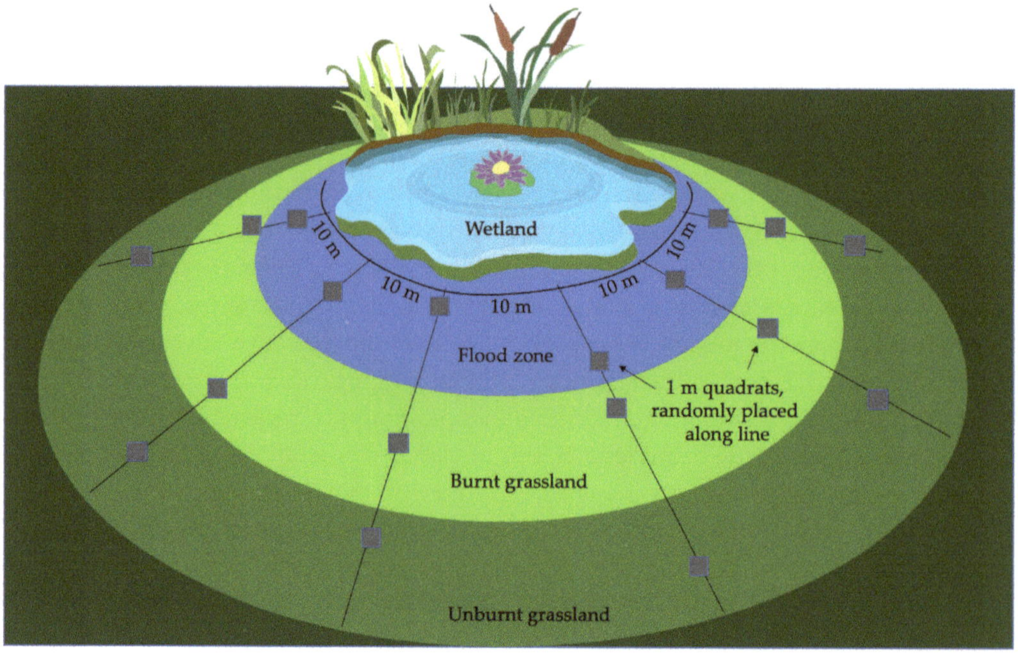

Figure 9.2 (Top) A schematic of a systematic sampling protocol using quadrat frames. Dividing a large area into smaller sampling units makes the survey task much more feasible. The survey can be performed in the field, or photos such as these can be taken for analysis once back at the office. CC BY 4.0. (Bottom) A quadrat frame divided into 10x10 cm squares, set out to monitor the species richness and abundance of plants in a grassland recovering from a fire. Photograph by Yohan Euan, https://commons.wikimedia.org/wiki/file:quadrat_sample.jpg, CC BY-SA 3.0.

9.1.3 Demographic studies

Demographic studies monitor individuals of different ages and sizes over time (Figure 9.3) to obtain a more comprehensive dataset than would be produced by population censuses. Most demographic studies use the same methods that what would be used for a population census; however, in addition to counting and marking, individuals would also be aged, measured for size and body condition, and sexed, when possible. The best demographic studies involve collecting these data from the same individuals over time, which is easiest when working with sedentary species (e.g. plants), populations in an enclosed space (e.g. in a small fenced reserve), animals that are fairly resident and/or habituated to human presence, or individuals carrying **biologging devices** (Kays et al., 2015). This may not always be possible, in which case biologists may obtain data from different individuals during each field session, to serve as an index for larger population trends.

Mark-recapture survey worksheet

Mark–recapture survey, basic formula (Caughley 1977):

$$N = \frac{M \times n}{m}$$

where
N = Population size;
M = Number of individuals marked (first visit);
n = Number of individuals seen (second visit);
m = Number of marked individuals seen (second visit).

Assume 21 individuals were marked with ear tags on the first visit. With 13 marked individuals recaptured out of a total of 30 individuals (both marked and unmarked) captured on a second visit, the estimated population size is:

$$N = \frac{21 \times 30}{13} = 48 \text{ individuals}$$

Figure 9.3 A biologist gathering biometric data from a juvenile central African slender-snouted crocodile (*Mecistops leptorhynchus*, CR) in the DRC. The crocodile will be tagged with a permanent marker before release so it can be recognised when caught again. Accompanying the photo is an example of mark-recapture survey worksheet to estimate population size. Photograph by Terese Hart, CC BY 4.0.

The data obtained from demographic studies are often used in combination with mathematical modelling to guide and refine conservation strategies. For example, researchers frequently compare the age structure (i.e. the percentage of juveniles, reproductively active adults, and older post-reproductive-age adults) of a declining population to that of a stable population to identify causes of decline, and the population parameters that are most sensitive to disturbances. This information can then be used to predict population sizes at different points in the future, and how those populations may respond to different management scenarios. The aim of many demographic studies is to predict, and identify strategies to reduce, extinction risk (see Section 9.2).

9.1.4 Recent progress in collecting survey data

Conservation activities are regularly impeded by insufficient information. This is especially true in tropical regions of the world, where most threatened species lack demographic data, and some species lack reliable data altogether. Faced with these gaps, biologists have started using several innovative methods to fill data gaps. Prominent examples include using market surveys (e.g. Kümpel et al., 2010, Ingram et al., 2015) and interviews with local people (e.g. Edwards and Plagányi, 2008) to obtain much-needed baseline survey data. It is important to note that such datasets, obtained second-hand rather than directly, can be unreliable and biased, especially if data are collected from harvesters unwilling to report on their own illegal activities. It is thus important to combine potentially unreliable datasets with reliable datasets, or obtain independent verification, before using such data to make important decisions. One such example comes from West Africa, where researchers wanted to quantify extinction risk for the Nigeria-Cameroon chimpanzee (*Pan troglodytes ellioti*, EN). Here, biologists related unreliable market survey data to two reliable datasets—orphan intake rate at wildlife sanctuaries and the number of young in wild groups—to estimate that the region's chimpanzees might be extinct within the next 20 years because hunting was two to 13 times higher than the population could sustain (Hughes et al., 2011).

> Camera traps, hair snares, and faecal samples all provide non-invasive sampling techniques to obtain baseline data needed for conservation assessments.

Collecting genetic material on elusive and rarely-seen animals with **non-invasive techniques** such as hair snares and faecal sampling are also becoming increasingly popular means of collecting survey data. Researchers in Gabon did just that, by using genetic material obtained from dung to estimate the population size, gender ratio, age distribution, breeding status, relatedness, and dispersal patterns of the region's forest elephant (*Loxodonta cyclotis*) population (Eggert et al., 2013). These non-invasive techniques reduce the need for researchers to be in the field, thereby reducing both the researchers' exposure to dangerous conditions and disturbances to the populations they are trying to monitor.

Camera traps represent another non-invasive survey technique whose popularity has greatly increased in recent years. These special cameras, often placed at supplemental food or next to wildlife paths, are activated automatically when an animal passes into the area covered by the camera's motion sensors (Figure 9.4). This photographic record of movement can then be used to obtain biodiversity inventories, population size estimations, or even to compile demographic datasets (Steenweg et al., 2017). Creative researchers at South Africa's Robben Island even successfully combined camera trapping with human facial recognition technology—more generally associated with law enforcement—to automate monitoring of individual African penguins (*Spheniscus demersus*, EN) (Sherley et al., 2010).

Figure 9.4 (Left) A nature conservation student sets a camera trap in northern South Africa to monitor leopard (*Panthera pardus*, VU) and brown hyena (Parahyaena brunnea, NT) populations on a privately protected area. Photograph by Kelly Marnewick, CC BY 4.0. (Right) Congo peafowl (*Afropavo congensis*, VU)—a highly elusive species—investigating a camera trap in the DRC. Photograph by Lukuru Foundation, CC BY 4.0.

9.2 Estimating Extinction Risk

Biologists often use the positive relationship between population size and likelihood of persistence (Section 8.7) to predict the probability that a population may go extinct at some point in the future. One of the most popular tools for making such predictions is **population viability analysis (PVA)**. A PVA can be thought of as a type of extinction risk assessment; it uses demographic data and mathematical methods to predict at what point in the future a population or species is likely to perish. In addition, by considering a species' resource requirements and the availability of limiting resources, biologists can use the results of a PVA to identify a species' most vulnerable life stages, and to estimate how management techniques may influence population size and extinction risk. In this way, PVAs can guide conservation decisions by highlighting the need to, for example, modify harvesting regulations, perform translocations (Section 11.2), or provide and protect a greater amount of suitable habitat. Even the IUCN's Red List Criteria (Section 8.5) uses PVA as a criterion to help prioritise conservation targets: populations with low extinction risk may not require immediate attention, while those approaching extinction thresholds will gain higher priority.

9.2.1 A word of warning

The purpose of this chapter is to provide a brief introduction on the usefulness of quantitative population biology methods such as PVA in conservation. While the methods for studying population sizes, fluctuations, and demographics are very powerful, they are also highly technical, and require specialist knowledge of mathematical procedures. Erroneous predictions from using incorrect methods, violating assumptions, and/or using inadequate data would run counter to well-intentioned objectives; and so, the increased popularity and use of PVA by insufficiently trained conservation scientists is of serious concern. For many people, the methods highlighted in this chapter are best learned by studying under the supervision of an expert, to better understand each model's assumptions and the newest developments in the field. For people with advanced mathematical skills who might want to study more on their own, texts such as *Quantitative Conservation Biology* (Morris and Doak, 2002) and *Bayesian Methods for Ecology* (McCarthy, 2007) may help. The development of user-friendly software packages, such as VORTEX and RAMAS (reviewed in Brook et al., 2000), have also expanded the PVA user-group in recent years. Nonetheless, when obtaining results—even from seasoned demographic modellers—it is important to remember that we cannot account for all future possibilities. Interpreting the results of a PVA, as any other model predicting the future, requires a great deal of caution and a healthy dose of common sense.

9.2.2 Probability of extinction

> Population viability analysis (PVA) uses demographic data and mathematical methods to predict if a population or species is likely to persist or perish.

The main purpose of a PVA is to estimate the viability (or time to extinction) of a species or population from observed population sizes and growth rates. Consider a population with 100 individuals that loses 50% of its individuals each year. A simple model will suggest that this population will lose 50 individuals the first year, 25 individuals the second year, and so on, until no individuals are left in the seventh year. The probability of extinction for this population is thus 100%, and the time to extinction is seven years. But how do we deal with the more realistic complex variations in population sizes we see in nature?

A more realistic PVA begins by constructing a mathematical model representing the population of interest using data obtained from a demographic study, which may include the current age (or size) structure of the population, average birth rates, and average survival rates of each age class. This dataset would be organised in a format suitable for PVA modelling using a database package, and then analysed using the methods of matrix algebra. Because results from this initial model have only one outcome—a population that is either stable or growing/declining at a fixed rate—it is called a **deterministic model**. Typically, deterministic models are then tailored to include a variety of independent environmental parameters, such as food availability, storm frequency, or invasive competitors. Variability can also be added into the model by allowing some or all the model elements (e.g. survival rate or habitat availability) to vary within their observed ranges of values. Catastrophic events, such as a fire that kills a large proportion of the population, can also be programmed to occur at random points in time. Hundreds or even thousands of simulations of this complex model can then be run to determine changes in population size over time, the probability of population extinction within a certain period, and the median time to extinction. Because of the variability built into this more complex model, each iteration's output will vary from the next; for that reason, it is called a **stochastic model**. The choice of models and the parameters included depend on the goals of the analysis and the management options under consideration.

9.2.3 Minimum viable population

When a PVA shows that a population has a relatively high risk of extinction, a logical next step would be to determine what could be done to prevent the extinction from happening. In general, protecting larger populations reduces extinction probability (Figure 9.5). To understand exactly how large is large enough, a PVA can also be used to estimate a **minimum viable population** (MVP). As the name implies, an MVP is the smallest number of individuals necessary for a population to have a chance of long-term persistence, despite the potential effects of demographic, environmental, and genetic stochasticity, and natural catastrophes faced by small populations (Section 8.7). This is well illustrated in the influential paper by Shaffer (1981), who compared setting MVP targets to planning for floods; engineers cannot rely on the average annual rainfall when designing flood control systems near rivers and wetlands. Instead, they must design systems that can also handle extreme rainfall and flooding events. These extreme events may occur rarely, perhaps once every 50 years, but they will likely occur during the lifetime of a flood control system. Similarly, to maximise the long-term protection of a threatened species, we must take actions that

> A minimum viable population (MVP) is an estimate of the smallest number of individuals necessary for a population to have a good chance of long-term persistence.

protect them in both average and extreme years characterised by catastrophic events such as cyclones/hurricanes, forest fires, and disease epidemics (Anderson et al., 2017). This is especially true considering future climate change scenarios, where every year may be uncharacteristically harsh, in effect a 50-year event.

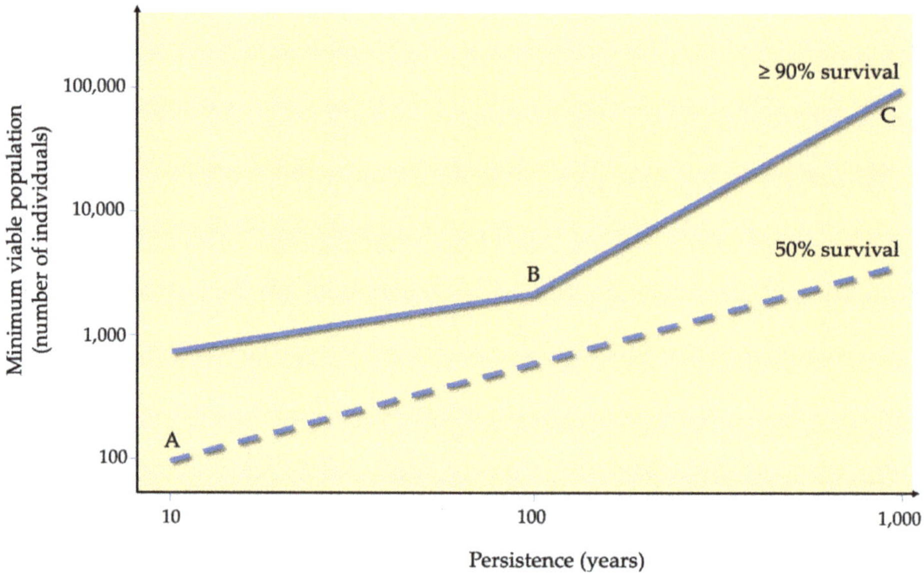

Figure 9.5 A graph (both axes on log scales) derived from a meta-analysis including 1,198 species showing how a larger minimum viable population (MVP) size translates to a higher likelihood of persistence over time. (A) If the goal is for 50% chance of persistence after 10 years, 100 individuals are required; (B) If the goal is for 90% chance of persistence after 100 years, 3,000 individuals are required; (C) If the goal is for 90% chance of persistence after 1,000 years, 100,000 individuals are required. After Traill et al., 2010, CC BY 4.0.

Several studies have attempted to come up with a "universal" MVP value that could ensure that a population of any species has a reasonable chance of persistence. The estimates vary greatly. For example, a universal MVP estimate from the 1980s, the "50/500 rule", suggested that at least 50 individuals are necessary to prevent inbreeding, and 500 to prevent genetic drift (Frankham et al., 2014). While this 50/500 rule is currently used to guide the IUCN Red List Criteria for small populations (see e.g. Table 8.1, Criteria D), more recent studies suggested that this estimate is much too low. For example, one study that considered over 1,000 species calculated that 1,377 individuals must be protected to ensure the survival of the population and species (Brooke et al., 2006). Another study argued that 4,169 adults needed to be protected (Traill et al., 2007), while a third study identified 7,316 adults as the universal MVP (Reed et al., 2003). The reason why these estimates are highly variable is because MVPs are context specific, with the results varying greatly by species, location, and degree of threat (Flather et al., 2011). For some species, it might be necessary to protect large numbers of individuals—maybe thousands or tens of thousands for invertebrates and annual plants with that can experience large population size fluctuations. For other species, such as those that are long-lived and reproduce regularly, protecting only a few hundred individuals may suffice. Unfortunately, many threatened species have population sizes much smaller than any of these recommended minimums. For example, half of the 23 surveyed elephant populations remaining in West Africa have fewer than 200 individuals (Bouché et al., 2011), a number considered to be vastly inadequate for their long-term survival, especially in the absence of strong conservation management.

While a universal MVP value will probably never be agreed upon, species and location specific MVP estimates have great value for guiding conservation efforts. For example, it can suggest the minimum number of individuals that need to be released to improve chances of reintroduction success (Section 11.2). MVP estimates can also be combined with a species' home range requirements to determine a **minimum dynamic area (MDA)**, which is the smallest area of suitable habitat required to sustain the MVP. The use of MVP and MDA, and factors influencing it, were well illustrated in several studies on South Africa's fragmented cheetah (*Acinonyx jubatus,* VU) population. Here, researchers originally found that translocations every 1–5 years can greatly improve the likelihood of persistence for 20 subpopulations with at least 10 cheetahs each, or for 10 subpopulations with at least 15 cheetahs each (Lindsey et al., 2009). But a follow-up study then showed that these results were context specific, and highly dependent on the presence of other predators that compete for the same prey (Lindsey et al., 2011). With no competitors, a minimum dynamic area of 200 km^2 would be sufficient to support 10 cheetahs. However, a reserve of at least 700 km^2 would be needed for 15 cheetahs and 15 lions, and even more if other competitor carnivores are present.

Because of the close relationship between population viability and habitat availability, these two factors are often considered together in **population and habitat viability assessments** (**PHVA**, http://www.cpsg.org/our-approach/workshop-processes/phva-workshop-process). Such an assessment was recently performed for Sierra Leone's western chimpanzees (*Pan troglodytes verus*, CR), where 53 conservation partners came together to develop a recovery plan for this highly threatened species (Carlsen et al., 2012). As an illustration of how conservation projects can bring people from different walks of life together, the participants for this PHVA came from universities, government, NGOs, and the private sector, and included Paramount Chiefs, representatives from the UN and Sierra Leone's government ministries, and well as experts in tourism, communications, population modelling, and wildlife rehabilitation.

> Because of the close relationship between population viability and habitat availability, these two factors are often considered together in population and habitat viability assessments (PHVA).

9.2.4 Effective population size

One of the most important considerations when estimating MVPs is deciding which individuals to include in the calculations. Because population viability depends greatly on a population's ability to produce young to increase in size (or at least balance out mortality), it makes sense that reproductive status is important in MVP estimations. For that reason, biologists often calculate MVPs using the **effective population size (N_e)**, an estimate of how many individuals or pairs in a population are actively breeding. Consider, for example, a school of 1,000 dolphins; it might have 990 immature individuals and only 10 mature dolphins (five males and five females) that are actively breeding. Even though the full population consists of 1,000 dolphins, the effective population size is only 10—just the mature breeding dolphins.

It is worth noting that the effective population size may sometimes be even smaller than the number of individuals capable of breeding at any one time. Factors that cause such a scenario include unequal sex ratios, variation in reproductive output, or an inability to find mates. Health status may also play a role; for example, many long-lived seabird species will forego breeding in years where adults did not attain a necessarily healthy body condition, or years when food is scarce (Crawford et al., 2008). Such reduced effective population sizes can lead to drastic population declines, especially when unsuitable conditions persist over consecutive years.

A population's effective size is often much smaller than the total population size because not all individuals are capable of breeding at any one time.

It is also important to remember that the individuals included in effective population size are not the only ones deserving conservation attention. For example, while young animals may not immediately contribute to population growth and stability, they remain a conservation priority for their potential to contribute to population viability in future. Protecting non-reproductive individuals is also important to avoid having cooperative breeders such as African wild dogs (*Lycaon pictus*, EN) succumb to Allee effects (Section 8.7.2). Individuals that forego reproduction because of poor body condition (e.g. malnourished individuals) can easily become reproductive, and contribute to population viability, if their stressors are mitigated. Lastly, for many species (e.g. many plants, fungi, bacteria, and protists), many (sometimes all) individuals may be dormant for long periods in the soil as seeds, spores, tubers, or other structures. While these dormant individuals may not be part of the effective breeding population, they still contribute to population viability in the long term.

9.2.5 Maximum sustainable yield

A population's maximum sustainable yield provide an estimate of the greatest number of individuals that can be harvested without detriment to the population.

An important but under-utilised benefit of PVAs is the ability to help conservation managers estimate sustainable harvest rates for wildlife populations at risk of overharvesting (Milner-Gulland and Rowcliffe, 2007). Many threatened species can withstand some level of harvesting, so long as harvest rates are lower than **recruitment** rates. To estimate the sustainable level of harvesting, biologists may use PVA to estimate a population's **maximum sustainable yield**—the greatest number of individuals that can be harvested without detriment to the population (Box 9.3). When estimating maximum sustainable yields for overharvested taxa, it is important to consider not only the total population size (or effective population size), but also harvesting biases produced by harvester preferences and techniques. For example, when estimating hunting quotas, hunter preferring larger animals (Lindsey et al., 2013; Barthold et al., 2016) and animal behaviour (Caro et al., 2009) can significantly influence model output. Also, in fisheries management, it is important to consider the outsized role older and larger fish play in recruitment rates, or the indirect damage fishing does the environment or to juvenile individuals (De Leo and Micheli, 2015). Lastly, it is important to consider how harvesting right at maximum sustainable yield levels may leave those populations less buffered to future disturbances (Cumming and Cumming, 2015)—it is thus advisable to maintain harvest quotas well below maximum levels.

Box 9.3 Sustainably Harvesting Fruit Bats Through Better Understanding of Life Histories

David T. S. Hayman

Molecular Epidemiology and Public Health Laboratory, Hopkirk Research Institute, Massey University, Palmerston North, New Zealand.

d.t.s.hayman@massey.ac.nz

> "Full-time hunters are employed to shoot them [...] Nothing is known about the natural factors that encourage or repress population growth in the straw-coloured fruit bat, or on what age category these factors have maximum effects. [...] No laws, customs or taboos protect the straw-coloured fruit bat from exploitation [...].
>
> Funmilayo, 1978

Funmilayo's comments from the 1970s still ring true about the problems currently facing straw-coloured fruit bats (*Eidolon helvum*, NT) throughout their African distribution range. The species is hunted widely in West and Central Africa (Figure 9.C), with scientists estimating that over 128,000 and up to 306,000 individuals are killed annually in Ghana (Kamins et al., 2011) and Côte D'Ivoire (Niamien et al., 2015). In Ghana, the population "is hunted far beyond maximum sustainable yield", concluded an initial study that predicted maximum sustainable yield based on population sizes and estimated intrinsic rate of increase (Kamins et al., 2011). This overhunting has been going on for a long time. Notes from a 1909 field trip to DRC reported them hunted in the "hundreds" (Allen et al., 1917).

Figure 9.C (Top) Smoked straw-coloured fruit bats, an important source of protein in many parts of Africa, for sale at a local market in Ghana. (Bottom) Straw-coloured fruit bats at their daytime roost. Photographs by D. Hayman, CC BY 4.0.

Despite these reports of intensive harvesting, straw-coloured fruit bats remain abundant with colonies comprised of several million individuals often reported. The species is also highly mobile, migratory, and panmictic, breeding freely across its continental distribution (Peel et al., 2017). These life history traits make it difficult to determine how hunting is impacting the population because presence and size of colonies are highly variable in space and time (Hayman and Peel, 2016).

What can the natural history of the species tell us about their possible vulnerability to the pressures inherent of being hunted? Straw-coloured fruit bats exhibit classic life history traits of long-lived species. These bats invest time and energy into single, well-developed pups that they nurse and carry, as Funmilayo noted, "until they are capable of independent existence, which gives [the pup] a high chance of survival". This investment in individual young and single annual breeding events means that straw-coloured fruit bats are susceptible to over-harvesting.

Good information on birth and death rates are required to accurately assess the impact of harvesting. Hayman et al. (2012) estimated birth rate and survival probability parameters in a single colony of up to 1 million straw-coloured fruit bats that roost in trees in Accra, Ghana, demonstrating the feasibility of obtaining such information. Histological examination of tooth growth layers allowed age estimation and life-table analyses to estimate an annual survival probability for juveniles of 43% and adults of 83%. Mark-recapture data using radio-collar telemetry and multi-state models to address confounding emigration estimated lower annual adult survival probability, c. 63%. True survival probabilities likely exist between these estimates, as follow up studies from four further locations suggest (Hayman and Peel, 2016), because permanent emigration may underestimate capture–recapture estimates and population decline may bias life table estimates. Birth rates for the species are high (0.96 young per female per year). Improved estimation of these key parameters will allow for critical analyses of harvest sustainability of straw-coloured fruit bat populations in future.

9.2.6 Sensitivity analysis

A particularly useful feature of PVA—and models in general—is that model parameters can be individually evaluated to better understand the implications of different management strategies. This is usually accomplished with a **sensitivity analysis**, a method that determines which parameter or combination of parameters has the biggest influence on population viability. Obviously, parameters that greatly influence population viability should become the focus of conservation efforts, whereas parameters that have a minimal effect can be given less attention. Some of the most popular model parameters to investigate are demographic parameters by age class, which can identify which life stages are most sensitive to conservation management. Such a sensitivity analysis might reveal that slight changes in adult mortality rates greatly affect population viability, whereas relatively large changes in juvenile recruitment rates have a minimal impact. Crouse et al. (1987) obtained such a result in their classic study on loggerhead turtles (*Caretta caretta* VU) living off the USA's Atlantic coast. At the time, great effort was invested in improving hatching success and ensuring that hatchlings to reach the sea. However, Crouse's study showed that, even if 100% egg and hatchling success was achieved, sea turtles will remain threatened unless adult survival were also improved. Results from this study subsequently played a significant role in initiating global efforts to reduce sea turtle bycatch during fisheries operations (see e.g. Fennessy and Isaksen, 2007; Ayinla et al., 2011).

9.3 Challenges to PVA Implementation

9.3.1 Lack of adequate data

Population biologists often require several years of survey data to distinguish long-term population trends from "model noise"—short-term population fluctuations caused by weather and other unpredictable events (Figure 9.6). For that reason, general guidelines suggest that at a minimum, six (Morris and Doak, 2002) to 10 (McCarthy et al., 2003) years' worth of population data are required before a PVA is attempted.

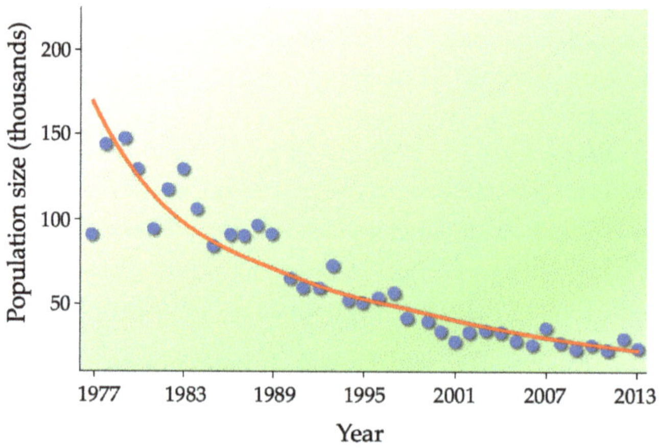

Figure 9.6 It often takes several years of data to distinguish long-term population trends from the "noise" caused by short-term fluctuations. In this example, it appears as if Kenya's topi (*Damaliscus lunatus jimela*, VU) population size is relatively stable, and sometimes even increasing, between 1977 and 1989. However, the 82% decline is unmistakable when long-term trends are considered. After Ogutu et al., 2016, CC BY 4.0.

In recent years, considerable effort has been invested in collating, summarising, and making available demographic datasets. One example is the Demographic Species Knowledge Index (Conde et al., 2019) meant to summarise demographic data obtained from ex situ conservation facilities (Section 11.5). Nevertheless, most African species continue to lack multi-year datasets, while many threatened species lack reliable survey data altogether. Because the enormous task of filling these data gaps is impractical, there is a need to be strategic as to which populations to consider for PVA purposes. For example, it does not make sense to conduct a PVA on each species in a threatened ecosystem when a few carefully selected indicator species will suffice to monitor ecosystem health (McGeoch et al., 2002). Other priorities for PVA efforts include (1) species harvested by humans, (2) species most sensitive to ecosystem changes, (3) species with the greatest uncertainty regarding viability, and (4) species that are the focus of current management efforts (Wilson et al. 2015).

Population biologists often require several years of survey data to distinguish long-term population trends from short-term population fluctuations.

But even in the absence of reliable and complete datasets, PVAs can still be useful. For example, sensitivity analysis can inform future data collection efforts, particularly to fill gaps that lead to high levels of uncertainty, or to verify data accuracy for particularly sensitive parameters.

9.3.2 Data reliability

While strategically filling data gaps should be a priority, it should not come at the expense of data quality and reliability. Many—perhaps most—population monitoring programmes are poorly designed (Buckland and Johnston, 2017), leading to biased data, poor survey precision, and misleading results. Poorly designed surveys not only waste valuable time and resources, but the erroneous results also seriously hamper conservation efforts.

To overcome these shortcomings, there are five criteria that a well-designed monitoring programme should satisfy (Buckland and Johnston, 2017). First, survey sites should represent the region or species of interest. Second, a sufficiently large number of monitoring sites should be chosen. Third, monitoring programmes should be set up that every target species—whether common or rare—is adequately counted. Fourth, species selected for monitoring should represent the community of interest, rather than charismatic species that are easily detected. Fifth, multiple surveys need to be conducted over time to detect long-term population trends. Given resource constraints, some compromises in survey design may at times be required. It may also be worth considering the use of citizen scientists and new technologies such as camera traps (Section 9.1.4) to improve data collection efficiency and to provide back-up evidence of reported species for follow-up expert review, if needed.

9.3.3 Model reliability

While PVAs can provide reasonably accurate predictions when based on reliable data (Brook et al., 2000; McCarthy et al., 2003), many conservationists continue to be sceptical of PVA results and their ability to predict future population changes over time (Crone et al., 2013). Part of the reason is our inability to accurately account for unanticipated future events, such as unusual weather events or the arrival of a new invasive species. There are also mechanistic challenges to PVA modelling, including their sensitivity to model assumptions and slight changes in model parameters i.e. slight changes in model input generate vastly different results. For this reason, some biologists have started to discourage the use of PVAs in conservation management, especially when faced with inadequate data (Ellner et al., 2002).

While this scepticism is important and model interrogation should always be welcomed (both aspects usually lead to model improvements), PVA will continue to play a crucial role in conservation in the foreseeable future. It is however important for biologists using PVA to be familiar with the challenges associated with model reliability, as well as the assumptions and limitations of each PVA model. It always helps to begin any PVA model with a clear understanding of the ecology of the target population, the threats it faces, and its demographic characteristics, which in turn enables the modellers to better evaluate model results.

9.4 Summary

1. Protecting a threatened species requires a firm grasp of its population biology. Long-term monitoring using biodiversity inventories, population censuses, and demographic studies can reveal temporal changes in population size and distribution and help to distinguish short-term fluctuations from long-term decline.
2. Biologists are increasingly relying on innovative methods to track wildlife populations and demographics. Among the most popular are market surveys, hair snares and faecal sampling, while photos taken by tourists and camera traps have also been used to obtain population-level data.
3. Population viability analysis (PVA) uses demographic, genetic, and environmental data to predict changes in population sizes and extinction risk over time. Sensitivity analysis can be used to guide conservation action by estimating how different management actions will affect a population's extinction probability.
4. Minimum viable population estimates can be used to determine how many individuals are needed to reduce the threat of extinction, while maximum sustainable harvest estimates can be used to set harvest limits on species threatened by overharvesting.
5. Many surveys are poorly designed, leading to biased data, poor survey precision, and misleading results, which hamper our ability to halt biodiversity losses. To overcome these challenges, surveys should be representative, sufficiently large, and conducted repeatedly over time.

9.5 Topics for Discussion

1. Read the manuscript by Pfab and Witkowski, (2000), which is a PVA study that is relatively easy to understand. Can you identify some strengths and weaknesses of this PVA? Which assumptions did this study make? What parameters were used? Are there any other model parameters you think could have been useful?
2. For this exercise, you are going to construct a simple PVA for a threatened frog species on a sheet of paper. This frog formerly occupied an expansive lowland forest, which over time was disturbed and degraded. A recent survey was able to find only ten frogs (five males and five females), all in

one small, isolated forest patch that can accommodate up to 20 frogs. In the spring, males and females form mating pairs; each pair typically produce zero, one, two, three, or four, offspring that survive to breeding age the following year (to create this demographic dataset, flip four coins for each mated pair; the number of heads is the number of offspring). The sex of the offspring is assigned at random (flip a coin for each young animal, with heads for males and tails for females. Individuals not mated because of uneven sex ratios do not breed. After the breeding season, all the adult frogs die. (A) Run five different population simulations for five generations each, and chart population size over time. What percentage of populations would go extinct within the 10 generations? (B) Perform a sensitivity analysis by making the frogs' living conditions more severe. For example, lower the number of frogs found during the survey to six, or impose 50% mortality on offspring every year due to introduced rats. (C) Perform another sensitivity analysis by making the frogs' living conditions more accommodating. For example, examine the impact of supplying extra food to the frogs, which would allow more offspring to be produced each year. Examine the results of all your different models to determine which factor is most important to address to ensure the frog species does not go extinct.

9.6 Suggested Readings

Anderson, S.C., T.A. Branch, A.B. Cooper, et al. 2017. Black-swan events in animal populations. *Proceedings of the National Academy of Sciences* 114: 3252–57. https://doi.org/10.1073/pnas.1611525114 Ignoring extreme weather events may severely underestimate extinction risk.

Buckland, S.T., and A. Johnston. 2017. Monitoring the biodiversity of regions: Key principles and possible pitfalls. *Biological Conservation* 214: 23–34. https://doi.org/10.1016/j.biocon.2017.07.034 Five principles for reliable surveys.

Danielsen, F., N.D. Burgess, P.M. Jensen, et al. 2010. Environmental monitoring: The scale and speed of implementation varies according to the degree of peoples' involvement. *Journal of Applied Ecology* 47: 1166–68. https://doi.org/10.1111/j.1365-2664.2010.01874.x Involving the local community in environmental monitoring increases the speed of conservation actions.

Guschanski, K., L. Vigilant, A. McNeilage, et al. 2009. Counting elusive animals: Comparing field and genetic census of the entire mountain gorilla population of Bwindi Impenetrable National Park, Uganda. *Biological Conservation* 142: 290–300. https://doi.org/10.1016/j.biocon.2008.10.024 Genetic techniques are providing new opportunities for studying populations.

Jewell, A. 2013. Effect of monitoring technique on quality of conservation science. *Conservation Biology* 27: 501–08. https://doi.org/10.1111/cobi.12066 The methods that researchers use to tag and monitor species can affect and sometimes even harm the species being studied.

Mascia, M.B., S. Pailler, M.L. Thieme, et al. 2014. Commonalities and complementarities among approaches to conservation monitoring and evaluation. *Biological Conservation* 169: 258–67.

https://doi.org/10.1016/j.biocon.2013.11.017 Describes different objectives in monitoring, with a goal of developing standard procedures for evaluating projects.

Sebastián-González, E., J.A. Sánchez-Zapata, F. Botella, et al. 2011. Linking cost efficiency evaluation with population viability analysis to prioritize wetland bird conservation actions. *Biological Conservation* 144: 2354–61. http://doi.org/10.1016/j.biocon.2011.06.015 Different management approaches are evaluated for their cost effectiveness on bird populations in Spain.

One of the following two texts:

Bibby, C., M. Jones, and S. Marsden. 1998. *Expedition Field Techniques: Bird Surveys* (London: Royal Geographic Society). http://www.conservationleadershipprogramme.org/media/2014/09/Bird_Surveying_Manual.pdf Methods for conducting biological surveys. Written on birds, but applicable to other taxa.

White, L., and A. Edwards. 2000. *Conservation Research in the African Rain Forests: A Technical Handbook* (New York: WCS). http://apes.eva.mpg.de/eng/pdf/documentation/WhiteEdwards2000 Methods for obtaining data on animals and their environment. Written for forest work, but also applicable in other ecosystems.

Bibliography

Allen, J.A., H. Lang, and J.P. Chapin. 1917. *The American Museum Congo Expedition Collection of Bats* (New York: Order of the Trustees, American Museum of Natural History). http://digitallibrary.amnh.org/handle/2246/1068

Anderson, S.C., T.A. Branch, A.B. Cooper, et al. 2017. Black-swan events in animal populations. *Proceedings of the National Academy of Sciences* 114: 3252–57. https://doi.org/10.1073/pnas.1611525114

Ayinla, O.A., A.B. Williams, D.A. Bolaji, et al. 2011. *Development of turtle excluder device (TED) and its adoption in Nigeria* (Lagos: Nigerian Institute for Oceanography and Marine Research). http://hdl.handle.net/1834/5337

Barthold, J.A., A.J. Loveridge, D.W. Macdonald, et al. 2016. Bayesian estimates of male and female African lion mortality for future use in population management. *Journal of Applied Ecology* 53: 295–304. https://doi.org/10.1111/1365-2664.12594

Bongers, F., L. Poorter, W.D. Hawthorne, et al. 2009. The intermediate disturbance hypothesis applies to tropical forests, but disturbance contributes little to tree diversity. *Ecology Letters* 12: 798–805. https://doi.org/10.1111/j.1461-0248.2009.01329.x

Bouché, P., I. Douglas-Hamilton, G. Wittemyer, et al. 2011. Will elephants soon disappear from West African savannahs? *PLoS ONE* 6: e20619. https://doi.org/10.1371/journal.pone.0020619

Bouley P., M. Poulos, R. Branco, et al. 2018. Post-war recovery of the African lion in response to large-scale ecosystem restoration. *Biological Conservation* 227: 233–42. https://doi.org/10.1016/j.biocon.2018.08.024

Brook B.W., J.J. O'Grady, A.P. Chapman, et al. 2000. Predictive accuracy of population viability analysis in conservation biology. *Nature* 329: 512–19. https://doi.org/10.1038/35006050

Brook, B., M. Burgman, and R. Frankham. 2000. Differences and congruencies between PVA packages: The importance of sex ratio for predictions of extinction risk. *Conservation Ecology* 4: 6.

Brook, B.W., L.W. Traill, and C.J.A. Bradshaw. 2006. Minimum viable population sizes and global extinction risk are unrelated. *Ecology Letters* 9: 375–82. https://doi.org/10.1111/j.1461-0248.2006.00883.x

Buckland, S.T., and A. Johnston. 2017. Monitoring the biodiversity of regions: Key principles and possible pitfalls. *Biological Conservation* 214: 23–34. https://doi.org/10.1016/j.biocon.2017.07.034

Carlsen, F., K. Leus, K. Traylor-Holzer, et al. 2012. *Western chimpanzee population and habitat viability assessment for Sierra Leone: Final report*. IUCN/SSC CBSG—Europe (Copenhagen: CBSG Europe). http://www.cpsg.org/sites/cbsg.org/files/documents/Sierra%20Leone%20Chimpanzee%20PHVA%20Final%20Report.pdf

Caro, T.M., C.R. Young, A.E. Cauldwell, et al. 2009. Animal breeding systems and big game hunting: Models and application. *Biological Conservation* 142: 909–29. https://doi.org/10.1016/j.biocon.2008.12.018

Caughley, G. 1977. *Analysis of Vertebrate Populations* (New York: Wiley).

Conde, D.A., J. Staerk, F. Colchero, et al. 2019. Data gaps and opportunities for comparative and conservation biology. *Proceedings of the National Academy of Sciences* 116: 9658–64. https://doi.org/10.1073/pnas.1816367116

Crawford, R.J.M., L.G. Underhill, J.C. Coetzee, et al. 2008. Influences of the abundance and distribution of prey on African penguins *Spheniscus demersus* off western South Africa. *African Journal of Marine Science* 30: 167–75. https://doi.org/10.2989/AJMS.2008.30.1.17.467

Crone, E.E., M.M. Ellis, W.F. Morris, et al. 2013. Ability of matrix models to explain the past and predict the future of plant populations. *Conservation Biology* 27: 968–78. https://doi.org/10.1111/cobi.12049

Crouse, D.T., L.B. Crowder, and H. Caswell. 1987. A stage-based population model for loggerhead sea turtles and implications for conservation. *Ecology* 68: 1412–23. https://doi.org/10.2307/1939225

Cumming, D.H.M., and G.S. Cumming. 2015. One Health: An ecological and conservation perspective. In: *One Health: The Theory and Practice of Integrated Health Approaches*, ed. by J. Zinsstag, et al. (Wallingford: CAB International).

De Leo, G.A., and F. Micheli. 2015. The good, the bad and the ugly of marine reserves for fishery yields. *Philosophical Transactions of the Royal Society B* 370: 20140276. https://doi.org/10.1098/rstb.2014.0276

Eggert, L.S., R. Buij, M.E. Lee, et al. 2014. Using genetic profiles of African forest elephants to infer population structure, movements, and habitat use in a conservation and development landscape in Gabon. *Conservation Biology* 28: 107–18. https://doi.org/10.1111/cobi.12161

Ellner, S.P., J. Fieberg, D. Ludwig, et al. 2002. Precision of population viability analysis. *Conservation Biology* 16: 258–61. https://doi.org/10.1046/j.1523-1739.2002.00553.x

Fennessy, S.T., and B. Isaksen. 2007. Can bycatch reduction devices be implemented successfully on prawn trawlers in the Western Indian Ocean? *African Journal of Marine Science* 29: 453–63. https://doi.org/10.2989/AJMS.2007.29.3.12.342

Flather, C.H., G.D. Hayward, S.R. Beissinger, et al. 2011. Minimum viable populations: Is there a 'magic number' for conservation practitioners? *Trends in Ecology and Ecolution* 26: 307–16. https://doi.org/10.1016/j.tree.2011.03.001

Formia, A., B.J. Godley, J.F. Dontaine, et al. 2006. Mitochondrial DNA diversity and phylogeography in West and Central African green turtles (*Chelonia mydas*). *Conservation Genetics* 7: 353–69. https://doi.org/10.1007/s10592-005-9047-z

Frankham, R., C.J.A. Bradshaw, and B.W. Brook. 2014. Genetics in conservation management: Revised recommendations for the 50/500 rules, Red List criteria and population viability analyses. *Biological Conservation* 170: 56–63. https://doi.org/10.1016/j.biocon.2013.12.036

Funmilayo, O. 1978. Fruit bats for meat: are too many taken? *Oryx* 14: 377–78. https://doi.org/10.1017/S0030605300016008

Hayman, D.T.S., and A.J. Peel. 2016. Can survival analyses detect hunting pressure in a highly connected species? Lessons from straw-coloured fruit bats. *Biological Conservation* 200: 131–39. https://doi.org/10.1016/j.biocon.2016.06.003

Hayman, D.T.S., R. McCrea, O. Restif, et al. 2012. Demography of straw-colored fruit bats in Ghana. *Journal of Mammalogy* 93: 1393–404. https://doi.org/10.1644/11-MAMM-A-270.1

Hughes, N., N. Rosen, N. Gretsky, et al. 2011. Will the Nigeria-Cameroon chimpanzee go extinct? Models derived from intake rates of ape sanctuaries. In: *Primates of Gashaka*, ed. by V. Sommer and C. Ross (New York: Springer). http://doi.org/10.1007/978-1-4419-7403-7

Ingram, D.J., L. Coad, B. Collen, et al. 2015. Indicators for wild animal offtake: Methods and case study for African mammals and birds. *Ecology and Society* 20: 40. http://doi.org/10.5751/ES-07823-200340

Kamins, A.O., O. Restif, Y. Ntiamoa-Baidu, et al. 2011. Uncovering the fruit bat bushmeat commodity chain and the true extent of fruit bat hunting in Ghana, West Africa. *Biological Conservation* 144: 3000–08. https://doi.org/10.1016/j.biocon.2011.09.003

Kays, R., M.C. Crofoot, W. Jetz, et al. 2015. Terrestrial animal tracking as an eye on life and planet. *Science* 348: aaa2478. http://doi.org/10.1126/science.aaa2478

Kümpel, N.F., E.J. Milner-Gulland, G. Cowlishaw, et al. 2010. Assessing sustainability at multiple scales in a rotational bushmeat hunting system. *Conservation Biology* 24: 861–71. https://doi.org/10.1111/j.1523-1739.2010.01505.x

Lindsey, P., C.J. Tambling, R. Brummer, et al. 2011. Minimum prey and area requirements of the vulnerable cheetah *Acinonyx jubatus*: Implications for reintroduction and management of the species in South Africa. *Oryx* 45: 587–99. https://doi.org/10.1017/S003060531000150X

Lindsey, P., K. Marnewick, H. Davies-Mostert, et al. 2009. *Cheetah (*Acinonyx *jubatus) population habitat viability assessment workshop report* (Johannesburg: IUCN CBSG and EWT). http://www.cbsg.org/sites/cbsg.org/files/documents/South%20African%20Cheetah%20PHVA%202009.pdf

Lindsey, P.A., G.A. Balme, P. Funston, et al. 2013. The trophy hunting of African lions: Scale, current management practices and factors undermining sustainability. *PLoS ONE* 8: e73808. https://doi.org/10.1371/journal.pone.0073808

Marnewick, K., S.M. Ferreira, S. Grange, et al. 2014. Evaluating the status of and African wild dogs *Lycaon pictus* and cheetahs *Acinonyx jubatus* through tourist-based photographic surveys in the Kruger National Park. *PloS ONE* 9: e86265. https://doi.org/10.1371/journal.pone.0086265

Maxwell, S.M., G.A. Breed, B.A. Nickel, et al. 2011. Using satellite tracking to optimize protection of long-lived marine species: Olive ridley sea turtle conservation in Central Africa. *PloS ONE* 6: e19905. https://doi.org/10.1371/journal.pone.0019905

McCarthy, M.A. 2007. *Bayesian Methods for Ecology* (Cambridge: Cambridge University Press). https://doi.org/10.1017/CBO9780511802454

McCarthy, M.A., S.J. Andelman, and H.P. Possingham. 2003. Reliability of relative predictions in population viability analysis. *Conservation Biology* 17: 982–89. https://doi.org/10.1046/j.1523-1739.2003.01570.x

McGeoch, M.A., B.J. van Rensburg, and A. Botes. 2002. The verification and application of bioindicators: A case study of dung beetles in a savanna ecosystem. *Journal of Applied Ecology* 39: 661–72. https://doi.org/10.1046/j.1365-2664.2002.00743.x

Metcalfe, K., P.D. Agamboué, E. Augowet, et al. 2015. Going the extra mile: Ground-based monitoring of olive ridley turtles reveals Gabon hosts the largest rookery in the Atlantic. *Biological Conservation* 190: 14–22. https://doi.org/10.1016/j.biocon.2015.05.008

Milner-Gulland, E.J., and M.J. Rowcliffe. 2007. *Conservation and Sustainable Use: A Handbook of Techniques* (Oxford: Oxford University Press).

Morris, W.F., and D.F. Doak. 2002. *Quantitative Conservation Biology* (Sunderland: Sinauer).

Niamien, M., J. Coffi, B. Kadjo, et al. 2015. Initial data on poaching of *Eidolon helvum* (Kerr, 1792) near-threatened species in Côte D'ivoire, West Africa. *European Journal of Scientific Research* 35: 219–27.

Ogutu, J.O., H.-P. Piepho, M.Y. Said, et al. 2016. Extreme wildlife declines and concurrent increase in livestock numbers in Kenya: What are the causes? *PLoS ONE* 11: e0163249. https://doi.org/10.1371/journal.pone.0163249

Parker, L. 2017. New ocean reserve, largest in Africa, protects whales and turtles. *National Geographic.* http://on.natgeo.com/2samx3a

Peel, A.J., J.L.N. Wood, K.S. Baker, et al. 2017. How does Africa's most hunted bat vary across the continent? Population traits of the straw-coloured fruit bat (*Eidolon helvum*) and its interactions with humans. *Acta Chiropterologica* 19: 77–92. https://doi.org/10.3161/15081109 ACC2017.19.1.006

Péron, G., and R. Altwegg. 2015. Twenty-five years of change in southern African passerine diversity: Nonclimatic factors of change. *Global Change Biology* 21: 3347–55. https://doi.org/10.1111/gcb.12909

Pfab, M.F., and E.T.F. Witkowski. 2000. A simple population viability analysis of the critically endangered *Euphorbia clivicola* R.A. Dyer under four management scenarios. *Biological Conservation* 96: 263–70. http://doi.org/10.1016/S0006-3207(00)00088-4

Pikesley S.K., P.D. Agamboue, J.P. Bayet, et al. 2018. A novel approach to estimate the distribution, density and at-sea risks of a centrally-placed mobile marine vertebrate. *Biological Conservation* 221:246256. https://doi.org/10.1016/j.biocon.2018.03.011

Reed, D.H., J.J. O'Grady, B.W. Brook, et al. 2003. Estimates of minimum viable population sizes for vertebrates and factors influencing those estimates. *Biological Conservation* 113: 23–34. https://doi.org/10.1016/S0006-3207(02)00346-4

Shaffer, M.L. 1981. Minimum population sizes for species conservation. *BioScience* 31: 131–34. https://doi.org/10.2307/1308256

Sherley, R.B., T. Burghardt, P.J. Barham, et al. 2010. Spotting the difference: Towards fully-automated population monitoring of African penguins *Spheniscus demersus*. *Endangered Species Research* 11: 101–11. https://doi.org/10.3354/esr00267

Stalmans M., T.J. Massad, M.J.S. Peel, et al. 2019. War-induced collapse and asymmetric recovery of large-mammal populations in Gorongosa National Park, Mozambique. *PLoS ONE* 14: e0212864. https://doi.org/10.1371/journal.pone.0212864

Steenweg, R., M. Hebblewhite, R. Kays, et al. 2017. Scaling-up camera traps: Monitoring the planet's biodiversity with networks of remote sensors. *Frontiers in Ecology and the Environment* 15: 26–34. https://doi.org/10.1002/fee.1448

Tinley, K.L. 1977. *Framework of the Gorongosa ecosystem*. Ph.D. thesis (Pretoria: University of Pretoria). http://hdl.handle.net/2263/24526

Traill, L.W., B.W. Brook, R.R. Frankham, et al. 2010. Pragmatic population viability targets in a rapidly changing world. *Biological Conservation* 143: 28–34. https://doi.org/10.1016/j.biocon.2009.09.001

Traill, L.W., C.J.A. Bradshaw, and B.W. Brook. 2007. Minimum viable population size: A meta-analysis of 30 years of published estimates. *Biological Conservation* 139: 159–66. https://doi.org/10.1016/j.biocon.2007.06.011

White, L., and A. Edwards. 2000. *Conservation Research in the African Rain Forests: A Technical Handbook* (New York: WCS). http://apes.eva.mpg.de/eng/pdf/documentation/WhiteEdwards2000.pdf

Wilson, H.B., J.R. Rhodes, and H.P. Possingham. 2015. Two additional principles for determining which species to monitor. *Ecology* 96: 3016–22. https://doi.org/10.1890/14-1511.1

Witt, M.J., B. Baert, A.C. Broderick, et al. 2009. Aerial surveying of the world's largest leatherback turtle rookery: A more effective methodology for large-scale monitoring. *Biological Conservation* 142: 1719–27. https://doi.org/10.1016/j.biocon.2009.03.009

Witt, M.J., E.B. Augowet, A.C. Broderick, et al. 2011. Tracking leatherback turtles from the world's largest rookery: Assessing threats across the South Atlantic. *Proceedings of the Royal Society B* 278: 2338–47. https://doi.org/10.1098/rspb.2010.2467

10. Conserving Ecosystems

Rising 500 m above the Sahelian plains of Mali, the sandstone cliffs of the Bandiagara Escarpment are home to the Dogon people, whose unique homes are carved into the cliff's walls. The Escarpment is a World Heritage Site, known for its outstanding cultural and natural value. However, both nature and humans are suffering from environmental degradation due to climate change and unsustainable land use. Photograph by Timm Guenther, https://commons.wikimedia.org/wiki/File:Les_Falaises_de_Bandiagara.jpg, CC BY-SA 3.0.

https://doi.org/10.11647/OBP.0177.10

Habitat loss (and its associated degradation) is currently the most important threat facing Africa's wildlife (Figure 10.1). When an ecosystem is destroyed or degraded, its ability to sustain wildlife is compromised, and the individuals that depend on that ecosystem for survival either need to adapt or move elsewhere, or they will die. Conversely, preventing ecosystem degradation and destruction is one of the single most important actions we can take to protect biodiversity. In the process, we also improve our own well-being, given that natural ecosystems are our first line of defence against natural disasters, and provide us with food, clean water, and other ecosystem services.

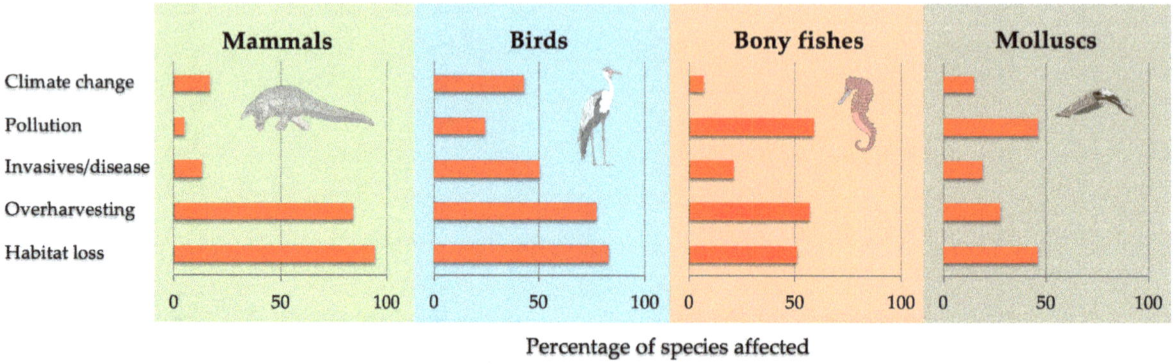

Figure 10.1 Habitat loss and degradation, much of which are driven by agriculture, are the most important threats to Africa's wildlife, followed by overharvesting, invasive species and disease, and pollution. Groups of species often face similar threats: mammals and birds are more likely to be threatened by habitat loss, while fish and molluscs are more likely to be threatened by pollution. Percentages add up to more than 100% because many species face multiple threats. The influence of climate change is under-estimated because its impact on most species still need to be assessed. Source: IUCN, 2019, CC BY 4.0.

Broadly speaking, ecosystem conservation involves three different activities: (1) monitoring ecosystems, (2) maintaining ecosystems, and (3) restoring damaged ecosystems. While many books have been written on each of these three activities, the broad overview this chapter provides will hopefully enable readers to gain a basic understanding of the tools and methods used in ecosystem conservation.

10.1 Ecosystem Monitoring

A complex and adaptive ecosystem in which all the chemical, physical, and biological components, functions, and processes are intact and functioning normally is considered a **healthy ecosystem** (but see Cumming and Cumming (2015), for a discussion on this value-based term). In contrast, disturbing any of an ecosystem's components, functions, and/or processes will, by definition, alter them to some degree (Table 10.1). In many cases, ecosystems that have been exposed to certain forms and levels of disturbances remain healthy because there is redundancy in the roles performed by different ecosystem components (Section 4.2.1). This ability of an ecosystem to withstand certain forms and levels of disturbances is referred to as ecosystem stability. Ecosystem stability could be the result of one or both of two qualities: resistance and resilience. **Resistance** is the ability of an ecosystem to retain the same characteristic communities and natural cycles throughout and after a disturbance event, while **resilience** is the ability of an ecosystem to rapidly recover or adapt after a disturbance event. For example, if the number of native aquatic insect species decline after non-native fishes are introduced to previously fish-free ponds, the pond's ecosystem has low resistance. But if the native insect community recovers rapidly after the non-native fishes were removed, the ecosystem is resilient.

Table 10.1 Three ways how humans have changed the natural world.

Natural function	Changes attributed to human activities
Land surface	As much as half of the world's ice-free land surface has been transformed to cater to people's need for natural resources. Much of these changes are driven by agricultural activities.
Nitrogen cycle	Human activities release massive amounts of nitrogen into natural ecosystems on a daily basis. Much of this occurs through the use of nitrogen fertilisers, burning fossil fuels, and cultivating nitrogen-fixing crops.
Atmospheric carbon cycle	Scientists estimate that humans would have doubled levels of carbon dioxide in the Earth's atmosphere by the middle of this century. This is primarily the result of fossil fuel use and deforestation.
Wildlife populations	Between 1970 and 2014, Sub-Saharan Africa have lost three quarters of its freshwater vertebrates; the rate of these declines shows no sign of reducing.
Pollutants	Pollution from human activities have become so omnipresent that it is hard to escape its impacts. Microplastics have been found in drinking water and the food we eat (Chapter 7).

Sources: MEA, 2005; Kulkarni et al., 2008; http://www.livingplanetindex.org

Many forms of ecosystem disturbances are easy to observe. Consequently, monitoring these visible forms of disturbances—such as the outright destruction of a forest or plastic pollution on a beach—focuses less on detection and more on developing systematic survey protocols (Section 9.1) that can provide information on whether a disturbance is spreading and increasing in intensity, or whether conservation action is successful in containing the threat. However, some disturbances are subtler, unobtrusive, and thus difficult to detect; examples include pesticide drift and agricultural runoff (Section 7.1). Adopting a "wait and see" approach to detecting these invisible forms of disturbances can be particularly damaging, since that approach generally ends at a point where the harm will either be impossible to reverse or will require significantly more resources and time than would have been the case if the problem was addressed earlier. In this way, there are many similarities between monitoring ecosystem health and human health—some ailments are easier to diagnose than others, but we avoid the worst-case scenarios by screening regularly for diseases and treating the threatening ones promptly.

Figure 10.2 Biologists sampling aquatic macroinvertebrates in the Okavango Delta, Botswana, as part of a biomonitoring project of the Freshwater Research Centre. Photograph by Helen Dallas/FRCSA, CC BY 4.0.

Perhaps the most popular method conservation biologists use to monitor ecosystem health is known as **biomonitoring**. By monitoring the abundance and/or fitness of sensitive species (Box 10.1), biologists can sometimes detect ecosystem degradation before it becomes apparent to the human eye or escalates to a point where it starts impacting human lives (Bornman and Bouwman, 2012). Monitoring environmental indicators such as macroinvertebrates (Figure 10.2) is particularly popular when examining the ecological condition of aquatic ecosystems; mayflies, caddisflies, and stoneflies—specialists of undisturbed streams—are often replaced by flies and midges in polluted and disturbed environments. Sometimes however, when plants or animals are not easily monitored, certain aspects of those species can still be monitored. One example is monitoring total plant biomass as a proxy for soil nutrients or intensity of herbivory. Another option is to perform a **bioassay**, during which a sensitive organism (typically water fleas or plankton) is released into a potentially contaminated environment to see if death or declining health occurs.

By monitoring the abundance and/or fitness of sensitive species, biologists can detect threats to biodiversity before it becomes apparent to the human eye.

Box 10.1 Using Insects to Monitor Environmental Health

Rosina Kyerematen

Department of Animal Biology and Conservation Science,
University of Ghana,
Legon, Ghana.

✉ *rkyerematen@ug.edu.gh*

Insects are important to nearly every terrestrial food web in the world and serve a multitude of different purposes: some insects are responsible for pollination of plants while others are scavengers that clean up dead plant and animal material. In some cases, our understanding of an insect species' ecological role can make it suitable as an indicator of environmental health. Biomonitoring looks at the presence and abundance of organisms within their natural communities to assess the impact of environmental disturbances; this knowledge can then be used to guide **ecosystem management**. An indicator taxon is one whose impact can be specifically and precisely measured; its abundance serves as a measure of the overall health of an ecosystem. Understanding how the presence and abundance of indicator species, and the relative abundance of tolerant and intolerant species, reflects the relative health of an environment can allow for rapid surveys of impaired ecosystems to assess trends as well as to track changes following remediation and restoration efforts.

Over the past few years, biomonitoring with insects as indicator taxa has become increasingly popular in Ghana. Butterflies are especially popular because they show varying relative sensitivities to environmental change; the abundance of certain butterfly species can for example be used to study the impact of habitat loss, fragmentation, and climate change (Kyerematen et al., 2018). The presence and abundance of butterflies more characteristic of open and disturbed ecosystems (Figure 10.A) can, for example, be used as an indicator of forest degradation.

Aquatic insects, particularly benthic macroinvertebrates, are also useful bioindicators. Freshwater resources, such as lakes and rivers, provide water for drinking and washing to local people, and a home for economically important taxa, such as fish and shellfish. Protecting these water sources is therefore important for safeguarding people's health and livelihoods. The presence, absence, and diversity of certain benthic macroinvertebrates, even at the order level, can provide valuable information about whether a waterbody is being degraded or not (Kyerematen et al., 2014; Nnoli et al., 2019). A recent study showed that dragonfly and damselfly diversities and populations along the coastal Densu River in Ghana vary widely depending on the physical condition of the river and surrounding area (Acquah-Lamptey et al., 2013).

Figure 10.A Because the green-banded swallowtail (*Papilio nireus*) prefers open woodlands, researchers in Ghana are using their abundances in forests as a measure of habitat degradation. Photograph by Celesta von Chamier, CC BY 4.0.

With their high diversity and varying tolerances for ecosystem conditions, insects are extraordinarily suited as ecological indicators in environmental monitoring. Each insect species is also part of a wider biological community with important ecological roles. If lost, not only will an abundance of other life be affected, such a loss may also hint at a looming crisis facing people living in those compromised ecosystems.

At times, conservation biologists may need to measure the physical environment to assess environmental health. This approach is particularly common when tracking pollution, for example by monitoring for changes in **biochemical indicators**. For example, measuring total phosphorus, nitrogen, and dissolved oxygen load in streams and other surface water can help scientists track eutrophication (Section 7.1.1). Measuring these and other biochemical indicators is usually accomplished directly via chemical analysis of environmental samples, such as soil and water. Sometimes however, biochemicals indicators are tracked indirectly via biological samples obtained from plants and animals. Because they bio-accumulate heavy metals and other pollutants, filter feeders such as clams and mussels (e.g. Bodin et al., 2013) are particularly useful in this regard as they can be used to detect very low concentrations of harmful chemicals in the environment.

10.1.1 Monitoring ecosystems with geospatial analysis

A persistent challenge facing biologists who monitor ecosystems—and other aspects of biodiversity—is achieving consistency across space and time. Consider a survey of a sensitive bird community to track ecosystem change; not only will different observers have varying levels of experience, but they will almost certainly see different birds during an early morning census compared to one later in the afternoon due to differences in biology and behaviour between species. These factors introduce error into monitoring data, which in turn can mask the effect a biologist tries to measure (Buckland and Johnston, 2017). Laboratory scientists' control for these **confounding factors** by making multiple measurements under strictly controlled conditions. But for conservation biologists working outside in the wind and rain, repeated observations under similar conditions can be near impossible.

Geospatial analysis offers a variety of tools that allow biologists to overcome some of these traditional field monitoring challenges. These tools use **geographic information systems (GIS)** computer software packages to store, display, and manipulate a wide variety of data representing the natural environment, biodiversity, and human land-use patterns as they relate to one another on Earth's surface. GIS thereby allows biologists to easily visualise and analyse spatial relationships between mapped data, which may include aspects such as vegetation types, climate, soils, topography, geology, water availability, species distributions, existing protected areas, human settlements, and human resource use (Figure 10.3). Understanding such relationships helps conservation biologists to prioritise their actions, for example by identifying areas where data are lacking, where an environmental change requires further investigation, or where gaps in regional protected areas network exist.

Remote sensing is a special branch of geospatial analysis directed at obtaining ecosystem data without making physical contact (i.e. boots on the ground) with the observation site. Before the turn of the 20th century, the most popular form of remote sensing was aerial photography from airplanes. These aerial photographs facilitated geographers' ability to draw maps of landscape features, including human infrastructure and natural vegetation patterns. Remote sensing opportunities greatly expanded from 1960 onward, with the launch of the National Aeronautical Space Administration's (NASA) first **Earth observation satellites**, to take photographs of Earth from space for weather forecasting. Subsequent satellite programmes expanded their scope to also collect additional data of Earth's surface and atmosphere. While much of this data would have been useful to conservation, early satellite data products were very expensive and thus largely out of reach of the larger conservation community. This all changed in 2008 when NASA started distributing their Earth observation products for free to the public, heralding an era in which remote sensing became a standard tool in the conservation field.

> Remote sensing offers a variety of tools that allow biologists to monitor biodiversity beyond the abilities of traditional field monitoring techniques.

Today, hundreds of Earth observation satellites circle the planet, offering near real-time access to unbiased and consistent environmental datasets of nearly all terrestrial surfaces, oceanic surfaces and floor depths, and the atmosphere, all from the comfort of a computer connected to the internet (Wilson et al., 2013). Scientists use these products in ecosystem monitoring efforts, including monitoring water quality (Dube et al. 2015), forest loss (Laporte et al.,

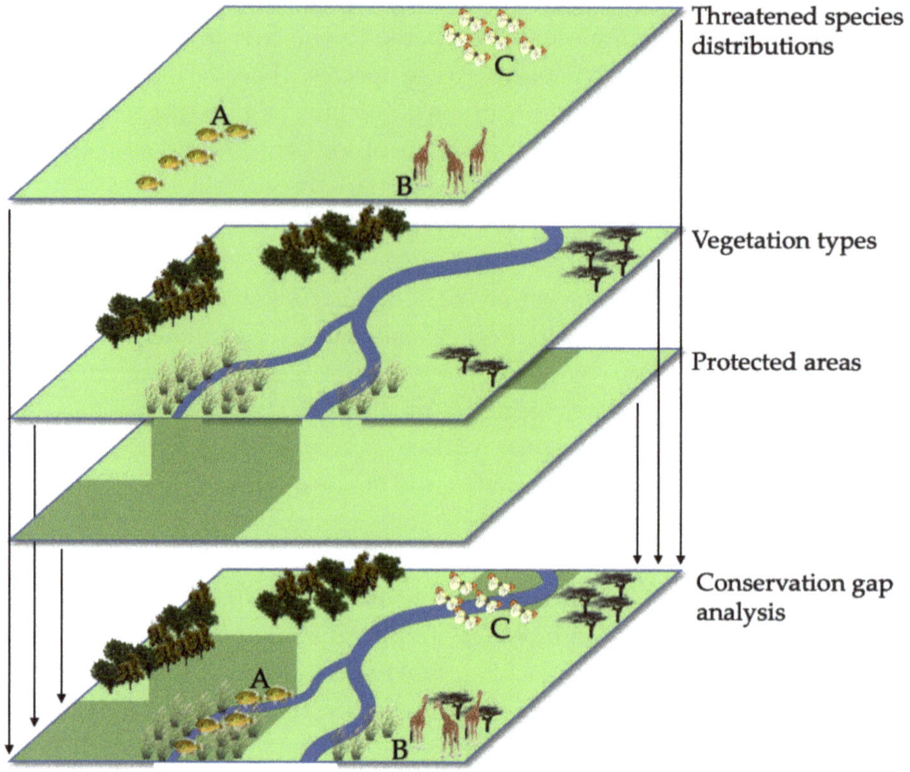

Figure 10.3 Distribution maps for three threatened species are overlaid onto maps of protected areas and vegetation types to identify potential conservation gaps (discussed in Chapter 13). This GIS analysis shows that Species A is well protected, Species B is protected to some extent, and Species C is not protected at all. Species C would thus be considered a high conservation priority. After Scott et al., 1991, CC BY 4.0.

2007), coral reef health (McClanahan et al., 2011), desertification (Symeonakis et al., 2004), and fire regimes (Archbald et al., 2010). Linking the information obtained by Earth observation satellites to biological information collected on the ground has proved invaluable in monitoring species' threat statuses (Di Marco et al., 2014), **ecosystem connectivity** (Wegmann et al., 2014), and habitat suitability (Torres et al., 2010), as well as understanding how biodiversity responds to environmental changes (Box 10.2).

Conservation biologists are often faced with the shifting baseline syndrome, where the reference points they use to measure their progress may be vastly different from earlier states.

The popularity and utility of these products have preceded and facilitated the expansion of other remote sensing applications in ecosystem monitoring. Among the most popular are radar products (Figure 10.4), which have become the standard method for obtaining elevation and other terrain data (NASA, 2009, 2013), as well as estimates of carbon stocks (Carreiras et al., 2012). Advances have also been made in using hyper-spectral imagery to monitor soil properties (Mashimbye et al., 2012) and even individual trees (Naidoo et al., 2012). LiDAR has enabled biologists to

Box 10.2 Remote Sensing and Spatial Analysis for African Conservation

Barend F. N. Erasmus

Global Change Institute (GCI), University of the Witwatersrand, Johannesburg, South Africa.

Barend.Erasmus@wits.ac.za

Remote sensing is the art and science of observing objects or landscapes from a distance, without being in direct contact with the environment. Although you can think of wildlife photography as a type of remote sensing, the term usually refers to aerial photography (Figure 10.B), or images taken via satellite. For both cases, there is a trade-off between the area covered by each photo (the footprint), how much digital storage space is available, and how often a satellite takes a picture of the same area. For satellites that take pictures of the earth systematically, along a pre-defined path, cloud cover determines how often you can get a usable image. Recent improvements in technology, now allow for the deployment of constellations of satellites that have a collective point-and-track capability to observe an area almost continuosly.

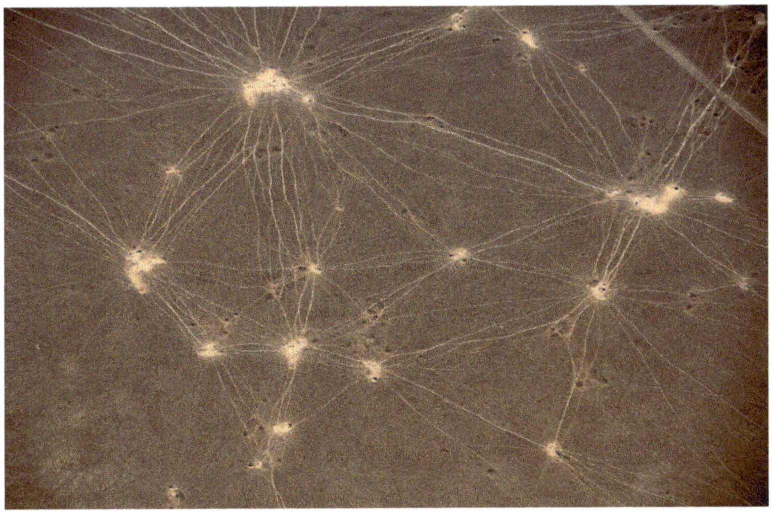

Figure 10.B Aerial photograph of Kaa pan in the southwestern Kalahari region of Botswana. Gemsbok (*Oryx gazella*, LC), red hartebeest (*Alcelaphus buselaphus caama*, LC), and eland (*Tragelaphus oryx*, LC) excavate holes for lekking (eating soil with high salt content), and the footpaths between the different lekking sites clearly show preferences for certain sites. In the same way, on a much broader scales, ungulates move across landscapes to access other resources, such as water and grass, whilst avoiding human hunters and predators. Photograph by B.F.N. Erasmus, CC BY 4.0.

Our eyes are sensitive to the colours red, green, and blue, together called the visible part of the electromagnetic spectrum. This is actually a very small part of the entire spectrum, and scientists have found, for example, that vegetation seen as uniformly green shows a lot of variation in the infrared part of the spectrum. For this reason, many satellites carry cameras that can "see" infrared light, and by proxy, measure vegetation health, biomass, and sometimes even structure. This capability, together with frequent revisit rates, allows for a unique view of how African landscapes change, whether through habitat loss, seasonal changes, or drought. Analyses of 13 years of remotely sensed data show that the most arid parts of southwest Botswana now experience typical summer vegetation conditions later in the year (Dubovyk et al., 2015), so herbivores must cope with a much longer dry season now than they experienced in 2000.

African landscapes are always changing, and sometimes in unpredictable ways—localised thunderstorms at the beginning of the rainy season can rapidly change a dry dustbowl landscape into green grazing. These "wet footprints" can cover areas as small as 1 km × 3 km; in contrast, a large frontal weather system can cover dozens or even hundreds of kilometres. An ungulate looking for green grazing not too far from drinking water, while trying to avoid predators and hunters, needs the ability to detect such green patches, and the strength and knowledge to move there. Obtaining this knowledge from an ungulate's perspective of the landscape is no small task. Hopcraft et al. (2014) show how common wildebeest and plains zebra in the Serengeti have different migration strategies: wildebeest move to green grass as quickly as they can, with little effort to avoid predators, whereas zebra gauge predation risk and forage quality concurrent to their trek.

Our ability to understand ungulate movements and other ecosystem patterns has been greatly enhanced over the last few years, as animal-tracking technology (where GPS positions are logged and stored using radio-frequency tracking, mobile phone networks, satellite systems, or any combination) and the resulting analyses became more sophisticated. It is now possible to distinguish between locations where animals often spend a little time (for example, a preferred shady tree for resting during the heat of the day), or the same location where they infrequently spend a lot of time (for example, once during an oestrus cycle, a lactating lioness will spend a lot of time at her den, with cubs). It is also possible to "see" when such an animal changes "mode" of movement. For example, researchers have found that springbok (*Antidorcas marsupialis*, LC) rams around the Etosha pan in Namibia show sedentary behaviour (small movements around a specific area) when grazing is good, which transitions to searching behaviour during dry seasons (long, relatively fast movements in a straight line) as they move to areas that had water or grass during previous drought periods (Lyons et al., 2013).

If we combine this animal tracking capability with regular remotely sensed images of vegetation quality, then we can start to answer questions about why certain animals move to certain areas. It also highlights the fact that it is very seldom good enough to put a fence around an area and call it preserved. Due to the changing nature of Africa's savannahs, animals need to be able to move to areas with water or forage, often outside reserves, when they become available. In the face of a changing climate and changing rainfall patterns, this ability to move long distance to reach vital resources remains one of the best adaptation responses that African ungulates may have to cope with climate change. However, fences, roads, and farmland may block these migrations across the landscape, putting those populations at risk of extinction (Section 5.1.1).

African conservation areas need to make provision for these animal movements, or risk conservation areas without sustainable animal populations. This presents a problem—how do we investigate options for large animals to move across a transformed rural landscape and minimise human-wildlife conflict while, on the other hand, still providing access to green grass or drinking water at a specific protected area? Both resources change in location and time of year, and only regular, detailed remote sensing of vegetation, combined with detailed animal movement studies, will provide the necessary picture in time and space.

map three-dimensional vegetation, which can be used to explain animal movements (Loarie et al., 2013; Davies et al., 2016) and measure carbon stocks and forest loss (Burton et al., 2017).

Despite the opportunities presented by remote sensing, it is critical to remember that it is not a substitute for traditional field monitoring methods. Most importantly, remote sensing applications cannot be considered reliable without verification using field data (see Burton et al., 2017). Most remote sensing products are also relatively new, which does not allow for enough opportunities to compare across time. In the absence of historical remotely sensed data, geospatial analysts usually choose the best **reference site** currently available; this may expose those analysts to **shifting baseline syndrome**, because the chosen reference site may be vastly different from earlier states the scientists are actually interested in studying (Bunce et al., 2008; Papworth et al., 2009). Remote sensing is, therefore, not a cure-all for ecosystem monitoring challenges; it is simply a powerful tool to supplement traditional field-based monitoring.

10.2 Maintaining Complex and Adaptive Ecosystems

Even when monitoring data show that an ecosystem is healthy, management is often required to maintain those desired conditions. That is because very few ecosystems are completely free of human influences. For example, rivers and streams carry

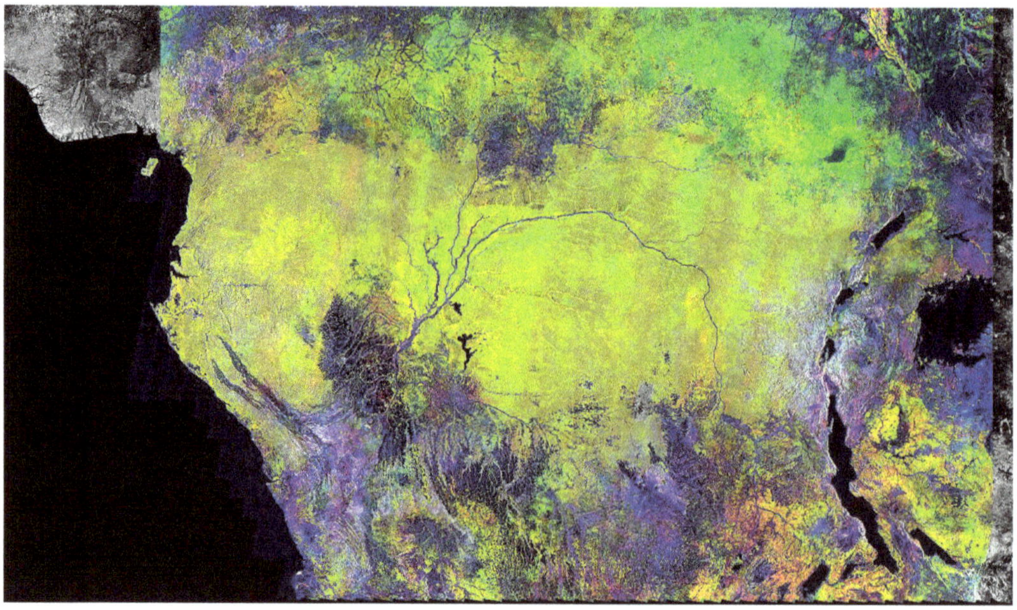

Figure 10.4 A radar composite of equatorial Africa obtained in 1996 by Japan's Earth Resources Satellite. The area shown covers about 7.4 million km^2. Due to the filters applied to the image, yellow represents flooded forests, palm plantations, and urban areas; green represents forest, and black represents surface water. Savannahs may be black, blue, purple, or green, depending on the ecoregion type. Displays such as these allow researchers to map ecosystems, and ecosystem loss at a much larger scale than if only field observations were used. Image courtesy of NASA, https://images.nasa.gov/details-PIA01348.html, CC0.

pollutants far beyond the point of contamination (Section 7.1) and roads acting as firebreaks suppress natural fire regimes. Today, even the most isolated patches of habitats may not be completely protected from the influences of global processes such as climate change. To maintain complex and adaptive ecosystems, conservation biology is guided by four complementary management principles: (1) maintain ecosystem processes, (2) minimise external threats, (3) be adaptive, and (4) be minimally intrusive.

10.2.1 Maintaining critical ecosystem processes

Ecologists generally divide **ecosystem processes** into four disparate yet interdependent categories: water cycling, nutrient cycling (which include the carbon and nitrogen cycle), energy flow, and community dynamics. The linkages between these processes create feedback loops, where changes in one factor may be amplified elsewhere. Maintaining ecosystem processes is thus very important because small, seemingly small, changes can have major impacts on biological communities.

The water cycle

The water cycle refers to the distribution of water through an ecosystem, and includes the absorption and distribution of water vapour, rainwater, and surface water in lakes, rivers, and oceans. Since much of the water cycle happens out of sight and is generally associated with large-scale phenomena, such as weather patterns and anthropogenic climate change, land managers sometimes fail to recognise how local factors influence the water cycle. This is a grave mistake; many deadly ecological disasters (e.g. desertification, flooding, and landslides) can be attributed to disturbances to the water cycle at the local scale.

Outside of ensuring sustainable use of water resources, maintaining vegetation cover arguably plays the most important role in preserving the water cycle at local scales. Plants and their roots enable soil to store and release water, and make these water reserves available for soil organisms, which in turn aid in decomposition of dead plants and animals (Section 4.2.2). In contrast, a loss of vegetation cover increases surface runoff, which leads to deteriorating soil conditions through nutrient **leaching** and erosion of fertile topsoil. An increasing number of studies have also shown how forest loss can change a region's climate by reducing rainfall which in turn exacerbates drought conditions (Lawrence and Vandecar, 2015). For example, forest clearing for agriculture has reduced rainfall by 50% over much of West Africa (Garcia-Carreras and Parker, 2011). Many forest restoration programmes thus focus on reversing these losses.

When restoring degraded forests and other ecosystems to repair the water cycle and other ecosystem services, it is important to remember that complex ecosystems with locally-adapted plants are generally the most effective in maintaining the water cycle and other ecosystem services (Burton et al., 2017). There have been cases across Africa where well-intended restoration efforts used fast-growing timber species such as gum (*Eucalyptus* spp.) and pine (*Pinus* spp.). While these single-crop plantations may superficially resemble a forest in structure and may even provide some of the same ecosystem services as native plants, some of these fast-growing exotic plants also bring significant environmental harm and negative externalities passed onto local people (van Wilgen and Richardson, 2014). Of particular concern is their role in disrupting local water cycles (Section 7.4.2), ironically the very aspect these forest restoration efforts aim to rehabilitate. The choice of species used for restoration should this be carefully considered to avoid unintended consequences later on.

Complex and adaptive ecosystems provide more opportunities for people to benefit from ecosystem services than uniform plantations with single species

The nutrient cycle

The nutrient cycle involves the cycling of essential nutrients such as carbon, nitrogen, sulphur, and phosphorus through the ecosystem. Like the water cycle, natural vegetation cover plays an important role in maintaining the nutrient cycle. That is because plant roots slow water runoff which, in turn, help soil to retain nutrients dissolved in water. Plants also form a major component of above-ground and below-ground biomass. When dead plant biomass is decomposed along with animal waste products, nutrients previously absorbed through plant roots are released back into the soil and water, where they can once again be absorbed by living plants and other consumers.

Unfortunately, vegetation cover, decomposition, and fire dynamics (discussed below) alone cannot ensure a healthy nutrient cycle. Much of Africa is nutrient impoverished because soil nutrients are lost quicker than they are replaced. One of the main causes is unsustainable agricultural practices (Sanchez, 2010), such as farming on sandy soils and in tropical forests. These areas are nutrient-poor, so crop yields are typically low. Because these areas are prone to leaching, a large proportion of synthetic fertilisers added to supplement impoverished soils leaches into groundwater or washes into nearby streams and lakes, threatening water supplies by causing harmful algae blooms and eutrophication (Section 7.1.1). To compensate, the failing farmers may resort to even more unsustainable land conversions (Wallenfang et al., 2015). Careful management of the nutrient cycle is thus critical for both biodiversity conservation and socio-economic well-being, particularly given its importance to food security (Drechsel et al., 2001). To achieve this, there is an urgent need to adopt more sustainable land management practices (Chapter 14).

> Ecosystem processes are linked into multiple feedback loops, so changes in one factor are amplified elsewhere.

The energy cycle

Energy flow—a crucial component of ecosystem productivity (Section 4.2.2)—refers to the capture and storage of solar energy by primary producers (photosynthetic plants, algae, and some bacteria), and the distribution of that energy to consumers, detritivores, and decomposers. Although solar energy can appear as an unlimited resource in many ecosystems, the energy available to consumers (i.e. herbivores and carnivores) is limited because only about 10% of the energy obtained at one trophic level is passed on to the next (Figure 10.5). Being at the top of the food chain, apex predators are in a particularly vulnerable position because seemingly small disruptions at lower trophic levels will have a cumulative impact on the energy available to them. Such disruptions may include reduced prey populations (e.g. overharvesting of herbivores, Section 7.2) or foraging disruptions (e.g. a predator needing to walk further to find prey). Research from Southern Africa's Kalahari Desert has showed that such disruptions, which are amplified by predators' high-energy lifestyles (it takes a lot of energy to bring down a large ungulate!), may put apex predators such as cheetahs on a downward spiral of energy deficits (Scantlebury et al., 2014). While such impacts may not always lead to direct mortality, these insidious, subtle, and easily-overlooked sublethal impacts compromise individuals' ability to reproduce, with extinction being the end result. To avoid such a scenario, maintaining energy flow generally involves maintaining complex, species-rich ecosystems so that consumers have ample opportunities to fulfil their energy needs for finding prey, growth, reproduction, and other activities.

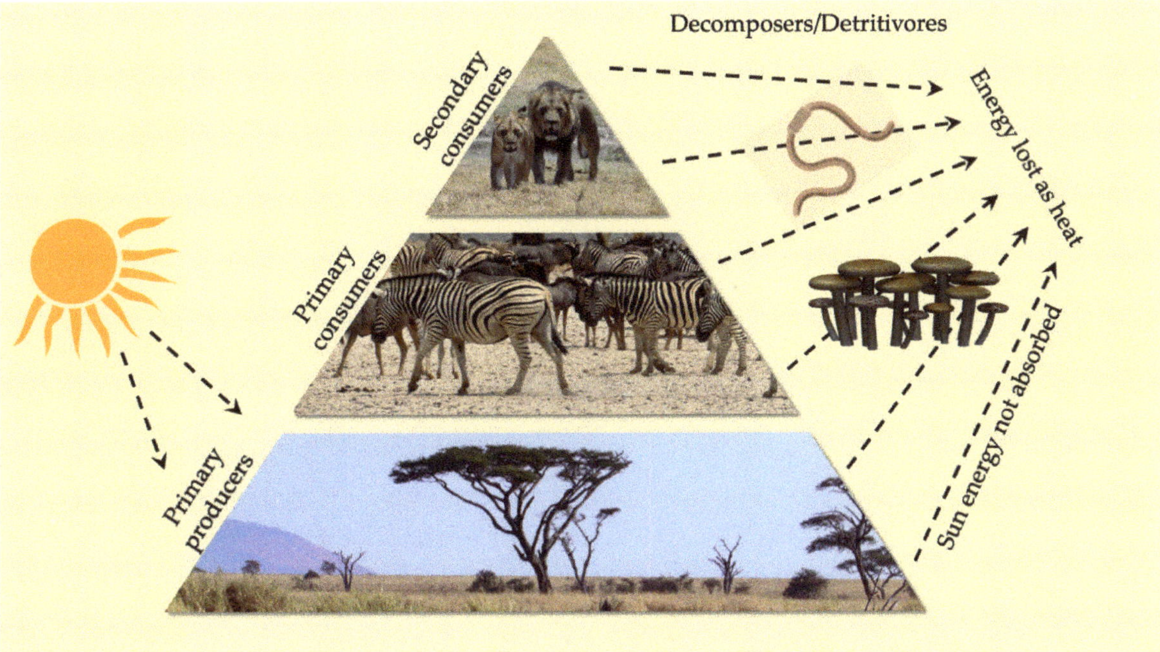

Figure 10.5 A food pyramid of a model savannah ecosystem, showing the various trophic levels and energy pathways. About 90% of energy is lost at each trophic level through respiration and animal waste excretion. CC BY 4.0.

Community dynamics

In ecosystem conservation, maintaining viable populations of different interacting species is as important as maintaining important ecosystem processes, such as ecosystem productivity and ecological succession. This focus often falls on maintaining populations that form part of important mutualistic relationships such as pollination and seed dispersal (Section 4.2.5), predator-prey interactions, and even healthy levels of competitive and parasitic interactions (which allow more species to persist). Of interest is the preservation of keystone species and ecosystem engineers, which has an outsized effect on community dynamics (Section 4.2.1). As illustrated in this, and other, chapters, disrupting community dynamics through pollution, overharvesting, or any other threat facing biodiversity (Chapter 5–7), generally leads to impoverished natural communities. Impoverished communities may in turn provide opportunities for invasive species to colonise an area, further perpetuating biodiversity losses. In Chapter 11, we discuss further how populations and species can be maintained.

Fire Dynamics

Although fire is generally not considered one of the four fundamental ecosystem processes, it plays such an important role in African biodiversity management, including maintaining the four fundamental ecosystem processes, that it deserves its own discussion. African farmers understand the importance of setting fire to keep cropland and grazing pastures productive; burning existing vegetation releases carbon and other essential nutrients beneficial for plant growth into the environment. Similarly, fire also plays a critical role in the flow of energy, community dynamics, and overall maintenance of **fire-dependent ecosystems**, such as grasslands, savannahs, and Mediterranean communities. Suitably low intensity fires seldom kill living plants; rather, they encourage seed germination and seedling growth by reducing dead material that may crowd new growth, by exposing bare mineral soil (the substrate required for many seeds to germinate), and by releasing vital nutrients into the soil. This periodic removal of dead material also prevents fuel load accumulation, thereby preventing future fires from becoming destructive. In contrast, without fire, fire-dependent ecosystems will slowly transform into unproductive scrublands suffocated by encroaching woody vegetation (Smit and Prins, 2015). Then, when wildfires do occur (e.g. through human negligence or lightning) the resultant accumulated fuel loads increase the intensity and heat of fires, creating very dangerous and difficult to control scenarios.

Obviously, given the potentially destructive force of fire, land managers who use fire as a management tool must consider many aspects before setting a **prescribed burn**, also known as a controlled burn. Foremost, to prevent a fire from becoming destructive to natural communities and nearby human developments, burning must be done in a well-planned manner with careful consideration given to the area's ecology, weather forecasts, and fire-readiness of the site (Goldammer and de Ronde, 2004; Kelly and Brotons, 2017). It is also recommended that prior to burning (or any other conspicuous management operation for that matter), land managers develop a public outreach plan to explain to local people the importance of fire in ecosystems management, and the steps taken to keep them and their properties safe. To further improve community relations and education, South Africa's Working on Fire programme (Figure 10.6) provides scholarships, fire training, and employment opportunities to local youths.

Fire **management plans** that match natural fire regimes produce the best results for effective **ecosystem management**. Land managers accomplish this by ensuring that their burn plans mimic the local area's natural fire season, fire frequency, and flame intensity, while also accounting for management goals and local ecological factors such as rainfall and geology (see e.g. van Wilgen et al., 2010, 2014). The size of each burn area must also be considered. Best practices suggest not burning the entirety of a community at a time; rather, burning only portions of an area allows for more habitat heterogeneity, provides opportunities for non-burrowing animals to take refuge in unburned areas, and maximises ecosystem diversity. Bringing all of these aspects together, scientists working in Tanzania's Ngorongoro Conservation Area determined that the area would respond best if land managers burn up to 20% of their grasslands annually or biannually (Estes et al., 2006). Similarly, field **experiments** in certain South African grasslands have found that plant diversity is highest when burns occur every second year, in winter or autumn (Uys et al., 2004). Other ecosystems, such as the Cape Floristic Region's fynbos, may need to burn only once every decade (Kraaij et al.,

Figure 10.6 A fire crew from South Africa's Working on Fire programme keeps a close eye on a controlled fire set to keeps the native savannah ecosystem healthy. Photograph by Working on Fire, CC BY 4.0.

2013). However, because of the high density of houses in some fynbos ecosystems, the periodic fires needed for locally-adapted vegetation to persist are often extinguished because of the threat to human settlement (van Wilgen et al., 2012).

While fire plays an important role in many African landscapes, it is important to note that overly frequent fires can be a threat even to fire-dependent communities. For example, habitat degradation resulting from too many fires in quick succession can leave a natural community vulnerable to invasions by harmful species (Masocha et al., 2011). Overly frequent fires can also prevent seedling recruitment by directly killing vulnerable young plants, and by depleting the seed bank because seedlings do not have sufficient time to mature and set seed.

Fire-sensitive ecosystems (e.g. tropical forests, high mountains, and peat bogs) must also be managed carefully to avoid fire disturbance, which can lead to habitat loss and edge effects (Chapter 5). One way to accomplish this is to educate farmers living adjacent to fire-sensitive ecosystems on how to safely manage their land with fire. Conservationists also need to be considerate when managing fire-dependent ecosystems adjacent to fire-sensitive ecosystems, as is the case with the unique patches of fire-dependent savannahs—remnants from the last Ice Age 15,000 years ago—that are surrounded by forest within Gabon's Lobé National Park (Jeffery et al., 2014). Careful fire management, led by good science, is bound to become increasingly important in the future, given that wildfires are expected to become more frequent and more intense under climate change (Pricope and Binford, 2012).

10.2.2 Minimising external threats

Human activity cannot and does not need to be eliminated from nature; in fact, the structure and diversity of many of today's natural landscapes—and to which today's wildlife are adapted to—are in part the result of past human activities (e.g. Garcin et al., 2018). Today, there are over 7 billion people on Earth, so our impacts are more pervasive than for the majority of history. There is, thus, an urgent need to utilise natural resources in such a way that future generations will also benefit from the ecosystem services that previous generations have left us. This requires a concerted effort from every sector in human society to minimise those threats we impose on the ecosystems around us. This includes preventing pollution, large-scale human disturbances, overharvesting, and habitat destruction (Chapter 5–7).

Major strides have been made in recent years towards achieving these goals. Governments are updating laws to safeguard the environment, industries are refining recycling and waste disposal methods, new techniques are being developed to remove pollutants from the environment, and individual citizens are becoming more aware of their individually small but collectively significant impacts on the environment. We should be proud of the progress being made and continue to strive for improvements. But one external threat that requires greater attention and understanding is invasive species.

Controlling invasive species

Invasive species degrade and destroy natural ecosystems by outcompeting native species, disturbing ecosystem processes, and altering the physical environment (Section 7.4). Limiting these harmful impacts can be particularly challenging since exotic species that establish themselves in a new area can build up such large numbers, become so widely dispersed, and be so thoroughly integrated into ecosystems (i.e. **naturalised**) that eradicating them entirely would be extraordinarily difficult and expensive, or as in the case of tickberry (*Lantana camara*) perhaps even impossible (Bhagwat et al., 2012). This is not only a problem facing conservation biology, but also agriculture, where invasive species often spread from one farm to another, forestry, where invasive species are spread between saw mills and along logging routes, and fisheries, where native resources are outcompeted, sometimes up-ending an entire local industry. The impact of invasive species on farming communities is particularly severe—they lose tens of billions of dollars each year while trying to combat deteriorating grazing lands, reduced crop yields, and escalating pest control expenses. One study from South Africa calculated that invasive plants result in financial losses of US $646 million each year—this figure would have been US $5 billion if invasive species control measures already implemented were absent (de Lange and van Wilgen, 2010). Consequently, a range of stakeholders have invested considerable resources in combatting invasive species.

Because invasive species are often very hard to eradicate once established (Figure 10.7), the foremost step in avoiding invasive species' harmful impacts is to avoid opportunities for new invasions (Section 7.4.1). This requires raising awareness across all levels of society about the dangers posed by invasive species, both to the natural world and to agricultural and natural resources systems. There is also a need for citizens, scientists, and industry to monitor for potential and known invasive species, and promptly implement intensive control efforts to stop establishment and spread. The Global Register of Introduced and Invasive Species (http://www.griis.org) is a free, online searchable source to facilitate these tasks, by providing information about the impact and control of invasive species. Governments can also partake in efforts to control invasive species. While most African countries screen agricultural imports for pests, countries, such as Australia and New Zealand, take this task particularly seriously, with trained officials screening each visitor (and returning residents) and package for hitchhiking species before they cross those countries' borders. Lastly, it would require increased dialogue between conservation biologists and land managers to make a careful and thorough assessment prior to the deliberate introduction of a new species, even if thought of as beneficial.

> The most important step in preventing biological invasions is to prevent the initial establishment of problem species. This requires educating people about the dangers posed by invasive species.

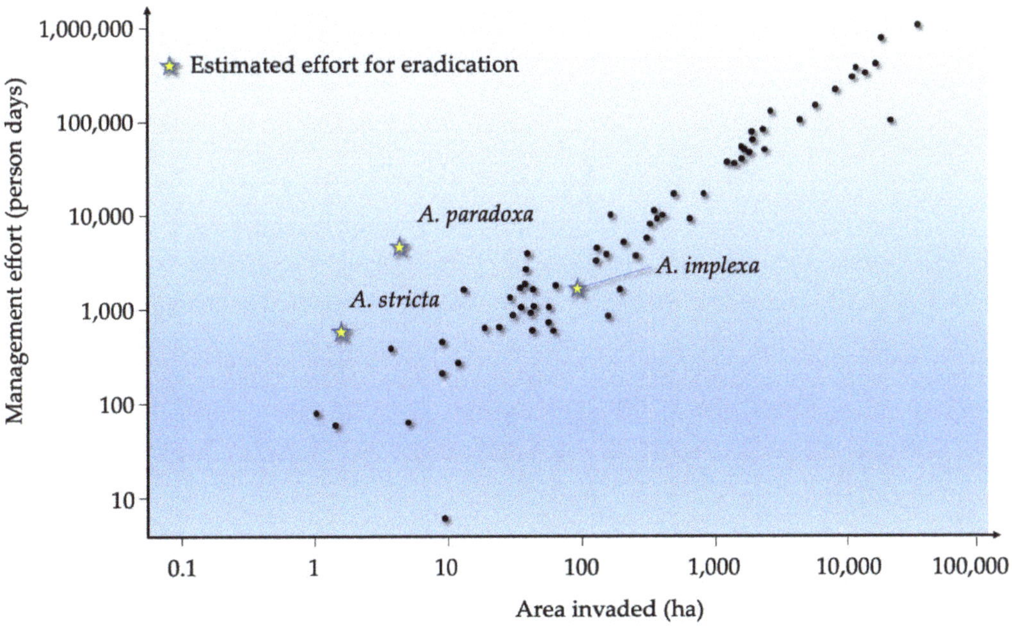

Figure 10.7 Data on the control of invasive species in South Africa have shown that the larger the infestation (a function of time passed since establishment), the more resources are required for eradication. For that reason, it is critical to avoid establishment of invasive species in the first place, and to respond promptly once new invasions have been discovered. Also shown is the effort required to eradicate three species of Australian wattle (*Acacia* spp.), based on extent of invasion. After Wilson et al., 2013, CC BY 4.0.

Despite best practices, not all invasions can be prevented, and for those that do occur, an early detection and rapid response strategy offers the best chance to limit harm. This usually involves raising awareness of potential invasive species to ensure biologists and other stakeholders will recognise a new invasion, efforts to screen for such species on a regular basis, and implementing direct attack approaches, such as using herbicides, pesticides, or mechanical control once detected. While addressing a new invasion as soon as possible, it is also important to consider and contain the risks each direct attack approach carries. For example, herbicides and pesticides carry a risk of killing non-target native species via pesticide drift (Section 7.1), while mechanical control may cause disturbances such as trampling, undue soil disturbance, and even pollution.

Controlling invasive species that have become established will require substantially more resources and manpower to combat than those detected early on. Even so, it is still worth initiating control measures as early as possible since harm and resource needs will only escalate over time. To overcome these impediments to invasive species control, the South African government established an exemplary model, the Working for Water programme, in 1995. Working for Water has three goals—poverty relief, water conservation, and invasive species control—which it accomplishes by combining invasive species control with job creation and social upliftment (van Wilgen and Wannenburgh, 2016). Specifically, the programme hires and trains unemployed people to eradicate water-thirsty invasive shrubs and trees across South Africa; some of the removed plants are subsequently sold as firewood at local markets at a profit for the participants. In its first 20 years, the programme has created over 227,000 person years of employment and treated over 28,000 km^2 of invasive species (Figure 10.8). Hopefully programmes, such as these, will inspire more governments to act to restore degraded ecosystems to their previously healthy state.

Figure 10.8 Members of South Africa's Working for Water programme mechanically removing invasive Australian wattles (*Acacia* spp.). To make sure the wattles don't grow again, the stumps are also treated with a herbicide. Photograph by Christo Marais/Department of Environmental Affairs, CC BY 4.0.

Another method to manage established pests is biological control, also called biocontrol. Biocontrol typically relies on one or more natural enemies from an invasive species' original range to control the pest in its introduced range (Section 4.2.7). One of the main benefits of biocontrol is that it ensures cost-effective, long-term, area-wide control of an invasive species, beyond the capabilities of chemical pesticides and mechanical control. Biocontrol also allows for opportunities to control invasive species that are hard to manage with chemical pesticides and mechanical control (at least without significant additional harm to the environment), such as submerged aquatic weeds (Coetzee et al., 2011). Third, biocontrol agents are highly host-specific, thereby eliminating the impact that chemical pesticides and mechanical control have on non-target organisms. Lastly, an effective biocontrol agent ideally eliminates the need for chemical pesticides and mechanical control, thereby reducing threats such as pesticide pollution (Section 7.1) and ecosystem degradation (Section 5.1).

Biocontrol does have some drawbacks. Primary among them is the significant upfront investment required, as candidate species first need to be found, and then extensively tested for host specificity and potential interactions with native wildlife before being released. Biocontrol also requires careful monitoring after release to determine effectiveness as well as to carefully check for impacts on non-target native species. This monitoring needs to be conducted over the long term, because biocontrol agents typically require several years before they establish self-sufficient colonies in the wild and might only then show signs of unintended impacts. Because alternative methods for controlling invasive species can also kill biocontrol agents (causing conflicting results and wasted resources), additional coordination is required before applying biocontrol and alternative pest management strategies simultaneously in the same area. Lastly, there is no guarantee that a biocontrol agent will be effective. For example, the tickberry continues to thrive despite the release of over 40 biocontrol agents (Zalucki et al., 2007). But when successful, the long-term savings from these upfront investments are generally well worth it. One study in Benin found that biocontrol of water hyacinth required a US $2 million upfront investment, but the resultant water quality improvements increased local incomes by US $84 million per year (de Groote et al., 2003). Another study from South Africa estimated a net gain of 50–3,500 times the investment, depending on the specific biocontrol agent used (de Lange and van Wilgen, 2010).

Several very successful biological control programmes have been implemented in Africa over the past century. Most famous is the rescue of the cassava crop (see Box 4.3). Another successful biocontrol programme, implemented across much of the continent, has reduced Kariba weed (*Salvinia molesta*) by over 95% within just a few years (e.g. Mbati and Neuenschwander, 2005; Diop and Hill, 2009; Martin et al, 2018). South Africa has been particularly active in the research and introduction of biocontrol agents. From 1913, when South Africa started controlling invasive cacti with hemipterans, to 2017, a total of 93 biocontrol agents have been released for the control of 59 invasive species (Zachariades et al., 2017).

While the most popular biocontrol agents generally involve insects, disease-causing pathogens can also be used for biocontrol. Feline panleukopenia virus (also known as feline distemper) was highly effective in managing a feral cat population that caused the extirpation of seabirds on Sub-Antarctic Marion Island; some birds are now even returning as breeders (Bester et al., 2002). In an effort to reduce the use of chemical pesticides on food crops, efforts are also currently underway to find fungal pathogens to control introduced pests impacting African crops, including pea leafminers (*Liriomyza huidobrensis*), originally from South America (Akutse et al., 2013), and banana weevils (*Cosmopolites sordidus*), originally from Southeast Asia (Akello et al., 2008).

Some of the most effective pest control programmes use an integrated pest management (IPM) approach that relies on using multiple pest control methods described above either simultaneously or in succession (van Wyk and van Wilgen, 2002). Strategic planning to coordinate best practices can also help offset some of the costs of invasive species control (Rahlao et al., 2010) and ensure that important pest sources are not missed (van Wilgen et al., 2007). When considering the best method to control an invasive species, it may also help to consider how our own actions inadvertently encourage invasive species. For example, an over-reliance on synthetic fertiliser has been shown to cause eutrophication (Chislock et al., 2013) and encourage growth of aquatic invasive plants (Coetzee and Hill, 2012; Bownes et al., 2012).

> The most effective pest control programmes use an integrated pest management (IPM) approach that relies on using multiple control methods either simultaneously or in succession.

Even though the impacts of invasive and other exotic species are generally considered negative, they do occasionally provide some benefits. For example, Australian pines (*Casuarina equisetifolia*) have been planted widely throughout Africa for timber, charcoal, and to stabilise eroding lands. Some people harvest invasive species to eat or to sell; examples include water hyacinth (Figure 10.9), prickly pears (*Opuntia* spp.), and Mediterranean mussels (*Mytilus galloprovincialis*). The latter has also become an important food source for the African black oystercatcher (*Haematopus moquini*, NT) (Kohler et al., 2009). Similarly, nearly 18 species of diurnal raptors, including four globally threatened species, nest and roost—sometimes in colonies numbering thousands of individuals—in stands of invasive Australian gum (*Eucalyptus* spp.) trees (Allan et al., 1997; Jenkins, 2005). Many birds also favour fruits produced by invasive plants, such as the tickberry and syringa (*Melia azedarach*); this however allows those plants to spread even further. Because they reproduce so fast, water hyacinth has been investigated as a resource for producing bioenergy (Güereña et al., 2015). A ground spider (*Prodida stella*, CR), endemic to the Seychelles and threatened by sea level rise (Gerlach, 2014), was first described from specimens collected on Australian pine, and may thus persist on trees further inland as their original range is submerged by rising sea levels. Nevertheless, assessments generally show that costs incurred from invasive species outweigh the benefits (Mwangi and Swallow, 2008). It is thus important to consider whether native species can fulfil the same functions in those cases where benefits of invasive species are touted.

Figure 10.9 Turning an undesirable weed into a cash crop: entrepreneurs from South Africa to Lake Victoria and Nigeria (pictured) have developed alternative income streams by training members of their communities in how to make handcrafts from invasive water hyacinth. Photograph by MitiMeth, CC BY 4.0.

10.2.3 Adaptive management

In decades past, ecosystem management in Africa was generally conducted following *a laissez-faire* (i.e. hands-off) approach where natural processes were allowed to follow their own course, with interventions only implemented when subjectively deemed absolutely necessary. While this passive management style may have worked in an era when ecosystems were less fragmented by roads and fences, and the impact of pollution and invasive species transported along rivers and streams were limited. But maintaining healthy ecosystems (especially small ones) through limited action is becoming increasingly difficult in today's human-dominated world. Leaving concerns unattended may seem fine in the short term, but such neglect can create problems that are very difficult to contain later on. For this reason, it is increasingly necessary to actively manage ecosystems not only to achieve conservation goals, but even to just avoid degradation of current conditions.

A major challenge of active (and reactive) conservation management is that nature consists of thousands of interacting components and feedback loops. Because it is impossible to fully understand how these different components interact, management strategies are typically implemented with an incomplete understanding of how interventions may impact broader ecosystem processes.

Management actions that draw from experimentation and prior experiences often deliver the desired results. However, despite good intentions and careful consideration, some interventions may later give rise to a cascade of unintended consequences that run counter to overall conservation goals. South Africa's Kruger National Park provides a good example. Conservation managers here once thought they could better protect the local wildlife by fencing the Park and constructing artificial waterholes at regular intervals, a policy that was implemented primarily in the 1960s and 1970s (Smith et al., 2007; Venter et al., 2008; van Wilgen and Biggs, 2011). But instead of providing conservation benefits, the fences blocked dispersal routes, which caused grazing herbivores to be increasingly sedentary around waterholes, leading to overgrazing. The increased availability of surface water also allowed elephant populations to increase to a point where they became a threat to other taxa, particularly large fruit-bearing trees such as baobabs (*Adansonia digitata*). The construction of artificial waterholes in arid sections of the Park saw an expansion of plains zebra (*Equus quagga*, NT) and common wildebeest (*Connochaetes taurinus*, LC) distributions, which in turn also attracted more lions into these arid areas. This increasing grazing competition and predation pressure caused populations of four locally rare antelopes to fall by 73–88% between 1986 and 2006, prompting fears of imminent extirpations of these iconic species in one of Africa's premier protected areas.

While it is hard to escape the looming threat of unintended consequences harming biodiversity, their impacts can be mitigated. One of the first steps involves setting management goals and objectives through close working relationships with a wide variety of stakeholders, including local people (Box 10.3) and scientists who can provide additional insights into the local social-ecological context within which land managers operate. Also important is developing a monitoring protocol that enables land managers to assess whether conservation goals are being met. When monitoring shows that management actions are ineffective or detrimental, it is absolutely critical to be willing and able to integrate this improved understanding into revised management strategies. This process, whereby new knowledge gained through repeated cycles of learning is used to revise and refine conservation strategies and management goals, is known as **adaptive management** (Venter et al., 2008; van Wilgen and Biggs, 2011). Rather than punishing management errors, adaptive management embraces the idea that we live in a complex world, and that management strategies frequently need be revised to account for new knowledge, shifting priorities, and even evolving societal values. Some of the best adaptive management plans explicitly mandate regular (e.g. every 5 years) reviews, where management strategies, goals, and objectives are formally reviewed and updated.

Despite good intentions, conservation activities may give rise to unintended consequences that run counter to overall conservation goals. Adaptive management can reduce further harm.

Box 10.3 Environmental Governance in the Serengeti Ecosystem

Alex Wilbard Kisingo

College of African Wildlife Management, Mweka, Tanzania.

akisingo@mwekawildlife.ac.tz

The greater Serengeti ecosystem of Tanzania (Figure 10.C) represents one of the most biologically productive ecoregions in the world. In recognition of its biological importance, much of this ecosystem is safeguarded under a mosaic of government, private, and co-managed protected areas, as well as community conserved areas (Kisingo, 2013a). Among these protected areas is the world-famous Serengeti National Park, as well as the Ngorongoro Conservation Area.

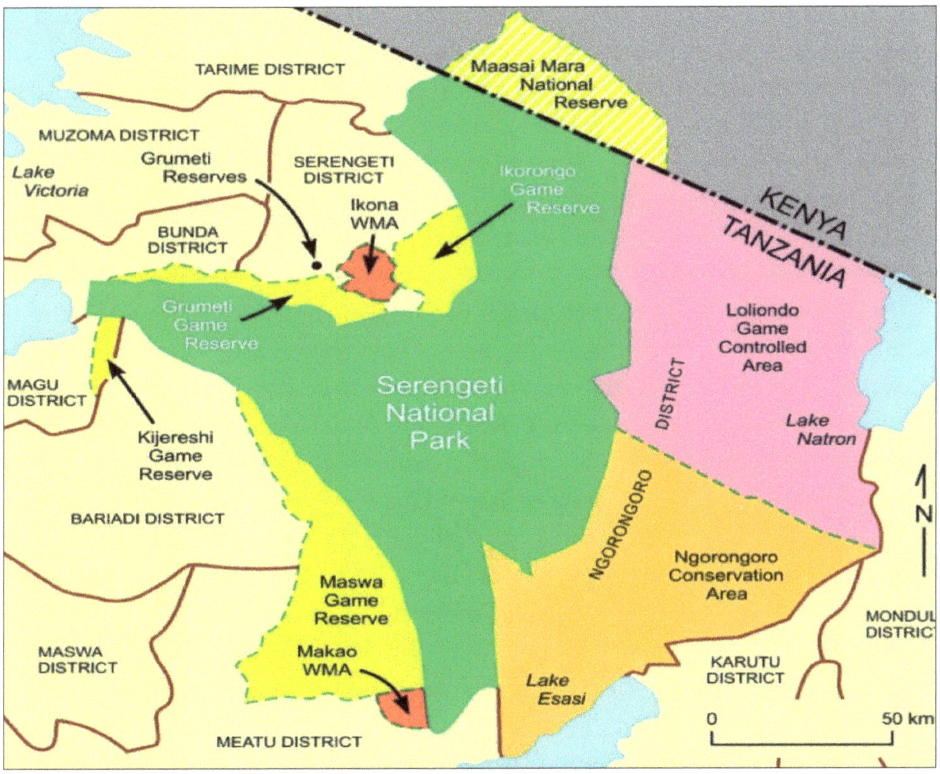

Figure 10.C A map of the greater Serengeti ecosystem, showing the location of the major protected areas in the region. Map by Alex Kisingo, CC BY 4.0.

The various governance models that dictated human actions in the colonial, post-colonial, and contemporary eras have had a profound effect on the ecosystem's conservation outcomes (Polasky et al., 2008; Sinclair et al., 2008). During most of the colonial period and immediately after Tanzania gained independence in 1961, management of the ecosystem was in the hands of government, which followed a fortress conservation approach (i.e. "fences and fines"). This included evicting traditional peoples from their ancestral lands, restricting the use of wildlife and other natural resources, and imposing heavy penalties on those on the wrong side of wildlife laws. This approach severed the link between people and nature, creating an atmosphere of hostility and resistance to conservation initiatives amongst the local communities. The results have been increased subsistence and commercial poaching, human encroachment into wildlife habitats, and a general lack of community support for conservation (Kisingo 2013b). As a result, Tanzania has seen its wildlife populations reduced and migratory corridors blocked, while invasive plant species are spreading and human-conservation conflicts are intensifying.

In contrast to this counter-productive approach, involving local people in decision making and implementation can enhance attainment of conservation and social outcomes. Such an involvement can occur through training and employment in conservation enterprises, developing social infrastructure through conservation-related financing, and improving a democratic governance space through increased awareness and capacity building (Kisingo 2013a). This socio-economic transformation, where local communities experience the benefits of conservation activities first-hand, can even create a positive feedback loop where local people not only reduce their dependence on protected wildlife resources for survival, but also become inspired to initiate their own grassroots conservation initiatives.

In 1989, nearly 30 years after Tanzania gained independence, some attempts were made to empower local people in the governance of the Serengeti ecosystem. The first major step in this regard involved developing the Serengeti Regional Conservation Strategy (SRCP), which encouraged the creation of integrated conservation and development projects (ICDPs, see Section 14.3) as a means for local people to gain direct benefit from conservation (Kisingo, 2013a). This was followed by the establishment of the Ngorongoro Pastoral Council in 2000, to include pastoralist communities living in the Ngorongoro Conservation Area (NCA) in decision making and protected areas management. Then, in 2003, the Ikona and Makao Community Wildlife Management Areas (WMAs) were established as a first step towards the co-management of wildlife in the Serengeti ecosystem.

Unfortunately, despite good intentions, these initiatives only had a limited impact in empowering local communities. An important reason for this failure was a general incompatibility between local policies and national legislations, notably a legal framework that favoured a top-down approach to decision making over the wishes and values of local communities. In the NCA, communities were more involved in the provisioning of scholarships and socio-economic assistance, in effect surrendering decision making power to government authorities. Local communities thus either never had a real seat at the table during the development and implementation of management strategies, or lost what little they had over time (Kisingo, 2013a). It should thus not come as a surprise that management of the Serengeti ecosystem generally presents mixed results in terms of conservation and social outcomes. Although wildlife populations are still better off in government protected areas when compared to WMAs, neglecting these community conserved areas (which function as critical wildlife corridors) undoubtedly also negatively impacts wildlife in government protected areas.

To be more effective at meeting conservation goals, there is an urgent need to adapt and reengineer ecosystem governance structures in the greater Serengeti ecosystem. Involving more stakeholders, particularly local people, would be an important first step. This involvement should be honest and transparent. Instead of calling one or two community representatives to a workshop to secure a rubber-stamp, start by gathering input from a wide variety of stakeholders and make an effort afterwards to show how that input was considered in final plans. In this way, even those who do not get what they want will at least know that they were heard, and hopefully understand why their desires were not met. There is also an urgent need for capacity building, to develop conservation management expertise among members of local communities, rather than the current over-reliance on outside experts who do not understand local dynamics. This could start small: community members can, for example, be involved in anti-poaching efforts through the use of Village Game Scouts (VGS), a strategy that reduces operational costs in tandem with boosting local incomes. The most promising and eager scouts can then be provided with additional professional development opportunities. Empowering local communities in this way, and enabling them to benefit from conservation activities, will provide opportunities for a wider variety of people, beyond just foreign tourists, to understand how conservation actions can support and benefit livelihoods. This will result in social outcomes that are desired by local communities, and in return, conservation authorities who then gain local support for future actions.

10.2.4 Being minimally intrusive

While adaptive management often necessitates active management, it does not embrace a philosophy of intervention for its own sake. If managers try to maintain their land by intervening in every aspect in nature, conservation will become inhibitively expensive, especially as the costs of managing unintended consequences escalate. Rather, effective adaptive management plans generally embrace a philosophy of "management by exception" (Venter et al. 2008), whereby interventions are implemented only when certain pre-determined thresholds of concern are exceeded, or when monitored trends suggest that those thresholds will soon be exceeded. (For examples of such thresholds, see van Wilgen and Biggs, 2011).

Thinking back to the example from Kruger National Park, discussed above, when it became clear the four species of antelopes were on their way to extirpation, a research programme was initiated to identify the causes for these population declines. Based on this research, park managers decided that the best way forward was to close the majority of the artificial waterpoints, and move some antelopes to a predator-proof enclosure where they can safely breed (van Wilgen and Biggs, 2011). To reverse the damage caused fences, park managers also prioritised re-establishing free movement of animals by removing these fences between the park and some adjacent properties (Venter et al., 2008). While lion and zebra populations subsequently moved out of the Park's arid region as hoped, and elephant populations stabilised, numbers of the four antelope species remained stubbornly low. This prompted park managers to evaluate whether the resources invested in these locally rare but globally common ungulates could be better spent on globally rare species occurring in the park, such as black rhinoceros (*Diceros bicornis*, CR). In other words, they adapted their threshold of concern from focussing on locally rare species to prioritising globally rare species (van Wilgen and Biggs, 2011). This decision remains controversial, even among park managers concerned about the potential loss of genetic diversity within the four ungulates. But adaptive management allows for shifting priorities; park managers may very well revisit their decisions if and when the park's rhinoceros are secured, and funds are freed up for other activities.

10.3 Restoring Damaged Ecosystems

Ecosystems are regularly disturbed by natural phenomena such as floods from cyclones/hurricanes, or wildfires started by lightning. Nevertheless, natural disturbances typically lead to succession and a return to ecological conditions that can sustain high levels of biodiversity. In contrast, ecosystems that humans have damaged or destroyed through activities such as unsustainable agriculture and deforestation, overgrazing, or pollution tend to lose their ability to rebound without human intervention.

Ecological restoration is the practice of restoring damaged ecosystems to a point where their ecosystem functions and species composition resemble their original or near-original state. **Restoration ecology,** in turn, is the scientific study of restoring damaged ecosystems, communities, and populations. Ecological restoration is often the best method for providing for the long-term use of degraded sites, whether considered from the perspectives of ecology, social benefit, or even economic benefit. Consequently, and unsurprisingly, this practice and the science originally developed in response to attempts to restore economically valuable ecosystem functions: creating wetlands to prevent flooding, reclaiming mining sites to prevent pollution and soil erosion, revegetating overgrazed rangelands to increase grass production, and planting trees on cleared areas to improve agroforestry.

> Ecological restoration aims to restore damaged ecosystems to a point where their ecosystem functions and species composition resemble their original or near-original state.

Best practices in ecological restoration have undergone major advances in recent decades. In the past, restoration methods mostly aimed for quick economic benefit, which resulted in simplified ecosystems that either failed to establish or degraded after a short time. To avoid such costly mistakes, restoration plans of today increasingly aim for the permanent re-establishment of healthy ecosystems that could support sustainable industries such as ecotourism, wildlife management, carbon sequestration, and low-level grazing by livestock. Ecological restoration often also makes economic sense; a study from South Africa found that every US $1 invested in restoring ecosystem services would generate US $8.30 for the local economy (de Wit et al., 2008).

Many grassroots conservation groups are at the forefront of initiatives that use ecosystem restoration to help make the connection between healthy ecosystems and socio-economic well-being. One prime example is the Green Belt Movement, a Kenyan initiative led by rural women to combat deforestation and restore degraded forests. They do this by helping rural women work together to grow and plant trees. Since its founding in 1977, the organisation has overseen the planting of over 51 million trees, which has helped restore forests on Mount Kenya, the Aberdares, and the Mau Complex. The planted trees have prevented erosion, stored rainwater, and provided firewood, timber, and food. In addition, over 30,000 women have been trained in sustainable trades such as forestry, beekeeping, and food processing.

10.3.1 Ecological restoration approaches

There are four main approaches to ecological restoration (Figure 10.10):

- *Natural regeneration*. Degraded areas, such as abandoned fields or logged areas, are allowed to naturally reseed and return to grasslands or forests. Land managers often choose this approach when active restoration is too expensive, when earlier restoration attempts have failed, or when experience has shown that the ecosystem is resilient and can recover on its own (e.g. Crouzeilles et al., 2017).

- *Rehabilitation*. Land managers improve conditions of a degraded ecosystem by transitioning it to another, different ecosystem type. For example, land managers could rehabilitate a degraded forest by transitioning it to a tree plantation. Rehabilitation could involve replacing just a few species or many species.

- *Partial restoration*. Land managers restore some ecosystem functions and some of the species that were dominant or characteristic of the ecosystem.

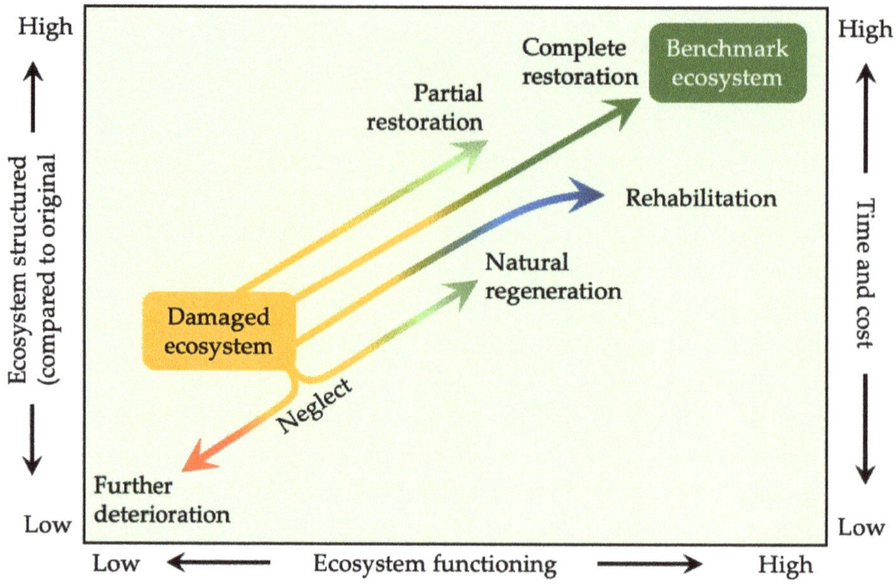

Figure 10.10 Several approaches can be followed when restoring an ecosystem, ranging from taking no action and letting the ecosystem regenerate naturally to completely restoring a degraded site. The best course of action will depend on a project's end goals and the resources available. After Bradshaw, 1990, CC BY 4.0.

For example, as a part of a grassland restoration, land managers might initially replant a few key species that are hardy and contribute to ecosystem functioning; they could delay restoration of rare species until later phases.

- *Complete restoration*. Land managers restore an area to benchmark ecosystem structure, mix of species, and ecosystem functioning. Complete restoration usually requires an active programme to modify the site, reintroduce native species, and eliminate or reduce the factors that were degrading the ecosystem.

Before a restoration project is initiated, and the type of approach is decided upon, land managers must consider how quickly the ecosystem can recover, resource needs and availability, the availability of locally adapted taxa, and the work that might be required to allow the restored community to persist over the long term. Examples of specific considerations include how to prepare soils, how to handle translocated organisms, when and how much fertiliser and water to add, and how to prevent invasions by unwanted species (Galatowitsch and Richardson, 2005; Zabbey and Tanee, 2016). It is also important to remember that ecosystems generally fail to recover if the factor that caused them to become degraded in the first place is not removed or reduced. For instance, efforts to reverse desertification (Section 5.3.4) would require a reduction of grazing pressure and unsustainable agricultural practices.

To measure restoration success, biologists often aim to restore degraded areas to conditions (ecosystem functions or species composition) comparable to a chosen benchmark or **reference site**. Reference sites provide practical targets for restoration and can be used to quantitatively assess of the success of a restoration project. Comparing restoration progress against a reference site also allows land managers to intervene or adjust their methods if restoration goals are not being met. This approach, in which land managers monitor conditions and adjust their protocols as and when needed, is known as **adaptive restoration**. (For a general discussion on adaptive management, see Section 10.2.3.)

10.3.2 Major restoration targets

Many human-altered ecosystems in Africa have proven to be good candidates for ecological restoration. These include tropical rainforests, wetlands, rangelands, and coral reefs. In addition, restoration projects in urban areas (Box 10.4) have become popular in recent years in part due to the enhanced quality of life for people living in the area.

Box 10.4 Sustainable Forest Restoration Using Natural Vegetation

Samuel Kiboi

School of Biological Sciences, University of Nairobi, Nairobi, Kenya.

✉ *samuel.kiboi@uonbi.ac.ke*

Deforestation is one of the main driving forces of biodiversity loss in Africa. Many rural and urban communities rely on wood biomass for energy in the form of either charcoal or firewood. This means that they must continuously source for the firewood or charcoal by harvesting living or dead trees. In many cases, the available energy source is live trees on farmlands which are planted as border trees, or random remnants of pre-existing vegetation within the farm. In some instances, farmers who have land in less densely populated areas have portions of forested areas or woodlots which are under continuous disturbance from wood harvesting. This is more common in rangelands or areas that have lower agricultural productivity. In other areas, such as urban settlements bordering forests, such as Kibera in Nairobi Kenya, there has been extensive harvesting of firewood and sometimes selectively for medicinal purposes or wood carving (Furukawa et. al., 2011).

Given the known benefits of intact forests, including improving food security and climate change mitigation, there are currently several efforts aimed at increasing Kenya's forest cover both in protected and unprotected areas. The Kenya Forest Service has always been at the forefront of restoration in protected areas, particularly in gazetted forest areas. Despite the general enthusiasm to increase forest cover, many structural and informational challenges remain. Most reforestation programmes classify seedlings as either "exotic" or "indigenous", but do not consider which species are best suited to local conditions. In addition, despite the numerous reforestation programmes initiated by individuals, government entities, and corporations, there is generally minimal follow-up maintenance after planting, which can jeopardise an entire project. The first three years after planting are especially crucial for proper seedling establishment and require intensive management, including weeding, mulching, and protection from herbivores. Perhaps the biggest challenge to the sustainability of these reforestation initiatives is the slow growth rate of many valuable indigenous trees that does not meet short-term harvest demands while also allowing for longer-term forest regeneration.

Figure 10.D (Top) June 2016: project participants planting trees on the University of Nairobi, Chiromo campus, following the Miyawaki method; (Bottom). January 2019: less than three years later, the trees have successfully grown to provide ecosystem services to the Chiromo campus. Photographs by Samuel Kiboi, CC BY 4.0.

At the University of Nairobi, successful urban forest islands with potential natural vegetation have been established using the "Miyawaki method" (Miyawaki, 2004). This method uses native trees to restore indigenous forests at timelines shorter than if natural regeneration was allowed to take its course (Figure 10.D). To create an urban green space on the university property, we selected 16 native tree species using a vegetation science study of remnant forests around Nairobi. Within 16 months, many of the trees had established well, with the best performing species, *Ehretia cymosa*, growing to over 2 m (Kiboi et. al., 2014). This study illustrates the importance of selecting locally adapted species in forest restoration initiatives.

Sustainable restoration practices can alleviate the short-term pressure from restored ecosystems while they mature to a self-sustaining structure. Not only should locally adapted species be promoted, but also native species that can be continuously coppiced, where new shoots rapidly replace harvested branches and portions of branches. Exotic species, such as Australian gum, pine, and mesquite (*Prosopis juliflora*) often display these characteristics, but those species are often invasive with detrimental effects on native ecosystems and communities. Fortunately, many African plant species are also good candidates for sustainable restoration initiatives, including camphor bush (*Tarchonanthus camphoratus*), sickle-leaved false-thorn (*Albizia harveyi*), silver clusterleaf (*Terminalia sericea*), and weeping wattle (*Peltophorum africanum*) (Kennedy, 1998; Kaschula et al., 2005). Although coppices may have more variable increases in biomass compared to initial planted stands, it is a sustainable way of biomass management especially in areas that experience high demand for harvestable wood. In addition, planting native trees and shrubs in farmlands typically provides beneficial ecosystem services through increasing the abundance and diversity of native insectivorous birds and pollinators of crops. In these various ways, the right management practices can lead to benefits for local people, biodiversity, and sustainable conservation practices.

Tropical forests: Tropical forests cover less than 10% of Earth's land surface; yet, they contain more than half of all terrestrial species (Cortlett and Primack, 2011). When these forests are lost, we lose substantial biodiversity and ecosystem services. For this reason, tropical forest restoration initiatives in Africa and elsewhere have received much attention in recent years. Towards the end of this chapter we will discuss a major global effort focussed on restoring degraded tropical forests, known as REDD+.

> Africa has already lost over 40% of its wetlands through human activity, with current loss rates among the highest in the world.

Wetlands: Africa has already lost over 40% of its wetlands through human activity, with current loss rates among the highest in the world (Davidson, 2014).

Because of the recognised importance of wetlands in providing flood control and other ecosystem services (Section 5.5.3), damaged wetlands are frequently targeted in restoration efforts. Wetlands are defined by their hydrology; therefore, wetland restoration projects often the focus on restoring a site's original hydrology. One such example comes from South Africa, where authorities (with support from the **World Bank**) have been working on restoring Africa's largest estuarine lake at iSimangaliso Wetland Park—a multistep process that involves restoration of the estuary's hydrology, controlling invasive plants around the wetland, and improving farming practices in the surrounding area (Whitfield et al., 2013). Wetland restoration can also occur through activities like dam removals (Section 11.3.2) or replacing exotic vegetation that deplete groundwater with native vegetation to promote groundwater retention (Sirami et al., 2013). Importantly, true wetland restorations are notoriously difficult to accomplish. It can be relatively easy to replant a wetland to look as it previously looked, but to restore the foundational hydrology often requires sophisticated engineering. In many cases, partial wetland rehabilitation is the best that can be achieved.

Mangrove swamps (Figure 10.11) provide nursery grounds for many economically important fisheries, protect coastal communities against powerful storms, and prevent saltwater from intruding into freshwater systems (van Bochove et al., 2014). They are also among the world's most important carbon sinks, storing four times more carbon per hectare than other types of tropical forests (Donato et al., 2011). Yet, over 35% of the world's mangrove swamps have already been degraded by agriculture, urban expansion, pollution, and commercial shellfish farming (MEA, 2005; Giri et al., 2011). To regain these lost services, several communities are now restoring their mangroves, while also adopting more sustainable practices to reduce damage to these important habitats (Feka et al., 2009). One of Africa's most ambitious mangrove restoration projects have been initiated in Senegal, where more than 300,000 local citizens planted more than 150 million mangrove trees across 140 km^2 between 2006 and 2013 (Cormier-Salem and Panfili, 2016). Mangrove (as any other) restoration projects do need to be planned carefully to ensure success. For example, it is important to choose ecologically-appropriate species to plant, rather than the fastest growing species that promises quick (but not necessarily optimal) results. Studies from Eritrea have also shown how fertiliser runoff caused by wave action could reduce lead to project failure (Sato et al., 2005). Another concern is that mangroves are often exploited, restored, and managed as forests, while the primary determinants of their function and structure—hydrology, soils, and nutrients—are neglected (Lewis, 2005; Gopal, 2013). Recent work showed that natural regeneration of mangrove swamps may produce more diverse, resilient, and productive ecosystems compared to planting efforts (Wetlands International, 2016). These issues will need to be addressed to ensure the long-term sustainability of mangrove restoration efforts.

Seasonal drylands: Through extensive land mismanagement (primarily overgrazing and unsustainable agriculture), a large portion of Africa's seasonal drylands are undergoing desertification, the conversion of once-productive land to desolate man-made deserts—large dry unproductive dust bowls with no vegetation.

Figure 10.11 (Left) Classic air-breathing roots of a mangrove tree in Senegal. Photograph by Ji-Elle, https://commons.wikimedia.org/wiki/File:Carabane-Mangrove.JPG, CC0. (Right) A woman collecting oysters among mangroves in Senegal's Saloum Delta. Photograph by Julien Saison, https://commons.wikimedia.org/wiki/File:Cueilleuse_traditionnelle_d%27Hu%C3%AEtres_de_mangrove,_Sine_Saloum,_femme_du_village_de_Soucouta,_S%C3%A9n%C3%A9gal.jpg, CC by-SA 4.0.

The degradation of these lands has crippled agriculture, obliterated natural biological communities, and displaced millions of people. While many drylands seem to regenerate naturally when pressures associated with land mismanagement are removed at an early stage, extended periods of mismanagement hamper recovery by leading to a loss of natural seed banks, nutrients, and microsites that allow for seedling establishment.

Somalia is home to one of the world's most effective desertification reversal programmes. Since the early 1990s, when Somalia's national government collapsed, Somalis have been tormented by warlords and civil war. The lack of effective governance also saw the rise of an unregulated charcoal trade; groves of thorn trees hundreds of years old were set ablaze, before the so-called "black gold" was exported to Arabia. The resultant wildfires and removal of trees caused an erosion crisis, turning grazing lands that once supported a diverse pastoralist community into unproductive wastelands. The resulting famine, exacerbated by droughts, caused even more Somalis to turn to a life of crime, piracy, and terrorism in a desperate effort to support their families. To reverse this decline, the humanitarian NGO Adeso successfully persuaded a regional government to create and enforce a ban on charcoal exports. Adeso also started educating local people about the links between the environment and their own lives, and introduced sustainable alternatives to the charcoal trade, such as promoting the use of solar cookers to reduce the need for charcoal fuel. To reverse desertification and prevent further erosion, Adeso showed local communities how to construct small and simple rock dams; the dams also provide a microenvironment suitable for thorn tree seeds to germinate. Adeso has been so successful in these ventures that they subsequently expanded their work to Kenya and South Sudan.

Coral Reefs: Coral reefs are one of the world's most important marine ecosystems, both ecologically and economically. They provide food to local communities, support ecotourism industries, and protect coasts by reducing wave energy by as much as 97% (Ferrario et al., 2014). Yet, coral reefs are also one of the most threatened marine ecosystems, impacted heavily by overharvesting, pollution, sedimentation, and climate change. Nevertheless, restoring coral reefs is well worth it; a meta-analysis found that it is nearly 20 times cheaper to restore coral reefs than to construct artificial systems for coastal protection (Ferrario et al., 2014). As such, several initiatives are now in progress to restore coral reefs, ranging from transplanting corals and boosting sea urchin populations for seaweed control to creating artificial reefs that can act as substrate for coral settlements (Lindahl, 2003; Edwards and Gomez, 2007).

10.3.3 The future of ecological restoration

Research in restoration ecology has grown rapidly in recent years. Many reviews (e.g. Suding, 2011) and books (e.g. Falk et al., 2016) have recently been published on the topic. The Society for Ecological Restoration (SER) was established in 1988 to support the field, and two scientific journals (*Restoration Ecology* and *Ecological Restoration*) publish hundreds of papers each year on the topic, in addition to the papers published in other ecological and conservation journals. The growth in research provides scientists and land managers more studies and evidence to inform planning and improvement of restoration projects.

A recent development in the field involves **biodiversity offsets** (ten Kate et al., 2004; MacFarlane et al., 2016). A system generally used by developers, biodiversity offsets aim to achieve no net loss of biodiversity during economic development; some projects even aim for a net overall biodiversity gain. Developers accomplish this by compensating for the ecosystem damage (or loss of threatened species populations, Kormos et al., [2014]) that may be incurred during a development project. This compensation usually follows one or more of three main strategies: (1) reducing the extent of damage at the development site, (2) restoring or protecting natural communities at a different "receptor site" as compensation for what is being lost, and (3) enhancement of the remaining natural communities after development.

> Because restoring damaged environments takes considerable time and resources, preserving intact ecosystems should be prioritised.

While biodiversity offsets (and other restoration initiatives in general) sound good in theory, it is important to remember that the most effective biodiversity conservation strategy remains protecting and managing intact ecosystems. Studies and practical experience have shown that ecological restoration efforts often fail to recreate key characteristics of their reference sites, including species composition or ecosystem functioning, even after years of effort and investment. It is also important to remember than some African ecosystems regenerate very slowly—tropical forests require more than

100 years to develop (Bonnell et al., 2011)—so even effective restorations may take decades to provide the full range of benefits. In cases where biodiversity offsets are pursued, it is critical to ensure that these initiatives indeed offer true conservation gains by mitigating the various associated risks (Coralie et al., 2015; Gordon et al., 2015; Maron et al., 2016).

10.4 Combatting Climate Change Through Ecosystem Conservation

Complex natural ecosystems play an important role in mitigating the destructive effects of climate change. Prominently, living plants sequester greenhouse gases from the atmosphere (Zarin et al., 2015); in contrast, their loss due to habitat loss increases greenhouse gas emissions (Section 6.1). Studies have also shown how ecosystems with high complexity (Betts et al. 2018) and species diversity (Mokany et al., 2014; Isbell et al., 2015) are better buffered against climate change. Lastly, by maintaining and restoring carbon-sequestering ecosystems, we also provide opportunities for climate-sensitive to persist despite the threat of climate change pace (Section 11.4).

One of the foremost initiatives aimed at combatting climate change through ecosystem conservation and restoration is known as the **Reducing Emissions from Deforestation and Forest Degradation (REDD+**, http://www.un-redd.org) programme. Set up by the UN, REDD+ provides financial incentives to local communities and landowners that make conservation of carbon-sequestering ecosystems worth more than destroying them. Funding for REDD+ is obtained through **carbon trading** programmes, in which individuals and organisations looking to offset their emissions buy **carbon credits**. The funds obtained through REDD+ are then invested in initiatives that promote ecological restoration and reduce local dependence of intact ecosystems by creating alternative income streams such as sustainable crop, timber, honey, milk, and meat production.

The original aim of REDD+ was to safeguard primary old-growth forests, but the diversity of goals set out by REDD+ also include improving ecosystem connectivity, protecting threatened species, and preventing further loss and degradation of carbon-sequestering ecosystems other than forests. Africa has been a major beneficiary of this programme. Since its inception in 2007, REDD+ projects have been funded in 27 Sub-Saharan African countries (http://www.reddprojectsdatabase.org), affording protection for over 1.6 million ha of forest (Panfil and Harvey, 2016), and providing opportunities for thousands of Africans who would not have had access to these funds otherwise.

> REDD+ provides financial incentives to local communities and landowners by making conservation of carbon-sequestering ecosystems worth more than destroying them.

Yet the future of these opportunities is not secure. Concerns exist regarding the effectiveness of REDD+ programmes, much of which is based on problematic implementation, long-term funding security, lack of monitoring, and lack of concrete conservation goals (Phelps et al., 2011; Panfil and Harvey. 2015; Fletcher et al., 2016). For example, there are concerns that REDD+ programmes can develop into a form of perverse subsidies, such as when native vegetation is cleared to establish plantations (Figure 10.12) with trees that have a high risk of becoming invasive (Lindenmayer et al., 2012). Similarly, there are concerns about the strong emphasis on forests, possibly at the expense of other important ecosystems and ecosystem services (Bond, 2016). Conservation biologists continue to be hopeful that REDD+, as with the range of approaches described in this chapter, will provide opportunities for land managers and scientists to successfully protect and restore biodiversity now and into the future.

Figure 10.12 The Reducing Emissions from Deforestation and Forest Degradation (REDD+) programme aims to promote forest conservation by paying landowners to protect forests on their lands. However, there are concerns that some REDD+ funds have become a form of perverse subsidies, for example when native vegetation is cleared to establish plantations of invasive trees. To ensure programme sustainability, it is important to strike a balance between meeting short-term development goals and ensuring the protection of a complex and adaptable biological landscape that can provide a variety of ecosystem services over the long term. Photographs by Johnny Wilson, CC BY 4.0.

10.5 Summary

1. Ecosystem conservation and management involves three different activities: (1) monitoring ecosystem components, (2) maintaining healthy ecosystems, and (3) restoring damaged ecosystems.
2. An ecosystem in which all the chemical, physical, and biological components and processes are functioning normally is considered healthy. Ecosystems that remain healthy through disturbance are resistant, while ecosystems that rapidly recover after disturbance are resilient.
3. Ecosystems can be monitored using direct observation, environmental or biochemical indicators, and remote sensing analysis. It is important that any monitoring method be consistent and repeatable across space and time.

4. To maintain ecosystems that can support diverse ecological communities, conservationists are guided by three complementary management principles: (1) maintain critical ecosystem processes (water cycling, nutrient cycling, energy flow, community dynamics), (2) minimise external threats, and (3) be adaptive yet minimally intrusive.
5. Ecological restoration is the practice of restoring damaged ecosystems to an agreed-upon benchmark. This can be accomplished via rehabilitation, partial restoration, complete restoration, or taking no action. The strategy followed will depend on each project's goals and resource availability.

10.6 Topics for Discussion

1. Read the articles by Bunce et al. (2008) and/or Papworth et al. (2009) about shifting baselines. Then think of a natural ecosystem in your region; it could even be an ecosystem in a protected area. How do you think that ecosystem looked 50 years ago? What about 100 years ago? And 1,000 years ago? At what time do you think the ecosystem was able to support the most diverse ecological community? What would you do to restore (or maintain) the ecosystem to this state? Would such a restoration project impact some species negatively? Is that a problem?
2. Consider all the aquatic communities in your region (ponds, marshes, streams, rivers, lakes, estuaries, coastal waters, etc.). Who is responsible for managing these ecosystems, and how do they balance the need for protecting biodiversity with the needs of society for natural resources? What additional conservation projects would you implement to help protect those ecosystems in the coming decades?
3. Imagine that the last population of a threatened bird species (which draws birdwatchers to the area) lives along a river nearby. This river also has numerous endemic species of fish, shellfish, and insects. A foreign company recently obtained permission to dam the river for hydropower generation. Beyond the impact of flooding, the dam will also cause various forms of pollution which will destroy the threatened birds' food source and nesting area. Upon writing about the challenge in the local newspaper, you receive US $1 million from an anonymous donor to save the bird. The company is willing to forego the development in exchange for the US $1 million. It will cost an additional $750,000 to implement an effective ecological restoration programme that can reverse the threatened birds' population declines. Is it better to buy out the company and not devote additional resources to ecological restoration and researching the bird? Or would you rather spend the money on finding alternative ways to protect the bird and the other endemic species? Explain your answers.

10.7 Suggested Readings

Crouzeilles, R., M.S. Ferreira, R.L. Chazdon, et al. 2017. Ecological restoration success is higher for natural regeneration than for active restoration in tropical forests. *Science Advances* 3: e1701345. https://doi.org/10.1126/sciadv.1701345 Natural regeneration is an appropriate restoration strategy under the right conditions.

Maron, M., C.D. Ives, H. Kujala, et al. 2016. Taming a wicked problem: Resolving controversies in biodiversity offsetting. *BioScience* 66: 489–98. https://doi.org/10.1093/biosci/biw038 Biodiversity offsets offer conservation opportunities as well as challenges.

Miller, B.P., E.A. Sinclair, M.H.M. Menz, et al. 2017. A framework for the practical science necessary to restore sustainable, resilient, and biodiverse ecosystems. *Restoration Ecology* 25: 605–17. https://doi.org/10.1111/rec.12475 Practical guidelines for ecological restoration success.

Panfil, S.N., and C.A. Harvey. 2015. REDD+ and biodiversity conservation: A review of the biodiversity goals, monitoring methods, and impacts of 80 REDD+ projects. *Conservation Letters* 9: 143–50. https://doi.org/10.1111/conl.12188 The UN's REDD+ programme shows promise, but several shortcomings need to be addressed.

van Wilgen, B.W., and A. Wannenburgh. 2016. Co-facilitating invasive species control, water conservation and poverty relief: Achievements and challenges in South Africa's Working for Water programme. *Current Opinion in Environmental Sustainability* 19: 7–17. https://doi.org/10.1016/j.cosust.2015.08.012 Ecosystem conservation, poverty relief, and job creation.

van Wilgen, B.W., and D.M. Richardson. 2014. Challenges and trade-offs in the management of invasive alien trees. *Biological Invasions* 16: 721–34. https://doi.org/10.1007/s10530-013-0615-8 Is the benefits gained from planting invasive species worth the costs?

Waldram, M.S., W.J. Bond, and W.D. Stock. 2008. Ecological engineering by a mega- grazer: White rhino impacts on a South African savanna. *Ecosystems* 11: 101–12. https://doi.org/10.1007/s10021-007-9109-9 Restoring ecosystem engineer populations results in a more diverse landscape.

Zachariades, C., I.D. Paterson, L.W. Strathie, et al. 2017. Assessing the status of biological control as a management tool for suppression of invasive alien plants in South Africa. *Bothalia* 47: 1–19. http://dx.doi.org/10.4102/abc.v47i2.2142 Biocontrol has many benefits

One of the following two manuscripts:

van Wilgen, B.W., and H.C. Biggs. 2011. A critical assessment of adaptive ecosystem management in a large savanna protected area in South Africa. *Biological Conservation* 144: 1179–87. https://doi.org/10.1016/j.biocon.2010.05.006 A general overview of an adaptive management programme.

van Wilgen, B.W., N. Govender, I.P.J. Smit, et al. 2014. The ongoing development of a pragmatic and adaptive fire management policy in a large African savanna protected area. *Journal of Environmental Management* 132: 358-368. http://dx.doi.org/10.1016/j.jenvman.2013.11.003 A specific overview of an adaptive management programme.

Bibliography

Acquah-Lamptey, D., R. Kyerematen, and E.O. Owusu. 2013. Using odonates as markers of the environmental health of water and its land related ecotone. *International Journal of Biodiversity and Conservation* 5: 761–69

Akello, J., T. Dubois, D. Coyne, et al. 2008. Endophytic *Beauveria bassiana* in banana (*Musa* spp.) reduces banana weevil (*Cosmopolites sordidus*) fitness and damage. *Crop Protection* 27: 1437–41. https://doi.org/10.1016/j.cropro.2008.07.003

Akutse, K.S., N.K. Maniania, K.K.M. Fiaboe, et al. 2013. Endophytic colonization of *Vicia faba* and *Phaseolus vulgaris* (Fabaceae) by fungal pathogens and their effects on the life-history parameters of *Liriomyza huidobrensis* (Diptera: Agromyzidae). *Fungal Ecology* 6: 293–301. https://doi.org/10.1016/j.funeco.2013.01.003

Allan, D.G., J.A. Harrison, R.A. Navarro, et al. 1997. The impact of commercial afforestation on bird populations in Mpumalanga Province, South Africa—Insights from bird atlas data. *Biological Conservation* 79:173–85. https://doi.org/10.1016/S0006-3207(96)00098-5

Archibald, S., R.J. Scholes, D.P. Roy, et al. 2010. Southern African fire regimes as revealed by remote sensing. *International Journal of Wildland Fire* 19: 861–78. https://doi.org/10.1071/WF10008

Bester, M.N., J.P. Bloomer, R.J. van Aarde, et al. 2002. A review of the successful eradication of feral cats from sub-Antarctic Marion Island, Southern Indian Ocean. *South African Journal of Wildlife Research* 32: 65–73. https://hdl.handle.net/10520/EJC117137

Betts, M.G., B. Phalan, S.J.K. Frey, et al. 2018. Old-growth forests buffer climate-sensitive bird populations from warming. *Diversity and Distributions* 24: 439–47. https://doi.org/10.1111/ddi.12688

Bhagwat S.A., E. Breman, T. Thekaekara, et al. 2012. A battle lost? Report on two centuries of invasion and management of *Lantana camara* L. in Australia, India and South Africa. *PLoS ONE* 7: e32407. https://doi.org/10.1371/journal.pone.0032407

Bodin, N., R. N'Gom-Kâ, S. Kâ, et al. 2013. Assessment of trace metal contamination in mangrove ecosystems from Senegal, West Africa. *Chemosphere* 90: 150–57. https://doi.org/10.1016/j.chemosphere.2012.06.019

Bond, W.J. 2016. Ancient grasslands at risk. *Science* 351: 120–22. https://doi.org/10.1126/science.aad5132

Bonnell, T.R., R. Reyna-Hurtado, and C.A. Chapman. 2011. Post-logging recovery time is longer than expected in an East African tropical forest. *Forest Ecology and Management* 261: 855–64. https://doi.org/10.1016/j.foreco.2010.12.016

Bornman, M.S., and H. Bouwman. 2012. Environmental pollutants and diseases of sexual development in humans and wildlife in South Africa: Harbingers of impact on overall health? *Reproduction in Domestic Animals* 47: 327–32. https://doi.org/10.1111/j.1439-0531.2012.02094.x

Bownes, A., M.P. Hill, and M.J. Byrne. 2013. The role of nutrients in the responses of water hyacinth, *Eichhornia crassipes* (Pontederiaceae) to herbivory by a grasshopper *Cornops aquaticum* Brüner (Orthoptera: Acrididae). *Biological Control* 67: 555–62. https://doi.org/10.1016/j.biocontrol.2013.07.022

Bradshaw, A.D. 1990. The reclamation of derelict land and the ecology of ecosystems. In: *Restoration Ecology: A Synthetic Approach to Ecological Research*, ed. by W.R. Jordan III et al. (Cambridge: Cambridge University Press).

Buckland, S.T., and A. Johnston. 2017. Monitoring the biodiversity of regions: Key principles and possible pitfalls. *Biological Conservation* 214: 23–34. https://doi.org/10.1016/j.biocon.2017.07.034

Bunce, M., L.D. Rodwell, R. Gibb, et al. 2008. Shifting baselines in fishers' perceptions of island reef fishery degradation. *Ocean and Coastal Management* 51: 285–302. https://doi.org/10.1016/j.ocecoaman.2007.09.006

Burton, M.E.H., J.R. Poulsen, M.E. Lee, et al. 2017. Reducing carbon emissions from forest conversion to oil palm agriculture in Gabon, *Conservation Letters* 10: 297–307. https://doi.org/10.1111/conl.12265

Carreiras, J.M.B, M.J. Vasconcelos, and R.M. Lucas. 2012. Understanding the relationship between aboveground biomass and ALOS PALSAR data in the forests of Guinea-Bissau (West Africa). *Remote Sensing of Environment* 121: 426–42. https://doi.org/10.1016/j.rse.2012.02.012

Chislock, M.F., E. Doster, R.A. Zitomer, et al. 2013. Eutrophication: Causes, consequences, and controls in aquatic ecosystems. *Nature Education Knowledge* 4: 10.

Coetzee, J.A., A. Bownes, and G.D. Martin. 2011. Prospects for the biological control of submerged macrophytes in South Africa. *African Entomology* 19: 469–88. https://doi.org/10.4001/003.019.0203

Coetzee, J.A., and M.P. Hill. 2012. The role of eutrophication in the biological control of water hyacinth, *Eichhornia crassipes*, in South Africa. *BioControl* 57: 247–61. https://doi.org/10.1007/s10526-011-9426-y

Coralie, C., O. Guillaume, and N. Claude. 2015 Tracking the origins and development of biodiversity offsetting in academic research and its implications for conservation: A review. *Biological Conservation* 192: 492–503. https://doi.org/10.1016/j.biocon.2015.08.036

Corlett, R., and R.B. Primack. 2011. *Tropical Rain Forests: An Ecological and Biogeographical Comparison* (Hoboken: Wiley-Blackwell). https://doi.org/10.1002/9781444392296

Cormier-Salem, M.-C., and J. Panfili. 2016. Mangrove reforestation: Greening or grabbing coastal zones and deltas? Case studies in Senegal. *African Journal of Aquatic Science* 41: 89–98. https://doi.org/10.2989/16085914.2016.1146122

Crouzeilles, R., M.S. Ferreira, R.L. Chazdon, et al. 2017. Ecological restoration success is higher for natural regeneration than for active restoration in tropical forests. *Science Advances* 3: e1701345. https://doi.org/10.1126/sciadv.1701345

Cumming, D.H.M., and G.S. Cumming. 2015. One Health: An ecological and conservation perspective. In: *One Health: The Theory and Practice of Integrated Health Approaches*, ed. by J. Zinsstag, et al. (Wallingford: CAB International).

Davidson, N.C. 2014. How much wetland has the world lost? Long-term and recent trends in global wetland area. *Marine and Freshwater Research* 65: 934–41. https://doi.org/10.1071/MF14173

Davies, A.B., C.J. Tambling, G.I.H. Kerley, et al. 2016. Effects of vegetation structure on the location of lion kill sites in African thicket. *PloS ONE* 11: e0149098. https://doi.org/10.1371/journal.pone.0149098

de Groote, H., O. Ajuonu, S. Attignon, et al. 2003. Economic impact of biological control of water hyacinth in Southern Benin. *Ecological Economics* 45: 105–17. https://doi.org/10.1016/S0921-8009(03)00006-5

de Lange, W.J., and B.W. van Wilgen. 2010. An economic assessment of the contribution of biological control to the management of invasive alien plants and to the protection of ecosystem services in South Africa. *Biological Invasions* 12: 4113–24. https://doi.org/10.1007/s10530-010-9811-y

de Wit, M., H. van Zyl, D. Crookes, et al. 2009. *Investing in natural assets: A business case for the environment in the City of Cape Town* (Cape Town: City of Cape Town).

Di Marco, M., G.M. Buchanan, Z. Szantoi, et al. 2014. Drivers of extinction risk in African mammals: The interplay of distribution state, human pressure, conservation response and species biology. *Philosophical Transactions of the Royal Society B* 369: 20130198. https://doi.org/10.1098/rstb.2013.0198

Donato, D.C., J.B. Kauffman, D. Murdiyarso, et al. 2011. Mangroves among the most carbon-rich forests in the tropics. *Nature Geoscience* 4: 293–97. https://doi.org/10.1038/ngeo1123

Drechsel, P., L. Gyiele, D. Kunze, et al. 2001. Population density, soil nutrient depletion, and economic growth in sub-Saharan Africa. *Ecological Economics* 38: 251–58. https://doi.org/10.1016/S0921-8009(01)00167-7

Dube, T., O. Mutanga, K. Seutloali, et al. 2015. Water quality monitoring in sub-Saharan African lakes: A review of remote sensing applications. *African Journal of Aquatic Science* 40: 1–7. https://doi.org/10.2989/16085914.2015.1014994

Dubovyk, O., T. Landmann, B.F.N. Erasmus, et al. 2015. Monitoring vegetation dynamics with medium resolution MODIS-EVI time series at sub-regional scale in southern Africa. *International Journal of Applied Earth Observation and Geoinformation* 38: 175–83. https://doi.org/10.1016/j.jag.2015.01.002

Edwards, A.J., and E.D. Gomez. 2007. *Reef Restoration Concepts and Guidelines: Making Sensible Management Choices in the Face of Uncertainty* (St. Lucia: Coral Reef Targeted Research and Capacity Building for Management Programme).

Estes, R.D., J.L. Atwood, and A.B. Estes. 2006. Downward trends in Ngorongoro Crater ungulate populations 1986–2005: Conservation concerns and the need for ecological research. *Biological Conservation* 131: 106–20. https://doi.org/10.1016/j.biocon.2006.02.009

Falk, D.A., M.A. Palmer, and J.B. Zedler. 2016. *Foundations of Restoration Ecology* (Washington: Island Press).

Feka, N.Z., G.B. Chuyong, and G.N. Ajonina. 2009. Sustainable utilization of mangroves using improved fish-smoking systems: A management perspective from the Douala-Edea wildlife reserve, Cameroon. *Tropical Conservation Science* 2: 450–68. https://doi.org/10.1177/194008290900200406

Ferrario, F., M.W. Beck, C.D. Storlazzi, et al. 2014. The effectiveness of coral reefs for coastal hazard risk reduction and adaptation. *Nature Communications* 5: 3794. https://doi.org/10.1038/ncomms4794

Fletcher, R., W. Dressler, B. Büscher, et al. 2016. Questioning REDD+ and the future of market-based conservation. *Conservation Biology* 30: 573–675. https://doi.org/10.1111/cobi.12680

Furukawa, T., K. Fujiwara, S. Kiboi, et al. 2011. Can stumps tell what people want: Pattern and preference of informal wood extraction in an urban forest of Nairobi, Kenya. *Biological Conservation* 144: 3047–54. https://doi.org/10.1016/j.biocon.2011.09.011

Galatowitsch, S., and D.M. Richardson. 2005. Riparian scrub recovery after clearing of invasive alien trees in headwater streams of the Western Cape, South Africa. *Biological Conservation* 122: 509–21. https://doi.org/10.1016/j.biocon.2004.09.008

Garcia-Carreras, L., and D.J. Parker. 2011. How does local tropical deforestation affect rainfall? *Geophysical Research Letters* 38: L19802. https://doi.org/10.1029/2011GL049099

Garcin, Y., P. Deschamps, G. Ménot, et al. 2018. Early anthropogenic impact on West Central African rainforests 2,600 years ago. *Proceedings of the National Academy of Sciences* 115: 3261–66. https://doi.org/10.1073/pnas.1715336115

Gerlach, J. 2014. *Prodida stella. The IUCN Red List of Threatened Species* 2014: e.T196234A2443094. http://doi.org/10.2305/IUCN.UK.2014-1.RLTS.T196234A2443094.en

Giri, C., E. Ochieng, L.L. Tieszen, et al. 2011. Status and distribution of mangrove forests of the world using Earth observation satellite data. *Global Ecology and Biogeography* 20: 154–59. https://doi.org/10.1111/j.1466-8238.2010.00584.x

Goldammer, J.G., and C. de Ronde. 2004. *Wildland Fire Management Handbook for Sub-Sahara Africa* (Freiburg: GFMC). http://gfmc.online/latestnews/GFMC-Wildland-Fire-Management-Handbook-Sub-Sahara-Africa-2004.pdf

Gopal, B. 2013. Mangroves are wetlands, not forests: Some implications for their management. In: *Mangrove Ecosystems of Asia*, ed. by I. Faridah-Hanum, et al. (New York: Springer). https://doi.org/10.1007/978-1-4614-8582-7

Gordon, A., J.W. Bull, C. Wilcox, et al. 2015. Perverse incentives risk undermining biodiversity offset policies. *Journal of Applied Ecology* 52: 532–37. https://doi.org/10.1111/1365-2664.12398

Güereña, D., H. Neufeldt, J. Berazneva, et al. 2015. Water hyacinth control in Lake Victoria: Transforming an ecological catastrophe into economic, social, and environmental benefits. *Sustainable Production and Consumption* 3: 59–69. https://doi.org/10.1016/j.spc.2015.06.003

Diop, O., and M.P. Hill, M.P. 2009. Quantitative post-release evaluation of biological control of floating fern, *Salvinia molesta* DS Mitchell (Salviniaceae), with *Cyrtobagous salviniae* Calder and Sands (Coleoptera: Curculionidae) on the Senegal River and Senegal River Delta. *African Entomology* 17: 64–70. https://doi.org/10.4001/003.017.0108

Hopcraft, J.G.C., J.M. Morales, H.L. Beyer, et al. 2014. Competition, predation, and migration: Individual choice patterns of Serengeti migrants captured by hierarchical models. *Ecological Monographs*, 84: 355–72. https://doi.org/10.1890/13-1446.1

Isbell, F., D. Craven, J. Connolly, et al. 2015. Biodiversity increases the resistance of ecosystem productivity to climate extremes. *Nature* 526: 574–77. https://doi.org/10.1038/nature15374

IUCN. 2019. *The IUCN Red List of Threatened Species.* http://www.iucnredlist.org

Jeffery, K.J., L. Korte, F. Palla, et al. 2014. Fire management in a changing landscape: A case study from Lopé National Park, Gabon. *Parks* 20: 39–52. https://doi.org/10.2305/IUCN.CH.2014.PARKS-20-1.KJJ.en

Jenkins, A.R. 2005. Lesser Kestrel *Falco naumanni*. In: *Roberts Birds of Southern Africa*, ed. by P.A.R. Hockey, et al. (Cape Town: Trustees of the John Voelcker Bird Book Fund).

Kaschula, S.A., W.E. Twine, and M.C. Scholes. 2005. Coppice harvesting of fuelwood species on a South African common: Utilizing scientific and indigenous knowledge in community based natural resource management. *Human Ecology* 33: 387–418. https://doi.org/10.1007/s10745-005-4144-7

Kelly, L.T., and L. Brotons. 2017. Using fire to promote biodiversity. *Science* 355: 1264–65. https://doi.org/10.1126/science.aam7672

Kennedy, A.D. 1998. Coppicing of *Tarconanthus camphoratus* (Compositae) as a source of sustainable fuelwood production: An example from the Laikipia Plateau, Kenya. *African Journal of Ecology* 36: 148–58. https://doi.org/10.1046/j.1365-2028.1998.00115.x

Kiboi, S., K. Fujiwara, and P. Mutiso. 2014. Sustainable management of urban green environments: Challenges and opportunities. In: *Sustainable Living with Environmental Risks*, ed. by N. Kaneko, et al. (Tokyo: Springer). https://doi.org/10.1007/978-4-431-54804-1

Kohler, S., M. Connan, J. Hill, et al. 2011. Geographic variation in the trophic ecology of an avian rocky shore predator, the African Black Oystercatcher, along the southern African coastline. *Marine Ecology Progress Series* 435: 235–49. https://doi.org/10.3354/meps09215

Kormos, R., C.F. Kormos, T. Humle, et al. 2014. Great apes and biodiversity offset projects in Africa: The case for national offset strategies. *PloS ONE* 9: e111671. https://doi.org/10.1371/journal.pone.0111671

Kraaij, T., R.M. Cowling, B.W. Wilgen, et al. 2013. Proteaceae juvenile periods and post-fire recruitment as indicators of minimum fire return interval in eastern coastal fynbos. *Applied Vegetation Science* 16: 84–94. https://doi.org/10.1111/j.1654-109X.2012.01209.x

Kyerematen, R., D. Acquah-Lamptey, E.H. Owusu, et al. 2014. Insect diversity of the Muni-Pomadze Ramsar site: An important site for biodiversity conservation in Ghana. *Journal of Insects* 2014: 985684. http://doi.org/10.1155/2014/985684

Kyerematen, R., S. Adu-Acheampong, D. Acquah-Lamptey, et al. 2018. Butterfly diversity: An indicator for environmental health within the Tarkwa Gold mine, Ghana. *Environment and Natural Resources Research* 8: 69–83. https://doi.org/10.5539/enrr.v8n3p69

Laporte, N.T., J.A. Stabach, R. Grosch, et al. 2007. Expansion of industrial logging in Central Africa. *Science* 316: 1451–1451. https://doi.org/10.1126/science.1141057

Lawrence, D., and K. Vandecar. 2015. Effects of tropical deforestation on climate and agriculture. *Nature Climate Change* 5: 27–36. https://doi.org/10.1038/nclimate2430

Lewis, R.R. 2005. Ecological engineering for successful management and restoration of mangrove forests. *Ecological Engineering* 24: 403–18. https://doi.org/10.1016/j.ecoleng.2004.10.003

Lindahl, U. 2003. Coral reef rehabilitation through transplantation of staghorn corals: Effects of artificial stabilization and mechanical damages. *Coral Reefs* 22: 217–23. https://doi.org/10.1007/s00338-003-0305-6

Lindenmayer, D.B., K.B. Hulvey, R.J. Hobbs, et al. 2012. Avoiding bio-perversity from carbon sequestration solutions. *Conservation Letters* 5: 28–36. https://doi.org/10.1111/j.1755-263X.2011.00213.x

Loarie, S.R., C.J. Tambling, and G.P. Asner. 2013. Lion hunting behaviour and vegetation structure in an African savanna. *Animal Behaviour* 85: 899–906. https://doi.org/10.1016/j.anbehav.2013.01.018

Lyons, A.J., W.C. Turner, and W.M. Getz. 2013. Home range plus: A space-time characterization of movement over real landscapes. *Movement Ecology* 1: 2. https://doi.org/10.1186/2051-3933-1-2

MacFarlane, D.M., S.D. Holness, A. von Hase, et al. 2016. *Wetland offsets: A best-practice guideline for South Africa* (Pretoria: SANBI and Department of Water and Sanitation). http://biodiversityadvisor.sanbi.org/wp-content/uploads/2014/09/Wetland-Offset-Guidelines-Version-7-For-stakeholder-comment.pdf

Maron, M., C.D. Ives, H. Kujala, et al. 2016. Taming a wicked problem: Resolving controversies in biodiversity offsetting. *BioScience* 66: 489–98. https://doi.org/10.1093/biosci/biw038

Martin, G.D., J.A. Coetzee, P.S.R. Weyl, et al. 2018. Biological control of *Salvinia molesta* in South Africa revisited. *Biological Control* 125: 7480. https://doi.org/10.1016/j.biocontrol.2018.06.011

Mashimbye, Z.E., M.A. Cho, J.P. Nel, et al. 2012. Model-based integrated methods for quantitative estimation of soil salinity from hyperspectral remote sensing data: A case study of selected South African soils. *Pedosphere* 22: 640–49. https://doi.org/10.1016/S1002-0160(12)60049-6

Masocha, M., A.K. Skidmore, X. Poshiwa, et al. 2011. Frequent burning promotes invasions of alien plants into a mesic African savanna. *Biological Invasions* 13: 1641–48. https://doi.org/10.1007/s10530-010-9921-6

Mbati, G., and P. Neuenschwander. 2005. Biological control of three floating water weeds, *Eichhornia crassipes, Pistia stratiotes,* and *Salvinia molesta* in the Republic of Congo."BioControl 50: 635–45. https://doi.org/10.1007/s10526-004-5863-1

McClanahan, T., J.M. Maina, and N.A. Muthiga. 2011. Associations between climate stress and coral reef diversity in the western Indian Ocean. *Global Change Biology* 17: 2023–2032. https://doi.org/10.1111/j.1365-2486.2011.02395.x

MEA (Millenium Ecosystem Assessment). 2005. *Ecosystems and Human Well-Being* (Covello: Island Press). https://www.millenniumassessment.org

Miyawaki, A. 2004. Restoration of the living environment based on vegetation ecology: Theory and practice. *Ecological Research* 19: 83–90. https://doi.org/10.1111/j.1440-1703.2003.00606.x

Mokany, K., S. Prasad, and D.A. Westcott. 2014. Loss of frugivore seed dispersal services under climate change. *Nature Communications* 5: 4971. https://doi.org/10.1038/ncomms4971

Mwangi, E., and B. Swallow. 2008. *Prosopis juliflora* invasion and rural livelihoods in the Lake Baringo area of Kenya. Con*servation and Society* 6: 130–40. https://doi.org/10.4103/0972-4923.49207

Naidoo, L., M.A. Cho, R. Mathieu, et al. 2012. Classification of savanna tree species, in the Greater Kruger National Park region, by integrating hyperspectral and LiDAR data in a Random Forest data mining environment. *ISPRS Journal of Photogrammetry and Remote Sensing* 69: 167–79. https://doi.org/10.1016/j.isprsjprs.2012.03.005

NASA. 2009. *ASTER Global Digital Elevation Model v. 2 (GDEM V2)* (Pasadena: NASA JPL). https://doi.org/10.5067/ASTER/ASTGTM.002

NASA. 2013. *NASA Shuttle Radar Topography Mission Global 3 Arc Second Sub-Sampled (SRTM3)* (Sioux Falls: NASA LP DAAC). https://doi.org/10.5067/MEASURES/SRTM/SRTMGL3S.003

Nnoli, H., R. Kyerematen, S. Adu-Acheampong, et al. 2019. Change in aquatic insect abundance: Evidence of climate and land-use change within the Pawmpawm River in southern Ghana. *Cogent Environmental Science* 5: 1594511. https://doi.org/10.1080/23311843.2019.1594511

Panfil, S.N., and C.A. Harvey. 2015. REDD+ and biodiversity conservation: A review of the biodiversity goals, monitoring methods, and impacts of 80 REDD+ projects. *Conservation Letters* 9: 143–50. https://doi.org/10.1111/conl.12188

Papworth, S.K., J. Rist, L. Coad, et al. 2009. Evidence for shifting baseline syndrome in conservation. *Conservation Letters* 2: 93–100. https://doi.org/10.1111/j.1755-263X.2009.00049.x

Phelps, J., E.L. Webb, and L.P. Koh. 2011. Risky business: an uncertain future for biodiversity conservation finance through REDD+. *Conservation Letters* 4: 88–94. https://doi.org/10.1111/j.1755-263X.2010.00155.x

Pricope, N.G., and M.W. Binford. 2012. A spatio-temporal analysis of fire recurrence and extent for semi-arid savanna ecosystems in southern Africa using moderate-resolution satellite imagery. *Journal of Environmental Management* 100: 72–85. https://doi.org/10.1016/j.jenvman.2012.01.024

Rahlao, S.J., S.J. Milton, K.J. Esler, et al. 2010. The distribution of invasive *Pennisetum setaceum* along roadsides in western South Africa: The role of corridor interchanges. *Weed Research* 50: 537–43. https://doi.org/10.1111/j.1365-3180.2010.00801.x

Sanchez, P.A. 2010. Tripling crop yields in tropical Africa. *Nature Geoscience* 3: 299–300. https://doi.org/10.1038/ngeo853

Sato, G., A. Fisseha, S. Gebrekiros, et al. 2005. A novel approach to growing mangroves on the coastal mud flats of Eritrea with the potential for relieving regional poverty and hunger. *Wetlands* 25: 776. https://doi.org/10.1672/0277-5212(2005)025[0776:ANATGM]2.0.CO;2

Scantlebury, D.M., M.G.L. Mills, R.P. Wilson, et al. 2014. Flexible energetics of cheetah hunting strategies provide resistance against kleptoparasitism. *Science* 346: 79–81. https://doi.org/10.1126/science.1256424

Scott, J.M., B. Csuti, and F. Davis. 1991. Gap analysis: An application of Geographic Information Systems for wildlife species. In: *Challenges in the Conservation of Biological Resources: A Practitioner's Guide*, ed. by D.J. Decker, et al. (Boulder: Westview Press).

Sirami, C., S.S. Jacobs, and G.S. Cumming. 2013. Artificial wetlands and surrounding habitats provide important foraging habitat for bats in agricultural landscapes in the Western Cape, South Africa. *Biological Conservation* 164: 30–38. https://doi.org/10.1016/j.biocon.2013.04.017

Smit, I.P.J., and H.H.T. Prins. 2015. Predicting the effects of woody encroachment on mammal communities, Grazing biomass and fire frequency in African savannas. *PloS ONE* 10: e0137857. https://doi.org/10.1371/journal.pone.0137857

Smit, I.P.J., C.C. Grant, and B.J. Devereux. 2007. Do artificial waterholes influence the way herbivores use the landscape? Herbivore distribution patterns around rivers and artificial water sources in a large African savanna park. *Biological Conservation* 136: 85–99. https://doi.org/10.1016/j.biocon.2006.11.009

Suding, K.N. 2011. Toward an era of restoration in ecology: Successes, failures and opportunities ahead. *Annual Reviews in Ecology, Evolution, and Systematics* 42: 465–87. https://doi.org/10.1146/annurev-ecolsys-102710-145115

Symeonakis, E., and N. Drake. 2004. Monitoring desertification and land degradation over sub-Saharan Africa. *International Journal of Remote Sensing* 25: 573–92. https://doi.org/10.1080/0143116031000095998

ten Kate, K., J. Bishop, and R. Bayon. 2004. *Biodiversity offsets: Views, experience, and the business case* (Gland: IUCN; London: Insight Investment). https://www.iucn.org/sites/dev/files/import/downloads/bdoffsets.pdf

Torres, J., J.C. Brito, M.J. Vasconcelos, et al. 2010. Ensemble models of habitat suitability relate chimpanzee (*Pan troglodytes*) conservation to forest and landscape dynamics in Western Africa. *Biological Conservation* 143: 416–25. http://doi.org/10.1016/j.biocon.2009.11.007

Uys R.G., W.J. Bond, and T.M. Everson. 2004. The effect of different fire regimes on plant diversity in southern African grasslands. *Biological Conservation* 118: 489–99. https://doi.org/10.1016/j.biocon.2003.09.024

van Bochove, J., E. Sullivan, and T. Nakamura. 2014. *The Importance of Mangroves to People: A Call to Action* (Cambridge: UNEP). http://wedocs.unep.org/handle/20.500.11822/9300

van Wilgen, B.W., and A. Wannenburgh. 2016. Co-facilitating invasive species control, water conservation and poverty relief: Achievements and challenges in South Africa's Working for Water programme. *Current Opinion in Environmental Sustainability* 19: 7–17. http://doi.org/10.1016/j.cosust.2015.08.012

van Wilgen, B.W., and H.C. Biggs. 2011. A critical assessment of adaptive ecosystem management in a large savanna protected area in South Africa. *Biological Conservation* 144: 1179–87. https://doi.org/10.1016/j.biocon.2010.05.006

van Wilgen, B.W., and D.M. Richardson. 2014. Challenges and trade-offs in the management of invasive alien trees. *Biological Invasions* 16: 721–34. https://doi.org/10.1007/s10530-013-0615-8

van Wilgen, B.W., N. Govender, I.P.J. Smit, et al. 2014. The ongoing development of a pragmatic and adaptive fire management policy in a large African savanna protected area. *Journal of Environmental Management* 132: 358–68. http://dx.doi.org/10.1016/j.jenvman.2013.11.003

van Wilgen, B.W., G. Forsyth, and P. Prins. 2012. The management of fire-adapted ecosystems in an urban setting: The case of Table Mountain National Park, South Africa. *Ecology and Society* 17: 8. http://doi.org/10.5751/ES-04526-170108

van Wilgen, B.W., G.G. Forsyth, H. de Klerk, et al. 2010. Fire management in Mediterranean-climate shrublands: A case study from the Cape fynbos, South Africa. *Journal of Applied Ecology* 47: 631–38. https://doi.org/10.1111/j.1365-2664.2010.01800.x

van Wilgen, B.W., J.L. Nel, and M. Rouget. 2007. Invasive alien plants and South African rivers: A proposed approach to the prioritization of control operations. *Freshwater Biology* 52: 711–23. https://doi.org/10.1111/j.1365-2427.2006.01711.x

van Wyk, E., and B.W. van Wilgen. 2002. The cost of water hyacinth control in South Africa: A case study of three options. *African Journal of Aquatic Science* 27: 141–49. https://doi.org/10.2989/16085914.2002.9626585

Venter, F.J., R.J. Naiman, H.C. Biggs, et al. 2008. The evolution of conservation management philosophy: Science, environmental change and social adjustments in Kruger National Park. *Ecosystems* 11: 173–92. https://doi.org/10.1007/s10021-007-9116-x

Wallenfang, J., M. Finckh, J. Oldeland, et al, 2015. Impact of shifting cultivation on dense tropical woodlands in southeast Angola. *Tropical Conserververvation Science* 8: 863–92. https://doi.org/10.1177/194008291500800402

Wegmann, M., L. Santini, B. Leutner, et al. 2014. Role of African protected areas in maintaining connectivity for large mammals. *Philosophical Transactions of the Royal Society of London B* 369: 20130193. https://doi.org/10.1098/rstb.2013.0193

Wetlands International. 2016. *Mangrove restoration: To plant or not to plant*? (Wageningen: Wetlands International). https://www.wetlands.org/publications/mangrove-restoration-to-plant-or-not-to-plant

Whitfield, A.K., G.C. Bate, T. Forbes, et al. 2013. Relinkage of the Mfolozi River to the St. Lucia estuary system—urgent imperative for the long-term management of a RAMSAR and World Heritage Site. *Aquatic Ecosystem Health and Management* 16: 104–10. https://doi.org/10.1080/14634988.2013.759081

Williams, A.E., R.E. Hecky, and H.C. Duthie. 2007. Water hyacinth decline across Lake Victoria—Was it caused by climatic perturbation or biological control? A reply. *Aquatic Botany* 87: 94–96. http://doi.org/10.1016/j.aquabot.2007.03.009

Wilson, J.R.U., P. Ivey, P. Manyama, et al. 2013. A new national unit for invasive species detection, assessment and eradication planning. *South African Journal of Science* 109: 0111. http://doi.org/10.1590/sajs.2013/20120111

Wilson, J.W., J.O. Sexton, R.R. Jobe, et al. 2013. The relative contribution of terrain, land cover, and vegetation structure indices to species distribution models. *Biological Conservation* 164: 170–76. https://doi.org/10.1016/j.biocon.2013.04.021

Zabbey, N., and F.B.G. Tanee. 2016. Assessment of asymmetric mangrove restoration trials in Ogoniland, Niger Delta, Nigeria: Lessons for future intervention. *Ecological Restoration* 34: 245–57. http://doi.org/10.3368/er.34.3.245

Zachariades, C., I.D. Paterson, L.W. Strathie, et al. 2017. Assessing the status of biological control as a management tool for suppression of invasive alien plants in South Africa. *Bothalia* 47: 1–19. http://dx.doi.org/10.4102/abc.v47i2.2142

Zalucki, M.P., M.D. Day, and J. Playford. 2007. Will biological control of *Lantana camara* ever succeed? Patterns, processes & prospects. *Biological Control* 42: 251–61. https://doi.org/10.1016/j.biocontrol.2007.06.002

Zarin, D.J., N.L. Harris, A. Baccini, et al. 2016. Can carbon emissions from tropical deforestation drop by 50% in 5 years? *Global Change Biology* 22: 1336–47. https://doi.org/10.1111/gcb.13153

11. Preventing Extinctions

Tourists appreciating a southern right whale (*Eubalaena australis*, LC) from a whale charter boat operating from Hermanus, South Africa. Considered the "right" whale by 19th century whalers, the species was hunted to near extinction by the early 1900s, when only 300 individuals were left in the world (Reilly et al., 2013). Following the international ban on whale hunting, their populations have steadily recovered. Today it is considered safe from extinction, enabling towns such as Hermanus to base their thriving tourism industry on whale watching. Photograph by Southern Right Charters, CC BY 4.0.

https://doi.org/10.11647/OBP.0177.11

There are many examples in this textbook illustrating how species have been saved from the brink of extinction. For some, the solution was simple: halt the threats that caused their populations to decline. In other cases, more drastic steps were required, like moving the last remaining individuals into captivity until the threats have been reversed. Many species that persist with low population sizes would likely not have survived without human intervention (Figure 11.1).

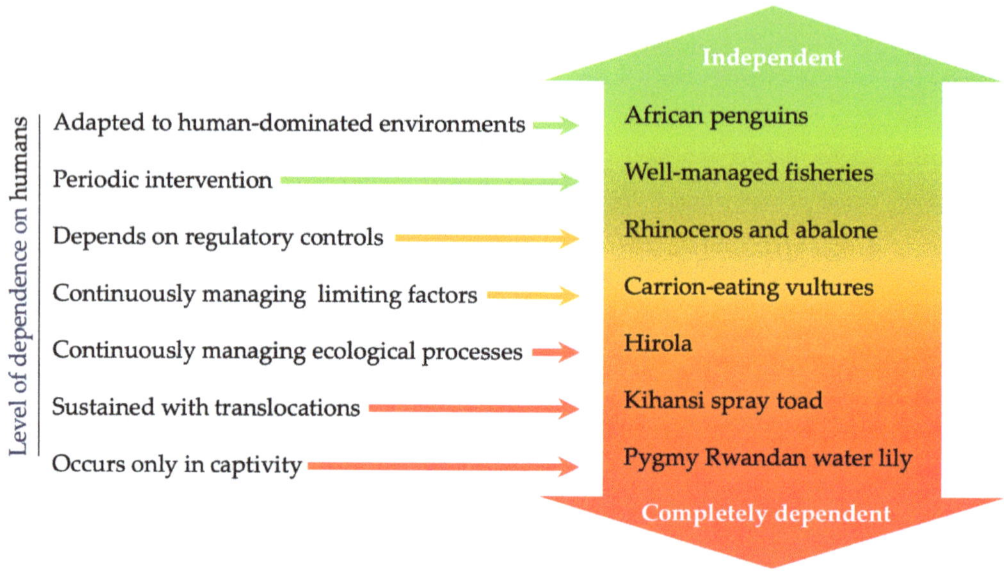

Figure 11.1 The continuum of species management approaches. Some threatened species exist under such low population sizes that they depend on active human intervention for recovery, while others can persist with minimal intervention. Each of the examples have been discussed elsewhere in the book. After Scott et al., 2005, CC BY 4.0.

In each of these success stories, the most important first steps involved determining the ecological needs of the species at risk and understanding the factors that made that species vulnerable to extinction (Section 8.5). This chapter reviews some of the most important concepts for understanding and managing those needs and risks. The concepts reviewed in this chapter include methods to study species and populations, actions that can be taken to increase population sizes, and strategies that can help maintain evolutionary processes such as genetic exchange. This chapter also considers how to manage for climate change and discuss the importance of ex situ conservation strategies.

11.1 Studying Species and Populations

To save a species from extinction, it is vital to have a firm grasp on the species' distinctive characters, in other words its **natural history**. To obtain this natural history information, 10 important factors need to be considered:

- *Population biology*: How many individuals are there in the population? How many males, females, juveniles, breeding adults, and individuals past breeding age are there? What is the species' life expectancy? How have these aspects changed over time? (see also Chapter 9)
- *Habitat*: In what kind of environment can the species be found? How do these ecosystems change over time and space? Does the species have a complex life history that requires multiple habitats (e.g. frogs that live on land generally need water for breeding)? What factors are important to maintain suitable habitat?

> To save a species from extinction, it is vital to have a firm grasp on the species' distinctive characters, in other words its natural history.

- *Distribution*: Where in the world can the species of concern be found? At what rate is its distribution increasing/decreasing? What factors drive these increases/decreases?
- *Morphology*: What are the defining traits, or range of traits, of the species' appearance? How do the species' unique morphological characteristics help it survive? Are there closely-related species that appear similar (i.e. cryptic species) and with which it can be misidentified?
- *Limiting resources*: What types of resources does the species need to survive? Are any of these resources in short supply? Does the distribution of these important resources change over time and space?
- *Physiology*: Are there any special requirements the species' physical and biochemical processes need for it to grow, survive, and reproduce? What are the conditions under which meeting these requirements is especially hard?
- *Behaviour*: How do individuals act or behave (Box 11.1)? Is the species sedentary, nomadic, or migratory? Do individuals group together, disperse at random throughout landscapes, or space themselves out at regular distances? How do these behaviours help it survive?
- *Genetics*: How much do genes vary within the species? How are the species' genetics linked to its morphology, physiology, and behaviour? Are there local genetic adaptations? Is the genetic variation in key traits sufficient to allow the species to adapt to environmental changes? Are there any deleterious genetic concerns? (Section 8.7.1)
- *Biological interactions*: In what ways do individuals of the species interact with each other and with other species? Which of these interactions are critical for survival? Are there any competitors, predators, parasites, or diseases affecting the species?
- *Interactions with humans*: How sensitive is the species to human activity? Do humans use the species in any way? Is the species sustainably harvested? Is the species associated with human-wildlife conflict (Section 14.4)?

Box 11.1 The Overlooked Role of Behavioural Ecology in the Conservation of African Mammals

Adrian M Shrader

Mammal Research Institute, Department of Zoology and Entomology, University of Pretoria, South Africa.

✉ *adrian.shrader@up.ac.za*

When considering the management and conservation of wild animals, topics linked to population and community ecology (e.g. carrying capacity, Hayward et al., 2007a) often come to mind. This is not surprising, as these disciplines consider broad patterns of population dynamics (e.g. birth rates and mortality rates), which are key to achieving management and conservation goals. While this information is necessary, in many instances, it fails to explain the mechanisms behind the patterns observed and answer key questions. For example, why do species prefer specific habitats? Why do some herbivores adjust their home ranges with the seasons? To answer these sorts of questions, we need to understand an animal's behavioural ecology.

Take for example the challenge of understanding the impacts that elephants cause within protected areas. A standard way to assess these impacts is to record which tree species are damaged and how many trees are affected (e.g. broken branches, bark stripping) (Boundja and Midgley, 2010). While this provides information on the trees most vulnerable to elephant damage, it does not explain why elephants are damaging the trees. Is it because the trees are a key part of the elephants' diet, or are these trees just abundant across the landscape and in the way of a moving herd? To answer these questions, we turn to behavioural ecology. By observing foraging elephants, or by walking down their feeding paths after they have left, we can determine the animals' diet, and generate an acceptability index (number eaten ÷ number available) of each tree species (Shrader et al., 2012). These data allow us to better understand the reasons behind elephant damage.

Other situations where behavioural ecology can help include reintroductions, population management, and human-animal conflict mitigation. For example, in South Africa, oribi are locally threatened by habitat loss and poaching. One conservation strategy is to relocate individuals away from known threats. Oribi are grassland specialists (Figure 11.A) that require both short and tall grasslands—therefore, release sites require a mosaic of these habitats. Moreover, within grasslands oribi perceive woodland patches to be dangerous, and tend to avoid feeding within 15 m of them (Stears and Shrader, 2015). If we do not consider how oribi utilise their environment, our estimate of available habitat at a release site may be greater than the area utilised. This mistake could reduce relocation success.

Figure 11.A The oribi is a grassland specialist that requires both short and tall grasslands and tends to forage at least 15 m from wooded patches. Photograph by K. Stears, CC BY 4.0.

With regards to population management, behavioural ecology is central to the conservation of southern white rhinoceros (*Ceratotherium simum simum*, NT) in the Hluhluwe-iMfolozi Park, South Africa. Within the park, the management policy incorporates space use and social ecology of the rhinos to facilitate population regulation (i.e. dispersal). To do this, the population can grow in the central core of the park. When rhino numbers get too high in the core, individuals naturally disperse into surrounding low-density areas, at which point they are captured by wildlife officers and transported to other areas. Thus, rhino behaviour itself is used to indicate when there are too many individuals within the fenced park (Linklater and Shrader, 2017).

Finally, behavioural ecology has helped reduce human-elephant conflict through the understanding that elephants are afraid of bees and will avoid feeding close to them. To capitalise on this fear, fences that incorporate beehives were designed and constructed around agricultural fields in northern Kenya, which helped reduce crop damage from raiding elephants. Of 32 raids recorded in the area, only one was at a farm with a beehive fence (King et al., 2011). These examples showcase how behavioural ecology can support, expand, and strengthen management and conservation of wildlife. These same principles can be applied to protect a wide range of animals across Africa, and elsewhere.

Understanding the natural history of a species directly informs conservation strategies. For example, if we know where a species occurs and what its habitat needs are, we are in a better position to prioritise which areas need to be protected or how ecosystems need to be restored. Similarly, if we know that an important food resource is missing, perhaps during a drought or due to human activities, conservationists could provide supplemental feeding until the limiting resource has recovered (Figure 11.2). Depending on the species in question, some factors play a more important role than others. For example, managing a disease outbreak may play a more important role in the conservation of a widespread migratory bird (that can spread diseases to other species), while managing for genetic diversity may play a more prominent role in the conservation of a small population of fishes restricted to only one lake. For many widespread species, different factors affect different subpopulations. In such cases each subpopulation might need to be managed as its own **evolutionary significant unit** (**ESU**; see e.g. Dubach et al., 2013) to retain unique local adaptations and genetic markers.

Figure 11.2 In some areas where diminishing food supplies threaten vulture populations, conservationists are supplementing their diets by placing carcasses at "vulture restaurants". These vulture restaurants often depend on cooperation with local farmers who donate livestock that have died. Photograph by Hoedspruit Endangered Species Centre, CC BY 4.0.

11.1.1 Obtaining natural history data

Conservationists rely on several resources and techniques to obtain natural history information. Initial steps often involve reviewing published and unpublished literature to understand what is known (and not known) about a species. Literature reviews do have some drawbacks: they can take a long time, may uncover contradictory information, and may lack critical information relevant to a local area or specific population. For this reason, and especially when decisions need to be made under tight schedules, conservation biologists may need to speed up their initial species review by sourcing natural history information from subject matter experts who are familiar with the species or ecosystem of concern.

Conservation biologists are also increasingly recognising the importance of **traditional ecological knowledge (TEK)**—detailed insights that rural people have on the ecology, behaviour, and distribution of the species around where they live (Shackeroff and Campbell, 2007; Brook and McLachlan, 2008). For example, while termites are often considered a pest by people living in urban settings, scientists are increasingly relying on TEK to understand the important contributions of termites to food security to human health, as well as to learn about ecological sustainable methods for their control when needed (Sileshi et al., 2009).

Conservation biologists are increasingly recognising the importance of traditional ecological knowledge—detailed insights that rural people have on the species around them.

While literature reviews, expert opinions, and traditional ecological knowledge are important first steps to collect natural history information, the most reliable method remains fieldwork, where multiple individuals from the population of concern in the area of interest are observed repeatedly over time. Indeed, most of natural history information we have today was obtained during detailed notetaking by naturalists—biologists who dedicate much of their time to better understand the natural world—in the field.

Unfortunately, there are still major gaps in our understanding of the living world. Consequently, a very large number of threatened species, including better-known groups (e.g. reptiles, Tolley et al., 2016), lack the kinds of data necessary to ensure that we can give them the best chance of survival. Filling these gaps is also becoming harder since it is costly and sometimes logistically impossible (or dangerous) for naturalists to spend an extended period in the field. There is also a trade-off in the breadth and depth of data collection possible: the more area one covers, the less detailed the data; conversely, when one collects more detailed data, the scope of the study is reined in for logistical constraints. Further, there is also a limit to the number of organisms any one individual observer can study at any one time.

Recent technological advances have greatly increased our ability to overcome the logistical constraints that impede conservation fieldwork. One of the most useful developments involves the miniaturisation (and reduced costs) of animal-borne **biologging devices**, such as radio telemetry and GPS tags (Kays et al., 2015). Previously reserved for projects with large grants that focused on large animals, the big clunky devices of a few decades ago have made way for devices small enough to fit comfortably on animals as small as beetles and frogs. Some biologging devices are now also solar-powered and transmit data through Earth observation satellites in real time, allowing researchers to track the behaviours of several organisms at a time from the comfort of their offices. Even better, some tracking technologies also collect environmental data and movement data simultaneously, allowing us to better understand how wildlife responds to changing environmental conditions. These new and sophisticated datasets can then be used to better understand threats to species (e.g. Scantlebury et al., 2014; Childress et al., 2016) and inform management of protected areas (e.g. Maxwell et al., 2011).

Species distribution modelling (SDM), also known as environmental niche modelling, is becoming increasingly popular for determining a species' distribution and habitat needs. SDMs overlay species location data, obtained during field work or using biologging devices, onto a selection of relevant environmental variables (e.g. forest cover, elevation, soil type) using GIS software, after which special modelling algorithms estimate the species' ecological niche and distribution (Figure 11.3, see also Figure 10.3). This information enables conservation biologists to identify previously unknown habitat patches (which may represent undiscovered and unprotected populations) or empty habitats (which may be used in translocations, see Section 11.2). The appeal of SDMs lies in the availability of user-friendly software packages that can use very limited datasets. For example, one study from West Africa successfully combined market survey data and SDM to determine the potential for sustainable extraction of 12 medicinal plant species (van Andel et al., 2015). Another study used SDMs to develop a holistic picture of diversity and endemism patterns of nearly all 250 African bat species (Herkt et al., 2016). While distribution modelling offers very useful conservation tools, it is important to learn about the different techniques under the guidance of an expert to avoid making costly mistakes (McPherson et al., 2006; Pearson et al., 2006).

Species distribution modelling, also known as environmental niche modelling, is becoming increasingly popular for determining a species' distribution and habitat needs.

Experimentation offers powerful methods to better understand competing theories and hypotheses, and to gain insight into how specific management actions may influence population dynamics. Experimentation is often associated with controlled environments such as laboratories; however, this is often impossible and sometimes even unethical to perform laboratory experiments on threatened species. Instead, conservation researchers may opt for natural experiments, which allows for the target species or population to be studied in its natural ecosystem.

A chronosequence study is a special type of natural experiment that overcome the long-term commitment some studies require to attain meaningful results. Also called space-for-time experiments, chronosequence studies allow us to infer long-term trends over a short study period using study systems that share similar qualities but are differently aged. Chronosequence studies are particularly popular when studying ecological restoration projects (Section 10.3) since some ecological processes often require many decades to develop (Bonnell et al., 2011). In one such example, conservation biologists needed only three summers worth of vegetation surveys to

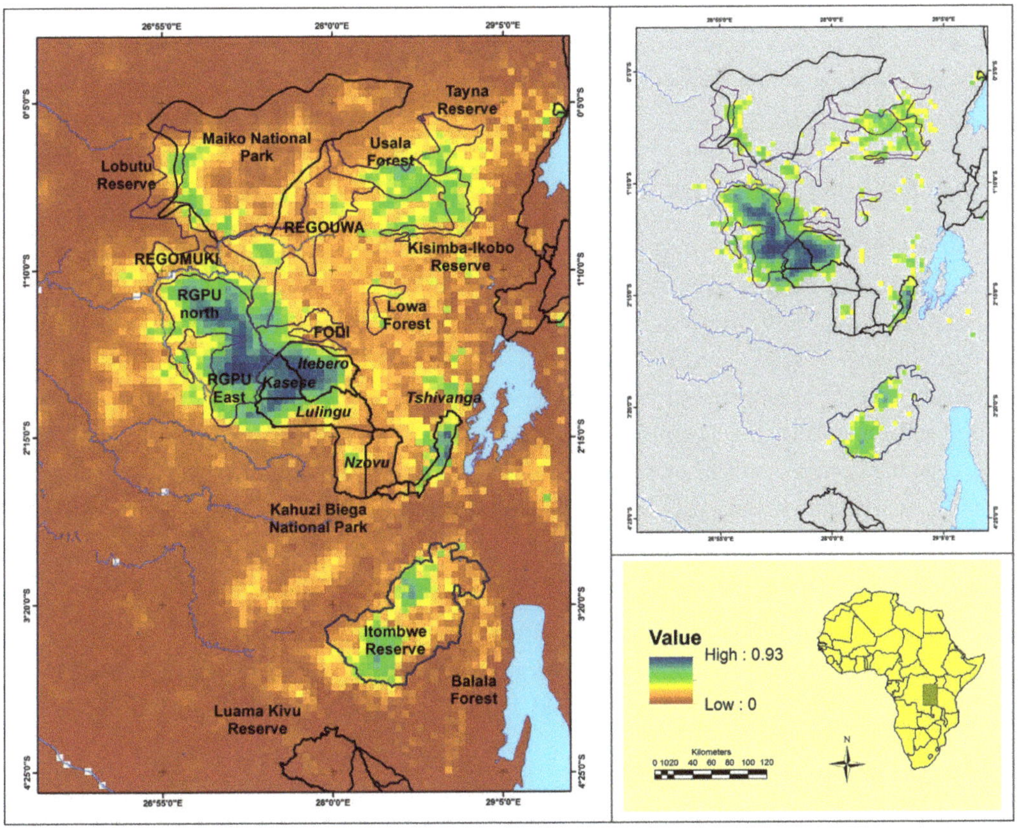

Figure 11.3 A species distribution model over the global range of the Grauer's gorilla. Purple and green areas indicate potentially suitable habitat while yellow and red areas indicate unsuitable habitat. The analysis highlighted that the gorilla is found in high-altitude forests far from deforestation activity. The map also shows which areas should be safeguarded to secure the species' survival. Source: Plumptre et al., 2016, CC BY 4.0.

show that some species recolonise coastal dune forests in the Maputaland-Podoland-Albany Biodiversity Hotspot only after 100 years since disturbance (Wassenaar et al., 2005).

Sometimes, despite their best efforts, biologists may still fail (or may not have enough time) to obtain much needed natural history information during a critical period. To overcome such a challenge, biologists have, at times, used natural history information of a **substitute species** (which is different from **surrogate species**, Section 13.3.5) to fill data gaps for a rare species (Caro et al., 2005). An example of this application comes from the USA where researchers used behavioural observations of a common butterfly to predict dispersal of another closely related butterfly that was too rare to properly study (Hudgens et al., 2012). It is important to note that using information from substitute species does have serious limits (Henry et al., 2019). For example, considering that different populations of a *single species* may have very different environmental needs and adaptations, using data from a *different species* may be even less useful. Care must therefore be taken when using data from substitute species with proper acknowledgement of the assumptions and uncertainty this approach adds to one's research.

11.2 Saving Species Through Translocations

Because the probability of extinction increases rapidly for small populations (Section 8.7), conservation biologists often invest considerable energy into increasing the size of small and declining populations. Often, these projects involve improving the extent and quality of suitable habitat (Chapter 10) or mitigating threats such as overharvesting (Chapter 12). When appropriate, conservation biologists may sometimes resort to translocations—moving individuals from sites where they are threatened (e.g. unprotected lands or a paper park) or overabundant (e.g. a well-managed protected area or ex situ conservation facility) to sites where they can offer a larger contribution to conservation efforts.

Conservation biologists generally recognise four basic translocation approaches:

- **Restocking** (also called augmentation) occurs when wildlife managers increase the size and genetic diversity of existing populations, by releasing individuals that have been raised in captivity or that have been obtained from other wild populations.
- **Reintroduction** occurs when wildlife managers release individuals into areas where they occurred in the past but not at present. The areas must be ecologically suitable and the factors that caused the extirpation must have been reduced or eliminated for a reintroduction to be successful.
- **Introduction** involves creating new populations by moving individuals to suitable areas outside that species' historical range. Introductions are usually considered when reintroductions are impossible because the species' historical range has been degraded too severely or because persistent threats will lead to reintroduction failure.
- **Assisted colonisation** (also called assisted migration) is a special class of introduction where biologists "assist" species with poor dispersal capabilities to adapt their ranges in response to environmental changes. It is anticipated that this strategy will become an important conservation tool in preventing extinctions where climate change outpaces the speed of natural migration.

> Understanding a species' ecological needs is critically important for translocations, because it influences the choice of release site and type of preparations needed.

11.2.1 Important considerations for translocations

Section 11.1 broadly discussed the importance of understanding the ecological and other natural history needs when protecting threatened species. Understanding a species' ecological needs is equally, if not more, important for translocations, because it influences the choice of release site and type of preparations needed (Figure 11.4). Complementing the 10 factors mentioned in Section 11.1, the next section briefly introduces some of the most important considerations during translocations.

Figure 11.4 A team of wildlife rehabilitators release a group of Cape vultures (*Gyps coprotheres*, EN) near a vulture restaurant in South Africa. Releasing the vultures near a supplemental food source greatly enhances their chances for survival after release. The vultures have wing tags to enable monitoring of each individual after release. Photograph by VulPro, CC BY 4.0.

Determining need and feasibility

Perhaps the most important factor to consider before starting a translocation is to determine whether it is necessary. Translocations carry risks, not only for the target population to be moved, but also the individuals left behind and for the recipient ecosystem. These risks expose translocation projects to a high risk of failure, particularly if preparations are inadequate and essential resources (e.g. funding, trained staff) are in short supply. Translocations also demand considerable resources—resources that can at times be better spent mitigating the threats the target population face. While these considerations may seem obvious, a recent review found that *most* translocations projects are initiated without proper cost-benefit analyses (Pérez et al., 2012). To improve translocation practices, conservationists seriously considering a translocation project are encouraged to review the 10 criteria outlined in Pérez et al. (2012), some of which also overlap with the considerations mentioned below.

Support from local stakeholders

It is also important to consider, at an early stage, how the public will view the translocation project. Some people may feel the resources used in a translocation are better invested elsewhere; others dislike translocations because they view it as a threat to their livelihood—this is especially true when carnivores are involved (Gusset et al., 2008a). Because of these and other potential conflicts and emotions, it is crucial that translocation projects (like any conservation activity) obtain the support from local stakeholders at an early stage. It is helpful to be transparent from the outset and to explain the project's goals, as well as the benefits the local community may gain (e.g. attract more tourists, restore a degraded ecosystem service). Good public outreach also provides opportunities to address the public's concerns and misconceptions about the project and about biodiversity conservation in general.

Identifying suitable habitat

It goes without saying that the probability for success is greatly improved when the translocated individuals are released in good quality habitat. This is particularly true for species with poor dispersal capabilities, such as plants that reproduce through vegetative propagation: the plants could die in an environment that is too sunny, shady, wet, or dry. While this point may seem obvious, many translocations fail because individuals are released in inferior habitats (Armstrong and Seddon, 2007). One of the reasons for this potential habitat mismatch is because wildlife may perceive the environment differently than humans, so a site that may look good to the human eye may lack one or more overlooked limiting resource. **Refugee species**—species forced to live in suboptimal habitat due to threats present in their preferred habitat (e.g. Ali et al., 2017)—also present a challenge to biologists who may unwittingly view inferior habitat as optimal and base conservation decisions on essentially bad information. The same challenge presents itself at **ecological traps**—unsuitable environments that an organism mistakenly perceives as optimal habitat (e.g. Sherley et al., 2017). These are some of the most important reasons why biologists need to be cautions when using species distribution models (SDM) when identifying areas suitable for translocations. To mitigate costly translocation failures, it is advisable that releases start small, and have multiple phases, to assess how released individuals respond to their new environment. Conducting experimental and adaptive releases can also reduce uncertainty by evaluating different release scenarios (Menges et al., 2016).

> Species forced to live in suboptimal habitat due to threats present in their preferred habitat may lead biologists to unwittingly view inferior habitat as optimal.

It is also important to ensure that any habitat identified as suitable is free from threats such as pollution and invasive species that may lead to declining health or even death for released individuals. A project in the Cape Floristic Region in South Africa provides a good example of how alert conservation biologists mitigated a threat that could have caused a translocation failure. The Clanwilliam sandfish (*Labeo seeberi*, EN) was once widespread in the region's Olifants-Doring River system. However, recent surveys indicated that the species had gone extinct in the Olifants River. Although biologists did find some juvenile fish in the Doring River and some of its tributaries, they also noticed that invasive fish predated on most of those juveniles before they reached adulthood (Jordaan et al., 2017). These ill-fated individuals were thus dispersing from the last remaining reproductive subpopulation persisting in the headwaters of one single Doring tributary to other parts of the river, which acted as a population sink. To prevent the species' extinction, biologists initiated a habitat restoration plan involving restoring natural stream flow regimens and eradicating predatory invasive fish in the headwaters of a second Doring tributary. They then installed barriers that prevented invasive fish from reaching the restored area before translocating 338 juvenile fish (Figure 11.5) there. With this habitat restoration plan, the biologists hope to establish a second viable population, and to improve the juveniles' chances of surviving to adulthood before they disperse back to areas where the invasive fishes occur (Jordaan et al., 2017).

Figure 11.5 (Top) Conservation biologists collecting threatened Clanwilliam sandfish for a reintroduction project in the Cape Floristic Region, South Africa. Photograph by John Lucas/explore4knowledge®, CC BY 4.0. (Bottom) A close-up view of Clanwilliam sandfish. Photograph by Gustav Klotz/explore4knowledge®, CC BY 4.0.

Considering genetics and behaviour

Translocation projects also need to consider the genetic makeup, social organisation, and behaviour of a species that is being released. It is preferable to use individuals from the same genetic stock as individuals that already occur (or have occurred) in the release area to avoid outbreeding depression and to capture local adaptations (Sections 8.7.1). Such efforts simultaneously also contribute to conservation of genetic diversity, as opposed to the pollution thereof if individuals from different genetic stock are mixed.

Group-living species, particularly those vulnerable to Allee effects (Section 8.7.2), need to be released in sufficient numbers so they can maintain their natural social organisation and behaviour. For species that need to be released in groups, it is preferable to release socially integrated animals rather than individuals unfamiliar with each other (Gusset et al., 2008b). Releasing groups of animals does have its own set of challenges. For example, social groups abruptly released from captivity may disperse explosively, possibly leading to project failure. This happened with African buffalo (*Syncerus caffer*, NT) herds translocated to South Africa's Addo Elephant National Park which fragmented into smaller groups after release, making them more vulnerable to lion (*Panthera leo*, VU) predation (Tambling et al., 2013). Fortunately, in this case, the buffaloes underwent several behavioural modifications over time, which eventually allowed their numbers to stabilise (Box 11.2). This contrasts with failed rock hyrax (*Procavia capensis*, LC) reintroductions in South Africa, where group disintegration post release exposed the animals to unsustainable predation levels (Wimberger et al., 2009).

How many individuals to release

The ultimate aim of translocation projects is to establish **ecologically relevant** populations, meaning populations that are self-sustaining, free from inbreeding, and an interactive participant of its community and ecosystem. The probability of achieving this goal increases as more individuals are being released. Because translocation projects typically do not have an unlimited supply of individuals to release, wildlife managers often rely on quantitative models (Section 9.2) to estimate the minimum number of individuals that should be released and how many times releases should occur. For example, a population viability analysis (PVA) on western lowland gorillas (*Gorilla gorilla gorilla*, CR) reintroduced to Gabon and the Republic of the Congo showed that the probability of persistence of an apparently established population could be increased significantly if more individuals were released (King et al., 2014).

> The ability to establish new populations through translocations does not reduce the need to protect threatened species still in their natural habitats.

While releasing more individuals certainly improves the likelihood of establishing a self-sustaining population, it is also important to determine how many individuals

Box 11.2 Large Predator Reintroductions: A Balancing Act

Craig J. Tambling

Department of Zoology and Entomology, University of Fort Hare, Alice, South Africa.

✉ *ctambling@ufh.ac.za*

Large predator numbers are declining, and African carnivores are no exception (Ripple et al. 2014). How to conserve African carnivores are a hotly debated topic now, with "fortress" type conservation areas considered the most viable option by many (Packer et al., 2013). In South Africa, this conservation model is the norm, and many small protected areas are now translocating large carnivores for ecotourism. However, these large carnivore translocations have repercussions for resident prey species. Understanding the ecological and biodiversity consequences of these translocations is thus important for the management of these small protected areas (Tambling et al., 2014).

In 2003, lions and spotted hyenas (*Crocuta crocuta*, LC) were reintroduced into the Addo Elephant National Park Main Camp Section after being absent from the area for over 100 years. Post-release monitoring of the six reintroduced lions indicated that at least 50% of their diet in the first two years following reintroduction was African buffalo. This was especially concerning to South African National Parks as this resident buffalo population contributes substantially to game auction sales each year, with the money raised being used to expand the national park system in South Africa (SANParks, 2009).

Following high predation rates of buffalo by lion and a 2007 buffalo census suggesting low juvenile recruitment, the coexistence of lion and buffalo in Addo was questioned. These concerns lead to a detailed assessment of buffalo behaviour and demographics between 2008 and 2011 (Tambling et al., 2012), which showed that by 2008–2009, juvenile buffalo recruitment (Figure 11.B) had rebounded to levels reminiscent of those prior to the lion reintroduction. Direct observations of the buffalo population showed drastic behavioural alteration following the high initial predation rates by lions. These behavioural changes included: (1) increased breeding herd sizes, (2) a reduction in nocturnal movement, and (3) greater use of open habitats at night and early morning when lions are hunting. These behavioural adjustments enabled the active defence of the breeding herds, reducing successful predation by lions and ensuring an increase in buffalo recruitment. Although this study suggests that prey populations are capable of behavioural adjustments to reduce predation, this is not always the case, with some species unable to respond, leading to precipitous declines in prey populations such as eland (*Tragelaphus oryx*, LC) (Leaver, 2014).

Figure 11.B After lions were reintroduced into the Addo Elephant National Park Main Camp Section, high levels of predation of buffalo (in particular, juveniles) prompted an investigation into the demographics and behaviour of the buffalo population. Results showed that buffalo were adjusting their behaviour to make greater use of open habitats, which subsequently led to improved juvenile buffalo survival. Photograph by C. Tambling, CC BY 4.0.

Case studies of predator-prey interactions following large predator reintroductions highlight the management challenges faced by small reserves where ecotourism, biodiversity, and financial goals each need to be met. Due to the small size of these "fortress" reserves, a local overabundance of predators can have severe ecological effects on prey populations. However, in many reserves, the high demand for large predators for ecotourism often results in costly reactive, rather than scientifically sound proactive, management. There is, however, a growing body of research on the proactive management of large carnivores, where wildlife managers aim to replicate ecological processes (i.e. lion inter-birth intervals) to limit management interventions required to control large predator numbers (Ferreira and Hofmeyr, 2014). In small reserves, lion inter-birth intervals are shorter than in large ecosystems, and so lengthening the inter-birth intervals to that observed in large ecosystems can reduce lion population growth rate in these small reserves (Miller et al., 2015). Understanding predator-prey interactions is important regardless of the conservation model employed to protect these large charismatic species.

The ultimate aim of most translocation projects is to establish populations that are self-sustaining, free from inbreeding, and interactive participants of their communities and ecosystems.

the target community *can* sustain. In other words, the release area should contain enough suitable habitat to support the territories of *all* the released individuals. To determine how many individuals can be sustained, wildlife managers may calculate the release area's **carrying capacity**—an estimate of the maximum number of individuals an ecosystem can support. The carrying capacity concept has its roots in the livestock trade, where farmers wanted to maximise the number of animals on their land without risking overgrazing. While the concept has gained popularity in conservation biology in recent decades, calculating the carrying capacity for wildlife is very complex because of all the multi-faceted interactions that characterise healthy ecosystems. For example, the carrying capacity for a wild population can depend on factors such as food, water, shelter, soil nutrients, and sunlight availability, as well as more species-specific natural history factors such as habitat quality, home range, sex ratios (Tambling et al., 2014), and interactions with other species (Lindsey et al., 2011).

Over the past few decades, through trial and error, adaptive management (Section 10.2.3), and the collection of vast amounts of demographic data, scientists have made significant progress in calculating carrying capacity for wildlife populations. Perhaps the most progress has been made in calculating carrying capacities for large ungulates, by monitoring vegetation biomass, which in turn is affected by soil nutrients and rainfall (Fritz and Duncan, 1994). Much progress has also been made in calculating carrying capacities of predators by monitoring prey densities (Hayward et al., 2007a). For most populations, however, carrying capacity isn't explicitly calculated, but implicitly estimated based on intuition. Refining existing carrying capacity models and developing new methods for other taxa remain an active area of research that will hopefully reduce conservation biologists' over-reliance on intuition in future years. But even in the absence of carrying capacity calculations, wildlife managers can track a population's health and overall fitness. When the health of a particularly successful population or its environment starts declining, a root cause may be that too many individuals have been released, or the population is being sustained above carrying capacity.

Preparing individuals for release

Translocation projects using individuals obtained from the wild are generally much more successful than those using captive-bred individuals, given that wild individuals are already adapted to a life where they must fend for themselves. Nevertheless, some projects may have to use captive-bred individuals, particularly when the target species is extinct in the wild, or when individuals were brought to an ex situ conservation facility because it is easier to breed them under human care in controlled conditions. In such cases, a great amount of effort may be required to prepare the captive-bred individuals for releases.

> A great amount of effort may be required to prepare captive-bred individuals for translocation because they may have lost adaptations required for survival and reproduction in the wild.

A major drawback when using captive-bred individuals is that they may have lost the important adaptations required for survival and successful reproduction in the wild. Pre-release training, which varies according to the species, can sometimes overcome this drawback. For predators, it may involve providing low risk prey, such as chickens and domestic rabbits in holding facilities until their hunting skills are better developed (Houser et al., 2011). For plants propagated indoors, it may involve hardening them off by placing them outside for increasingly longer periods to gradually introduce them to sun, wind, and temperature changes during the day. To help young birds disassociate humans from food, human trainers sometimes use puppets or wear costumes (Figure 11.6) during feeding time to mimic the appearance and behaviour of wild individuals (Valutis and Marzluff, 1999). Another method, which may promote behavioural enrichment, involves **cross-fostering**, in which unrelated parents helps raise the offspring of a threatened species. In carnivore conservation, this technique has shown much promise to augment litter size and encourage gene flow using orphaned African wild dog (*Lycaon pictus*, EN) pups (McNutt et al., 2008). Interspecific cross-fostering has also been used in bird conservation, where biologists use common species to incubate eggs abandoned by threatened species (e.g. Powell and Cuthbert, 1993). However, cross-fostering using different species may lead to a new set of problems, like behavioural changes and hybridisation, if the young subsequently associate with the wrong species. A great amount of care and research are thus needed before such strategies are attempted.

Figure 11.6 Wattled cranes (*Grus carunculate*, VU) sometimes lay two eggs, but always abandon the second egg. Conservation biologists in South Africa are collecting the discarded eggs, which are then hatched in a captive breeding programme. To avoid the captive chicks associating humans with food and safety after release, handlers use special crane costumes when interacting with the birds. Photograph by Daniel Dolpire, CC BY 4.0.

Whether using captive-bred or wild individuals for translocations, individuals may have to be fed, sheltered, trained, or otherwise cared for after release to give them time to become more familiar with their new surroundings. This approach, known as **soft release**, involves keeping the released individuals in pre-release holding facilities for a period; it may also include some form of assistance after release to increase opportunities for success. Soft releases also provide an opportunity to introduce captive-bred organisms to wild individuals of the same species that can act as "instructors" for survival in the new environment, or for unfamiliar individuals to bond into cohesive units (Gusset et al., 2006).

The alternative to soft release is a **hard release**—an abrupt release of individuals from captivity without assistance such as food supplementation. While hard releases are popular (because they are relatively easy to perform), it is a risky strategy that faces a high risk to failure (Brown et al., 2007; Wimberger et al., 2009). Hard releases can however be appropriate under the right conditions (Hayward et al., 2007b). For example, hard releases are often use in **head-starting** programmes (Figure 11.7) for reptiles and amphibians (Scheele et al., 2014), where conservation biologists collect wild individuals and raise them past their most vulnerable life stages before releasing them again where they were collected.

Figure 11.7 Local children releasing leatherback sea turtles (*Dermochelys coriacea*, VU) as part of an environmental education project in Equatorial Guinea. Known as head-starting, conservationists would sometimes collect sea turtle eggs in the wild, hatched in captivity, and raise the offspring past their most vulnerable stage before releasing them back where they were found. Photograph by Katharine Clukey/TOMAGE-INDEFOR, CC BY 4.0.

Post-release monitoring

A translocation project does not end after the last individual was released. Rather, ongoing monitoring should be implemented to determine whether a translocation was successful, what degree of success was achieved, whether adaptive management is needed, whether additional releases should be conducted, or whether the project should be aborted. A well-designed monitoring plan can also highlight the consequences of translocation on the broader ecosystem, such as the impact that predators introduced to a new area may have on prey populations (Box 11.2) and competing species (Groom et al., 2017). Because some responses in translocated populations can be rather subtle and take many years to show or subside, post-release monitoring should ideally be a long-term endeavour. For example, by monitoring seemingly successful elephant reintroductions across five protected areas in South Africa, researchers found that stress hormones in released animals continued to decline 24 years post release (Jachowski et al., 2013). Long-term monitoring will also help wildlife managers better understand the ultimate fate of the released individuals. Many apparently successful translocations fail because the released individuals die after several years without ever reproducing. Highlighting the importance of post-release monitoring, one study from South Africa found that 70% of captive-bred oribi (*Oribia oribi*, LC) died within two months of release, mostly due to predation (Grey-Ross et al., 2009). Another study found that reintroduced cheetahs were all killed within a year of release (Houser et al., 2011). These were expensive lessons, but post-release monitoring ensured that the reason for failures are known and can be addressed ahead of future releases.

Helping other translocation projects

Strategies used in successful translocation projects were nearly always informed by releases conducted by other wildlife managers who circulated their experiences to the wider conservation community. It is important to pay this effort forward; new translocation projects should make every effort to track and publish their results to inform others. While it is always easier to present the results of successful projects, publishing the lessons from failed projects is also important (Wimberger et al., 2010; Godefroid et al., 2011). Equally important is the publication of project costs, to enable wildlife managers to better determine under which conditions translocations represent a cost-effective conservation strategy. For example, a large African wild dog reintroduction programme in South Africa achieved their initial goal of establishing nine self-sustaining packs much more quickly than expected—five years rather than 10—yet reintroducing all these populations cost 20 times more than if the funds were used to enhance protection of existing packs within protected areas (Lindsey et al., 2005). With more information available, future conservationists would hopefully be able to have better guidelines to maximise cost-effectiveness and the likelihood of project success.

11.3 Managing and Facilitating Movement Dynamics

Some ecosystems are transient in nature—their character is temporary and will change because of disturbance and succession. Consequently, species that occupy those transient habitats are bound to be naturally extirpated at one time or another. Consider, for example, a small population of wildflowers occurring in a river's floodplain; at some stage, there is going to be a flood that will wash away those flowers. But the flooding also disperses seeds downstream, allowing for new wildflower populations to establish in suitable habitat elsewhere. These shifting populations linked by movements between them are better characterised as a **metapopulation** (a "population of populations") (Figure 11.8) consisting of several subpopulations. For some metapopulations, every subpopulation is transient: their distribution changes dramatically with each generation. Other metapopulations involve relatively permanent subpopulations with only a few individuals dispersing each generation. Some metapopulations consist of one or more **source populations** whose sizes are stable or increasing, and several **sink populations** whose sizes fluctuate depending on environmental conditions. Some sink subpopulations may undergo such dramatic fluctuations that they would be extirpated in unfavourable years were it not for **population rescue** by immigrants from source populations.

A metapopulation (a "population of populations") consists of several subpopulations linked by movements of individuals between them.

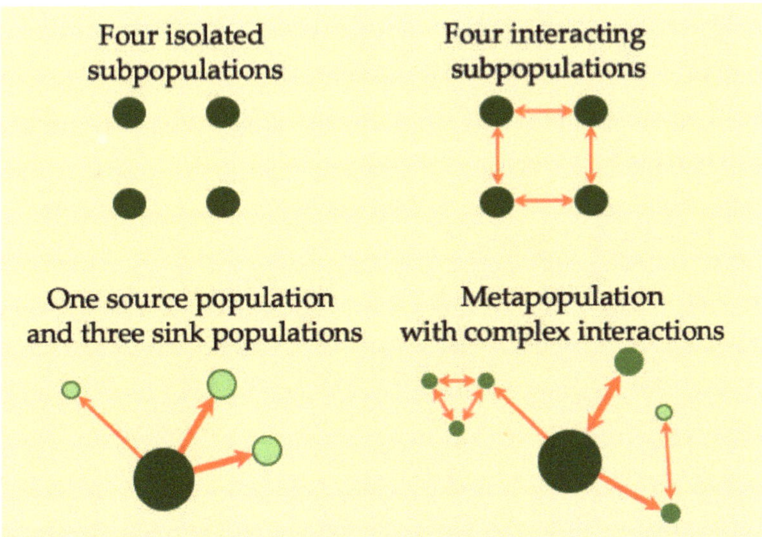

Figure 11.8 A range of metapopulation patterns is possible in nature. In this illustration, population size is represented by the size of the circle, while movement direction and intensity are indicated by the direction and thickness of the arrows. After White, 1996, CC BY 4.0.

Habitat fragmentation threatens metapopulation dynamics by reducing opportunities for dispersal across the landscape (Chapter 5). When there is too little movement of individuals between habitat fragments, the dwindling subpopulations within those fragments are at risk of extirpation or even extinction (Section 8.7). In contrast, well-connected subpopulations maintain themselves by colonising empty niches, exchanging genetic material, and adapting to changing environments. Dispersal also maintains critical ecosystem processes, such as pollination and seed dispersal (Section 4.2.5). Consequently, conservation biologists have invested significant resources in recent years to maintain and restore wildlife movements within fragmented ecosystems.

11.3.1 Connectivity in terrestrial ecosystems

Maintaining and restoring **ecosystem connectivity**—the ability of ecosystems to facilitate the dispersal of individuals between different areas—involves maintaining and restoring wildlife movements that are (at risk of being) impeded by human activities. The most popular method to maintain (or restore) connectivity in a fragmented landscape is to maintain (or restore) **habitat linkages**, also called wildlife corridors, habitat corridors, dispersal corridors, or movement corridors. All these terms refer to continuous tracts of suitable habitat with little to no dispersal barriers that connect otherwise isolated habitat patches and populations.

> Maintaining and restoring ecosystem connectivity is an important strategy for conserving wildlife whose movements are impeded by human activities.

Some of the most prominent efforts to restore habitat linkages involve habitat restoration. For example, plans are currently underway to use forest regeneration to reconnect nine forest fragments in Tanzania's East Usambara Mountain; if successful, this project would establish the largest contiguous forest block (over 3,000 km^2) in the Eastern Arc Mountain Biodiversity Hotspot (Newmark, 2008). The positive impact of this project is expected to be immense. It has been estimated that the restoration of just 80 km^2 of forest would stave off the first fragmentation-induced extinctions by over 2,000 years, compared to an estimated seven years until the first extinction if these forest fragments were to remain unconnected (Newmark et al., 2017).

Connectivity is important in every ecosystem on Earth. However, given the linear characteristic of **riparian zones** along rivers and stream—and hence a larger proportional impact of edge effects (Section 5.1.2)—we might consider connectivity in these spatially restricted systems to be particularly important (Figure 11.9). Protecting and restoring riparian zones as habitat linkages resonates with a variety of people because these areas provide a range of important ecosystem services, including flood control and water purification (Section 4.2.4). Conservationists can tap into this energy by lobbying for laws that prohibit activities such as logging, housing, and industrial developments within a certain distance from a river or stream. By protecting ecosystem services associated with riparian zones, these laws simultaneously also maintain wildlife refuges (Monadjem and Reside, 2008), source populations (Vosse et al., 2008), and habitat linkages (Bentrup et al., 2012; McLennan and Plumptre, 2012). In contrast, inadequate protection of riparian ecosystems not only compromises connectivity, but also negatively affect species not overtly dependent on these buffer areas. For example, research from Southeast Asia has shown that losing riparian ecosystems in an otherwise palm oil dominated landscape reduced stream quality, which in turn reduced local fish diversity by up to 36% (Giam et al., 2015). In contrast, protecting riparian zones were found to increase palm oil yields (Horton et al., 2018). With so many riparian areas currently being degraded and destroyed, there is an urgent need for stronger riparian protection laws (Chapter 12), and for more effective enforcement of those laws.

Figure 11.9 Protecting riparian zones such as this one along the Turkwel River in northern Kenya is an effective strategy for maintaining connectivity and securing a range of ecosystem services. Photograph by Bernard Dupont, https://www.flickr.com/photos/berniedup/17966234205, CC BY-SA 2.0.

Restoring connectivity may also involve removing or otherwise mitigating human constructs that block wildlife dispersal. This is a major aim of TFCAs, which aim to restore dispersal between protected areas (Jones et al., 2012) by removing fences and other human constructs while still maintaining sustainable land tenures (Andersson et al., 2013). These efforts, accomplished through partnerships with local communities, are re-establishing historical mass migration routes, which in turn will hopefully also boost those areas' ecotourism potential (Box 11.3). Efforts to revive extinct mass migrations also seem to be paying off! For example, in Botswana, the removal of veterinary fences—meant to prevent spread of diseases from wildlife to livestock, but also cutting off the world's second largest wildebeest migration—have seen several hundred plains zebras (*Equus quagga*, NT) returning to old migration routes within four years (Bartlam-Brooks et al., 2011).

Section 5.1.1 discussed how inconsiderate fence placements threaten wildlife, while the paragraph above explained how removing fences can improve connectivity. Ironically, and illustrating the difficulties conservationists face when dealing with conflicting demands, strategically placed fences can sometimes also be used as a conservation tool. For example, researchers working on a fragmented lion population in Botswana found that the most effective way to improve this population's viability was through strategic placement of fences to direct dispersal between protected areas (Cushman et al., 2016). Strategically placed predator-proof fences may at times also be

Box 11.3 Transfrontier Conservation Areas: Managing Biodiversity Across International Boundaries

Simon M. Munthali

Kavango-Zambezi Transfrontier Conservation Area Programme, Kasane, Botswana.

http://www.kavangozambezi.org

TFCAs are components of a larger ecosystems that straddles the border between two or more countries, encompassing one or more protected areas as well as multiple-resource areas used by communities and private landholders. They are also managed for sustainable use of natural resources (Singh, 1998). The concept recognises that borders are political rather than ecological (Dallimer and Strange, 2015), and aims to ensure that key ecological processes continue to function where political borders have divided ecosystems, river basins, or wildlife corridors (Cumming, 1999).

TFCAs are widely being established in Africa. One of these is the 520,000 km^2 Kavango-Zambezi TFCA (KAZA)—a conservation and development initiative of Angola, Botswana, Namibia, Zambia, and Zimbabwe.

The benefits of the KAZA include:

- Re-establishment of the seasonal wildlife migration routes and connectivity among the many protected areas (national parks, community conservancies, and wildlife and forest reserves) within the region (Figure 11.C). The primary wildlife focus is the savannah elephant (*Loxodonta africana*), whose population of about 250,000 is predominantly concentrated in Chobe National Park (Botswana), Hwange National Park (Zimbabwe), and Bwabwata National Park (Namibia). Elephants need unimpeded movement to protected areas where population densities are much lower, such as Luengue-Luiana and Mavinga National Parks (Angola), and Sioma Ngwezi and Kafue National Parks (Zambia). This movement would reduce pressure on the ecosystems that are currently overpopulated and enable elephants and other species to better coexist—especially grazing herbivores that depend on the same habitats as the elephant.

- Expanding the wildlife-based economy, primarily ecotourism, into agricultural marginal areas (with predominantly Kalahari sand soils), through community-private partnerships. Through these partnerships, local communities would benefit from employment and business opportunities in ecotourism activities.

- Opportunities for local communities to participate in decision-making, and influencing policies and legislation related to natural management such as coordination of the fishing closed season between Namibia and Zambia during the fish breeding season (December–March) in the Zambezi River.

- Formation of alliances among different stakeholders (governments, private sector, NGOs, and local communities) to maximise skills and resources in promoting sustainable land use, conserving biodiversity and alleviating poverty.

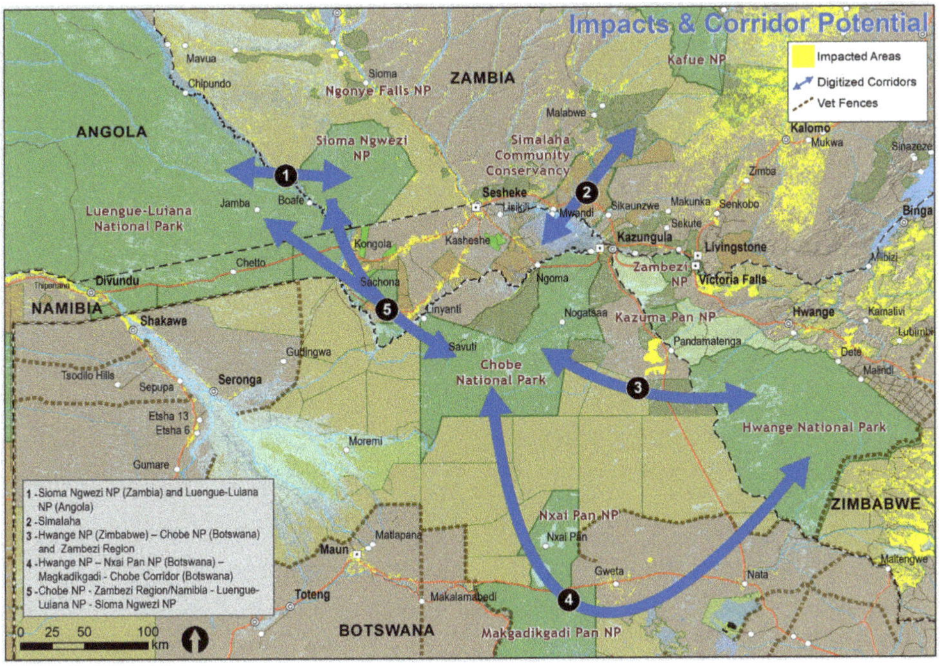

Figure 11.C Location of priority wildlife dispersal corridors between the various national parks of the KAZA TFCA. Map by Peace Parks Foundation, CC BY 4.0.

Despite these benefits, there are obstacles to progress in attaining the benefits of the KAZA. Notable among these are social and political factors, such as increasing human population density, increasing cultivation of land, and expanding human settlements in wildlife corridors. Many of these factors trigger human-wildlife conflicts and poaching both for local consumption of bushmeat and for the illegal sale of elephant ivory. To mitigate these threats, the following strategies are being implemented:

- A Master Integrated Development Plan for the KAZA has been developed, which provides initial zoning. Its key feature is spatially allocating land into various uses (human settlement, agriculture, and protected wildlife areas, including wildlife dispersal corridors). The Master Integrated Development Plan also assists in creating awareness about the value of the wildlife corridors, which traverse communal areas.

- Promotion of conservation agriculture as a tool for improving land stewardship, intensification of agriculture, and improving crop yields per unit area of land, and therefore decreasing the likelihood of cutting down forested areas in and around wildlife corridors to plant new agricultural fields. Currently, within the KAZA, conservation agriculture is being piloted in Angola, Namibia, and Zambia. Conservation agriculture is crop production that strives to achieve acceptable profits together with high and sustained production levels while concurrently conserving the environment.
- Promotion of community-private partnerships in ecotourism development. Over the past four years, Ngoma safari lodge (Botswana), and Machenje sport fishing lodge (Zambia) have been developed specifically in support of securing wildlife corridors. They also provide incentives to the local communities for adopting wildlife conservation as a supplement to their land use practices. These lodges are in addition to the numerous existing tourist resorts in the KAZA.
- A law enforcement and anti-poaching strategy for the KAZA is being developed to coordinate transboundary law enforcement surveillance and fines to prevent poaching of protected wildlife. In addition, KAZA partner countries are integrating other security agencies, such as the military, police, immigration, and customs officials, to prevent the illegal export of wildlife products such as elephant ivory and bushmeat out of the KAZA.
- Reducing human-wildlife conflicts (Section 14.4) through improved land use planning, solar-powered electrified fencing encircling clusters of village fields and facilities and use of chilli-pepper-based olfactory repellents to deter elephants from entering crop fields.

The KAZA has made considerable progress to date in coordinating conservation efforts among the wildlife agencies and national parks across five countries in Southern Africa. The principal success has been measures to allow the continued migration of elephants along existing migration routes across international borders. The challenges ahead—from inadequate funding for wildlife patrolling and anti-poaching activities to increasing populations of rural people outside the protected areas and across migration routes—remain significant.

required to avoid human-wildlife conflict (Packer et al., 2013, but see Creel et al., 2013), and to facilitate the recovery of threatened species, as is the case for Africa's rarest antelope, the hirola (*Beatragus hunter* CR) (Ng'weno et al., 2017). The final word here is that management must remain responsive to both positive and negative impacts of tools, such as fences, rather than relegating them to bins, such as good or bad. (See also Dupuis-Desormeaux et al., [2018] for the use of fence-gaps and exclusionary fences to mitigate some negative fence impacts.)

At times, when it is impractical to establish or restore continuous habitat linkages, biologists may opt to protect and restore **stepping stone habitats** (Figure 11.10). As the name implies, stepping stone habitats are a special type of habitat linkage that facilitate dispersal along a patchwork of isolated habitat patches within a matrix of unsuitable or inhospitable habitat. Stepping stones thereby divide long dispersal events through a long stretch of inhospitable terrain up into shorter, and thus more manageable, sections. Stepping stone habitats are particularly important for migratory species that rest and refuel at stop-over sites between the end-points of their migratory route (Runge et al., 2015)—each stop-over site can be viewed as a stepping stone habitat. Prominent examples of stepping stone habitats that deserve protection include sacred forests which can act as stop-over sites for migratory forest birds; wetlands and estuaries (see Box 5.3), which can act as stop-over sites by migratory waterbirds; and small forest reserves, which can act as stepping stones between a network of other protected areas (Riggio and Caro, 2017).

Protecting and restoring stepping stone habitats can maintain connectivity in areas where it is impractical to establish or restore continuous habitat linkages.

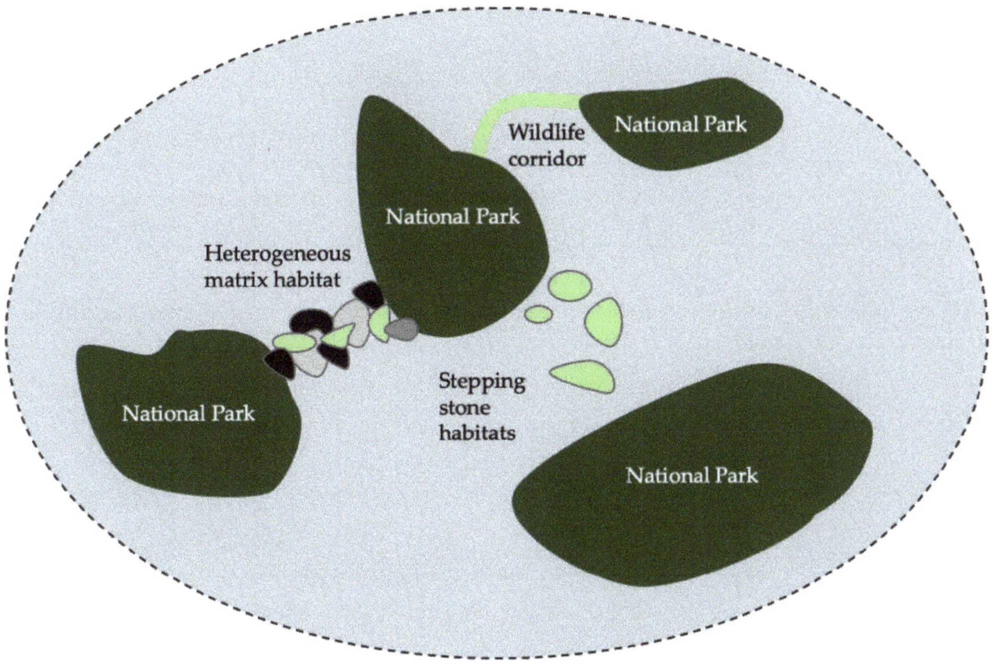

Figure 11.10 Methods to reconnect fragmented metapopulations (or maintain connectivity) can take many forms. The three main strategies are to maintain or restore wildlife corridors (e.g. to link two isolated forest patches), maintain or restore stepping stone habitats (e.g. a patchwork of wetlands or sacred forests), or facilitating movement through the matrix with sustainable land use tenures (e.g. removing fences). After Bennett, 2004, CC BY 4.0.

11.3.2 Connectivity in freshwater ecosystems

> While dams play an important role in hydropower generation and securing a year-round supply of water, recent evidence suggests that they create many environmental problems, including blocking species dispersal.

Dams have always played an important role in hydropower generation and securing a year-round supply of water for farms, industries, and cities. Unfortunately, recent evidence suggests that reservoirs may create more problems than they solve (Section 5.3.2). Of concern is their contribution to greenhouse gases (Deemer et al., 2016), as well as their role in blocking dispersal of aquatic organisms. To counter these negative impacts, governments across the world are decommissioning and removing dams and other types of artificial water impoundments. For instance, over the past 30 years more than 1,174 dams were removed in the USA; the 72 dams removed in 2016 alone restored more than 3,000 km of streams (Thomas-Blate, 2016). Similar efforts are also underway in Europe (http://www.ecrr.org), where river restoration efforts have been initiated at over 1,100 locations across 31 countries. Unfortunately, not only are efforts to restore freshwater connectivity lagging across Africa; in many cases, even more rivers are currently being dammed (Winemiller et al., 2016).

11.3.3 Connectivity in marine ecosystems

Ecosystem connectivity is also important in marine ecosystems. Many marine organisms, including economically important species, breed and feed in different areas at different times of the year, and use established dispersal routes to move between those areas. It is thus important to protect these dispersal routes so we can maintain these marine ecosystems and ecosystem services.

> Maintaining movement dynamics in marine seascapes involves protecting and restoring marine corridors, estuarine linkages, and coastal habitat linkages.

There are three main strategies to maintain and restore movement dynamics of marine seascapes. First, marine corridors—zones used by whales and other marine species to move between feeding and breeding grounds—should be protected. Marine biologists in several countries successfully reduced collisions between whales and ocean-faring vessels with minor adjustments to shipping lanes that previously crossed marine corridors (Silber et al., 2012). Second, estuarine linkages should be protected, and restored where needed. For example, biologists in South Africa restored the natural flow regime of the St Lucia Estuary, Africa's largest estuarine lake, by removing dredge spoil in the estuary mouth (Nunes et al., 2018). Third, coastal habitat linkages—beaches and littoral shallows used by wildlife for dispersal, breeding, and feeding—need to be maintained. Studies from South Africa have highlighted how poor protection of connectivity pathways between coastal habitats can compromise these areas' high levels of species richness and endemism (von der Heyden, 2009; Harris et al., 2014).

11.3.4 Mimicking connectivity

In the absence of habitat linkages, wildlife managers may be able to mimic dispersal dynamics by sporadically translocating a few individuals between subpopulations. Managing populations in this way may be a good alternative in cases where areas earmarked for translocations are too small to sustain a single viable population. Such is the case in South Africa, where conservation biologists occasionally move threatened predators within a small and fragmented protected areas network, where none of the areas are large enough to host a viable population on their own (see Box 8.3). Managing isolated and small populations so intensively nearly always requires sound underlying principles and extensive quantitative analyses (Chapter 9) for guidance.

11.3.5 Management considerations in connectivity conservation

While intuitively appealing, there are a few potential drawbacks to connectivity that conservation planners should consider when planning to establish new habitat linkages (reviewed in Haddad et al., 2014). Prominently, connecting historically isolated populations may lead to outbreeding depression, for example when populations with different local adaptations are connected. Habitat linkages may also act as bottlenecks that expose dispersing animals to greater risks of predation and enable pests and diseases to spread easier. Care must be taken to ensure that wildlife do indeed perceive the landscape "connected"; a habitat linkage that may look good to the human eye may in fact be perceived as inhospitable habitat to wildlife (Newmark, 2008). A recent study from the Americas has shown that the habitat quality of a single stepping stone habitat can determine whether a migration is successful or not (Gómez et al., 2017).

Although the benefits for connecting landscapes for conservation generally outweigh the drawbacks (Haddad et al., 2014), it is important to carefully plan to avoid those drawbacks. Genetic studies can be useful in both determining connectivity among populations (von der Heyden, 2009; Godley et al., 2010) and help researchers detecting potential deleterious factors, such as outbreeding depression (Figure 11.11, see also Frankham et al., 2011; Ralls et al., 2018). Modelling approaches that combine a target species' movement limitations with radio tracking technologies (e.g. Godley et al., 2010) or remotely sensed environmental variables (e.g. Wegmann et al., 2014) could help to estimate whether a landscape is indeed connected. Much effort has also been invested in finding the optimal width of habitat corridors. For example, one study in lowland forests suggested that corridors that are 30–40 m wide might be adequate for migration of most species while corridors that are 200 m wide will be adequate for all species (Laurance and Laurance, 1999). This is useful guidance, but ecosystems vary, as do target species (Wilson et al., 2010; Pryke and Samways, 2012) and, thus, some corridors may need to be even wider.

Although the benefits for reconnecting fragmented landscapes generally outweigh the drawbacks, it is important to carefully plan to avoid those drawbacks.

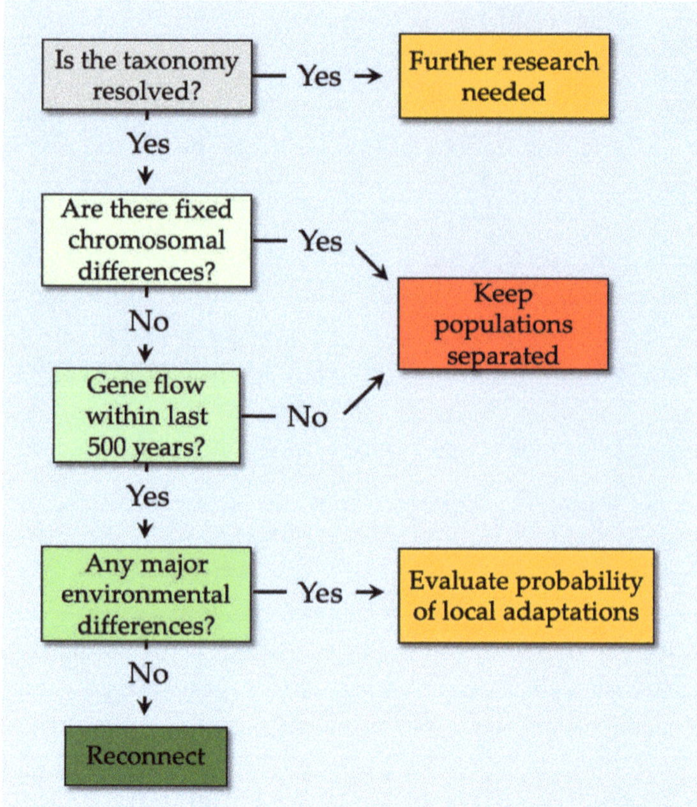

Figure 11.11 An example of a decision tree to avoid outbreeding depression, which can be used guide decisions for reconnecting fragmented landscapes. After Frankham et al., 2011, CC BY 4.0.

11.4 Managing Species Sensitive to Climate Change

Earth's temperature is well on its way to exceed the 2°C increase cap set by global authorities in 2016 (***Paris Agreement***, Section 12.2.1). Many species that need to adapt to these changes are unable to do so, either because of their limited dispersal capabilities or because of human-induced habitat fragmentation (Section 6.3.5). Others that can disperse may risk decoupling of important symbiotic relationships, as the species involved may not disperse at the same speed, or the same distance (Section 6.3.2). While slowing habitat loss could slow the overall impacts of climate change (Section 10.4), preventing the extinction of many climate-sensitive species will require a range of pro-active conservation management strategies that allow species to adapt at their own pace as and when needed.

One of the most important strategies for protecting climate-sensitive species is to identify and protect their likely future habitats. This task of predicting where suitable habitats may be found in future is generally accomplished by identifying and projecting a species' climatic niche (or bioclimatic envelope) using species distribution models (SDM, Pearson and Dawson, 2003). Section 11.1.1 described how SDM use location data overlaid onto environmental variables to estimate a species' environmental niche, and how this information can then be used to predict where else a species may occur in a landscape. A similar strategy is followed when predicting a species' future climate-adapted range. Here, location data are overlaid onto present-day climate variables (e.g. average temperature and rainfall) to define the species' climatic niche; these niche limits are then projected onto the landscape of interest using future climate scenarios (Section 6.2). Much effort has also been made in recent years to incorporate aspects, such as physiology (Kearney and Porter, 2009) and biological interactions (e.g. Araújo and Luoto, 2007), in predicting future ranges.

> Preventing the extinction of climate-sensitive species will require a range of pro-active conservation strategies that allow those species to adapt at their own pace as and when needed.

Once future ranges have been identified, the next task is to recognize and protect/restore critical dispersal pathways (Section 11.3). While a general strategy of increasing ecosystem-wide connectivity will certainty also benefit climate-sensitive species, conservationists could specifically target climate adaption, by maintaining and restoring **climate corridors**—dispersal pathways between the current and future ranges (Mawdsley et al., 2009). Several efforts (e.g. Williams et al., 2005; Phillips et al., 2008; Ayebare et al., 2013) are currently underway to establish and protect species-specific and community-specific climate corridors, as predicted using advanced distribution modelling techniques. These and other studies have shown that likely climate corridors often include north–south river valleys, ridges, and coastlines to facilitate poleward distribution shifts, while habitat linkages that cross gradients of elevation, rainfall, and soil types will help climate adaptation across more complex landscapes.

Species with dispersal limitations and specialised interactions may not always benefit from increased connectivity. Instead, those species may rely on **climate refuges**—areas that are resilient to climate change and thus able to continue to support climate-sensitive communities in future. Africa offers two good examples that illustrate how climate refuges can be identified. The first study, on South African birds, identified climate refuges as areas where temperatures seldom rise above the threshold known to negatively impact a specific species' fitness (Cunningham et al., 2013). The second study, on northern Mozambique's coral reefs (McClanahan and Muthiga, 2017), identified two kinds of climate refuges: (a) areas where temperatures never reached a point where it would kill the corals, and (b) areas situated in deeper and cooler water but with the full spectrum of light, which allowed corals to thrive while avoiding heat stress. Both these studies highlight why protecting and restoring complex natural ecosystems (see also Betts et al. 2018) is so important for climate change mitigation.

> Climate-sensitive species that are dispersal-limited may not benefit from increased connectivity. Instead, they will rely on climate refuges—areas that are resilient to climate change.

Assisted colonisation is an alternative conservation strategy to save species with dispersal limitations and specialised interactions. Also called assisted migration, assisted colonisation involves the pro-active translocation of climate-sensitive species from their present ranges to their future ranges. Sometimes, even species able to self-disperse may require assisted colonisation. For example, African penguins (*Spheniscus demersus*, EN) are currently undergoing population declines because of climate change-induced shifts in fish populations on which they depend for food (Sherley et al., 2017). To re-establish this important biological interaction, conservationists are currently using assisted colonisation to establish two new penguin colonies further east from existing colonies (Birdlife South Africa, 2019), in an area where fish populations have remained healthy (Figure 11.12).

Figure 11.12 BirdLife South Africa, in partnership with CapeNature, are introducing rehabilitated African penguins to two new sites several hundred kilometres east of existing colonies. The hope is that the translocated penguins will establishing colonies that is buffered from the negative effects of climate change and fluctuating fish populations. Here members of the public are witnessing the first releases. Photograph by Michael Bridgeford, CC BY 4.0.

As with any translocation project, introducing climate-sensitive species to new areas carries significant risks, including decoupling them from critical limiting resources and symbiotic relationships. It is thus imperative to start small, by translocating only a few well-monitored individuals. If monitoring shows that the initial releases were successful, one can then plan for further releases over time. Because this strategy is still new, it is also important to disseminate your experiences to the broader conservation community, for example by presenting results at conferences or in scientific journals.

11.5 Ex Situ Conservation Strategies

The best strategy for protecting biodiversity over the long term is to protect existing wild populations in their natural ecosystems. This strategy, known as on-site, or **in situ conservation**, not only protect entire ecological communities—including thousands of species and their interactions—but also natural processes and ecosystem services. However, if the last populations of a threatened species are too small to remain viable, if they continue to decline despite conservation efforts, or if their threats do not subside, then in situ conservation may prove ineffective. In such cases, sometimes the only option left to prevent an imminent extinction is to capture those last remaining individuals and transfer them to a facility where they can be cared for under artificial, human-controlled conditions. This strategy is known as off-site, or **ex situ conservation**, and may involve individuals that were collected in the wild, orphaned, confiscated, or displaced and have nowhere else to go. Thanks to ex situ efforts, several African plants and animals that are extinct in the wild continue to survive in zoos, aquaria, and botanical gardens. Examples include four to seven species of ancient cycads (*Encephalartos* spp.) from Southern Africa, and the pygmy Rwandan water lily (*Nymphaea thermarum*, EW), which is the world's smallest water lily (IUCN, 2019).

> For species facing imminent extinction, sometimes the only option left may be to capture the remaining individuals and transfer them to captivity.

Ex situ and in situ conservation are complementary strategies (Figure 11.13; see also Conde et al., 2011). For example, many ex situ conservation programmes aim to raise enough healthy individuals to support translocation projects when appropriate habitats are available. Ex situ conservation efforts were instrumental in preventing the extinction of the live-bearing Kihansi spray toad (*Nectophrynoides asperginis*, EW). Populations of this Tanzanian endemic declined precipitously following the establishment of a hydroelectric dam, which caused the toad's waterfall spray-zone habitat to dry up. The species was subsequently declared *Extinct in the Wild* in 2009. Tanzanian conservationists, however, demonstrated good foresight by inviting zoos from the USA to collect adults for a captive breeding effort even before the dam was built. This effort is now yielding positive results: after a decade of captive breeding, the erection of an artificial sprinkler system for habitat restoration, and experimental releases (Vandvik et al., 2014), nearly 10,000 toads were released to their former range in May 2018 (Anon, 2018).

Safeguarding a well-represented sample of the world's biodiversity play only a small role in ex situ conservation efforts. Maintaining self-sustaining wildlife populations under human care not only reduce the need to collect individuals for research from the wild; it also allows researchers to study aspects such as physiology, genetics, and demographics of threatened species (Conde et al., 2019) using methods that might not be possible without animals in captivity. These studies can then provide knowledge and experience to help protect both ex situ and in situ populations. For example, the establishment of the Demographic Species Knowledge Index (Conde et al., 2019), summarise demographic data obtained from ex situ conservation facilities, play a crucial role in filling gaps in datasets for population viability analyses (Section 9.2) Ex situ facilities also play a critical role in captive breeding, head-starting, public outreach, education, and fundraising for in situ conservation. Many ex situ facilities have also become directly involved—and sometimes even taking leading roles—in

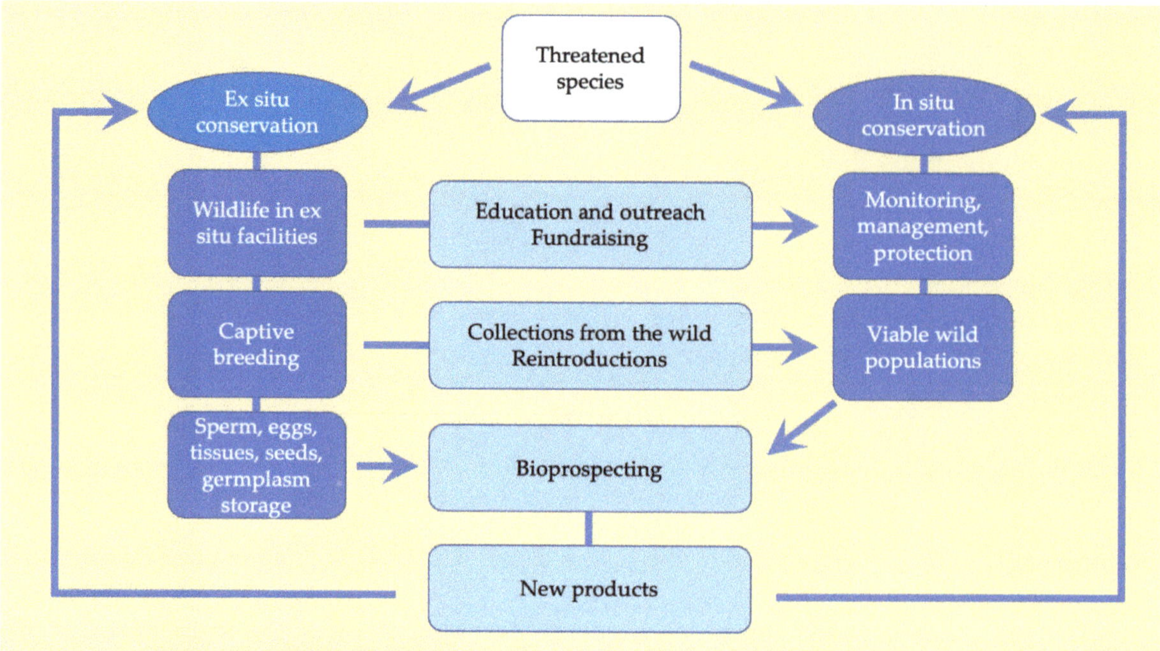

Figure 11.13 There are several ways in which in situ (on site) and ex situ (off-site) conservation can complement each other. No species conforms exactly to this idealised model, but nearly all species present some of these elements. After Maxted, 2001, CC BY 4.0.

field conservation efforts (Wilson et al., 2019). Lastly, many ex situ facilities directly connect conservation to social and economic progress through off-site education, employment, and implementation of a range of different community development activities (Ferrie et al., 2013).

Recent efforts to increase knowledge transfer among ex situ facilities has greatly enhanced their contribution to overall conservation efforts. Facilitated by organisations such as the IUCN's Conservation Planning Specialist Group (CPSG), ex situ facilities now regularly share information on best practices for care and handling of species in human care, including aspects such as nutritional requirements, optimal housing conditions, and veterinary techniques to anaesthetize, immobilise, and reduce stress for animals when they are being moved or during medical treatments (see http://www.cpsg.org). Much of this information is stored in a central database called the Zoological Information Management System (ZIMS). Maintained by Species360, ZIMS keeps track of animal **husbandry**, medical, and breeding information on over 6.8 million animals belonging to more than 21,000 species for over 1,000 member institutions in 90 countries. Ex situ facilities that maintain these records and comply with operations standards in animal welfare, conservation, education, and research can also apply to become an accredited institution with the Pan-African Association for Zoos and Aquaria (PAAZA), or its parent organisation, the World Association of Zoos and Aquariums (WAZA). As of mid-2019, four Sub-Saharan African ex situ facilities were accredited by WAZA, and 19 by PAAZA.

11.5.1 Types of ex situ facilities

Many types of facilities help to preserve ex situ populations. Here we describe some of the most common, including zoos and aquaria for animals, and botanical gardens and seed banks for plants.

Zoos around the world currently contribute to the conservation of nearly 7,000 species of terrestrial vertebrates (mammals, birds, reptiles, and amphibians) by caring for more than 500,000 individual animals. They do not do this alone; they often work with government agencies, universities, and a variety of other organisations who use zoo animals for research, education, and other conservation activities. While zoos traditionally focussed on displaying charismatic animals that draw visitors, many zoos are now also investing in the conservation of small threatened vertebrates, as well as invertebrates, such as butterflies, beetles, dragonflies, spiders, and molluscs (many of which are also cheaper to maintain). South Africa's National Zoological Gardens, which houses more than 9,000 individual animals belonging to 705 species, is Africa's largest zoo by variety of captive species and individuals. The zoo also hosts a variety of daily school programmes meant to inspire kids to a career in conservation; these include holiday courses, a zoo club, and guided tours at night.

> Ex situ conservation facilities compliment field conservation efforts through captive breeding, public outreach, education, knowledge generation, and fundraising.

Aquaria are the aquatic version of zoos, specialised in caring, displaying, and conserving marine and freshwater biodiversity, such as fishes, corals, molluscs, and crustaceans (Figure 11.14). One such institution is South Africa's uShaka Marine World, the world's fifth largest aquarium and home to more than 390 marine species—most from the Western Indian Ocean—held in 11 million litres of seawater. Most organisms currently in aquaria have been obtained from the wild, but conservationists are constantly refining techniques to breed more species in captivity to limit wild collecting. Recent and dramatic increases in aquaculture, which currently accounts for roughly a third of fish and shellfish production globally, have made ex situ conservation of aquatic species even more important. The hope is that these ex situ populations will help maintain genetic stocks and act as insurances against disease outbreaks introduced by domestic fish, molluscs, and crustaceans.

Botanical gardens (and **arboretums,** which specialise on trees and other woody plants) are dedicated to the collection, cultivation and educational curation of living plant species. Botanical gardens across the world house more than 6 million living plants, representing over 80,000 species—approximately 25% of the world's vascular flora (Wyse Jackson, 2001). The world's oldest and largest botanical garden—the Royal Botanic Gardens in London, UK—maintains over 28,000 plant taxa, nearly 10% of plant taxa in the world. In Sub-Saharan Africa, there are at least 153 botanical gardens in 33 countries, which range from small community-organised centres to world-famous conservation hubs, such as South Africa's Kirstenbosch Botanical

Figure 11.14 (Top) Many aquaria host tours and children's programmes, some involving opportunities to touch the organisms, for additional enrichment to visitors. Photograph by Karen Schermbrucker, courtesy of Two Oceans Aquarium, CC BY 4.0. (Bottom) Aquaria also provide opportunities to observe species such as this black musselcracker (*Cymatoceps nasutus*, VU) at South Africa's Two Oceans Aquarium, which many people would not have experienced otherwise. Photograph by Geoff Spiby, courtesy of Two Oceans Aquarium, CC BY 4.0.

Garden. Like zoos and aquaria, botanical gardens play a critical role in conservation efforts through public outreach and education. For example, Ghana's Aburi Botanical Garden established a model medicinal plant garden where the public can gain first-hand knowledge on how to combine conservation, cultivation, and sustainable use of medicinal plants (Gillett et al. 2002).

A few botanical gardens and research institutes have developed collections of seeds, known as **seed banks**, which take advantage of the fact that seeds of most plants can survive for long periods when stored in cold, dry conditions. The seeds deposited in seed banks may be obtained from the wild, or from cultivated specimens.

When gathering material from the wild, botanists generally target populations from across a species' geographical and habitat ranges so their collections can capture as much of each species' genetic diversity as possible. In this way, seed banks play a crucial role not only in conservation of plant species richness, but also genetic diversity. Seed banks may even be the only means some plant species are protected. Because many seeds of each species are usually collected, seed banks also provide a convenient opportunity for translocation projects. That is because safeguarded seed collections can be used to propagate not just large numbers of seedlings but, in some cases, custom-developed genetic mixtures to maximise local adaptations. The world's largest and most diverse seed bank is the Millennium Seed Bank, UK. At the end of 2018, the Millennium Seed Bank catalogued over 2.25 billion seeds from over 39,000 species; its billionth seed, from an African bamboo, was deposited in April 2007. In addition to safeguarding a portion of plant diversity, the Millennium Seed Bank has also benefitted countries, such as Botswana, Burkina Faso, and Mali through the redistribution of banked seeds to aid ecological restoration efforts.

Seed banks contribute to conservation of genetic diversity of plants by collecting material across target species' geographical and habitat ranges.

11.5.2 Challenges facing ex situ facilities

While the contribution of ex situ conservation facilities to overall biodiversity conservation strategies is significant (Conde et al., 2011), there are some drawbacks that need to be considered. For example, due to the limited number of individuals that can be maintained under human care, especially for larger animals, there is an increased risk that captive populations may suffer from threats facing small populations, such as inbreeding depression and demographic stochasticity (Section 8.7). There is also a concern that ex situ conservation can contribute to hybridisation concerns, for example if different cryptic species are accidentally managed as a single species. To avoid these threats, many ex-situ facilities manage their captive populations jointly as a single interbreeding metapopulation. They do this through studbooks which track the origin, pedigree, and demographic history of each individual in participating facilities. By maintaining and referring to these studbooks, ex situ conservation facilities can make informed decisions regarding transfer and breeding recommendations. The establishment of a European studbook for African dwarf crocodiles (*Osteolaemus* spp.) even addressed concerns about potential hybridisation between cryptic species (Schmidt et al., 2015).

Ex-situ facilities often manage captive populations as a single metapopulation using studbooks to track the origin and demographic history of breeding individuals.

Funding also remains an obstacle, given that ex situ facilities typically require large, long-term, funding commitments, in comparison to many in situ conservation activities. One consequence of funding limitations is that ex situ facilities mostly focus on showy or charismatic species that attract visitors, so small and less charismatic species are not always afforded equal protection (Brooks et al., 2009). Many ex situ facilities are also more inclined to house non-threatened species that are easier and less costly to care for, rather than threatened species with specialised needs (Table 11.1). For example, despite the fear of looming mass amphibian extinctions due to a disease caused by the chytrid fungus (*Batrachochytrium dendrobatidis*) (Alroy, 2015), 75% of ex situ amphibian collections consist of non-threatened species, with only 6.2% of all threatened amphibians afforded ex situ protection (Dawson et al., 2016). Neglecting threatened species in ex situ conservation efforts also creates a feedback loop, by maintaining a limited understanding on how to care for the species most in need.

Table 11.1 Number and percentages of terrestrial vertebrate species from Sub-Saharan Africa currently maintained in the world's ex situ facilities. Values in parenthesis represent percentage of all species[a], threatened species[b], and CITES-listed species[c] for each taxon class, respectively.

	Mammals	Birds	Reptiles	Amphibians	Total
Worldwide[a]	659 (55%)	1,470 (65%)	197 (27%)	44 (5%)	2,370 (47%)
Africa	110 (9%)	234 (10%)	34 (4%)	6 (1%)	384 (8%)
Asia	136 (11%)	327 (14%)	22 (3%)	2 (0%)	487 (10%)
Oceania	37 (3%)	61 (3%)	6 (1%)	1 (0%)	105 (2%)
Europe	191 (12%)	465 (20%)	73 (10%)	19 (2%)	748 (15%)
North America	145 (12%)	311 (14%)	53 (7%)	14 (2%)	523 (10%)
South America	40 (3%)	72 (3%)	197 (27%)	44 (5%)	353 (7%)
Threatened species[b]	45 (23%)	42 (20%)	22 (21%)	8 (4%)	117 (16%)
Extinct in the Wild	1 (100%)	0 (0%)	0 (0%)	1 (100%)	2 (100%)
Critically Endangered	7 (26%)	4 (19%)	5 (25%)	3 (5%)	19 (15%)
Endangered	13 (16%)	12 (15%)	2 (5%)	3 (3%)	30 (10%)
Vulnerable	24 (27%)	26 (23%)	15 (33%)	1 (2%)	66 (23%)
CITES-listed species[c]	95 (50%)	121 (62%)	45 (25%)	1 (6%)	262 (45%)
Appendix I species	30 (58%)	4 (44%)	8 (80%)	1 (6%)	43 (49%)
Appendix II species	58 (44%)	112 (62%)	37 (22%)	0 (0%)	207 (43%)
Appendix III species	7 (100%)	5 (100%)	0 (0%)	0 (0%)	12 (100%)

Source: https://zims.species360.org, current as of April-2019. Compiled by Johanna Staerk (Species360).

Fortunately, ex situ facilities have responded to these concerns by developing several innovative mechanisms that enables them to contribute more to the conservation of threatened species. For example, ex situ facilities all agree that attracting more visitors attracts more funding. To attract more visitors, zoos and aquaria are increasingly keeping animals in enclosures that are representative of their natural environments; this keeps the animals heathier and providing more opportunities to exhibit natural behaviours which, in turn, leave visitors more satisfied. Some zoos and aquaria have also established special displays where visitors can feed, touch, or otherwise interact with animals. Many ex situ facilities have also started inviting local artists to display sculptures and other artwork, which adds to the experience for visitors and attracting people that might not otherwise have visited. A rather unusual—but very successful—attempt to increase foot traffic comes from the USA, where the California Academy of Sciences hosts dance parties with laser shows, food, and drinks every Thursday night (http://www.calacademy.org/nightlife), which visitors can enjoy while visiting the Academy's aquarium and other conservation exhibits.

While the contribution of ex situ facilities to species conservation is significant, many rare species are ill-suited for ex situ efforts. Some species simply do not adapt or reproduce in captivity, while others that do relatively well in captivity experience behavioural and physiological changes or acquire diseases (Brossy et al., 1999) that prevent releases in the wild. Even so, the conservation biologists working at ex situ facilities constantly try to find ways to overcome these challenges. For example, staff at ex situ facilities sometimes use assisted reproductive techniques such as **artificial incubation** of bird and reptile eggs, or **artificial insemination** (Box 11.4) to overcome reproductive challenges (e.g. if individuals cannot mate because they are in different locations). Others use cryopreservation and **genome resource banks** for the long-term storage of embryos, eggs, sperm, or purified DNA, at least until those tissues can be used to increase a species' genetic diversity, or perhaps even to resurrect an extinct species (see de-extinction, Section 8.8). However, many ex situ conservation techniques are difficult and expensive to implement. When possible, it is almost always preferable to preserve species in situ where they can be self-sustaining, free from inbreeding, and an interactive participant of their community and ecosystem.

11.6 Thoughts on Neglected Taxa

Most of today's species-centric conservation initiatives are biased towards species that are showy, charismatic, or economically important. Consequently, conservation efforts for the vast majority of taxa are neglected, particularly in Africa where conservation funding is often more limited than elsewhere. One well-known example is known as **plant blindness**, the perception that animals take precedence above plants in conservation efforts. This isn't just a case of hurt feelings among botanists: there are likely significantly more plant than animal species that should be considered as threatened (see Table 2.1); however, thorough threat assessments are hampered because, as a group, plants receive significantly less funding compared to animals

Box 11.4 Saving the Northern White Rhinoceros with Assisted Reproduction Technologies

Morné de la Rey

Embryo Plus,
Brits, South Africa.
http://www.embryoplus.com

A few centuries ago, Earth's wildernesses enabled animals to roam and breed relatively freely. Today, sprawling cities, agriculture, and fences not only restrict animals' ability to forage, but also limit reproduction between differing gene pools. These stresses create smaller and more isolated populations which are being edged toward extinction.

There are several landscape-scale conservation initiatives to counteract these imbalances. But some species and populations are so rare that they depend on intensive management to remain viable. Assisted reproductive techniques (ART) provide promise for helping such species. Over the past 30 years, ARTs have greatly enhanced how the livestock industry preserve, improve, and proliferate genetic stock. Now, efforts are also underway to use ARTs to ensure the preservation of biodiversity.

The many types of ARTs

ARTs include a wide array of medical procedures to address infertility, and to make reproduction possible between individuals unable to do so naturally (e.g. animals in different protected areas). In this way, biologists can ensure genetic exchange while eliminating the risks inherent in translocation such as spread of diseases, adaptation to new environments, and disruption of group dynamics.

ARTs have various levels from relatively simple to very complex. The most basic technique is artificial insemination. A major advantage of this technique is that it can multiply male genetic contributions by inseminating more females than would be possible in nature. Much progress has also been made in improving viability of cryopreserved semen to overcome challenges with timing of female reproductive cycles and other logistical constraints.

As for multiplying female genetic contributions, methods involve embryo transfer and in vitro fertilisation (IVF). With multiple ovulation embryo transfer (MOET) egg fertilisation occurs naturally; with IVF, it occurs in a laboratory incubator. In both cases, an embryo is eventually transferred to a surrogate mother which will carry it until birth. Scientists are currently working on improving viability of stored germplasm, so that embryos can be cryopreserved until a suitable surrogate mother is ready.

The third technique is nuclear transfer, also known as cloning. This very delicate procedure involves replacing the haploid DNA of an unfertilised egg with diploid DNA of another; cells are then cultured, after which the embryo is transferred to a surrogate mother.

Using ARTs to save the northern white rhino

Once widespread across Central Africa, poaching has pushed the northern white rhinoceros (*Ceratotherium simum cottoni*, EW) to the brink of extinction. Today, only two females remain, both in a semi-captive setting at Ol Pejeta Conservancy, Kenya. Incapable of natural reproduction, this species is committed to extinction without drastic intervention.

A cutting-edge initiative is currently underway to use ARTs to save this iconic species. While the project's exact trajectory is still being developed, likely steps include optimising procedures for harvesting, maturing, and fertilising eggs, followed by embryo transfer into surrogate southern white rhinoceros. Some preliminary successes have also been achieved to generate stem cells from skin biopsies (Ben-Nun et al., 2011), which could be used in cloning. Genetic material (tissue samples and semen) of several northern white rhinos has been cryopreserved at various places around the world. However, there is a limited amount of sperm available (there are no males left), and so artificial insemination and IVF with northern white rhinoceros depends on embryo transfer successes. Many partners have been assembled to pool resources and ideas in support of this initiative, including Ol Pejeta Conservancy, Embryo Plus, Fauna & Flora International (FFI), Back to Africa, Dvur Karlove Zoo, Leibniz-IZW, Avantea, San Diego Zoo, and Kenya Wildlife Service.

Refining ARTs on other species

Before ARTs are implemented on the near-extinct rhinoceros, it is advisable to optimise procedures on another species. A logical choice would be the closely related southern white rhinoceros (*C. simum simum*, NT). However, the southern subspecies is also threatened, so we should look for more common mammals first.

Veterinarians at Embryo Plus routinely perform ARTs on domestic cattle, so efforts are currently focussed on building from this experience to work with wild bovines (Figure 11.D). For example, Embryo Plus recently produced the world's first African buffalo (*Syncerus caffer*, NT) through IVF; the healthy calf named Pumelelo (meaning success in isiZulu) was born in June 2016. Embryo Plus has also produced several western Zambian sables (*Hippotragus niger kirkii*) from southern sable (*H. niger niger*) surrogates using embryo transfer. Plans are also underway to investigate the viability of using eland (*Tautragus oryx*, LC) and domestic horses (*Equus ferus*) as surrogate mothers for mountain bongo (*T. eurycerus isaaci*, CR) and Grevy's zebra (*E. grevyi*, EN), respectively.

Figure 11.D (Top) The world's first African buffalo calf conceived by in vitro fertilisation. (Bottom) The world's first western Zambian sable born from a southern sable surrogate mother. Both species also breed successfully on their own, but scientists are refining their techniques on more common species before attempting them on highly threatened species. Photographs by Morné de la Rey/ Embryo Plus,

From dream to reality

The long-term objective of the northern white rhinoceros project is to establish a viable breeding herd which can be reintroduced into secure habitats. But much work remains for this dream to become reality. While there was one successful attempt in producing a healthy bongo calf by transferring an embryo to an eland mother (Woolf, 1986), inter-species embryo transfer remains challenging. Due to a rhinoceros' size, we also need to ensure ART procedures can be performed safely without placing undue stress on the patient. Lastly, because each species' embryos have different requirements in the laboratory, extensive research is necessary before ARTs can be attempted on a new species.

Although ARTs in wildlife management is still in its infancy, we are confident that early breakthroughs hold promise for the survival of the northern white rhinoceros, as well as other threatened species that may one day benefit from these procedures.

(Negron-Ortiz, 2014). One explanation for this disparity is that plants are often seen as the backdrop of the environment rather than the critical foundation (as primary producers) of every food web on Earth. While showy plant species may indeed have highly visible roles in maintaining the environment and regional economies, neglected species may play an equally—sometimes even more—important role in maintaining ecosystems and ecosystem services (Schleuning et al., 2016).

Fortunately, the number of professional and amateur societies interested in protecting neglected taxa, such as reptiles, amphibians, invertebrates, fungi, and plants are rising. Some groups of experts are also organised into Specialist Groups (https://www.iucn.org/ssc-groups) by the IUCN. These societies and expert groups highlight the plight of neglected taxa and are willing to provide in-house expertise on best practices for protecting those species.

11.7 Summary

1. A species may be threatened by a combination of many factors, all of which must be addressed in a comprehensive conservation plan that considers its natural history.
2. New populations of threatened species can be established in the wild using either captive-raised or wild-caught individuals. Animals used in translocation projects sometimes require special care and behavioural training before release as well as care and monitoring after release.
3. Maintaining and facilitating movement dynamics is very important for protecting wildlife in their natural ecosystems. To do this, connectivity must be preserved by ensuring that habitat linkages such as wildlife corridors and stepping stone habitats that are intact, functional, and free from human-made obstacles.
4. Preventing biodiversity losses under climate change requires ecosystem preservation, maintaining and restoring climate corridors and refugia, and assisted colonisation for species unable to adapt their ranges quick enough.
5. Some species that are in danger of going extinct in the wild can be maintained in zoos, aquaria, botanical gardens, and seed banks; this strategy is known as ex situ conservation. Ex situ conservation contributes to field conservation through research, skills development, public outreach, conservation education, fundraising, captive breeding, and head-starting.

11.8 Topics for Discussion

1. How do you judge whether a reintroduction project is successful? Develop simple and then increasingly detailed criteria to evaluate a project's success.
2. Tying concepts from different chapters together, what are the biggest challenges standing in the way of conserving Africa's migratory birds?
3. Use the advanced search functions on the IUCN Red List website (https://www.iucnredlist.org) to pick one species occurring in your country that is threatened by climate change. Referring to Chapter 6, how does climate change threaten this species? What strategies can be used to prevent this species' extinction?
4. What roles do ex situ facilities play in the conservation of threatened species in Africa? Discuss two or three different roles. Do you think there are certain aspects in conservation that they can make a larger contribution to than is currently the case?
5. Find two or three examples of wild or semi-wild populations of African species maintained on other continents? Are those species threatened in their natural distribution ranges? Does maintaining populations of African species on other continents represent a successful conservation strategy? Explain your answer.

11.9 Suggested Readings

Hoffmann, M., J.W. Duckworth, K. Holmes, et al. 2015. The difference conservation makes to extinction risk of the world's ungulates. *Conservation Biology* 29: 1303–13. https://doi.org/10.1111/cobi.12519 Conservation action does make a difference, as illustrated by ungulate conservation.

IUCN/SSC. 2013. *Guidelines for Reintroductions and Other Conservation Translocations* (Gland: IUCN/SSC). https://portals.iucn.org/library/efiles/documents/2013-009.pdf Guidelines for species reintroductions and translocations.

Mawdsley, J.R., R. O'Malley, and D.S. Ojima. 2009. A review of climate-change adaptation strategies for wildlife management and biodiversity conservation. *Conservation Biology* 23: 1080–89. https://doi.org/10.1111/j.1523-1739.2009.01264.x A review of methods to combat climate change.

Menges, E.S., S.A. Smith, and C.W. Weekley. 2016. Adaptive introductions: How multiple experiments and comparisons to wild populations provide insights into requirements for long-term introduction success of an endangered shrub. *Plant Diversity* 38: 238–46. https://doi.org/10.1016/j.pld.2016.09.004 Experimentation can reduce uncertainty in population translocation.

Miller, B., W. Conway, R.P. Reading, et al. 2004. Evaluating the conservation mission of zoos, aquariums, botanical gardens, and natural history museums. *Conservation Biology* 18: 86–93. https://doi.org/10.1111/j.1523-1739.2004.00181.x Eight tough questions directed at ex situ facilities.

Nogués-Bravo, D., D. Simberloff, C. Rahbek, et al. 2016. Rewilding is the new Pandora's box in conservation. *Current Biology* 26: R87–R91. https://doi.org/10.1016/j.cub.2015.12.044 Are there limits to wildlife translocation dreams?

Pérez, I., J.D. Anadón, M. Díaz, et al. 2012. What is wrong with current translocations? A review and a decision-making proposal. *Frontiers in Ecology and the Environment* 10: 494–501. https://doi.org/10.1890/110175 Important considerations for translocations.

Sherley, R.B., K. Ludynia, B.M. Dyer, et al. 2017. Metapopulation tracking juvenile penguins reveals an ecosystem-wide ecological trap. *Current Biology* 27: 563–68. https://doi.org/10.1016/j.cub.2016.12.054 Detecting and dealing with ecological traps requires holistic thinking.

Bibliography

Ali, A.H., A.T. Ford, J.S. Evans, et al. 2017. Resource selection and landscape change reveal mechanisms suppressing population recovery for the world's most endangered antelope. *Journal of Applied Ecology* 54: 1720–29. https://doi.org/10.1111/1365-2664.12856

Alroy, J. 2015. Current extinction rates of reptiles and amphibians. *Proceedings of the National Academy of Sciences* 112: 13003–08. https://doi.org/10.1073/pnas.1508681112

Andersson, J., M. de Garine-Wichatitsky, D. Cumming, et al. 2013. *Transfrontier Conservation Areas: People Living on the Edge* (New York: Routledge).

Anon. 2018. 9,873 Kihansi spray toads return to their environment. *IPP Media*. https://go.shr.lc/2B6N7wB

Araújo, M.B., and M. Luoto. 2007. The importance of biotic interactions for modelling species distributions under climate change. *Global Ecology and Biogeography* 16: 743–53. https://doi.org/10.1111/j.1466-8238.2007.00359.x

Armstrong, D.P., and P.J. Seddon. 2007. Directions in reintroduction biology. *Trends in Ecology and Evolution* 23: 20–25. https://doi.org/10.1016/j.tree.2007.10.003

Ayebare, S., R. Ponce-Reyes, S.B. Segan, et al. 2013. *Identifying climate resilient corridors for conservation in the Albertine Rift* (Chicago: MacArthur Foundation).

Bartlam-Brooks, H.L.A., M.C. Bonyongo, and S. Harris. 2011. Will reconnecting ecosystems allow long-distance mammal migrations to resume? A case study of a zebra *Equus burchelli* migration in Botswana. *Oryx* 45: 210–16. https://doi.org/10.1017/S0030605310000414

Ben-Nun, I.F., S.C. Montague, M.L. Houck, et al. 2011. Induced pluripotent stem cells from highly endangered species. *Nature Methods* 8: 829–31. https://doi.org/10.1038/nmeth.1706.

Bennett, G. 2004. *Linkages in practice: A review of their conservation value* (Gland: IUCN). https://portals.iucn.org/library/node/8412

Bentrup, G., M. Dosskey, G. Wells, et al. 2012. Connecting landscape fragments through riparian zones. In: *Forest Landscape Restoration*, ed. by J. Stanturf, et al. (Dortrecht: Springer). https://doi.org/10.1007/978-94-007-5326-6

Betts, M.G., B. Phalan, S.J.K. Frey, et al. 2018. Old-growth forests buffer climate-sensitive bird populations from warming. *Diversity and Distributions* 24: 439–47. https://doi.org/10.1111/ddi.12688

BirdLife South Africa. 2019. *Creating new penguin colonies*. https://www.birdlife.org.za/what-we-do/seabird-conservation/what-we-do/coastal-seabird-conservation/creating-penguin-colonies/

Bonnell, T.R., R. Reyna-Hurtado, and C.A. Chapman. 2011. Post-logging recovery time is longer than expected in an East African tropical forest. *Forest Ecology and Management* 261: 855–64. https://doi.org/10.1016/j.foreco.2010.12.016

Boundja, R.P., and J.J. Midgley. 2010. Patterns of elephant impact on woody plants in the Hluhluwe-iMfolozi Park, Kwazulu-Natal, South Africa. *African Journal of Ecology* 48: 206–14. https://doi.org/10.1111/j.1365-2028.2009.01104.x

Brook, R.K., and S.M. McLachlan. 2008. Trends and prospects for local knowledge in ecological and conservation research and monitoring. *Biodiversity and Conservation* 17: 3501–12. https://doi.org/10.1007/s10531-008-9445-x

Brooks, T.M., S.J. Wright, and D. Sheil. 2009. Evaluating the success of conservation actions in safeguarding tropical forest biodiversity. *Conservation Biology* 23: 1448–57. https://doi.org/10.1111/j.1523-1739.2009.01334.x

Brossy, J.J., A.L. Plös, J.M. Blackbeard, et al. 1999. Diseases acquired by captive penguins: What happens when they are released into the wild? *Marine Ornithology* 27: 185–86

Brown, M., M. Perrin, and B. Hoffman. 2007. Reintroduction of captive-bred African Grass-Owls *Tyto capensis* into natural habitat. *Ostrich* 78: 75–79. https://doi.org/10.2989/OSTRICH.2007.78.1.11.55

Caro, T.J. Eadie, and A. Sih. 2005. Use of substitute species in conservation biology. *Conservation Biology* 19: 1821–26. https://doi.org/10.1111/j.1523-1739.2005.00251.x

Childress, B., D. Harper, B. Hughes, et al. 2004. Satellite tracking lesser flamingo movements in the Rift Valley, East Africa: Pilot study report. *Ostrich* 75: 57–65. https://doi.org/10.2989/00306520409485413

Conde, D.A., J. Staerk, F. Colchero, et al. 2019. Data gaps and opportunities for comparative and conservation biology. *Proceedings of the National Academy of Sciences* 116: 9658–64. https://doi.org/10.1073/pnas.1816367116

Conde, D.A., N. Flesness, F. Colchero, et al. 2011. An emerging role of zoos to conserve biodiversity. *Science* 331: 1390–91. https://doi.org/10.1126/science.1200674

Creel, S., M.S. Becker, S.M. Durant, et al. 2013. Conserving large populations of lions–the argument for fences has holes. *Ecology Letters* 16: 1413. https://doi.org/10.1111/ele.12145

Cumming, D.H.M. 1999. *Study on development of transboundary natural resources management areas in Southern Africa — environmental context: Natural resources, land use and conservation* (Washington: Biodiversity Support Programme).

Cunningham, S.J., A.C. Kruger, M.P. Nxumalo, et al. 2013. Identifying biologically meaningful hot-weather events using threshold temperatures that affect life-history. *PLoS ONE* 8: e82492. https://doi.org/10.1371/journal.pone.0082492

Cushman, S.A., N.B. Elliot, D.W. Macdonald, et al. 2016. A multi-scale assessment of population connectivity in African lions (*Panthera leo*). *Landscape Ecology* 31: 1337–53. https://doi.org/10.1007/s10980-015-0292-3

Dallimer, M., and N. Strange. 2015. Why socio-political borders and boundaries matter in conservation. *Trends in Ecology and Evolution* 30: 132–39. https://doi.org/10.1016/j.tree.2014.12.004

Dawson, J., F. Patel, R.A. Griffiths, et al. 2016. Assessing the global zoo response to the amphibian crisis through 20-year trends in captive collections. *Conservation Biology* 30: 82–91. https://doi.org/10.1111/cobi.12563

Deemer, B.R., J.A. Harrison, S. Li, et al. 2016. Greenhouse gas emissions from reservoir water surfaces: A new global synthesis. *BioScience* 66: 949–64. https://doi.org/10.1093/biosci/biw117

Dubach, J.M., M.B. Briggs, P.A. White, et al., 2013. Genetic perspectives on "Lion Conservation Units" in Eastern and Southern Africa. *Conservation Genetics* 14: 741–55. https://doi.org/10.1007/s10592-013-0453-3

Dupuis-Desormeaux, M., T.N. Kaaria, M. Mwololo, et al. 2018. A ghost fence-gap: Surprising wildlife usage of an obsolete fence crossing. *PeerJ* 6: e5950. https://dx.doi.org/10.7717%2Fpeerj.5950

Ferreira, S.M., and M. Hofmeyr. 2014. Managing charismatic carnivores in small areas: Large felids in South Africa. South African *Journal of Wildlife Research* 44: 32–42. https://doi.org/10.3957/056.044.0102

Ferrie, G.M., K.H. Farmer, C.W. Kuhar, et al. 2014. The social, economic, and environmental contributions of Pan African Sanctuary Alliance primate sanctuaries in Africa. *Biodiversity and Conservation* 23: 187–201. http://doi.org/10.1007/s10531-013-0592-3

Frankham, R., J.D. Ballou, M.D.B. Eldridge, et al. 2011. Predicting the probability of outbreeding depression. *Conservation Biology* 25: 465–75. https://doi.org/10.1111/j.1523-1739.2011.01662.x

Fritz, H., and P. Duncan. 1994. On the carrying capacity for large ungulates of African savanna ecosystems. *Proceedings of the Royal Society B* 256: 77–82. https://doi.org/10.1098/rspb.1994.0052

Giam, X., R.K. Hadiaty, H.H. Tan, et al. 2015. Mitigating the impact of oil-palm monoculture on freshwater fishes in Southeast Asia. *Conservation Biology* 29: 1357–67. https://doi.org/10.1111/cobi.12483

Gillett, H. 2002. *Conservation and sustainable use of medicinal plants in Ghana* (Cambridge: UNEP-WCMC). https://wedocs.unep.org/bitstream/handle/20.500.11822/7487/Conservation_sustainable_use_medicinal_plants_Ghana.pdf

Godefroid, S., C. Piazza, G. Rossi, et al. 2011. How successful are plant species reintroductions? *Biological Conservation* 144: 672–82. https://doi.org/10.1016/j.biocon.2010.10.003

Godley, B.J., C. Barbosa, M. Bruford, et al. 2010. Unravelling migratory connectivity in marine turtles using multiple methods. *Journal of Applied Ecology* 47: 769–78. https://doi.org/10.1111/j.1365-2664.2010.01817.x

Gómez, C., N.J. Bayly, D.R. Norris, et al. 2017. Fuel loads acquired at a stopover site influence the pace of intercontinental migration in a boreal songbird. *Scientific Reports* 7: 3405. https://doi.org/10.1038/s41598-017-03503-4

Grey-Ross, R., C.T. Downs, and K. Kirkman. 2009. Reintroduction failure of captive-bred oribi (*Ourebia ourebi*). *South African Journal of Wildlife Research* 39: 34–38. https://doi.org/10.3957/056.039.0104

Groom, R.J., K. Lannas, and C.R. Jackson. 2017. The impact of lions on the demography and ecology of endangered African wild dogs. *Animal Conservation* 20: 382–90. https://doi.org/10.1111/acv.12328

Gusset, M., A.H. Maddock, G.J. Gunther, et al. 2008a. Conflicting human interests over the re-introduction of endangered wild dogs in South Africa. *Biodiversity and Conservation* 17: 83–101. http://doi.org/10.1007/s10531-007-9232-0

Gusset, M., R. Slotow, and M.J. Somers. 2006. Divided we fail: The importance of social integration for the re-introduction of endangered African wild dogs (*Lycaon pictus*). *Journal of Zoology* 270: 502–11. https://doi.org/10.1111/j.1469-7998.2006.00168.x

Gusset, M., S.J. Ryan, M. Hofmeyr, et al. 2008b. Efforts going to the dogs? Evaluating attempts to re-introduce endangered wild dogs in South Africa. *Journal of Applied Ecology* 45: 100–08. https://doi.org/10.1111/j.1365-2664.2007.01357.x

Haddad, N.M., L.A. Brudwig, A.I. Damschen, et al. 2014. Potential negative ecological effects of corridors. *Conservation Biology* 28: 1178–87. https://doi.org/10.1111/cobi.12323

Harris, L., E.E. Campbell, R. Nel, et al. 2014. Rich diversity, strong endemism, but poor protection: Addressing the neglect of sandy beach ecosystems in coastal conservation planning. *Diversity and Distributions* 20: 1120–35. https://doi.org/10.1111/ddi.12226

Hayward, M.W., J. Adendorff, L. Moolman, et al. 2007b. The successful reintroduction of leopard *Panthera pardus* to the Addo Elephant National Park. *African Journal of Ecology* 45: 103–04. https://doi.org/10.1111/j.1365-2028.2006.00673.x

Hayward, M.W., J. O'Brien, and G.I.H. Kerley. 2007a. Carrying capacity of large African predators: Predictions and tests. *Biological Conservation* 139: 219–29. https://doi.org/10.1016/j.biocon.2007.06.018

Henry, E., E. Brammer-Robbins, E. Aschenough, et al. 2019. Do substitute species help or hinder endangered species management? *Biological Conservation* 232: 127–30. https://doi.org/10.1016/j.biocon.2019.01.031

Herkt, K.M.B., G. Barnikel, A.K. Skidmore, et al. 2016. A high-resolution model of bat diversity and endemism for continental Africa. *Ecological Modelling* 320: 9–28. http://doi.org/10.1016/j.ecolmodel.2015.09.009

Horton, A.J., E.D. Lazarus, T.C. Hales, et al. 2018. Can riparian forest buffers increase yields from palm oil plantations? *Earth's Future* 6: 1082–96. https://doi.org/10.1029/2018EF000874

Houser, A., M. Gusset, C.J. Gragg, et al. 2011. Pre-release hunting training and post-release monitoring are key components in the rehabilitation of orphaned large felids. *South African Journal of Wildlife Research* 41: 11–20. https://doi.org/10.3957/056.041.0111

Hudgens, B.R., W.F. Morris, N.M. Haddad, et al. 2012. How complex do models need to be to predict dispersal of threatened species through matrix habitats? *Ecological Applications* 22: 1701–10. https://doi.org/10.1890/11-1048.1

IUCN. 2019. *The IUCN Red List of Threatened Species.* http://www.iucnredlist.org

Jachowski, D.S., R. Slotow, and J.J. Millspaugh. 2013. Delayed physiological acclimatization by African elephants following reintroduction. *Animal Conservation* 16: 575–83. http://doi.org/10.1111/acv.12031

Jones, T., A.J. Bamford, D. Ferrol-Schulte, et al. 2012. Vanishing wildlife corridors and options for restoration: A case study from Tanzania. *Tropical Conservation Science* 5: 463–74. https://doi.org/10.1177/194008291200500405

Jordaan, M., A. Lubbe, C. Bragg, et al. 2017. *Labeo seeberi. The IUCN Red List of Threatened Species* 2017: e.T11071A100162293. http://doi.org/10.2305/IUCN.UK.2017-3.RLTS.T11071A100162293.en

Kays, R., M.C. Crofoot, W. Jetz, et al. 2015. Terrestrial animal tracking as an eye on life and planet. *Science* 348: aaa2478. http://doi.org/10.1126/science.aaa2478

Kearney, M., and W. Porter. 2009. Mechanistic niche modelling: Combining physiological and spatial data to predict species' ranges. *Ecology Letters* 12: 334–50. https://doi.org/10.1111/j.1461-0248.2008.01277.x

King, L.E., I. Douglas-Hamilton, and F. Vollrath. 2011. Beehive fences as effective deterrents for crop-raiding elephants: Field trials in northern Kenya. *African Journal of Ecology* 49: 431–39. https://doi.org/10.1111/j.1365-2028.2011.01275.x

King, T., C. Chamberlan, and A. Courage. 2014. Assessing reintroduction success in long-lived primates through population viability analysis: Western lowland gorillas *Gorilla gorilla gorilla* in Central Africa. *Oryx* 48: 294–303. https://doi.org/10.1017/S0030605312001391

Laurance, S.G., and W.F. Laurance. 1999. Tropical wildlife corridors: Use of linear rainforest remnants by arboreal mammals. *Biological Conservation* 91: 231–39. https://doi.org/10.1016/S0006-3207(99)00077-4

Leaver, J. 2014. *Options for eland: A multi-scale assessment of antipredatory responses of a vulnerable prey species to their major predator in the Eastern Cape, South Africa.* MSc thesis (Port Elizabeth: NMMU). http://hdl.handle.net/10948/6570

Lindsey, P., C.J. Tambling, R. Brummer, et al. 2011. Minimum prey and area requirements of the Vulnerable cheetah *Acinonyx jubatus*: implications for reintroduction and management of the species in South Africa. *Oryx* 45: 587–99. https://doi.org/10.1017/S003060531000150X

Lindsey, P.A., R. Alexander, J.T. du Toit, et al. 2005. The cost efficiency of wild dog conservation in South Africa. *Conservation Biology* 19: 1205–14. https://doi.org/10.1111/j.1523-1739.2005.00088.x

Linklater, W., and A.M. Shrader. 2017. Rhino challenges: Spatial and social ecology for habitat and population management. In: *Conserving Africa's Mega-diversity in the Anthropocene: The Hluhluwe-iMfolozi Park Story*, ed. by J.P.G.M. Cromsigt, et al. (Cambridge: Cambridge University Press). https://doi.org/10.1017/9781139382793

Mawdsley, J.R., R. O'Malley, and D.S. Ojima. 2009. A review of climate-change adaptation strategies for wildlife management and biodiversity conservation. *Conservation Biology* 23: 1080–89. https://doi.org/10.1111/j.1523-1739.2009.01264.x

Maxted, N. 2001. Ex situ, in situ conservation. In: *Encyclopedia of Biodiversity*, ed. by S.A. Levin (San Diego: Academic Press).

Maxwell, S.M., G.A. Breed, B.A. Nickel, et al. 2011. Using Satellite tracking to optimize protection of long-lived marine species: Olive ridley sea turtle conservation in Central Africa. *PLoS ONE* 6: e19905. https://doi.org/10.1371/journal.pone.0019905

McClanahan, T.R., and N.A. Muthiga. 2017. Environmental variability indicates a climate-adaptive center under threat in northern Mozambique coral reefs. *Ecosphere* 8: e01812. https://doi.org/10.1002/ecs2.1812

McLennan, M.R., and A.J. Plumptre. 2012. Protected apes, unprotected forest: Composition, structure and diversity of riverine forest fragments and their conservation value in Uganda. *Tropical Conservation Science* 5: 79–103. https://doi.org/10.1177/194008291200500108

McNutt, J.W, M.N. Parker, M.J. Swarner, et al. 2008. Adoption as a conservation tool for endangered African wild dogs (*Lycaon pictus*). *African Journal of Wildlife Research* 38: 109–13. https://doi.org/10.3957/0379-4369-38.2.109

McPherson, J.M., W. Jetz, and D.J. Rogers. 2006. Using coarse-grained occurrence data to predict species distributions at finer spatial resolutions—possibilities and limitations. *Ecological Modelling* 192: 499–522. http://doi.org/10.1016/j.ecolmodel.2005.08.007

Menges, E.S., S.A. Smith, and C.W. Weekley. 2016. Adaptive introductions: How multiple experiments and comparisons to wild populations provide insights into requirements for long-term introduction success of an endangered shrub. *Plant Diversity* 38: 238–46. https://doi.org/10.1016/j.pld.2016.09.004

Miller, S.M., C.J. Tambling, and P.J. Funston. 2015. GrowLS: Lion (*Panthera leo*) population growth simulation for small reserve management planning. *African Journal of Wildlife Research* 45: 169–77. https://doi.org/10.3957/056.045.0169

Monadjem, A., and A. Reside. The influence of riparian vegetation on the distribution and abundance of bats in an African savanna. *Acta Chiropterologica* 10: 339–48. http://doi.org/10.3161/150811008X414917

Negrón-Ortiz, V. 2014. Pattern of expenditures for plant conservation under the Endangered Species Act. *Biological Conservation* 171: 36–43. https://doi.org/10.1016/j.biocon.2014.01.018

Newmark, W.D. 2008. Isolation of African protected areas. *Frontiers in Ecology and the Environment* 6: 321–28. https://doi.org/10.1890/070003

Newmark, W.D., C.N. Jenkins, S.L. Pimm, et al. 2017. Targeted habitat restoration can reduce extinction rates in fragmented forests. *Proceedings of the National Academy of Sciences* 114: 9635–40. https://doi.org/10.1073/pnas.1705834114

Ng'weno, C.C., N.J. Maiyo, A.H. Ali, et al. 2017. Lions influence the decline and habitat shift of hartebeest in a semiarid savanna. *Journal of Mammalogy* 98: 1078–87. https://doi.org/10.1093/jmammal/gyx040

Nunes, M., J.B. Adams, and G.M. Rishworth. 2018. Shifts in phytoplankton community structure in response to hydrological changes in the shallow St Lucia Estuary. *Marine Pollution Bulletin* 128: 275–86. https://doi.org/10.1016/j.marpolbul.2018.01.035

Packer, C., A. Loveridge, S. Canney, et al. 2013. Conserving large carnivores: Dollars and fence. *Ecology Letters* 16: 635–41. https://doi.org/10.1111/ele.12091

Pearson, R.G., and T.P. Dawson. 2003. Predicting the impacts of climate change on the distribution of species: Are bioclimate envelope models useful? *Global Ecology and Biogeography* 12: 361–71. https://doi.org/10.1046/j.1466-822X.2003.00042.x

Pearson, R.G., W. Thuiller, M.B. Araújo, et al. 2006. Model-based uncertainty in species range prediction. *Journal of Biogeography* 33: 1704–11. https://doi.org/10.1111/j.1365-2699.2006.01460.x

Pérez, I., J.D. Anadón, M. Díaz, et al. 2012. What is wrong with current translocations? A review and a decision-making proposal. *Frontiers in Ecology and the Environment* 10: 494–501. https://doi.org/10.1890/110175

Phillips, S.J., P. Williams, G. Midgley, et al. 2008. Optimizing dispersal corridors for the Cape Proteaceae using network flow. *Ecological Applications* 18: 1200–11. https://doi.org/10.1890/07-0507.1

Plumptre, A.J., S. Nixon, D.K. Kujirakwinja, et al. 2016. Catastrophic decline of world's largest primate: 80% loss of Grauer's gorilla (*Gorilla beringei graueri*) population justifies Critically Endangered status. *PloS ONE* 11: e0162697. https://doi.org/10.1371/journal.pone.0162697

Powell, A.N., and F.J. Cuthbert. 1993. Augmenting small populations of plovers: An assessment of cross-fostering and captive-rearing. *Conservation Biology* 7: 160–68. https://doi.org/10.1046/j.1523-1739.1993.07010160.x

Pryke, J.S., and M.J. Samways. 2012. Conservation management of complex natural forest and plantation edge effects. *Landscape Ecology* 27: 73–85. https://doi.org/10.1007/s10980-011-9668-1

Ralls, K., J.D. Ballou, M.R. Dudash, et al. 2018. Call for a paradigm shift in the genetic management of fragmented populations. *Conservation Letters* 11: 1–6. https://doi.org/10.1111/conl.12412

Reilly, S.B., J.L. Bannister, P.B. Best, et al. 2013. *Eubalaena australis. The IUCN Red List of Threatened Species* 2013: e.T8153A44230386. http://doi.org/10.2305/IUCN.UK.2018-1.RLTS.T8153A50354147.en

Riggio, J., and T. Caro. 2017. Structural connectivity at a national scale: Wildlife corridors in Tanzania. *PLoS ONE* 12: e0187407. https://doi.org/10.1371/journal.pone.0187407

Ripple, W.J., J.A. Estes, R.L. Beschta, et al. 2014. Status and ecological effects of the world's largest carnivores. *Science* 343: 12341484. https://doi.org/10.1126/science.1241484

Runge, C.A., J.E.M. Watson, S.H.M. Butchart, et al. 2015. Protected areas and global conservation of migratory birds. *Science* 350: 1255–58. http://doi.org/10.1126/science.aac9180

SANParks. 2009. Sought-after disease-free Addo buffalo to go on auction. *SANParks News*. https://www.sanparks.org/about/news/?id=1246

Scheele, B.C., D.A. Hunter, L.F. Grogan, et al. 2014. Interventions for reducing extinction risk in Chytridiomycosis-threatened amphibians. *Conservation Biology* 28: 1195–205. https://doi.org/10.1111/cobi.12322

Schleuning, M., J. Fründ, O. Schweiger, et al. 2016. Ecological networks are more sensitive to plant than to animal extinction under climate change. *Nature Communications* 7: 13965. https://doi.org/10.1038/ncomms13965

Schmidt, F., F.A. Franke, M.H. Shirley, et al. 2015. The importance of genetic research in zoo breeding programmes for threatened species: The African dwarf crocodiles (genus *Osteolaemus*) as a case study. *International Zoo Yearbook* 49: 125–36. https://doi.org/10.1111/izy.12082

Scott, J.M., D.D. Goble, J.A. Wiens, et al. 2005. Recovery of imperilled species under the Endangered Species Act: The need for a new approach. *Frontiers in Ecology and the Environment* 3: 383–89. https://doi.org/10.1890/1540-9295(2005)003[0383:ROISUT]2.0.CO;2

Shackeroff, J.M., and L.M. Campbell. 2007. Traditional ecological knowledge in conservation research: Problems and prospects for their constructive engagement. *Conservation and Society* 5: 343

Sherley, R.B., K. Ludynia, B.M. Dyer, et al. 2017. Metapopulation tracking juvenile penguins reveals an ecosystem-wide ecological trap. *Current Biology* 27: 563–68. https://doi.org/10.1016/j.cub.2016.12.054

Shrader, A.M., C. Bell, L. Bertolli, et al. 2012. Forest or the trees: At what scale do elephants make foraging decisions? *Acta Oecologica* 42: 3–10. http://doi.org/10.1016/j.actao.2011.09.009

Silber, G.K., A.S.M. Vanderlaan, A.T. Arceredillo, et al. 2012. The role of the International Maritime Organization in reducing vessel threat to whales: Process, options, action and effectiveness. *Marine Policy* 36: 1221–33. http://doi.org/10.1016/j.marpol.2012.03.008

Sileshi, G.W., P. Nyeko, P.O.Y. Nkunika, et al. 2009. Integrating ethno-ecological and scientific knowledge of termites for sustainable termite management and human welfare in Africa. *Ecology and Society* 14: 48. http://www.ecologyandsociety.org/vol14/iss1/art48

Singh, J. 1998. The lessons learnt: The development and management of transboundary parks world-wide. In: *Study on the Development and Management of Trans-boundary Conservation Areas in Southern Africa* (Gabarone: RCSA).

Stears, K., and A.M. Shrader. 2015. Increases in food availability can tempt oribi antelope into taking greater risks at both large and small spatial scales. *Animal Behaviour* 108: 155–64. http://doi.org/10.1016/j.anbehav.2015.07.012

Tambling, C.J., D.J. Druce, M.W. Hayward, et al. 2012. Spatial and temporal changes in group dynamics and range use enable anti-predator responses in African buffalo. *Ecology* 93: 1297–304. https://doi.org/10.1890/11-1770.1

Tambling, C.J., J.W. Wilson, P. Bradford, et al. 2014. Fine-scale differences in predicted and observed cheetah diet: Does sexual dimorphism matter? *South African Journal of Wildlife Research* 44: 90–94. https://doi.org/10.3957/056.044.0109

Tambling, C.J., S.M. Ferreira, J. Adendorff, et al. 2013. Lessons from management interventions: Consequences for lion-buffalo interactions. *South African Journal of Wildlife Research* 43: 1–11. https://doi.org/10.3957/056.043.0116

Thomas-Blate, J. 2017. Celebrating a great year for dam removals in 2016. *American Rivers*. https://www.americanrivers.org/2017/02/celebrating-great-year-dam-removal-2016

Tolley, K.A., G.J. Alexander, W.R. Branch, et al. 2016. Conservation status and threats for African reptiles. *Biological Conservation* 204: 63–71. http://doi.org/10.1016/j.biocon.2016.04.006

Valutis, L.L., and J.M. Marzluff. 1999. The appropriateness of puppet-rearing birds for reintroduction. *Conservation Biology* 13: 584–91. https://doi.org/10.1046/j.1523-1739.1999.97443.x

van Andel, T.R., S. Croft, E.E. van Loon, et al. 2015. Prioritizing West African medicinal plants for conservation and sustainable extraction studies based on market surveys and species distribution models. *Biological Conservation* 181: 173–81. https://doi.org/10.1016/j.biocon.2014.11.015

Vandvik, V., I.E. Måren, H.J. Ndangalasi, et al. 2014. Back to Africa: Monitoring post-hydropower restoration to facilitate reintroduction of an extinct-in-the-wild amphibian. *Ecosphere* 5: 1–16. https://doi.org/10.1890/ES14-00093.1

von der Heyden, S. 2009. Why do we need to integrate population genetics into South African marine protected areas planning? *African Journal of Marine Science* 31: 263–69. https://doi.org/10.2989/AJMS.2009.31.2.14.886

Vosse, S., K.J. Esler, D.M. Richardson, et al. 2008. Can riparian seed banks initiate restoration after alien plant invasion? Evidence from the Western Cape, South Africa. *South African Journal of Botany* 74: 432–44. https://doi.org/10.1016/j.sajb.2008.01.170

Wassenaar, T.D., R.J. van Aarde, S.L. Pimm, et al. 2005. Community convergence in disturbed subtropical dune forests. *Ecology* 86: 655–66. https://doi.org/10.1890/03-0836

Wegmann M, L. Santini, B. Leutner, et al. 2014. Role of African protected areas in maintaining connectivity for large mammals. *Philosophical Transactions of the Royal Society B* 369: 20130193. https://doi.org/10.1098/rstb.2013.0193

White, P.S. 1996. Spatial and biological scales in reintroduction. In: *Restoring Diversity: Strategies for Reintroduction of Endangered Plants*, ed. by D.A. Falk, et al. (Washington: Island Press).

Williams, P., L. Hannah, S. Andelman, et al. 2005. Planning for climate change: Identifying minimum-dispersal corridors for the Cape proteaceae. *Conservation Biology* 19: 1063–74. https://doi.org/10.1111/j.1523-1739.2005.00080.x

Wilson, J.W., R. Bergl, L.J. Minter, et al. 2019. The African elephant *Loxodonta* spp. conservation programmes of North Carolina Zoo: Two decades of using emerging technologies to advance in situ conservation efforts. *International Zoo Yearbook* 53: *in press*. https://doi.org/10.1111/izy.12216

Wilson, J.W., R.L. Stirnemann, Z.S. Schaikh, et al. 2010. The response of small mammals to natural and human-altered edges associated with Afromontane forests of South Africa. *Forest Ecology and Management* 259: 926–31. http://doi.org/10.1016/j.foreco.2009.11.032

Wimberger, K., C.T. Downs, and M.R. Perrin. 2009. Two unsuccessful reintroduction attempts of rock hyraxes (*Procavia capensis*) into a reserve in the KwaZulu-Natal Province, South Africa. *South African Journal of Wildlife Research* 39: 192–201. https://doi.org/10.3957/056.039.0213

Wimberger, K., C.T. Downs, and R.S. Boyes. 2010. A survey of wildlife rehabilitation in South Africa: Is there a need for improved management? *Animal Welfare* 19: 481–99.

Winemiller, K.O., P.B. McIntyre, L. Castello, et al. 2016. Balancing hydropower and biodiversity in the Amazon, Congo, and Mekong. *Science* 351: 128–29. http://doi.org/10.1126/science.aac7082

Woolf, N.B. 1986. New hope for exotic species. *BioScience* 36: 594–97. https://doi.org/10.2307/1310192.

Wyse Jackson, P.S. 2001. International review of the ex situ plant collections of the botanic gardens of the world. *Botanic Gardens Conservation News* 3: 22–33.

12. Biodiversity and the Law

One hundred and five tonnes of confiscated ivory and one tonne of confiscated rhino horn ablaze in Nairobi National Park, Kenya. The Presidents of the Republic of the Congo, Kenya, and Chad have personally set such stockpiles of seized contraband ablaze as a symbolic illustration of their support for efforts to stamp out wildlife crimes. Photograph by Mwangi Kirubi, https://commons.wikimedia.org/wiki/File:Nairobi-Ivory-Burn-by-Mwangi-Kirubi-7.jpg, CC BY-SA 4.0.

 https://doi.org/10.11647/OBP.0177.12

The negative impact of human activities on the natural environment is apparent wherever you look. Some impacts are an unavoidable consequence of human activities; vast resources are currently invested in finding ways to mitigate those impacts. Other impacts, often entirely preventable, are rooted in greed. Consider how the worst polluters are corporations that prioritise profits over environmental and human health. Similarly, many threatened species continue to be illegally exploited in an unsustainable manner; in the worst cases, the profits from poaching are funding human-rights atrocities and organised criminal networks. Because society pays the price for environmental crimes—which generally benefit only a few people—there is broad interest in preventing environmental abuse, and to punish the perpetrators.

Environmental crimes are generally divided into two categories: wildlife crimes—the illegal exploitation of biodiversity (including but not restricted to **wildlife trafficking** and biopiracy), and pollution crimes—the illegal trade and disposal of waste and hazardous substances. As with other crimes, environmental crimes are generally defined by legislative action, when governments pass environmental laws and regulations that restrict certain kinds of activities. The effectiveness of these laws and regulations in protecting the environment relies on three main factors: (1) identifying conservation priorities, (2) establishing regulations that addresses those needs, and (3) enforcing environmental laws and regulations.

Because society pays the price for environmental crimes—which benefit only a few people—there is broad interest in preventing environmental abuse, and to punish the perpetrators.

12.1 Identifying Legislative Priorities

Humans have always depended on the environment to fulfil their most basic needs. Before the Industrial Revolution, fulfilling those basic needs generally occurred at sustainable levels. Over the last few centuries, however, exponential human population growth and rates of resource extraction have put enormous pressure on the environment. Today, many wildlife populations and ecosystems are unable to cope with these pressures. Increased globalisation has exacerbated many of these problems. For example, with most Asian rhinoceros and pangolin populations on the brink of extinction (IUCN, 2019), Asian traders are increasingly filling their orders for elephant (Figure 12.1), rhinoceros, and pangolin body parts from African suppliers (Biggs et al., 2013; Wasser et al., 2015; Heinrich et al., 2016).

Identifying which species and ecosystems need to be prioritised for legislative action can be confusing, and sometimes even seems in conflict with more readily available information at hand. For example, many hunters believe that the animals they target persist in healthy numbers despite claims to the contrary from conservation biologists. In other areas, logging companies claim they operate sustainably, yet tropical forests continue to shrink. In the face of conflicting information, it is critical for conservation biologists to rely on consistent, repeatable, and transparent methods

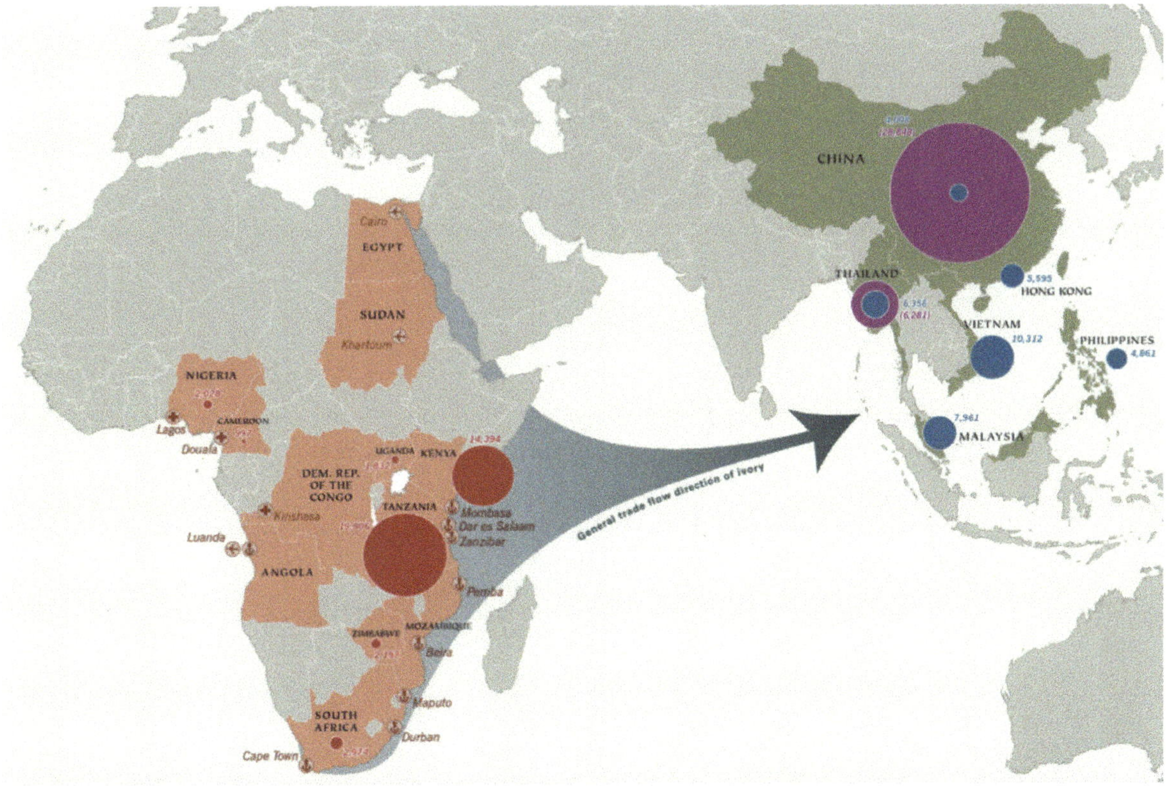

Figure 12.1 Key global ivory smuggling routes from 2009–2011, based on seizure data. More recent work has shown that poached elephants continue to originate from Tanzania, Mozambique, and Cameroon, while several seizures were now also from Gabon and the Republic of the Congo (Wasser et al., 2015). Map by CIA, https://www.flickr.com/photos/ciagov/30885483595, CC0.

to identify those populations, species, and ecosystems that may need (additional) regulatory protections.

Currently, the most popular method to identify legislative priorities is to use the IUCN's Red List criteria, developed to reflect a taxon's risk of extinction (Section 8.5). Following these criteria (which can be applied on a global or local scale), species that are considered *Extinct in the Wild*, *Critically Endangered*, *Endangered*, and *Vulnerable* are officially considered "threatened with extinction" and would thus receive higher priority than species that are *Near Threatened* or *Least Concerned*.

Although coarse filter approaches, which focus on groups of species and threatened ecosystems (Section 8.5.1), have been a catalyst for many international treaties and protected areas, legal mechanisms at the national and regional level do not always allow for its use. Through lobbying and education, these legislative branches will hopefully improve their receptiveness for coarse filter approaches in setting future legislative agendas.

12.2 Environmental Laws and Policies

When conservation priorities have been identified, there are several options available to preserve biodiversity. One option could involve the establishment of protected areas where ecological restoration (Section 10.3) and species conservation projects (Chapter 11) can be carried out. Conservation biologists could also start an environmental education programme (Section 15.5) that would help people live more sustainably on unprotected lands (Chapter 14). Under certain conditions, however, especially when control and protection measures fail, restrictions or outright bans of some human activities may be necessary (Keeley and Scoones, 2014). The most effective restrictions and bans involve legislative actions that also establish mechanisms to enforce environmental laws and regulations, and mechanisms that reduce consumer demand (Challender and MacMillan, 2014).

Environmental laws and regulations are implemented at three different levels: international treaties, national laws, and local laws. While the scope of each of these levels differs, they are intricately connected with one another. International treaties influence national laws, but also depend on their enforcement to succeed, while national laws are guided by local needs as well as customary laws that have been in place for generations. Ideally speaking, international and national laws set minimum benchmarks, which regional and local governments adopt and enforce. Local and regional laws may sometimes set stricter standards in areas where the environment is more sensitive, more damaged, or more important for human well-being. Local and national legislatures may also choose to ignore broader legislation, through non-cooperation and non-enforcement. But this is not advisable as it may lead to further environmental deterioration, loss of funding, and even trade embargoes and sanctions (Section 12.4.4) that could harm local economies.

12.2.1 International agreements

International agreements provide frameworks that allow countries to work together to protect biodiversity (Sands and Peel, 2012). These international agreements, called treaties or conventions, are needed for five important reasons: (1) many species migrate and disperse across administrative borders, (2) ecosystems do not follow administrative boundaries, (3) pollution spreads by air and water across regions and around the globe, (4) many biological products are traded internationally, and (5) some environmental problems (e.g. climate change and pollution) require global cooperation and coordination. To pass international treaties, agreements are negotiated at international conferences under the authority of international bodies such as the UN, UNEP, or IUCN and come into force when they are ratified by an agreed-upon number of countries. These treaties are then implemented at the local level when signatory countries pass national laws to enforce them.

> International agreements provide frameworks that allow countries to work together to protect biodiversity.

One of the most important international environmental treaties is the ***Convention on Biological Diversity*** (**CBD**, https://www.cbd.int). The CBD formulated and signed following the UN **Earth Summit** (also called Rio Summit) held in Rio de Janeiro, Brazil in 1992, has played a major role in raising awareness of the value of biodiversity to humanity. At this meeting, representatives from 178 countries formulated and eventually signed the CBD, obligating signatory countries to protect biodiversity through careful management of nature for the benefit of humans. The CBD was expanded in 2010 to also include recommendations for the protection of IUCN Red Listed species and ecosystems, as part of the **Aichi Biodiversity Targets** (Table 12.1).

Table 12.1 The UN, with governments across the world, have agreed to work on five strategic goals and 20 specific targets (collectively known as Aichi Biodiversity Targets) to halt the loss of biodiversity and protect and restore what remains.

CBD strategic goal	Aichi Target
A. Address underlying causes of biodiversity losses	1. Improve awareness of biodiversity values
	2. Integrate biodiversity values into development
	3. Eliminate perverse subsidies; incentivise sustainability
	4. Implement plans for sustainable consumption and production
B. Reduce pressures on biodiversity	5. Reduce the rate of habitat loss by at least 50%
	6. Ensure sustainable use of marine resources
	7. Ensure sustainable agriculture, aquaculture, and forestry
	8. Reduce pollution to non-detrimental levels
	9. Identify and control priority invasive species
	10. Reduce pressures on climate-sensitive ecosystems
C. Safeguard ecosystems, species, and genetic diversity	11. Increase coverage of well-managed protected areas
	12. Prevent the extinction of threatened species
	13. Prevent genetic erosion of biodiversity
D. Enable more people to enjoy the benefits of biodiversity	14. Restore and safeguard ecosystems and essential services
	15. Restore and enhance resilience of degraded ecosystems
	16. Ensure fair and equitable sharing of ecosystem services
E. Implement participatory biodiversity strategies	17. Implement participatory national biodiversity strategies
	18. Respect and conserve traditional knowledge
	19. Improve, share, and apply biodiversity knowledge
	20. Mobilise resources to address Aichi Targets

Source: https://www.cbd.int/sp/targets

There are also several international agreements seeking the direct protection of targeted threatened species. One of the most important treaties of this nature is CITES (***Convention on International Trade in Endangered Species of Wild Fauna and Flora***, https://cites.org), agreed upon in 1973 in Washington, DC. This treaty, ratified by 175 countries, establishes lists (known as Appendices) of species for which member nations agree to ban, restrict, control, and monitor international trade. Over 35,000 species of plants and animals appear on these appendices, many also listed as threatened by the IUCN. With a few exceptions, the international trade of wild-caught specimens on Appendix I is prohibited; trade in Appendix II species is strictly regulated to ensure sustainability, while trade in Appendix III species require a certification of origin. Once member countries pass local laws to comply with CITES, police, customs inspectors, wildlife officers, and governmental agents appointed for that purpose can arrest individuals possessing or trading in products from the listed species. The World Conservation Monitoring Centre (WCMC), which operates within UNEP, is tasked with managing the CITES database and monitoring whether member countries are enforcing recommendations.

The *Convention on the Conservation of Migratory Species of Wild Animals* (sometimes shortened to *Bonn Convention*, http://www.cms.int) is another important treaty that seeks the protection of specifically targeted species. The *Bonn Convention* came into force in 1983, and has over 120 Parties, including 37 from Sub-Saharan Africa. As with CITES, the *Bonn Convention* categorises species under Appendices. Species on Appendix I are threatened with extinction; "Range States" to Appendix I species are obliged to afford those species' strict protections. Appendix II lists species whose populations would significantly benefit from international cooperation. Three important agreements that involve Sub-Saharan species have been concluded under the *Bonn Convention*: (1) the *African-Eurasian Waterbird Agreement* (http://www.unep-aewa.org), which, amongst others, things bans the use of lead shot around aquatic ecosystems; (2) the *Gorilla Agreement* (http://www.cms.int/gorilla), which binds Parties to protect gorillas in their habitats; and the *Agreement on the Conservation of Albatrosses and Petrels* (https://acap.aq), which coordinates international efforts to mitigate known threats to seabirds.

Several international agreements seek the protection of important ecosystems. Perhaps the most prominent is *Convention Concerning the Protection of the World's Cultural and Natural Heritage* (http://whc.unesco.org), which protects natural (and cultural) areas of international significance. As of mid-2019, **UNESCO** (the organisation managing the list of **World Heritage Sites**) recognised 35 natural World Heritage Sites in Sub-Saharan Africa; this includes some of the world's most famous conservation areas, such as Serengeti National Park in Tanzania, Bwindi Impenetrable Park in Uganda, and the Aldabra Atoll of the Seychelles. In addition, five World Heritage Sites are recognised for their natural *and* cultural significance: this includes Gabon's Ecosystem and Relict Cultural Landscape of Lopé-Okanda, Tanzania's Ngorongoro Conservation Area, and the Maloti-Drakensberg Park—a transboundary site composed of South Africa's Drakensberg National Park and Lesotho's Sehlathebe National Park (Figure 12.2).

Figure 12.2 The Maloti-Drakenberg Park World Heritage Site, on the borders of South Africa and Lesotho, protects globally significant natural and cultural heritage. Photograph by Diriye Amey, https://commons.wikimedia.org/wiki/File:South_Africa_-_Drakensberg_(16261357780).jpg, CC BY 4.0.

Another important treaty that seems ecosystem protection is the ***Ramsar Convention on Wetlands*** (http://www.ramsar.org), which recognises the ecological, scientific, economic, cultural, and recreational value of freshwater, estuarine, and coastal marine ecosystems. All but three Sub-Saharan African countries have signed the *Ramsar Convention*; this binds each member country to conserve and sustainably utilise its wetlands (particularly those that support migratory waterfowl), and to officially declare at least one internationally significant wetland as protected. As of mid-2019, 252 Sub-Saharan African wetlands, covering over 1 million km^2, were declared internationally significant under Ramsar guidelines. South Africa and Burkina Faso have the most Ramsar Wetlands (23 and 22, respectively), while the Republic of the Congo has the largest area (138,138 km^2) designated. The world's largest Ramsar wetland, the DRC's Ngiri-Tumba-Maidombe, is 65,696 km^2 in size (over twice the size of Lesotho!).

International treaties are particularly important to the marine environment, since about two-thirds of the world's oceans (50% of the planet) are considered international waters—that is, being outside any country's **exclusive economic zone (EEZ)**, all states have the freedom to fish, travel, do research, etc. in these areas. Three examples of international agreements protecting such marine ecosystems are (1) the 1972 *Convention on the Prevention of Marine Pollution by Dumping of Wastes and Other Matter* (http://www.imo.org/en/OurWork/Environment/LCLP) which regulates pollutants into the marine environment, (2) the 1982 *Convention on the Law of the Sea* (http://www.un.org/Depts/los) which establishes guidelines for management of marine natural resources, and (3) the 2009 *Agreement on Port State Measures* (http://www.fao.org/port-state-measures) which sanctions monitoring for illegal, unreported, and unregulated fishing at shipping ports.

> International treaties are particularly important to the marine environment, since about two-thirds of the world's oceans (50% of the planet) fall outside any country's jurisdiction.

In addition to being party to these and other global treaties, several African countries are also members of agreements that address regional environmental concerns.

Among the most prominent is the 2003 *Revised African Convention on the Conservation of Nature and Natural Resources* (or *Maputo Convention*). The most progressive reforms of the *Maputo Convention* include the recognition that nature is a finite resource, that the needs of future generations and traditional peoples must be considered, and that the harmful impacts of civil strife on the environment must be mitigated.

The 2016 ***Paris Agreement***, which deals with greenhouse gas emission reduction and climate change mitigation, serves to illustrate the difficult political negotiations (Figure 12.3) involved in the adoption of an international treaty. Although the negative effects of climate change have been known for several decades (Section 6.1), until recently there has been a distinct lack of action to curb global greenhouse gas emissions. For example, as an early call to action on reducing greenhouse gas emissions, representatives from 154 countries signed the UN *Framework Convention on Climate Change* (UNFCCC) at the Earth Summit in May 1992. In the following years, negotiations during annual UNFCCC conferences (formally known as "Conference of the Parties", or COPs) led in the ***Kyoto Protocol***, adopted in Japan in 1997, which marked the first attempt to set legally binding emission reduction targets. Despite broad appeal among its 192 parties, the *Kyoto Protocol* faced an uphill battle from the start because the USA (the world's biggest greenhouse gas emitter at the time) refused to ratify it, and China (which recently overtook the USA as the biggest emitter) was exempted from compliance. While this has left the *Kyoto Protocol* largely a failure, it provided important lessons that contributed to the successful passing of the *Paris Agreement* (http://unfccc.int), which was negotiated and adopted through consensus by 195 countries (this time including the USA and China) in December 2015. The *Paris Agreement* went into effect on 4 November 2016 after the minimum 55 countries ratified it, marking a breakthrough in the decades-long battle to curb global greenhouse gas emissions. By mid-2019, all but one country in the world (the non-signatory being the Holy See, who as UNFCCC observer nation that cannot sign but strongly support the Agreement) have signed and/or ratified the Agreement. Most relevant to African member states are the mechanisms set up to provide developing countries with large amount of aid for climate change mitigation and adaption, much of which involves ecosystem conservation (see REDD+, Sections 15.3).

While it is still too early to judge the effectiveness of the *Paris Agreement*, the 1987 *Montreal Protocol* on *Substances that Deplete the Ozone Layer* (http://ozone.unep.org) illustrates how international cooperation *can* be effective in preventing environmental disasters. In the 1970s, scientists discovered that a range of chemicals (primarily chlorofluorocarbons, or CFCs) commonly used in agriculture, energy production, and even common household items (such as refrigerators and aerosol spray canisters) were depleting the atmospheric **ozone layer**. The ozone layer is critical for human life; by cutting the amount of harmful ultraviolet radiation from the sun that reaches the Earth's surface, protection from the ozone layer reduces skin cancer, cataracts, and crop damage. In response to this threat, the *Montreal Protocol* aimed to phase out those substances that were responsible for ozone depletion. Since then, the ozone layer has steadily recovered; current projections suggest that the ozone layer will return

Figure 12.3 A small group of COP21 delegates, led by UNFCCC Executive Secretary Christiana Figueres, negotiating the final terms of the *Paris Agreement* before its adoption on 12 December 2015. Photograph by Benjamin Géminel, https://www.flickr.com/photos/cop21/23596677582, CC0.

to 1980 levels in the second half of the 21st century. Towards the end of his tenure as Secretary General of the UN (1997–2006), Ghana's Kofi Annan declared, "Perhaps the single most successful international agreement to date has been the *Montreal Protocol*". The *Montreal Protocol*'s success is directly due to this widespread adoption and implementation.

12.2.2 National and local laws

Traditional African societies have long recognised that preserving the environment is important for human well-being. Consequently, many African cultures had mechanisms in place before the arrival of European colonists that allowed these historical societies to exploit **communal resources** on a long-term, sustainable basis. These mechanisms included mystical beliefs, local customs, and cultural taboos that ensured the protection of wildlife and land with cultural and spiritual significance. While sacred forests are prominent examples, not all sacred sites are/were forested. For example, the sandy beaches on Guinea-Bissau's Poilão Island was also regarded as sacred by the people of the Bijagós Archipelago, ensuring the protection of one of the world's most important green turtle (*Chelonia mydas*, EN) nesting sites (Catry et al., 2002). These mechanisms, referred to as customary laws, also limited access to certain territories and imposed restrictions on harvesting methods, harvest times, and types of individuals that may be harvested. Strict sanctions for violations ensured that customary laws were generally followed, often through self-policing. In some ways, this traditional approach to natural resource management was not so different from certain wildlife management systems in Europe and elsewhere at the time—or even from today's more formal law systems—which place restrictions on how we utilise nature. While some customary laws continue to regulate activities in certain regions of Africa (e.g. Walters et al., 2015), in many areas, they were lost when European authorities replaced traditional authorities during colonisation.

Today, an increasing number of international treaties and environmental organisations are achieving their conservation goals by promoting respect for and inclusion of the cultural and spiritual values that traditional peoples attach to the environment. This includes the CBD, UN, IUCN, and African Union, all promoting the integration of traditional ecological knowledge (TEK) in conservation activities and regulations (Mauro and Hardison, 2000). A growing number of national governments are also institutionalising these efforts by passing laws recognising traditional rights, providing traditional peoples with land titles, and declaring areas of spiritual and cultural significance as protected. Conservation scientists are also increasingly relying on TEK to better understand ecological networks (Sileshi et al., 2009; Gómez-Baggethun et al., 2013), to ensure sustainable utilisation of natural resources (Mbata et al., 2002; Terer et al., 2012), and to secure the continued survival of severely threatened species such as the Cross River gorilla (*Gorilla gorilla diehli*, CR), of which fewer than 300 individuals remain (Etiendem et al., 2011).

While governments are becoming increasingly respectful of customary laws and traditional lifestyles, in many areas the customs inherent to them have fallen by the wayside under increased industrialisation, urbanisation, and globalisation. An increasing number of traders of traditional products are also using more effective collection and harvesting techniques, thereby pushing many species to extinction (Section 7.2). To fill these regulatory voids and to ensure sustainable utilisation of natural resources, statuary (passed by legislatures); regulatory (passed by regulatory agencies; and case (passed by judicial bodies) laws are playing an important role in protecting Africa's natural heritage.

Laws that protect the environment (and which can be passed by local or national branches of government) can generally be divided into three categories:

- Natural resource management laws, which define the limits of fair and sustainable use of land, water, minerals, and biodiversity.
- Pollution laws, which regulate dumping of waste and other harmful substances into the environment.
- Tax incentives, which encourage environmentally responsible behaviours.

Environmental laws that address natural resource use are well known because they impact the activities of the public and some businesses. These include hunting, trapping, and fishing regulations that limit the size and number of animal and plant products that can be collected, and the equipment that can be used for harvesting. Such regulations are typically enforced through licencing requirements, harvest reporting, and law enforcement patrols. Authorities may also set up mechanisms to restrict the sale, transport, and killing of sensitive species, including restricting the sale of firearms and ammunition.

Many people have also been exposed to restrictions that control the ways in which land is used to protect biodiversity. For example, uncontrolled fires may severely damage natural communities, so practices (such as building campfires) that contribute to accidental fires are often rigidly controlled. In some areas, vehicles and even foot traffic may be restricted to protect ecosystems and resources that are sensitive to disturbance, such as bird and turtle nesting areas on beaches, or sources of drinking water. One of the most popular methods of restricting activities in sensitive ecosystems and around sensitive resources is to pass laws that establish protected areas (Chapter 13).

Commercial operations are also subject to laws that govern natural resource use. Zoning laws, for example, prevent development of sensitive areas, such as riparian forests, beaches, wetlands, and floodplains. In areas where development is permitted, national laws typically require environmental impact assessments (EIAs, see Dana et al. 2012; Biamah et al. 2013) prior to development (Figure 12.4). Construction sites are surveyed during these assessments to ensure that damage is not done to threatened species or sensitive ecosystems. For major regional and national projects such as dams, mines, oil extraction, and highway construction, environmental impact statements must often be prepared that describe a project's potential damage, and mediatory actions taken.

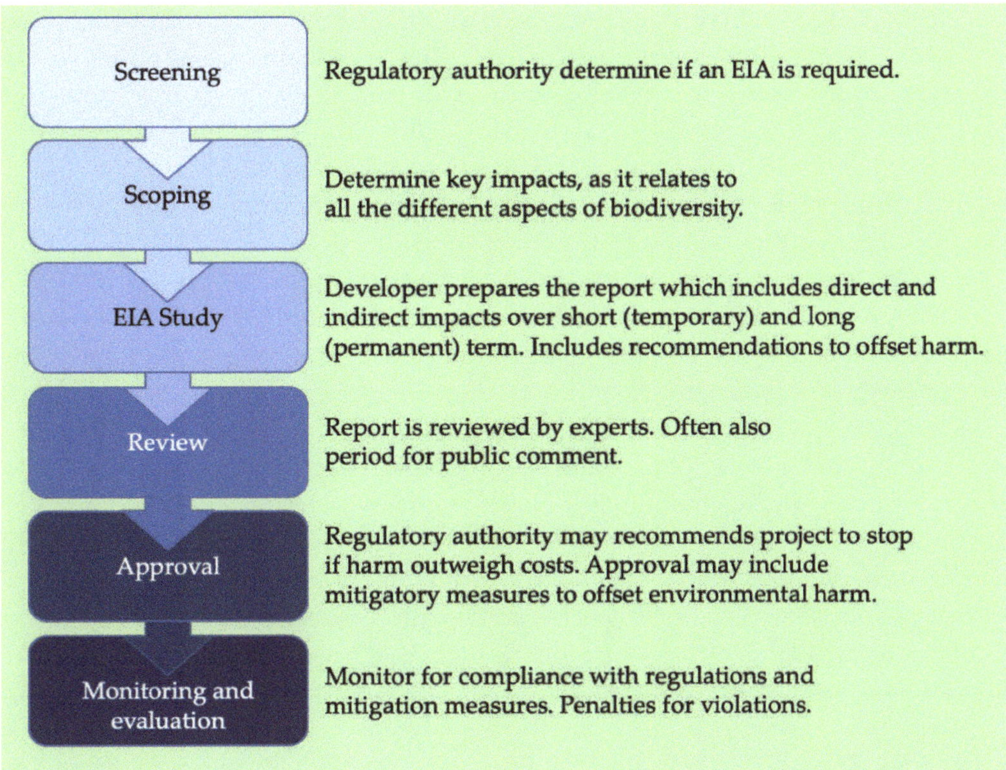

Figure 12.4 The steps required in a typical environmental impact assessment (EIA). EIAs are generally performed prior to a new development to assess potential environmental damage the development may cause, and to identify steps that can be taken to mitigate the damage. After Biamah et al., 2013, CC BY 4.0.

For industries that exploit threatened species and ecosystems, certification of a product's origin is increasingly being used as a mechanism to ensure that wild populations are not depleted by illegal collections (Poole and Shepherd, 2016). These certifications may state that environmental regulations, sustainable practices, and socially responsible methods have been followed, or that products were farmed, captive reared, or horticulturally derived rather than collected in the wild. To offset the damage caused by deforestation, various governments have recently made a concerted effort to minimise threats to their forests, including announcing timber export bans and moratoriums on commercial logging. Further afield, the USA, European Union, and Australia have also started placing bans on imported timber that was illegally harvested, some of which was sourced in Africa. Such bans are very effective in reducing the market value of unsustainably sourced products, while also increasing the market share for responsible businesses.

In recognising the immense harm invasive species inflict on the environment (Section 7.4), some countries have also enacted laws aimed at combatting invasive species. One example is South Africa, where over 500 current and potential invasive species are classified under three categories (http://www.invasives.org.za): Category 1 (destroy immediately, may not be owned), Category 2 (kept only with permit, no trade), and Category 3 (no trade, no breeding, but no need to remove) (Zengeya et al., 2017). Category 2 includes popular pets, such as mallards (*Anas platyrhynchos*, LC), that can hybridise with native waterfowl, as well as plants, such as gum trees (*Eucalyptus* spp.) that reduce local water availability (Section 7.4.2). Complementing this effort, the city of Cape Town's local government launched a competition (http://www.capetowninvasives.org.za) (with prizes) during the first half of 2017 for people who report the location for any of 28 priority invasive species.

Laws that regulate waste management and prevent pollution (Section 7.1) deal with aspects such as air emissions, sewage treatment, hazardous waste, solid waste, and wastewater dumping. In the unfortunate event that pollution ends up in the environment, such laws may also sanction contaminant clean-up. The primary aim of most pollution laws is to protect human health, property, and natural resources such as drinking water, forests, and commercial and sport fisheries. At the same time, they also protect biological communities that would otherwise be destroyed by pollution. For example, air pollution that exacerbates respiratory disease (in humans and animals) also damages commercial forests. Similarly, drinking water pollution which sickens people also kills aquatic species, such as turtles, amphibians, and fish. These examples once again show how intricately human health and economic well-being are linked to the health of the environment.

Most laws meant to protect biodiversity are restrictive in nature, but some regulations take a different tact by rewarding individuals who contribute to biodiversity conservation. Although under-utilised in Africa, perhaps the most popular regulatory reward mechanisms are subsidies and tax incentives. For example, several industrialised countries provide subsidies and tax rebates for citizens and industries that install sustainable energy alternatives, such as solar panels, acquire greener transport options, such as hybrid and electric vehicles, and invest in **green infrastructure,** such as green roofs and permeable surfaces (Section 14.2). South Africa took its first step of this kind in 2016, when BirdLife South Africa's Fiscal Benefits Project influenced the introduction of a new tax incentive into national legislation that rewards citizens for making conservation commitments on their land (Stevens, 2017). This tax incentive allows landowners to pay reduced taxes based on the value of their land they have formally declared and manage as a protected area. (For a detailed financial analysis of a similar incentive in Canada, see Schuster et al., 2017). By financially rewarding responsible citizens, national governments can put a smile on their citizens' faces, while also saving money over the long term given that it is often cheaper to protect intact ecosystem services than restoring damaged ecosystems.

> Tax incentives can encourage environmentally responsible behaviours and reward individuals who contribute to biodiversity conservation.

12.3 Environmental Law Enforcement

A single unlawful act—whether negligent or on purpose, by one single person or business—can harm countless ecosystems over a very wide geographic area. Such harm may persist for long periods of time (years, decades, and longer), and impact the lives of thousands of people. For that reason, mutual respect dictates that people and corporations alike abide by the environmental laws and regulations governing their activities. Unfortunately, while most people and businesses comply with environmental laws, it seems that there will always be those who take more than their fair share, corrupt government officials who facilitate smuggling, and greedy corporations that ignores the laws or searches for ways around them for profit. Consequently, there is a constantly need for government structures to evaluate whether environmental laws and regulations are enforced, whether violators are prosecuted, and whether amendments or new laws are needed.

> Mutual respect dictates that people and corporations alike abide by the environmental laws and regulations.

Environmental laws can be enforced in several ways. In general, the system works when offences are investigated, and violators are apprehended by law enforcement officers, such as the police (Figure 12.5). Vigilant citizens can also play a role by reporting offences to authorities; financial rewards are increasingly being offered as an incentive for citizens to report environmental crimes. Some districts, environmental agencies, and protected areas may also employ dedicated environmental compliance officers, such as game rangers and anti-poaching units, to monitor human activities that may negatively impact biodiversity. Sometimes, the mere presence of environmental compliance officers is enough to deter illicit activities. With adequate enthusiasm, training, support, and equipment, these teams can have a positive impact on an area's biodiversity and its people in a short period of time.

Figure 12.5 Rangers at Garamba National Park, DRC, found 73 kg of giant ground pangolin (*Smutsia gigantea*, VU) scales (from about 20 animals) and two elephant tusks in this handcuffed poacher's possession. Considered the world's most trafficked animals, Africa's four pangolin species (and Asia's four species) are threatened with extinction (IUCN, 2019). Photograph by Naftali Honig/African Parks, CC BY 4.0.

When caught, violators are usually punished by being charged fines and/or civil damages, and/or being sentenced to time in prison. Serve severe penalties can act as strong deterrent to those who consider engaging in environmental crimes. For example, Zimbabwe recently sentenced a rhinoceros poacher to 35 years in prison (Rademeyer, 2016), Nigeria sanctioned 26 mining companies for not complying with environmental laws (NAN, 2015), and Cameroon fined two ivory traffickers US $500,000 plus five years in prison (WWF, 2017). While such severe fines are usually reserved for major offences involving charismatic species, a South African court recently signalled that all biodiversity matters by sentencing a Spanish couple in possession of illegally collected plants to 12 years in prison, in addition to a US $150,000 fine (Steyn, 2015). As always, it is critically important that every violator is treated equally under the law, whether the violator is the owner of a company that dumps noxious chemicals into a river, a corrupt government official who facilitates smuggling of illegal wildlife products, or an individual caught hunting illegally in a protected area.

12.3.1 New technologies in environmental law enforcement

While environmental law enforcement is the single best predictor of conservation success across Africa (Hilborn et al., 2006; Tranquilli et al., 2012), catching and prosecuting perpetrators can be a difficult and dangerous task. Over the last few years, an increasing number of law enforcement officers have died while protecting the environment (WWF, 2016). Journalists reporting on environmental crimes are also increasingly persecuted, kidnapped, and even murdered (RSF, 2015). Well-organised environmental crime syndicates linked to drug smuggling, terrorism, and other human-rights abuses use increasingly sophisticated tools and tactics to evade detection. Moreover, armed poachers frequently outnumber law enforcement officials. Consequently, refining old and developing new strategies in environmental law enforcement are increasingly necessary.

One of the most promising developments in wildlife conservation has been the rapid development of molecular and other analytical tools and increased data processing capacity, leading to better detection, tracking, and prosecution environmental crimes. One promising development has been the increased use of genetic analysis to aid law enforcement. For example, **DNA barcoding**—a genetic analysis method that can identify the species of unknown tissue samples—helped expose illegal trade in five species of cycad (*Encephalartos* spp.), each of them threatened and listed on CITES Appendix I (Williamson et al., 2016). Elsewhere, biologists have started using stable isotope analysis—a technique that analyses an animal's diet—to determine the origin (captive-bred or wild-caught) of parrots that are for sale (Alexander et al., 2019), and whether rare cycads were wild-collected before or after the practice was banned (Retief et al., 2014). To fully harness the power of molecular methods, wildlife agencies in South Africa and Kenya have even set up dedicated wildlife crime forensic laboratories (Wasser et al., 2007, 2015), where conservationists work closely with forensic scientists to solve wildlife crimes (Box 12.1). These initiatives have already paid off in Kenya, where molecular methods have helped increase conviction rates for environmental crimes from 43% in 2013 to over 90% in 2016 (ODPP, 2017)!

> In Kenya, molecular methods have helped increase conviction rates for environmental crimes from 43% in 2013 to over 90% in 2016.

Conservationists have also become more mindful of the strategies they use to plan and conduct law enforcement monitoring. For example, park managers in Chad now use sophisticated mapping technologies to plan and monitor vulnerable wildlife as well as anti-poaching patrols (Box 12.2), while conservationists working in Cameroon are using acoustic sensors which identify times and areas of increased poaching activity (Astaras et al., 2017). Biologists in the Albertine Rift in turn use a spatial planning software package called Marxan (http://marxan.org)—generally used to identify the locations of new protected areas—to ensure law enforcement activities are more cost effective (Plumptre et al., 2014). To keep staff out of harm's way and to cover more ground, environmental agencies have also started using unmanned aerial vehicles (UAVs) for law enforcement monitoring (see Box 15.1).

> Conservationists are becoming more mindful of the strategies they use to plan and conduct law enforcement monitoring.

Box 12.1 Insect Biodiversity Helps Solve African Wildlife Crimes

Martin H. Villet

Southern African Forensic Entomology Research Laboratory, Department of Zoology and Entomology, Rhodes University, Grahamstown, South Africa.

m.villet@ru.ac.za

Poaching and pollution are crimes akin to murder and poisoning and forensic biologists have a set of vitally important tools to convict perpetrators of such crimes: insect biodiversity.

While police detectives sometimes evaluate insects found at crime scenes to help solve murders, forensic entomologists and anti-poaching investigators can use the biodiversity associated with the decomposition of carcasses to solve poaching crimes. The flies and beetles involved in decomposition are like two hands of a clock, flies ticking along in days and beetles indicating weeks.

At least 14 families of flies, including blow flies (Calliphoridae, Figure 12.A), flesh flies (Sarcophagidae), house flies (Muscidae), cheese skippers (Piophilidae), and soldier flies (Stratiomyidae), breed on carcasses in Africa (Villet, 2017), some of them arriving within an hour of the death of the animal to start the decomposition clock. They lay eggs, which hatch into larvae that eat the carcass and grow at a steady pace. The age of these larvae can be estimated by measuring their size when a carcass is found, providing a minimum time between death and discovery. The larvae eventually mature into pupae that give rise to adult flies; this process can also be calibrated to inform the timeline of evidence.

Over 90 species of beetles from at least 10 families also breed on carcasses (Villet, 2011), and can be used to cross-validate evidence from fly larvae; their longer life cycles provide a record that spans a longer period of weeks. Beetles also arrive in a sequence linked to the decomposition process. This pattern of ecological succession starts with clown beetles (Histeridae) and rove beetles (Staphylinidae) that prey on fly larvae, followed by hide beetles (Dermestidae) and carrion beetles (Silphidae) that feed on the dried tissues left by the fly larvae and, finally, by spider beetles (Anobiidae) and hair beetles (Trogidae) that eat the hair, feathers, scales, skin and cartilage left at the end of decomposition (Villet, 2011). The ecological succession clock covers a month or more, depending on the weather and the characteristics of the carcass.

Insect biodiversity can reveal other forensic details, too (Villet, 2015). For instance, insects that feed on drugged or poisoned animal tissue can bioaccumulate contaminants and provide samples for analysis even after the

Figure 12.A Adult blow flies (*Chrysomya marginalis*) emerging from a savannah elephant carcass, with barn swallows (*Hirundo rustica*, LC) feeding in the background. Photograph by Cameron Richards, CC BY 4.0.

carcass has become too decomposed to analyse directly. Insects may even indicate the presence of these contaminants through their behaviour. Animal remains that have been transported from elsewhere and dumped may harbour insects that indicate the route that was travelled. For example, poached parts of African animals bearing insects from Asia have almost certainly travelled through those areas.

The diversity of insects can also provide evidence of environmental crimes involving pollution, a field called environmental forensics. Lethal levels of pollution will change the structure of insect communities, affecting the most sensitive species first. This insight underlies the certified South African Scoring System for aquatic biomonitoring and related scoring systems developed in other African countries for rating the health of rivers based on the biodiversity of their invertebrate inhabitants (Villet, 2015). Sub-lethal levels of pollutants affect insect reproduction and development, which can be detected in impaired reproduction in adult insects and developmental anomalies in insect larvae (Villet, 2017), including increased asymmetry between the left and right sides of the body (termed *fluctuating asymmetry*) and peculiar developmental patterns. Such research is called environmental forensic entomotoxicology, and it is an exciting new field linking the study of insect biodiversity, environmental damage, and conservation biology.

Box 12.2 Protecting Elephants in a Hostile Region

Lorna Labuschagne

Previous Address:
African Parks, Zakouma National Park,
N'Djaména, Chad.

Current address:
Frankfurt Zoological Society, Serengeti Conservation Project,
Arusha, Tanzania.

✉ *lorna.labuschagne@fzs.org*

"Extinction is forever" is a phrase we hear often, but perhaps don't consider deeply enough. The passenger pigeon (*Ectopistes migratorius*, EX) in North America is a prime example. Early naturalist accounts describe how this species were once so numerous that flocks blackened the sky, and yet it was possible to kill each one. Africa's elephants are currently under similarly huge pressure, especially in Central and West Africa. The well documented story of the elephants of Zakouma National Park in Chad is a good example, where an estimated 4,000 elephants lost their lives between 2002 and 2010 to feed the insatiable demand for ivory (https://www.africanparks.org/the-parks/zakouma).

In the past, a densely-packed elephant herd was an effective defence against horsemen with spears, whose hunting method centred on isolating an individual. With armed groups coming from as far afield as Darfur, Sudan, the modus operandi of the poachers on horseback has not changed much over the past 200 years, except that the spear has been replaced with an automatic rifle. With today's poachers shooting indiscriminately into a tightly packed herd, the result is a devastating massacre. In the past, as many as 60 elephants of all ages were killed in a single attack in Zakouma, with many dying later from festering bullet wounds and small calves ending up lost or orphaned. The trauma of such slaughter on these intelligent animals is hard to imagine and is perhaps best understood by the fact that the Zakouma herds stopped breeding for almost five years. So how does one endeavour to stop such carnage on a free-roaming population, and allow elephants to live a normal life again, especially in an open system where herds range widely?

Each area in Africa is different, and it is important to remember that what works in one area will not always work in another. To address a poaching problem, the situation must be carefully assessed, historical information evaluated, and a "feel" for the threats acquired. It is also important to remember that no anti-poaching team can function without the support staff that keeps them equipped and mobile (mechanics, buyers, bookkeepers, etc.).

Below is a list of key initiatives forming the basis of an efficient protection system for a conservation area (Figure 12.B).

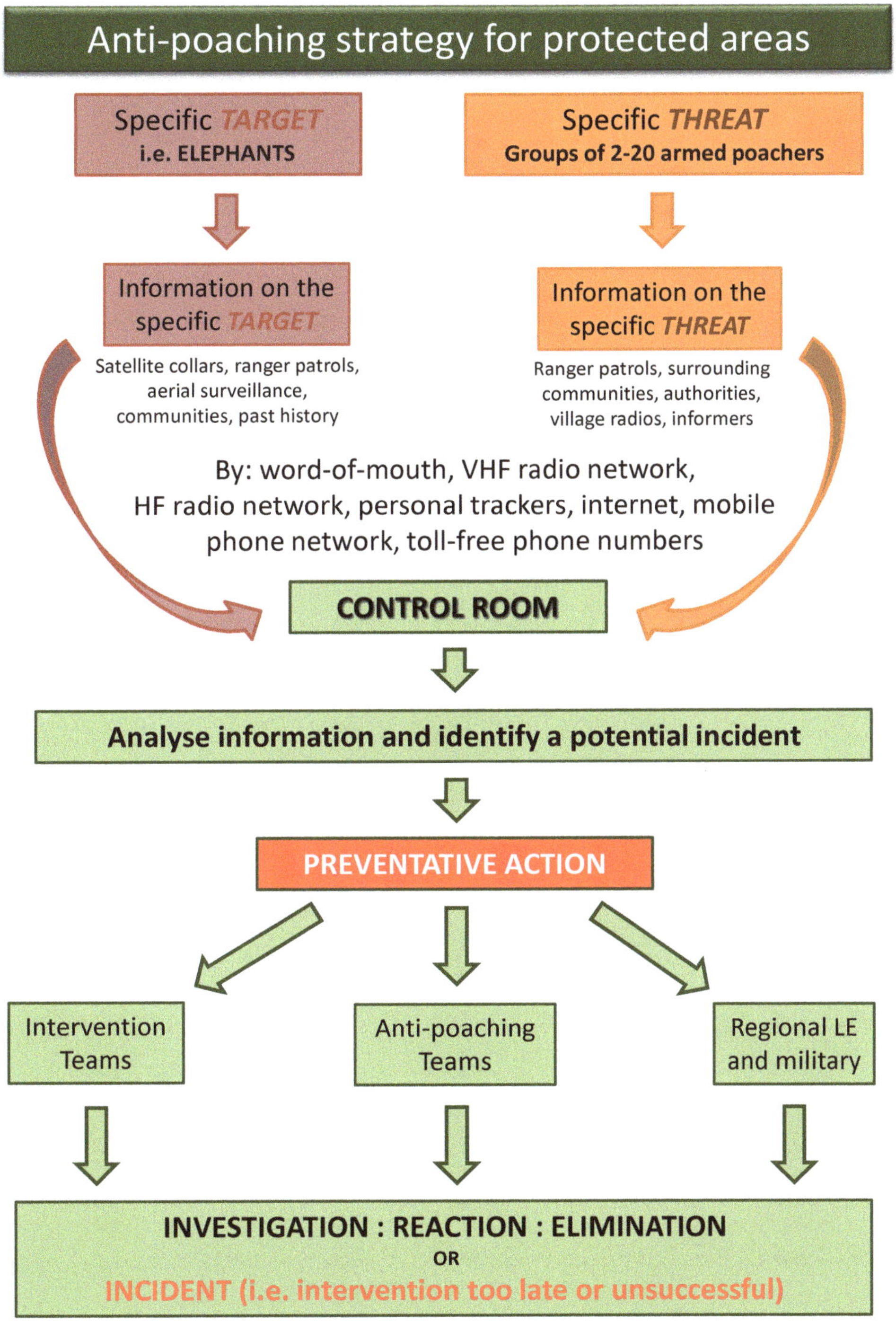

Figure 12.B (2) A flow diagram example of an efficient anti-poaching strategy for protected areas. Diagram by Parc National de Zakouma/MEP-AP, CC BY 4.0.

- *Finding and Tracking Animals*: To protect a species, a good understanding of its movements is needed. Several parks in Africa achieve this by fitting satellite GPS collars on individual elephants in different herds. Animal tracking has in the past been primarily used for research purposes, but today the data is also used to monitor elephant movements and adapt anti-poaching patrols accordingly.

- *Communication*: It is impossible to stop poaching without good communication—be it by mobile or satellite phone, a radio network, or personal trackers with a messaging function. Where the terrain allows it, a digital VHF radio network should ideally be put in place with linked relay stations and portable radios to ensure communication throughout most of the protected area.

- *Central Control Room*: A Central Control Room (CCR) (Figure 12.C) is where all anti-poaching activities are coordinated day and night. Ideally the park should work on a *predictable but unpredictable* anti-poaching system; rangers and their families know when they will be on patrol again (*predictable*), but the day-to-day deployment is *unpredictable* and coordinated by the CCR using all information available, such as real-time elephant movements, in their decision making. Where to deploy patrols should ideally be made by at least three people and the command then given to the Patrol Leaders, who are trained and equipped with GPS units. Where ranger posts or Forward Operational Bases (FOBs) are used, the unpredictable component can also include not knowing which ranger post or FOB they will be sent to, or with whom. A good rotational policy among rangers plays an important part in keeping rangers alert and motivated.

- *Anti-Poaching Monitoring Technology*: Today sophisticated mapping techniques are especially helpful for planning and reporting. Not only that, many are available at no cost on the internet. These tools allow conservation practitioners to monitor, record, and display the movements of animals, anti-poaching patrols, aircraft, and poaching incidents, and plot all of them in different layers on a map or satellite image. This is a key component to monitoring the patrol effort and coverage of an area and should be coordinated by the team in the CCR.

- *Accessibility Throughout the Year*: Although it can be difficult based on the area or budget, conservationists must be inventive in adapting to changing weather and field conditions throughout the year. Airstrips, for example, should be carefully placed to support rangers year-round; accessibility throughout the year is important for logistics but also for evacuation of ill or wounded rangers if needed.

- *Intelligence Gathering*: Not much happens in rural Africa without somebody knowing about it. The key is to get that information to your CCR. Cell phones are increasingly common, and you might also consider having a toll-free phone number. In areas without good GSM coverage, another option is to put in place a "Village Radio" system, where digital VHF radios are programmed in such a way that private calls can be made, allowing for a radio to be installed in a village and still protect the sensitive communications of a park. Having communication in key villages around a park, which speak with the CCR about any illegal activity that they have picked up in the surrounding communities, helps provide much needed security to local people and an important link with the park management.

Figure 12.C An example of a Central Control Centre, which usually operates for 24 hours every day of the year. Photograph by Vanessa Stephen/Parc National de Zakouma/MEP-AP, CC BY 4.0.

As protectors of elephants and other wildlife, park managers must assess the situation, decide what can be done in an area, and try it, but most importantly, park managers must employ an adaptive management strategy (Section 10.2.3) and *continue adapting over time to a changing situation*. Poachers change their strategies, and therefore so must park managers. Ultimately, the goal is to reduce the number of poaching incidents to allow wildlife populations to recover. This can be a daunting task with pitfalls among successes; always remember, keeping field ranger morale high is a key component to ensuring success.

In recent years, some of the world's biggest conservation organisations banded together to form the SMART (Spatial Monitoring and Reporting Tool) partnership. The main goal of the partnership is improving protected areas management, particularly environmental law enforcement, by enabling law enforcement officials and biologists to more easily collate and process information collected during monitoring and patrols. The partnership accomplishes this through the development of a freely available and fully customisable software package that includes real-time mapping, basic analysis tools, and automatic report generation abilities (Wilson et al., 2019). These features allow park managers to be more strategic in their conservation work by allowing them to better plan, evaluate, and implement their activities. SMART is rapidly becoming the standard in environmental law enforcement across the developing world, and several national governments in Africa have already adopted SMART as its environmental crime monitoring platform.

Despite this progress, older technologies are still being used very effectively in law enforcement. To name a few examples, conservationists continue to rely on tools, such as **passive integrated transponder (PIT) tags** (Gibbons and Andrews, 2004) and embedding GPS transmitters (e.g. Christy and Stirton, 2015) to identify and track stolen wildlife products such as rhinoceros horns, elephant tusks, valuable timber, and expensive ornamental plants. Lastly, environmental law enforcement officials continue to rely so heavily on well-trained domestic dogs to detect trafficked wildlife products and apprehend environmental criminals that several organisations now specialise in training dogs for conservation purposes.

12.4 The Limits of Environmental Laws and Regulations

Despite all the efforts to protect biodiversity through laws and regulations, the scale of environmental crimes continues to increase year after year. Today, the US $91–258 billion environmental crime industry is the world's fourth largest illegal enterprise, after drug smuggling, counterfeiting, and human trafficking (Nellemann et al., 2016). Increased financial support for environmental law enforcement could certainty help: current spending to combat environmental crimes, globally estimated at US $20–30 million a year, is a mere drop in the bucket compared to the losses incurred from these same crimes, which are 10,000 times greater (Nelleman et al., 2016). There is also a need to prevent environmental crimes before they happen, given that the damages incurred cannot always be undone by punishing the offenders.

> Environmental crime is the world's fourth largest illegal enterprise, after drug smuggling, counterfeiting, and human trafficking.

An important step towards reducing the scale of environmental crimes is to address the ineffectiveness of environmental regulations. That includes addressing the range of tactics that criminals use to facilitate non-compliance (Chapron et al., 2017), but also ensuring there are mechanisms that remove the incentives for people to engage in environmental crimes. In the following section, we look at some of the most prominent challenges that complicate environmental law enforcement.

12.4.1 Lack of capacity

The foremost reason why environmental laws fail is that authorities often lack the capacity for effective monitoring and enforcement. Lack of capacity is a major problem in the marine fisheries industry due to the size of the oceans and the cost of patrolling them. This is particularly prominent in West Africa's oceanic waters (Figure 12.6), which experience the world's highest levels of illegal and unregulated fishing (Agnew et al., 2009; Gremillet et al., 2015). Apprehending the environmental criminals operating in these and other areas requires resources and manpower, both in short supply.

Figure 12.6 Fishermen from Tanji fishing village in The Gambia fixing their nets after a day out at sea. Intensive harvesting has reached crisis level off West Africa, where a largely unregulated fisheries industry threatens not only fish populations and the people relying on fish, but also seabirds, marine turtles, whales, and dolphins. Photograph by Jan Kruithof, https://www.flickr.com/photos/jankruithof/30426220994, CC BY 2.0.

Lack of capacity is exacerbated by law enforcement officers that turn a blind eye to actions they deem innocuous, or when prosecutors fail to indict criminals out of fear of reprisals. Regulatory controls may also no longer exist or be enforced in regions that experience substantial political instability, economic hardship, civil unrest, or war (Hanson et al., 2009; Beyers et al., 2011). But even in areas where regulatory controls exist, prosecution can be complex, and thus very hard, especially when the illegal activities cross international boundaries and different legal jurisdictions. Such a breakdown of legal mechanisms often leaves natural resources vulnerable to whoever can exploit them.

12.4.2 Conflicting government priorities

Clashing priorities between different government structures complicate the enforcement of environmental laws. We see this when agencies overseeing mining activities issue inappropriate permits because of pressure for economic development, or because of corrupt agreements between businesses and government officials (Mascia and Pailler, 2011). Another example of mixed priorities occurs when a national government gives permission to extractive companies to exploit protected areas (Section 13.7.3) or communal lands without first consulting and obtaining local input and consent. Such government-sanctioned violations are generally very difficult to prosecute and require an active and caring citizenry to take their governments to task.

One of the most popular methods for citizens to make themselves heard is activist activities, such as public protests. Concerned individuals can also launch petitions on websites such as https://www.change.org, http://www.greenpeace.org, and https://www.avaaz.org. There is even a website for whistle-blowers (https://wildleaks.org) who want to report environmental crimes anonymously. Another positive development is the growing number of successful lawsuits that concerned citizens and environmental justice organisations have brought against their governments for environmental violations (e.g. CER, 2017; Yende, 2017). The Kenyan conservation organisation Wildlife Direct has taken this a step further; they are keeping citizens informed about lawsuits involving environmental crimes through a website dedicated to tracking and reporting on such cases (https://wildlifedirect.org/legal-program-3).

12.4.3 Informal economies, traditional activities, and the law

Law enforcement can at times be counterproductive. This is true especially in areas where the separation between informal/unreported and illegal activities is blurred. For example, traditional people who graze their livestock, collect medicinal plants, or hunt and trap animals in protected areas that were established on ancestral land seldom have criminal intent. But because formal law systems seldom account for these informal activities, those people are engaging in illegal activities. Similarly, confusing terminology may also lead to unintended conflict. For example, in some parts of Cameroon, the cultural definition of a hunter describes someone who owns a gun and makes a living from hunting animals (Hofner et al., 2018). In this context, some people consider it within the law to trap animals with snares, or even to make sporadic "hunting" trips into a protected area where hunting is forbidden, given that it is not for commercial purposes.

When dealing with vulnerable peoples whose livelihoods are threatened, an approach that involves sensitivity and compassion generally offers more effective and enduring resolutions. Many conservation initiatives have not only failed but have also created long-lasting negative attitudes by preventing traditional peoples from sustaining their livelihoods. Before implementing new regulations, governments should carefully consider if they would disrupt livelihoods. If so, it might be wise to consider if some form of sustainable utilisation isn't possible. For example, while pastoralist activities in Tanzania's

> When dealing with vulnerable people, conservation initiatives must also consider impacts on livelihoods and the potential for sustainable utilisation

Ngorongoro Conservation Area may lead to wildlife declines (Boone et al., 2002, but see Ogutu et al., 2016), conservation authorities decided on a suitable compromise by allowing Maasai herders to graze their livestock in the area on the condition that they exit daily. (See also Section 13.5.2 for discussion on **zoning**.)

Despite best intentions, sustainable utilisation is not always possible, and the actions of people engaging in environmentally detrimental activities are not compatible with conservation goals. In such cases, it is important to implement controls that enable those people to transition toward sustainable activities. Failing that, conservation activities may unintentionally force the affected people to resort to illegal activities such as poaching out of desperation to obtain food and income. When banning previously-allowed activities that have become unsustainable, it helps to provide affected people with start-up resources and market access to help them comply with new restrictions while also meeting broader societal needs. For example, to fulfil income and nutritional needs when bushmeat harvesting is banned, it might be necessary to help hunters transition to farming with animals that reproduce quickly in captivity. Raising poultry can be a good alternative to bushmeat because chickens grow quickly, provide eggs, feed on insect pests, and need little land for maintenance. Farming with locally-adapted wildlife, such as large snails (e.g. Carvalho et al., 2015) and cane rats (*Thryonomys swinderianus*, LC) (e.g. van Vliet et al., 2016), has also proven to be a profitable and sustainable alternative to the bushmeat trade (for a review on wildlife farming for conservation, see Tensen, 2016). Initially, many people may resist the risks involved in leaving behind familiar activities. It is, therefore, important to explain carefully the reasoning behind those changes (e.g. "bushmeat hunting drives away tourist dollars", Rogan et al., 2017). It may also be beneficial to enable the affected individuals to travel to areas where they can see first-hand how more sustainable activities can benefit local people.

> When banning activities which have become unsustainable, it helps to provide start-up resources and market access to help affected people comply with new restrictions.

12.4.4 Trade embargoes and sanctions

The basic premise of CITES is that, so long as participating countries abide by agreed-upon regulations, trade involving species of concern will not be stopped, only monitored. However, when agreements are not met, or compliance falls short, then trade is banned in part or in whole. For example, in early 2016, following failures to comply with international trade regulations, CITES instituted blanket suspensions on trade of all CITES-regulated products against 14 African countries (CITES, 2016). Another example from 2016: after various high-profile environmental crimes that involved CITES-regulated species, the USA afforded protection to lions and elephants under their *Endangered Species Act* (https://www.fws.gov/endangered). Trophy hunters from the USA now face significant regulatory and logistical barriers which has all but eliminated hunting of these species for wealthy American hunters, threatening a US

$500 million per year industry that supports over 53,000 jobs and protects over 1.4 million km^2 of land (SCIF, 2015). In both these examples, businesses operating within the law are unfortunately also impacted.

> Many formerly destructive companies are now voluntarily pursuing opportunities to prove, through special certifications, that their products are harvested responsibly and sustainably.

To avoid scenarios such as these, it is much more advantageous for pressure to mount from *within* non-complying countries and industries before outside pressure takes effect. To that end, many formerly destructive companies are now voluntarily pursuing opportunities to prove, through special certifications, that their products are harvested responsibly and sustainably. Four prominent certification agencies operating in Africa are the Forest Stewardship Council (FSC) which sets guidelines for the responsible management of forests, the Marine Stewardship Council (MSC) which sets standards for sustainable fisheries, the Roundtable on Sustainable Palm Oil (RSPO) which promotes sustainable production of palm oil, and the Rainforest Alliance which promotes sustainable agriculture (For a more complete treatment of sustainability standards, see https://www.isealalliance.org, https://www.evidensia.eco, and https://www.iisd.org/ssi). Given that most certification schemes were established relatively recently, they are not without their flaws, but collaborations with conservation biologists (e.g. Christian et al., 2013; WWF, 2013) play a major role in ensuring continued improvements.

12.5 Conclusion

Challenging problems are often solved when a diverse group of people from different backgrounds and viewpoints come together for mutual benefit. Solving environmental crime is no different: history shows that building trust and respect for fellow human beings and future generations is more powerful than the threat of force. Effective law enforcement efforts, which require multi-level cooperation from international structures down to individual people (Box 12.3), are often characterised by partnerships between wildlife agencies and local people (Biggs et al., 2016). These partnerships may take the form of environmental educational campaigns to incentivise conservation action and sustainable resource use (Abensperg-Traun, 2009). Cooperation between different individuals at the grassroots level is another very effective means to ensure sustainable resource use. These efforts may take many forms, but they begin with individual and group decisions to prevent the destruction of habitats and species to preserve something of perceived economic, cultural, biological, scientific, or recreational value. Through collaboration and cooperation, both from the grassroots level up, and governments down, conservation biologists can achieve their goals, by ensuring free and fair treatment of all citizens regardless of their diverse and sometimes opposing viewpoints on natural resource management.

Box 12.3 Thoughts on Poaching and Illegal Wildlife Trafficking in Sub-Saharan Africa

Tamar Ron

Biodiversity Conservation consultant.

✉ *tamarron@bezeqint.net*

The two main causes of the alarmingly rapid wildlife loss in Africa today are: (1) unsustainable use of land and natural resources, mostly related to decision making that does not prioritise conservation considerations; and (2) overharvesting of wild animals and plants through poaching and illegal logging.

Poaching and illegal logging can be locally driven for subsistence use, resulting from poverty; for lack of other protein, energy, and income sources; and, at times, by the intensifying impact of armed conflict or post-conflict situations. In a country like Angola, for example, where war has significantly diminished wildlife populations, the continuous impact of intensive bushmeat poaching may well lead to the extinction of remnant core populations of species that have initially survived the armed conflict. It is severely impacting even the iconic and endemic giant black sable (*Hippotragus niger variani*, CR). Bushmeat poaching is typically unselective; it targets mainly large and medium sized mammals, but also smaller mammals, birds, reptiles and freshwater fish. Similarly, illegal logging for wood, charcoal, or slash-and-burn-based cultivation, results in rapid and irreversible biodiversity degradation and loss.

In contrast, commercial poaching is often driven by international trafficking, whether of live animals (Figure 12.D) or animal and plant products, from source countries to destination markets. Illegal traffickers are often well financed, sophisticated, and involved in other forms of serious crime, at times even in terrorism (e.g. Nellemann et al., 2014). Illegal wildlife trade is selective and forms an imminent threat to iconic species with commercial value, such as elephants, rhinoceros, big cats, great apes, pangolins, sea turtles, parrots, and rosewood, to name a few.

While the core causes, nature, and impacts of subsistence and commercial poaching are different, local community members are a centrepiece of both. Their intimate acquaintance with local wildlife and their habitats is vital; therefore, effective wildlife protection is based on their active engagement. After many generations of alienation, local community members must be included in the decision-making process for the sustainable management of natural resources. The Namibian conservancies (Section 14.3) offer a model of success in engaging communities in conservation, by protecting their rights, securing their fair benefits as the resource owners and not merely as workers, and providing them with adequate training.

Figure 12.D In 2004, Angolan authorities confiscated Massamba, an orphaned chimpanzee (Pan troglodytes, EN), from poachers, as part of a crackdown on the illegal wildlife trade. Photograph by Tamar Ron, CC BY 4.0.

Wildlife crime is a serious threat to biodiversity and while local influences should be recognized and addressed, these crimes should be treated as a global enforcement priority. Continuous and coordinated national efforts of all relevant sectors, and with global cooperation, are essential to success. Such efforts must include: (1) improving awareness at all levels; (2) adequate legislation and policies; (3) realistically deterring punishments and forfeiture of wildlife crime revenues; (4) strengthening enforcement and intelligence capacities in all source, destination, and transit countries; (5) addressing governance challenges; and (6) trying to eradicate the markets, or at least to reduce the demand for illicit wildlife products. The poaching drivers, international crime syndicates, and middlemen, must be targeted. Enforcement focused mostly at the poachers' level can never achieve the desired results. Often, they are no less victims than their target species. Further, if the world wants to protect iconic species, it cannot be expected that the burden of their conservation, and human-wildlife-conflict damages, should fall solely on those communities that happen to share their habitat. The global effort and substantive support required should not be viewed as a contribution, but as mutual responsibility for achieving a global goal.

Lastly, we may have to accept that total eradication of wildlife crime may not be achievable. There is no magic remedy, nor a single perpetrator. Nevertheless, integrated efforts to reduce these crimes must be strengthened at all levels. We simply cannot give up on our fellow species.

12.6 Summary

1. Environmental laws and regulations are implemented at three different levels: international treaties, national laws, and local laws. Each of these levels is intricately connected: international treaties influence—but also depend on—national laws to succeed, while national laws are guided by local needs and customary laws that have been in place for generations.
2. International treaties and conventions provide frameworks for countries to cooperate on protecting species, ecosystems, and other levels of biodiversity. International agreements are important because: (1) many species migrate and disperse across borders, (2) ecosystems do not follow administrative boundaries, (3) pollution spreads by air and water across regions and around the globe, (4) many biological products are traded internationally, and (5) some environmental problems require global cooperation and coordination.
3. National governments protect biodiversity by regulating natural resource use and preventing pollution. Subsidies and tax incentives can also be used to reward citizens and businesses that engage in environmentally-responsible behaviours.
4. There is a constant need to evaluate environmental laws and regulations to ensure that they are enforced, violators are prosecuted, and new laws and amendments are passed as and when needed. Law enforcement agencies and scientists are constantly looking for new ways to address enforcement shortcomings.
5. When banning environmentally detrimental activities that are not compatible with conservation goals, it is critical to help the affected people to transition toward sustainable activities. Failing that, conservation activities may unintentionally force those people to resort to illegal activities such as poaching out of desperation to obtain food and income.

12.7 Topics for Discussion

1. Identify two or three environmental laws regulating human activities in your country. Discuss how these national laws relate to international laws and local laws.
2. A wide range of international and national laws protect threatened species and ecosystems. Why do species and ecosystems covered by such laws not recover?
3. Trapping birds and small mammals are traditional activities for boys and young men across Africa, and traditional hunting is seen as a competitive

sport to sharpen the mind. These activities are also important to meet local income and nutritional needs. How should we deal with the hunting of threatened species outlawed by national governments, but encouraged by local traditions? Should people be allowed to hunt for bushmeat, even if it includes great apes and other rare species, to pay for necessities, such as schooling and medicine?

12.8 Suggested Readings

Biggs, D., R. Cooney, D. Roe, et al. 2016. Developing a theory of change for a community-based response to illegal wildlife trade. *Conservation Biology* 31: 5–12. https://doi.org/10.1111/cobi.12796 Four pathways to community-level action against wildlife crimes.

Chapron, G., Y. Epstein, A. Trouwborst, et al. 2017. Bolster legal boundaries to stay within planetary boundaries. *Nature Ecology and Evolution* 1: 0086. https://doi.org/10.1038/s41559-017-0086 Various tactics are used to make biodiversity laws less effective.

Christian, C., D. Ainley, M. Baileyet al. 2013. A review of formal objections to Marine Stewardship Council fisheries certifications. *Biological Conservation* 161: 10–17. https://doi.org/10.1016/j.biocon.2013.01.002 Strengths and weaknesses of schemes certifying sustainability.

Christy, B., and B. Stirton. 2015. How killing elephants finances terror in Africa. *National Geographic.* http://on.natgeo.com/1I5N2aO. A riveting account of two journalists tracking poaching routes.

Etiendem, D., L. Hens, and Z. Pereboom. 2011. Traditional knowledge systems and the conservation of Cross River gorillas: A case study of Bechati, Fossimondi, Besali, Cameroon. *Ecology and Society* 16: 22. http://doi.org/10.5751/ES-04182-160322 Traditional belief systems can be used to promote conservation.

Harfoot, M., S.A.M. Glaser, D.P. Tittensor, et al. 2018. Unveiling the patterns and trends in 40 years of global trade in CITES-listed wildlife. *Biological Conservation* 223: 47–57. https://doi.org/10.1016/j.biocon.2018.04.017 Even legal wildlife trade is having a massive impact on biodiversity.

Hilborn, R., P. Arcese, M. Borner, et al. 2006. Effective enforcement in a conservation area. *Science* 314: 1266. https://doi.org/10.1126/science.1132780 Enforcement of environmental laws really makes a difference.

Plumptre, A.J., R.A. Fuller, A. Rwetsiba, et al. 2014. Efficiently targeting resources to deter illegal activities in protected areas. *Journal of Applied Ecology* 51: 714–25. https://doi.org/10.1111/1365-2664.12227 Prioritising law enforcement efforts can save money.

Tulloch, A.I.T., N. Auerbach, S. Avery-Gomm, et al. 2018. A decision tree for assessing the risks and benefits of publishing biodiversity data. *Nature Ecology and Evolution* 2: 1209–17. https://doi.org/10.1038/s41559-018-0608-1 Publishing biodiversity data has risks that should be considered

Wasser, S.K., L. Brown, C. Mailand, et al. 2015. Genetic assignment of large seizures of elephant ivory reveals Africa's major poaching hotspots. *Science* 349: 84–87. https://doi.org/10.1126/science.aaa2457 Genetics can be used to identify poaching hotspots.

Bibliography

Abensperg-Traun, M. 2009. CITES, sustainable use of wild species and incentive-driven conservation in developing countries, with an emphasis on southern Africa. *Biological Conservation* 142: 948–63. http://doi.org/10.1016/j.biocon.2008.12.034

Agnew, D.J., J. Pearce, G. Pramod, et al., 2009. Estimating the worldwide extent of illegal fishing. *PLoS ONE* 4: e4570. https://doi.org/10.1371/journal.pone.0004570

Alexander, J., C.T. Downs, M. Butler, et al. 2019. Stable isotope analyses as a forensic tool to monitor illegally traded African grey parrots. *Animal Conservation* 22: 134–43. https://doi.org/10.1111/acv.12445

Astaras, C., J.M. Linder, P. Wrege, et al. 2017. Passive acoustic monitoring as a law enforcement tool for Afrotropical rainforests. *Frontiers in Ecology and the Environment* 15: 233–34. https://doi.org/10.1002/fee.1495

Beyers, R.L., J.A. Hart, A.R.E. Sinclair, et al. 2011. Resource wars and conflict ivory: The impact of civil conflict on elephants in the Democratic Republic of Congo-the case of the Okapi Reserve. *PloS ONE* 6: e27129. https://doi.org/10.1371/journal.pone.0027129

Biamah, E.K., J. Kiio, and B. Kogo. 2013. Chapter 18–Environmental Impact assessment in Kenya. *Developments in Earth Surface Processes* 16: 237–64. https://doi.org/10.1016/B978-0-444-59559-1.00018-9

Biggs, D., F. Courchamp, R. Martin, et al. 2013. Legal trade of Africa's rhino horns. *Science* 339: 1038–39. http://doi.org/10.1126/science.1229998

Biggs, D., R. Cooney, D. Roe, et al. 2016. Developing a theory of change for a community-based response to illegal wildlife trade. *Conservation Biology* 31: 5–12. https://doi.org/10.1111/cobi.12796

Boone, R.B., M.B. Coughenour, K.A. Galvin, et al. 2002. Addressing management questions for Ngorongoro Conservation Area, Tanzania, using the SAVANNA modelling system. *African Journal of Ecology* 40: 138–50. https://doi.org/10.1046/j.1365-2028.2002.00357.x

Carvalho, M., F. Rego, J.M. Palmeirim, et al. 2015. Wild meat consumption on São Tomé Island, West Africa: Implications for conservation and local livelihoods. *Ecology and Society* 20: 27. http://doi.org/10.5751/ES-07831-200327

Catry, P., C. Barbosa, B. Indjai, et al. 2002. First census of the green turtle at Poilão, Bijagós Archipelago, Guinea-Bissau: The most important nesting colony on the Atlantic coast of Africa. *Oryx* 36: 400–03. https://doi.org/10.1017/S0030605302000765

CER (Centre for Environmental Rights). 2017. Victory in SA's first climate change court case. *CER News*. https://cer.org.za/news/victory-in-sas-first-climate-change-court-case

Challender, D.W.S., and D.C. MacMillan. 2014. Poaching is more than an enforcement problem. *Conservation Letters* 7: 484–94. https://doi.org/10.1111/conl.12082

Chapron, G., Y. Epstein, A. Trouwborst, et al. 2017. Bolster legal boundaries to stay within planetary boundaries. *Nature Ecology and Evolution* 1: 0086. https://doi.org/10.1038/s41559-017-0086

Christian, C., D. Ainley, M. Bailey, et al. 2013. A review of formal objections to Marine Stewardship Council fisheries certifications. *Biological Conservation* 161: 10–17. https://doi.org/10.1016/j.biocon.2013.01.002

Christy, B., and B. Stirton. 2015. How killing elephants finances terror in Africa. *National Geographic*. http://on.natgeo.com/1I5N2aO

CITES. 2016. A good week for wildlife — CITES meeting takes bold decisions in fight against illicit wildlife trafficking and on ensuring sustainability. *CITES News*. https://cites.org/eng/news/pr/cites_meeting_takes_bold_decisions_in_fight_against_illicit_wildlife_trafficking_and_on_sustainability_19012016

Dana, G.V., A.R. Kapuscinski, and J.S. Donaldson. 2012. Integrating diverse scientific and practitioner knowledge in ecological risk analysis: A case study of biodiversity risk assessment in South Africa. *Journal of Environmental Management* 98: 134–46. https://doi.org/10.1016/j.jenvman.2011.12.021

Etiendem, D., L. Hens, and Z. Pereboom. 2011. Traditional knowledge systems and the conservation of Cross River gorillas: A case study of Bechati, Fossimondi, Besali, Cameroon. *Ecology and Society* 16: 22. http://doi.org/10.5751/ES-04182-160322

Gibbons, W.J., and K.M. Andrews. 2004. PIT tagging: Simple technology at its best. *Bioscience* 54: 447–54. https://doi.org/10.1641/0006-3568(2004)054[0447:PTSTAI]2.0.CO;2

Gómez-Baggethun, E., E. Corbera, and V. Reyes-García. 2013. Traditional ecological knowledge and global environmental change: Research findings and policy implications. *Ecology and Society* 18: 72. http://doi.org/10.5751/ES-06288-180472

Gremillet, D., C. Peron, P. Provost, et al. 2015. Adult and juvenile European seabirds at risk from marine plundering off West Africa. *Biological Conservation* 182: 143–47. https://doi.org/10.1016/j.biocon.2014.12.001

Hanson, T, T.M. Brooks, G.A.B. da Fonseca, et al. 2009. Warfare in Biodiversity Hotspots. *Conservation Biology* 23: 578–87. https://doi.org/10.1111/j.1523-1739.2009.01166.x

Heinrich, S., T.A. Wittmann, T.A.A. Prowse, et al. 2016. Where did all the pangolins go? International CITES trade in pangolin species. *Global Ecology and Conservation* 8: 241–53. https://doi.org/10.1016/j.gecco.2016.09.007

Hilborn, R., P. Arcese, M. Borner, et al. 2006. Effective enforcement in a conservation area. *Science* 314: 1266. https://doi.org/10.1126/science.1132780

Hofner, A.N., C.A. Jost Robinson, and K.A.I. Nekaris. 2018. Preserving Preuss's red colobus (*Piliocolobus preussi*): An ethnographic analysis of hunting, conservation, and changing perceptions of primates in Ikenge-Bakoko, Cameroon. *International Journal of Primatology* 39: 895–917. https://doi.org/10.1007/s10764-018-0020-3

IUCN. 2019. *The IUCN Red List of Threatened Species*. http://www.iucnredlist.org

Keeley, J., and I. Scoones. 2014. *Understanding Environmental Policy Processes: Cases from Africa*. New York: Earthscan).

Mascia, M.B., and S. Pailler. 2011. Protected area downgrading, downsizing, and degazettement (PADDD) and its conservation implications. *Conservation Letters* 4: 9–20. https://doi.org/10.1111/j.1755-263X.2010.00147.x

Mauro, F., and P.D. Hardison. 2000. Traditional knowledge of indigenous and local communities: International debate and policy initiatives. *Ecological Applications* 10: 1263–69. https://doi.org/10.1890/1051-0761(2000)010[1263:TKOIAL]2.0.CO;2

Mbata, K.J., E.N. Chidumayo, and C.M. Lwatula. 2002. Traditional regulation of edible caterpillar exploitation in the Kopa area of Mpika district in northern Zambia. *Journal of Insect Conservation* 6: 115–30. https://doi.org/10.1023/A:1020953030648

NAN (News Agency of Nigeria). 2015. Nigeria: Govt sanctions 26 mining company for non-compliance with environmental laws. *All Africa Media*. http://allafrica.com/stories/201506180356.html

Nellemann, C., R. Henriksen, A. Kreilhuber, et al. 2016. *The rise of environmental crime—A growing threat to natural resources, peace, development and security* (Cambridge: UNEP). http://wedocs.unep.org/handle/20.500.11822/7662

Nellemann, C., R. Henriksen, P. Raxter, et al. 2014. *The environmental crime crisis—Threats to sustainable development from illegal exploitation and trade in wildlife and forest resources* (Nairobi: UNEP; Arendal: GRIDArendal). http://www.grida.no/publications/178

ODPP (Office of the Director of Public Prosecution). 2017. New act leads to reduction in wildlife crimes. *ODPP News*. http://www.odpp.go.ke/new-act-leads-to-reduction-in-wildlife-crimes

Ogutu, J.O., H.-P. Piepho, M.Y. Said, et al. 2016. Extreme wildlife declines and concurrent increase in livestock numbers in Kenya: What are the causes? *PLoS ONE* 11: e0163249. https://doi.org/10.1371/journal.pone.0163249

Plumptre, A.J., R.A. Fuller, A. Rwetsiba, et al. 2014. Efficiently targeting resources to deter illegal activities in protected areas. *Journal of Applied Ecology* 51: 714–25. https://doi.org/10.1111/1365-2664.12227

Poole, C.M., and C.R. Shepherd. 2016. Shades of grey: The legal trade in CITES-listed birds in Singapore, notably the globally threatened African grey parrot *Psittacus erithacus*. *Oryx* 51: 411–17. https://doi.org/10.1017/S0030605314000234

Rademeyer, J. 2012. *Killing for Profit: Exposing the Illegal Rhino Horn Trade* (Johannesburg: Zebra Press).

Retief, K., A.G. West, and M.F. Pfab. 2014. Can stable isotopes and radiocarbon dating provide a forensic solution for curbing illegal harvesting of threatened cycads? *Journal of Forensic Sciences* 59: 1541–51. https://doi.org/10.1111/1556-4029.12644

Rogan, M.S., P.A. Lindsey, C.J. Tambling, et al. 2017. Illegal bushmeat hunters compete with predators and threaten wild herbivore populations in a global tourism hotspot. *Biological Conservation* 210: 233–42. https://doi.org/10.1016/j.biocon.2017.04.020

RSF (Reporters Sans Frontiers) 2015. *Hostile climate for environmental journalists* (Paris: RSF). https://rsf.org/sites/default/files/rapport_environnement_en.pdf

Sands, P., and J. Peel. 2012. *Principles of International Environmental Law* (Cambridge: Cambridge University Press). https://doi.org/10.1017/9781108355728

Schuster, R., E.A. Law, A.D. Rodewald, et al. 2017. Tax shifting and incentives for biodiversity conservation on private lands. *Conservation Letters* 11: e12377. https://doi.org/10.1111/conl.12377

SCIF (Safari Club International Foundation). 2015. *The economic contributions of hunting-related tourism in Eastern and Southern Africa* (Tuscon: SCIF). http://safariclubfoundation.org/wp-content/uploads/2016/06/Southwick-Associates-2015_FINAL.pdf

Sileshi, G., P. Nyeko, P. Nkunika, et al. 2009. Integrating ethno-ecological and scientific knowledge of termites for sustainable termite management and human welfare in Africa. *Ecology and Society* 14: 48. https://www.ecologyandsociety.org/vol14/iss1/art48

Stevens, C. 2017. South Africa gets first biodiversity tax incentive. *Birdlife South Africa News*. https://www.birdlife.org/africa/news/south-africa-gets-first-biodiversity-tax-incentive

Steyn, P. 2015. Big illegal market for little critters. *National Geographic*. http://on.natgeo.com/1ROc68B

Tensen, L. 2016. Under what circumstances can wildlife farming benefit species conservation? *Global Ecology and Conservation* 6: 286–98. https://doi.org/10.1016/j.gecco.2016.03.007

Terer, T., A. Muthama Muasya, F. Dahdouh-Guebas, et al. 2012. Integrating local ecological knowledge and management practices of an isolated semi-arid papyrus swamp (Loboi, Kenya) into a wider conservation framework. *Journal of Environmental Management* 93: 71–84. https://doi.org/10.1016/j.jenvman.2011.08.005

Tranquilli, S., M. Abedi-Lartey, F. Amsini, et al. 2012. Lack of conservation effort rapidly increases African great ape extinction risk. *Conservation Letters* 5: 48–55. https://doi.org/10.1111/j.1755-263X.2011.00211.x

van Vliet, N., D. Cornelis, H. Beck, et al. 2016. Meat from the wild: Extractive uses of wildlife and alternatives for sustainability. *Current Trends in Wildlife Research* 1: 225–65. http://doi.org/10.1007/978-3-319-27912-1_10

van Wilgen, B.W., B. Reyers, D.C. le Maitre, et al. 2008. A biome-scale assessment of the impact of invasive alien plants on ecosystem services in South Africa. *Journal of Environmental Management* 89: 336–49. https://doi.org/10.1016/j.jenvman.2007.06.015

Villet, M.H. 2011. African carrion ecosystems and their insect communities in relation to forensic entomology. *Pest Technology* 5: 1–15.

Villet, M.H. 2015. History, accomplishments, and challenges of forensic entomology in Africa. In: *Forensic Entomology: International Dimensions and Frontiers*, ed. by J.K. Tomberlin and M.E. Benbow (Boca Raton: Taylor and Francis).

Villet, M.H. 2017. Forensic significance. In: *Manual of Afrotropical Diptera*, ed. by A.H. Kirk-Spriggs and B. Muller (Pretoria: SANBI). http://afrotropicalmanual.org

Walters, G., J. Schleicher, O. Hymas, et al. 2015. Evolving hunting practices in Gabon: Lessons for community-based conservation interventions. *Ecology and Society* 20: 31. http://doi.org/10.5751/ES-08047-200431

Wasser, S.K., C. Mailand, R. Booth, et al. 2007. Using DNA to track the largest ivory seizure since the 1989 trade ban. *Proceedings of the National Academy of Sciences* 104: 4228–33. https://doi.org/10.1073/pnas.0609714104

Wasser, S.K., L. Brown, C. Mailand, et al. 2015. Genetic assignment of large seizures of elephant ivory reveals Africa's major poaching hotspots. *Science* 349: 84–87. https://doi.org/10.1126/science.aaa2457

Williamson, J., O. Maurin, S.N.S. Shiba, et al. 2016. Exposing the illegal trade in cycad species (Cycadophyta: Encephalartos) at two traditional medicine markets in South Africa using DNA barcoding. *Genome* 59: 771–81. https://doi.org/10.1139/gen-2016-0032

Wilson, J.W., R. Bergl, L.J. Minter, et al. 2019. The African elephant *Loxodonta* spp. conservation programmes of North Carolina Zoo: Two decades of using emerging technologies to advance in situ conservation efforts. *International Zoo Yearbook* 53: *in press*. https://doi.org/10.1111/izy.12216

WWF. 2013. *WWF FAQ on the RSPO P&C review*. http://wwf.panda.org/rspo/pcreview

WWF. 2016. *Ranger insurance* (Washington: WWF). http://tigers.panda.org/reports/ranger-insurance-report-2016

WWF. 2017. Court slams half a million dollar fine on ivory traffickers. *WWF News*. http://www.wwf-congobasin.org/news/?301593/Court-Slams-half-a-million-dollar-fine-on-ivory-traffickers

Yende, S.S. 2017. Concourt strikes a blow for nature. *Fin24*. https://www.fin24.com/economy/concourt-strikes-a-blow-for-nature-20170818

Zengeya, T., P. Ivey, D.J. Woodford, et al. 2017. Managing conflict-generating invasive species in South Africa: Challenges and trade-offs. *Bothalia* 47: 1–11. https://doi.org/10.4102/abc.v47i2.2160

13. The Importance of Protected Areas

Sea anemones and cold-water corals are among the species that enjoy protection in the 1000 km^2 Table Mountain National Park Marine Protected Area (MPA), South Africa. The MPA is divided into several no-take zones which act as breeding and nursery areas for marine life, as well as zones where harvesting is allowed under certain conditions. Photograph by Andrew Beard, https://www.flickr.com/photos/andrewbeard/13268749044, CC BY 2.0.

https://doi.org/10.11647/OBP.0177.13

With its rich biological diversity, Africa plays a critical role in global conservation efforts. Yet, many of the continent's most threatened species and ecosystems continue to face an uncertain future. In light of increasing human populations that need an increasing amount of natural resources each year, safeguarding the region's biodiversity is a major challenge. One of the best ways to meet this challenge is to designate protected areas—regions where human activities are regulated or, at times, even prohibited by law.

Biodiversity conservation is most effective when we maintain healthy, functioning, and intact ecosystems. Although it is true that many species and populations live outside protected areas, and some wildlife populations (Craigie et al., 2010) and natural communities (Lindsey et al., 2014) are declining even when protected, well-managed protected areas continue to be the most effective method to safeguard biodiversity (Brooks et al., 2009; Ihwagi et al., 2015). Illustrating the point, a global meta-analysis, which included 952 locations across Sub-Saharan Africa, found that wildlife populations are 15% larger and species richness is 11% higher inside protected areas compared to populations directly outside (Gray et al., 2016). Differences may be even starker at individual sites: tea fields on Tanzania's East Usambara Mountains held only 8% of the bird species present in the adjacent protected forest (Newmark, 2008), while some vultures in Eswatini now exclusively breed in protected areas (Monadjem and Garcelon, 2005). Studies from Tanzania have also shown how wildlife in protected areas are more resilient to climate change (Beale et al., 2013a), because habitat loss and fragmentation occur at four times their respective rates outside protected areas relative to inside them (see also Potapov et al., 2017). Consequently, until such a time that we can live more sustainably on unprotected lands, protected areas will remain an important cornerstone in our efforts to protect biodiversity. But how do we know what or where to protect, how much to protect, or how to effectively manage a protected area?

> Protecting existing wild populations in their natural ecosystems not only protects ecological communities and interactions, but also natural processes and ecosystem services.

13.1 Establishing Protected Areas

A **protected area** is "a clearly defined geographical space (Figure 13.1), recognised, dedicated and managed through legal or other effective means to achieve the long-term conservation of nature with associated ecosystem services and cultural values" (Dudley, 2008). Given this broad definition, it comes as no surprise that governments, organisations, and local communities use a variety of mechanisms to establish protected areas. The most popular of these mechanisms are:

- Government action, which can occur at a national, regional, or local level.
- Community-based initiatives by local people and traditional groups.

- Land purchases and holdings by private individuals and organisations.
- Protected areas established through co-management agreements.
- Development of biological field stations or marine laboratories.

Figure 13.1 Land clearing and agricultural development pushes right up to the eastern edge of Bwindi Impenetrable National Park, Uganda. It is important for protected areas—and zones within those areas—to have clearly defined boundaries to avoid confusion on where and how human activities are regulated. Photograph by Jason Houston/USAID, https://www.flickr.com/photos/usaid-biodiversity-forestry/38484053220, CC0.

13.1.1 Government protected areas

Government actions are generally considered the most secure form of protection because they involve passage of laws and buy-in from multiple levels of society. Of course, legislation establishing a protected area does not guarantee that the species and ecosystems therein are adequately preserved. Small populations, especially those living in small protected areas, often require active management (Section 8.7.5) to ensure their continued survival. Another concern is that laws protecting national parks and other wildlife sanctuaries are not strictly enforced, leading to so-called paper parks—parks that appear on official government lists, but with wildlife monitoring, law enforcement, and ecosystem management lacking on the ground (Laurance et al., 2012). However, government-sanctioned protected areas do lay a solid foundation for partnerships among governments, international conservation organisations, multinational banks, research institutes, and educational organisations. Such partnerships can bring together funding, training, and scientific and management expertise to maximise the potential value of those protected areas.

13.1.2 Community conserved areas

In many areas, local people already protect biological communities, forests, wildlife, rivers, and coastal waters in the vicinity of their homes. Protection on these **community conserved areas** is enforced by village elders and councils to ensure the sustainable use of natural resources such as food supplies and drinking water. Natural areas have also been set aside by royal families and churches to provide a space for spiritual activities (see Box 2.1) and sustainable harvesting of medicinal plants (see Box 5.2). Because human activities are highly restricted in these sacred spaces, they provide an important refuge for biodiversity. Today, an increasing number of traditional communities link cultural advocacy directly to conservation through the establishment of protected areas on their lands as a safeguard against developments that would compromise their way of living. Other communities establish protected areas to attract tourists and ensure the protection of special wildlife. One such example is the Iyondji Bonobo Community Reserve in the DRC, which protects bonobos (*Pan paniscus*, EN), forest elephants (*Loxodonta cyclotis*), as well as one of the world's most enigmatic birds, the Congo peafowl (*Afropavo congensis*, VU) (Dupain et al., 2013).

Traditional communities may link cultural advocacy to conservation by establishing protected areas as a safeguard against developments that would compromise their way of living.

13.1.3 Privately protected areas

Over the last few decades, many African countries have adopted a more Western form of land tenure under private ownership. Wealthy individuals or groups of people have taken advantage of this opportunity by acquiring large tracts of land for ecotourism purposes (de Vos et al., 2019). Because the ecotourism potential of these privately protected areas depends on how well the property is managed (Clements et al., 2016), private landowners often invest considerable effort to maintain and even increase wildlife populations on their land. Privately protected areas have unique advantages over government-protected areas. For example, they have local buy-in from landowners and their employees by design; this is often a significant stumbling block for government-protected areas. Private sites could also employ innovative funding mechanisms that allow them to fast-track land acquisition, perhaps in response to threats such as development. In some areas, privately protected areas may even employ more people, pay better wages, and contribute more to local economies that government protected areas (Sims-Castley et al., 2005). Privately protected areas can, therefore, play a significant role in overall conservation efforts (see Box 2.3), particularly in areas where threatened species (Cousins et al., 2010) and ecosystems (Gallo et al., 2009) are underrepresented in government-protected areas.

Because the ecotourism potential of private protected areas depends on how they are managed, landowners prioritize maintaining and even increasing wildlife populations on their land.

Despite the advantages of privately protected areas, we must also consider the drawbacks. Like many community conserved areas, privately protected areas are not permanently protected by the same mechanisms and oversight as government protected areas are. Ownership and management style can also change at the whim of the landowner, or perhaps the heirs. At times, management practices may be detrimental to the species and ecosystems these privately protected areas claim to protect, for example, through introduction of invasive species and harmful breeding practices (Milner et al., 2007), and by resisting regulatory controls (Cousins et al., 2010). Innovative strategies will thus be required to ensure that these areas do contribute to biodiversity protection, which include education, support, and methods that balance financial gains with conservation goals.

13.1.4 Co-managed protected areas

Local people who support conservation and the protection of their local natural resources are often inspired to take the lead in protecting their local biodiversity. Governments and conservation organisations can assist such initiatives by allowing local people to access specialist expertise and obtain financial assistance to develop conservation and ecotourism infrastructure. These conservation areas, characterised by partnerships between different levels of society that share decision-making responsibilities and consequences of management actions, have been termed co-managed protected areas. Tanzania, where the management of more than two million hectares of forests and woodlands have been transferred to local groups (Blomley et al., 2019), has been particularly active in this regard. One of the biggest strengths of **co-management** is that, with proper consultation and engagement, it avoids **eco-colonialism**—the unfortunate practice by some governments and conservation organisations of disregarding the rights and practices of local people during the establishment and management of new conservation areas or environmental laws and regulations.

Contractual parks offer a good model on how to avoid eco-colonialism. These protected areas are established and managed through agreements with private or communal landowners whose land forms part of a protected area (usually a national park). This not only allows a larger area to be protected, but also allows local people to benefit from biodiversity conservation through benefit sharing and job generation initiatives. Contractual parks play an important role, especially in South Africa, where it is used as a tool to meet both conservation goals and restitution of previously dispossessed land. One such example is the |Ai-|Ais/Richtersveld TFCA (Figure 13.2), which protects a huge number of succulent plant species and a variety of desert ecosystems at the border between Namibia and South Africa. Much of this national park is make up of communal lands, with the landowners—the local Nama people—having co-management and benefit-sharing agreements with the South African government (Reid et al., 2004). Incorporating activities of the landowners in sections of the park enriches tourism experiences, such as boating, hiking, and birdwatching, and contributes to preserving the Namas' cultural identity, pastoral lifestyle, and threatened local languages (Chennels, 1999).

Figure 13.2 A lone giant quiver tree (*Aloidendron pillansii,* CR) stands guard over a desert valley in ǀAi-ǀAis/Richtersveld TFCA, on the border between South African and Namibia. This TFCA is special in that it is an agreement park, established through the cooperation between governments and private landowners. Photograph by Vincent van Oosten, https://pixabay.com/en/richtersveld-south-africa-desert-758235, CC0.

13.1.5 Field stations and marine laboratories

Biological field stations and marine laboratories are a special kind of protected area that provide a dedicated stable space for scientists, students, and even the general public to pursue research projects on all kinds of natural phenomena in an intact environment (Tydecks et al., 2016). By facilitating collaboration and long-term observation, work done at field stations in Africa has led to several fundamental scientific advances, including improved understanding of environmental responses to climate change and acid rain, as well as advances in social development through conservation activities. Today, there are biological field stations in at least 24 Sub-Saharan African countries (Tydecks et al., 2016). Among them are Namibia's Gobabeb Research and Training Centre which focuses on desert conservation, Kenya's Mpala Research Centre (Box 13.1) which investigates the potential for wildlife and livestock to coexist, Nigeria's A.P. Leventis Ornithological Research Institute (see Box 15.4) which focuses on bird conservation, and Uganda's Makerere University Biological Field Station which has a long, distinguished record of primate research.

13.2 Classification of Protected Areas

Protected areas vary greatly in how they are managed. For some, particularly those that protect very sensitive and/or recovering wildlife populations and ecosystems, human activity—even activities, such as photography, hiking, or bird watching (which can cause trampling and disturb shy animals)—may at times need to be forbidden except for specially arranged guided tours. For others, extraction of natural resources may be permitted albeit regulated.

To distinguish how protected areas are managed, the IUCN developed six categories to classify protected areas based on how the land is used (Table 13.1). Of

Box 13.1 Mpala Research Centre: A Living Laboratory for (More than Just) Scientists

Anchal Padukone and Dino J. Martins

Mpala Research Centre,
Nanyuki, Kenya.
http://www.mpala.org

In the heart of Kenya's Laikipia district, Mpala Conservancy stretches over 200 km^2 of semi-arid savannah, acacia bushland, wooded grassland, rocky escarpments and riverine communities along the Ewaso Nyiro and Ewaso Narok rivers. The area is home to an abundance of wildlife, including all the classic savannah mammals: impala (*Aepyceros melampus*, LC), Grant's gazelles (*Nanger granti*, LC), reticulated giraffe (*Giraffa camelopardalis reticulata*, EN), leopards (*Panthera pardus*, VU), lions (*P. leo*, VU), spotted hyenas (*Crocuta crocuta*, LC), and some of the largest savannah elephant (*Loxodonta africana*, VU) and African wild dog (*Lycaon pictus*, EN) populations in Kenya. There are also a few species typical of the northern regions of the Somali-Maasai centre of endemism, such as Grevy's zebra (*Equus grevyi*, EN) and gerenuk (*Litocranius walleri*, NT). Mpala also functions as a working cattle ranch, with upwards of 2,000 cattle, camels, and sheep that are available for use by researchers.

This "multiple use" landscape and its neighbouring ranches provide exceptional opportunities for researchers to study interactions among humans, their domestic herds and wildlife in an area where they coexist. Since much of East Africa's wildlife is found in similar areas outside formal protection, such research could provide essential and widely applicable knowledge for conservation efforts. They will be particularly important as conservation managers will increasingly have to balance wildlife and rangeland management needs to remain effective, in a context of human population growth and economic development. The Mpala Research Centre, established here in 1994, attracts hundreds of scientists every year who use this "living laboratory" to pursue projects varying in scope from the population biology of individual species to community-level dynamics and ecosystem functioning (Rubenstein and Rubinoff, 2014).

Having a research station in this area facilitates long-term and large-scale field experiments, including the Kenya Long-term Exclosure Experiment (KLEE). The 18 KLEE plots are designed to keep out different groups of animals: some plots only exclude megaherbivores (e.g. elephants and giraffes); others exclude all large herbivores; still others only exclude domestic cattle, among other combinations. This allows for controlled studies of the effects of different groups of herbivores on the vegetation and on each other. This research reveals that while domestic stock and wild grazers compete for forage during the dry season, the presence of zebras enhances cattle weight gain during the wet season—perhaps because zebras consume dead grass parts, improving forage quality for cattle (Riginos et al., 2012). Other studies at Mpala have also shown that wildlife and livestock can coexist and facilitate each other's success, given the right approaches in management (Odadi et al., 2011; Ogutu et al. 2016).

Another long-time focus of research at Mpala is the threatened Grevy's zebra. Dr. Daniel Rubenstein (Princeton University) and his research group examine the influence of environmental features on competitive behaviour and reproductive patterns in plains zebra and Grevy's zebra. In turn, they are interested in how these social processes influence zebra population size. Their findings have the potential to inform management strategies in areas where Grevy's numbers are too low to be self-sustaining (Rubenstein, 2010).

Involving the non-scientific community, especially those living around conservancies, is crucial to the long-term success of conservation efforts. Recognising this, Mpala has hosted several citizen science initiatives. For example, The Great Grevy's Rally was a photographic census that relied on inputs from both scientists and members of the public, who travelled to conservancies, such as Mpala, to take pictures of every Grevy's zebra they could find. Researchers processed these images using the Image Based Ecological Information System (IBEIS, http://ibeis.org) to differentiate individuals using their stripe patterns. This allows them to determine population size and structure, and assess whether zebra numbers are stable, increasing, or decreasing.

Figure 13.A Participants in the Kid's Twiga Tally trying to differentiate individual giraffes using photos to understand how the IBEIS software works. Photograph by Danielle Martin, CC BY 4.0.

Also hosted at Mpala, the Kids Twiga Tally (Kahumbu et al., 2016) was a similar "sight-resight" survey of reticulated giraffes that relied on IBEIS software to distinguish between individuals and determine population structure. Its 70 young participants (Figure 13.A) came from both city schools and nearby pastoralist communities, spanning a range of socio-economic backgrounds. After spending two days taking pictures of giraffes on GPS-enabled cameras, the children returned to their schools having contributed meaningfully to conservation science.

these categories, the first five can be defined as true protected areas, because the environment is managed primarily for biological diversity. The sixth category, *Protected area with sustainable use of natural resources*, refers to **extractive reserves** that are managed primarily for the sustainable production of natural resources, such as timber and grazing lands. Nevertheless, extractive reserves can play an important role in conservation: (1) they frequently protect much larger areas than do other types of protected areas; (2) they still provide habitat for many species that were present pre-extraction; and (3) they often border and can thus provide a buffer around, and wildlife linkage between, category I–V protected areas.

Managers of extractive reserves must seek balance between the harvest of natural resources and risking environmental degradation from unsustainable practices.

It is important to note that not all protected areas are covered under the IUCN's six-category system. Prominently are RAMSAR wetlands (Section 12.1.2) which are not incorporated under formal protected areas, but still protected under international law. Other examples include locally-managed marine areas and indigenous reserves, some of which are as effectively managed as formal protected areas. The IUCN is currently working on a new classification, called 'other effective area-based conservation measures' (OECM; IUCN WCPA, 2018), to officially recognise the contribution of areas falling outside formal protected area networks to biodiversity conservation efforts.

13.3 Prioritisation: What Should be Protected?

Historically, the boundaries of protected areas was often determined through pragmatic considerations, such as the availability of funds and land, and political influence, rather than ecological considerations. Many conservation areas were thus established on "lands that nobody wants": marginal areas with little agriculture and development potential, or areas that were too remote to have high commercial value (a trend that continues even today: Venter et al., 2018). Other protected areas were established in locations with charismatic megafauna, so ecosystems without those species remained unprotected. Consequently, some of Africa's most threatened species and ecosystems remain under-protected (Beresford et al., 2011).

In a crowded world with finite natural resources and limited funding, it is becoming increasingly important to be strategic about where protected areas are established. To

Table 13.1 Description of Categories I–VI of the IUCN's classification of protected areas.

Category		Description
Ia	Strict nature reserve	Managed strictly for biodiversity conservation. Serves as reference sites for research and monitoring. Human visitations and impacts highly regulated.
Ib	Wilderness area	Generally large and relatively unmodified natural areas without significant human habitations. Managed to preserve the area's natural character and ecological integrity.
II	National Park	Large natural areas set aside for protection of biodiversity and ecosystem processes. Also managed to support human activities (spiritual, education, scientific, recreation) compatible with biodiversity protection.
III	National monument of feature	Managed to protect a natural feature (e.g. seamount, geological feature, ancient grove) with outstanding cultural and/or natural significance. Can cover a small area, and often have high visitor value.
IV	Habitat/species management area	Protected area dedicated to the protection of a specific species of habitat. May at times required regular and active intervention to ensure primary management goals are met
V	Protected landscape/ seascape	An area with a significant natural or cultural value, created by the interaction between people and nature. Managed to safeguard the interactions that sustains the area's value. Often act as model for sustainability
VI	Managed-resource protected area	Managed primarily for the low-level, non-industrial, sustainable use of natural resources. Generally large, with most of its ecosystems intact.

Source: After Dudley, 2008

do this, conservation biologists and policy makers must answer three key questions: (1) What is most important to protect? (2) Where would it be best protected? (3) How could it be most effectively protected? Three criteria can be used to answer the first two of these questions:

- *Distinctiveness* (*or irreplaceability*): Ecosystems with species that are distinct in their taxonomy (e.g. ecosystems that contain the only species in a taxonomic group) or geographic distribution (e.g. endemic species), or ecosystems with unique attributes (e.g. scenic landscapes, unusual geological features).
- *Endangerment* (*or vulnerability*): Areas that contain concentrations of species threatened with extinction, or ecosystems in danger of being destroyed.
- *Utility*: Species and ecosystems that people value, including culturally

significant species, economically valuable species or ecosystems, or areas that can contribute to combating climate change.

Using these criteria, scientists have developed several broadly complementary methods to prioritise areas for protection. The approaches differ more in what traits they emphasise rather than in their fundamental principles. Thus, although some people may argue about which approach is better, each approach contributes to the protection of biodiversity.

13.3.1 Species approach

Many protected areas are created to protect (e.g. threatened, culturally significant, or keystone) species. Species that provide the motivation to establish a protected area are known as **focal species**. As a prominent example using the focal species concept, the Alliance for Zero Extinction (http://www.zeroextinction.org) identified 67 priority sites across Sub-Saharan Africa (853 sites globally) that contain the last remaining populations of one or more *Endangered* or *Critically Endangered* species. **Flagship species**, such as gorillas, are a special kind of focal species because they capture public attention, have symbolic value, and are important for ecotourism purposes. Many flagship species and focal species are also **umbrella species**, because their protection indirectly benefits other species and ecosystem components with which they share their landscape.

> Protected areas are often established to protect threatened or charismatic species, unique ecosystems, and or wilderness areas.

13.3.2 Ecosystem approach

There is debate among conservation biologists over whether ecosystems rather than individual species should be the primary target of conservation efforts. Supporters of an ecosystem approach argue that protecting and managing ecosystems can preserve more species and provide more value to people than spending the same amount of money to protect individual species. Focusing on ecosystems also allows for greater flexibility in justifying conservation efforts, because it can be easier to demonstrate the economic value of ecosystems for helping to control floods, filtering water, and providing opportunities for recreation. To that end, the WWF has identified 238 ecoregions across the globe (the "Global 200")—57 of them in Sub-Saharan Africa—that are most crucial to the biodiversity conservation (Olson et al., 2002). This Global 200 analysis formed the basis of a more recent global assessment that identified 41 at-risk ecoregions—areas of high conservation priorities because they are undergoing high levels of habitat conversion and have low protected areas coverage (Watson et al., 2016). Africa has several at-risk ecoregions, particularly in Angola, South Africa, the DRC, and West Africa's Sahel region. The IUCN Red List of Ecosystems (RLE, Section 8.5.1) is another example of an ecosystem-focused prioritisation for conservation. While the ecosystems approach overcomes several limitations of the species approach, some conservationists argue that focussing on distinct ecosystems may, in itself, be detrimental, and that the scope of conservation should be expanded, for example by also including biogeographic transition zones (van Rensburg et al., 2013).

13.3.3 Wilderness approach

Wilderness areas are large areas where people have had little influence on the environment (relative to other areas), they have few people living in them, and are unlikely places for human development in the short term. These areas are conservation priorities because they may be the only places where animals that require large home ranges can continue to survive in the wild. Further, wildernesses can serve as controls or benchmarks for researchers to measure the effect of human disturbance on nature. The most popular way to identify wilderness areas is to identify areas without roads. While very few roadless areas remain, many of the world's most important roadless wildernesses, some larger than 10,000 km^2, are in Africa (Ibisch et al., 2016). Of concern is that, second to South America, Africa also leads the world in wilderness losses over the past decade (Potapov et al., 2017). It is worth emphasising that even wilderness areas have had a long history of human activity (Roberts et al., 2017). It is not always necessary or even possible to eliminate all human activity from such areas, if those activities do not obstruct conservation goals.

13.3.4 Hotspot approach

Multiple prominent initiatives have prioritised conservation in areas where large concentrations of species can be protected in a relatively small area. Perhaps the most prominent example is the **Global Biodiversity Hotspots** initiative. Combining a species approach with an ecosystem approach, Global Biodiversity Hotspots are areas with exceptionally high levels of biological diversity and endemism—that is, irreplaceable biodiversity—that are threatened with imminent habitat destruction (Table 13.2). Norman Myers, a British biologist who launched his conservation career as a wildlife photographer in Kenya, originally proposed the Biodiversity Hotspot concept (Myers, 1988). Working with a team of prominent scientists, Myers identified 25 Hotspots (five of them in Sub-Saharan Africa), which contained 44% of all vascular plant species and 35% of all terrestrial vertebrate species on only 1.4% of the Earth's land surface (Myers et al., 2000). More recently, Conservation International (CI) identified an expanded set of 36 Biodiversity Hotspots (Mittermeier et al., 2005), eight of which are in Sub-Saharan Africa (Figure 13.3). This expanded set of Biodiversity Hotspots covers only 2.3% of Earth's surface yet contains over 50% of all plant species and over 40% of all terrestrial vertebrate species.

While the Global Biodiversity Hotspots highlight some of the most important global conservation priorities, none of these Hotspots are small enough to be contained in a single protected area—in fact, most of these Hotspots identify whole regions, not

Table 13.2 A natural history comparison of Sub-Saharan Africa's eight Global Biodiversity Hotspots.

Location	Original extent (× 1,000 km²)	Remaining undisturbed vegetation (%)	Number of species		
			Plants	Birds	Mammals
Guinean Forests of West Africa	620	15	9,000	917	390
Succulent Karoo	103	29	6,356	225	75
Cape Floristic Region	90	20	9,000	320	127
Maputaland-Pondoland-Albany	274	25	8,100	631	202
Coastal Forests of Eastern Africa	291	10	4,050	633	198
Eastern Afromontane	1,018	11	7,600	1,300	490
Indian Ocean Islands[a]	601	10	13,500	503	211
Horn of Africa	1,659	5	5,000	697	220

Source: Mittermeyer et al., 2004; https://www.cepf.net/our-work/biodiversity-hotspots.

[a] *Includes Madagascar and Mascarene islands*

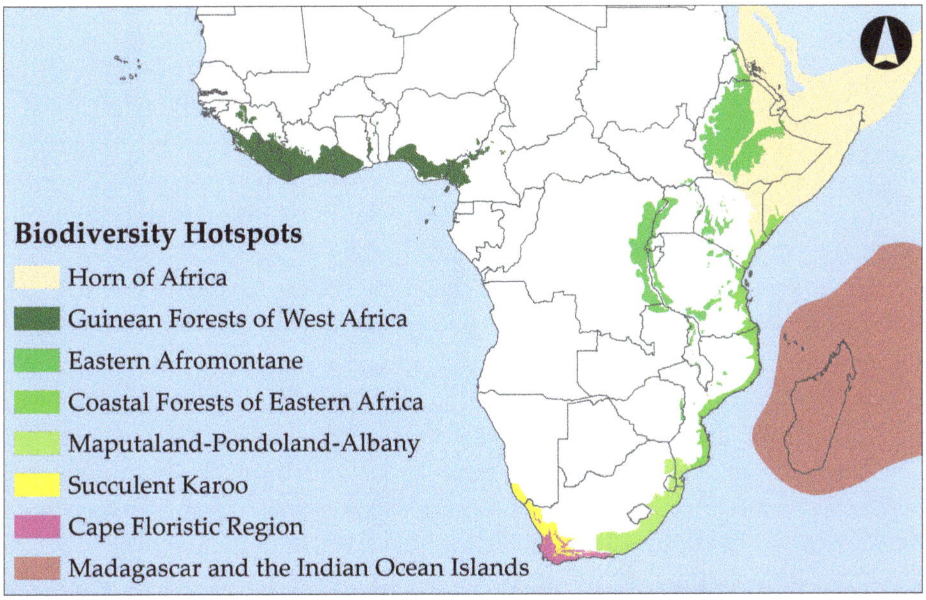

Figure 13.3 Sub-Saharan Africa's eight Global Biodiversity Hotspots. These areas are targets for protection because of their high biodiversity, endemism, and significant threat of imminent extinctions. After Mittermeier et al., 2005. Map by Johnny Wilson, CC BY 4.0.

projects, requiring conservationists to still make decisions for prioritising protection within them. To create actionable priorities from within regional hotspots, several initiatives aim to identify local hotspots of species richness that can be conserved as one protected area of a manageable size. One such approach is the **Key Biodiversity Areas (KBA)** programme (Eken et al., 2004), which identifies conservation priorities using standardised criteria and thresholds that account for concentrations of threatened species and/or globally significant population aggregations. The KBA programme functions as an umbrella designation for several taxon-specific approaches, most prominently BirdLife International's Important Bird and Biodiversity Areas (IBA) programme (Fishpool and Evans, 2011). Other KBA programmes include PlantLife International's Important Plant Areas programme (e.g. Smith and Smith 2004), as well as the Important Sites for Freshwater Biodiversity programme (Darwall et al., 2005). One example from Guinea used KBA criteria and thresholds regarding threatened mammals to provide suggestions for expanding the country's protected areas network (Brugiere and Kormos, 2009).

13.3.5 Gap analysis approach

Assessing the performance of existing protected areas can be done by spatially comparing their footprint to prioritised conservation areas (as above). Such an assessment offers not only an assessment of existing protected areas performance, but also offers a means to identify conservation gaps—important areas that still need to be protected to meet broader conservation goals. Such assessments, which systematically evaluate whether different aspects of biodiversity are adequately protected, are collectively known as **systematic conservation planning** assessments (McIntosh et al., 2017). Perhaps the most popular systematic conservation planning method is **gap analysis**, during which scientists overlay maps of species (or ecosystem) distributions with maps of protected areas to identify species (called gap species, see also Figure 10.3) or ecosystems that are not adequately protected in existing protected areas networks (Box 13.2).

> Gap analysis enables conservation planners to identify species or ecosystems that are not adequately protected in existing protected areas networks.

When identifying conservation gaps, it is important to think carefully about the taxa or ecosystem used to make the assessment. Many conservation assessments assume that one well-known species group can act as a **biodiversity indicator** (also known as a biodiversity surrogate or **surrogate species**) for lesser-known taxa, so establishing a protected area to protect one gap species will also afford protection to other under-protected taxa. While this is true to some level, several studies have shown that this may not always be the case (Rodrigues and Brooks, 2007; Carwardine et al., 2008; Jones et al., 2016).

Box 13.2 Identifying Key Sites for Conservation in the Albertine Rift

Andrew J. Plumptre[12]

[1]*Albertine Rift Program, Wildlife Conservation Society, Kampala, Uganda.*

[2]*Current Address: Key Biodiversity Area Secretariat, c/o BirdLife International, Cambridge, UK.*

✉ *aplumptre@keybiodiversityareas.org*

The Albertine Rift is one of the richest regions on Earth for vertebrate diversity (Figure 13.B). Spanning about 100 km either side of the international border of the eastern DRC, it includes forests, wetlands and savannahs from eastern DRC and western Uganda, Rwanda, Burundi, and Tanzania, and runs from the northern end of Lake Albert to the southern end of Lake Tanganyika. It contains more than 40% of Africa's mammals, 52% of Africa's birds, as well as 19% of its amphibians and plants, in only 1% of the continent's surface area.

Figure 13.B (Left) Mubwindi Swamp, in Bwindi Impenetrable National Park, an important site for mountain gorillas and the Albertine Rift endemic Grauer's Rush Warbler (*Bradypterus grayeri*, EN). (Right) A Grauer's gorilla, the largest of the four gorilla subspecies and a flagship for conservation efforts in the Albertine Rift. Photographs by A.J. Plumptre/WCS, CC BY 4.0.

It also contains more endemic and globally threatened species than any other ecoregion in Africa (Plumptre et al., 2007). Endemic large charismatic species include the eastern gorilla (*Gorilla beringei*, CR), golden monkey (*Cercopithecus kandti*, EN), Congo bay owl (*Phodilus prigoginei*, EN), and Ruwenzori turaco (*Ruwenzorornis johnstoni*, LC). The lakes in the Albertine Rift each also contain several hundred unique fish species. Unfortunately, this rich biodiversity also occurs in one of the most densely populated parts of Africa, and the threats to existing protected areas are high.

The Wildlife Conservation Society (WCS) has been working to support the conservation of six key landscapes in the Albertine Rift (ARCOS, 2004), particularly focusing on (a) identifying critical areas for conservation of threatened and endemic species; (b) undertaking research and monitoring of species and key landscapes; and (c) supporting the conservation of critical sites and the creation of new protected areas to conserve large and small mammals, birds, reptiles, amphibians and plants in all protected areas, as well as sites where new protected areas might be established. These surveys have identified critically important areas in eastern DRC, such as the Itombwe and Kabobo Massifs where new species have been identified and some species were rediscovered, having been last seen more than 50 years ago. Working with local communities, the surveys have been used to design the boundaries of newly established protected areas to ensure that they capture as much of the biodiversity as feasible. Once the local people in the area are presented with survey results and options for protection discussed, they often realise the importance of their site and propose more stringent protection measures than conservationists initially thought possible.

Using species distribution models (SDM) of the region's endemic and globally threatened species, WCS gained an understanding of where these species should occur both now and under future climate change scenarios (Ayebare et al., 2018). Using Marxan software (Possingham et al., 2000), WCS then identified those areas that would conserve all the species of conservation interest at minimum cost (Plumptre et al., 2019). This procedure identified the Itombwe and Kabobo Massifs together with the Sitebi Hills east of Mahale Mountains National Park in western Tanzania as being critical for conservation of species that are currently not adequately protected (Figure 13.C).

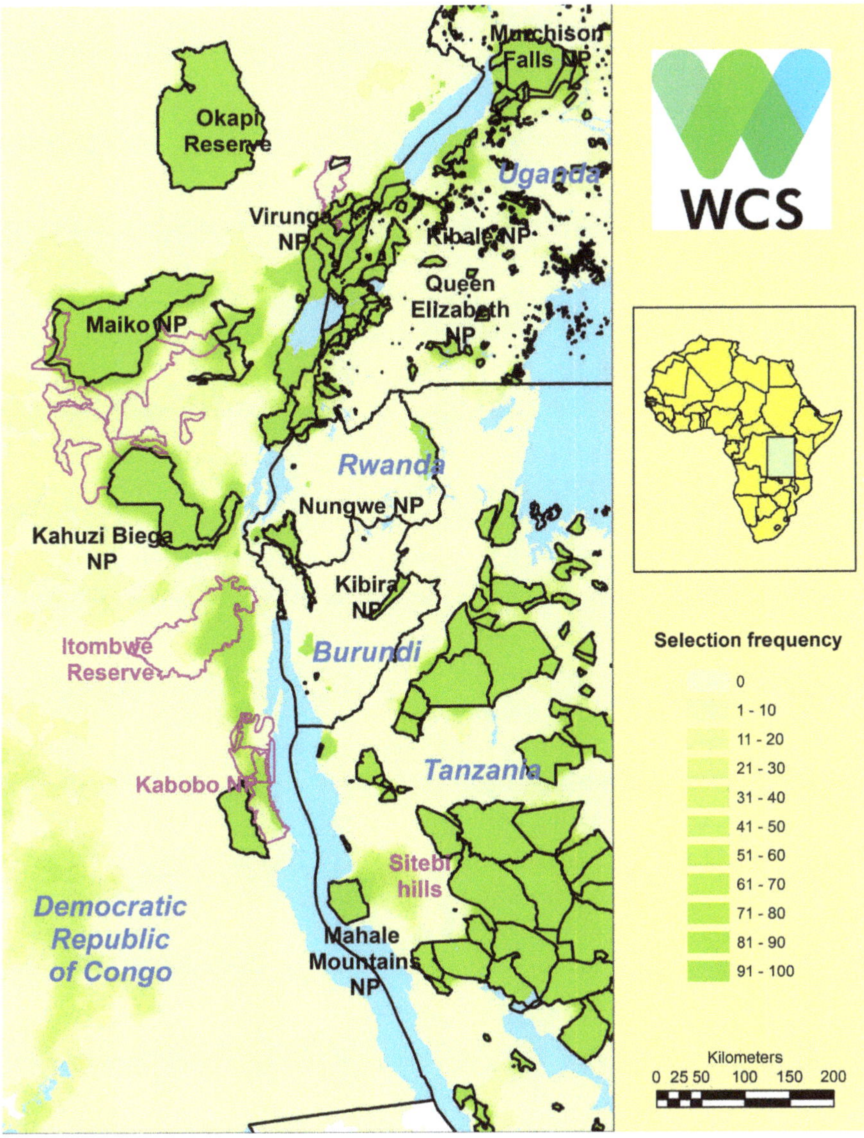

Figure 13.C Selection frequency of 5 km^2 cells in the Albertine Rift from Marxan analysis, indicating priority areas for the conservation of endemic and threatened mammals, birds, reptiles, amphibians, and plants. Existing protected areas (all highlighted) were locked in but proposed protected areas such as Itombwe and Kabobo and community reserves (purple boundary) were not. Darker green areas indicate priority conservation sites. Image courtesy of WCS Albertine Rift Program, CC BY 4.0.

These results were used to develop an Albertine Rift Action Plan (Plumptre et al., 2016), together with detailed conservation action plans for the preservation of the six core landscapes and their unique and threatened species, both inside and outside of protected areas, now and into future.

13.3.6 Optimisation approach

Prioritisation efforts typically need to consider multiple factors in addition to biodiversity, such as cost-effectiveness, socio-economics, site condition, and potential threats that may impact a proposed protected area. Technical computer software known as "decision support tools" are providing a new way to identify conservation priorities that meet a suite of conservation objectives. One of the most popular packages is Marxan (http://marxan.org), a freely available program that identifies the optimal location for protected areas based on flexible user-defined criteria (Watts et al., 2009). The user-defined criteria can be complex; for example, one can set the model parameters to choose the areas that best protect certain aspects of biodiversity (e.g. protect at least 25% of each vegetation type) while reducing costs and minimising impact on other stakeholders; model input can include measured data, as well as expert input. In one such example, conservation biologists from South Africa, Eswatini, and Mozambique used Marxan to identify potential locations for new protected areas in the Maputaland Centre of Endemism which the three countries share. They found that adding 4,291 km^2 to the existing protected areas network could generate US $18.8 million in revenues *while* fulfilling their conservation objectives: protecting 44 landcover types, 53 species, and 14 ecological processes (Smith et al., 2008).

> Decision support tools help identify conservation priorities that meet a suite of objectives, including cost-effectiveness, socio-economics, and site condition.

Regardless which prioritisation approach one follows, it is important to remember that prioritising species and ecosystems in need of protection does not amount to "doing conservation". Real conservation only happens when a conservation plan that will implement those suggestions is drawn up and put in place. A review of eight different systematic conservation assessments in South Africa provides a good foundation to guide conservation biologists in the process from prioritisation to implementation (Knight et al., 2006).

13.4 How Much Land Should We Protect?

As of mid-2019, there were just over 7,500 protected areas covering over 4.5 million km^2 of land and ocean surface (UNEP-WCMC, 2019) scattered across Sub-Saharan Africa (Figure 13.4). The country with the largest number of protected areas is South Africa with over 1,500 protected areas, while the country with the largest total area under protection is Tanzania, with over 360,000 km^2. While these statistics may seem impressive, seeing these numbers in perspective is important before performance is judged. Currently, one of the most prominent sets of targets used to measure conservation progress is laid out in the international **Aichi Biodiversity Targets** (https://www.cbd.int/sp/targets). The global conservation area target reads:

> By 2020, at least 17 per cent of terrestrial and inland water areas and 10 per cent of coastal and marine areas, ... are conserved ... and integrated into the wider landscape and seascape.

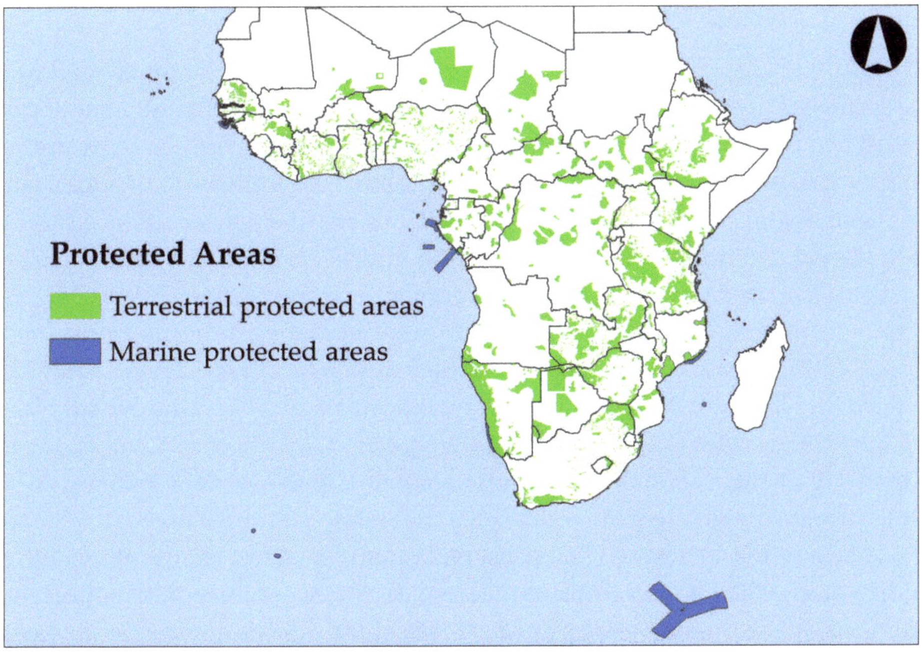

Figure 13.4 The location of Sub-Saharan Africa's terrestrial and marine protected areas (MPAs), which falls under the IUCN's categories I–VI classification for protected areas. Note that many small protected areas do not clearly show up at this scale. Source: UNEP-WCMC, 2019. Map by Johnny Wilson, CC BY 4.0.

The good news is that as a region, Sub-Saharan Africa is well on its way to achieving the Aichi terrestrial target, since just under 17% of the region's total land and inland water surfaces were protected as of mid-2019 (UNEP-WCMC, 2019). Further good news is that 22 Sub-Saharan African countries have protected more than 17% of their land area, with Seychelles (42%), Republic of the Congo (41%), and Tanzania (38%) leading the way. Sub-Saharan Africa's protected areas network is also one the best performers globally in affording protection to migratory birds (Runge et al., 2015) and terrestrial megafauna (Lindsey et al., 2017).

> While Sub-Saharan Africa as a region is well on its way to achieving its goal of protecting 17% of terrestrial areas, the percentage of land protected is very uneven among countries.

Despite this progress, some notable gaps remain. Foremost, the percentage of land protected is very uneven among countries. While a few countries have reached the Aichi protected areas target, there were also 10 countries with less than 5% of their land protected, and an additional six countries which protect less than 10%. Furthermore, the amount of land protected does not necessarily translate to adequate protection for all ecosystems (Watson et al., 2016). For example, despite having the most protected areas, South Africa protects only 8% of its land, well below the Aichi target. Many protected areas also qualify as paper parks (Tranquilli et al., 2012, 2014), with a questionable contribution towards achieving conservation goals.

13.4.1 A neglected system: marine protected areas

When thinking about conservation in Africa, many people's minds will wander towards images of a charismatic terrestrial mammal, like an elephant, lion, or gorilla. But what about all the whales, dolphins, sea urchins, starfish, nudibranchs, and other wonderful marine creatures? Perhaps due to the outsized influence of Africa's famous land mammals on the ecotourism sector, Africa's marine conservation efforts have always lagged behind terrestrial conservation efforts. In total, just over 700,000 km^2 (7%) of Sub-Saharan Africa's marine environment is protected (UNEP-WCMC, 2019). The gaps in marine conservation are even more obvious when one considers that as of mid-2019, only six countries have achieved the 10% Aichi Target, with Gabon (29%) and St. Helena (28%) leading the way. Marine protection is particularly lacking along the Atlantic coast (Klein et al., 2015), where many of 15 coastal countries protect less than 1% of their coastal and oceanic waters. It is also worth keeping in mind that the 10% coverage target (a modest goal that many countries may fail to achieve), may not be enough to achieve key conservation and sustainable development goals (Spalding et al., 2008). For example, to reverse declining commercially important fish populations, it is estimated that as much as 30% of the marine environment may need to be protected (O'Leary et al., 2016).

There is clearly an urgent need to establish more **marine protected areas** (MPAs), protected areas within oceanic and coastal environments (Box 13.3). There is also an urgent need to scale up law enforcement in the marine environment (Brashares et al., 2004). Increasing our marine protection efforts—which even local communities can initiate (Rocliffe et al., 2014)—is well worth it: it strengthens local fisheries (Kerwath et al., 2009; Lester et al. 2009) and offers educational and recreational opportunities, such as swimming and diving, which in turn generates ecotourism revenue. For example, Africa' oldest MPA, Tsitsikamma National Park in South Africa (established in 1964), attracts over 170,000 visitors each year (Chadwick et al., 2014); the tourism revenues support numerous jobs and are a major stimulant of the local economy (Oberholzer et al., 2010). This is in stark contrast to the marine environment off West Africa, where unregulated fisheries are putting tremendous strain on local economies amid a lack of ecotourism infrastructure (Agnew et al., 2009; Gremillet et al., 2015).

13.5 Designing Protected Areas

The unplanned way in which protected areas have historically been established means that their design may at times impede rather than aid their goals. For example, many protected areas are too small to sustain viable populations of the species they

Box 13.3 Marine Protected Areas in East Africa and the Western Indian Ocean

Abraham J. Miller-Rushing

Acadia National Park, US National Park Service, Bar Harbor, ME, USA.

How can MPAs in the Western Indian Ocean best enhance the preservation of biodiversity and the economies in this Global Biodiversity Hotspot? The ecosystems of the East African coast and nearby islands are diverse—mangrove forests, river deltas, coastal lagoons, rocky shores, sandy beaches, coral reefs, mud flats, seagrass beds, and open water. These areas are also economically important, with millions of people dependent on these waters' shrimp, fish, and other natural resources for their livelihoods.

How effective are these MPAs, both in protecting biodiversity and people's livelihoods? In 2006, an assessment of eight MPAs in Kenya, Tanzania, and Seychelles found several shortcomings, including inadequacies in staffing, funding, stakeholder engagement, and articulation of goals and management practices. Also, there needed to be additional monitoring and research to inform management and policy (Hockings et al., 2006). Despite these faults, the abundance and size of fish increased dramatically in several MPAs within 10 years of implementing fishing restrictions (McClanahan et al., 2007). The size and quality of fish caught in surrounding fishing grounds also increased substantially, probably due to fish dispersing from the MPAs.

Following these successes, the number and management of MPAs in the area have steadily increased and improved, at least, in part, due to cultivating better relationships with local stakeholders. One such example comes from the Quirimbas archipelago, just off the coast of northern Mozambique, where the Quirimbas National Park (over 1,000 km^2) is managed through a cooperative effort of 40 villages, the government of Mozambique, and WWF. At the northern end of the Quirimbas archipelago, a few kilometres north of Quirimbas National Park, the Vamizi Conservation Project (Figure 13.D) protects an additional 230 km^2 around the islands of Vamizi, Rongui and Macaloe. The Vamizi Project was initiated in 2002 as an innovative community-based management project involving local communities, international NGOs, and a group of individual investors. After protection, fish populations quickly began rebounding and had positive spill-over effects on fish around the reserve (da Silva et al., 2015). The stories of the abundant fish have contributed to a challenge for the project—attracting commercial fishermen from outside the area. To help ensure the financial and scientific sustainability of the project, partners developed a luxury ecotourism site and a research centre on Vamizi Island.

Figure 13.D Vamizi Island has some of the world's richest and most pristine coral reefs, as well as the last population of the grey reef shark (*Carcharhinus amblyrhynchos*, NT) in Mozambique. The reefs are now protected thanks to a collaborative conservation effort that includes the local community. Photograph by Isabel Marques da Silva, CC BY 4.0.

Other protected areas have met variable degrees of success, as conservation managers and communities in the region test different approaches and figure out how best to sustain MPAs in a challenging environment. Different approaches are likely to work in different situations, depending on availability of resources, local stakeholders, and other constraints. As MPAs in the region continue to develop, coordination among countries could improve the value of the MPAs to biodiversity conservation. Already there are examples of multiple pathways to improving and expanding MPAs to protect biodiversity and achieve sustainable fisheries in this region (McClanahan et al., 2016). The future is hopeful.

are meant to protect. To avoid and mitigate such mistakes, conservation biologists are increasingly exploring methods to design more efficient protected areas networks.

Conservation biologists often start the process of designing protected areas networks by considering "the four Rs":

- *Representation*: A network of protected areas should protect as much of the biodiversity (including species, ecosystems, genetic diversity, ecosystem processes, etc.) of a region, country, or subcontinent (depending on the scale of planning) as possible.

- *Resiliency*: Protected areas should be large enough that they can maintain biodiversity (including species, ecosystems, genetic diversity, etc.) for the foreseeable future, including in the face of climate change and natural disasters such as cyclones/hurricanes and uncontrollable wildfires.
- *Redundancy*: A network of protected areas should not rely on a single protected area to conserve key aspects of a region's biodiversity; rather important aspects of biodiversity should be included in multiple protected areas to ensure their long-term existence.
- *Reality*: Each protected area requires sufficient funding, political will, defensibility, and local buy-in to support biodiversity over the long term.

In addition to the four Rs (which can also be applied to species protection), the following questions can also help guide planning of protected area networks (Figure 13.5):

- How large of an area must be protected and what landscape features must it include to effectively and sustainably protect biodiversity that may not be able to persist outside it?
- Is a single large protected area better, or are multiple smaller reserves more effective?
- What shape should a protected area be?
- When creating multiple protected areas, should conservation managers create them near one another or far apart? Should they be connected in some way, or should they be isolated from one another?
- How should human activities be accommodated? What activities should be allowed?

To prepare readers for discussions with land managers, conservation planners, and policy makers who are in the process of developing new protected areas, the next section introduces some of the most important principles related to protected areas design. It is important to note that this introduction is not meant to serve as a universal set of rules for the design of protected areas. Because every project presents a special and unique set of circumstances, a single set of simplistic or overly general guidelines will not suffice. Also, the principles discussed below have been explored mainly in terms of protecting terrestrial vertebrates, vascular plants, and large invertebrates, so it is still uncertain how they apply to freshwater and marine nature protected areas.

13.5.1 What size should a protected area be?

The design of protected areas, and their size, is often addressed through the lens of the island biogeography model that states that large islands can accommodate more species and larger populations than small islands (Section 5.1). Research on extinction rates of populations (Newmark, 1996; Woodroffe and Ginsberg, 1998) and species richness (Harcourt et al., 2001; Brashares et al., 2001) has shown that protected areas function very much like islands. Specifically, because large protected areas contain greater habitat diversity than small protected areas, larger protected areas can accommodate (a) more species, (b) a larger range of ecosystem processes, and (c) viable populations of large species that range over large areas and live at low densities.

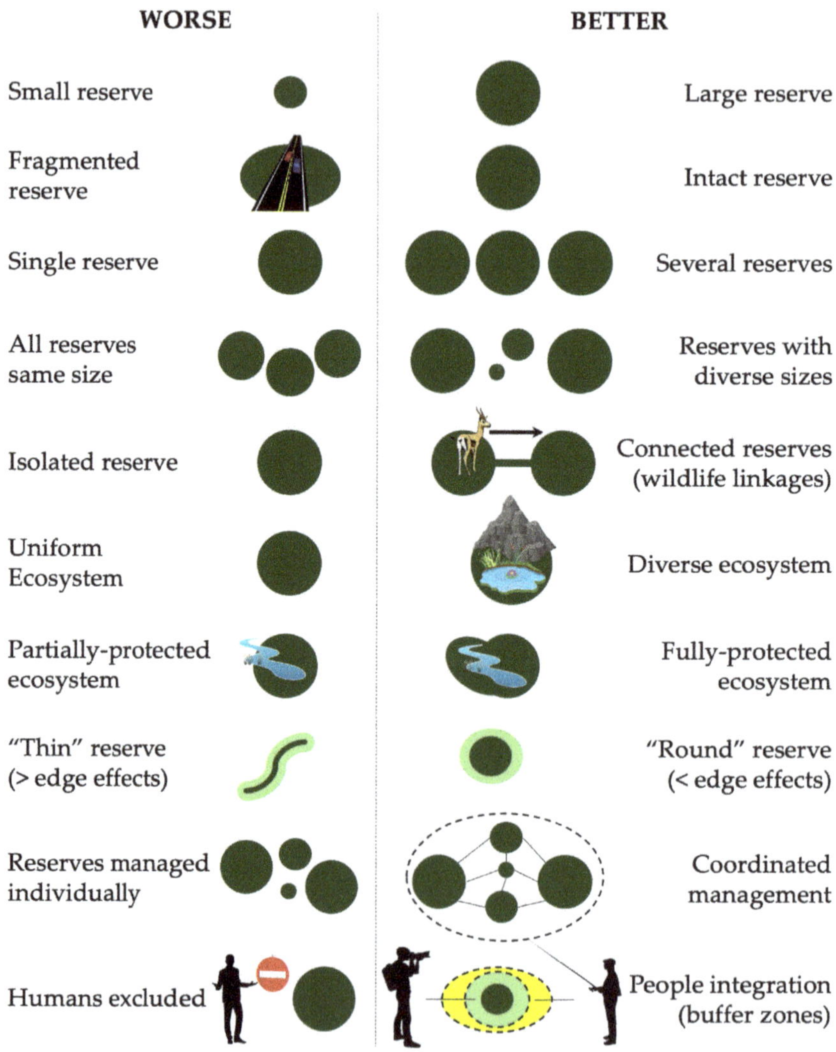

Figure 13.5 There are several major principles of reserve design to consider when establishing a new protected area or redrawing the boundaries of an existing protected area. While addressing all these principles is not always possible, the designs on the right are generally considered preferable to those on the left. After Shafer, 1997, CC BY 4.0.

Given the range of costs and benefits of establishing large protected areas, conservation biologists have debated whether creating a single large reserve or several small reserves of the same total area—known as the **SLOSS (Single Large Or Several Small) debate**—is better. As discussed in Section 5.1.1, habitat fragmentation is currently one of the main drivers of species extinctions; it divides large populations into more vulnerable subpopulations, leads to undesirable edge effects, creates barriers to dispersal, and provides entry points for invasive species. These negative impacts are also of concern for protected areas, especially those that are small and fragmented (leading to larger perimeter:area ratios). For example, fragmentation concentrates elephants (Vanak et al., 2010) and apex predators (Cozzi et al., 2013) in the core of protected areas, greatly limiting the effective protected area for these taxa. However, these same impacts do not alter ungulate foraging (Kiffner et al., 2013), leading, potentially, to overgrazing near reserve borders. Studies have also shown how wildlife experience higher levels of mortality near protected area boundaries (Balme et al., 2010). Ignoring such edge effects could disrupt the long-term conservation value of a protected area, particularly small ones that could effectively function as edge habitat in its entirety. Because one big fragmented reserve has many of the characteristics of several small protected areas, conservation planners should aim to establish properly-placed large protected areas and to keep them as intact as possible. It is thus good practice to restrict and even remove highways, fences, farms, logging operations, and other human activities inside protected areas because of how they fragment habitats and reduce habitat availability overall.

> Large protected areas are generally preferred over small ones because they can support a greater variety of ecosystems and larger wildlife populations.

But how do we know when a protected area is big enough? Ultimately, optimal size depends on the area over which important natural processes take place, which varies depending on the ecosystem. In some cases, the functional size may be quite small, such as a desert spring, a mountain bog, or a rocky outcrop. In contrast, the functional size of tropical forests, seasonal drylands, and desert communities are typically quite large, possibly spanning across country borders. Understanding and planning for protecting these different targets thus requires a familiarity with the functioning and ecology of each ecosystem.

When considering the size of a proposed protected area, conservation managers must also consider how well the area can be monitored and defended from threats. In some instances, an entire community may be incorporated into a relatively small protected area that is easy to monitor and defend against pollution, invasive species, and so forth. More often however, only a portion of the target community can be protected. In such cases it is important to consider how secure the conservation target will ultimately be. For example, if an aquatic organism needs protection, clearly the protection of its immediate habitat is critical. However, if a major threat is upstream from its habitat, then protection of the immediate habitat alone will be insufficient. Instead, managers would need to find ways to prevent outside threats from impacting populations inside the protected area. One option could be to discuss the threats and how to mitigate them with surrounding landowners, perhaps by facilitating their adoption of sustainable land-use practices. If the magnitude of the threats cannot be reduced to acceptable levels, a prioritisation programme might be used to identify critical sub-components of a larger ecosystem that will still accomplish the necessary protection. These kinds of considerations can become very complex and involved. But they are also very important to consider as options, especially when dealing with ecosystems situated between a variety of stakeholders.

13.5.2 Zoning as a solution to conflicting demands

While the general consensus seems to be that larger protected areas are better than smaller ones, establishing a properly-placed large protected area can be an imposing challenge. In a few special cases, large protected areas may be established through cooperation between multiple levels of society. More often, however, conservation biologists are faced with limited resources, and stakeholders can reasonably ask why a large park is required in an area that can otherwise be used to support agriculture or other types of businesses that may provide quick profits and jobs.

One way to deal with such conflicting demands while still achieving the target of protecting a large area is through a method called **mixed-use zoning**. Mixed-use zoning prioritises the overall conservation objectives for a protected area but also sets aside designated areas where certain regulated human activities are permitted (Box 13.4). In this way, some areas (or zones) may be designated for subsistence agriculture, shade-grown crops, timber production, hunting, ecotourism, or water management. Other areas are designated are dedicated to recovery of threatened species, ecotourism, ecosystem restoration, and scientific research. This is the case at the W-Arly-Pendjari (WAP) Complex, which straddles the border zone between Benin, Burkina Faso, and Niger. The core of the complex consists of three national parks covering 14,948 km^2, set aside for strict biodiversity conservation. These national parks are surrounded by as many as 16 additional reserves, partial reserves, and hunting concessions, bringing the total area of protected Sudano-Sahelian savannah to 26,000 km^2 (WHC, 2018).

> Mixed-use zones sets aside areas for certain regulated human activities within a larger conservation area. This approach helps abate conflicting land use pressure.

Through its Biosphere Reserves programme, UNESCO has pioneered a formal zoning approach that integrates human activities, scientific research, biodiversity conservation, and tourism at a single location (Coetzer et al., 2014). A **biosphere reserve** is divided into three zones to delineate different levels of human use (Figure 13.6). The core of a typical biosphere reserve is a **no-take zone** (also called a core zone), strictly protected for biodiversity and ecosystem functioning. Around the core area is a restricted-use buffer zone, where people can conduct traditional, low-impact activities, such as collecting edible plants and small amounts of wood for fuel, and scientists can conduct non-destructive research. Outside of the buffer zone is a transition zone

Box 13.4 Zoning: Something for Everyone in the Forests of Dzangha-Sangha

Richard Carroll[1,2]

[1]*World Wildlife Fund, Washington DC, USA.*

[2]*Current address: The Pimm Group, Nicholas School of the Environment, Duke University, Durham, NC, USA.*

✉ *richardwcarroll@hotmail.com*

Located in the dense forests of southwestern Central African Republic (CAR), in a wedge between neighbouring Cameroon to the West and the Republic of the Congo to the East, the Dzanga-Sangha Project (DSP) aims to conserve CAR's last lowland tropical forest by integrating conservation and regional development. The DSP, which formally began in 1988 with the establishment of a cooperative agreement between WWF and the CAR government, is an integrated conservation and development project (ICDP); its activities are focused on protected area management, rural development, tourism, and project administration, as well as sustainable use of natural resources and applied ecological and social research. The focal area of the DSP is the Dzanga-Sangha Complex of Protected Areas (Figure 13.E), an area of 4,589 km^2 comprising the Dzanga-Sangha Special Reserve (3,359 km^2) and Dzanga Ndoki National Park (1,143 km^2). The Complex is home to healthy populations of forest elephant (Figure 13.F), western lowland gorilla (*Gorilla gorilla gorilla*, CR), chimpanzee, and other wildlife characteristic of the Northwest Congolian Moist Lowland Forest (Carroll, 1992). The forest also shelters the BaAka Pygmies, a hunter-gatherer community whose traditional livelihood is directly linked to the forest and its plant and wildlife resources (Robinson and Remis, 2014).

Many of the WWF-supported programmes in Central Africa have sought to create the conditions for traditional peoples, such as Pygmies to maintain their lifestyles, and to adapt to changing social conditions should they choose. In the case of the DSP, two-thirds of the Complex area is classified as a "Special Reserve", a designation that the CAR government created to accommodate traditional peoples' use of the forest. While traditional hunting and gathering are broadly allowed in the Special Reserve, national laws specifically prohibit hunting of "integrally protected species", such as gorillas, chimpanzees and elephants, in the Complex and elsewhere in CAR.

To establish a "safe zone" where wildlife can reproduce away from human pressures (Blom et al., 2004), and to accommodate tourism, one third of the

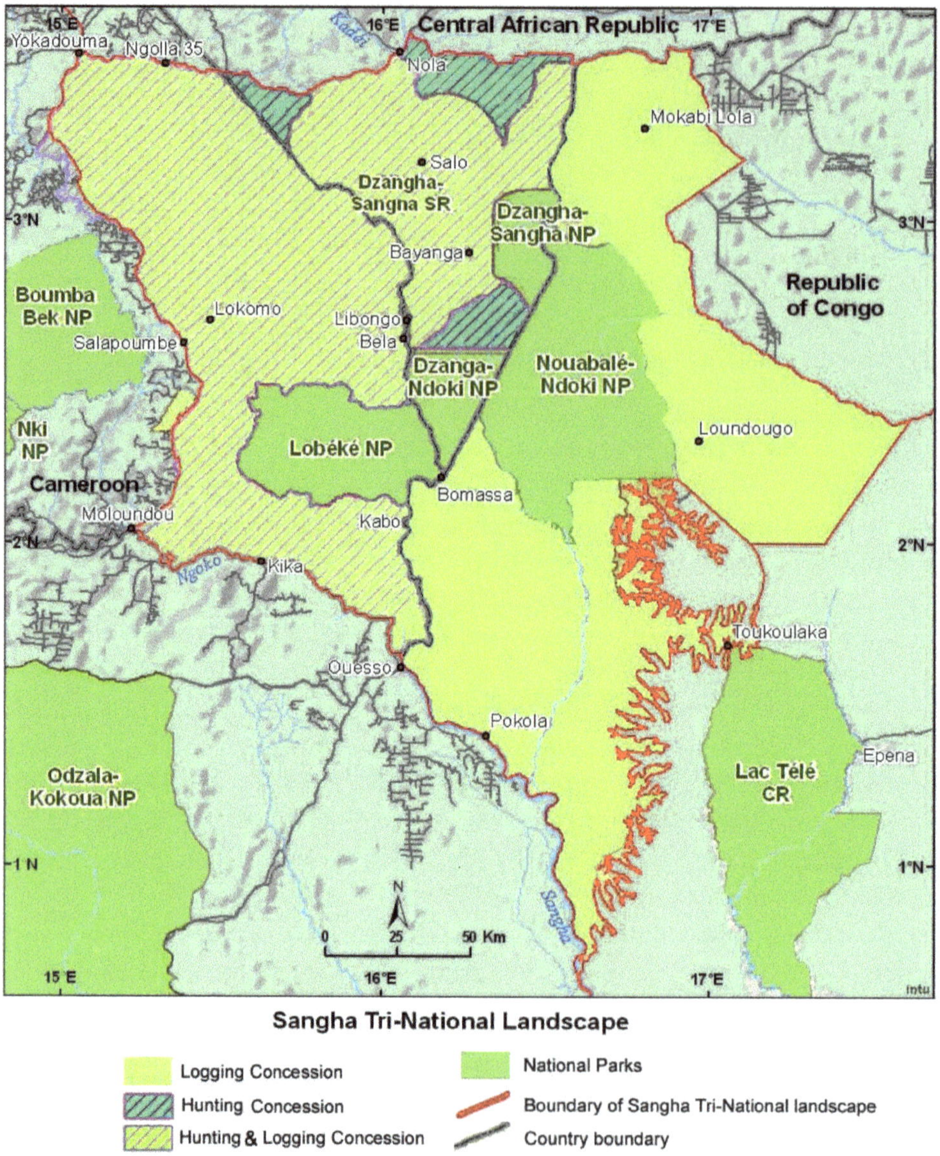

Figure 13.E The location of Dzanga-Sangha Special Reserve and Dzanga-Ndoki National Park, CAR, in relation to the Sangha Tri-National Landscape. Source: Endamana et al., 2010, CC BY 3.0.

Complex is designated as a national park. Hunting is not allowed in the national park; as compensation, 40% of all tourist receipts go to a village association, which includes BaAka, and 50% pays salaries for local employees of the park and special reserve. In other words, 90% of the dividends earned from tourism activities goes to the local people affected by conservation activities. The local community by and large supports the designation of this no-hunting zone, both to sustain their traditional activities and those of tourists.

Figure 13.F Forest elephants burrow for nutrients in Dzanga-Ndoki National Park's mineral rich pools. Photograph by Ana Verahrami/Elephant Listening Project, CC BY 4.0.

Building on the successes in CAR, the DSP is also an active partner in the 36,000 km^2 transboundary Sangha Tri-National (STN) initiative. Reflecting the Peace Park concept, the initiative is a multi-national effort to protect a large block of contiguous forests, the heart of which lies at the meeting point of the Congo-CAR-Cameroon boundaries. This initiative includes CAR's Dzanga-N'Doki National Park, as well as two adjacent national parks: Cameroon's Lobéké National Park (430 km^2) and Republic of the Congo's Nouabalé-Ndoki National Park (4,190 km^2). These three national parks are surrounded by extensive buffer zones that include the Dzanga-Sangha Special Reserve, forests around Lobeke (700 km^2) and the peripheral zone in Republic of the Congo with almost 12,000 km^2 of logging concessions. STN was declared as the first landscape level World Heritage Site in 2012.

In summary, the Dzanga-Sangha Project is an ambitious, long-term effort of the CAR government, WWF, and other participating partners to save the largest and most biologically diverse tract of forest remaining in the region. Moreover, the evolution of the STN initiative demonstrates the shift from site-focused conservation to a more eco-regional or landscape strategy that incorporates the impact of human activities and the movement of animal populations across international boundaries.

that allows some sustainable development (such as small-scale farming) and some medium-impact natural resource extraction (such as selective logging and fishing). As of mid-2019, there were 73 UNESCO Biosphere Reserves in 26 different Sub-Saharan African nations (http://www.unesco.org/new/en/natural-sciences); new reserves are regularly being added.

Zoning is also proving effective in resolving conflicting demands over marine environments. Like terrestrial biosphere reserves, zoned MPAs consist of core zones where marine organisms can escape and recover from human disturbances, and multiple-use zones where activities such as harvesting of natural resources are permitted. Of course, harvesting fish and other marine species is not the only human activity that needs to be regulated. For example, many marine organisms are sensitive to anthropogenic noise, which interferes with communication and other important behaviours (Shannon et al., 2015). Creating multiple types of multiple-use areas can allow for different intensities of human activities. The is well illustrated at Eritrea's Sheik Said Marine National Park; here, only approved scientific research is allowed in the restricted zone, low-impact ecotourism activities such as snorkelling and spiritual activities are allowed in the sanctuary zone, while noisy motorboats and limited take are allowed in the habitat protection zone (Habtemariam and Fang, 2016).

> Zoned marine protected areas include core zones where marine organisms can escape human disturbance, and multiple-use zones where certain activities are permitted.

While resolving conflicting demands for space, zoning also provides benefits to biodiversity and people. For example, when compared to nearby unprotected sites, zoned MPAs typically have greater total weight of commercially important fish, greater numbers of individual fish, and greater coral reef cover (Lester et al., 2009). Conditions that allow marine organisms within MPAs to thrive, in turn, create opportunities for fish and other sea creatures to spill from the MPA into adjacent unprotected areas where they can be caught by local fishers, with a goal of a more sustainable harvest overall. A study from South Africa evaluated this hypothesis by attaching radio transmitters onto several white stumpnose (*Rhabdosargus globiceps* VU), an important fish for both commercial and recreational fisheries (Kerwath et al., 2009). This study showed that tagged fish spent 50% of their time outside the MPA, which would make them theoretically available to fishermen half of the time, while fish that did not leave protected waters produced offspring that could later disperse into multiple-use areas.

Despite the clear benefits of zoning, enforcing restrictions remains a major challenge. Even with good public outreach efforts and the threat of fines, harvesters of natural resources may still move toward and sometimes even into restricted zones to access more abundant or accessible natural resources. The resultant overharvesting at the margins of protected areas may prevent wildlife from dispersing into a wider area, which then make it hard for people that abide by the rules to access natural resources. The primary challenge in zoning is thus to find a compromise that the

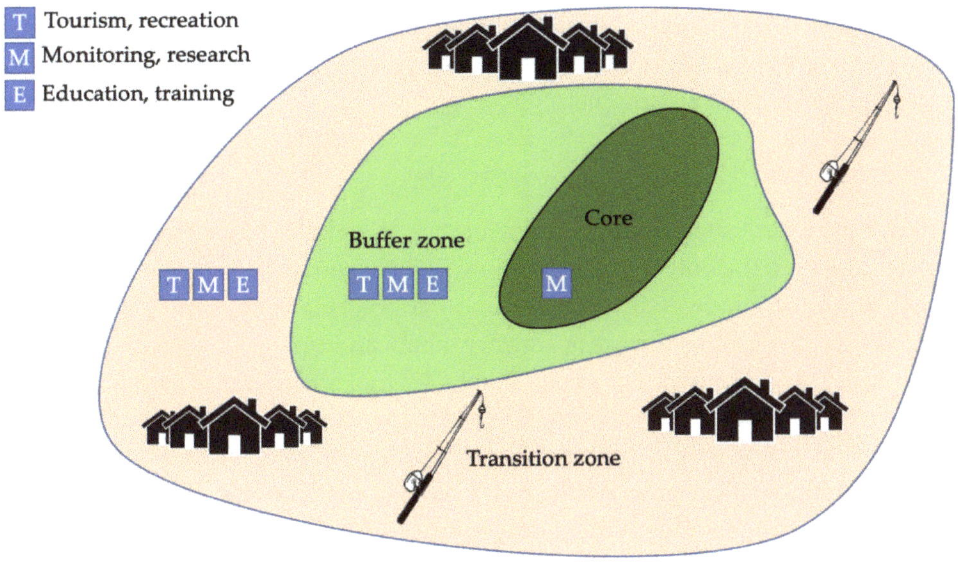

Figure 13.6 (Top) The general zones of a biosphere reserve: a core area set aside strictly for biodiversity conservation; a restricted-use buffer zone where human activities compatible with conservation are carried out; and a buffer zone dedicated to sustainable development. (Bottom) Fishermen on their traditional fishing boats in the buffer zone of Ethiopia's Lake Tana Biosphere Reserve. Photograph by Alan Davey, https://www.flickr.com/photos/adavey/2260748777, CC BY 2.0.

various stakeholders are willing to accept, and that provides opportunities for the long-term sustainable use of natural resources. Once those compromises have been agreed upon, a combination of local involvement, public outreach, clear posting of information signs, and visible enforcement of zoning restrictions can significantly increase the success of a zoning plan.

13.5.3 Connectivity among protected areas

Although large protected areas are preferable to smaller ones, sometimes small protected areas are the only available option, and conservation biologists must figure out how to protect biodiversity in these small areas. This is important in an African context, where most protected areas are very small, and only very few are sufficiently large to truly fulfil biodiversity needs (Table 13.3). To help conservation biologists meet this challenge, there is a growing body of evidence showing that small protected areas, even ones less than a hectare, can in fact be effective at maintaining viable wildlife populations. But how can that be? Does it suggest that small conservation areas are also useful for conservation purposes?

Table 13.3 A size comparison of Sub-Saharan Africa's 10 largest protected areas.

Name	Location	Size (km^2)	Established
Prince Edward Island Marine Protected Area	South Africa	181,230	2013
Termit & Tin Toumma National Nature and Cultural Reserve	Niger	97,000	2012
Ouadi Rimé-Ouadi Achim Faunal Reserve	Chad	77,950	1969
Air and Ténéré Reserves	Niger	77,360	1988
Central Kalahari Game Reserve	Botswana	52,800	1961
Namib-Naukluft National Park	Namibia	49,768	1979
Borana Controlled Hunting Area	Ethiopia	45,366	1973
Selous Game Reserve	Tanzania	44,000	1905
Ngiri-Tumba-Maindombe*	DRC	65,696	2008
Okavango Delta system*	Botswana	55,374	1996

Source: https://www.protectedplanet.net
** Ramsar wetlands*

One of the main reasons why some wildlife populations can persist in small protected areas is that these areas violate an important assumption—that protected areas are isolated from one another. But we now know that wildlife populations often disperse between protected areas through the surrounding **habitat matrix** (Pryke et al., 2015). This dispersal maintains both metapopulation dynamics (Section 11.3) and reduces the risk of deleterious genetic effects (Section 8.7.1), allowing a network of small protected areas to effectively function as one large conservation area (Wegmann et al., 2014). In contrast, reserve isolation create population sinks for wildlife meant to be protected (Newmark, 2008). Consequently, re-establishing or maintaining connectivity within protected areas networks, and particularly among small reserves, has become an important strategy for enhancing their conservation value

> Landscape connectivity may enable a network of small protected areas to effectively function as one large conservation area.

Many of the strategies used to maintain and restore ecosystem connectivity (Section 11.3) can be applied to protected areas management. However, this can be challenging given that administrative boundaries seldom consider natural ecosystem boundaries (Dallimer and Strange, 2015). Consequently, many ecosystems are artificially divided between different countries, each with its own development needs and management styles. Furthermore, many border barriers meant to restrict movement of people also restrict wildlife movement.

Bioregional management seeks to conserve such large ecosystems that cross political borders. One way to accomplish this is to establish a **transfrontier conservation area** (TFCA) (also known as Peace Park or transboundary protected area), in which two or more countries collaboratively manage a shared ecosystem for mutual benefit (Hanks, 2008; see also Box 2.2 and Box 11.3). In addition to pooling scarce resources, this cooperative management style often includes removal of human-made physical barriers such as fences to allow free movement of animals (and sometimes also people, such as pastoralists) within the TFCA (Section 11.3.1). Sub-Saharan Africa first transboundary protected area was created in 1954, with the establishment of W National Park in Benin, Burkina Baso, and Niger, so named because the River Niger is shaped like letter "W" in this area. But it was only after the creation of the Peace Parks Foundation in 1997, and the Kgalagadi Transfrontier Park 2000, on the border between Botswana and South Africa (Anderson et al., 2013), that the concept gained widespread popularity in the region.

> Transfrontier conservation areas enable two or more countries to collaboratively manage a shared ecosystem for mutual benefit.

13.5.4 What about small isolated reserves?

At times, there will be no other choice than to accept that a small reserve is the only option available to achieve in situ conservation. In those cases, it is certainly better to accept the challenge. For many species, especially plants, a small protected area is the only buffer they have against extinction (Wintle et al., 2019). Biologists in South Africa have also pioneered an initiative to maintain species that require large home ranges in small, isolated protected areas by artificially managing dispersal dynamics (see Box 8.3). Small reserves, especially those located in or near populated areas (see

Box 14.2), can also serve as locations for public outreach, conservation education, recreation, and citizen science that can improve public engagement with nature and awareness of conservation issues (Miller and Hobbs, 2002). Lastly, in addition to serving as stepping stones (Section 11.3.1), even small protected areas in urban areas provide various ecosystem services, including mitigating the urban heat island effect and reducing flooding (Feyisa et al., 2014, see also Section 7.1.6). In each of these cases, conservation biologists must creatively consider how to replicate natural processes across a small and/or fragmented protected areas network to ensure that they function on a scale that will maintain the target populations and communities.

13.6 Managing Protected Areas

Many people today have a misconception that the job of a conservation manager is done once a protected area is established. This might have been true if nature were "in balance" (a flawed concept in today's human-dominated world, see e.g. Pimm, 1991). However, reality is very different. In many cases, humans have modified the environment so much that important populations and ecosystem processes cannot be maintained without at least some intervention, even inside protected areas. It is also important to regulate the activities of people who enter protected areas, particularly those who feel that reserves and national parks are shared public spaces that should be open to hunting, fishing, logging, farming, or mining activities. If we ignore these threats by leaving protected areas unmanaged and regulations unenforced, the biodiversity they are supposed to protect will almost certainly be lost over time.

> Protected areas management should ideally be guided by a carefully-designed management plan assembled and regularly reviewed by a team of experts.

Every single protected area on Earth requires some form of management to be effective. Ideally, a protected area's management is guided by a carefully-designed management plan assembled and reviewed by a team of experts (Henschel et al., 2014). While the details of each protected area's management plan will be different, important aspects to address include monitoring and maintaining complex and adaptive ecosystems (Chapter 10), managing threatened species (Chapter 11), and providing resources, training, and memorable experiences to local people and visitors (discussed below). Management plans should also address which activities are prohibited (e.g. hunting or campfires) which activities are encouraged (e.g. wildlife photography, citizen science projects), and how rules and regulations will be enforced (Chapter 12). Lastly, the best management plans have a system in place to ensure that goals and activities are regularly reviewed and updated to account for new knowledge and experiences, and changing priorities.

In some protected areas, particularly small ones, it may be necessary to artificially maintain conditions that enable local wildlife to persist. One such example is the maintenance of natural fire regimes in fire-adapted ecosystems (Section 10.2.1). Another example is the temporary (or sometimes permanent) supply of limiting resources, such as exposed mineral licks, carcasses for scavengers, and nest boxes for bats and birds. Conservation managers might also establish artificial water sources or plant native fruit trees to support local (or translocated) wildlife. When taking such steps, it is important to strike a balance between establishing protected areas free from human influence and creating semi-natural areas in which plants and animals become so dependent on people that their persistence is not sustainable over the long term.

Management actions are generally implemented without completely understanding how the action will influence local ecosystem processes and wildlife populations. In light of this uncertainty, and despite good intentions, it should come as no surprise that some management actions may not achieve conservation goals. Some management actions may even later show to have unintended consequences that harm local biodiversity. While some actions are easy to reverse, some may put conservation managers on a cycle of reactionary management that is hard to escape. For that reason, it is important to carefully consider both the benefits and drawbacks of a management action before implementation. It is also important to be ready and willing to adapt management protocols as and when needed (see adaptive management, Section 10.2.3).

13.6.1 The importance of monitoring

An important aspect of a protected area management plan involves setting up a well-designed, long-term monitoring plan to assess whether conservation goals are being met. The exact types of information gathered will depend on the goals and objectives of each protected area, but can include tracking threatened wildlife populations, monitoring ecosystem health, or evaluating whether a threat is increasing or decreasing. These assessments may involve a wildlife survey (Section 9.1), taking regular measurements of various ecosystem indicators (Section 10.1), and/or conducting regular law enforcement monitoring (Section 12.3). In recognition of the linkages between the wellbeing of people and success of conservation (Oberholzer et al., 2010; Oldekop et al., 2016; Hauenstein et al., 2019), many conservation biologists are now also combining biodiversity monitoring with monitoring local peoples' well-being.

> A protected area management plan should include a long-term monitoring plan to assess whether conservation goals are being met.

Monitoring may highlight uncomfortable realities for conservation managers. An example could be management actions that prove to harm biodiversity (discussed above). Another uncomfortable reality is when one species needs to be prioritised over another. This is the case on protected islands off Southern Africa's west coast, where biologists have resorted to selectively culling Cape fur seals (*Actocephalus pusillus*, LC) that predate on three species of threatened seabirds; in one case, this predation led to the abandonment of an entire seabird breeding colony (Makhado et al., 2009). Even more problematic is when one threatened species causes significant harm to another. This is the case in Uganda's Kibale National Park, where chimpanzees kill as much as 12% of the area's Ugandan red colobus monkeys (*Procolobus tephrosceles*, EN) each year (Watts and Mitani, 2002; Lwanga et al., 2011). It is however important to not confuse sustainable levels of predation with real threats that can lead to extinction. For example, in Ethiopia, the big-headed African mole rat (*Tachyoryctes microcephalus*, EN) is the favoured prey of the similarly-threatened Ethiopian wolf (*Canis simensis*, EN). However, rather than predation by the wolves, habitat loss from agriculture and overgrazing is the most important threat to the survival of the mole rat (Lavrenchenko and Kennerley, 2016), as well as the wolf (Marino and Sillero-Zubiri, 2011).

The control of any wildlife population, even invasive species in protected areas, can become very emotional for the public. It may even give rise to animal rights advocacy groups that oppose or even impede conservation. Such is the case in South Africa, where a well-organised group of local citizens opposed the eradication of invasive Himalayan tahrs (*Hemitragus jemlahicus*, NT), relatives of goats, which threatened imperilled Fynbos plants in a World Heritage Site (Gaertner et al., 2016). To avoid unnecessary conflict with such citizen groups, which can quickly turn into a public relations nightmare, it is important to consider whether drastic management actions are necessary. If so, it is wise to involve and educate the public for the need of such actions at an early stage.

Because monitoring can be resource-intensive, it is important to ensure the scale and methods of monitoring are appropriate for management needs. For small reserves, tracking only a few ecosystem components during periodic site visits might be sufficient. In contrast, for large or remote protected areas, geospatial analysis with environmental data obtained through remote sensing methods (Section 10.1.1) may be a more feasible way to monitor legal and illegal human impacts, such as logging (Figure 13.7), shifting cultivation, hunting, and mining. Many protected areas are also increasingly reliant on local people, researchers, tourists, and other groups of people to contribute to monitoring, particularly through citizen science projects (Section 15.4.1).

13.6.2 The importance of working with local people

The future of a protected area almost always depends on the degree of support, neglect, or hostility it receives from people who may be living inside the protected area, or in the surrounding area. Local people are unlikely to support conservation areas where there is a history of mistrust or disagreement between them and conservation authorities, or where park managers have not worked with and/or discussed conservation goals with them. This is particularly true when local people have been displaced by conservation actions (Cross, 2015; Baker et al., 2012) or are victims of human-wildlife conflict (Section 14.4). Such victims will understandably be angry and frustrated and may even reject conservation regulations altogether. Escalating cycles of hostility due to

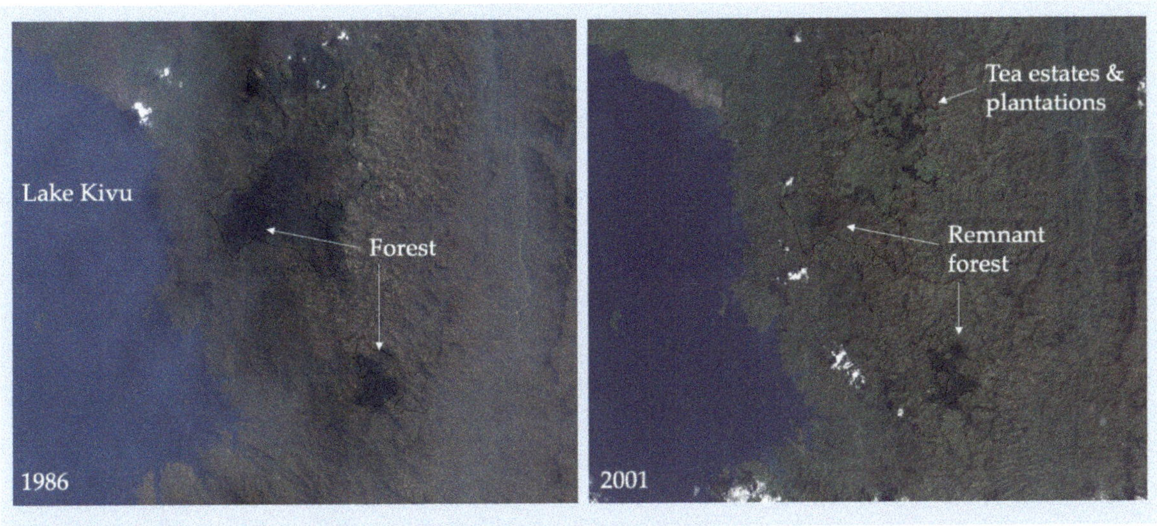

Figure 13.7 Satellite imagery provides a cost-effective method for monitoring ecosystem conditions, both inside and outside protected areas. These freely available NASA Landsat images show how Rwanda's Gishwati Forest lost 99.4% of its 1,000 km^2 forest cover between 1986 (left) and 2001 (right). The area was declared a national park in 2016, and wildlife populations have started increasing thanks to habitat protection and restoration efforts. Photographs by NASA, https://earthobservatory.nasa.gov/images/38644/gishwati-forest-rwanda, CC BY 4.0.

enforcement efforts can even lead to outright violence, during which protected areas staff, residents, and tourists can be threatened, hurt, or even killed.

To avoid such an ugly scenario, a central part of any protected area's management plan must be a policy to ensure that local communities value, and benefit from, conservation activities. The ultimate goal of such a policy should not only be to ensure that local people become strong supporters of conservation efforts, but that they later also willingly contribute to them. At a very basic level, this can be accomplished by developing a range of ecotourism opportunities, particularly those that encourage participation in citizen science projects (Section 15.4.1), and those that afford opportunities where the goals and benefits of a protected area can be explained to local people. South Africa's SANParks does this by encouraging school visits and accommodating a variety of income groups through a multi-tiered fee system (Beale et al., 2013b). When conservation displaces local people or limits activities previously allowed, it might also be worth investigating whether there is room to practice traditional activities in a sustainable way. Such is the case in South Africa, where the regional conservation authority Ezemvelo KZN Wildlife allows local people to sustainably harvest plant resources, such as thatching grass and medicinal plants, from protected areas they manage (Beale et al., 2013b; see also Section 13.5.2).

> The future of a protected area depends on the degree of support, neglect, or hostility it receives from people who live inside the protected area, or in the surrounding area.

The next level of involvement includes benefit sharing. This often takes the form of compensatory payments for people who have lost assets due to conservation actions (Hall et al., 2014; see also Section 14.4). Some park managers also provide educational and employment support to local communities. One example comes from Botswana's Okavango Delta region, where employment opportunities generated through ecotourism ventures at Moremi Game Reserve greatly improved relationships between local communities and park managers (Mbaiwa and Strongza, 2011; see also Section 14.3). African Parks, who manages 15 national parks across 10 African countries, have made local involvement (Figure 13.8) and community development a core part of their mission, which they accomplish by constructing schools, facilitating entrepreneurship, and funding healthcare services. The third level of involvement involves co-management partnerships, where local people directly participate in park management and planning (discussed in Section 13.1.4).

Figure 13.8 Wildlife experts working with African Parks fitting an elephant in Garamba National Park, DRC, with a satellite tracking device. Garamba's management staff sometimes invites chiefs and other local villagers to take part in park events; touching a live elephant and seeing how biologists, veterinarians, and other experts operate allows the visitors to connect to conservation on a very personal level. Photograph by Naftali Honig/African Parks, CC BY 4.0.

13.6.3 The importance of accommodating visitors

Developing plans that accommodate outside visitors is also an important aspect of protected areas management. Tourists are some of the most important outside visitors to attract. Their spending stimulates local economies, and provides funds for salaries, maintenance, and other conservation initiatives (Ferraro and Hanauer, 2014). When tourism activities are combined with citizen science projects (Section 15.4.1), visitors can also contribute to monitoring, further expanding the capacity of protected areas staff. Accommodating university students and other researchers is also important, as they could provide valuable information to park managers and training to staff at a steeply discounted price, compared to work by expensive outside consultants who may not always understand local dynamics.

While visitors provide significant benefits, it is important to monitor harmful elements they may knowingly or unknowingly introduce (Buckley et al., 2016). For example, visitors may introduce invasive species (Spear et al., 2013; Foxcroft et al., 2019) or induce behavioural changes in the animals they observe (Geffroy et al., 2015). Visitors may also directly damage protected ecosystems: frequent boating and diving among reefs can degrade reef communities when divers' flippers, boat hulls, and anchors crush fragile corals. Visitors may even kill wildlife directly when they trample wildflowers, disrupt nesting birds, collide into animals that are crossing roads, or spread diseases to wildlife (Ryan and Walsh, 2011). When visitor activities are restricted, especially previously-allowed activities, park managers need to be able to explain reasons for the current policies and ensure that reasonable alternatives are available. For example, if the number of tourists visiting a special wildlife spot must be restricted to prevent damage to a site, the tourists could be offered the chance to visit a different site or participate in another activity.

> While ecotourism provides opportunities for employment, income, and monitoring, it is important to manage the multiple threats introduced by visitors.

13.6.4 The IUCN Green List of Protected Areas

A challenge that park managers frequently face is objectively determining how well their protected areas are managed. While profit margins, tourist numbers, species diversity, and population indices offer some form of evaluation criteria, it is not a fool-proof system: some well-managed protected areas are not very accessible to tourists, while carelessly increasing species richness or wildlife populations will likely have detrimental consequences. Tools such as the Management Effectiveness Tracking Tool (Stolton et al., 2007), Spatial Monitoring and Reporting Tool (Moreto, 2015), and Rapid Assessment and Prioritisation of Protected Areas Management (Ervin, 2003) have helped park managers assess whether the goals of their management plans were being achieved. But with no global standard of best practices against which protected areas are objectively assessed, park managers are mostly left to evaluate success based on their own subjective criteria and goals.

To fill this gap, the IUCN recently established the Green List of Protected Areas (http://www.iucn.org/greenlist) which aims to increase the number of protected areas that are effectively and fairly managed (Figure 13.9). Nominated protected areas will be evaluated against a set of standards which attest to management structures that can achieve long-term positive impacts on biodiversity and people. This list of standards, adapted to reflect local contexts within which evaluated protected areas operate, is divided into four higher level components: (1) good governance, (2) sound design and planning, (3) effective management, and (4) successful conservation outcomes (Figure 13.10). There are even plans, through a "Fair Finance" initiative, to reward protected areas that receive Green List status by making resources available to further strengthen their accomplishments.

Figure 13.9 Ethiopia's Simien Mountains National Park, where Gelada baboons (*Theropithecus gelada*, LC) roam in packs of hundreds, and globally threatened species such as the Walia ibex (*Capra walie*, EN) and Ethiopian wolf (*Canis simensis*, EN) hang on to the edge of existence. The continued persistence of these and other species endemic to this World Heritage Site depends on effective management of protected areas such as this. Photograph by Hulivili, https://en.wikipedia.org/wiki/File:Semien_Mountains_13.jpg, CC BY 2.0.

The Green List has only recently been established; hence, not many protected areas have been evaluated by the time this book was written. Sub-Saharan Africa's first Green List sites were Kenya's Lewa Wildlife Conservancy and Ol Pejeta Conservancy, both which formed part of the 2014 initial trial period. Both sites were re-certified in 2018, when Kenya's Ol Kinyei Conservancy was also added to the Green List. Hopefully many more sites will follow suit in the near future.

13.7 Challenges for Protected Areas

The biggest challenges that park managers will face in the coming decades stem from a growing human population. When key natural resources, such as firewood

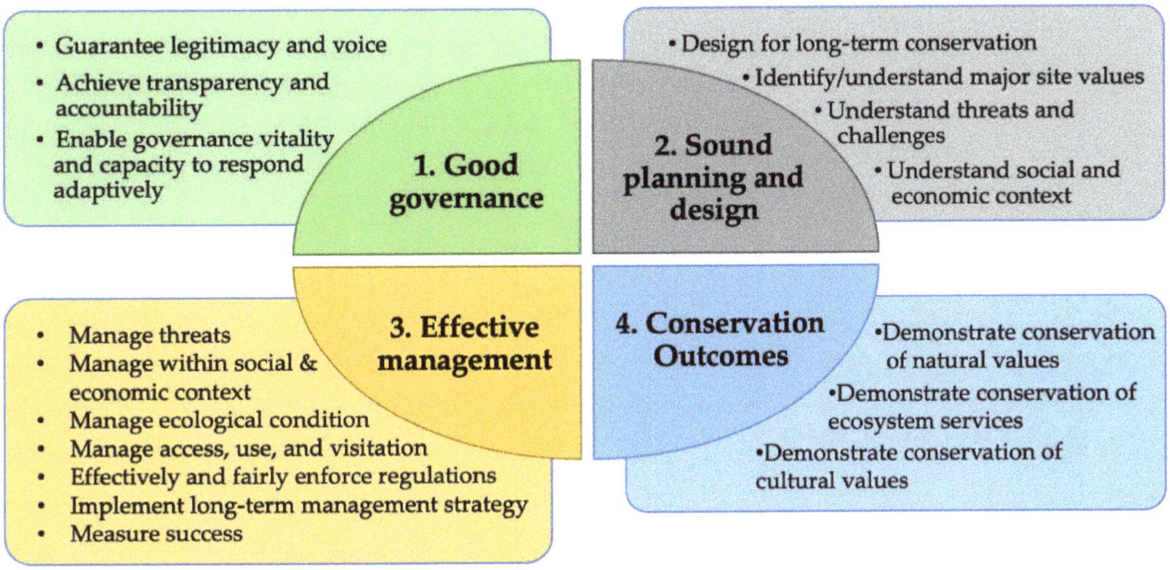

Figure 13.10 The list of generic standards, to be adapted for local contexts, against which protected areas will be evaluated before achieving IUCN Green List of Protected Areas status. After IUCN and WCPA, 2017, CC BY 4.0.

and bushmeat, become harder to find, conflict is inevitable as more people look for new lands where they can fulfil their needs. As more people encroach into protected areas, so too will habitat loss, pollution, invasive species, and diseases. Despite conservationists' best efforts to build collaborations with nearby communities, park managers need to anticipate that this ever-greater demand for space and natural resources will add additional challenges to their work plans. Below, we discuss three challenges that will likely continue to pose threats in future, and for which there are not always easy solutions.

13.7.1 Funding limitations

To enable protected areas to achieve their full potential, there must be adequate funding to support a team of well equipped, properly trained, and motivated staff (James et al. 2001; Gill et al., 2017). There is also a need for buildings, vehicles, communications equipment, and other appropriate infrastructure and resources to enable the staff to fulfil their duties, and for tourists to have a memorable time. The cost of these resources can quickly add up; for example, researchers estimated that more than $1 billion is needed each year to manage Africa's protected areas that include lion populations (Lindsey et al., 2018). Yet, Africa's protected areas are frequently understaffed, lack basic equipment, and face funding shortages (Tranquilli et al., 2014; Watson et al., 2014). Without the means to travel, communicate, and protect themselves, even motivated staff may find themselves stuck at their duty stations, unaware of what is happening elsewhere in their park. Some of these challenges can be solved with an adequate ecotourism plan, which can be facilitated from the grassroots level up or government level down. A growing number of funding mechanisms, including private and international donors, have also started to fill funding gaps (Section 15.3) which, in turn, has allowed more NGOs to assist in conservation areas management (Tranquilli et al., 2012; Lindsey et al., 2014). Above all, a carefully assembled management and monitoring plan, which is adequately funded, is key to the success of protected areas.

13.7.2 Planning for climate change

Because protected areas are fixed in space and time, many species that are currently protected will adjust their ranges beyond the borders of existing protected areas due to climate change. One study from South Africa found that 62% of bird species will lose some degree of protection over the next few decades, with five species losing at least 85% of their protected ranges (Coetzee et al., 2009). Studies in West Africa yielded remarkably similar results, where 63% of amphibians, 63% of mammals, and 55% of bird species face decreased protection due to changing climate (Baker et al., 2015). The situation is even worse for taxa with too little protection as it is. For example, suitable habitat for only 5% of African bat species is currently protected; due to climate change, it will further decrease by 2050 (Smith et al., 2016).

To ensure the future protection of species vulnerable to climate change, we must incorporate species' predicted distribution ranges into the planning of protected areas networks. For species that disperse easily, this requires protecting gaps in their current and future ranges (Hole et al., 2011), as well as protecting, maintaining, and restoring potential dispersal pathways (Section 11.3). For poor dispersers, conservationists could start experimenting with assisted colonisations, or identify and protect their climate refugia (Section 11.4). For many species, however, establishing protected areas in their future ranges will be nearly impossible simply because no land is available. These species will greatly depend on conservation efforts outside protected areas, which we will discuss in Chapter 14.

13.7.3 Facing degazettement

It may be reasonable to assume that protected areas (especially government protected areas, established by law) afford permanent protection to biodiversity on those lands. Unfortunately, that is not the case—between 1950 and 2017, at least 227 different protected areas in Sub-Saharan Africa lost (partially or fully) lost their legal protected status (WWF and CI, 2016), in a process formally known as **protected area downgrading, downsizing, and degazettement** (**PADDD**, http://www.padddtracker.org). There are a variety of reasons behind PADDDs. For example, some protected areas have been PADDDed because of environmental degradation caused by conflicting land uses, including

> Mining pressure is currently the leading cause for downgrading and degazettement of African protected areas.

illegal logging, illegal agriculture, and land invasions; in such cases governments (in consultation with conservation managers) may determine that the resources needed for land rehabilitation are better spent protecting other sites (Fuller et al., 2010). Others have been PADDDed because incorrect procedures were followed during establishment—in such cases, it might be prudent to carefully consider if a compromise could be reached that combines the goals of conservation and development (Section 14.3). However, the vast majority of African PADDDs are enacted because of more sinister motives, such as to undercut conservation restrictions (Mascia and Pailler, 2011). For example, when examining each threat individually, data from WWF and CI (2016) suggest that mining pressure was the leading cause of previous African PADDDs. Considering that nearly 30% of African protected areas are still earmarked for oil and gas exploration (Leach et al., 2016), the threat from mining will likely also continue in the foreseeable future (Durán et al., 2013; Edwards et al., 2013).

Most conservationists consider the PADDD process a bad precedent that should be avoided unless necessary. While there are legitimate reasons behind some PADDDs (Fuller et al., 2010), few are enacted with conservation goals in mind. In many cases, government officials remove the protected status of lands without even consulting conservation scientists and park managers. Such decisions are particularly frustrating when important areas that protect threatened species and ecosystems are affected. Combatting the continuing threat of PADDDs will depend on national and international conservation organisations partnering with vigilant citizens who take ownership of their natural treasures. Until citizenry can trust that government officials have the interests of their natural heritage at heart, protected areas PADDDs will remain a highly controversial topic.

13.8 Summary

1. Establishing protected areas is the most effective method for safeguarding biodiversity. Seventeen percent of Sub-Saharan Africa's land surface is included in over 7,500 protected areas, with new reserves and parks regularly designated. In contrast, only 7% of the region's marine and coastal environments are protected, with protection highly uneven among countries.
2. Government agencies and conservation organisations set priorities for establishing new protected areas based on the relative distinctiveness, endangerment, and utility of a species or ecosystems. Many protected areas are established to preserve species of special significance, unique ecosystems, wilderness areas, and concentrations of threatened species. Gap analysis is used to identify elements of biodiversity not accommodated in existing protected area networks.
3. While protected areas have previously been designed haphazardly, conservation biologists are developing guidelines for designing more

effective protected areas. As general guidelines, protected areas should be large whenever possible, and should not be fragmented. Conservation planners should also aim to create linked networks of conservation areas to encourage wildlife dispersal.

4. Protected areas must be actively managed to maintain biodiversity. Monitoring provides much needed information to evaluate whether management activities are achieving their intended objectives or need to be adapted.
5. Managing interactions with local people and visitors is critical to the success of protected areas and should be part of a management plan. To obtain and maintain local support, managements plans should consider benefit sharing and co-management partnerships.

13.9 Topics for Discussion

1. Obtain a map of your region's protected areas (e.g. nature reserves and national parks) and multiple-use managed areas (e.g. hunting and logging concessions). (https://protectedplanet.net is a good source.) If you could designate an additional protected area, where would it be? What shape would your protected area be? What would the management goals be for these additional areas? Explain all your answers.
2. Think about a protected area that you have visited. What is the main goal of this protected area? Do you think park management is succeeding in the goal? What are they doing particularly well? What could they do to manage the protected area better?
3. Think of a protected area near you that safeguards an aquatic environment, such as a beach, estuary, or lake. What unique challenges do you think the people managing that protected area face that managers of a terrestrial protected area do not face?
4. Many countries are developing protected areas that cross international borders. What are the main goals of these parks? Are they achieving their goals? What are the main challenges?
5. How can national parks continue to function optimally in countries where the central governments have largely ceased to function, and where corruption is rampant?

13.10 Suggested Readings

de Vos, A., H.S. Clements, D. Biggs, et al. 2019. The dynamics of proclaimed privately protected areas in South Africa over 83 years. *Conservation Letters* 12: e12644. https://doi.org/10.1111/

conl.12644 Tracking the growth of privately protected areas in South Africa as a function of national legislation.

Ferro, P.J., M.M. Hanauer, and K.R.E. Sims. 2011. Conditions associated with protected area success in conservation and poverty reduction. *Proceedings of the National Academy of Sciences* 108: 13913–18. https://doi.org/10.1073/pnas.1011529108 Protected areas can provide many benefits for poor people living nearby.

Fox, H.E., M.B. Mascia, X. Basurto, et al. 2012. Reexamining the science of marine protected areas: Linking knowledge to action. *Conservation Letters* 5: 1–10. https://doi.org/10.1111/j.1755-263X.2011.00207.x Biological, social, and policy issues are all important in effective marine protected areas.

Gross, J.E., S. Woodley, L.A. Welling, et al. 2016. *Adapting to Climate Change: Guidance for Protected Area Managers and Planners* (Gland: IUCN). https://portals.iucn.org/library/node/46685 Best practices guidelines for managing protected areas under climate change.

Ihwagi, F.W., T. Wang, G. Wittemyer, et al. 2015. Using poaching levels and elephant distribution to assess the conservation efficacy of private, communal and government land in northern Kenya. *PLoS ONE* 10: e0139079. https://doi.org/10.1371/journal.pone.0139079 A comparison of different types of protected areas in protecting biodiversity

Knight, A.T., A. Driver, R.M. Cowling, et al. 2006. Designing systematic conservation assessments that promote effective implementation: Best practice from South Africa. *Conservation Biology* 20: 739–50. https://doi.org/10.1111/j.1523-1739.2006.00452.x An overview of eight systematic conservation plans from design to implementation.

Mascia, M.B., and S. Pallier. 2011. Protected areas downgrading, downsizing, and degazettement (PADDD) and its conservation implications. *Conservation Letters* 4: 9–20. https://doi.org/10.1111/j.1755-263X.2010.00147.x In many areas of the world, the official protection of national parks and other conservation areas is being withdrawn.

Mbaiwa, J.E., and A.L. Stronza. 2011. Changes in resident attitudes towards tourism development and conservation in the Okavango Delta, Botswana. *Journal of Environmental Management* 92: 1950–59. https://doi.org/10.1016/j.jenvman.2011.03.009 Local people being positive about conservation.

Olds, A.D., R.M. Connolly, K.A. Pitt, et al. 2012. Habitat connectivity improves reserve performance. *Conservation Letters* 5: 56–63. https://doi.org/10.1111/j.1755-263X.2011.00204.x Networks of protected areas benefit from connectivity.

Survival International. 2014. *Parks Need People* (London: Survival International). https://assets.survivalinternational.org/documents/1324/parksneedpeoples-report.pdf Integrating cultural protection with biodiversity conservation.

Bibliography

Agnew, D.J., J. Pearce, G. Pramod, et al., 2009. Estimating the worldwide extent of illegal fishing. *PLoS ONE* 4: e4570. https://doi.org/10.1371/journal.pone.0004570

Andersson, J., M. de Garine-Wichatitsky, D. Cumming, et al. 2013. *Transfrontier Conservation Areas: People Living on the Edge* (New York: Routledge).

ARCOS (Albertine Rift Conservation Society) 2004. *A framework for conservation in the Albertine Rift: 2004–30* (Kampala: ARCOS).

Ayebare, S., A.J. Plumptre, D. Kujirakwinja, and D. Segan. 2018. Conservation of the endemic species of the Albertine Rift under future climate change. *Biological Conservation* 220: 67–75. https://doi.org/10.1016/j.biocon.2018.02.001

Baker, D.J., A.J. Hartley, N.D. Burgess, et al. 2015. Assessing climate change impacts for vertebrate fauna across the West African protected area network using regionally appropriate climate projections. *Diversity and Distributions* 21: 991–1003. https://doi.org/10.1111/ddi.12337

Baker, J., E.J. Milner-Gulland, and N. Leader-Williams. 2012. Park gazettement and integrated conservation and development as factors in community conflict at Bwindi Impenetrable Forest, Uganda. *Conservation Biology* 26: 160–70. http://doi.org/10.1111/j.1523-1739.2011.01777.x

Balme, G.A., R.O.B. Slotow, and L.T.B. Hunter. 2010. Edge effects and the impact of non-protected areas in carnivore conservation: Leopards in the Phinda–Mkhuze Complex, South Africa. *Animal Conservation* 13: 315–23. https://doi.org/10.1111/j.1469-1795.2009.00342.x

Beale, C.M., S. van Rensberg, W.J. Bond, et al. 2013b. Ten lessons for the conservation of African savannah ecosystems. *Biological Conservation* 167: 224–32. https://doi.org/10.1016/j.biocon.2013.08.025

Beale, C.M., N.E. Baker, M.J. Brewer, et al. 2013a. Protected area networks and savannah bird biodiversity in the face of climate change and land degradation. *Ecology Letters* 16: 1061–68. https://doi.org/10.1111/ele.12139

Beresford, A.E., G.M. Buchanan, P.F. Donald, et al. 2011. Poor overlap between the distribution of protected areas and globally threatened birds in Africa. *Animal Conservation* 14: 99–107. https://doi.org/10.1111/j.1469-1795.2010.00398.x

Blake, S., S.L. Deem, S. Strindberg, et al. 2008 Roadless wilderness area determines forest elephant movements in the Congo Basin. *PloS ONE* 3: e3546. https://doi.org/10.1371/journal.pone.0003546

Blom, A., R. van Zalinge, E. Mbea, et al. 2004. Human impact on wildlife populations within a protected Central African forest. *African Journal of Ecology* 42: 23–31. https://doi.org/10.1111/j.0141-6707.2004.00441.x

Blomley, T., F. Nelson, H. Doulton, et al. 2019. Scaling up community forest enterprises in Tanzania. *IIED Briefing*, April 2019. https://pubs.iied.org/pdfs/17701IIED.pdf

Brashares, J.S., P. Arcese, and M.K. Sam. 2001. Human demography and reserve size predict wildlife extinction in West Africa. *Proceedings of the Royal Society B* 268: 2473–78. https://doi.org/10.1098/rspb.2001.1815

Brashares, J.S., P. Arcese, M.K. Sam, et al. 2004. Bushmeat hunting, wildlife declines, and fish supply in West Africa. *Science* 306: 1180–83. https://doi.org/10.1126/science.1102425

Brooks, T.M., S.J. Wright, and D. Sheil. 2009. Evaluating the success of conservation actions in safeguarding tropical forest biodiversity. *Conservation Biology* 23: 1448–57. https://doi.org/10.1111/j.1523-1739.2009.01334.x

Brugiere, D., and R. Kormos. 2009. Review of the protected area network in Guinea, West Africa, and recommendations for new sites for biodiversity conservation. *Biodiversity and Conservation* 18: 847–68. https://doi.org/10.1007/s10531-008-9508-z

Buckland, S.T., and A. Johnston. 2017. Monitoring the biodiversity of regions: Key principles and possible pitfalls. *Biological Conservation* 214: 23–34. https://doi.org/10.1016/j.biocon.2017.07.034

Buckley, R.C., C. Morrison, and J.G. Castley. 2016. Net effects of ecotourism on threatened species survival. *PloS ONE* 11: e0147988. https://doi.org/10.1371/journal.pone.0147988

Carroll, R. 1992. Central African Republic. In: *The Conservation Atlas of Tropical Forests Africa*, ed. by J.A. Sayer, et al. (London: Palgrave Macmillan). https://portals.iucn.org/library/sites/library/files/documents/1992-063.pdf

Carwardine, J., K.A. Wilson, M. Watts, et al. 2008. Avoiding costly conservation mistakes: The importance of defining actions and costs in spatial priority setting. *PLoS ONE* 3: e2586. https://doi.org/10.1371/journal.pone.0002586

Chadwick. P., J. Duncan, and K. Tunley. 2014. *State of management of South Africa's marine protected areas* (Cape Town: WWF South Africa). http://awsassets.wwf.org.za/downloads/final_wwf_marine_report_02_dec_2014_web_1.pdf

Chennels, R. 1999. *What have we achieved?* (Cape Town: South African San Institute).

Clements, H., J. Baum, and G.S. Cumming. 2016. Money and motives: An organizational ecology perspective on private land conservation. *Biological Conservation* 197: 108–15. https://doi.org/10.1016/j.biocon.2016.03.002

Coetzee, B.W.T., M.P. Robertson, B.F.N. Erasmus, et al. 2009. Ensemble models predict Important Bird Areas in southern Africa will become less effective for conserving endemic birds under climate change. *Global Ecology and Biogeography* 18: 701–10. https://doi.org/10.1111/j.1466-8238.2009.00485.x

Coetzer, K.L., E.T.F. Witkowski, and B.F.N. Erasmus. 2014. Reviewing Biosphere Reserves globally: Effective conservation action or bureaucratic label? *Biological Reviews* 89: 82–104. https://doi.org/10.1111/brv.12044

Cousins, J., J. Sadler, and J. Evans. 2010. The challenge of regulating private wildlife ranches for conservation in South Africa. *Ecology and Society* 15: 28. https://www.ecologyandsociety.org/vol15/iss2/art28

Cozzi, G., F. Broekhuis, J.W. McNutt, et al. 2013. Comparison of the effects of artificial and natural barriers on large African carnivores: Implications for interspecific relationships and connectivity. *Journal of Animal Ecology* 82: 707–15. https://doi.org/10.1111/1365-2656.12039

Craigie, I.D., J.E.M. Baillie, A. Balmford, et al. 2010. Large mammal population declines in Africa's protected areas. *Biological Conservation* 143: 2221–28. https://doi.org/10.1016/j.biocon.2010.06.007

Cross, H. 2016. Displacement, disempowerment and corruption: Challenges at the interface of fisheries, management and conservation in the Bijagós Archipelago, Guinea-Bissau. *Oryx* 50: 693–701. https://doi.org/10.1017/S003060531500040X

da Silva, I.M., N. Hill, H. Shimadzu, et al. 2015. Spillover effects of a community-managed marine reserve. *PloS ONE* 10: e0111774. https://doi.org/10.1371/journal.pone.0111774

Dallimer, M., and N. Strange. 2015. Why socio-political borders and boundaries matter in conservation. *Trends in Ecology and Evolution* 30: 132–39. https://doi.org/10.1016/j.tree.2014.12.004

Darwall, W., K. Smith, T. Lowe, et al. 2005. *The status and distribution of freshwater biodiversity in Eastern Africa* (Gland: IUCN). https://www.iucn.org/downloads/the_status_and_distribution_of_freshwater_biodiversity_in_eastern_africa.pdf

de Vos, A., H.S. Clements, D. Biggs, et al. 2019. The dynamics of proclaimed privately protected areas in South Africa over 83 years. *Conservation Letters* 12: e12644. https://doi.org/10.1111/conl.12644

Dudley, N. 2008. *Guidelines for applying protected area management categories* (Gland: IUCN). https://www.iucn.org/theme/protected-areas/about/protected-area-categories

Dupain, J., A. Fowler, P. Kasalevo, et al. 2013. The process of creation of a new protected area in the Democratic Republic of Congo: The case of the Iyondji Community Bonobo Reserve (DRC). *Pan Africa News* 20: 10–13.

Durán, A.P., J. Rauch, and K.J. Gaston. 2013. Global spatial coincidence between protected areas and metal mining activities. *Biological Conservation* 160: 272–78. https://doi.org/10.1016/j.biocon.2013.02.003

Edwards, D.P., S. Sloan, L. Weng, et al. 2014. Mining and the African environment. *Conservation Letters* 7: 302–11. https://doi.org/10.1111/conl.12076

Eken, G., L. Bennun, T.M. Brooks, et al. 2004. Key biodiversity areas as site conservation targets. *BioScience* 54: 1110–18. https://doi.org/10.1641/0006-3568(2004)054[1110:KBAASC]2.0.CO;2

Endamana, D., A.K. Boedhihartono, B. Bokoto, et al. 2010. A framework for assessing conservation and development in a Congo Basin Forest Landscape. *Tropical Conservation Science* 3: 262–81. https://doi.org/10.1177%2F194008291000300303

Ervin, J. 2003. Rapid assessment of protected area management effectiveness in four countries. *BioScience* 53: 833–41. https://doi.org/10.1641/0006-3568(2003)053[0833:RAOPAM]2.0.CO;2

Ferraro, P.J., and M.M. Hanauer. 2014. Quantifying causal mechanisms to determine how protected areas affect poverty through changes in ecosystem services and infrastructure. *Proceedings of the National Academy of Sciences* 111: 4332–37. https://doi.org/10.1073/pnas.1307712111

Feyisa, G.L., K. Dons, and H. Meilby. 2014. Efficiency of parks in mitigating urban heat island effect: An example from Addis Ababa. *Landscape and Urban Planning* 123: 87–95. https://doi.org/10.1016/j.landurbplan.2013.12.008

Fishpool, L.D.C., and M.I. Evans. 2001. *Important Bird Areas in Africa and Associated Islands* (Cambridge: BirdLife International).

Foxcroft, L.C., D. Spear, N.J. van Wilgen, et al. 2019. Assessing the association between pathways of alien plant invaders and their impacts in protected areas. *NeoBiota* 43: 1–25. https://doi.org/10.3897/neobiota.43.29644

Fuller, R.A., E. McDonald-Madden, K.A. Wilson, et al. 2010. Replacing underperforming protected areas achieves better conservation outcomes. *Nature* 466: 365–67. https://doi.org/10.1038/nature09180

Gaertner, M., B.M.H. Larson, U.M. Irlich, et al. 2016. Managing invasive species in cities: A framework from Cape Town, South Africa. *Landscape and Urban Planning* 151: 1–9. https://doi.org/10.1016/j.landurbplan.2016.03.010

Gallo, J.A., L. Pasquini, B. Reyers, et al. 2009. The role of private conservation areas in biodiversity representation and target achievement within the Little Karoo region, South Africa. *Biological Conservation* 142: 446–54. https://doi.org/10.1016/j.biocon.2008.10.025

Geffroy, B., D.S.M. Samia, E. Bessa, et al. 2015. How nature-based tourism might increase prey vulnerability to predators. *Trends in Ecology and Evolution* 30: 755–65. https://doi.org/10.1016/j.tree.2015.09.010

Gill, D.A., M.B. Mascia, G.N. Ahmadia, et al. 2017. Capacity shortfalls hinder the performance of marine protected areas globally. *Nature* 543: 665–69. https://doi.org/10.1038/nature21708

Gray, C.L., S.L. Hill, T. Newbold, et al. 2016. Local biodiversity is higher inside than outside terrestrial protected areas worldwide. *Nature Communications* 7: 12306. https://doi.org/10.1038/ncomms12306

Gremillet, D., C. Peron, P. Provost, et al. 2015. Adult and juvenile European seabirds at risk from marine plundering off West Africa. *Biological Conservation* 182: 143–47. https://doi.org/10.1016/j.biocon.2014.12.001

Habtemariam, B.T., and Q. Fang. 2016. Zoning for a multiple-use marine protected area using spatial multi-criteria analysis: The case of the Sheik Seid Marine National Park in Eritrea. *Marine Policy* 63: 135–43. https://doi.org/10.1016/j.marpol.2015.10.011

Hall, J.M., N.D. Burgess, S. Rantala, et al. 2014. Ecological and social outcomes of a new protected area in Tanzania. *Conservation Biology* 28: 1512–21. https://doi.org/10.1111/cobi.12335

Hanks, J. 2003. Transfrontier Conservation Areas (TFCAs) in Southern Africa: Their role in conserving biodiversity, socioeconomic development and promoting a culture of peace. *Journal of Sustainable Forestry* 17: 127–48. https://doi.org/10.1300/J091v17n01_08

Harcourt, A.H., S.A. Parks, and R. Woodroffe. 2001. Human density as an influence on species/area relationships: Double jeopardy for small African reserves? *Biodiversity and Conservation* 10: 1011–26. https://doi.org/10.1023/A:1016680327755

Hauenstein, S., M. Kshatriya, J. Blanc, et al. 2019. African elephant poaching rates correlate with local poverty, national corruption and global ivory price. *Nature Communications* 10: 2242. https://doi.org/10.1038/s41467-019-09993-2

Henschel, P., L. Coad, C. Burton, et al. 2014. The lion in West Africa is critically endangered. *PLoS ONE* 9: e83500. https://doi.org/10.1371/journal.pone.0083500

Hockings, M., S. Stolton, F. Leverington, et al. 2006. *Evaluating Effectiveness: A Framework for Assessing Management Effectiveness of Protected Areas* (Gland: IUCN). https://portals.iucn.org/library/efiles/documents/PAG-014.pdf

Hole, D.G., B. Huntley, J. Arinaitwe, et al. 2011. Toward a management framework for networks of protected areas in the face of climate change. *Conservation Biology* 25: 305–15. https://doi.org/10.1111/j.1523-1739.2010.01633.x

Ibisch, P.L., M.T. Hoffmann, S. Kreft, et al. 2016. A global map of roadless areas and their conservation status. *Science* 354: 1423–27. https://doi.org/10.1126/science.aaf7166

Ihwagi, F.W., T. Wang, G. Wittemyer, et al. 2015. Using poaching levels and elephant distribution to assess the conservation efficacy of private, communal and government land in northern Kenya. *PLoS ONE* 10: e0139079. https://doi.org/10.1371/journal.pone.0139079

IUCN WCPA (World Commission on Protected Areas). 2017. *IUCN Green List of Protected and Conserved Areas: Standard, v. 1.1* (Gland:IUCN).

IUCN WCPA (World Commission on Protected Areas). 2018. *PARKS* 24: *Special Issue: Standard* (Gland:,IUCN). https://doi.org/10.2305/IUCN.CH.2018.PARKS-24-SI.en

James, A., K.J. Gaston, and A. Balmford. 2001. Can we afford to conserve biodiversity? *BioScience* 51: 43–52. https://doi.org/10.1641/0006-3568(2001)051[0043:CWATCB]2.0.CO;2

Jones, K.R., A.J. Plumptre, J.E.M. Watson, et al. 2016, Testing the effectiveness of surrogate species for conservation planning in the Greater Virunga Landscape, Africa. *Landscape and Urban Planning* 145: 1–11. https://doi.org/10.1016/j.landurbplan.2015.09.006

Kahumbu, P., D. Martins, J. Schieltz, et al. 2016. Transforming the next generation through citizen science in Kenya: The kids twiga tally *Swara* April–June 2016: 52–56. https://dir.princeton.edu/pdf_dir/Kids%20Twiga%20Tally.pdf

Kerwath, S.E., E.B. Thorstad, T.F. Næsje, et al. 2009. Crossing invisible boundaries: The effectiveness of the Langebaan Lagoon Marine Protected Area as a harvest refuge for a migratory fish species in South Africa. *Conservation Biology* 23: 653–61. https://doi.org/10.1111/j.1523-1739.2008.01135.x

Kiffner, C., C. Stoner, and T. Caro. 2013. Edge effects and large mammal distributions in a national park. *Animal Conservation*: 97–107. https://doi.org/10.1111/j.1469-1795.2012.00577.x

Klein, C.J., C.J. Brown, B.S. Halpern, et al. 2015. Shortfalls in the global protected area network at representing marine biodiversity. *Scientific Reports* 5: 17539. https://doi.org/10.1038/srep17539

Knight, A.T., A. Driver, R.M. Cowling, et al. 2006. Designing systematic conservation assessments that promote effective implementation: Best practice from South Africa. *Conservation Biology* 20: 739–50. https://doi.org/10.1111/j.1523-1739.2006.00452.x

Laurance, W.F., D.C. Useche, J. Rendeiro, et al. 2012. Averting biodiversity collapse in tropical forest protected areas. *Nature* 489: 290–94. https://doi.org/10.1038/nature11318

Lavrenchenko, L., and R. Kennerley. 2016. *Tachyoryctes microcephalus*. *The IUCN Red List of Threatened Species* 2016: e.T21293A115161321. http://doi.org/10.2305/IUCN.UK.2016-3.RLTS.T21293A22276163.en

Leach, K, S.E. Brooks, and S. Blyth. 2016. *Potential threat to areas of biodiversity importance from current and emerging oil and gas activities in Africa* (Cambridge: UNEP-WCMC). https://www.unep-wcmc.org/system/dataset_file_fields/files/000/000/394/original/African_threat_mapping_270716.pdf

Lester, S.E., B.S. Halpern, K. Grorud-Colvert, et al. 2009. Biological effects within no-take marine reserves: A global synthesis. *Marine Ecology Progress Series* 384: 33–46. https://doi.org/10.3354/meps08029

Lindsey P.A., V.R. Nyirenda, J.L. Barnes, et al. 2014. Underperformance of African protected area networks and the case for new conservation models: Insights from Zambia. *PLoS ONE* 9: e94109. https://doi.org/10.1371/journal.pone.0094109

Lindsey, P.A., G. Chapron, L.S. Petracca, et al. 2017. Relative efforts of countries to conserve world's megafauna. *Global Ecology and Conservation* 10: 243–52. https://doi.org/10.1016/j.gecco.2017.03.003

Lindsey, P.A., J.R.B. Miller, L.S. Petracca, et al. 2018. More than $1 billion needed annually to secure Africa's protected areas with lions. *Proceedings of the National Academy of Sciences*: E10788-E10796. https://doi.org/10.1073/pnas.1805048115

Lwanga, J.S., T.T. Struhsaker, P.J. Struhsaker, et al. 2011. Primate population dynamics over 32.9 years at Ngogo, Kibale National Park, Uganda. *American Journal of Primatology* 73: 997–1011. http://doi.org/10.1002/ajp.20965

Makhado, A.B., M.A. Meÿer, R.J.M. Crawford, et al. 2009. The efficacy of culling seals seen preying on seabirds as a means of reducing seabird mortality. *African Journal of Ecology* 47: 335–40. https://doi.org/10.1111/j.1365-2028.2008.00966.x

Marino, J., and C. Sillero-Zubiri. 2011. *Canis simensis*. *The IUCN Red List of Threatened Species* 2011: e.T3748A10051312. http://doi.org/10.2305/IUCN.UK.2011-1.RLTS.T3748A10051312.en

Mascia, M.B., and S. Pailler. 2011. Protected area downgrading, downsizing, and degazettement PADDD and its conservation implications. *Conservation Letters* 4: 9–20. https://doi.org/10.1111/j.1755-263X.2010.00147.x

Mbaiwa, J.E., and A.L. Stronza. 2011. Changes in resident attitudes towards tourism development and conservation in the Okavango Delta, Botswana. *Journal of Environmental Management* 92: 1950–59. https://doi.org/10.1016/j.jenvman.2011.03.009

McClanahan, T.R., J.M. Maina, N.A. Graham et al. 2016. Modeling reef fish biomass, recovery potential, and management priorities in the western Indian Ocean. *PloS ONE* 11: e0154585. https://doi.org/10.1371/journal.pone.0154585

McClanahan, T.R., N.A.J. Graham, J.M. Calnan, et al. 2007. Toward pristine biomass: Reef fish recovery in coral reef marine protected areas in Kenya. *Ecological Applications* 17: 1055–67. https://doi.org/10.1890/06-1450

McIntosh, E.J., R.L. Pressey, S. Lloyd, et al. 2017. The impact of systematic conservation planning. *Annual Reviews of Environment and Resources* 42: 677–97. https://doi.org/10.1146/annurev-environ-102016-060902

Miller, J.R., and R.J. Hobbs. 2002. Conservation where people live and work. *Conservation Biology* 16: 330–37. https://doi.org/10.1046/j.1523-1739.2002.00420.x

Milner, J.M., E.B. Nilsen, and H.P. Andreassen. 2007. Demographic side effects of selective hunting in ungulates and carnivores. *Conservation Biology* 21: 36–47. https://doi.org/10.1111/j.1523-1739.2006.00591.x

Mittermeier, R.A., P. Robles-Gil, M. Hoffman, et al. 2005. *Hotspots Revisited: Earth's Biologically Richest and Most Endangered Terrestrial Ecoregions* (Chicago: University of Chicago Press).

Monadjem, A., and D.K. Garcelon. 2005. Nesting distribution of vultures in relation to land use in Swaziland. *Biodiversity and Conservation 14*: 2079–93. https://doi.org/10.1007/s10531-004-4358-9

Moreto, W.D. 2015. Introducing intelligence-led conservation: Bridging crime and conservation science. *Crime Science* 4: 15. https://doi.org/10.1186/s40163-015-0030-9

Myers, N. 1988. Threatened biotas: "Hot spots" in tropical forests. *Environmentalist* 8: 187–208. https://doi.org/10.1007/BF02240252

Myers, N., R.A. Mittermeier, C.G. Mittermeier, et al. 2000. Biodiversity hotspots for conservation priorities. *Nature* 403: 853–58. https://doi.org/10.1038/35002501

Newmark, W.D. 1996. Insularization of Tanzanian parks and the local extinction of large mammals. *Conservation Biology* 10: 1549–56. https://doi.org/10.1046/j.1523-1739.1996.10061549.x

Newmark, W.D. 2008. Isolation of African protected areas. *Frontiers in Ecology and the Environment* 6: 321–28. https://doi.org/10.1890/070003

O'Leary, B.C., M. Winther-Janson, J.M. Bainbridge, et al. 2016. Effective coverage targets for ocean protection. *Conservation Letters* 9: 398–404. https://doi.org/10.1111/conl.12247

Oberholzer, S., M. Saayman, A. Saayman, et al. 2010. The socio-economic impact of Africa's oldest marine park. *Koedoe* 52: 879. https://doi.org/10.4102/koedoe.v52i1.879

Odadi, W.O., M.K. Karachi, S.A. Abdulrazak, et al. 2011. African wild ungulates compete with or facilitate cattle depending on season. *Science* 333: 1753–55. https://doi.org/10.1126/science.1208468

Ogutu, J.O., H.-P. Piepho, M.Y. Said, et al. 2016. Extreme wildlife declines and concurrent increase in livestock numbers in Kenya: What are the causes? *PLoS ONE* 11: e0163249. https://doi.org/10.1371/journal.pone.0163249

Oldekop, J.A., G. Holmes, W.E. Harris, et al. 2016. A global assessment of the social and conservation outcomes of protected areas. *Conservation Biology* 30: 133–41. https://doi.org/10.1111/cobi.12568

Olson, D.M., and E. Dinerstein. 2002. The Global 200: Priority ecoregions for global conservation. *Annals of the Missouri Botanical Garden* 89: 199–24. https://doi.org/10.2307/3298564

Pimm, S.L. 1991. *The Balance of Nature?* (Chicago: University of Chicago Press).

Plumptre, A.J., S. Ayebare, D. Kujirakwinja, and D. Segan. 2019. Conservation planning for Africa's western Rift: Conserving a biodiverse region in the face of multiple threats. *Oryx* 53: *in press*.

Plumptre, A.J., S. Ayebare, D. Segan, et al., 2016. *Conservation action plan for the Albertine Rift*. WCS Report. https://albertinerift.wcs.org/About-Us/News/articleType/ArticleView/articleId/10950/Conservation-Action-Plan-for-Albertine-Rift-identifies-key-sites-and-species.aspx

Plumptre, A.J., T.R.B. Davenport, M. Behangana, et al. 2007. The biodiversity of the Albertine Rift. *Biological Conservation* 134: 178–94. https://doi.org/10.1016/j.biocon.2006.08.021

Possingham, H.P., I.R. Ball, and S. Andelman. 2000. Mathematical methods for identifying representative reserve networks. In: *Quantitative Methods for Conservation Biology*, ed. by S. Ferson and M. Burgman (New York: Springer). https://doi.org/10.1007/b97704

Potapov, P., M.C. Hansen, L. Laestadius, et al. 2017. The last frontiers of wilderness: Tracking loss of intact forest landscapes from 2000 to 2013. *Science Advances* 3: e1600821. https://doi.org/10.1126/sciadv.1600821

Pryke, J.S., M.J. Samways, and K. de Saedeleer. 2015. An ecological network is as good as a major protected area for conserving dragonflies. *Biological Conservation* 191: 537–45. https://doi.org/10.1016/j.biocon.2015.07.036

Reid, H., D. Fig, H. Magome, et al. 2004. Co-management of contractual national parks in South Africa: Lessons from Australia. *Conservation and Society* 2: 377.

Riginos, C., L.M. Porensky, K.E. Veblen, et al. 2012. Lessons on the relationship between livestock husbandry and biodiversity from the Kenya Long-term Exclosure Experiment (KLEE). *Pastoralism* 2: 10. https://doi.org/10.1186/2041-7136-2-10

Roberts, P., C.O. Hunt, M. Arroyo-Kalin, et al. 2017. The deep human prehistory of global tropical forests and its relevance for modern conservation. *Nature Plants* 3: 17093. https://doi.org/10.1038/nplants.2017.93

Robinson, C.A.J., and M.J. Remis. 2014. Entangled Realms: Hunters and Hunted in the Dzanga-Sangha Dense Forest Reserve (APDS), Central African Republic. *Anthropological Quarterly* 87: 613–33. https://doi.org/10.1353/anq.2014.0036

Rocliffe, S., S. Peabody, M. Samoilys, et al. 2014. Towards a network of locally managed marine areas (LMMAs) in the Western Indian Ocean. *PLoS ONE* 9: e103000. https://doi.org/10.1371/journal.pone.0103000

Rodrigues, A.S.L., and T.M. Brooks. 2007. Shortcuts for biodiversity conservation planning: The effectiveness of surrogates. *Annual Review of Ecology, Evolution, and Systematics* 38: 713–37. https://doi.org/10.1146/annurev.ecolsys.38.091206.095737

Rubenstein, D.I. 2010. Ecology, social behavior, and conservation in zebras. *Advances in the Study of Behavior* 42: 231–58. https://doi.org/10.1016/S0065-3454(10)42007-0

Rubenstein, D.I., and I. Rubinoff. 2014. Mpala's Beginnings. *Mpala Memos* Apr: 5–6 2014. http://www.mpala.org/documents/Get_our_Newsletter_33_3158696297.pdf

Runge, C.A., J.E.M. Watson, S.H.M. Butchart, et al. 2015. Protected areas and global conservation of migratory birds. *Science* 350: 1255–58. https://doi.org/10.1126/science.aac9180

Ryan, S.J., and P.D. Walsh. 2011. Consequences of non-intervention for infectious disease in African great apes. *PloS ONE* 6: e29030. https://doi.org/10.1371/journal.pone.0029030

Shafer, C.L. 1997. Terrestrial nature reserve design at the urban/rural interface. In: *Conservation in Highly Fragmented Landscapes*, ed. by M.W. Schwartz (New York: Chapman and Hall). https://doi.org/10.1007/978-1-4757-0656-7_15

Shannon, G., M.F. McKenna, L.M. Angeloni, et al. 2015. A synthesis of two decades of research documenting the effects of noise on wildlife. *Biological Reviews* 91: 982–1005. https://doi.org/10.1111/brv.12207

Sims-Castley, R., G.I. Kerley, B. Geach, et al. 2005. Socio-economic significance of ecotourism-based private game reserves in South Africa's Eastern Cape Province. *Parks* 15: 6–18.

Smith, A., M.C. Schoeman, M. Keith, et al. 2016. Synergistic effects of climate and land-use change on representation of African bats in priority conservation areas. *Ecological Indicators* 69: 276–83. https://doi.org/10.1016/j.ecolind.2016.04.039

Smith, R.J., J. Easton, B.A. Nhancale, et al. 2008. Designing a transfrontier conservation landscape for the Maputaland centre of endemism using biodiversity, economic and threat data. *Biological Conservation* 141: 2127–38. https://doi.org/10.1016/j.biocon.2008.06.010

Smith, T.J., and G.F. Smith. 2004. Selecting important plant areas in southern Africa: News and views. *South African Journal of Science* 100: 434–35. https://hdl.handle.net/10520/EJC96312

Spalding, M.D., L. Fish, and L.J. Wood. 2008. Toward representative protection of the world's coasts and oceans—progress, gaps, and opportunities. *Conservation Letters* 1: 217–26. https://doi.org/10.1111/j.1755-263X.2008.00030.x

Spear, D., L.C. Foxcroft, H. Bezuidenhout, et al. 2013. Human population density explains alien species richness in protected areas. *Biological Conservation* 159: 137–47. https://doi.org/10.1016/j.biocon.2012.11.022

Stolton, S., M. Hockings, N. Dudley, et al. 2007. *Management effectiveness tracking tool: Reporting progress at protected area sites* (Gland: WWF). http://assets.panda.org/downloads/mett2_final_version_july_2007.pdf

Tranquilli, S., M. Abedi-Lartey, F. Amsini, et al. 2012. Lack of conservation effort rapidly increases African great ape extinction risk. *Conservation Letters* 5: 48–55. https://doi.org/10.1111/j.1755-263X.2011.00211.x

Tranquilli, S., M. Abedi-Lartey, K. Abernethy, et al. 2014. Protected areas in tropical Africa: Assessing threats and conservation activities. *PloS ONE* 9: e114154. https://doi.org/10.1371/journal.pone.0114154

Tydecks, L., V. Bremerich, I. Jentschke, et al. 2016. Biological field stations: A global infrastructure for research, education, and public engagement. BioScience 66: 164–71. https://doi.org/10.1093/biosci/biv174

UNEP-WCMC. 2019. *World Database on Protected Areas*. https://www.protectedplanet.net

van Rensburg, B.J., S. Hugo, N. Levin, et al. 2013. Are environmental transitions more prone to biological invasions? *Diversity and Distributions* 19: 341–51. https://doi.org/10.1111/ddi.12026

van Wilgen, B.W., and H.C. Biggs. 2011. A critical assessment of adaptive ecosystem management in a large savanna protected area in South Africa. *Biological Conservation* 144: 1179–87. https://doi.org/10.1016/j.biocon.2010.05.006

Vanak, A.T., M. Thaker, and R. Slotow. 2010. Do fences create an edge-effect on the movement patterns of a highly mobile mega-herbivore? *Biological Conservation* 143: 2631–37. https://doi.org/10.1016/j.biocon.2010.07.005

Venter, O., A. Magrach, N. Outram, et al., 2018. Bias in protected-area location and its effects on long-term aspirations of biodiversity conventions. *Conservation Biology* 32: 127–34. https://doi.org/10.1111/cobi.12970

Watson, J.E., N. Dudley, D.B. Segan, et al. 2014. The performance and potential of protected areas. *Nature* 515: 67–73. https://doi.org/10.1038/nature13947

Watson, J.E.M., K.R. Jones, R.A. Fuller, et al. 2016. Persistent disparities between recent rates of habitat conversion and protection and implications for future global conservation targets. *Conservation Letters* 9: 413–21. https://doi.org/10.1111/conl.12295

Watts, D.P., and J.C. Mitani. 2002. Hunting behavior of chimpanzees at Ngogo, Kibale National Park, Uganda. *International Journal of Primatology* 23: 1–28. https://doi.org/10.1023/A:1013270606320

Watts, M.E, I.R. Ball, R.R. Stewart, et al. 2009. Marxan with Zones: Software for optimal conservation-based land- and sea-use zoning. *Environmental Modelling and Software* 24: 1513–21. https://doi.org/10.1016/j.envsoft.2009.06.005

Wegmann M., L. Santini, B. Leutner, et al. 2014 Role of African protected areas in maintaining connectivity for large mammals. *Philosophical Transactions of the Royal Society B* 369: 20130193. https://doi.org/10.1098/rstb.2013.0193

WHC (World Heritage Committee). 2018. *W-Arly-Pendjari Complex*. http://whc.unesco.org/en/list/749

Wintle, B.A., H. Kujala, A. Whitehead, et al. 2019. Global synthesis of conservation studies reveals the importance of small habitat patches for biodiversity. *Proceedings of the National Academy of Sciences* 116: 909–14. https://doi.org/10.1073/pnas.1813051115

Woodroffe, R., and J.R. Ginsberg. 1998. Edge effects and the extinction of populations inside protected areas. *Science* 280: 2126–28. https://doi.org/10.1126/science.280.5372.2126

WWF, and CI (Conservation International). 2016. *PADDDtracker.org Data Release v. 1.1*, http://www.padddtracker.org

14. Conservation on Unprotected Lands

Farm workers tending rice fields near Monrovia, Liberia. Agricultural expansion is a major contributor to habitat loss, but is also a common cause of pollution, invasive species, and at scale, even climate change. There is an urgent need to adopt more sustainable agricultural practices and policies to ensure this sector continues to sustain food security and economic activity, but not at the expense of biodiversity. Photograph by blk24ga, https://commons.wikimedia.org/wiki/File:Liberia,_Africa_-_panoramio_(103).jpg, CC BY 3.0.

https://doi.org/10.11647/OBP.0177.14

Well-managed protected areas are essential tools for securing intact ecosystems and the biodiversity they sustain. However, it will not be sufficient to rely solely on protected areas to preserve biodiversity. Harming ecosystems such as rivers and streams on unprotected lands has repeatedly been shown to decrease biodiversity also *within* protected areas (Colvin et al., 2011; Woodborne et al., 2012). Furthermore, many species only occur on unprotected lands (Beresford et al. 2011), and some species even fare better outside protected areas (Murgatroyd et al., 2016). Other species need to move out of protected areas to access important seasonal resources: about two-thirds of Kenya's large animals regularly move from protected areas into unprotected rangelands in search of food and water (Young et al., 2005; Western et al. 2009a). Lastly, restoring damaged areas and maintaining intact ecosystems inside and outside of protected areas provides ecosystem services such as water and air purification. And so, while we must continue to pursue protected areas for the many benefits they do offer, we must not forget about the value of the spaces between protected areas. In this chapter, we will explore how efforts on unprotected lands can complement conservation efforts in protected areas.

14.1 Human-Dominated Landscapes

In every country on Earth, significant portions of unprotected lands still harbour some of their original biota (Figure 14.1). Consider, for example, remote regions that are considered "wilderness" by governments and the general public. Most of these areas are inhabited by low-density human societies that practice a traditional way of life. With relatively little outside influence from modern technology, these traditional peoples are often dependent on—and thus highly concerned with—the health of their environment. More importantly, traditional peoples have been an integral part of their environments for thousands of years. The present mixture and relative densities of wildlife in these "wildernesses" thus reflect the historical activities (e.g. fishing, hunting, fire management, land clearing, and planting of useful plant crops) of the people living in those areas (Roberts et al., 2017). These activities do not degrade the environment if human population densities remain low and natural resources are harvested sustainably. To regulate these activities, most traditional societies have an established system of rights to natural resources, known as customary laws, which an increasing number of governments recognise (Section 12.2.2). Conservation biologists should follow this example: rather than being considered a threat to the "pristine" environments in which they live, traditional peoples should be seen as important partners in conservation efforts because protecting their lifestyles also ensures the protection of biodiversity (Box 14.1).

> Traditional peoples are important partners in conservation efforts because protecting their lifestyles also ensures the protection of biodiversity.

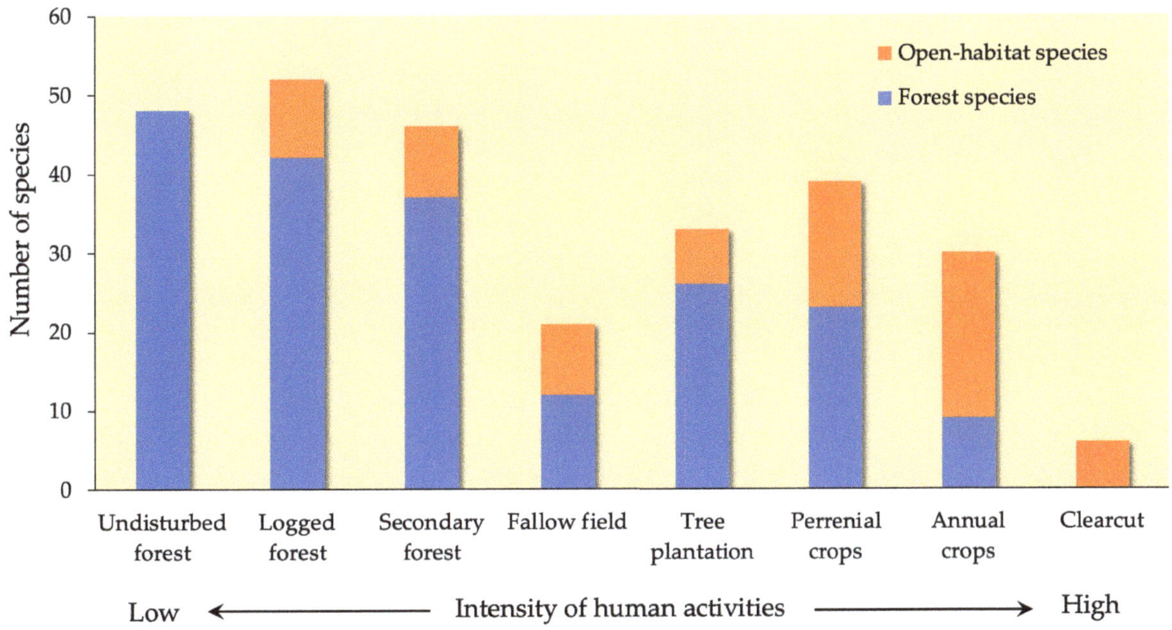

Figure 14.1 Vertebrate species richness in a variety of land-use systems in West Africa. As human impacts increase, the average number of forest species generally declines while open-habitat species increase. After Norris et al., 2010, CC BY 4.0.

Box 14.1 Traditional People and Conservation: Turning the Page

Abraham J. Miller-Rushing[1] *and John W. Wilson*

[1]*Acadia National Park, US National Park Service, Bar Harbor, ME, USA.*

How can we balance the conservation of biodiversity with the rights of traditional peoples? Conservation has a mixed history in this regard (Brockington et al., 2006). In many cases, conservation projects and traditional peoples have supported each other, but there are also many examples where vulnerable peoples have been abused and dispossessed of their ancestral lands. Here we briefly discuss three such examples from different regions of Africa: the "Pygmies" of Central Africa's forests, the "Bushmen" (also known as San, or First People) of the Kalahari Desert, and the Maasai of East Africa's savannahs.

Relocating local people from protected areas, either before or after establishment, is a relatively common strategy across the world. Amongst other reasons, these relocations are thought of as a strategy to reduce stress on wildlife from hunting and other forms of resource extraction which, in turn, enables park managers to reach their conservation and tourism goals quicker. However, in many cases relocated people are not adequately supported as they transitioned to new lifestyles, leading to increased poverty, declining health, and loss of identity. When vulnerable traditional peoples are relocated, humanity also loses a rich cultural heritage, including local knowledge about native wildlife and traditional medicines.

The Baka, a "Pygmy" tribe that lives in southern Cameroon, have greatly suffered from poorly executed conservation activities. This traditional hunter-gatherer community often identify themselves as "forest peoples" for their strong ties to the forests of Central Africa, which defines their history, culture, and livelihoods. Their ties are now under strain; over the past few decades, several hunting concessions and protected areas were established on ancestral Baka land without proper consultation. In addition to losing their ancestral land, there have also been multiple instances where law enforcement patrols arrested, abused, tortured, and even killed Baka community members that were suspected of illegal hunting (FPP, 2016). Regardless of the back story, it is inherently unjust for hunting to be legal for rich foreigners on ancestral lands where traditional peoples are now prohibited from doing so. This is all even truer when the people who forgo their traditional activities in the name of conservation do not receive compensation for their losses in return.

The second example involves the "Bushmen" (also known as San, or First People). For tens of thousands of years, the Bushmen have lived a nomadic or semi-nomadic hunting and gathering lifestyle in the Kalahari Desert of Southern Africa. Over the last few centuries however, the Bushmen have struggled to maintain their lifestyles as larger African tribes, and later European colonists have staked claims on their land. In 1961, the government of Botswana established the Central Kalahari Game Reserve to protect wildlife and dedicate a place for the Bushmen to practice their traditional lifestyles. However, the government also drilled wells, and established a school and health post which unintentionally encouraged the Bushmen to adopt a more sedentary lifestyle in which they grew crops and raised livestock in the reserve. In 1985, the government of Botswana decided that the Bushmen's lifestyles were no longer compatible with the goals of the reserve. After a process that most observers, including the UN (Anaya, 2010), considered inadequate and unethical, between 1997 and 2001 the Bushmen were banned from hunting within the reserve, and forced to relocate to settlements outside its boundaries.

In 2006, the High Court of Botswana ruled that the relocations were illegal, and that the government had to allow the Bushmen to return to the reserve. The government, however, allowed only a very limited number of people to return, prohibited traditional hunting, and took other actions that failed to meet their human rights obligations (Shaheed, 2016). The government defended these actions by saying that they were protecting biodiversity, and that the hunting pressure by Bushmen amounted to poaching of protected wildlife. To outside observers, it seemed that keeping them away from a lucrative diamond mine, established in the reserve in 2014, was the stronger motivation (Haines, 2016).

The Maasai are also very familiar with land conflict in the name of conservation. They occupy an area of 160,000 km^2 of seasonal lands in the Great Rift Valley in southern Kenya and northern Tanzania. Here they live a semi-nomadic, pastoralist lifestyle, and rely almost entirely on livestock for food and income. In the early 1900s, British colonists began taking lands away from the Maasai to create ranches and later also protected areas. This included the iconic Maasai Mara National Reserve and Serengeti National Park on either side of the Kenya-Tanzania border. Together these parks protect one of the last great mammal migrations in the world, attracting tourists from far and wide.

Maasai pastoralists who live adjacent to these and other protected areas regularly come in conflict with wildlife, particularly lions, which pose a threat to the herders and their livestock. The Maasai have a tradition of hunting lions, particularly after livestock predation, but also as a traditional rite of passage to manhood. Killing lions is now illegal in Kenya as the region's lion population is dwindling from the impacts of habitat loss and too many lion hunters. As an alternative to retaliatory killings, the Kenyan government financially compensates herders for predation losses; however, the government has not always been consistent in this compensation (Goldman et al., 2010). At times, corrupt officials have also exploited the Maasai by taking or misusing their share of tourism revenues (although corruption seems to be declining and compensation and revenue sharing have improved in recent years). Also, in Tanzania, tens of thousands of Maasai were evicted from their ancestral land in the Serengeti wildlife corridor in 2013. And yet, instead of strict protection, the land in question was earmarked for a foreign-owned hunting concession (Smith, 2014). (The Tanzanian government has since reversed this land grab and fired those behind it.)

Conservation conflicts with traditional peoples are unnecessary and avoidable. Research—done by conservation biologists—shows that traditional peoples have a relatively minor impact on the environment, especially when compared to that of commercial hunters and exploited migrant labourers who have little incentive to live sustainability (Thibault and Blaney, 2003; Poulsen et al., 2009; Fa et al., 2016). Conservation biologists and traditional peoples thus face similar concerns: habitat loss and fragmentation, human encroachment, commercial hunting and other capitalist ventures, and armed militias. Such is the case even in 2019, in southern Cameroon, where the clearing of the Meyomessala forests for Chinese-owned rubber plantations not only threatens many threatened and iconic species, but also the livelihoods of the local Baka community (Sixtus, 2008). As such, working together will benefit both conservation biologists and traditional peoples.

Fortunately, there are several models illustrating the compatibility of conservation and traditional lifestyles. One example comes from northern Kenya, where the community-run Northern Rangeland Trust sustainably manages 44,000 km^2 of land. Here, core areas are set aside for wildlife, while buffer areas are subjected to a rotational livestock grazing framework. This model benefits both livestock and wildlife by preventing overgrazing, increasing habitat heterogeneity, and maximising overall biodiversity, thereby allowing 480,000 people (from 15 ethnic groups) living in the area to enjoy peace, prosperity, and healthy ecosystems. Livestock and wildlife populations—which show limited competition under the right conditions (Kartzinel et al., 2015)—are also thriving, and at times they may even benefit one another (Odadi et al., 2011). A major accomplishment here is that the population size of Africa's rarest antelope, the hirola, doubled in only three and a half years. The area now boasts a growing tourism industry, which boosts employment, and provides revenue that is spent on health, education, and other societal needs.

Some challenges remain, most notably human-wildlife conflict, and overcoming a history of mistrust. Droughts and political instability have also contributed to tensions between conservationists and traditional peoples in some areas. Even so, conservation biologists have an obligation to follow best practices and ensure that human rights are not violated through conservation activities (Borrini-Feyerabend and Hill, 2015). There is also a need to abide by international laws and to recognise the rights of people who want to live a traditional lifestyle; this includes respecting their right to **free, prior, and informed consent** (**FPIC**, see UN-REDD, 2013) before mutually-beneficial conservation actions are implemented. With more African countries adopting a western form of land tenure that enables private land ownership, empowering traditional peoples by helping them to obtain **legal title** (the right to land ownership that is recognised by the government) can also help establish trust and locally managed protected areas where resources are harvested sustainably (Rai and Bawa, 2013). Recent studies have shown that deforestation rates decline on such legally designated traditional lands, particularly in places undergoing rapid land-use changes (Nolte et al., 2013). Gaining trust and working with traditional peoples is not always easy. But only under those circumstances can we have any hope of establishing a sustainable conservation model that will weather the test of the times.

Communal lands in East Africa that are dedicated to pastoralism provide a good example illustrating the compatibility between traditional peoples and conservation efforts (McGahey et al., 2007). In contrast to modern livestock farming systems that maintain livestock in a restricted area, **pastoralism** involves regularly herding livestock to new areas in search of fresh pasture and water. Pastoralists sometimes also use fire to enhance land productivity, which, together with intense but short-term grazing, maintains native grassland and savannah ecosystems by preventing encroachment of woody plants (Section 10.2.1). By keeping their grazing lands suitable for livestock, pastoralists also maintain those areas in a state that is suitable for native biodiversity and the natural resources they need for survival. In fact, levels of biodiversity on well-managed pastoral land may rival (Msuha et al., 2012) or even exceed (ILRI, 2006) that of adjacent protected areas where such activities are excluded.

However, when the rules governing pastoral systems break down due to agricultural developments and fences that impede pastoralist movements, the resultant breakdown of traditional grazing systems lead to overgrazing, negatively impacting people, livestock, and wildlife (Western et al., 2009b; Groom and Western, 2013). These are the kinds of threats facing the hirola (*Beatragus hunteri*, CR), also known as the Hunter's antelope, which shares much of its range with Somali pastoralists in north-eastern Kenya. With fire suppression, elephant extirpation, and a breakdown of traditional grazing systems, trees are encroaching on this facultative grazer's last remaining strongholds; the loss of grasslands is also harm cattle production in the region (Ali et al., 2017). On pastoralist lands, biodiversity conservation and human livelihoods truly operate hand in hand, but with a fragile balance that requires continued maintenance.

Biodiversity conservation and human welfare operate hand-in-hand on unprotected lands, but with a fragile balance that requires continued maintenance.

People who live in rural areas and sell natural resources that they extract from healthy ecosystems also play an important role in conservation by engaging in sustainable natural resource management. Consider all the unprotected estuaries and marine areas that support commercial fisheries for a moment. When fisheries are managed in a sustainable way, not only does this benefit the people that depend on these commercially important species, but other native species can also thrive. Such a win-win outcome was illustrated in Cameroon's mangrove swamps, where the unsustainable harvesting of mangrove trees for smoking fish was mitigated with fuel-efficient stoves. These new stoves reduced both the amount of time and firewood needed for smoking fish, benefiting the mangroves as well as the fishers' profit margins (Feka et al., 2009). Fuel-efficient stoves form an important role in reducing firewood harvesting pressure in Cameroon and many other African countries.

People living in urban centres can also contribute to conservation efforts by raising environmental awareness among fellow citizens, participating in activism, lobbying, and fundraising activities, and generating knowledge through citizen science projects (see Box 15.3). They can also help reduce the multiple pressures that their cities exert on the surrounding environment. Among the most exciting recent developments have been the development and installation of **green infrastructure**, such as urban forests, green roofs, urban wetlands, permeable sidewalks, urban farms, and rain gardens (Figure 14.2). Not only does green infrastructure save money by reducing energy consumption and pollution clean-up costs, it also reduces overall maintenance (Odefey et al., 2012) and improves overall well-being (Demuzere et al., 2014). Consequently, green infrastructure is increasingly being integrated in urban planning across the world, including North America (EPA, 2018), Europe (Natural England, 2009), Asia (e.g. Kennedy et al., 2016), and in South Africa (Culwick et al., 2016).

Figure 14.2 Green infrastructure enables people living in urban centres to reduce their pressure on natural ecosystems and live more comfortable lives. (Top) A green wall to provide cooling and air purification in Madrid, Spain. Photograph by Jean-Pierre Dalbéra, https://www.flickr.com/photos/dalbera/4657766022, CC BY 2.0. (Bottom) A constructed wetland in Harbin, China, built to provide flood control, a wildlife refuge, and a nature experience. Photograph by Richard Primack, CC BY 4.0.

Many urbanites are also eager to work with government agencies and conservation NGOs to make their cities more biodiversity friendly by restoring urban waterways and wetlands (Box 14.2), and replanting abandoned industrial sites and other damaged urban areas with native vegetation that can support pollinators, birds, and other wildlife. Such efforts foster neighbourhood pride, create a sense of community, and provide a sense of satisfaction to people who like to be close to nature. These restored areas, and other urban green spaces, can also serve to highlight the links between human well-being and nature, which may make those city dwellers who remain on the side-lines of conservation more receptive to the more challenging aspects of conservation, such as prescribed fire and invasive species management (Gaertner et al., 2016). Establishing and maintaining areas to protect biodiversity where people live and work, termed **reconciliation ecology** (Rosenzweig, 2003), will increase in importance as Africa's urban centres continue to expand over the next decades (Seto et al., 2011).

Box 14.2 Importance of Protected Areas in Cities: Insights from the City of Cape Town

Pippin M. L. Anderson

Department of Environmental and Geographical Science, University of Cape Town, South Africa.

✉ *pippin.anderson@uct.ac.za*

With urbanisation rates of 3% per year, Sub-Saharan Africa's sprawling cities are predicted to increase in area by seven times over coming decades (Anderson et al., 2014). Much of this growth will be associated with weak decentralised governance and limited resources for environmental management. Just under half of all Africans live below the poverty line, and African city-dwellers tend to make greater use of natural resources than citizens of other continents (Anderson et al., 2014). Meanwhile, biodiversity and conservation concerns tend to be eclipsed by, or be independent of, other pressures such as poverty, unemployment, and access to food, water, and housing. Despite the challenges of promoting biodiversity conservation under these conditions, sprawling cities will benefit greatly from conserving some green space within their boundaries.

The City of Cape Town in South Africa has exceptional biodiversity; the region around the city hosts some 9,000 plant species on just 90,000 km^2 of land. But as urban areas spread, natural areas are being developed to meet housing and economic needs. Elsewhere, native ecosystems are being suppressed by agricultural transformation, invasive species, annual floods and coastal erosion are intensifying, and altered fire regimes are hindering natural processes. These alterations to the natural landscape have significant economic consequences due to eroded ecosystem services (O'Farrell et al., 2012) conservatively valued at US $150–450 million per annum (de Wit et al., 2009). Similarly, urbanisation also reduces human well-being through loss of education and recreation opportunities afforded by green spaces (Goodness and Anderson, 2014).

The Edith Stephens Nature Reserve, a small (0.39 km^2) protected area in the City of Cape Town, illustrates the important role of green space in an urban environment (Figure 14.A). Despite the small size and isolated nature of the reserve, the site provides several benefits to the surrounding community. Originally established to protect a wetland that holds the threatened Cape quillwort (*Isoetes capensis*, EN), the reserve offers nature-watching opportunities to the public via a bird hide overlooking the wetland, and a quiet neutral space for socialising that is not linked to gang or political territories. The reserve also has an environmental education centre that regularly offers children's workshops and holiday programmes, as well as a teacher-training programme. Even so, the site faces considerable urban pressures, including heavy foot traffic across the reserve, nutrient run-off from the adjacent agricultural land and urban area, and informal activities, such as brick cleaning on the property's boundary.

Figure 14.A The Edith Stephens Wetland, South Africa, pictured here with Table Mountain in the background, is surrounded by urban settlement, making it an important reserve for the local communities as it provides green space and ecosystem services. The reserve's success lies in the fact that it is managed in close collaboration with the adjacent communities. Photograph by Pippin Anderson, CC BY 4.0.

The reserve was included in the Cape Flats Nature project initiated in 2002, which sought to link urban nature conservation with social justice in the city's most economically marginal areas (Katzschner, 2013). The idea was to establish conservation practices that integrate ecological sustainability with community empowerment and social well-being. Even though the Cape Flats Nature project ended in 2010, the principles of the project live on through ongoing efforts by the reserve's small staff and local volunteers recruited from adjacent communities and government public works programmes. The reserve has developed good relations with its neighbours, and the staff claim that they enjoy lower incidences of crime than many other protected areas in the city. The Edith Stephens Nature Reserve exemplifies the kind of hybrid conservation practice that is required in an urban African setting where social and biodiversity requirements need to be balanced, especially in small areas with significant development pressures.

14.1.1 The impact of agriculture

Habitat loss from agricultural expansion is arguably the biggest current challenge to biodiversity conservation in Africa (Balmford et al., 2012; Laurance et al. 2013; Maxwell et al., 2016). At the root of this problem is the need to supply food and other resources to a growing human population. Exacerbating the situation, much of Sub-Saharan Africa's arable land has already been degraded to such a degree that it cannot sustain viable food production anymore (Drechsel et al., 2001). Most of these losses are not due to natural factors, but to poor land management practices, such as overgrazing, continual ploughing of fields, and heavy use of fertilisers. These practices release nitrogen, carbon, and oxygen into the atmosphere and compromise the soil's ability to hold water, leading to erosion, soil salinisation, desertification, and even climate change (Vågen et al., 2005). This not only lead to collapsing ecosystem services, but also increased competition for space as even more land must be converted for agriculture, and to accommodate people and their activities. Such land conflicts are only going to become worse with climate change (Zabel et al., 2014).

In light of the seemingly irreconcilable conflict between agriculture and conservation, some conservation biologists have suggested that the only way in which we can secure a future for biodiversity is through a **land sparing** approach, in which agricultural investments are focussed on intensifying practices on land already dedicated to farming and no more. One of the main drawbacks of such a high-yield approach is that the impact of intensive agricultural practices degrades natural ecosystems even far from the immediate area, for example through nutrient and pesticide pollution (Section 7.1). For that reason, others support a **land sharing** approach that promotes biodiversity-friendly agricultural practices, even if that means agricultural lands continue to expand. One of the main drawbacks of this land-sharing approach is that it still alters ecosystem composition, which would threaten species that need large territories and habitats, in addition to leading to more human-wildlife conflict (Section 14.4).

While the land-sparing versus land-sharing frameworks make for good intellectual debate, the reality is that both are undesirable scenarios when carried out as opposite extremes. There is no denying that agriculture is important—it's the primary source of livelihood for millions of Africans, and critical for food security. But because there is a finite amount of land available for food production, we have no choice but to develop methods that will allow greater yields on existing agricultural lands without depleting the soil or damaging more ecosystems. In other words, we need to adopt a hybrid approach where some lands are dedicated to large protected areas where human activities are restricted, some lands are dedicated to wildlife-friendly agro-ecosystems, and some lands are used for intensive food production (Fischer et al., 2014; Law et al., 2017). Much of Africa has employed a similar structure historically and even today, where large areas that are relatively untouched or regenerating are interspersed with low-intensity traditional agricultural systems that continue to support a range of native species (Şekercioğlu, 2012). Traditional farming systems offer many strategies showing how natural ecosystem services can be used to improve yields, including the use of biocontrol and crop diversification to keep pests and diseases at bay, and planting of nitrogen-fixing legumes to improve soil fertility. This is in stark contrast to intensive modern agricultural practices dedicated to single crop specialisation; these impoverished ecosystems cannot maintain themselves but rather rely on continuous use of fertilisers and pesticides to remain productive.

With the increased realisation that biodiversity-friendly farming practices can also produce economic benefits, many government programmes have begun to promote and subsidise the adoption of **sustainable agricultural intensification** (http://www.fao.org/ag/ca/AfricaTrainingManual.html; see also Pretty et al., 2011; Garnett et al., 2013). Also known as **conservation agriculture**, this farming approach blends traditional agricultural practices with improved and locally adapted crops, as well as integrated crop and pest management strategies (Figure 14.3) to boost yields on existing farmland while creating cost and labour savings; it may even reduce the amount of land under cultivation (Stevenson et al., 2013). Some of these strategies include minimal tilling, crop rotation, intercropping, and terracing (to prevent agricultural runoff which, in turn, prevents erosion). Soil nutrient levels are enhanced through **fertiliser microdosing**, and by planting legumes, encouraging decomposition by termites, and using crop residues as mulch before composting it directly into the soil. Crop yields are further improved by maintaining windbreaks such as riparian buffer zones, which have the added benefit of enhancing the diversity of seed dispersers, biocontrol agents, and pollinators. In one study, fertiliser microdosing increased sorghum and millet yields and, thus, incomes for 25,000 smallholder farmers in Mali, Burkina Faso, and Niger by 50–130% (Tabo et al., 2011). Another region-wide study found that conservation agricultural techniques could increase average crop yields by nearly 400 kg per hectare

Biodiversity-friendly farming practices can also produce economic benefits; and so, many government programmes are now promoting sustainable agricultural intensification.

(Corbeels et al., 2014). It is important to remember that conservation agriculture is a departure from both traditional and intensive monocrop farming techniques. Adequately training farmers in best practices and new techniques is therefore crucial to programme success (Gatare et al., 2013).

Figure 14.3 A vegetable farm maintained by a local community living adjacent to Gorongosa National Park, Mozambique. The farm is maintained following several conservation agriculture principles. For example, to conserve water, the ground is covered in mulch obtained from a nearby grassy field, while a drip irrigation system was installed by park staff and a local women's association. To further strengthen local development and collaboration, part of the crop is also sold at the tourist restaurant inside the park. Photograph by Iñaki Abella Gutiérrez/Bio+, CC BY. 4.0.

To take advantage of the multiple benefits to be gained from integrated, biodiversity-friendly farming techniques, entire industries have started adopting such practices. Among the most prominent are cacao and coffee, where many growers now produce their crops under native shade trees (Box 14.3). While Africa has had a long history of shade-grown cacao and coffee production, recent decades have seen many farmers transitioning towards intensive farming in full sun, which allows for easier mechanisation. But crops grown in full sun are generally also of lower quality and more susceptible to pest outbreaks (Kellerman et al., 2008; Bisseleua et al., 2009; Tscharntke et al., 2011). In contrast, shade-grown cacao and coffee benefits both the farmers and biodiversity: a study from Ethiopia found that shade coffee farms had over double the number of bird species in comparison with nearby forest sites (Buechley et al., 2015). Shade farming could also be a strategy for farmers trying to cope with increasing temperatures due to climate change (Blaser et al., 2018). Considering the large global markets for cacao and coffee, reverting to traditional growing methods here would have a large positive impact on the environment simply due to the economy of scale.

Box 14.3 Preserving Biodiversity Through Shaded Agroforestry

Hervé D. Bisseleua

World Agroforestry Centre (ICRAF), Nairobi, Kenya.

hbissel@gmail.com

Chocolate is one of the most universal treats in the world, but could your sweet tooth be increasing biodiversity loss? The chocolate tree (*Theobroma cacao*) is traditionally grown in areas with dense and diverse canopies of shade trees, home to an abundant variety of plants and animals (Figure 14.B). The chocolate industry is strongly dependent on small-scale agriculture, but also highly vulnerable to pest and disease outbreaks, and climate change. These production challenges combined with increasing global demand for chocolate has increased economic and social pressures to achieve higher yields within a shorter timeframe. Higher yields could be achieved through reduced shade tree management and increased use of chemical pesticides and fertiliser. But these techniques lead to deforestation, biodiversity loss, and loss of ecosystem functioning. Higher yield techniques in the short term are also not sustainable over the long term: work in Cameroon and elsewhere showed that the promotion of high-yielding hybrid cacao varieties under direct full sun have contributed to more frequent outbreaks of pests and diseases (Kellerman et al., 2008; Bisseleua et al., 2009; Tscharntke et al., 2011).

To achieve more sustainable cacao production, and to benefit from ecosystem services, such as enhanced biological control of pests and diseases, and increased soil fertility, West Africa's cacao farmers are now gradually returning to agroforestry practices that embrace increased shade tree diversity. Farmers adopting these techniques are already reaping benefits. For instance, in Ghana and Cameroon, cacao yields from shaded cacao agroforestry systems are 12–23% higher compared to full sun systems (Bisseleua et al., 2009; Asare and Raebild, 2016). In eastern Côte d'Ivoire, the use of leguminous trees as shade in rehabilitated cacao plantations is also reported to increase the survival rate and yield of cacao trees (Smith Dumont et al., 2014). Cacao grown in shade may produce for 60–100 years, whereas production may only last for 20 years without shade (Obiri et al., 2007). In addition to environmental services, diversified shade trees may also provide additional income opportunities, such as timber and firewood production, medicine, local spices, and fruit, from native shade trees such as the njangsang tree (*Ricinodendron heudelotii*) and bush mango (*Irvingia gabonensis*) (Smith Dumont et al., 2014). Importantly, in all these multi-strata systems, a higher density and diversity of shade trees means higher

Figure 14.B (Top) Cacao grown as a monoculture, without shade trees, in Ghana. Biodiversity is greatly reduced and the cacao trees are more susceptible to pests and diseases in this framework. Photograph by Phillip Allman, CC BY 4.0. (Bottom) Coffee grown under a diverse canopy of trees in Ethiopia. Like cacao, coffee cultivated under shade provides a forest structure in which birds, insects, and other wildlife can flourish. Photograph by Evan Buechley, CC BY 4.0.

densities and diversity of pollinators and biological pest control agents such as ants and social wasps, which in turn increase cacao yields even more (Bisseleua et al., 2017).

In conclusion, better land management practices, such as allowing a diversity of shade trees to grow among the cacao crop, increases both biodiversity and revenue for farmers. Tropical agroforestry is thus a promising approach to reconcile biodiversity conservation and economic development. Educating farmers on shaded agroforestry systems and creating complimentary economic incentives and policies would help farmers adapt to better management practices faster, ultimately allowing agroforestry systems to contribute more to biodiversity conservation. A guilt-free sweet tooth, indeed!

Sustainable agricultural practices are much needed in the conservation portfolio of Africa, where farmers have been slow to adopt conservation agriculture. Two main challenges are noted. First, spreading ideas and innovations across different agricultural sectors has been challenging, and much work remains to implement them. Second, due to incredible ecosystem diversity across the region, specific practices are not equally suitable everywhere; there is thus need for more research into flexible practices that can be modified to meet local growing conditions (Giller et al., 2009). Strengthening coordination of agricultural research and cooperation at local and regional scales (Gonthier et al., 2014), as well as better land-use allocation (Law et al., 2015) may solve some food production challenges. Adapting biodiversity-friendly certification schemes to consider local dynamics might also encourage more African farmers to adopt biodiversity-friendly techniques (Gove et al. 2008; Buechley et al., 2015). Not only would biodiversity benefit, but these practices would enable farmers across Africa to receive higher prices for their crops and recover a large portion of the land lost to land degradation each year without the need for more land conversion.

14.1.2 The impact of logging, mining, and other extractive industries

As with intensive agriculture, high-impact resource extraction industries have not traditionally been compatible with conservation needs. These include mining, oil and gas extraction, dredging, quarrying, and logging, which have often been associated with complete ecosystem destruction. While it is easy to criticise these industries for their impact on nature, it is important for conservationists to remember that we are all dependent on those industries in some way or other, even to perform our conservation activities. Rather than criticise, it is more productive to partner with and influence with industries to contribute to conservation efforts.

There are many examples illustrating how partnerships between conservation biologists and extractive industries can benefit conservation. One of the best examples comes from the timber industry, which has the potential to greatly increase forest conservation opportunities (Clark et al., 2009). Traditionally known for leaving unsightly clear-cuts behind them, research has shown that improved logging techniques facilitate quicker ecosystem recovery after harvesting, which in turn also benefits biodiversity. For example, while the response of wildlife to logging differs depending on the harvesting method and forest type (Ofori-Boateng et al., 2013), primates (Stokes et al., 2010; Morgan et al., 2018), amphibians (Ofori-Boateng et al. 2013), and birds (Şekercioğlu, 2002) can all tolerate responsible-done light-touch logging techniques (but see also Bicknell et al., 2013; Gatti et al., 2015).

> Conservationists are also dependent on extractive industries. Rather than criticise, it is more productive to partner with and influence these industries to contribute to conservation efforts.

Guided by this research, some sectors of the timber industry have been keen to adopt more sustainable logging techniques (Figure 14.4) that focus on reducing damage to the soil, stream banks, and remaining trees, while removing just the largest trees. This approach reduces soil disturbance, erosion, waste, and carbon emissions. The Forest Stewardship Council (FSC) and other similar organisations are setting certification standards for sustainable logging, which enable certified logging operations to sell their products at higher prices on world markets. In addition to impact reduction within logged areas, certification schemes typically also require timber companies to avoid logging high conservation value forests, which is a good strategy for protecting ecosystem services and biodiversity in general. Some agroforestry companies also allow local people to cultivate rare medicinal and aromatic plants in the shaded areas on their concessions which reduces harvesting pressure on wild populations (Rao et al., 2004). Lastly, a key element for wildlife management in logged forests is to stop hunters, fishers, trappers, and plant collectors from entering the impacted area after timber harvest by closing unused logging roads. (For more discussion on the impact of the logging road on biodiversity, see Laurance et al., 2014 and Benítez-López et al., 2017.) (For an example in fisheries, see Box 7.2)

Many African mining companies have also become active partners in conservation. These partnerships include contributions like providing funding for conservation activities, participating in biodiversity offset programmes (Section 10.3.3), and subsidising conservation agriculture efforts. South Africa offers several examples illustrating how extractive industries can develop productive conservation partnerships. Two South African diamond trading companies, De Beers Group and E. Oppenheimer and Son, converted 2,500 km^2 of their properties earmarked for diamond mining and exploration into protected areas used for ecotourism and environmental research. Also, in South Africa, the petrochemical company SASOL supports a wide range of environmental programmes, including sponsoring natural history field guides, anti-poaching programs, and threatened species recovery projects.

Despite these and other examples of progress, many challenges associated with extractive industries remain unsolved. For example, pollution from these industries continues to threaten Africa's environment, and many extractive companies remain

Figure 14.4 Reduced impact logging techniques facilitate quicker ecosystem recovery. In this example from Mozambique, foresters took only the largest trees, and left some scattered logging slash (i.e. cut branches) to provide shelter for wildlife and to promote natural seed germination. Logged areas are also surrounded by stands of intact forest to promote wildlife and seed dispersal. Photograph by Johnny Wilson, CC BY 4.0.

unfriendly towards conservation activities. Many of these challenges stem from efforts to cut costs by ignoring legal and regulatory requirements, particularly when it involves stakeholders from outside Africa who prioritise short-term profits over long-term sustainability. Tardy monitoring and enforcement of regulatory requirements (sometimes driven by corruption) also remain as obstacles (Linder and Palkovitz, 2016). The best solutions for overcoming these challenges involve continued diplomacy to establish new (and strengthen existing) partnerships with such industries. Conservation biologists also need to educate industry workers and the general public to be on the lookout for environmental violations and inadequate law enforcement, which needs to be reported and addressed before more damage is done. Above all, it is important to remind legislators and other members of society that environmental damage can harm, sometimes irreversibly, our own ability to have fulfilling lives.

14.2 Smart Development Outside Conservation Areas

Infrastructure development poses a significant and escalating challenge to biodiversity conservation efforts. Dams and fences impede wildlife dispersal and migrations (Section 5.1.1), power distribution lines and high-rise buildings pose a collision hazard to birds and bats (Rushworth et al., 2014; Frick et al., 2017), and city expansions compete with biodiversity for space. Expanding road networks are particularly harmful because roads open new areas for deforestation, urban sprawl, agricultural expansion, and unsustainable hunting (Laurence et al., 2006; Benítez-López et al., 2017). A recent review found that 75% of Sub-Saharan Africa's development corridors—large-scale infrastructure developments meant to stimulate economic growth—would cut through sparsely-populated and low-quality agricultural, range, and forest lands (Laurance et al., 2015; Sloan et al., 2016). These developments not only threaten the wildlife living within those areas (Benítez-López et al., 2017), but also the carbon-sequestering potential of large swathes of tropical forests (Laurance et al., 2015). Most alarming, the corridors' planners have shown very little regard for existing biodiversity conservation efforts, given that the proposed transportation network cuts directly through 408 existing protected areas, which includes 69 national parks, biosphere reserves, World Heritage Sites, and Ramsar wetlands (Sloan et al., 2016). In contrast, only five of the 33 planned and active development corridors would cross areas of low conservation priority and with promising agricultural potential (Laurance et al., 2015).

Conservation versus development is not a zero-sum game. Rather, biodiversity conservation improves our own well-being by enabling us to obtain the necessary resources to support our livelihoods and our industries' profit margins (Chapter 4). One way to maintain these benefits while also promoting economic development (Section 15.1) is to focus on improving existing infrastructure in disturbed and populated areas, rather than creating new developments that bisect marginal lands, protected areas, and wilderness areas. Developing marginal lands and wildernesses seldom makes sense, not only because these areas are sparsely populated, but also because many are low-nutrient environments that would never support sustainable agriculture (Balmford et al., 2001; Laurance et al., 2015).

> Improving existing infrastructure in disturbed and populated areas makes more economic sense than creating new developments in wilderness areas.

When new developments are necessary, there are usually opportunities to balance diverging interests. For example, to offset the large land footprint of renewable energy, new wind farms and transmission cables could be directed to already degraded land. In many cases, the compromise might even contribute more to socio-economic developmental goals than the original plans. This was well illustrated in Tanzania, where a proposed road development would have disrupted the famous Mara-Serengeti migration route for large mammals, with potentially dire consequences to the area's ecotourism industry (Dobson et al., 2010; Holdo et al., 2011). To avoid such an impact, scientists used computer models to identify an alternative route that would not only minimise disturbance, but also achieve greater socio-economic development (Hopcraft et al., 2015). Studies, such as these, have provided important foundations for similar work to mitigate the impact of fences on wildlife (Durant et al., 2015), and by making minor adjustments to shipping lanes to reduce collisions between whales and ocean-faring vessels (Silber et al., 2012).

There are also opportunities to make existing infrastructure more wildlife friendly. Of interest is the maintenance of connectivity despite the presences of potential barriers such as fences and roads. For example, strategically placed fence-gaps and exclusionary fences, as well as tunnels placed under fences can be used to facilitate continued dispersal of selected species in fenced areas (Dupuis-Desormeaux et al., 2018). Similarly, warning signs (Figure 14.5), overpasses (e.g. Ford et al., 2009) and underpasses (e.g. Dell'Amore 2012) along paved roads can keep motorists safe from collisions with large animals. One study from Canada found that strategically placed **wildlife crossings** could reduce vehicle collisions involving large mammals by 96% (Ford et al., 2009), also reducing the chance of human injuries and damage to vehicles (Huijser et al., 2009). While this field of research is still relatively new, much headway has been made in making wildlife crossings cost-effective (https://arc-solutions.org) and determining their optimal placement (Bastille-Rousseau et al. 2018).

Figure 14.5 While road signs across Africa warn motorists of dangerous wildlife, the same strategy can be used to protect threatened species. Examples include signs warning motorists about (A) African dogs in Mozambique; (B) threatened frogs in Cape Town, South Africa; (C) threatened dung beetles in South Africa; and (D) rare turtles in South Africa. Photograph C by Vince Smith, https://www.flickr.com/photos/vsmithuk/2890596997, CC BY 2.0; otherwise by Johnny Wilson, CC BY 4.0.

14.3 Linking Conservation to Socio-Economic Development

Long-term project viability is critically linked to purposeful economic development. Therefore, conservationists are increasingly looking for ways to link conservation to sustainable development, particularly in areas that are impoverished. **Integrated conservation and development projects** (ICDP) are one of the most popular mechanisms by which this could be accomplished. ICDPs combine conservation activities and local customs with aspects of economic development, including poverty reduction, job creation, health care, and food security. A major goal of ICDPs is for local people to become involved in conservation efforts and have access to opportunities and markets for which sustainable use of natural resources is more valuable than its destructive use. Zambia's Community Markets for Conservation programme (COMACO) illustrates on how this goal can be achieved (Lewis et al., 2011). Working around the Luangwa Valley's national parks, COMACO helps food-insecure households and bushmeat hunters to meet their nutritional and income needs through sustainable production of honey, soy, Chama rice, groundnuts, and peanut butter (Figure 14.6). As additional incentive, COMACO connects participants to high-value markets where the villagers' locally crafted products and sustainably cultivated produce can earn significantly higher prices than locally. Through this project, the area's average household income has more than quadrupled, over 1,400 bushmeat hunters have adopted more sustainable lifestyles, and over 10,000 km^2 of land have been dedicated to community-conserved areas where wildlife populations are now thriving.

Figure 14.6 Many households face a stark choice between poaching/land clearing or food insecurity. As an alternative, COMACO helps rural households in Zambia meet their income and nutritional needs through several sustainable income streams. For example, groundnut farmers are provided opportunities to sell their produce at high-value markets, while groundnut shells (a waste product) are pressed into briquettes that can be used as a renewable fuel source. Photograph by COMACO, CC BY 4.0.

Community-based natural resource management (CBNRM) represents another approach in which local landowners and community groups can benefit economically from biodiversity and conservation. In previous years, government officials managed biodiversity both inside and outside protected areas through top-down mechanisms with little to no local input. Gaining little economic benefit from the wildlife on their lands, local communities had few incentives to participate in conservation efforts; in some cases, they even became hostile to conservation projects that impeded their activities (Section 13.6.2). To overcome this imbalance, centralised management systems are increasingly transitioning to CBNRM models that involve collaborative management of natural resources on private and communal lands. By empowering local communities and strengthening accountability, government officials and conservation organisations hope that CBNRM projects can simultaneously counterbalance pressures on local wildlife and contribute to economic development in ways that will have long-lasting positive impacts.

Namibia hosts one of the most ambitious CBNRM projects to date. With seed money from external funders such as the US Agency for International Development (USAID), the Namibian government granted community groups the opportunity to manage the wildlife on their own lands. To obtain these rights, interested community groups needed to form a management committee and determine the boundaries of its land, after which the government designated the group's land as a "community conservancy". Participating conservancies then worked with tourist operators—who employed members from the local communities—to provide opportunities for wildlife viewing and hunting (Naidoo et al., 2016), while also allowing tourists to learn about Namibia's cultural heritage at traditional villages. Revenues from these joint ventures were used to build and maintain even more tourist facilities, and train and pay game guards (also hired from the communal group) who monitor wildlife and human activities on the conservancies. These endeavours have been extremely successful (NACSO, 2015): from the programme's inception in 1996 to 2014, Namibia's terrestrial protected area coverage increased from 14% to 20%. Wildlife populations also rebounded: for example, Namibia's elephant population increased from 7,500 to 20,000. Local communities have since reaped the benefits (Störmer et al., 2019). For example, in just 2014, Namibia's CBNRM projects generated US $6 million in income and provided employment to 5,800 people (NACSO, 2015).

Unfortunately, maintaining programmes, even successful ones, remains challenging over the long term. Consequently, many previous ICDPs and CBNRM projects have only been partially successful. This includes Zimbabwe's iconic Communal Areas Management Programme for Indigenous Resources (CAMPFIRE) of the 1990s (Box 14.4), once considered a global model for conservation on unprotected lands. There are many reasons for these projects' partial successes and failures, including funding limitations, project over-complexity, and political instabilities (Pooley et al., 2014). Although disappointing, these failures have offered valuable lessons that enabled conservation groups to adapt to the challenges of maintaining similar projects over the long term. Today, ICDPs and CBNRM are regarded as worthy of serious consideration, with successful programmes across southern, East, West, and Central Africa (Roe et al., 2009). In addition to providing employment and food security, revenues from ICDPs and CBNRM projects have been used to build schools, clinics, and community centres; improve roads and sanitation; and establish crèches, community gardens, and nurseries (Arntzen et al., 2007; NACSO, 2015). In the end, ICDPs and CBNRM projects will be judged as successful when they can demonstrate that they can both protect wildlife and ensure improved livelihoods over the long term. To achieve these outcomes, a critical component of any ICDP or CBNRM project is the ongoing monitoring of biological, social, and economic factors to determine how effective the programmes are in meeting their goals. Involving local people in these monitoring efforts may increase information sharing and help to determine how aware the people themselves are of the benefits and challenges each project presents (Braschler, 2009).

Box 14.4 Confronting Human-Wildlife Conflict in Zimbabwe

Steven Matema[1,2]

[1]*African Conservation Trust, Applied Ecology Unit, Durban, South Africa.*

[2]*African Wildlife Economy Institute, Department of Animal Sciences, Stellenbosch University, South Africa.*

smatemah@gmail.com

In Zimbabwe, several agro-pastoral communities live at the edge of protected areas. They thus come into conflict with wildlife species such as elephants, lions, chacma baboon (*Papio ursinus*, LC), leopard, spotted hyena (*Crocuta crocuta*, LC), bushpig (*Potamochoerus africana*, LC), and common warthog (*Phacochoerus africanus*, LC) on a regular basis. Zimbabwean law does not provide compensation for crop and livestock losses due to wildlife damage; farmers, thus, develop negative attitudes towards wildlife. Fences and the use of unpalatable buffer crops have not been as successful in mitigating human-wildlife conflict as wildlife conservationists had envisioned (Parker and Osborne, 2006). Instead, lethal control has been the predominant method for managing human-wildlife conflict outside of protected areas, causing a rapid decline in native wildlife populations.

Starting in 1975, the government began to experiment with "people-centred" human-wildlife conflict management strategies (Table 14.1) by adopting the principle that good environmental stewardship is contingent on conferring use and management rights to those directly affected by wildlife depredation. This was the basis for the Communal Areas Management Programme for Indigenous Resources (CAMPFIRE). Under CAMPFIRE, smallholder agro-pastoralists, Rural District Councils (RDC, the land authorities in rural areas), and private safari operators co-manage wildlife outside protected areas and share income from controlled safari hunting and tourism (Murphree, 2009; Taylor, 2009). CAMPFIRE led to a dramatic increase in wildlife populations outside of Zimbabwe's protected areas: the elephant population increased, and the buffalo population stabilised or declined only slightly outside protected areas (Taylor, 2009). Many of the project's benefits have also been sustained, despite Zimbabwe's political volatility over time (Balint and Mashinya, 2008). Yet, in socio-economical terms, CAMPFIRE has largely failed: powerful politicians and local traditional leaders captured benefits, and natural resource governance arrangements have been politicised because political party-affiliated RDC councillors automatically chair local

Table 14.1 Policy and legislative changes for a people-centred approach to wildlife conservation in Zimbabwe, 1975–2005.

Year	Key event	Outcomes for conservation and human-wildlife conflict
1975	Parks and Wildlife Act enacted	Act gives authority to private landowners of white origin to exploit game for profit but leaves out black agro-pastoralists. Wildlife increases on private land. Human-wildlife conflict and negative attitudes toward wildlife by black agro-pastoralists persist.
1978	Wildlife Industries New Development for All (WINDFALL) programme	Culling of meat from parks and distribution to neighbouring communities as a strategy to mitigate human-wildlife conflict improves attitudes towards wildlife. Revenue sent to district councils with no local participation, decision making, or community ownership.
1982	Parks and Wildlife Act amended	The amendment makes provision for authority to be granted to district councils to manage wildlife in rural areas on behalf of the communities.
1984	CAMPFIRE conceived by Department of Parks and Wildlife Management	Target is: collective ownership with defined rights of access to natural resources, appropriate and legitimate institutions, technical and financial assistance.
1989	Authority granted to the first two Rural District Councils	Implementation of CAMPFIRE. Local participation but devolution stops at Rural District Council level.
2005	Direct payment system introduced in CAMPFIRE	Communities receive income due to them from the safari operator directly into a community bank account, bypassing the Rural District Councils (another level of elite capture of income).

Sources: Murphree, 2009; Taylor, 2009.

CAMPFIRE committees following the amendment of the Rural District Councils Act in 2002 (Matema and Andersson, 2015). Ongoing research in the Zambezi Valley also showed that trophy quality has declined since the early 2000s: the horn size of African buffalo (*Syncerus caffer*, NT) and elephant declined respectively by 42% (down from 1.35 m to 0.79 m) and 40% (down from 1.47 m to 0.91 m) between 2006 and 2014 (Matema et al., *unpublished litt.*). This suggests a decline in the number of adult animals and/or indiscriminate hunting of wildlife, indicating that CAMPFIRE may have also failed to reach its ultimate conservation goals.

Two major lessons were learnt from the CAMPFIRE experience. First, if community-based conservation is to be effective as a human-wildlife conflict mitigation strategy, attention needs to be paid to local and national political dynamics. Second, devolution—the transfer of decision power to local levels—is important. The enactment of Zimbabwe's *Indigenisation and Economic Empowerment Act* (2008), which makes provision for rural communities to form Community Share Ownership Trusts to exploit natural resources in their areas, provided a model that CAMPFIRE could have adopted to achieve complete devolution. However, the political elite has used the 2008 Act to demand shares in, or a complete take-over of, wildlife conservancies owned by ranchers of white origin. Communities living next to these conservancies have been excluded in these take-overs with the concomitant escalation of human-human conflict about wildlife (Nyahunzvi, 2014), and negative implications for local tolerance of wildlife species that kill livestock and damage crops. To curb elite capture of income, global compacts are needed, such as the recent ban of imports of wildlife trophies into the USA until there is evidence that local people are equitably sharing revenue from CAMPFIRE. The CAMPFIRE model can work and create greater tolerance for wildlife so long as it buffers local people from income losses. That means compensation in lieu of retaliation on species damaging crops and killing livestock.

14.4 Confronting Human-Wildlife Conflict

As a growing human population continues to encroach on the last remaining wildernesses, wildlife populations are facing increased competition for space and food. Inevitably, as animals are being displaced from degraded ecosystems, they will increasingly meet humans. Some of these interactions will be negative ranging from direct conflict (e.g. injury and even death to one or both participants) and indirect conflict (e.g. transmission of diseases) to opportunity costs (e.g. loss of income due to crop damage and livestock predation). Although **human-wildlife conflict** is not unique to Africa, Africans are generally very vulnerable due to high levels of poverty and dependence on land, which limits options for conflict mitigation. Managing human-wildlife conflict is thus an important issue to consider in the management of potentially dangerous species, especially near protected area borders.

14.4.1 Dealing with predators

When wildlife impedes human activities, the traditional solution is to either kill the animal or to exclude it from the area with a barrier such as a fence. Killing problem animals can take the form of pro-active lethal control to avoid losses, or retaliatory killings in response to losses. While there is a sense of instant gratification after killing a problem animal, it provides only a temporary solution at best, and may even give rise to a new set of challenges. For example, work on black-backed jackals (*Canis mesomelas*, LC) showed that killing territorial individuals may cause a breakdown in their local social structure, in turn allowing multiple roaming sub-adult animals to take advantage of the vacant territory (Minnie et al., 2016). Killing apex predators could also give rise **mesopredator release,** where medium-sized carnivores and omnivores (e.g. jackals and baboons) flourish in the absence of their natural enemies (Brashares et al., 2010). Indiscriminate poisoning and trapping also kills beneficial non-target animals that opportunistically scavenge, such as owls, vultures, and harmless ant-eating mammals (Brown, 2006; Ogada et al., 2015). Thus, while killing problem animals may seem an intuitive solution, it is seldom the best strategy.

Pastoralist communities are particularly vulnerable to predators. Because they are nomadic, pastoralists do not always have access to permanent or sturdy structures to protect their livestock and themselves. Consequently, conservation biologists are spending considerable energy on finding predatory-friendly approaches that offer lasting solutions for pastoralist communities. Among the most successful are schemes that provide compensation payments to pastoralists who forego retaliatory killings following livestock losses (Dickman et al. 2011). In Kenya, for example, compensation schemes reduced retaliatory killings of lions by 73–91% (Maclennan et al., 2009; Hazzah et al., 2014). Retaliatory killings can be reduced even more when compensation schemes are combined with other strategies; for example, one study that encouraged the use of mobile enclosures for livestock, communal herding, and "lion guardians" (the latter drawing on local knowledge and traditional values to mitigate conflict) saw a drop of 99% in retaliatory killings (Hazzah et al., 2014).

> Non-lethal control of problem animals involved in human-wildlife conflict may provide more benefits than lethal control.

Livestock on commercial and smallholder farms are also vulnerable to predation when foraging away from protective structures. Non-lethal options to reduce livestock losses under these circumstances include predator-proof fences using native thorny plants, corralling pregnant females and calves during their vulnerable periods (Schiess-Meier et al., 2007), and setting up visual, chemical, or acoustic repellents in predation hotspots. Eliminating poor livestock husbandry (Woodroffe and Frank, 2005; Gusset et al., 2009; Newsome et al., 2015) and tardy disposal of deceased animals (Humphries et al., 2015) can also avoid situations where predators are attracted to domestic animals in the first place. But perhaps one of the most successful programmes has been the use of livestock guarding animals, which could be dogs (Figure 14.7), donkeys, and other domesticated animals trained to protect livestock. A study from South Africa found that livestock depredation was eliminated on 91% of farms after the placement of guardian dogs, saving each of the 94 participant farms US $3,189 per year (Rust et al., 2013); Namibian farmers reported equally encouraging results with guardian animals (Marker et al., 2005). While there is an upfront cost involved in obtaining a guardian animal, recent work found that their deployment is generally more efficient and cost-effective than the cost of lethal options (McManus et al., 2015).

Figure 14.7 Placing large guard dogs with livestock is a highly effective, non-lethal strategy to reduce predator attacks. The mere presence of the dog is often enough to keep predators away. Several conservation organisations are now providing trained guard dogs to reduce instances of human-wildlife conflicts involving predators. Photograph by Cheetah Conservation Fund, https://en.wikipedia.org/wiki/File:Kangal_Shepherd_(livestock-guarding_dog)_and_flock_of_goats_in_Namibia.jpg, CC BY-SA 3.0.

The collaborations between farmers and conservation biologists to reduce livestock predation have benefited biodiversity conservation as well. Populations of African wild dogs (*Lycaon pictus*, EN) and lions are rebounding on some unprotected lands (Woodruffe, 2011; Blackburn et al., 2016), while farmers using guardian animals are also more tolerant of some predators on their properties (Rust et al., 2013). Not only do these farmers enjoy seeing native wildlife on their properties; some have even completely switched focus away from livestock to more profitable ecotourism (Sims-Castley et al., 2005) and wildlife ranching endeavours (Lindsey et al., 2013).

14.4.2 Dealing with crop raiders

Non-lethal management of crop-raiding animals is also a high priority. The traditional non-lethal method of dealing with potentially dangerous crop-raiding species (e.g. elephants) involves maintaining electric fences (Kioko et al., 2008), but this method is expensive and requires electricity. To overcome these challenges, conservationists and communities have developed several innovative strategies that may even supplement incomes. One such method is to establish buffer fences made of honey-producing beehives (Scheijen et al., 2019) or chilli plants (Parker and Osborn, 2006; Chang'a et al., 2016); tea plants have also been used successfully to keep crop-raiding gorillas (*Gorilla* spp.) at bay (Seiler and Robbins, 2016). Using a different approach, conservation biologists in Tanzania developed a harmless, low-cost alarm kit to deter elephants

(Bale, 2016). This four-step system involves first shining bright flashing lights at an approaching elephant, followed by loud air horns, then launching a grenade filled with chilli powder, sand, and a loud firecracker, and, as a last resort, launching exploding fireworks toward the approaching elephant.

14.4.3 Concluding thoughts on human-wildlife conflict

Whether dealing with dangerous animals or crop raiders, one of the most effective mechanisms for dealing with human-wildlife conflict is to develop awareness and opportunities for at-risk people to benefit from potentially harmful animals (Blackburn et al., 2016). Studies in northern Ethiopia found that most people—even those who have been victims of human-wildlife conflict—have positive attitudes towards wildlife and believe that they can co-exist (Eshete et al., 2015). The reason for such positive attitudes is that a substantial portion of the affected people are aware of benefits from ecosystem services, including ecotourism opportunities. Such positive attitudes toward wildlife play a crucial role in the protection of a range of endemic species in this Global Biodiversity Hotspot, including the Walia ibex (*Capra walie*, EN) and Ethiopian wolf (*Canis simensis*, EN).

> One of the most effective mechanisms for dealing with human-wildlife conflict is to develop awareness and opportunities for at-risk people to benefit from potentially harmful animals.

As discussed earlier, both ICDPs and CBNRM programmes offer opportunities for local people to gain direct benefits from local wildlife, even potentially dangerous species. There are also research opportunities to further human-wildlife conflict mitigation beyond direct benefits to local people. For example, much progress has been made in understanding how lion (Tuqa et al., 2014) and elephant (Granados et al., 2012; Chiyo et al., 2014) behaviours relate to human activities; a logical next step would be to use this information to reduce conflict (e.g. Packer et al., 2005). An increasing number of resources are available to aid these and other efforts. The IUCN Human-Wildlife Conflict Task Force has taken the lead to collate much of this information; their library (http://www.hwctf.org/resources/document-library) is sorted by species and topic. They also provide free training manuals (e.g. Parker et al., 2007) and host regular workshops.

14.5 Summary

1. Many species persist outside protected areas, in areas such as traditional farmland, sustainably logged forests, and unprotected rangelands. These areas can and must play a more important role in ongoing conservation efforts.

2. Traditional peoples that live on undeveloped land have beliefs that are compatible with biodiversity conservation. There are conservation strategies that can benefit traditional people and protect biodiversity.

3. Areas intensively used by humans can also contribute to conservation efforts. Biodiversity-friendly techniques are being developed and implemented for the agriculture, logging, and fisheries industries, many which have been adopted. Mines and other extractive industries can participate in biodiversity offset programmes, and partner with conservation organisations to contribute to local biodiversity protection. But there remains a need to monitor the activities of extractive industries to ensure that cost-cutting measures do not lead to biodiversity losses.

4. Integrated conservation and development projects (ICDPs) and community-based natural resource management (CBNRM) projects link biodiversity conservation with economic development. There is however a need to ensure these approaches remain resilient to challenges that may threaten their long-term success.

5. Human-wildlife conflict, such as livestock predation and crop raiding, remains a major conservation challenge. Multiple mechanisms have been developed to help victims overcome or mitigate such losses. Some of these mechanisms have even allowed victims to benefit from the animals they previously thought of as nuisances.

14.6 Topics for Discussion

1. Imagine that the government informs you that a highly threatened species lives on land that you planned to develop. Would you be happy, angry, confused, or proud? What are your options in terms of the planned development? What would be a fair compromise that would protect your rights and interests, the rights of the public, and the well-being of the species?

2. Imagine your country builds an expensive dam to provide hydroelectricity and water for irrigation. It will take decades to pay back the costs of construction and lost ecosystem services; some of those costs may never be recovered. Who are the winners of such a project, and who are the losers? How are each of these groups (consider both people and wildlife groups) affected? What do you think can be done to make the project more worthwhile?

3. Do you think that the purchase of "green" (environmentally-responsible) products is an effective way to promote biodiversity conservation? Would you be willing to spend more money for timber, fuelwood, coffee, chocolate, palm oil, and other products that have been produced in a sustainable way,

and if so, how much more? How could you determine whether the purchase of such products was really making a difference?

4. Think of a family (someone you know or heard of) that has been a victim of human-wildlife conflict or contracted a disease while being in nature. What happened? What did the family lose? Was the family compensated for their losses? How and by whom? If you had the opportunity to establish a plan to prevent or mitigate future conflicts, what would you do?

14.7 Suggested Readings

Balmford, A., R. Green, and B. Phalan. 2012. What conservationists need to know about farming. *Proceedings of the Royal Society B* 279: 2714–24. https://doi.org/10.1098%2Frspb.2012.0515 Farming is the basis of modern civilisation but can also be damaging to nature.

Cox, R.L., and E.C. Underwood. 2011. The importance of conserving biodiversity outside of protected areas in Mediterranean ecosystems. *PLoS ONE* 6: e14508. https://doi.org/10.1371/journal.pone.0014508 Unprotected lands have the potential to contribute to an overall conservation strategy.

Hassanali, A., H. Herren, Z.R. Khan, et al. 2008. Integrated pest management: The push-pull approach for controlling insect pests and weeds of cereals, and its potential for other agricultural systems including animal husbandry. *Philosophical Transactions of the Royal Society of London B* 363: 611–21. https://doi.org/10.1098/rstb.2007.2173 Benefits of integrated pest management strategies extend beyond pest control and increased crop yields.

Hopcraft, J.G.C., S.A.R. Mduma, M. Borner, et al. 2015. Conservation and economic benefits of a road around the Serengeti. *Conservation Biology* 29: 932–36. https://doi.org/10.1111/cobi.12470 Compromises between conservation and development might contribute more to socio-economic developmental goals than the original plans.

Laurance, W.F., S. Sloan, L. Weng, et al. 2015. Estimating the environmental costs of Africa's massive "development corridors". *Current Biology* 25: 3202–08. https://doi.org/10.1016/j.cub.2015.10.046 Il-conceived development wastes resources and harms biodiversity.

Lewis, D., S.D. Bell, J. Fay, et al. 2011. Community Markets for Conservation (COMACO) links biodiversity conservation with sustainable improvements in livelihoods and food production. *Proceedings of the National Academy of Sciences* 108: 13957–62. https://doi.org/10.1073/pnas.1011538108 An example of a programme that links conservation with socio-economic upliftment.

McManus, J.S., A.J. Dickman, D. Gaynor, et al. 2015. Dead or alive? Comparing costs and benefits of lethal and non-lethal human-wildlife conflict mitigation on livestock farms. *Oryx* 49: 687–95. https://doi.org/10.1017/S0030605313001610 Non-lethal methods provide more benefits than lethal methods in controlling predators of livestock.

Morgan, D., R. Mundry, C. Sanz, et al. 2018. African apes coexisting with logging: Comparing chimpanzee (*Pan troglodytes troglodytes*) and gorilla (*Gorilla gorilla gorilla*) resource needs and responses to forestry activities. *Biological Conservation* 218: 277–86. https://doi.org/10.1016/j.biocon.2017.10.026 Guidance for sustainable logging aimed at protecting apes.

Pretty, J., C. Toulmin, and S. Williams. 2011. Sustainable intensification in African agriculture. *International Journal of Agricultural Sustainability* 9: 5–24. https://doi.org/10.3763/ijas.2010.0583 Sustainable agricultural intensification benefits conservation and food security.

Bibliography

Ali, A.H., A.T. Ford, J.S. Evans, et al. 2017. Resource selection and landscape change reveal mechanisms suppressing population recovery for the world's most endangered antelope. *Journal of Applied Ecology* 54: 1720–29. https://doi.org/10.1111/1365-2664.12856

Anaya, J. 2010. *Report of the Special Rapporteur on the situation of human rights and fundamental freedoms of indigenous people. Addendum. The situation of indigenous peoples in Botswana* (New York: UNHRC). https://undocs.org/A/HRC/15/37/Add.2

Anderson, P.M.L., C. Okereke, A. Rudd, et al. 2014. Regional assessment of Africa. In: *Urbanization, Biodiversity and Ecosystem Services: Challenges and Opportunities: A Global Assessment*, ed. by T. Elmqvist, et al. (New York: Springer). https://doi.org/10.1007/978-94-007-7088-1

Arntzen, J., T. Setlhogile, and J. Barnes. 2007. *Rural livelihoods, poverty reduction and food security in Southern Africa: Is CBNRM the answer*? (Washington: USAID). http://unpan1.un.org/intradoc/groups/public/documents/cpsi/unpan026980.pdf

Asare, A., and A. Raebild. 2016. Tree diversity and canopy cover in cocoa systems in Ghana. *New Forests* 47: 287–302. https://doi.org/10.1007/s11056-015-9515-3

Bale, R. 2016. How chili condoms and firecrackers can help save elephants. *National Geographic.* http://on.natgeo.com/28SSeWj

Balint, P.J., and J. Mashinya. 2008. CAMPFIRE during Zimbabwe's national crisis: Local impacts and broader implications for community-based wildlife management. *Society and Natural Resources* 21: 783–96. https://doi.org/10.1080/08941920701681961

Balmford, A., J.L. Moore, T. Brooks, et al. 2001. Conservation conflicts across Africa. *Science* 291: 2616–19. https://doi.org/10.1126/science.291.5513.2616

Balmford, A., R. Green, and B. Phalan. 2012. What conservationists need to know about farming. *Proceedings of the Royal Society B* 279: 2714–24. https://doi.org/10.1098/rspb.2012.0515

Bastille-Rousseau, G., J. Wall, I. Douglas-Hamilton et al. 2018. Optimizing the positioning of wildlife crossing structures using GPS telemetry. *Journal of Applied Ecology* 55: 2055–63. https://doi.org/10.1111/1365-2664.13117

Benítez-López, A., R. Alkemade, A.M. Schipper, et al. 2017. The impact of hunting on tropical mammal and bird populations. *Science* 356: 180–83. https://doi.org/10.1126/science.aaj1891

Beresford, A.E., G.M. Buchanan, P.F. Donald, et al. 2011. Poor overlap between the distribution of protected areas and globally threatened birds in Africa. *Animal Conservation* 14: 99–107. https://doi.org/10.1111/j.1469-1795.2010.00398.x

Bicknell, J.E., M.J. Struebig, D.P. Edwards, et al., 2013. Improved timber harvest techniques maintain biodiversity in tropical forests. *Current Biology* 24: R1119–20. https://doi.org/10.1016/j.cub.2014.10.067

Bisseleua, H.B.D., A.D. Missoup, and S. Vidal. 2009. Biodiversity conservation, ecosystem functioning, and economic incentives under cocoa agroforestry intensification. *Conservation Biology* 23: 1176–84. https://doi.org/10.1111/j.1523-1739.2009.01220.x

Bisseleua, H.B.D., D. Begoude, H. Tonnang, et al. 2017. Ant-mediated ecosystem services and disservices on marketable yield in cocoa agroforestry systems. *Agriculture, Ecosystems and Environment* 247: 407–17. https://doi.org/10.1016/j.agee.2017.07.004

Blackburn, S., J.G.C. Hopcraft, J.O. Ogutu, et al. 2016. Human-wildlife conflict, benefit sharing and the survival of lions in pastoralist community-based conservancies. *Journal of Applied Ecology* 53: 1195–205. https://doi.org/10.1111/1365-2664.12632

Blaser, W.J., J. Oppong, S.P. Hart, et al., 2018. Climate-smart sustainable agriculture in low-to-intermediate shade agroforests. *Nature Sustainability* 1: 234–39. https://doi.org/10.1038/s41893-018-0062-8

Borrini-Feyerabend, G., and R. Hill. 2015. Governance for the conservation of nature. In: *Protected Area Governance and Management*, ed. by G.L. Worboys, et al. (Canberra: ANU Press). http://doi.org/10.22459/PAGM.04.2015

Braschler, B. 2009. Successfully implementing a citizen-scientist approach to insect monitoring in a resource-poor country. *BioScience* 59: 103–04. https://doi.org/10.1525/bio.2009.59.2.2

Brashares J.S., C.W. Epps, and C.J. Stoner. 2010. Ecological and conservation implications of mesopredator release. In: *Trophic Cascades*, ed. by J. Terborgh and J. Estes (Washington: Island Press).

Brockington, D., J. Igoe, and K. Schmidt-Soltau. 2006. Conservation, human rights, and poverty reduction. *Conservation Biology* 20: 250–52. https://doi.org/10.1111/j.1523-1739.2006.00335.x

Brown, C.J. 2006. *Historic distribution of large mammals in the Greater Fish River Canyon Complex, southern Namibia, and recommendations for re-introductions* (Windhoek: Namibia Nature Foundation). http://www.the-eis.com/data/literature/Greater%20Fish%20River%20Canyon%20Complex%20Historic%20distribution%20of%20mammals.pdf

Buechley, E.R., Ç. H. Şekercioğlu, A. Atickem, et al. 2015. Importance of Ethiopian shade coffee farms for forest bird conservation. *Biological Conservation* 188: 50–60. https://doi.org/10.1016/j.biocon.2015.01.011

Chang'a, A., N. de Souza, J. Muya, et al. 2016. Scaling-up the use of chili fences for reducing human-elephant conflict across landscapes in Tanzania. *Tropical Conservation Science* 9: 921–30. https://doi.org/10.1177%2F194008291600900220

Chiyo, P.I., J.W. Wilson, E.A. Archie, et al. 2014. The influence of forage, protected areas, and mating prospects on grouping patterns of male elephants. *Behavioral Ecology* 25: 1494–504. https://doi.org/10.1093/beheco/aru152

Clark, C.J., J.R. Poulsen, R. Malonga, et al. 2009. Logging concessions can extend the conservation estate for Central African tropical forests. *Conservation Biology* 23: 1281–93. https://doi.org/10.1111/j.1523-1739.2009.01243.x

Colvin, C., A. Burns, K. Schachtschneider, et al. 2011. *Coal and water futures in South Africa: The case for protecting headwaters in the Enkangala grasslands* (Cape Town: WWF-SA). http://awsassets.wwf.org.za/downloads/wwf_coal_water_report_2011_web.pdf

Corbeels M, R.K. Sakyi, R.F. Kühne, et al. 2014. Meta-analysis of crop responses to conservation agriculture in sub-Saharan Africa. *CCAFS Report No.* 12 (Copenhagen: CCAFS). https://ccafs.cgiar.org/publications/meta-analysis-crop-responses-conservation-agriculture-sub-saharan-africa

Culwick, C., and K. Bobbins. 2016. A framework for a green infrastructure planning approach in the Gauteng City-Region. *GCRO Research Report* 04 (Johannnesburg: GCRO). http://www.gcro.ac.za/media/reports/GCRO_Green_Assets_Report_Digital_version__book.pdf

de Wit, M., H. van Zyl, D. Crookes, et al. 2009. *Investing in Natural Assets: A business case for the environment in the City of Cape Town* (Cape Town: City of Cape Town).

Dell'Amore, C. 2012. Pictures: Elephant underpass reuniting Kenya herds. *National Geographic*. http://on.natgeo.com/Lkubxo

Demuzere, M., K. Orru, O. Heidrich, et al. 2014. Mitigating and adapting to climate change: Multi-functional and multi-scale assessment of green urban infrastructure. *Journal of Environmental Management* 146: 107–15. https://doi.org/10.1016/j.jenvman.2014.07.025

Dickman, A.J., E.A. Macdonald, and D.W. Macdonald. 2011. A review of financial instruments to pay for predator conservation and encourage human-carnivore coexistence. *Proceedings of the National Academy of Sciences* 108: 13937–44. https://doi.org/10.1073/pnas.1012972108

Dobson, A.P., M. Borner, A.R.E. Sinclair, et al. 2010. Road will ruin Serengeti. *Nature* 467: 272–73. https://doi.org/10.1038/467272a

Drechsel, P., L. Gyiele, D. Kunze, et al. 2001. Population density, soil nutrient depletion, and economic growth in sub-Saharan Africa. *Ecological Economics* 38: 251–58. https://doi.org/10.1016/S0921-8009(01)00167-7

Dupuis-Desormeaux, M., T.N. Kaaria, M. Mwololo, et al. 2018. A ghost fence-gap: Surprising wildlife usage of an obsolete fence crossing. *PeerJ* 6: e5950. https://dx.doi.org/10.7717%2Fpeerj.5950

Durant, S.M., M.S. Becker, S. Creel, et al. 2015. Developing fencing policies for dryland ecosystems. *Journal of Applied Ecology* 52: 544–51. https://doi.org/10.1111/1365-2664.12415

Edwards, D.P., S. Sloan, L. Weng, et al. 2014. Mining and the African environment. *Conservation Letters* 7: 302–11. https://doi.org/10.1111/conl.12076

EPA (Environmental Protection Agency). 2018. *Green Infrastructure*. https://www.epa.gov/green-infrastructure

Eshete, G., G. Tesfay, H. Bauer, et al. 2015. Community resource uses and Ethiopian wolf conservation in Mount Abune Yosef. *Environmental Management* 56: 684–94. https://doi.org/10.1007/s00267-015-0529-6

Fa, J.E., J. Olivero, M.A. Farfán, et al. 2016. Differences between Pygmy and non-Pygmy hunting in Congo Basin forests. *PLoS ONE* 11: e0161703. https://doi.org/10.1371/journal.pone.0161703

Feka, N.Z., G.B. Chuyong, and G.N. Ajonina. 2009. Sustainable utilization of mangroves using improved fish-smoking systems: A management perspective from the Douala-Edea wildlife reserve, Cameroon. *Tropical Conservation Science* 2: 450–68. https://doi.org/10.1177%2F194008290900200406

Fischer, J., D.J. Abson, V. Butsic, et al. 2014. Land sparing versus land sharing: Moving forward. *Conservation Letters* 7: 149–57. https://doi.org/10.1111/conl.12084

Ford, A.T., A.P. Clevenger, and A. Bennett. 2009. Comparison of methods of monitoring wildlife crossing-structures on highways. *Journal of Wildlife Management* 73: 1213–22. https://doi.org/10.2193/2008-387

FPP (Forest Peoples Programme). 2016. FPP Statement on survival international's complaint against WWF. *FPP Press Release*. https://www.forestpeoples.org/en/news-article/2016/fpp-statement-survival-internationals-complaint-against-wwf

Frick, W.F., E.F. Baerwald, J.F. Pollock, et al. 2017. Fatalities at wind turbines may threaten population viability of a migratory bat. *Biological Conservation* 209: 172–77. https://doi.org/10.1016/j.biocon.2017.02.023

Gaertner, M., B.M.H. Larson, U.M. Irlich, et al. 2016. Managing invasive species in cities: A framework from Cape Town, South Africa. *Landscape and Urban Planning* 151: 1–9. https://doi.org/10.1016/j.landurbplan.2016.03.010

Garnett, T., M. Appleby, A. Balmford, et al. 2013. Sustainable intensification in agriculture: Premises and policies. *Science* 341: 33–34. http://doi.org/10.1126/science.1234485

Gatere, L., J. Lehmann, S. DeGloria, et al. 2013. One size does not fit all: Conservation farming success in Africa more dependent on management than on location. *Agriculture, Ecosystems, and Environment* 179: 200–07. https://doi.org/10.1016/j.agee.2013.08.006

Gatti, R.C., S. Castaldi, J.A. Lindsell, et al. 2015. The impact of selective logging and clearcutting on forest structure, tree diversity and above-ground biomass of African tropical forests. *Ecological Research* 30: 119–32. https://doi.org/10.1007/s11284-014-1217-3

Giller, K.E., E. Witter, M. Corbeels, et al. 2009. Conservation agriculture and smallholder farming in Africa: The heretics' view. *Field Crops Research* 114: 23–34. https://doi.org/10.1016/j.fcr.2009.06.017

Goldman, N., J.R. de Pinho, and J. Perry. 2010. Maintaining complex relations with large cats: Maasai and lions in Kenya and Tanzania. *Human Dimensions of Wildlife* 15: 332–46. https://doi.org/10.1080/10871209.2010.506671

Gonthier, D.J., K.K. Ennis, S. Farinas, et al. 2014. Biodiversity conservation in agriculture requires a multi-scale approach. *Proceedings of the Royal Society B* 281: 20141358. https://doi.org/10.1098/rspb.2014.1358

Goodness, J., and P.M.L. Anderson. 2014. Local assessment of Cape Town. In: *Urbanization, Biodiversity and Ecosystem Services: Challenges and Opportunities: A Global Assessment*, ed. by T. Elmqvist, et al. (New York: Springer). https://doi.org/10.1007/978-94-007-7088-1

Gove, A.D., K. Hylander, S. Nemomisa, et al. 2008. Ethiopian coffee cultivation—Implications for bird conservation and environmental certification. *Conservation Letters* 1: 208–16. https://doi.org/10.1111/j.1755-263X.2008.00033.x

Granados, A., R.B. Weladji, and M.R. Loomis. 2012. Movement and occurrence of two elephant herds in a human-dominated landscape, the Bénoué Wildlife Conservation Area, Cameroon. *Tropical Conservation Science* 52: 150–62. https://doi.org/10.1177%2F194008291200500205

Groom, R., and D. Western. 2013. Impact of land subdivision and sedentarization on wildlife in Kenya's southern rangelands. *Rangeland Ecology and Management* 66: 1–9. https://doi.org/10.2111/REM-D-11-00021.1

Gusset, M., M.J. Swarner, L. Mponwane, et al. 2009. Human-wildlife conflict in northern Botswana: livestock predation by endangered African wild dog *Lycaon pictus* and other carnivores. *Oryx* 43: 67–72. https://doi.org/10.1017/S0030605308990475

Haines, G. 2016. Travellers urged to boycott Botswana over its treatment of Bushmen. *Telegraph.* http://tgr.ph/hid7nl

Hazzah, L., S. Dolrenry, L. Naughton, et al. 2014. Efficacy of two lion conservation programs in Maasailand, Kenya. *Conservation Biology* 28: 851–60. https://doi.org/10.1111/cobi.12244

Holdo, R.M., J.M. Fryxell, A.R.E. Sinclair, et al. 2011. Predicted impact of barriers to migration on the Serengeti wildebeest population. *PLoS ONE* 6: e16370. https://doi.org/10.1371/journal.pone.0016370

Hopcraft, J.G.C., S.A.R. Mduma, M. Borner, et al. 2015. Conservation and economic benefits of a road around the Serengeti. *Conservation Biology* 29: 932–36. https://doi.org/10.1111/cobi.12470

Huijser, M.P., J.W. Duffield, A.P. Clevenger, et al. 2009. Cost-benefit analyses of mitigation measures aimed at reducing collisions with large ungulates in the United States and Canada; a decision support tool. *Ecology and Society* 14: 15. https://www.ecologyandsociety.org/vol14/iss2/art15

Humphries, B.D., T.R. Hill, and C.T. Downs. 2015. Landowners' perspectives of black-backed jackals (*Canis mesomelas*) on farmlands in KwaZulu-Natal, South Africa. *African Journal of Ecology* 53: 540–49. https://doi.org/10.1111/aje.12247

ILRI (International Livestock Research Institute). 2006. *Pastoralism: The surest way out of poverty in East African drylands* (Nairobi: ILRI). https://cgspace.cgiar.org/handle/10568/2274

Kartzinel, T.R., P.A. Chen, T.C. Coverdale, et al. 2015. DNA metabarcoding illuminates dietary niche partitioning by African large herbivores. *Proceedings of the National Academy of Sciences* 112: 8019–24. https://doi.org/10.1073/pnas.1503283112

Katzschner, T. 2013. Cape Flats Nature: Rethinking urban ecologies. In: *Contested Ecologies: Dialogues in the South on Nature and Knowledge*, ed. by L.J.F. Green (Cape Town: HSRC Press).

Kellermann, J.L., M.D. Johnson, A.M. Stercho, et al. 2008. Ecological and economic services provided by birds on Jamaican Blue Mountain coffee farms. *Conservation Biology* 22: 1177–85. https://doi.org/10.1111/j.1523-1739.2008.00968.x

Kennedy, C., M. Zhong, and J. Corfee-Morlot. 2016. Infrastructure for China's Ecologically Balanced Civilization. *Engineering* 2: 414–25. https://doi.org/10.1016/J.ENG.2016.04.014

Kioko, J., P. Muruthi, P. Omondi, et al. 2008. The performance of electric fences as elephant barriers in Amboseli, Kenya. *South African Journal of Wildlife Research* 38: 52–58. https://doi.org/10.3957/0379-4369-38.1.52

Laurance, W.F., B.M. Croes, L. Tchignoumba, et al. 2006. Impacts of roads and hunting on central African rainforest mammals. *Conservation Biology* 20: 1251–61. https://doi.org/10.1111/j.1523-1739.2006.00420.x

Laurance, W.F., J. Sayer, and K.G. Cassman. 2014. Agricultural expansion and its impacts on tropical nature. *Trends in Ecology and Evolution* 29: 107–16. https://doi.org/10.1016/j.tree.2013.12.001

Laurance, W.F., S. Sloan, L. Weng, et al. 2015. Estimating the environmental costs of Africa's massive "development corridors". *Current Biology* 25: 3202–08. https://doi.org/10.1016/j.cub.2015.10.046

Law, E.A., B.A. Bryan, E. Meijaard, et al. 2017. Mixed policies give more options in multifunctional tropical forest landscapes. *Journal of Applied Ecology* 54: 51–60. https://doi.org/10.1111/1365-2664.12666

Law, E.A., E. Meijaard, B.A. Bryan, et al. 2015. Better land-use allocation outperforms land sparing and land sharing approaches to conservation in Central Kalimantan, Indonesia. *Biological Conservation* 186: 276–86. https://doi.org/10.1016/j.biocon.2015.03.004

Lewis, D., S.D. Bell, J. Fay, et al. 2011. Community Markets for Conservation (COMACO) links biodiversity conservation with sustainable improvements in livelihoods and food production. *Proceedings of the National Academy of Sciences* 108: 13957–62. https://doi.org/10.1073/pnas.1011538108

Linder, J.M., and R.E. Palkovitz. 2016. The threat of industrial oil palm expansion to primates and their habitats. In: *Ethnoprimatology*, ed. by M. Waller (Cham: Springer). https://doi.org/10.1007/978-3-319-30469-4

Lindsey, P.A., C.P. Havemann, R. Lines, et al. 2013. Determinants of persistence and tolerance of carnivores on Namibian ranches: Implications for conservation on southern African private lands. *PloS ONE* 8: e52458. https://doi.org/10.1371/journal.pone.0052458

Maclennan S.D., R.J. Groom, D.W. Macdonald, et al. 2009. Evaluation of a compensation scheme to bring about pastoralist tolerance of lions. *Biological Conservation* 11: 2419–27. https://doi.org/10.1016/j.biocon.2008.12.003

Marker, L.L., A.J. Dickman, and D.W. Macdonald. 2005. Perceived effectiveness of livestock-guarding dogs placed on Namibian farms. *Rangeland Ecology and Management* 58: 329–36. https://doi.org/10.2111/1551-5028(2005)058[0329:PEOLDP]2.0.CO;2

Matema, S., and J.A. Andersson. 2015. Why are lions killing us? Human-wildlife conflict and social discontent in Mbire District, northern Zimbabwe. *Journal of Modern African Studies* 53: 93–120. https://doi.org/10.1017/S0022278X14000664

Maxwell, S.L., R.A. Fuller, T.M. Brooks, et al. 2016. The ravages of guns, nets and bulldozers. *Nature* 536: 143–45. https://doi.org/10.1038/536143a

McGahey, D., J. Davies, and E. Barrow. 2008. Pastoralism as conservation in the Horn of Africa: Effective policies for conservation outcomes in the drylands of Eastern Africa. *Annals of Arid Zones* 46: 353–77.

McManus, J.S., A.J. Dickman, D. Gaynor, et al. 2015. Dead or alive? Comparing costs and benefits of lethal and non-lethal human-wildlife conflict mitigation on livestock farms. *Oryx* 49: 687–95. https://doi.org/10.1017/S0030605313001610

Minnie, L., A. Gaylard, and G.I.H. Kerley. 2016. Compensatory life-history responses of a mesopredator may undermine carnivore management efforts. *Journal of Applied Ecology* 53: 379–87. https://doi.org/10.1111/1365-2664.12581

Morgan, D., R. Mundry, C. Sanz, et al. 2018. African apes coexisting with logging: Comparing chimpanzee (*Pan troglodytes troglodytes*) and gorilla (*Gorilla gorilla gorilla*) resource needs and responses to forestry activities. *Biological Conservation* 218: 277–86. https://doi.org/10.1016/j.biocon.2017.10.026

Msuha, M.J., C. Carbone, N. Pettorelli, et al. 2012. Conserving biodiversity in a changing world: Land use change and species richness in northern Tanzania. *Biodiversity and Conservation* 21: 2747–59. http://doi.org/10.1007/s10531-012-0331-1

Murgatroyd, M., L.G. Underhill, L. Rodrigues, et al. 2016. The influence of agricultural transformation on the breeding performance of a top predator: Verreaux's Eagles in contrasting land use areas. *Condor* 118: 238–52. https://doi.org/10.1650/CONDOR-15-142.1

Murphree, M.W. 2009. The strategic pillars of communal natural resource management: Benefit, empowerment and conservation. *Biodiversity and Conservation* 18: 2551–62. https://doi.org/10.1007/s10531-009-9644-0.

NACSO (Namibian Association of CBNRM Support Organisations). 2015. *The state of community conservation in Namibia—A review of communal conservancies, community forests and other CBNRM initiatives* (Windhoek: NACSO). http://www.nacso.org.na/sites/default/files/2014-15_SoCC-Report.pdf

Naidoo, R., L.C. Weaver, R.W. Diggle, et al. 2016. Complementary benefits of tourism and hunting to communal conservancies in Namibia. *Conservation Biology* 30: 628–38. https://doi.org/10.1111/cobi.12643

Natural England. 2009. *Natural England's green infrastructure guidance* (NE176) (York: Natural England). http://publications.naturalengland.org.uk/publication/35033

Newsome, T.M., J.A. Dellinger, C.R. Pavey, et al. 2015. The ecological effects of providing resource subsidies to predators. *Global Ecology and Biogeography* 24: 1–11. https://doi.org/10.1111/geb.12236

Nolte, C., A. Agrawal, K.M. Silvus, et al. 2013. Governance regime and location influence avoided deforestation success of protected areas in the Brazilian Amazon. *Proceedings of the National Academy of Sciences* 110: 4956–5961. https://doi.org/10.1073/pnas.1214786110

Norris, K., A. Asase, B. Collen, et al. 2010. Biodiversity in a forest-agriculture mosaic—The changing face of West African rainforests. *Biological Conservation* 143: 2341–50. https://doi.org/10.1016/j.biocon.2009.12.032

Nyahunzvi, D.K. 2014. The resurgence in resource nationalism and private protected areas: Through the lens of Save Valley Conservancy's indigenisation. *Journal of Nature Conservation* 22: 343–46. https://doi.org/10.1016/j.jnc.2013.08.003

O'Farrell, P.J., P.M.L. Anderson, D. le Maitre, et al. 2012. Insights and opportunities offered by a rapid ecosystem service assessment in promoting a conservation agenda in an urban biodiversity hotspot. *Ecology and Society* 17: 8. http://doi.org/10.5751/ES-04886-170327

Obiri, B.D., G.A. Bright, M.A McDonald, et al. 2007. Financial analysis of shaded cocoa in Ghana. *Agroforestry Systems* 71: 139–49. https://doi.org/10.1007/s10457-007-9058-5

Odadi, W.O., M.K. Karachi, S.A. Abdulrazak, et al. 2011. African wild ungulates compete with or facilitate cattle depending on season. *Science* 333: 1753–55. https://doi.org/10.1126/science.1208468

Odefey, J., S. Detwiler, K. Rousseau, et al. 2012. *Banking on green: A look at how green infrastructure can save municipalities money and provide economic benefits community-wide* (Washington: American Rivers and others). https://www.americanrivers.org/conservation-resource/banking-on-green

Ofori-Boateng, C., W. Oduro, A. Hillers, et al. 2013. Differences in the effects of selective logging on amphibian assemblages in three west African forest types. *Biotropica* 45: 94–101. https://doi.org/10.1111/j.1744-7429.2012.00887.x

Ogada, D., A. Botha, and P. Shaw. 2015. Ivory poachers and poison: Drivers of Africa's declining vulture populations. *Oryx*: 1–4. https://doi.org/10.1017/S0030605315001209

Packer, C., D. Ikanda, B. Kissui, et al. 2005. Lion attacks on humans in Tanzania — understanding the timing and distribution of attacks on rural communities will help to prevent them. *Nature* 436: 927–28. https://doi.org/10.1038/436927a

Parker G.E., F.V. Osborn, R.E. Hoare, et al. 2007. *Human-elephant conflict mitigation: A training course for community-based approaches in Africa* (Livingstone: Elephant Pepper Development Trust; Nairobi: IUCN/SSC AfESG). https://www.iucn.org/sites/dev/files/import/downloads/heccombaptmen.pdf

Parker, G.E., and F.V. Osborn. 2006. Investigating the potential for chilli *Capsicum* spp. to reduce human-wildlife conflict in Zimbabwe. *Oryx* 40: 343–46. https://doi.org/10.1017/S0030605306000822

Pooley, S., J.A. Mendelsohn, and E.J. Milner-Gulland. 2014. Hunting down the chimera of multiple disciplinarity in conservation science. *Conservation Biology* 28: 22–32. https://doi.org/10.1111/cobi.12183

Poulsen, J.R., C.J. Clark, G. Mavah, et al. 2009. Bushmeat supply and consumption in a tropical logging concession in northern Congo. *Conservation Biology* 23: 1597–608. https://doi.org/10.1111/j.1523-1739.2009.01251.x

Pretty, J., C. Toulmin, and S. Williams. 2011. Sustainable intensification in African agriculture. *International Journal of Agricultural Sustainability* 9: 5–24. https://doi.org/10.3763/ijas.2010.0583

Rai, N.D., and K.S. Bawa. 2013. Inserting politics and history in conservation. *Conservation Biology* 27: 425–28. https://doi.org/10.1111/cobi.12026

Rao, M.R., M.C. Palada, and B.N. Becker. 2004. Medicinal and aromatic plants in agroforestry systems. *Afroforestry Systems* 61: 107–22. https://doi.org/10.1023/B:AGFO.0000028993.83007.4b

Roberts, P., C.O. Hunt, M. Arroyo-Kalin, et al. 2017. The deep human prehistory of global tropical forests and its relevance for modern conservation. *Nature Plants* 3: 17093. https://doi.org/10.1038/nplants.2017.93

Roe, D., F. Nelson, and C. Sandbrook. 2009. Community Management of Natural Resources in Africa: Impacts, Experiences and Future Directions. *Natural Resource Issues* 18 (London: IIED). http://pubs.iied.org/17503IIED

Rosenzweig, M.L. 2003. Reconciliation ecology and the future of species diversity. *Oryx* 37: 194–205. https://doi.org/10.1017/S0030605303000371

Rushworth, I., and S. Krüger. 2014. Wind farms threaten southern Africa's cliff-nesting vultures. *Ostrich* 8: 13–23. http://doi.org/10.2989/00306525.2014.913211

Rust, N.A., T.M. Whitehouse-Tedd, and D.C. MacMillan. 2013. Perceived efficacy of livestock-guarding dogs in South Africa: Implications for cheetah conservation. *Wildlife Society Bulletin* 37: 690–97. https://doi.org/10.1002/wsb.352

Scheijen, C.P.J., S.A. Richards, J. Smit, et al., 2019. Efficacy of beehive fences as barriers to African elephants: A case study in Tanzania. *Oryx* 52: 92–99. https://doi.org/10.1017/S0030605317001727

Schiess-Meier, M., S. Ramsauer, T. Gabanapelo, et al. 2007. Livestock predation—Insights from problem animal control registers in Botswana. *Journal of Wildlife Management* 71: 1267–74. https://doi.org/10.2193/2006-177

Seiler, N., and M.M. Robbins. 2015. Factors influencing ranging on community land and crop raiding by mountain gorillas. *Animal Conservation* 19: 176–88. https://doi.org/10.1111/acv.12232

Şekercioğlu, C.H. 2002. Effects of forestry practices on vegetation structure and bird community of Kibale National Park, Uganda. *Biological Conservation* 107: 229–40. http://doi.org/10.1016/S0006-3207(02)00097-6

Şekercioğlu, C.H. 2012. Bird functional diversity and ecosystem services in tropical forests, agroforests and agricultural areas. *Journal of Ornithology* 153: 153–61. https://doi.org/10.1007/s10336-012-0869-4

Seto, K.C., M. Fragkias, B. Günerapl, et al. 2011. A meta-analysis of global urban land expansion. *PloS ONE* 6: e23777. https://doi.org/10.1371/journal.pone.0023777

Shaheed, F. 2016. *Report of the Special Rapporteur in the field of cultural rights on her visit to Botswana* (New York: UNHRC). http://daccess-ods.un.org/access.nsf/Get?Open&DS=A/HRC/31/59/Add.1&Lang=E

Silber, G.K., A.S.M. Vanderlaan, A.T. Arceredillo, et al. 2012. The role of the International Maritime Organization in reducing vessel threat to whales: Process, options, action and effectiveness. *Marine Policy* 36: 1221–33. https://doi.org/10.1016/j.marpol.2012.03.008

Sims-Castley, R., G.I. Kerley, B. Geach, et al. 2005. Socio-economic significance of ecotourism-based private game reserves in South Africa's Eastern Cape Province. *Parks* 15: 6–18.

Sixtus, M. 2018. Indigenous communities at risk as Chinese rubber firm uses land. *Al Jazeera* http://aje.io/83rm4

Sloan, S., B. Bertzky, and W.F. Laurance. 2017. African development corridors intersect key protected areas. *African Journal of Ecology* 55: 731–37. https://doi.org/10.1111/aje.12377

Smith Dumont, E., G.M. Gnahoua, L. Ohouo, et al. 2014. Farmers in Côte d'Ivoire value integrating tree diversity in cocoa for the provision of ecosystem services. *Agroforestry Systems* 88: 1047–66. https://doi.org/10.1007/s10457-014-9679-4

Smith, D. 2014. Tanzania accused of backtracking over sale of Masai's ancestral land. *Guardian.* https://gu.com/p/43cxc

Stevenson, J.R., N. Villoria, D. Byerlee, et al. 2013. Green Revolution research saved an estimated 18 to 27 million hectares from being brought into agricultural production. *Proceedings of the National Academy of Sciences* 110: 8363–68. https://doi.org/10.1073/pnas.1208065110

Stokes, E.J., S. Strindberg, P.C. Bakabana, et al. 2010. Monitoring great ape and elephant abundance at large spatial scales: Measuring effectiveness of a conservation landscape. *PLoS ONE* 4: e10294. https://doi.org/10.1371/journal.pone.0010294

Störmer, N., L.C. Weaver, G. Stuart-Hill, et al. 2019. Investigating the effects of community-based conservation on attitudes towards wildlife in Namibia. *Biological Conservation* 233: 193–200. https://doi.org/10.1016/j.biocon.2019.02.033

Tabo, R, A. Bationo, B. Amadou, et al. 2011. Fertilizer microdosing and "Warrantage" or inventory credit system to improve food security and farmers' income in West Africa. In: *Innovations as Key to the Green Revolution in Africa*, ed. by A. Bationo, et al. (Dordrecht: Springer). https://doi.org/10.1007/978-90-481-2543-2

Taylor, R. 2009. Community based natural resources management in Zimbabwe: the experience of CAMPFIRE. *Biodiversity and Conservation* 18: 2563–83. https://doi.org/10.1007/s10531-009-9612-8

Thibault, M., and S. Blaney. 2003. The oil industry as an underlying factor in the bushmeat crisis in Central Africa. *Conservation Biology* 17: 1807–13. https://doi.org/10.1111/j.1523-1739.2003.00159.x

Tscharntke, T., Y. Clough, S.A. Bhagwat, et al. 2011. Multifunctional shade-tree management in tropical agroforestry landscapes—a review. *Journal of Applied Ecology* 48: 619–29. https://doi.org/10.1111/j.1365-2664.2010.01939.x

Tuqa, J.H., P. Funston, C. Musyoki, et al. 2014. Impact of severe climate variability on lion home range and movement patterns in the Amboseli ecosystem, Kenya. *Global Ecology and Conservation* 2: 1–10. https://doi.org/10.1016/j.gecco.2014.07.006

UN-REDD. 2013. *Guidelines on free, prior, and informed consent* (Geneva: UN-REDD). https://theredddesk.org/resources/guidelines-free-prior-and-informed-consent

Vågen, T.-G., R. Lal, and B.R. Singh. 2005. Soil carbon sequestration in sub-Saharan Africa: a review. *Land Degradation and Development* 16: 53–71. https://doi.org/10.1002/ldr.644

Western, D., R. Groom, and J. Worden. 2009b. The impact of subdivision and sedentarization of pastoral lands on wildlife in an African savanna ecosystem. *Biological Conservation* 142: 2538–46. https://doi.org/10.1016/j.biocon.2009.05.025

Western, D., S. Russell, and I. Cuthill. 2009a. The status of wildlife in protected areas compared to non-protected areas of Kenya. *PLoS One* 4: e6140. https://doi.org/10.1371/journal.pone.0006140

Woodborne, S., K.D.A. Huchzermeyer, D. Govender, Det al. 2012. Ecosystem change and the Olifants River crocodile mass mortality events. *Ecosphere* 3: 1–17. https://doi.org/10.1890/ES12-00170.1

Woodroffe, R. 2011 Demography of a recovering African wild dog (*Lycaon pictus*) population. *Journal of Mammalogy* 92: 305–15. https://doi.org/10.1644/10-MAMM-A-157.1

Woodroffe, R., and L.G. Frank. 2005. Lethal control of African lions (*Panthera leo*): Local and regional population impacts. *Animal Conservation* 8: 91–98. https://doi.org/10.1017/S1367943004001829

Young, T.P., T.M. Palmer, and M.E. Gadd. 2005. Competition and compensation among cattle, zebras, and elephants in a semi-arid savanna in Laikipia, Kenya. *Biological Conservation* 112: 251–59. https://doi.org/10.1016/j.biocon.2004.08.007

Zabel F, B. Putzenlechner, and W. Mauser. 2014. Global agricultural land resources—A high resolution suitability evaluation and its perspectives until 2100 under climate change conditions. *PLoS ONE* 99: e107522. https://doi.org/10.1371/journal.pone.0107522

15. An Agenda for the Future

A family in rural Tanzania celebrating solar power arriving in their village. New technologies provide great opportunities for more sustainable lifestyles. Photograph by Power Africa, https://www.flickr.com/photos/usaidafrica/26570196501, CC0.

https://doi.org/10.11647/OBP.0177.15

The field of conservation biology has set itself some imposing—but critical—goals: to describe Earth's biological diversity, to protect what remains, and to restore what is damaged. To understand what a significant undertaking this is, consider the Living Planet Index (http://www.livingplanetindex.org) which shows that, already in 2014, Sub-Saharan Africa's vertebrate populations were on average down 56% compared to 1970 levels (WWF, 2018). Declines were even more pronounced for freshwater vertebrates which showed a 75% decline. With wildlife declines showing no sign of halting, we are in a race against time to prevent catastrophic losses. Conservation biology is a truly crisis discipline (Soulé, 1985; Kareiva and Marvier, 2012), because decisions often need to be made under pressure, with limited resources, and constrained by tight deadlines. At the same time, the discipline needs to offer a long-term conservation vision that extends beyond the immediate crisis, despite unreliable commitments to seeing such plans through to completion.

Despite the challenges we face, there are many positive signs for cautious optimism. Some threatened species are recovering, the number of well-managed protected areas is increasing, and, in some cases, natural resources are being used more prudently on unprotected lands. We have also increased our capacity to restore degraded ecosystems to such a level that we are now reintroducing species that were once extinct in the wild. Our improved ability to protect biodiversity is in no small way attributable to the wide range of productive local, national, and international collaborative efforts that have been cultivated over the past few decades. It is also because the field of conservation biology has expanded for the better, by developing linkages with rural development, economics, the arts, social sciences, and government policy, to name a few.

Make no mistake, many challenges remain unaddressed and under-addressed, and new ones will surely also arise. These challenges all need to be faced head-on, because there is no "Planet B": Earth is our one and only planet. There will be times when the biodiversity crisis will feel insurmountable. When that happens, it is important to remember that every individual human can play a role in saving our natural heritage. If just one-tenth of Sub-Saharan Africa's 1 billion people use one less plastic item (e.g. plastic bags, drinking straws, food wrappers) a week, there would be a reduction of 100,000,000 plastic items each week. People operating at the regional and global scales, such as company executives and government officials, also have an important task—ensuring that mechanisms are in place for all citizens to contribute to ensuring that future generations will inherit a healthier environment. Below, we offer a few holistic strategies towards a sustainable future.

15.1 Achieving Sustainable Development

Efforts to preserve biological diversity are regularly perceived as in conflict with societal progress (Redpath et al., 2013). Perhaps the root of this conflict lies with the fact that most of the development we see today is *unsustainable*—that is, it risks depleting natural resources to a point where they will no longer be available for use or to provide ecosystem services. Moreover, governments and businesses often measure success in terms of **economic growth**, which occurs when an economy increasingly produces more goods and services (often measured as GDP). Economic policies that favour economic growth are generally based on an implicit but erroneous assumption that the supply of natural resources is unlimited. A society that aims for economic growth is therefore bound to fail at one point or another.

Economic policies that favour growth are based on the erroneous assumption that natural resources are unlimited. It is thus bound to fail at one point or another.

Figure 15.1 Sustainable development heals the rift between development and conservation; it aims to simultaneously meet conservation goals and human needs, CC BY 4.0.

To overcome these perceptions and conflicts, scientists, policy makers, and conservation managers are increasingly highlighting the need for **sustainable development**—economic activities that satisfy both present and future needs without compromising the natural world (Figure 15.1). Sustainable development is closely linked to **economic development**, a multi-dimensional concept that describes economic activities that aim to improve income and health without necessarily increasing consumption of natural resources. We should thus all strive for sustainable development, which emphasises economic development without unsustainable economic growth.

Sustainable development aims to satisfies present and future needs without compromising the natural world.

There are many good examples across Africa that illustrate the progress made towards sustainable development. For instance, many governments are investing in national parks and their infrastructure (such as staff and facilities) to protect biological diversity *and* provide economic opportunities for local communities. Similarly, stakeholders in large projects are increasingly engaging with one another to mitigate the negative impacts of infrastructure developments. One prime example was the 2015 Pan-African Business and Biodiversity Forum (http://www.panbbf.org), where representatives from business, governments, civil society, academia, development organisations, and financial institutions from across Africa came together to discuss how sustainable development can benefit nature, people, and business.

Unfortunately, there are also people and organisations that are taking advantage of this positive energy by misusing the term "sustainable development" to **greenwash** industrial activities that are harming the environment. For instance, a plan to establish a palm oil plantation that would damage a forested wilderness should not be considered sustainable development simply because the company agrees to protect a small plot of forest adjacent to the damaged area (see biodiversity offsets, Section 10.3.3). Similarly, many environmentally-destructive companies try to mislead customers with "environmentally-friendly" (often green-coloured) imagery on packages which are otherwise no better than the standard manufactured products. It is, therefore, critical for scientists, policy makers, and citizens to carefully study the issues, understand why different groups make arguments, and make thoughtful decisions about which actions or policies will best meet seemingly contradictory demands.

15.2 Dealing with Technological Advances

Over the past several decades, we have experienced a boom in new technologies to make our lives easier, our work more efficient, and our lifestyles more sustainable. Conservation biologists have adopted many of these new technologies to great success (Pimm et al., 2015). Consider, for example, the use of unmanned aerial vehicles (UAVs) to monitor environmental changes (Box 15.1), freely-available satellite imagery to monitor ecosystems (Section 10.1.1) and wildlife (Section 11.1.1), and molecular methods to monitor for wildlife crimes (Section 12.3.1). Hand-held devices that capture and send field data in real time are also increasing in popularity, as they enable conservation and law enforcement agencies to learn of and respond to threats much quicker than before (Wilson et al., 2019). To better streamline these efforts, there are groups such as Wildlabs which specialise in connecting the conservation community with engineers and entrepreneurs who develop such new technologies.

While conservation biologists certainly benefit from new technologies, these advances sometimes pose new challenges. Hunters now use powerful guns and motorised vehicles where historically they used bows and arrows and walked on foot. Sea fishing industries have transformed from using small wind-powered and hand-powered boats to large motorised fleets with freezers that can stay at sea for months at a time. Some technologies are so powerful that they allow for the alteration of entire ecosystems in a relatively short span of time. Some of these transformations are intentional, such as the creation of dams and the conversion of new agricultural land; others, such as pollution, are negative by-products from human activities. The impacts of these developments on ecosystems and wildlife are enormous and ominous; they have also stimulated the growth, expansion, and evolution of conservation biology.

Box 15.1 Not Just for War: Drones in Conservation

Meg Boeni and Richard Primack

Biology Department, Boston University, Boston, MA, USA.

Many of us have experienced the difficulty of following a moving herd of zebras, elephants, or any other large mammal from a vehicle or on foot. But what if this could be done from the sky? Efforts, such as mapping threatened species habitat, monitoring deforestation, and even fighting forest fires, have been aided for over 40 years with an "eye in the sky" using satellite and other aerial imagery (Pettorelli et al., 2013). The recent emergence of drones, or unmanned aerial vehicles (UAVs), has begun to make it even easier to facilitate conservation efforts from above.

Drone technology was originally developed for military applications but is fast becoming a vital resource to conservation biologists and natural resource managers. The increased popularity of drones in conservation is due to several distinct advantages. They are cheaper than airplanes or satellites; basic models that can fly up to 150 m high are available for around US $2,000. Because they operate from the ground, they are also less affected by weather conditions such as cloud cover. Drones can carry a range of sensors and equipment—video, thermal imaging, or sound—that allow them to detect organisms and ecological processes that would be impossible to study otherwise, especially at large scales. New organisations such as Conservation Drones have greatly facilitated discussions and innovations in this rapidly developing technology. Lastly, some governments are highly receptive to these new technologies. Leading the way is Rwanda, where regulators are setting the stage for an airport dedicated to civilian and commercial drones (Simmons, 2016).

While conservationists are just beginning to explore the flexibility and applicability of drones, they have already proven their worth in African conservation initiatives (Figure 15.A). With encouragement from national park officials, drones have been used to survey elephant populations in Burkina Faso (Vermuelen et al., 2013) and chimpanzee (*Pan troglodytes*, EN) nests and fruiting trees in Gabon (van Andel et al., 2015). In South Africa, drones assist anti-poaching patrols in remote areas of national parks (Mulero-Pázmány et al., 2014). There are even discussions of using drones to plant trees in reforestation efforts, and to directly manage wildlife, such as deploying noise-making drones to block an elephant herd from entering farming areas.

Despite progress, a range of obstacles still must be overcome. For example, drones are often prohibited from flying near government buildings (which often includes conservation infrastructure); many countries also continue to

Figure 15.A A researcher launching a drone to monitor an ecosystem restoration project in Namibia. Monitoring forests and other aspects of biodiversity with drones, also known as unmanned aerial vehicles, can be faster, cheaper, and safer, than from the ground or aircraft. Photograph by Miggan91, https://commons.wikimedia.org/wiki/File:A_female_researcher_flying_a_drone_in_the_field_in_Namibia.jpg, CC BY-SA 4.0.

uphold strict and arduous legal requirements for drone use. It is also important to remember that drones will never replace the need for rangers and researchers on the ground. They do however hold great promise in their potential to overcome some of the fundamental challenges that conservation biologists have always faced.

> Renewable energy sources are needed to create a sustainable society. They must also be evaluated for their environmental impact, with systems developed to mitigate those impacts.

Technologies developed to achieve sustainable development may sometimes also present new conservation challenges. For example, to combat climate change, scientists and engineers are rushing to reduce our dependence on fossil fuels by developing carbon-neutral and **energy efficient** alternatives. As these renewable energy sources have become more assimilated into our everyday lives, their unintended consequences on the environment have also become better understood. We now know that large wind farms (Figure 15.2) pose a significant collision hazard to birds (Rushworth and Krüger, 2014) and bats (Frick et al., 2017), while large solar-panel arrays that concentrate sunlight can also expose wildlife to burning temperatures (Walston et al., 2016). The impacts of hydroelectric dams are cause for even more concern: in addition to harming local fisheries and freshwater biodiversity (Section 5.3.2), these and other artificial reservoirs also generate large amounts of greenhouse gases that contribute to climate change (Deemer et al., 2016). Bioenergy also seems to create more problems than solutions, since it has become an important driver of habitat loss (Kleiner, 2007; see also Box 6.1). Similarly, hydrological fracturing for natural gas extraction—not in itself a carbon-neutral energy alternative but claimed to do less damage than coal and petroleum—has turned out to be so damaging to the environment and human health that several governments have now banned the practice (Section 7.1.1).

Figure 15.2 Wind power is become a popular technology representing a greener future. Yet, like other forms of carbon-neutral energy, wind power also has serious negative impacts on biodiversity that need to be mitigated to be sustainable. Photograph by Lollie-Pop, https://www.flickr.com/photos/lollie-pop/64839752, CC BY 2.0.

Despite the challenges posed by emerging technologies, none have yet posed an insurmountable threat. For example, we have already solved the ozone crisis by banning harmful chemicals such as chlorofluorocarbons (CFCs) (Section 12.2.1). We have also come a long way toward a sustainable fossil-fuel free world by setting guidelines for reducing the impact of wind power generation on wildlife (Reid et al., 2015; Martin et al., 2017), reducing the negative impacts of bioenergy production (Correa et al., 2017), safeguarding nuclear power stations and reusing nuclear waste (Heard and Brook, 2017), and developing more affordable solar power (Randall, 2016). It is important, however, to note that none of these emerging threats were solved by people who defended the status quo or resisted change, but by individuals who were alert and rapidly responded to new challenges before they reached a crisis point.

Environmental challenges are not solved by defending the status quo or resisting change, but by being alert and rapidly responding to new challenges before they reach a crisis point.

15.3 Funding Conservation Activities

Much of the Earth's biodiversity is concentrated in the tropics, a sizeable portion of which occurs in Africa. While people living in the tropics may be willing and eager to preserve the wildlife around them, they are often unable to accomplish the task due to funding constraints (James et al., 2001; McClanahan and Rankin, 2016). Because many of these areas experience high levels of poverty and rapid rates of population growth, the little aid these areas receive are generally diverted to short-term socio-economic programmes that ensure elected officials remain in power, rather than long-lasting sustainable solutions. This scenario is not limited to the tropics or to Africa. In fact, one of the biggest challenges facing conservation biologists across the world is inadequate funding—many areas lack basic operational funds for protected areas (Section 13.7.1), with even less for staff training, retaining top talent, keeping promises to local communities, and fulfilling the obligations set out in international treaties.

There are many organisations that continuously work to fill these funding deficits. Prominently active in Africa are multilateral organisations, such as the UN Environmental Programme (UNEP), as well as the World Bank in association with its partner organisations. A key World Bank partner organisation is the Forest Carbon Partnership Facility which helps countries in their REDD+ (discussed below) preparedness. Another is the Global Environment Facility (GEF), established to channel money from industrialised countries to conservation and environmental projects in developing countries. From 1991 to 2016, the World Bank-GEF partnership allocated over US $4 billion to more than 1,000 projects in Sub-Saharan Africa, with another $25 billion acquired through co-financing partnerships (http://www.thegef.org/projects). Prominent projects include a US $35 million project to reverse environmental damage at Central Africa's Lake Victoria, a US $16 million project to strengthen community conservancies in Mozambique, and a US $13 million project to bolster management effectiveness at Zambia's Kafue National Park.

Another significant development has been the rise of NGOs that directly fund and manage conservation activities. NGOs rely on several funding mechanisms to accomplish their goals, including membership dues, donations from wealthy individuals, sponsorships from corporations, and grants from foundations and multilateral consortiums. NGOs use these funds to advance scientific research and conservation training, to implement large-scale conservation projects, and to develop locally-adapted conservation strategies (Shackeroff and Campbell, 2007), often in collaboration with local communities (Rodríguez et al., 2007). For example, BirdLife International provides alternative environmentally friendly income streams by training local guides to help tourists find rare and elusive bird species (Biggs et al., 2011); other NGOs train park rangers and wildlife biologists, set up ecotourism lodges, and create opportunities to sell hand-made crafts.

> Multilateral consortiums and nongovernmental conservation organisations (NGOs) have emerged as important supporters of local conservation projects.

Another innovative funding approach, namely **debt-for-nature swaps**, leverages the huge international debt owed by developing countries to protect biodiversity. Major lenders (usually commercial banks or industrialised-country governments) have financed massive loans around the world, some of which they may never see repaid. One opportunity for the creditors to recoup some of this money is to restructure or sell the debt at a steep discount. Working with funders, investors, and development organisations, conservation groups may then buy a portion of these debts or help debtor country restructure this debt, in exchange for environmental commitments (in some cases, creditors may even directly engage with the debtor country). These commitments usually involve the debtor countries using the savings to annually fund, in their own currency, conservation activities, including enacting certain policies, acquiring lands for conservation, managing protected areas, and implementing conservation education programmes. In other words, freeing up money previously being spent to repay debt to now fund conservation activities. Some of the African countries that have benefitted from such debt swaps include Botswana, Cameroon, Ghana (Figure 15.3), Guinea Bissau, Mozambique, Seychelles, Tanzania, and Zambia (Sheikh, 2018). In one such example, The Nature Conservancy (TNC), the French government, and a group of creditors known as the Paris Club negotiated a US $22 million debt restructuring deal with the Seychelles in exchange for the creation of a climate adaptation trust fund and increased marine protection. As part of the deal, the Seychelles agreed to increase its marine protected areas (MPA) network from 1% to 30% coverage (400,00km^2), and to develop and implement a comprehensive spatial management plan for all its territorial waters (TNC, 2015).

Figure 15.3 To stimulate sustainable ecotourism, a debt-for-nature swap agreement facilitated the creation of Ghana's Kakum National Park to protect 375 km^2 of tropical forest that was destined for agriculture. Part of the agreement included development of local museums, interpretive trails, and a canopy walk to create income streams for local communities. Photograph by flowcomm, https://www.flickr.com/photos/flowcomm/42966954391, CC BY 2.0.

Another new strategy to obtain conservation funding is **payment for ecosystem services (PES)** schemes. Through these programmes, governments, conservation NGOs, and businesses develop markets from which landowners can receive direct payments for protecting and restoring ecosystems and ecosystems services. In a pilot project funded and coordinated by WWF and CARE Kenya, 514 farmers living upstream of Kenya's Lake Naivasha received US $20,000 in payments from water users downstream to restore and maintain riparian forests to improve flood control and water purification services (Chiramba et al., 2011).

To combat climate change, a major international initiative financially rewards communities for preserving their carbon stocks. This initiative, established by the UN in 2007 and called Reducing Emissions from Deforestation and Forest Degradation (REDD+, see also Section 10.4) receives its operational funds from individuals (such as people traveling on aeroplanes) and organisations seeking carbon credits to offset their carbon emissions. These funds are then used for results-based payments for conservation of carbon stocks such as forests and peatlands, the loss of which causes about 35% of Africa's greenhouse gas emissions (WRI, 2018). Today, REDD+ has already supported carbon conservation projects in over 30 Sub-Saharan African countries (http://www.reddprojectsdatabase.org). Being a major component of the *Paris Agreement* (Section 12.2.1), many more projects will hopefully be supported in coming years.

15.3.1 How effective is conservation funding?

Despite all these conservation resources, conservation activities continue to be underfunded due to a mismatch between funding needs and availability (Watson et al., 2014; McClanahan and Rankin, 2016; Gill et al., 2017; Lindsey et al., 2018). Exacerbating these shortfalls, conservation budgets continue to be dwarfed by spending from competing human activities and well-funded special-interest groups. For example, while the US $1.2–2.4 billion annually needed to secure Africa's protected areas with lions (Lindsey et al., 2018) is an enormous amount of money, it is much less than the US $26 billion in perverse subsidies that was paid to Africa's fossil fuel industry in 2015 (Whitley and van der Burg, 2015), which in turn is dwarfed by the whopping US $640 billion the USA budgets for military defence (DOD, 2017).

> While conservation funding is increasing, it continues to be dwarfed by perverse subsidies and spending by well-funded special-interest groups.

Many conservation projects are also constrained by weak institutional capacity, inappropriate nepotism, and even corruption in governments and NGOs (Section 2.4). There is sometimes a tendency for conservation organisations to compete, causing them to duplicate efforts in parallel rather than cooperating efficiently. Others spend a large percentage of their funds on maintaining extensive headquarters in expensive cities; these expenses are sometimes justifiable because of work on policy or advocacy, but they are sometimes wasteful and can come at a great cost to efforts in the field. Consequently, donors are increasingly worried about how funds earmarked for conservation will be spent—will funds be used to protect biodiversity and reducing poverty, or will they be diverted to other purposes? Thus, while new projects are often more effective, due in part to lessons learned from past experiences (Pooley et al., 2014), there is also a tendency to restrict funding to short-term cycles, and to add additional rules to prevent inappropriate spending. These additional constraints are making funding applications and accounting processes increasingly cumbersome and time-consuming, requiring even more time in the office than in the field. By focusing on short-term outcomes to meet reporting requirements, they also restrict grantees' ability to invest in organisational resilience and staff development, to adapt to changing circumstances, and to incorporate new ideas mid-cycle (Nelson et al., 2017).

Over the past few years, conservation groups have tried to develop several kinds of grassroots initiatives that can be low cost and self-sustaining. Among the most popular are privately protected areas, integrated conservation and development projects (ICDPs), and community-based natural resource management (CBNRM, Section 14.3) (Box 15.2). Other projects promote farming with native wildlife, such as snails (Carvalho et al., 2015) and cane rats (*Thryonomys swinderianus*, LC) (van Vliet et al. 2016) as a means to generate income while reducing pressure on wildlife targeted by the bushmeat trade (for a review on wildlife farming for conservation, see Tensen, 2016). To reduce human-wildlife conflict (Section 14.4), some communities have also found dual purpose in income-generating activities, such as beekeeping, and planting cash crops, such as tea and hot pepper plants, which also serve as barriers to nuisance animals.

Box 15.2 Supporting Self-Organised Action for Conservation in Africa

Duan Biggs[1,2]

[1]*Environmental Futures Research Institute, Griffith University, Nathan, Queensland, Australia.*

[2]*School of Public Leadership & Department of Conservation Ecology, Stellenbosch University, South Africa.*

https://www.resilientconservation.org

The conservation of biodiversity, especially outside of protected areas, faces ongoing budget constraints. One strategy to overcome such constraints is to facilitate and support individuals, communities, and organisations to self-organise to achieve positive conservation outcomes. Two terms are especially relevant in this regard: emergence (the coming about of new conservation initiatives and activities, McCay, 2002) and robustness (the durability and sustainability of these initiatives over time, Cox et al., 2010).

Central to the emergence of robust self-organised conservation activities is the particular composition of actors around a site or region of conservation interest, as well as a context that supports experimentation and learning (Figure 15.B). For example, where community conservancies are able to try different income-generating activities (e.g. photographic tourism, trophy hunting) and learn from each other through supported networks, the conditions for emergence will be strengthened (Child, 1996; Naidoo et al., 2016).

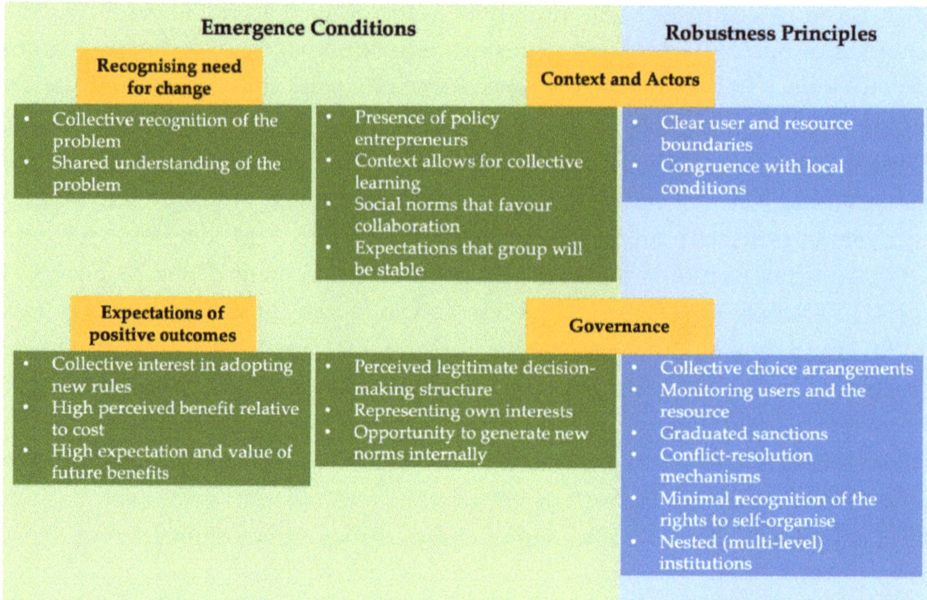

Figure 15.B The relationship between emergence and robustness as a support framework for the emergence of robust self-organised conservation activities. After Biggs et al., 2019, CC BY 4.0.

Also important are governance structures that enable communities and societies to have a central voice in the formulation of rules and policies. In this way, decision-making structures are perceived to be legitimate, and people are more empowered to take ownership of decisions that have important implications for their livelihoods (Cox et al., 2010; Biggs et al., 2019). For example, the recent ban on the import of elephant hunting trophies from Africa into the USA reduced benefit flows to communities. In addition, this ban weakened the perceived legitimacy of decision-making structures as affected communities did not have a voice in deliberations over the ban.

The final critical element is known as "nested enterprises", which means the presence of multiple overlapping institutions that support emerging conservation initiatives and activities. Successful nested enterprises include local community-based groups which are linked to national and international NGOs and have representation in local and national government (Biggs et al., 2019). For example, NGO support to community conservancies in Namibia plays an important role in aiding conservancies to access support for challenges such as human-wildlife conflict and finding partner organisations for tourism development.

Africa provides several notable examples where appropriate conditions have allowed for the emergence of self-organised conservation action on previously unprotected lands. A prominent example includes the development and expansion of privately protected area in Southern Africa (Box 2.3; Section 13.1.3). Another example is the development of community conservancy programmes, which have substantially extended the conservation estate and delivered socio-economic benefits in Kenya (Ihwagi et al., 2016) and Namibia (Naidoo et al., 2016; Störmer et al., 2019). Zimbabwe's CAMPFIRE program (Box 14.4) has also contributed to the expansion of conservation land on a large scale and remains partially successful despite Zimbabwe's current political crisis (Balint and Mashinya, 2008; Biggs et al., 2019). In each of these cases the conservation benefits have been substantial. For example, in Zimbabwe, elephant numbers on communal land increased from 4,000 to over 20,000 in just over a decade, while in Namibia, over 160,000 km^2 of land now has stronger protections due to conducive conditions for emergence of self-organized conservation.

Recent history has shown that the presence of structures that support the emergence of robust self-organised action for conservation can have substantial benefits to biodiversity and to people. But securing the future of such initiatives relies on striking a careful balance between letting local individuals, communities, and organisations "do their own thing", and providing external support and guidance when needed.

Further aiding these efforts is ecotourism, which has become a very lucrative market over the past few decades. Consequently, several private landowners and communal groups have converted their agricultural land into areas that maintain wildlife (Section 13.1). Some of these landowners cater to low-impact activities, such as bird watching (Figure 15.4) and guided safaris, while others offer hunting opportunities for wealthy individuals from North America, Europe, and Asia (Clements et al., 2016; Naidoo et al., 2016). The commercialisation of large, dangerous, and rare animals is particularly significant since more land in Africa is currently managed for regulated trophy hunting than national parks (IUCN/PACO, 2009; Flack, 2011). Because many rare and sought-after species targeted by trophy hunters require healthy ecosystems to thrive, other aspects of biodiversity also benefit, including the numerous birds, fish, insects, and plants that are not being commercially exploited in such game reserves.

> By reaping social and economic benefits from conservation, local communities have been inspired to take the lead in protecting biodiversity on their own lands.

Despite these conservation gains from the regulated hunting industry, legitimate concerns linger, including overcrowding and poor treatment of some animals, the ethics of trading and killing threatened species, and whether selective hunting and

Figure 15.4 The yellow-headed rockfowl (*Picathartes gymnocephalus*, VU) is endemic to the mountain forests of West Africa from Guinea to Ghana. As one of Africa's most sought-after birds, several protected areas and nature guides tailor their businesses to showing this species to traveling birders. Photograph by Nik Borrow, CC BY 4.0.

breeding complement or run counter to overall conservation objectives (Milner et al., 2007). The actual contribution of regulated hunting to society at large is also still being debated (IUCN/PACO, 2009; Murray, 2017), especially since some hunting concessions are established through land grabs and eco-colonialism (see Box 14.1). Similarly, there is also concerns that legal markets for threatened species may stimulate black markets and overharvesting (Lenzen et al., 2012; Hsiang and Sekar, 2016). Finding the balance between developing responsible trade opportunities in threatened species that can fund conservation activities, and risking overharvesting, is a highly emotional issue (e.g. Biggs et al., 2013a,b; Collins et al., 2013; Litchfield, 2013; Prince and Okita-Ouma, 2013) that conservation biologists will continue to grapple with in the coming years.

In the end, given the importance of nature to human well-being, it is unfortunate that conservationists continue to struggle to obtain funding and other resources. Research has shown that under-funded conservation activities run a high risk of failure (McCreless et al., 2013) while the rush to monetise nature risks weakening protection of species without immediate or realised value (Muradian et al., 2013; Balding and Williams, 2016). This contrasts with investments in protecting the natural world, which could save trillions of dollars and benefit millions of people (Costanza et al., 2014; Shindell et al., 2016). We look forward to the day when governments and individuals shift some funding from perverse subsidies to industries such as fossil fuels and unsustainable fisheries (Section 4.5.3) to supporting more conservation organisations and activities.

15.4 Building Lasting Partnerships

Productive partnerships are one of the most important components of any successful conservation undertaking. Throughout this textbook, we have seen how successful partnerships can ensure effective law enforcement, sustainable development, ecosystem protection, and threat mitigation. Yet, many conservation projects continue to fail due to a lack of collaboration between community groups, scientists, and government leaders. Other projects fail due to unproductive partnerships, such as those relying too much on foreign consultants who lack the necessary understanding of cultural intricacies and organisational objectives in recipient countries (Mcleod et al., 2015). When considering conservation's funding deficits, it is critical to wisely use what limited funds we have by maximising each project's prospects for success. Accomplishing this task starts with partnership composition.

15.4.1 Partnerships with local people

One of the most important groups to partner with is local people, particularly those individuals who are directly affected, positively and sometimes not so positively—hopefully only in the short term—by conservation projects (Redpath et al., 2013; Hall et al., 2014). Conservation projects are significantly more likely to achieve their long-term goals when they incorporate local histories and find ways to work with existing relationships between local people and their land (Waylen et al., 2010; Oldekop et al., 2016). When local people understand and buy into a project's goals and purposes, they may not only become partners in conservation, but also take on leadership roles in, or become activists for, environmental causes.

When local people buy into a project's goals and purposes, they may not only become partners in conservation, but also take on leadership roles in, or become activists for, environmental causes.

Environmental monitoring by volunteer citizen scientists provides one of the prominent success stories involving local partnerships (Figure 15.5). For example, using hand-held devices (e.g. smart phones) with GPS capabilities, local communities are now able to map natural resources in their forests (http://www.mappingforrights.org), wildlife distributions (Box 15.3), and poaching hotspots (Edwards and Plagányi, 2008), as well as forest loss (DeVries et al., 2016) and human-wildlife conflict (Larson et al., 2016). In Ethiopia, citizen scientists are empowered to perform tasks usually reserved for specialists, such as maintaining long-term demographic studies on birds (Şekercioğlu, 2011). Even people that lack confidence can contribute to these efforts, through platforms such as iNaturalist which have automated features to help users identify unknown organisms they may encounter.

There are many benefits to local involvement in biodiversity monitoring. For example, field data collected by citizen scientists—which are often as accurate as those collected by specialists (Danielsen et al., 2014; Schuttler et al. 2018)—allow biologists to obtain information from more areas more regularly and more cheaply than would be the case if specialists collected that same data. Local involvement also ensures that conservation decisions and actions are more effective and quicker to implement (Danielsen et al. 2010) and improves engagement, creating stronger advocates for conservation (Granek et al., 2008).

Box 15.3 Tracking Species in Space and Time: Citizen Science in Africa

Phoebe Barnard[1,2]

[1]*Biodiversity Futures Programme and Climate Change BioAdaptation, South African National Biodiversity Institute, Cape Town, South Africa.*

[2]*Current Address: Conservation Biology Institute, Corvallis, OR, & University of Washington, Bothell, Bothell, WA, USA.*

✉ *phoebe.barnard@consbio.org*

Planners and managers all know that keeping their eye on the world around them is crucial for good decision-making. But even in the richest nations, it's not always easy to gather enough data to get a detailed sense of environmental change in multiple dimensions—or even to keep track of what's happening on the far side of a large national park, reserve, or mountain range.

In Africa, perhaps even more than the rest of the world, the need for biodiversity monitoring data far outstrips the capacity of professional scientists to deliver it. And yet, in Namibia, South Africa, Eswatini, Lesotho, Kenya,

Figure 15.C Citizen science allows local people such as these birdwatchers from Limpopo, South Africa to make an important contribution to conservation biology. Photograph by Lisa Nupen, CC BY 4.0.

Tanzania, Zimbabwe and Botswana, the combination of public interest in biodiversity, technology, and recreation is giving rise to highly motivated "armies" of civil society volunteers (Figure 15.C). These citizen scientists not only help create remarkably detailed, high-quality datasets, but also make aspirations for ecological study a reality.

Bird data, as in so many regions, form the crux of dynamic citizen science in Africa. There are atlas projects such as the Second Southern African Bird Atlas Project, Tanzania Bird Atlas, Kenya Bird Map, and Nigerian Bird delivering important data on bird distributions in space and time across key parts of the African continent. The best of these are linked directly with academic research and applied conservation planning, policy and management, to enable adaptive responses to global change challenges (Barnard et al., 2016). In South Africa, IUCN Red Data books, environmental impact assessments (EIAs), systematic conservation plans, and national biodiversity assessments are now based partly on bird atlas data, as are dozens of high impact journal publications. These datasets can highlight places where bird ranges are shrinking or numbers are declining, such as the secretarybird (*Sagittarius serpentarius*, VU) across Southern Africa (Figure 15.D), or expanding rapidly, such as the invasive common mynah (*Acridotheres tristis*, LC).

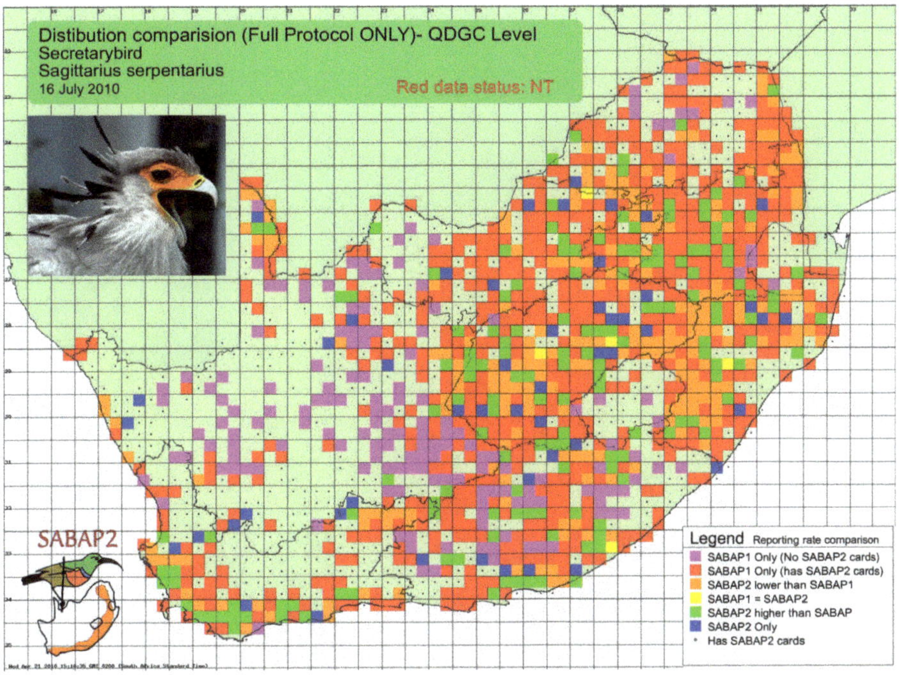

Figure 15.D Distribution data collected by citizen scientists during the second South African Bird Atlas Project (SABAP2) (ongoing since 2007) have shown that the secretarybird has disappeared in many areas where it was recorded during the SABAP1 survey (1987-1991). Red squares show population decline or disappearance, yellow squares show stable populations, and green square show population increase. Survey squares are approximately 25 km^2. Map courtesy of SANBI and University of Cape Town, CC BY 4.0.

Citizen science-based biodiversity monitoring works well in countries in which at least part of the population is mobile, interested, and moderately educated. Despite these being quite daunting obstacles in some areas, there are several important initiatives that enable new citizen scientists to contribute to biodiversity monitoring, even by those with very limited or no literacy. One such example is MammalMap, a major initiative that uses camera traps to track important and visible taxa across the continent.

Many of Africa's most dynamic and productive citizen-science projects supporting conservation biology arise from the University of Cape Town's Animal Demography Unit. The unit was founded in order to bring together civil society volunteerism, professional science, and conservation biology. The ADU, with its projects to monitor birds, frogs, butterflies, mammals, reptiles and other groups, deserves national and global investment as a powerful hub of cost-effective biodiversity monitoring.

Citizen science helps track biodiversity in space and time, providing important snapshots of the state of the environment during times of dizzying environmental change. It also builds love, knowledge, and custodianship of biodiversity among people who need to re-connect with nature and find meaning in their lives. These volunteers contribute their time, fuel and energy towards national, regional and global causes. This is a crucial cause for biodiversity in Africa, which needs investment in order to spread to all levels of society.

Figure 15.5 Under the guidance of a conservation biologist, a group of citizen scientists monitor lesser flamingo (*Phoeniconaias minor*, NT) and black crowned crane (*Balearica pavonina*, VU) at wetlands in Guinea. Photograph by Guinea Ecology, CC BY 4.0.

15.4.2 Partnerships among conservation professionals

Conservation biologists need to be more deliberate in fostering appropriate inter-organisational partnerships. Such partnerships enable new information to spread quicker and enable conservationists to learn from each other and to know whom to contact when advice is sought. Strategic partnerships also enable specialisation among organisations that they need not "do it all". It allows sharing of scarce resources (e.g. trained volunteers, temporary staff, and citizen scientists) from one organisation to another when not being utilised at a time. It also facilitates better coordination of activities, particularly at large scales, which improves project efficiency (Kark et al., 2015) organisational resilience (Maciejewski and Cumming, 2015), and conservation outcomes (Bonebrake et al., 2019). Lastly, research in Uganda showed that involving a variety of partners, especially governmental authorities, from the outset results in faster project implementation (Twinamatsiko et al., 2014).

Prospective collaborators are generally already familiar with each other. However, at times appropriate collaborators may be outside one's immediate network; this is especially true for conservation start-ups or people who have recently entered the field. In these cases, there are several effective strategies to foster new and effective partnerships. One of the best options is to attend professional meetings (Figure 15.6) such as those presented by the Society for Conservation Biology (SCB)'s Africa Section (https://conbio.org/groups/sections/africa). While this can be intimidating at first, it is worth thinking ahead of time how your own interests can be integrated with that of potential collaborators. At an organisational level, one can also contact a third party, such as the Africa Biodiversity Collaborative Group, which specialises in bringing appropriate partners together. Lastly, social media (e.g. Facebook, Twitter, ResearchGate) and biodiversity observation platforms (e.g. iNaturalist) serve to connect conservationists and naturalists from across the spectrum who wish to discuss their activities with other like-minded individuals in a more informal, less intimidating setting.

Like a marriage or friendship, professional partnerships also require constant maintenance (WWF, 2000). Project partners will invariably have different biases, objectives, and interests. They may also compete for the same funding sources, face historical legacies that complicate cooperation, or be confused about their roles in a project. It is therefore advisable for new partnerships to start small, and to take on little risk. For example, rather than initiating a project to save a high-priority species, it may be more conducive to gain experience by focussing on a less critical species or preparing a local sanctuary for a reintroduction. Once the foundation of the new partnership is set, steps can be taken towards expansion, for example by inviting new types of partners, and taking on more complex projects. More information on nurturing partnerships can be obtained by researching topics such as social-ecological system resilience, or by attending a course or workshop in organisational leadership fundamentals.

Figure 15.6 Conferences provide a good opportunity to meet other conservation biologists and to establish new collaborations. Here are members of the SCB's Africa Section after a business meeting at the 2015 International Congress for Conservation Biology, which was held in Montpellier, France. Photograph by Israel Borokini, CC BY 4.0.

15.5 Environmental Education and Leadership

Every year, conservation biologists acquire a vast body of knowledge from projects all over Africa and beyond. Yet, this information is often only communicated at small working groups and specialised meetings, published as technical papers in scientific journals with expensive subscription fees, or worse, not communicated at all. This leaves the general public detached from conservation work which, in turn, gives them (especially people living in urban centres) a sense that they live independent from nature and the knowledge gained by scientists. It also creates opportunities for **wilful ignorance**, where citizens can normalize the environmental damage caused by their activities. To avoid these scenarios, conservation biologists need to be more proactive in outreach and environmental education, which aims to raise the public's awareness and knowledge about the environment so they can adjust to live more sustainably.

One of the best ways to raise the public's environmental awareness is to involve them in local conservation projects, especially those that include fieldwork and site visits. Citizen science projects, as discussed above, present one of the most effective strategies. The public could also be invited to a guided tour where they are introduced to your organisation's activities or provided with volunteer opportunities for stewardship workdays at a local protected area. During such workdays, ordinary citizens might help with tasks, such as invasive plant control, nest box installation, and recording wildlife behaviours. An effective public relations programme can also connect people who want to engage with conservation; such a programme may involve conservation exhibits in public spaces, articles written by conservation biologists for local magazines and newspapers, or public presentations.

Children and youth are one of the most important audiences for environmental education and outreach efforts. Exposing children to the wonders of the natural world instils in them a personal sense of competence, ethics, and environmental awareness that will last a lifetime (Johnson et al., 2013). These children can also influence their parents' attitudes and behaviour towards environmental issues (Damerell et al., 2013). Ignoring children during outreach events, or recruiting ill-prepared teachers (Nkambwe and Essilfie, 2012), may however turn children against the environment, which they may see as a dangerous place detached from their own lives (Adams and Savahl, 2013). It could also lead to **nature deficit disorder**, a situation where spending less time in nature leads to behavioural problems (Louv, 2005). Consequently, many conservation organisations are now sponsoring and establishing schools to ensure young children are exposed to the importance of the environment. Others are working with children by hosting school groups, screening documentaries, publishing children's books, and offering field programmes and school outings to nearby protected areas.

Exposing children to the wonders of the natural world instils in them a personal sense of competence, ethics, and environmental awareness that will last a lifetime.

Environmental education and outreach cultivates the next cohort of conservation leaders. Today, young African conservationists can develop their leadership skills by pursuing funding opportunities to attend conferences and workshops, and fellowships to study at research institutes affiliated with local universities (Box 15.4). Some people might also be interested in the South African Wildlife College and College of African Wildlife Management, both which specialise in preparing students for a career in wildlife management. Several prestigious awards are also available that provide African youth conservation leaders with the resources they need to achieve their goals. Many conservation NGOs are also increasingly focussed on building leadership capacity through exposure to real-world conservation dilemmas. For example, the Zoological Society of London (ZSL), combat pangolin poaching in Central Africa through a specially designed mentoring programme in which young conservationists shadow experiences professionals to learn best practices in field assessments, legal protection, and demand reduction.

Reaching people who are not usually attracted to nature-based activities remains a challenge. One option is to blend conservation education and outreach with attractions and activities without an obvious conservation link. Sporting events have proven very successful in this regard. For example, an annual half marathon hosted inside South Africa's Kruger National Park has become an important opportunity to attract new people to conservation while also raising conservation funds. Another example is the Maasai Olympics, held every second year in Kenya's Amboseli-Tsavo ecosystem,

Box 15.4 The Contribution of Education Towards Conservation in Africa

Shiiwua Manu[1,2] *and Samuel Ivande*[1,2]

[1]*AP Leventis Ornithological Research Institute (APLORI), University of Jos Biological Conservatory, Laminga, Jos-East LGA, Nigeria.*

[2]*Department of Zoology, University of Jos, Nigeria.*

https://aplori.org

Improving the capacity of local people to appropriately manage natural resources in their domain is vital and fundamental for the successful conservation of biodiversity. This is usually a core objective of several environmental conservation organisations. Approaches to achieve this have often ranged from organising awareness campaigns, establishing sustainable livelihood programmes, delivering workshops to provide technical support and training to individuals, local groups, government agencies and policy officers.

One model to highlight is the A.P. Leventis Ornithological Research Institute (APLORI) model. APLORI, focused on academic training, founded a research institute and field station in 2001 to train graduate students at masters and doctorate levels in conservation biology, and to facilitate research in a tropical savannah environment (Figure 15.E). APLORI is in the Amurum Forest Reserve—one off Nigeria's key Important Bird Areas—and was established following an understanding between the Leventis Foundation, the University of Jos, Nigeria Conservation Foundation, and the Laminga community of Jos East—the reserve's host community.

One key vision of the institute is to train and equip the students who will eventually be in the driving seat of ecological and conservation research and policy in the region. To date, APLORI has trained 104 students at the master's level, with about 37 of these graduates going on to pursue doctorate degrees. APLORI is also host to many research projects needing a West African base; so far it has supported tropical ecological research for over 25 researchers from various leading universities across Europe and America. This also ensures that students at APLORI benefit from the expertise of visiting researchers.

After 14 years of APLORI's existence, its graduates have begun to occupy key positions working at the frontlines to advance ecological research in the region. Of the Institute's 104 graduates, 88% are actively engaged in teaching and research and are influencing policy at various universities, NGOs, and governmental agencies across Africa at various levels. At least four of these graduates are in leadership positions in important NGOs in the region including

Figure 15.E (Top) Dayo Osinubi, contributor to the African-Eurasian Migratory Landbirds Action Plan, participates in a workshop aimed at developing local capacity to influence policy at various levels across Africa. (Bottom) Shiiwua Manu fitting a unique ring onto the legs of a masked weaver (*Ploceus* spp.) during an APLORI mist netting session. Photographs by Will Cresswell, CC BY 4.0.

BirdLife Africa, Flora & Fauna International in Liberia, and A.G. Leventis Foundation.

The involvement of these graduates has greatly advanced the scope and quality of ecological research in the region. This is evidenced by the over a hundred published articles in international journals. A review of the research projects and publications from the institute indicates that the research scope is steadily advancing from simple biodiversity inventories and distribution updates to more detailed studies of population trends and dynamics, as well as aspects of animal behaviour, foraging, breeding, and genetic and molecular studies of tropical species and Palearctic migrants.

Much of APLORI's research uses birds to better understand the tropical environment. For example, observing breeding and migratory movements of some Afrotropical species like Abdim's stork (*Ciconia abdimii*, LC), black coucal (*Centropus grillii*, LC), rosy bee-eater (*Merops malimbicus*, LC), and the African cuckoo (*Cuculus gularis*, LC), have contributed to improve our knowledge of how seasonality influences their use of the Afrotropical landscape (Ivande et al., 2012; Cox et al., 2012, 2014). Similarly, studies of Palearctic migrants in the Afrotropics have revealed ecological flexibility in non-breeding habitat occupancy (Ivande and Cresswell, 2016) as well as high within-winter survival and site fidelity in species like whinchats (*Saxicola rubetra*, LC) which have returned to the very same winter territories every year (Wilson and Cresswell, 2006; Blackburn and Cresswell, 2015a,b). Constant Effort Site mist netting of birds, which was initiated at APLORI in 2002, has also improved our understanding of migratory passage times and survival in tropical environments (McGregor et al., 2007; Iwajomo et al., 2011) while other projects have used birds to highlight the effects of habitat fragmentation on biodiversity (Manu et al., 2005, 2007).

The location of APLORI in the Laminga community represents an effective model of successful community development projects associated with conservation projects in an area. For example, all APLORI's field assistants and support staff are employed from the community thus ensuring improved livelihoods as well as conservation skills for these individuals. This is in addition to the other community projects including: establishment of community woodlots, repair of access roads, construction of a community borehole for water, a police post, and a piggery, all of which contribute to livelihoods in the community.

Certainly, Africa with its increasing population and the attending anthropogenic pressures still needs more skilled personnel to adequately manage and conserve its vast natural resources. The APLORI model highlights the vital contribution that quality academic training and education can make.

which raises conservation awareness within the local community. Local NGOs such as the Korup Rainforest Conservation Society (KRCS) in Cameroon raises funds from membership fees; these fees are then used to host football games between local youths and park rangers, and to buy farm equipment awarded to the winners in exchange for environmental commitments. Music concerts at botanical gardens (Figure 15.7) and national parks (e.g. https://www.montybrett.com/baroque-in-the-bush) have also successfully exposed new audiences to environmental issues.

Africa is in desperate need of the next generation of conservation heroes who are up to the task of addressing a growing list of complex problems. We have learnt much over the past few decades about how to better protect the natural environment in the face of growing human populations, increased consumption, and socio-economic

Figure 15.7 Conservation facilities such as Kirstenbosch Botanical Gardens, South Africa, are attracting new people to their work by hosting concerts and other types of entertainment offers. Photograph by Ivan Hendricks, courtesy of Kirstenbosch Botanical Gardens, CC BY 4.0.

transformations. We have also developed strong foundations in environmental education and leadership that will help us reach more people and cultivate stronger leaders. But many ecosystems continue to be in a state of distress, many species are facing extinction, and many people continue to live indifferent to their environment. The time for action is now.

15.6 Summary

1. The field of conservation biology has set itself some imposing tasks: to describe Earth's biological diversity, to protect what remains, and to restore what is damaged. It is also a crisis discipline because decisions often need to be made under pressure, with limited resources, and under tight deadlines. A long-term conservation vision is also needed that extends beyond the immediate crisis.
2. Efforts to preserve biodiversity while overcoming conflicting human needs can be accomplished by striving towards sustainable development—economic development that satisfies both present and future needs without unsustainable economic growth that is compromising the natural world.
3. New technologies have greatly aided conservation efforts but have also created new challenges. Emerging threats are never solved by people who defend the status quo or resist change, but by individuals who rapidly respond to new challenges as soon as they arise.

4. One of the biggest challenges facing conservation biologists is inadequate funding. Fortunately, an increasing number of mechanisms are being established to fill funding voids, including multilateral funding consortiums, debt-for-nature swaps, payments for ecosystem services, and a range of grassroots initiatives.
5. To avoid leaving urban citizens detached from nature, conservation biologists need to reach out and educate the public, and particularly children, about their work. This can be achieved through citizen science projects, field programmes for the public, and writing materials suitable for adults and children for newspapers, magazines, and websites.

15.7 Topics for Discussion

1. Think of a very important conservation challenge facing your local area. How much funding do you think would be needed to address the problem? What types of funding sources would you pursue? What are the most important benefits you would highlight to the granting agency to convince them to fund the project?
2. Several initiatives have tried to generate rural income by offering trophy hunting and wildlife viewing opportunities. Do you think these two activities are compatible with each other? What ethical, economic, political, environmental, and social issues does each initiative raise?
3. The world is moving away from fossil fuels towards renewable, carbon-neutral energy solutions, prominently solar energy, wind energy, nuclear energy, hydropower, and bioenergy. Make a list of benefits and drawbacks of each renewable energy solution. Which renewable energy solution do you think is the best, and which is the worst? What do you think is the best way to generate energy in your region and why?
4. How has studying conservation biology changed your lifestyle or level of political activity? How do you think you can make the biggest difference in protecting biodiversity?
5. Which section of this textbook appealed to you the most and why?

15.8 Suggested Readings

Damerell, P., C. Howe, and E.J. Milner-Gulland. 2013. Child-orientated environmental education influences adult knowledge and household behaviour. *Environmental Research Letters* 8: 015016. https://doi.org/10.1088/1748-9326/8/1/015016 Environmental education focussed on children changes the behaviours of parents as well.

Granek, E.F., E.M.P. Madin, M.A. Brown, et al. 2008. Engaging recreational fishers in management and conservation: Global case studies. *Conservation Biology* 22: 1125–34. https://doi.org/10.1111/j.1523-1739.2008.00977.x Fishers can become strong advocates for conservation.

Pooley, S., J.A. Mendelsohn, and E.J. Milner-Gulland. 2014. Hunting down the chimera of multiple disciplinarily in conservation science. *Conservation Biology* 28: 22–32. https://doi.org/10.1111/cobi.12183 Projects combining conservation and development often fail due to their complexity, but it is important to learn from them so that mistakes are not repeated.

Waylen, K.A., A. Fischer, P.J.K. McGowan, et al. 2010. Effect of local cultural context on the success of community-based conservation interventions. *Conservation Biology* 24: 1119–29. https://doi.org/10.1111/j.1523-1739.2010.01446.x Conservation actions need to be tailored to local conditions.

Joseph, L.N., R.F. Maloney, and H.P. Possingham. 2009. Optimal allocation of resources among threatened species: A project prioritization protocol. *Conservation Biology* 23: 328–38. https://doi.org/10.1111/j.1523-1739.2008.01124.x Prioritising conservation spending can increase spending efficiency.

Kark, S., A. Tulloch, A. Gordon, et al. 2015. Cross-boundary collaboration: Key to the conservation puzzle. *Current Opinion in Environmental Sustainability* 12: 12–24. https://doi.org/10.1016/j.cosust.2014.08.005 Conservation collaborations have many benefits, but also drawbacks that need to be considered.

Muradian, R., M. Arsel, L. Pellegrini, et al. 2013. Payments for ecosystem services and the fatal attraction of win-win solutions. *Conservation Letters* 6: 274–79. https://doi.org/10.1111/j.1755-263X.2012.00309.x Innovative funding strategies also have their downsides.

Redpath, S.M., J, Young, A. Evely, et al. 2013. Understanding and managing conservation conflicts. *Trends in Ecology and Evolution* 28: 100–09. https://doi.org/10.1016/j.tree.2012.08.021 Many conservation conflicts can be solved through open dialogue.

Swaisgood, R.R., and J.K. Sheppard. 2010. The culture of conservation biologists: Show me the hope! *BioScience* 60: 626–30. https://doi.org/10.1525/bio.2010.60.8.8 While it is easy to feel hopeless about conservation, certain activities can turn that despair into hope.

Bibliography

Adams, S., and S. Savahl. 2015. Children's perceptions of the natural environment: A South African perspective. *Children's Geographies* 13: 196–211. https://doi.org/10.1080/14733285.2013.829659

Balding, M., and K.J.H. Williams. 2016. Plant blindness and the implications for plant conservation. *Conservation Biology* 30: 1192–99. https://doi.org/10.1111/cobi.12738

Balint, P.J., and J. Mashinya. 2008. CAMPFIRE during Zimbabwe's national crisis: Local impacts and broader implications for community-based wildlife management. *Society and Natural Resources* 21: 783–96. https://doi.org/10.1080/08941920701681961

Barnard, P., R. Altwegg, I. Ebrahim, et al. 2017. Early warning systems for biodiversity in southern Africa—How much can citizen science mitigate imperfect data? *Biological Conservation* 208: 183–88. https://doi.org/10.1016/j.biocon.2016.09.011

Biggs, D., F. Courchamp, R. Martin, et al. 2013a. Legal trade of Africa's rhino horns. *Science* 339: 1038–39. http://doi.org/10.1126/science.1229998

Biggs, D., F. Courchamp, R. Martin, et al. 2013b. Rhino poaching: Supply and demand uncertain—response. *Science* 340: 1168–69. https://doi.org/10.1126/science.340.6137.1168-b

Biggs, D., J. Turpie, C. Fabricius, et al. 2011. The value of avitourism for conservation and job creation—An analysis from South Africa. *Conservation and Society* 9: 80. https://doi.org/10.4103/0972-4923.79198

Biggs, D., N.C. Ban, J.C. Castilla, et al. 2019. Insights on fostering the emergence of robust conservation actions from Zimbabwe's CAMPFIRE program. *Global Ecology and Conservation* 17: e00538. https://doi.org/10.1016/j.gecco.2019.e00538

Blackburn, E., and W. Cresswell. 2015a. High within-winter and annual survival rates in a declining Afro-Palaearctic migratory bird suggest that wintering conditions do not limit populations. *Ibis* 158: 92–105. https://doi.org/10.1111/ibi.12319

Blackburn, E., and W. Cresswell. 2015b. Fine-scale habitat use during the non-breeding season suggests that winter habitat does not limit breeding populations of a declining long-distance Palearctic migrant. *Journal of Avian Biology* 46: 622–33. https://doi.org/10.1111/jav.00738

Bonebrake, T.C., F. Guo, C. Dingle, et al. 2019. Integrating proximal and horizon threats to biodiversity for conservation. *Trends in Ecology and Evolution* 34: *in press*. https://doi.org/10.1016/j.tree.2019.04.001

Carvalho, M., F. Rego, J.M. Palmeirim, et al. 2015. Wild meat consumption on São Tomé Island, West Africa: Implications for conservation and local livelihoods. *Ecology and Society* 20: 27. http://doi.org/10.5751/ES-07831-200327

Child, B. 1996. The practice and principles of community-based wildlife management in Zimbabwe: The CAMPFIRE programme. *Biodiversity and Conservation* 5: p. 369–98. https://doi.org/10.1007/BF00051780

Chiramba, T., S. Mogoi, I. Martinez, et al. 2011. *Payment for environmental services pilot project in Lake Naivasha basin, Kenya—A viable mechanism for watershed services that delivers sustainable natural resource management and improved livelihoods*. UN-Water International Conference (Zaragoza: UNEP). http://www.un.org/waterforlifedecade/green_economy_2011/pdf/session_4_biodiversity_protection_cases_kenya.pdf

Clements, H., J. Baum, and G.S. Cumming. 2016. Money and motives: An organizational ecology perspective on private land conservation. *Biological Conservation* 197: 108–15. https://doi.org/10.1016/j.biocon.2016.03.002

Collins, A., G. Fraser, and J. Snowball. 2013. Rhino poaching: supply and demand uncertain. *Science* 340: 1167–67. https://doi.org/10.1126/science.340.6137.1167-a

Correa, D.F., H.L. Beyer, H.P. Possingham, et al. 2017. Biodiversity impacts of bioenergy production: Microalgae vs. first generation biofuels. *Renewable and Sustainable Energy Reviews* 74: 1131–46. https://doi.org/10.1016/j.rser.2017.02.068

Costanza, R., R. de Groot, P. Sutton, et al. 2014. Changes in the global value of ecosystem services. *Global Environmental Change* 26: 152–58. https://doi.org/10.1016/j.gloenvcha.2014.04.002

Cox, D.T., M.J. Brandt, R. McGregor, et al. 2012. The seasonality of breeding in savannah birds of West Africa assessed from broodpatch and juvenile occurrence. *Journal of Ornithology* 154: 671–83. https://doi.org/10.1007/s10336-013-0930-y

Cox, D.T.C., and D.W. Cresswell. 2014. Mass gained during breeding positively correlates with adult survival because both reflect life history adaptation to seasonal food availability. *Oecologia* 174: 1197–204. https://doi.org/10.1007/s00442-013-2859-5

Cox, M., G. Arnold, and S.V. Tomas. 2010. A review of design principles for community-based natural resource management. *Ecology and Society* 15: 38. http://www.ecologyandsociety.org/vol15/iss4/art38

Damerell, P., C. Howe, and E.J. Milner-Gulland. 2013. Child-orientated environmental education influences adult knowledge and household behaviour. *Environmental Research Letters* 8: 015016. https://doi.org/10.1088/1748-9326/8/1/015016

Danielsen, F., N.D. Burgess, P.M. Jensen, et al. 2010. Environmental monitoring: the scale and speed of implementation varies according to the degree of peoples involvement. *Journal of Applied Ecology* 47: 1166–68. https://doi.org/10.1111/j.1365-2664.2010.01874.x

Danielsen, F., P.M. Jensen, N.D. Burgess, et al. 2014. A multicountry assessment of tropical resource monitoring by local communities. *BioScience* 64: 236–51. https://doi.org/10.1093/biosci/biu001

Deemer, B.R., J.A. Harrison, S. Li, et al. 2016. Greenhouse gas emissions from reservoir water surfaces: A new global synthesis. *BioScience* 66: 949–64. https://doi.org/10.1093/biosci/biw117

DeVries, B., A.K. Pratihast, J. Verbesselt, et al. 2016. Characterizing forest change using community-based monitoring data and Landsat time series. *PloS ONE* 11: e0147121. https://doi.org/10.1371/journal.pone.0147121

DOD (Department of Defense). 2017. *National defense budget estimates for FY 2018* (Washington: US DoD). https://comptroller.defense.gov/Portals/45/Documents/defbudget/fy2018/FY18_Green_Book.pdf

Edwards, C.T.T., and E.E. Plagányi. 2008. Participatory assessment of the South African abalone resource and its impact on predicted population trajectories. *South African Journal of Science* 104: 185–91. http://ref.scielo.org/8jnjw3

Flack, P. 2011. *The South African Conservation Success Story* (Cape Town: Rowland Ward Publications).

Frick, W.F., E.F. Baerwald, J.F. Pollock, et al. 2017. Fatalities at wind turbines may threaten population viability of a migratory bat. *Biological Conservation* 209: 172–77. https://doi.org/10.1016/j.biocon.2017.02.023

Gill, D.A., M.B. Mascia, G.N. Ahmadia, et al. 2017. Capacity shortfalls hinder the performance of marine protected areas globally. *Nature* 543: 665–69. https://doi.org/10.1038/nature21708

Granek, E.F., E.M.P. Madin, M.A. Brown, et al. 2008. Engaging recreational fishers in management and conservation: Global case studies. *Conservation Biology* 22: 1125–34. https://doi.org/10.1111/j.1523-1739.2008.00977.x

Hall, J.M., N.D. Burgess, S. Rantala, et al. 2014. Ecological and social outcomes of a new protected area in Tanzania. *Conservation Biology* 28: 1512–21. https://doi.org/10.1111/cobi.12335

Heard, B.P., and B.W. Brook. 2017. Closing the cycle: How South Australia and Asia can benefit from re-inventing used nuclear fuel Management. *Asia and the Pacific Policy Studies* 4: 166–75. https://doi.org/10.1002/app5.164

Hsiang, S., and N. Sekar. 2016. Does legalization reduce black market activity? Evidence from a global ivory experiment and elephant poaching data. *NBER Working Paper* 22314 (Cambridge: NBER). https://doi.org/10.3386/w22314

Ihwagi, F.W., T. Wang, G. Wittemyer, et al. 2015. Using poaching levels and elephant distribution to assess the conservation efficacy of private, communal and government land in northern Kenya. *PLoS ONE* 10: e0139079. https://doi.org/10.1371/journal.pone.0139079

IUCN/PACO. 2009. *Big game hunting in West Africa. What is its contribution to conservation?* (Ouagadougou: IUCN/PACO). https://portals.iucn.org/library/sites/library/files/documents/2009-074-En.pdf

Ivande, S.T., and W. Cresswell. 2016. Temperate migrants and resident bird species in Afro-tropical savannahs show similar levels of ecological generalism. *Ibis* 158: 496–505. https://doi.org/10.1111/ibi.12371

Ivande, S.T., S.A. Manu, Z.J. Wala, et al. 2012. Aspects of the breeding biology of Abdim's Storks *Ciconia abdimii* in Nigeria. *Malimbus* 34: 82–91. http://malimbus.free.fr/articles/V34/34082091.pdf

Iwajomo, S.B., U. Ottosson, Y. Barshep, et al. 2011. The stopover behaviour of the Garden Warbler *Sylvia borin* in Obudu, southeast Nigeria. *Ornis Svecica* 21: 29–36.

James, A., K.J. Gaston, and A. Balmford. 2001. Can we afford to conserve biodiversity? *BioScience* 51: 43–52. https://doi.org/10.1641/0006-3568(2001)051[0043:CWATCB]2.0.CO;2

Johnson, L.R., J.S. Johnson-Pynn, D.L. Lugumya, et al. 2013. Cultivating youth's capacity to address climate change in Uganda. *International Perspectives in Psychology* 2: 29–44. https://doi.org/10.1037/a0031053

Kareiva, P., and M. Marvier. 2012. What is conservation science? *BioScience* 62: 962–69. https://doi.org/10.1525/bio.2012.62.11.5

Kark, S., A. Tulloch, A. Gordon, et al. 2015. Cross-boundary collaboration: key to the conservation puzzle. *Current Opinion in Environmental Sustainability* 12: 12–24. https://doi.org/10.1016/j.cosust.2014.08.005

Kleiner, K. 2007. The backlash against biofuels. *Nature Reports Climate Change* 2: 9–11. https://doi.org/10.1038/climate.2007.71

Larson, L.R., A.L. Conway, S.M. Hernandez, et al. 2016. Human-wildlife conflict, conservation attitudes, and a potential role for citizen science in Sierra Leone, Africa. *Conservation and Society* 14: 205. https://doi.org/10.4103/0972-4923.191159

Lenzen, M., D. Moran, K. Kanemoto, et al. 2012. International trade drives biodiversity threats in developing nations. *Nature* 486: 109–12. https://doi.org/10.1038/nature11145

Lindsey, P.A., J.R.B. Miller, L.S. Petracca, et al. 2018. More than $1 billion needed annually to secure Africa's protected areas with lions. *Proceedings of the National Academy of Sciences*: E10788-E10796. https://doi.org/10.1073/pnas.1805048115

Litchfield, C.A. 2013. Rhino poaching: Apply conservation psychology. *Science* 340: 1168. https://doi.org/10.1126/science.340.6137.1168-a

Louv, R. 2008. *Last Child in the Woods: Saving Our Children from Nature-Deficit Disorder* (Chapel Hill: Algonquin Books).

Maciejewski, K., and G. Cumming. 2015. The relevance of socioeconomic interactions for the resilience of protected area networks. *Ecosphere* 6: 1–14. https://doi.org/10.1890/ES15-00022.1

Manu, S., W. Peach, and W. Cresswell. 2007. The effects of edge, fragment size and degree of isolation on avian species richness in highly fragmented forest in West Africa. *Ibis* 149: 287–97. https://doi.org/10.1111/j.1474-919X.2006.00628.x

Manu, S., W. Peach, C. Bowden, et al. 2005. The effects of forest fragmentation on the population density and distribution of the globally endangered Ibadan Malimbe *Malimbus ibadanensis*. *Bird Conservation International* 15: 275–85. https://doi.org/10.1017/S0959270905000444

Martin, C.M., E.B. Arnett, R.D. Stevens, et al. 2017. Reducing bat fatalities at wind facilities while improving the economic efficiency of operational mitigation. *Journal of Mammalogy* 98: 378–85. https://doi.org/10.1093/jmammal/gyx005

McCay, B.J. 2002. Emergence of institutions for the commons: Contexts, situations, and events. In: *The Drama of the Commons*, by National Research Council (Washington: National Academies Press). https://doi.org/10.17226/10287

McClanahan, T.R., and P.S. Rankin. 2016. Geography of conservation spending, biodiversity, and culture. *Conservation Biology* 30: 1089–101. https://doi.org/10.1111/cobi.12720

McCreless E., P. Visconti, J. Carwardine, et al. 2013. Cheap and nasty? The potential perils of using management costs to identify global conservation priorities. *PLoS ONE* 8: e80893. https://doi.org/10.1371/journal.pone.0080893

McGregor, R., M.J. Whittingham, and W. Cresswell. 2007. Survival rates of tropical birds in Nigeria, West Africa. *Ibis* 149: 615–18. https://doi.org/10.1111/j.1474-919X.2007.00670.x

Mcleod, E., B. Szuster, J. Hinkel, et al. 2015. Conservation organizations need to consider adaptive capacity: Why local input matters. *Conservation Letters* 9: 351–60. https://doi.org/10.1111/conl.12210

Milner, J.M., E.B. Nilsen, and H.P. Andreassen. 2007. Demographic side effects of selective hunting in ungulates and carnivores. *Conservation Biology* 21: 36–47. https://doi.org/10.1111/j.1523-1739.2006.00591.x

Mulero-Pázmány, M., R. Stolper, L.D. van Essen, et al. 2014. Remotely piloted aircraft systems as a rhinoceros anti-poaching tool in Africa. *PloS ONE* 9: e83873. https://doi.org/10.1371/journal.pone.0083873

Muradian, R., M. Arsel, L. Pellegrini, et al. 2013. Payments for ecosystem services and the fatal attraction of win-win solutions. *Conservation Letters* 6: 274–79. https://doi.org/10.1111/j.1755-263X.2012.00309.x

Murray, C.K. 2017. *The lion's share? On the economic benefits of trophy hunting* (Melbourne: Economists at Large). http://www.hsi.org/assets/pdfs/economists-at-large-trophy-hunting.pdf

Naidoo, R., L.C. Weaver, R.W. Diggle, et al. 2016. Complementary benefits of tourism and hunting to communal conservancies in Namibia. *Conservation Biology* 30: 628–38. https://doi.org/10.1111/cobi.12643

Nelson, F., L. Hazzah, J. Kasaona, et al. 2017. Rethinking conservation funding models in Africa (commentary). *Mongabay*. https://news.mongabay.com/2017/08/rethinking-conservation-funding-models-in-africa-commentary

Nkambwe, M., and V.N. Essilfie. 2012. Misalignment between policy and practice: Introducing environmental education in school curricula in Botswana. *Educational Research and Reviews* 7: 19–26.

Oldekop, J.A., G. Holmes, W.E. Harris, et al. 2016. A global assessment of the social and conservation outcomes of protected areas. *Conservation Biology* 30: 133–41. https://doi.org/10.1111/cobi.12568

Pettorelli, N., K. Safi, and W. Turner. 2014. Satellite remote sensing, biodiversity research and conservation of the future. *Philosophical Transactions of the Royal Society B* 369: 20130190. https://doi.org/10.1098/rstb.2013.0190

Pimm, S.L., S. Alibhai, R. Bergl, et al. 2015. Emerging technologies to conserve biodiversity. *Trends in Ecology and Evolution* 30: 685–96. https://doi.org/10.1016/j.tree.2015.08.008

Pooley, S., J.A. Mendelsohn, and E.J. Milner-Gulland. 2014. Hunting down the chimera of multiple disciplinarity in conservation science. *Conservation Biology* 28: 22–32. https://doi.org/10.1111/cobi.12183

Prins, H.H.T., and B. Okita-Ouma. 2013. Rhino poaching: Unique challenges. *Science* 340: 1167–68. https://doi.org/10.1126/science.340.6137.1167-b

Randall, T. 2016. World energy hits a turning point: Solar that's cheaper than wind. *Bloomberg* http://bloom.bg/2iWLc7q

Redpath, S.M., J, Young, A. Evely, et al. 2013. Understanding and managing conservation conflicts. *Trends in Ecology and Evolution* 28: 100–09. https://doi.org/10.1016/j.tree.2012.08.021

Reid, T., S. Krüger, D.P. Whitfield, et al. 2015. Using spatial analyses of bearded vulture movements in southern Africa to inform wind turbine placement. *Journal of Applied Ecology* 52: 881–92. https://doi.org/10.1111/1365-2664.12468

Rodríguez, J.P., A.B. Taber, P. Daszak, et al. 2007. Globalization of conservation: A view from the South. *Science* 317: 755–56. https://doi.org/10.1126/science.1145560

Rushworth, I., and S. Krüger. 2014. Wind farms threaten southern Africa's cliff-nesting vultures. *Ostrich* 8: 13–23. http://doi.org/10.2989/00306525.2014.913211

Schuttler, S.G., R.S. Sears, I. Orendain, et al. 2018. Citizen science in schools: Students collect valuable mammal data for science, conservation, and community engagement. *BioScience* 69: biy141. https://doi.org/10.1093/biosci/biy141

Şekercioğlu, Ç.H. 2011. Promoting community-based bird monitoring in the tropics: Conservation, research, environmental education, capacity-building, and local incomes. *Biological Conservation* 151: 69–73. https://doi.org/10.1016/j.biocon.2011.10.024

Shackeroff, J.M., and L.M. Campbell. 2007. Traditional ecological knowledge in conservation research: Problems and prospects for their constructive engagement. *Conservation and Society* 5: 343.

Sheikh, P.A. 2018. *Debt-for-nature initiatives and the tropical forest conservation act (TFCA): Status and implementation* (Washington: Congressional Research Services). http://www.policyarchive.org/handle/10207/1351

Shindell, D.T., Y. Lee, and G. Faluvegi. 2016. Climate and health impacts of US emissions reductions consistent with 2°C. *Nature Climate Change* 6: 503–07. https://doi.org/10.1038/nclimate2935

Simmons, D. 2016. Rwanda begins Zipline commercial drone deliveries. *BBC*. http://bbc.in/2tvnfq7

Soulé, M.E. 1985. What is conservation biology?: A new synthetic discipline addresses the dynamics and problems of perturbed species, communities, and ecosystems. *BioScience* 35: 727-734. https://doi.org/10.2307/1310054

Störmer, N., L.C. Weaver, G. Stuart-Hill, et al. 2019. Investigating the effects of community-based conservation on attitudes towards wildlife in Namibia. *Biological Conservation* 233: 193–200. https://doi.org/10.1016/j.biocon.2019.02.033

Tensen, L. 2016. Under what circumstances can wildlife farming benefit species conservation? *Global Ecology and Conservation* 6: 286–98. https://doi.org/10.1016/j.gecco.2016.03.007

TNC (The Nature Conservancy). 2015. *Debt swap to finance marine conservation in the Seychelles.* http://www.nature.org/newsfeatures/pressreleases/debt-swap-tofinance-marine-conservation-in-the-seychelles.xml

Twinamatsiko, M., J. Baker, M. Harrison, et al. 2014. *Linking conservation, equity and poverty alleviation: understanding profiles and motivations of resource users and local perceptions of governance at Bwindi Impenetrable National Park, Uganda* (London: IIED). http://pubs.iied.org/14630IIED

van Andel, A.C., S.A. Wich, C. Boesch, et al. 2015. Locating chimpanzee nests and identifying fruiting trees with an unmanned aerial vehicle. *American Journal of Primatology* 77: 1122–34. https://doi.org/10.1002/ajp.22446

van Vliet, N., D. Cornelis, H. Beck, et al. 2016. Meat from the wild: Extractive uses of wildlife and alternatives for sustainability. In: *Current Trends in Wildlife Research*, ed. by R. Mateo (Basel: Springer). https://doi.org/10.1007/978-3-319-27912-1

Vermeulen, C., P. Lejeune, J. Lisein, et al. 2013. Unmanned aerial survey of elephants. *PloS ONE* 8: e54700. https://doi.org/10.1371/journal.pone.0054700

Walston, L.J., K.E. Rollins, K.E. LaGory, et al. 2016. A preliminary assessment of avian mortality at utility-scale solar energy facilities in the United States. *Renewable Energy* 92: 405–14. https://doi.org/10.1016/j.renene.2016.02.041

Watson, J.E., Dudley, N., Segan, D.B., et al. 2014. The performance and potential of protected areas. *Nature* 515: 67–73. https://doi.org/10.1038/nature13947

Waylen, K.A., A. Fischer, P.J.K. McGowan, et al. 2010. Effect of local cultural context on the success of community-based conservation interventions. *Conservation Biology* 24: 1119–29. https://doi.org/10.1111/j.1523-1739.2010.01446.x

Whitley S., and L. van der Burg, 2015. *Fossil fuel subsidy reform in Sub-Saharan Africa: From rhetoric to reality* (London and Washington: New Climate Economy). https://newclimateeconomy.report/workingpapers/workingpaper/fossil-fuel-subsidy-reform-in-sub-saharan-africa-from-rhetoric-to-reality-2

Wilson, J., and W. Cresswell. 2006. How robust are Palearctic migrants to habitat loss and degradation in the Sahel? *Ibis* 148: 789–800. https://doi.org/10.1111/j.1474-919X.2006.00581.x

Wilson, J.W., R. Bergl, L.J. Minter, et al. 2019. The African elephant *Loxodonta* spp. conservation programmes of North Carolina Zoo: Two decades of using emerging technologies to advance in situ conservation efforts. *International Zoo Yearbook* 53: *in press*. https://doi.org/10.1111/izy.12216

WRI (World Resources Institute). 2018. *Climate Analysis Indicators Tool: WRI's climate data explorer*. http://cait2.wri.org

WWF. 2000. *Stakeholder collaboration: Building bridges for conservation* (Washington: WWF). http://wwf.panda.org/?4263/Stakeholder-Collaboration-Building-Bridges-for-

WWF. 2018. *Living Planet report 2018: Aiming higher* (Gland: WWF). https://wwf.panda.org/knowledge_hub/all_publications/living_planet_report_2018

Appendix A

Selected Sources of Information

Searchable databases provide a convenient way to find information on species, places, and topics. With the help of citizen scientists, these databases are rapidly expanding. Below are a few online databases that are free to use. Many also allow users to contribute their own data.

Biodiversity A-Z

- http://www.biodiversitya-z.org
- A thesaurus for biodiversity terminology.

Conservation Training

- https://www.conservationtraining.org
- Free conservation-based training materials, provided by TNC.

Copenhagen databases of African vertebrates

- https://macroecology.ku.dk/resources/african-vertebrates
- Distribution maps for Africa's mammals, birds, snakes, and amphibians.

eBird

- http://ebird.org
- Citizen science platform for the global birding community.

Encyclopaedia of Life

- http://www.eol.org
- Developing resource documenting the biology of all species known to science.

Evidensia

https://www.evidensia.eco

Comprehensive information on sustainability standards.

Global Biodiversity Information Facility

http://www.gbif.org

Free and open access to biodiversity data.

Global Register of Introduced and Invasive Species (GRIIS)

http://www.griis.org

Information about invasive species.

iNaturalist

http://www.inaturalist.org

A citizen science project that collects distribution data on all species.

Learning for Nature

https://learningfornature.org

e-Learning resource by the UNDP.

Mongabay

https://news.mongabay.com

A leading environmental news source.

Movebank

https://www.movebank.org

A free online database for animal tracking data

Protected Planet

https://www.protectedplanet.net

Comprehensive global spatial dataset on protected areas.

PADD tracker

http://www.padddtracker.org

Monitors protected area downgrading, downsizing, and degazettement.

Species+

- https://www.speciesplus.net
- Provides information on species covered by multilateral environmental agreements.

Vital Signs

- http://vitalsigns.org
- Collects and integrates data on agriculture, ecosystems, and human well-being.

Appendix B

Selected Environmental Organisations

Online search engines such as Google provide powerful tools to obtain information about conservation topics and opportunities. While much of the information obtained in this way is valuable, the growing popularity of the Internet has also allowed the rapid distribution of false and misleading information. You should, thus, carefully consider the source of the information you obtain online.

Similarly, it is also important to thoroughly research any conservation organisations with whom you are interested in working with. This task is particularly difficult in Africa, where most organisations have not yet been assessed for their effectiveness in carrying out conservation activities. As a starting point, you can see whether the organisation that interests you is a member of an international affiliate body, such as the IUCN or World Association for Zoos and Aquariums, which sets strict standards for organisational memberships. Online databases such as GuideStar (https://www.guidestar.org), Charity Navigator (https://www.charitynavigator.org), Better Business Bureau (https://www.bbb.org), and Great Nonprofits (https://greatnonprofits.org) are also good options for organisation vetting.

Below is a partial list of credible conservation organisations active on a regional scale in Africa.

Africa Biodiversity Collaborative Group (ABCG)

- Washington, DC, USA
- http://www.abcg.org
- Tackles conservation challenges by strengthening collaborations.

African Conservation Foundation (ACF)

- Nairobi, Kenya and Yaoundé, Cameroon
- https://www.africanconservation.org
- Saves Africa's endangered wildlife by building local capacity.

African World Heritage Fund

- Midrand, South Africa
- https://awhf.net
- Works to protect Africa's World Heritage Sites.

African Parks

- Johannesburg, South Africa
- https://www.african-parks.org
- Manages protected areas in collaboration with governments and communities.

African Wildlife Foundation (AWF)

- Nairobi, Kenya
- http://www.awf.org
- Works to ensure that wildlife and wild lands thrive.

Albertine Rift Conservation Society (ARCOS)

- Kampala, Uganda
- http://www.arcosnetwork.org
- Promotes biodiversity conservation in the Albertine Rift region.

Association for Tropical Biology and Conservation (ATBC)

- Lawrence, KS, USA
- http://tropicalbiology.org
- Fosters scientific understanding and conservation of tropical environments.

BirdLife International

- Nairobi, Kenya and Accra, Ghana
- http://www.birdlife.org/africa
- Strives to conserve birds and their habitats, with national partners across Africa.

Born Free Foundation

- Horsham, UK
- http://www.bornfree.org.uk
- Protects threatened species in the wild.

Botanical Gardens Conservation International (BGCI)

- Nairobi, Kenya
- http://www.bgci.org
- Guides, encourages, and supports botanical gardens.

Cambridge Conservation Initiative (CCI)

- Cambridge, UK
- http://www.cambridgeconservation.org
- A partnership of conservation leaders working towards a sustainable future.

Centre for International Forestry Research (CIFOR)

- Yaoundé, Cameroon and Nairobi, Kenya
- https://www.cifor.org
- Conducts research on forests and landscape management.

CGIAR (formerly Consultative Group for International Agricultural Research)

- Montpellier, France
- http://www.cgiar.org
- The world's largest agricultural innovation network.

CITES Secretariat of Wild Fauna and Flora

- Geneva, Switzerland
- https://cites.org
- The official UN body tasked with regulating the global trade in endangered species.

Conservation International (CI)

- Arlington, VA, USA
- http://www.conservation.org
- Saves nature through science, policy, and partnerships.

Conservation Leadership Programme

- Cambridge, UK
- http://www.conservationleadershipprogramme.org
- Supports leadership development of early career conservationists.

Convention on Biological Diversity (CBD) Secretariat

- Montreal, Canada
- https://www.cbd.int
- The official UN body tasked with promoting the goals of the CBD.

Critical Ecosystem Partnership Fund (CEPF)

- Arlington, VA, USA
- http://www.cepf.net
- Provides financial and technical support to conserve critical ecosystems.

Darwin Initiative

- London, UK
- http://www.darwininitiative.org.uk
- Assists developing countries implement biodiversity convention commitments.

Earthwatch Institute

- Boston, MA, USA
- http://earthwatch.org
- Helps citizen scientists contribute to field conservation projects.

East African Wild Life Society (EAWLS)

- Nairobi, Kenya
- https://eawildlife.org
- Promotes conservation and sustainable use of the environment.

EcoHealth Alliance

- New York, NY, USA
- https://www.ecohealthalliance.org
- Studies connections between humans, wildlife, and ecosystems.

The Environmental Foundation for Africa (EFA)

- Freetown, Sierra Leone
- http://www.efasl.org
- Protects and restores the environment in West Africa.

Environmental Investigation Agency (EIA)

- London, UK
- https://eia-international.org
- Activist organisation focussed on exposing environmental crimes.

Environmental Law Alliance Worldwide (ELAW)

- Eugene, OR, USA
- http://elaw.org
- Helps partners gain skills and build strong conservation organisations.

Fauna & Flora International (FFI)

- Cambridge, UK
- http://www.fauna-flora.org
- Africa's first conservation society; has been protecting African wildlife since 1903.

FitzPatrick Institute of African Ornithology

- Cape Town, South Africa
- http://www.fitzpatrick.uct.ac.za
- Promotes and undertakes scientific studies on African birds.

Forest Carbon Partnership Facility

- Washington. DC
- http://www.forestcarbonpartnership.org
- Assist countries with their REDD+ preparations to reduce emissions from forest loss.

Forest Stewardship Council (FSC)

- Bonn, Germany
- https://ic.fsc.org
- Sets the standards for responsibly managed forests.

Frankfurt Zoological Society (FZS)

- Frankfurt, Germany
- https://fzs.org
- Maintains wilderness areas and biodiversity.

Future for Nature

- Arnhem, The Netherlands
- http://futurefornature.org
- Provides mentoring and other assistance to young conservationists.

Game Rangers Association of Africa (GRAA)

- Johannesburg, South Africa
- http://www.gameranger.org
- Provides support, networks, and representation for rangers.

Global Environment Facility (GEF)

- Washington, DC, USA
- http://www.thegef.org
- Provide grants for biodiversity and sustainable development projects.

Global Forest Watch (GFW)

- Washington, DC, USA
- http://www.globalforestwatch.org
- Empower people to better protect forests.

Global Wildlife Conservation (GWC)

- Austin, TX, USA
- https://www.globalwildlife.org
- Protects species and habitats through science-based field action.

Goldman Environmental Foundation

- San Francisco, CA, USA
- http://www.goldmanprize.org
- Recognises environmental activists who have made an impact.

Greenpeace Africa

- Johannesburg, South Africa
- http://www.greenpeace.org/africa
- Activist organisation known for protests against environmental crime

High Seas Alliance

- Washington, DC, USA
- http://highseasalliance.org
- Facilitates cooperation for protection of high seas.

ICLEI Africa

- Cape Town, South Africa
- http://africa.iclei.org
- A network of governments committed to sustainable urban development.

International Fund for Animal Welfare (IFAW)

- Nairobi, Kenya and Cape Town, South Africa
- http://www.ifaw.org/africa
- Rescues and protects animals around the world.

Intergovernmental Panel on Climate Change (IPCC)

- Geneva, Switzerland
- http://www.ipcc.ch
- The UN's authority on climate change.

Intergovernmental Science-Policy Platform on Biodiversity and Ecosystem Services (IPBES)

- Bonn, Germany
- https://www.ipbes.net
- The UN's authority on nature's contributions to people (NCP), or ecosystem services.

The International Ecotourism Society (TIES)

- Washington, DC, USA
- http://www.ecotourism.org
- Promotes responsible tourism practices.

International Institute for Environment and Development (IIED)

- London, UK
- https://www.iied.org
- Promotes sustainable development to protect the environment.

International Institute of Tropical Agriculture (IITA)

- Ibadan, Nigeria
- http://www.iita.org
- Works to enhance crop quality and productivity.

International Tropical Timber Organization (ITTO)

- Yokohama, Japan
- http://www.itto.int
- Promotes sustainable management of tropical forest resources.

International Union for Conservation of Nature (IUCN)

- Gland, Switzerland
- https://www.iucn.org
- Coordinates international conservation efforts and produces Red Lists.

International Criminal Police Organisation (INTERPOL)

- Lyon, France
- https://www.interpol.int/Crime-areas/Environmental-crime
- Facilitate prosecution of international environmental crimes.

iSeal

- London, UK
- https://www.isealalliance.org
- A membership organisation for sustainability standards.

Jane Goodall Institute

- Vienna, VA, USA
- http://www.janegoodall.org
- Inspiring people to conserve the natural world.

Leadership for Conservation in Africa (LCA)

- Pretoria, South Africa
- http://lcafrica.org
- Influences business leaders to support investment in conservation.

Marine Stewardship Council (MSC)

- London, UK
- https://www.msc.org
- Promotes sustainable fishing practices.

National Geographic Society (NGS)

- Washington, DC, USA
- https://www.nationalgeographic.org
- One of the world's largest scientific and educational institutions.

Natural Capital Coalition

- London, UK
- https://naturalcapitalcoalition.org
- Collaboration of the global natural capital community.

The Nature Conservancy (TNC)

- Arlington, VA, USA
- https://www.nature.org
- Conserves threatened species and their habitats, emphasising land preservation.

Oxpeckers Centre for Investigative Environmental Journalism

- Johannesburg, South Africa
- https://oxpeckers.org
- Investigative journalists focusing on African environmental issues.

Pan-African Association of Zoos and Aquaria (PAAZA)

- Johannesburg, South Africa
- http://www.zoosafrica.com
- Guides and accredits African Zoos and Aquaria.

Peace Parks Foundation
- Stellenbosch, South Africa
- http://www.peaceparks.org
- Facilitates the establishment of transfrontier conservation areas.

The Pew Charitable Trusts
- London, UK
- http://www.pewtrusts.org
- Advances scientific understanding of environmental problems.

Project Aware
- Rancho Santa Margarita, CA, USA
- https://www.projectaware.org
- A movement of scuba divers protecting the planet's oceans.

Rainforest Alliance
- New York, NY, USA
- http://www.rainforest-alliance.org
- Advances sustainable forestry, agriculture, and ecotourism.

Rainforest Trust
- London, UK
- https://www.rainforesttrust.org
- Protecting forests by aquiring land for conservation.

Rapid Response Facility (RRF)
- Cambridge, UK
- http://www.rapid-response.org
- Provides emergency support to natural World Heritage sites.

Regional Partnership for Coastal and Marine Conservation (PRCM)
- Dakar, Senegal
- http://www.prcmarine.org
- Working on marine conservation in West Africa.

Roundtable on Sustainable Palm Oil (RSPO)

- Kuala Lumpur, Malaysia
- http://www.rspo.org
- Advances sustainable palm oil production.

Royal Botanic Gardens, Kew

- Richmond, Surrey, UK
- https://www.kew.org
- A leading botanical research institute with an enormous plant collection.

Rufford Foundation

- London, UK
- https://www.rufford.org/
- Funds conservation projects across the developing world.

Sahara Conservation Fund

- St. Louis, MO, USA
- https://www.saharaconservation.org
- Conserves biodiversity of the Sahara Desert and bordering Sahelian grasslands.

SEED

- Berlin, Germany
- https://www.seed.uno
- A global partnership that promotes sustainable development.

Society for Conservation Biology (SCB)

- Arlington, VA, USA
- http://conbio.org
- The leading scientific society for conservation biology.

Society for Ecological Restoration (SER)

- Washington, DC, USA
- http://www.ser.org
- Scientific society that promotes ecological restoration.

Species360 (formerly International Species Information System)

- Bloomington, MN, USA
- https://www.species360.org
- Gathers and shares information about animals kept in zoos and aquaria.

Tropical Biology Association

- Nairobi, Kenya
- http://www.tropical-biology.org
- Help scientists manage and conserve natural resources in tropical regions.

Tusk

- New York, NY, USA
- http://www.tusk.org
- Supports and connects conservation initiatives and expertise.

United Nations Environment Programme (UNEP)

- Nairobi, Kenya
- http://www.unep.org
- Coordinates the UN's environmental activities.

West Africa Biodiversity and Climate Change (WA BiCC)

- Accra, Ghana
- https://www.wabicc.org
- Improve conservation and climate-resilient growth across West Africa.

Western Indian Ocean Marine Science Association (WIOMSA)

- Zanzibar, Tanzania
- http://www.wiomsa.org
- Scientific society that promotes marine sciences.

Wetlands International

- Dakar, Senegal
- http://africa.wetlands.org
- Dedicated to the conservation and restoration of wetlands.

Whiteley Fund for Nature

London, UK

http://whitleyaward.org

Funds conservation leaders and projects in developing countries.

WildAid

San Francisco, CA, USA

http://wildaid.org

Working to end the illegal wildlife trade.

WILDLABS

Cambridge, UK

https://www.wildlabs.net

Platform that promotes technology-enabled conservation.

WildLeaks

Los Angeles, CA, USA

https://wildleaks.org

An online whistleblower platform for biodiversity crimes.

Wildlife Conservation Network (WCN)

San Francisco, CA, USA

https://wildnet.org

Supports community-based conservation projects.

Wildlife Conservation Society (WCS)

Bronx, NY, USA

http://www.wcs.org

One of the world's leaders in biodiversity conservation and research.

Wildlife Trade Monitoring Network (TRAFFIC)

Cambridge, UK

http://www.traffic.org

Promotes sustainable wildlife trade and combats wildlife crime.

World Bank

- Washington, DC, USA
- http://www.worldbank.org
- Provides loans to developing countries for economic development.

Worldwatch Institute

- Washington DC, USA
- http://www.worldwatch.org
- Highlights links between the economy and environment.

World Association of Zoos and Aquariums (WAZA)

- Gland, Switzerland
- http://www.waza.org
- Guides, encourages, and supports zoos and aquaria.

World Conservation Monitoring Centre (UNEP-WCMC)

- Cambridge, UK
- https://www.unep-wcmc.org
- An UN agency that supports biodiversity assessments and policy.

World Resources Institute (WRI)

- Washington, DC, USA
- http://www.wri.org
- Promotes sustainable development with sound environmental management.

World Wide Fund For Nature (WWF)

- Gland, Switzerland
- https://www.panda.org
- One of the world's largest conservation organisations.

Zoological Society of London (ZSL)

- London, UK
- https://www.zsl.org
- Manages several projects to protect threatened species and ecosystems.

Appendix C

Obtaining Conservation Funding

Funding limitations often hamper conservation activities. Because conservation funding is limited, there is much competition for the few options available. Below are 15 tips to make the writing of funding proposals less tedious, time-consuming, and depressing. The list is not meant to be exhaustive, and by no means a guarantee for funding—no tip can ever do that. But these generalities should give early-career conservationists a better chance for success.

1. **Start early**. Obtaining funding is a highly competitive endeavour, one you are more likely to fail in with a rushed job. It generally takes several months to put together a proposal that can convince assessors that your proposed work is well planned and feasible, and that your team is up to the task. To get there, you need to allow for enough time to put together a well-functioning team, develop and refine all your ideas, design a well-polished proposal, adapt it to specific grant requirements, conduct pilot studies, obtain external advice, address comments and concerns, and navigate institutional bureaucracy.
2. **Be a team player**. Assembling a good team is perhaps your most important decision towards funding success. Remember, your team will be your main support network during this process. They will brainstorm with you, look for funding opportunities, and help develop, edit, and critique your proposal. Make sure you assemble a team willing to contribute to these tasks—it is no fun doing the work alone, only for others to claim the funds and fame. Second, a carefully selected team confers reputation. As unfair as it may seem, funders invest in projects that maximise returns with minimal risk. They do this by funding established experts with a track record of successful grant management. This poses a significant barrier to early-career conservationists—how can you obtain funding without a track record, and vice versa? The best way to overcome this barrier is to assemble a team that includes reputable collaborators where each member provides a different

set of skills to assure success. (Note that established researchers are also increasingly relying on collaborations due to the interdisciplinary nature of conservation.) Make sure you state somewhere in your proposal (generally in a personnel section) why your team is the best to do this work, and how each team member's skills complement the others. Instead of viewing this as an impediment, see this requirement as an opportunity to learn from and network with experts—your project will most likely also be better off as a result.

3. **Focus on the funder's priorities**. Funders will have set priorities from which they will not deviate. Thus, while you and your colleagues may believe that your idea is truly ground-breaking, trying to convince funders to adapt their priorities to fit your grand idea simply will not happen. Instead, either find a funder whose priorities align with yours, or adapt your proposal to fit within the funder's stated priorities. In some cases, funders require that you state how your priorities align with theirs—make sure you do it, using the exact wording the funders used in their call for proposals.

4. **Your assessor is not an expert**. Funders usually appoint a small panel of assessors with a general understanding of the funder's priorities to quickly and efficiently adjudicate and rank funding proposals against each other. Having assessments done by non-experts has implications for how a proposal is written. First, do not assume that the assessor has specific knowledge of your field, or that s/he will just "get" the value of your project. Your proposal needs to clearly explain your plan in simple terms so that a lay person on the street will also care. Second, while technical terms (i.e. jargon) may be fine in specialist journals, they should be avoided at all costs in funding proposals. That also includes abbreviations, which can frustrate an assessor who needs to remind him/herself of the abbreviation's meaning.

5. **Follow the guidelines**. Before starting to write the proposal, read through the guidelines. While doing this, draw up a checklist documenting every requirement (e.g. budgets, timelines, margin sizes, fonts) that needs to be addressed and adhered to. Follow this checklist while writing the proposal. Then, when you are done, go over the guidelines again to make sure you did not miss a "hidden" requirement. While it may be tempting to make a small tweak, say to fit within the page limit, even minor deviations to the guidelines will stand out to assessors who look at hundreds of proposals in quick succession.

6. **Keep it simple**. As mentioned earlier, funders like to invest in projects that maximum returns for minimum risk. One way to meet this requirement is to have a carefully selected team of collaborators in place. Equally important is to propose projects that are realistic, with simple and obtainable goals. Remember, most grants run on one-year cycles, and there is only so much

one can accomplish in that timeframe. While you may think your overly ambitious project will impress assessors, more likely it will be viewed as a money drain and too risky to fund.

7. **Be exciting**. A grant is a reward for promising exciting work. Getting that award letter is undeniably an exciting moment in anyone's career. But before that excitement, you are going to have to think hard about ways to first make the assessors excited. This is difficult, because there are many constraints to proposals. Foremost is the challenge of finding a balance between simplicity and excitement. It is also difficult to excite an anonymous assessor with a limited understand of your work. But this situation is hardly unique: businesses all over the world constantly work on strategies to impress anonymous customers who are also considering competitor products. Remember, you, as the salesman, have only one opportunity to sell your project—through that piece of paper your proposal is printed on. While a proposal should remain formal, a marketing strategy that includes a memorable title that provokes curiosity, and an attractive layout that shows thoughtfulness and organisation, can do wonders for making your proposal stand out.
8. **Get to the point**. Another way to provoke excitement is to make sure you keep the assessor's attention from the start. Because you have only seconds to make an impression, this effort starts with a memorable title. Also, do not start the proposal like a journal article with a long background overview. Instead, use those first few sentences to immediately draw the assessor's attention to the significance of your work. As a good rule of thumb, use that first paragraph to point out what major societal problem you are addressing, why addressing it now is essential, and how you are proposing to solve it. Putting the most thought-provoking information upfront shows your assessor that you are confident and organised.
9. **Develop testable hypotheses**. You have a much better chance of success if your aims/objectives are immediately visible. **So write them in bold text, in their own line.** They also need to be written in a way to show they are objectively testable. Consider the aim of solving pesticide pollution. How would you define "solved"? Nobody using pesticides anymore? Nobody getting sick from pesticides? You see, lofty and ill-defined aims provide opportunities for confusion, a risk of appearing unrealistic, and probably a funding denial. To give the assessor assurance that your conclusions will be valid, there is an expectation (especially among assessors who are scientists) for applicants to state their main aims as testable hypotheses, followed by likely testable outcomes. It may require some thinking to frame an objective in an exciting way.

10. **Be exact and specific**. Science and research are about discovering objective facts and testable outcomes. It is important for you to show assessors that you grasp these concepts. Use your methods section to address each of your hypotheses, one at a time. As you do this, detail exactly how you will collect data free from bias, and what models/statistics you will use to ensure your results are reliable. To show clarity and understanding, either spell out potentially subjective and context-specific terms such as "larger", "amazing", and "plenty", or better yet, avoid them altogether. Also avoid vague throw-away statements like "we will model the population"; those will only hurt your cause. Instead, use that space to describe in detail how you will model the population.
11. **State your impact**. Some of the greatest discoveries of our time originated from pure scientific studies (i.e. those without obvious and immediate practical benefits). Even so, funders and scientists are increasingly debating the merits of funding pure over applied scientific studies (i.e. studies that directly and immediately benefit the public). While there is undoubtedly a need for better balance in funding allocations, there currently seems to be a strong bias towards funding applied research. Hence, unless grant guidelines explicitly state not to mention it, you should use some space to explain how your work will benefit society at large. It is important to note that the assessors may not share your background or values. Thus, do not assume the value of your work is self-evident—you really need to spell it out.
12. **State your outreach strategy**. While funding agencies generally support the cause they fund, they also want to attach their name to that cause and be recognised for their contributions. Funding agencies attached to governments in turn want tax-funded projects to be publicly accessible rather than restricted to the collective memories of specialists. A good outreach campaign also prevents the public from feeling detached from science and conservation. It is thus becoming increasingly important (and sometimes mandated) to state what steps you will take to communicate your project's results to the broader public.
13. **You are not alone**. As discussed in point 1, you should have a team of collaborators willing to help you. Do not be shy asking them for help; after all, they will also benefit from the funding and fame. It is also worth talking to co-workers who were previously successful getting the funds you target, as there are often unwritten nuances in how proposals should be framed. BUT you should also remember that your proposal is not the only one being assessed. There are likely hundreds of others. They will be ranked, and the most exciting proposals will be funded. You should think very carefully, every step of the way, how to make your proposal stand out from the crowd.

14. **Call on external help**. Once you and your team finished writing the proposal, ask friends and family who are not part of your team to read and comment on it. First prize is if you can get input from lay people who are not familiar with your work. Ask them if the proposed work excites them, and which parts they do not understand. If your proposal bores or confuses them, then you have more work to do to avoid boring and confusing the assessors. Every extra person willing to read your proposal provides an extra opportunity to test your message and improve your work.

15. **Do not give up**. Obtaining funding is not easy. It is increasingly the case that funding cuts forces more conservationists to compete for the same, if not smaller, pot of money. Funding success also depends on factors out of your control (e.g. quality and number of other proposals), leaving the chance of success to an element of luck. That does not mean applying for funding is a waste of time. Foremost, you will not succeed if you do not try. Funders may also provide comments on proposals, which enables you to improve it for the next round. Lastly, obtaining funding really is a numbers game. Do not put all your eggs in one basket by submitting your proposal to only one funder. Rather, identify several potential funders, tweak your proposal to fit their guidelines and priorities, and submit *to every one of them*. If you have a worthy idea, and you use every failure as an opportunity to refine your message, you will eventually achieve success.

Appendix D

Environmental Calendar

Several decades ago, the UN initiated a global outreach effort to mark the anniversary dates of key environmental treaties as an opportunity for us to pause and reflect on the natural environment's importance in our lives. Following this example, some environmental organisations has started devoting additional days to celebrate environmental issues not pertinently covered by UN treaties. Perhaps the most well-known being WWF's Earth Hour, held every year or 29 March, during which businesses and the public turn off non-essential lights for one hour, from 8:30–9:30pm, as a symbol of their commitment to the environment. These celebrations have become an important tool to help raise public awareness of the plight of the natural world, and many organisations are taking actions to promote environmental issues through newspaper articles, radio interviews, festivals, important announcements, seminars, and guided walks. Below is a list of some prominent celebrations in the annual environmental calendar. You, your friends, and your organisation may celebrate only some of these days, or all of them; it's all up to personal choices.

Celebration	Date	Inaugural year
International Zebra Day	31 January	2016
World Wetlands Day	2 February	1997
World Pangolin Day	Third Saturday in February	2012
*World Wildlife Day	3 March	2014
International Day of Action for Rivers	14 March	1997
World Frog Day	20 March	2014?
*International Day of Forests	21 March	2013
*World Water Day	22 March	1993
Earth Hour	29 March	2008
Earth Day	22 April	1970

World Penguin Day	25 April	Unclear
*World Migratory Bird Day	Second Saturday in May	2006
*International Day for Biological Diversity	22 May	2000
World Turtle Day	23 May	2000
*World Environmental Day	5 June	1974
*World Oceans Day	8 June	1992
World Sea Turtle Day	16 June	2005
*World Day to Combat Desertification and Drought	17 June	1995
World Albatross Day	19 June	2020
World Giraffe Day	21 June	2014
*World Population Day	11 July	1989
World Chimpanzee Day	14 July	2018
World Snake Day	16 July	2013
World Ranger Day	31 July	2007
World Lion Day	10 August	2013
World Elephant Day	12 August	2012
World Lizard Day	14 August	Unclear
International Vulture Awareness Day	First Saturday in Sept.	2009
*International Day for the Preservation of the Ozone Layer	16 September	1995
World Rhino Day	22 September	2010
World Gorilla Day	24 September	2017
World Environmental Health Day	26 September	2011
World Animal Day	4 October	1925
*International Day for Preventing the Exploitation of the Environment in War and Armed Conflict	6 November	2002
World Fisheries Day	21 November	1998
International Cheetah Day	4 December	2011
*World Soil Day	5 December	2014
*International Mountain Day	11 December	2003

**Officially celebrated by the UN*

Glossary

acid rain Rain with a low pH that develops when moisture in the atmosphere combines with oxides to produce acidic compounds like nitric and sulphuric acids.

adaptive management A management plan that is monitored for effectiveness, and adjustments are made when management goals are not met.

adaptive restoration A restoration project that uses adaptive management to achieve its goals.

agricultural runoff Water that collects and carries pollutants in its flow from agricultural lands into lakes, rivers, and oceans.

Aichi Biodiversity Targets A set of 5 strategic goals and 20 achievable targets for 2020 that was agreed upon by politicians to measure progress in biodiversity conservation.

albedo The ratio of solar radiation (i.e. sunlight energy) a body reflects or absorbs. Pale surfaces (e.g. pale sand) has a high albedo and generally reflect more sunlight, while dark surfaces (e.g. forests) has a low albedo and absorb more sunlight.

Allee effect Describes the correlation of fitness and population size, whereby the average fitness of individuals are reduced when their population drops below a certain number or density of individuals.

alleles Different forms of a gene, which arise through **mutations** that change DNA sequences. One example is different blood types on humans, produced by different alleles of the genes for specific blood proteins.

alpha diversity The total number of species found in a biological community, such as a lake or a forest. Also called species richness.

amenity value The intangible but desirable values people attach to certain aspects of nature. Includes ecotourism and other recreational values of biodiversity.

Anthropocene The current geological age marked by human activities that exert a dominant influence on Earth's climate and environment.

anthropogenic climate change *See* climate change

anthroponotic disease Diseases such as measles that can be transmitted from humans to animals. Compare to zoonotic disease.

arboretum A specialised botanical garden that focuses on collecting and conserving trees and other woody plants.

artificial incubation A captive breeding strategy that involves humans placing eggs in an incubator until hatching.

artificial insemination Human-assisted introduction of sperm into a receptive female animal to better manage her reproductive output.

assisted colonisation The establishment of populations of climate-sensitive species at new, suitable locations outside of their natural distribution range. Also called assisted migration.

augmentation programme *See* restocking programme.

background extinction rate The natural rate of extinctions that can be expected without the influence of humans as the primary driver of extinctions.

bequest value The perceived benefit people receive from preserving a natural resource or species for future generations. Also known as beneficiary value.

beta diversity Describes the rate at which species composition changes across a region, or along a gradient or transect.

Big Five The five species that big-game hunters consider the most difficult to hunt on foot — elephant, black rhinoceros, buffalo, leopard, and lion. Recently adopted by the safari industry to reflect the same five species tourists most like to see.

binomial An exclusive two-part name taxonomists give when they formally describe a species. Usually in italic font when typed; underlined when hand-written, e.g. *Panthera leo* (lion) or Homo sapiens (humans).

bioaccumulation *See* biomagnification.

bioassay Using the response of living plants or animals exposed to certain environmental conditions to evaluate an ecosystem's condition.

bioblitz A period of intense biological surveying where experts across a range of taxa come together to record all the living species within a designated area and time.

biochemical indicator A chemical substance used to evaluate ecosystem condition.

biodegradation Natural decomposition by bacteria and other living organisms.

Biodiversity Hotspot *See* Global Biodiversity Hotspots.

biodiversity indicators A species or groups of species that can be used to provide a measure of the total biodiversity in an area. Also known as surrogate species or biodiversity surrogates.

biodiversity inventory An attempt to document which species are present (and presumably absent) in some defined locality.

biodiversity offset When developers compensate for the loss of biodiversity during a development by proposing to protect or restore ecosystems elsewhere.

biodiversity Shortened form of biological diversity, which describes the range of species, genetic diversity within each species, and the multitude of complex biological communities with their associated interactions and ecosystem processes.

bioenergy Renewable energy products, such as ethanol and biodiesel (collectively called biofuels), derived from plants and/or waste products over a short period of time, rather than through long-term geological processes. Compare to fossil fuels.

biogeographic transition zone A region where different ecosystems meet and overlap. Also called ecotone.

biogeography The study of factors that shape organisms' distribution over space and time.

biologging device Data-recording devices (e.g. GPS tags, accelerometers) that are deployed on an animal to collect information such as movement, speed, and temperature. Also called biologgers.

biological community All the species of a locality that interact with one another.

biological control The use of natural predators, parasites, and pathogens to manage or eliminate pests and the damage they cause. Also called biocontrol.

biological definition of species A group of individuals that breed (or could breed) with each other in the wild, but do not breed with members of other groups. Compare to morphological and evolutionary definition of species.

biological diversity *See* biodiversity.

biological interaction The effects that living organisms have on one another. Interactions such as competition may be negative, while others, such as cooperation, may be positive. See also symbiotic relationships.

biomagnification Describes the process through which pesticides and other toxins accumulate and become more concentrated in animals at higher levels of the food chain. Also called bioaccumulation.

biome A large distinct biological community that evolved in response to a shared climatic region. All grasslands on Earth are an example of a biome.

biomimicry An approach by which scientists and engineers turn to nature to solve challenges or develop new technologies.

biomonitoring Using the presence, abundance, and health of organisms to infer the ecological condition of an ecosystem.

biopiracy The collection and use of biological materials for scientific, commercial, or personal benefit without appropriate permission or permits.

bioprospecting The continuous search for valuable or useful natural products.

bioregional management A management system that focuses on conservation across a single large ecosystem, particularly those that cross political borders.

biosorpsion The removal of heavy metals and toxic organic compounds from the environment by plants, microorganisms, and fungi.

biosphere reserve A protected areas model established by the UN to promote compatibility between biodiversity conservation, sustainable development, and the well-being of local people.

biota All the plants, animals, and other wildlife of a region or ecosystem.

biotic attrition The net loss of local biodiversity as species immigrate in response to climate change.

bushmeat crisis The sharp decline in wild animal populations caused by humans hunting for food. It is a crisis because it leads to impoverished natural communities and declining food security.

bushmeat Wild sources of protein obtained on land by hunting and collecting birds, mammals, snails, and caterpillars.

bycatch Animals that are incidentally caught, injured, or killed during fishing operations.

catchment area An area of land in which all surface water (from rain, melting snow, and natural springs) drains off into a common outlet at a lower elevation. In this way, a trickle drains into rivulets, then into a stream, then a river, and eventually into a lake or the sea.

carbon credit A permit that allows the holder to produce a certain amount (usually one tonne) of carbon emissions without additional fines or penalties.

carbon neutral A lifestyle, industry, or activity with a net zero carbon footprint, achieved by balancing carbon emissions with carbon offsets, often in the form of buying carbon credits.

carbon sequestration The capture and long-term storage of atmospheric carbon dioxide.

carbon sink Natural environments such as oceans and forests that are characterised by their ability to absorb carbon dioxide from the atmosphere.

carbon trading The buying and selling of carbon credits.

carcinogenic compound A substance that causes cancer.

carnivore An animal whose diet consists primarily of meat, which can be obtained by scavenging or hunting.

carrying capacity The maximum number of individuals or quantity of biomass of a species that an ecosystem can sustainably support.

census A repeatable sampling protocol to estimate the abundance or density of a population or species.

chromosome Components in the cells of living organisms that carry genetic information.

circadian rhythms The inherent physiological and behavioural responses of living organisms that roughly follow light and darkness patterns in the 24-hour day cycle.

CITES *See Convention on International Trade in Endangered Species of Wild Fauna and Flora.*

citizen scientist A public volunteer participating in science projects.

climate change The complete set of climate characteristics — temperature; precipitation; pressure systems; wind patterns; and oceanic currents — that are changing both locally and regionally due to human influences.

climate corridor A habitat linkage specifically aimed at protecting the dispersal routes that species will use during climate adaptation.

climate refuge (plural: climate refugia) Areas that are resilient to climate change and, thus, able to continue to support climate-sensitive species in future.

climatic envelope The suitable climatic range within which an organism can live and reproduce.

climax species Species that are characteristic of ecosystems in the last stages of succession.

cloning The process of producing genetically identical individuals (called clones).

co-management A conservation strategy characterised by partnerships between different levels of society that share decision-making responsibilities and consequences of management actions.

coarse-filter assessments Methods to identify communities and ecosystems that are threatened, rather than evaluating each individual species in a community or ecosystem.

colonise The process whereby a population establishes itself in a new area.

committed to extinction Species that are so rare that they are virtually guaranteed of extinction in the near future. Also called functionally extinct.

commodity value *See* direct use value.

communal resources Common property resources that belong to the community rather than single individuals.

community conserved area A protected area managed by local people.

community-based natural resource management (CBNRM) A conservation model that involves transferring authority of natural resources and land to local communities.

confounding factor An unmeasured variable that influences other variables of interest, thereby causing erroneous results.

conservation advocacy Describes the roles that conservation biologists adopt to guide social, political, and economical systems towards a personally-preferred outcome — adopting environmentally-friendly practices. Compare to conservation science.

conservation agriculture Environmentally friendly agricultural practices that place an emphasis on ecosystem services such as natural pollination and biocontrol. Also known as sustainable agricultural intensification.

conservation biology An integrated, multidisciplinary subject that aims to ensure the long-term preservation of biodiversity.

conservation science Describes activities that conservation biologists undertake to objectively describe biodiversity and measure biodiversity's response to stressors and safeguards. Compare to conservation advocacy.

conservation refugee A person whose life was uprooted by conservation activities. While conservation may involve restricting some human activities and, at times, even relocation, it is critical to assess whether it is necessary. If so, it is important to ensure opportunities exist so that those affected have other viable opportunities to sustain their livelihoods afterwards.

consumptive use value The value of natural resources consumed near where they are collected. Compare to productive use value.

contractual park Protected areas established and managed through agreements with private or communal landowners whose land forms part of the protected area (usually a national park).

controlled burn *See* prescribed burn.

convention International laws that that are negotiated at conferences under the authority of international bodies such as the UN. Also called treaty or international agreement.

***Convention on Biological Diversity* (CBD)** A treaty that obligates signatory countries to protect biodiversity through careful management of nature for the benefit of humans.

***Convention on International Trade in Endangered Species of Wild Fauna and Flora* (CITES)** The treaty that establishes lists (known as Appendices) of species for which member nations agree to ban, restrict, control, and monitor international trade.

cooperative breeder A social breeding system where only a few individuals in a group breed, while additional group members, called helpers, provide additional care for the offspring.

coral bleaching The breakdown of important symbiotic relationships between algae and coral when water is too warm, causing coral to die, during which they turn completely white.

crisis discipline Describes the reality that conservation biologists often face when they need to take creative steps to respond to imminent threats without complete knowledge of the systems requiring attention.

cross-fostering Conservation strategy in which a closely-related common species helps raise the offspring of a rare species.

cryopreservation Long-term preservation of purified DNA, eggs, sperm, embryos, and other tissue by freezing it at very low temperatures, usually in liquid nitrogen.

cryptic species An undescribed species that has been wrongly classified and grouped with a similar-appearing species.

customary law Customs and standards that have existed in a particular place or particular human society for generations, and that many formal law systems continue to regard as legal practice.

de-extinction Creating an organism that is genetically or visually like an extinct species.

debt-for-nature swap An agreement in which a developing country commits to fund conservation activities in exchange for cancellation of some of its debt.

decomposer Organisms (mainly bacteria, fungi, and protists) that break complex organic tissues and wastes into simple compounds by releasing enzymes, after which they absorb the nutrients. Compare to detritivores.

deep ecology The ethical view that species and biodiversity have an existence value independent from human needs, and that humans have an inherent responsibility to protect species and biodiversity.

deforestation The destruction of forests.

degazettement *See* PADDD.

demographic stochasticity Refers to variation in demographic traits (e.g. sex ratios, birth rates, death rates) of populations across years that cause population sizes to fluctuate. Also called demographic variation.

demographic study Monitors individuals of different ages and sizes over time to obtain a more comprehensive dataset than would be produced by a population census.

demographic variation *See* demographic stochasticity.

desertification The conversion of once-productive land to man-made deserts — large, dry unproductive dust bowls with no vegetation.

deterministic model A model with only one possible outcome. Compare to stochastic model.

detritivore Organisms such as earthworms, millipedes, slugs, and sea cucumbers that obtain nutrients by consuming decaying tissue and organic waste products. Compare to decomposers.

direct use value Values derived from the first-hand use of natural goods or services. Also known as commodity value.

DNA Acronym for deoxyribonucleic acid, which is the heredity material that stores genetic instructions for growth, reproduction, and functioning in all known living organisms.

DNA barcoding A technique used to rapidly identify unknown organisms or parts of organisms by comparing the unknown organism's DNA with a database of DNA sequences to see where it matches.

drylands Ecosystems such as those in the Sahel that are characterised by water scarcity. In general, there is a balance between evaporation and precipitation, in contrast to deserts where there is more evaporation.

Earth observation satellite A satellite designed to collect information on Earth's environment.

Earth Summit A major international conference, hosted by the UN in Rio de Janeiro in 1992, that resulted in several new high-profile environmental agreements. Also known as the Rio Summit.

eco-colonialism The unfortunate practice by some governments and conservation organisations of disregarding the rights and practices of local people during the establishment and management of new conservation areas or environmental laws and regulations.

ecological footprint A measure of the impact of humans in the environment, often expressed as the amount of land required to sustain all human activities.

ecological restoration The practice of restoring damaged ecosystems to their original or near-original state.

ecological trap A low-quality habitat that an organism mistakenly prefers to a high-quality habitat.

ecologically extinct A species that persists at such low numbers that its role in an ecosystem is negligible. Also called functionally extinct.

ecologically naïve Occurs when a species has not evolved with, and therefore does not recognise, a new danger such as a predator, and consequently does not defend itself from that threat.

ecologically relevant When a population or species is a self-sustaining, free from inbreeding, and an interactive participant of its community and ecosystem, it is considered ecologically relevant.

economic development Economic activities that aims to improve aspects such as income, health, and life expectancy, without necessarily increasing consumption of natural resources.

economic growth Economic activity based on an implicit but erroneous assumption that the supply of natural resources is unlimited.

ecoregion A relatively large geographical area that contains distinct natural communities that are separated from other ecoregions by vast oceans, broad deserts (e.g. Sahara in North Africa), or high mountains (e.g. Himalayas in Asia) that act as major barriers to movement.

ecosystem A community of interacting living organisms, together with its associated non-living chemical and physical environment. Compare to habitat.

ecosystem connectivity The ability of an ecosystem to facilitate dispersal of individuals between different areas.

ecosystem diversity The full variety of components that make up an ecosystem — i.e., assemblages of species and the physical environments in which they live. *See also* gamma diversity.

ecosystem engineer Organisms whose activities create, maintain, or modify ecosystems in such a way that they create or maintain suitable habitat for other species.

ecosystem management Activities that aim to preserve ecosystem components and processes.

ecosystem process The geochemical, physical, and biological processes and components that enable ecosystems to persist.

ecosystem productivity The ability of ecosystems to generate living biomass, starting with plants utilising the sun's energy.

ecosystem services All the benefits people gain from ecosystems and other components of biodiversity. Compare to nature's contributions to people (NCP).

ecotone *See* biogeographic transition zone.

ecotourism Tourism directed towards animals, plants, and other aspects of biodiversity.

edge effects Altered biological and environmental conditions associated with the edges of fragmented habitats.

effective population size (*Ne*) The number of individuals in a population that can breed with each other.

emigration The act of moving away from an area to settle in another. Compare to immigration.

endemic (species) A species native to one area and nowhere else on Earth.

energy efficient Products designed specifically to use less energy, especially from fossil fuels.

environmental crime An illegal act that directly harms the environment. These crimes are unique in that they have explicit laws and regulations that forbid them.

environmental economics A subdiscipline of economics that examines the contribution of ecosystem services to global economies, including the environmental costs of economic transactions and environmental policies.

environmental education Efforts to raise the public's awareness and knowledge about the environment, so they can manage their behaviours to live sustainably.

environmental ethics A discipline within philosophy that emphasises the ethical value of biodiversity.

environmental impact assessment (EIA) Assessments performed prior to a new development to assess potential environmental damage the development may cause, and to identify steps that can be taken to mitigate the damage. Also called ecological risk analysis.

environmental justice A movement that aims to empower poor and marginalised people to protect the environment around them.

environmental stochasticity Describes environmental conditions that vary unpredictably, which in turn cause population sizes to fluctuate.

environmentalism A movement that aims to protect the natural environment for its own sake.

eutrophication The process during which aquatic environments are degraded by nutrient pollution. Often characterised by algal blooms, oxygen depletion, and dead zones.

evolution The process by which organisms develop new traits in response to selective pressures such as mate choice and environmental changes.

evolutionary definition of a species A group of individuals that share unique similarities in their genetic makeup and, hence, their evolutionary past. Compare to biological and morphological definition of species.

evolutionary significant unit (ESU) A population that is considered distinct for conservation purposes. ESUs are generally geographically isolated and thus have unique local adaptations and genetic markers that should be maintained to ensure persistence. Also called "stocks" in fisheries management.

ex situ conservation Caring for biodiversity under artificial, human-controlled conditions, such as in zoos, aquariums, and botanical gardens. Compare to in situ conservation.

exclusive economic zones (EEZ) The oceanic waters and floor within a certain distance (generally 200 nautical miles or 370 km) from a country's coast to which that country claims exclusive rights to marine resources.

existence value The benefit people receive from simply knowing that an ecosystem or species exists.

exotic species A species that has been introduced to areas outside of its natural distribution range by human activity. Also called a non-native species or alien species. Compare to endemic species.

experiment A procedure undertaken to support or refute a hypothesis.

extant species A species that is presently alive; the opposite of extinct.

externalities Hidden costs and benefits of economic activities that are passed on to people not directly involved in the transactions, or to society at large.

extinct (species) A species that has no living individuals; the opposite of extant.

extinction cascade A series of linked extinction events following one another.

extinction debt Describes the time lag between harmful activities and species extinctions.

extinction vortex Describes a process whereby the factors that affect small populations can drive its size progressively downward towards extinction.

extirpated *See* locally extinct.

extractive reserve A protected area that is managed primarily for the sustainable production of natural resources, such as timber.

extrapolation Estimating unknown trends or patterns from observations in another area or time.

feedback loop Occurs when a system's outputs are routed back as input for that same system. Positive feedback loops amplify in outputs, while negative feedback loops reduce the outputs or buffer the system against changes.

feral (species) An escaped domestic species that has become wild.

fertiliser microdosing The application of very small quantities of fertiliser at the root of young crop plants. This lowers operational costs by reducing the amount of fertiliser required later and improving the efficiency of nutrient use by plants and microorganisms.

fire-dependent ecosystems Ecosystems that require periodic fires to persist.

fitness The relative ability of an individual to survive and reproduce.

flagship species A species that capture public attention, have symbolic value, and are important for ecotourism purposes.

focal species A species that provide the motivation to establish a protected area.

food chain The linear relationship between organisms at different trophic levels, where organisms at higher levels obtain nutrients and energy by feeding on organisms at lower levels.

food web A interconnected network of food chains that represent the feeding relationships among different organisms.

fortress conservation A school of thought that believes that conservation is best achieved by setting aside protected areas where nature can and should exist largely in isolation from human activities.

fossil fuels Energy sources, such as coal, natural gas, and oil, that formed over millions of years from the remains of living organisms buried in the Earth's crust. Compare to bioenergy.

founder effect The situation where a new population established ("founded") by only a few individuals have much less genetic diversity than the original population that the founders left behind.

fracking *See* hydrological fracturing.

free, prior, and informed consent (FPIC) A formal process, meant to protect traditional people's rights, that establishes bottom-up participation in activities on ancestral land. Consent sets a much higher threshold than consultation and includes veto power against projects that affect traditional lifestyles.

functionally extinct Describes species that are ecologically extinct (i.e. so rare that they do not contribute to ecosystem processes anymore), or species that are committed to extinction (i.e. virtually guaranteed of going extinct in the near future).

Gaia hypothesis The idea that all the biological, physical, and chemical properties on Earth interacts to form a complex, self-regulating superorganism, and that these interactions maintain the conditions necessary for life to persist.

gamma diversity The total number of species that occur across an entire region, such as a mountain range or continent, that includes many ecosystems.

gap analysis An analysis during which scientists overlay maps of species (or ecosystem) distributions with maps of protected areas to identify those species or ecosystems that are not covered under existing protected areas networks.

gene pool The total diversity of genes and alleles in a population or species.

gene The functional units of hereditary information that provide the blueprint of an organisms

general circulation models (GCM) The most popular group of mathematical models used to predict the impact of climate change.

generalist species A species that can live in a variety of different environments. Compare to specialist species.

genetic diversity The full range of variability in genetic material within a species. Compare to genetic variation.

genetic drift A random reduction in the relative abundance of alleles in small populations.

genetic pollution The uncontrolled flow of genetic material from one species or population to another during hybridisation. Also called genetic swamping or genetic mixing.

genetic variation Genetic differences between different individuals of a population. Compare to genetic diversity.

genetically modified organism (GMO) An organism that can provide useful or improved products and services after its genetic material has been altered using genetic engineering techniques.

genome resource bank Frozen collection of DNA, eggs, sperm, embryos, germplasms, and other kinds of genetic materials that are preserved for scientific research and breeding programmes.

genotype The particular mix of genes and alleles in an individual.

genus (plural: genera) A taxonomic rank in the biological classification system that comes above species and below family. Often abbreviated to the first letter on second use. For example, the genus elephant (*Loxodonta*) contains two species, the savannah elephant (*Loxodonta africana*) and the forest elephant (*L. cyclotis*).

geographic information systems (GIS) Computer software packages used to store, display, manipulate, and analyse data representing the natural environment, biodiversity, and human land-use patterns.

geological epoch A major subdivision of Earth's geological recent history that ends with -cene. Important epochs include the Pleistocene (also known as the Ice Age), Holocene, and Anthropocene (the current human-dominated era).

geospatial analysis Data analysis techniques that use GIS software to better understand spatial relationships between different GIS datasets.

ghost fishing A term used when fishing gear that was lost, dumped, or abandoned continues to catch (and kill) aquatic organisms.

glacial period Periods of time in Earth's history, known as ice ages, noted for colder temperatures and moving glaciers — massive, heavy ice sheets constantly moving under its own weight. The most recent glacial period ended about 15,000 years ago.

Global Biodiversity Hotspots Thirty-six (36) regions with extraordinary high number of species, many of which are endemic, that are also under immediate and intense threat from human activity.

global warming The general trend of increasing global temperatures due to increased greenhouse gas concentrations in the atmosphere.

globalisation The increased integration, interaction, and interdependence among different economies, governments, and organisations across the world.

globally extinct No living individuals of that species remain anywhere in the world.

governance The formal and informal rules and norms that guide societal functioning.

green infrastructure Urban infrastructure (i.e. green roofs, wetlands, and permeable surfaces) that is constructed in such a way that it harnesses free ecosystem services. Green infrastructure is more effective, attractive, and cheaper than conventional infrastructure.

greenhouse effect Warming caused by heat trapped near the Earth's surface by increased concentrations of greenhouse gases in the atmosphere.

greenhouse gases Transparent gases in the atmosphere that function much like the glass covering a greenhouse by allowing sunlight to pass through the atmosphere but trapping the reflected heat energy so that it stays close to Earth's surface.

greenwashing Misuse of terms, such as sustainable development, environmentally friendly, or "green" as it pertains to environmentally sound choices in order to hide activities that are harming the environment.

groundwater Subsurface reservoirs of freshwater, held in underground aquifers, in rock fissures, and in the pores between sand, dirt, and gravel particles.

gross domestic product (GDP) The value of all goods produced, and services provided, in a country over the course of one year.

habitat A species-specific term that refers to the suitable area within which an organism can find food, shelter, and mates for reproduction. Compare to ecosystem, the term used to describe a suitable area for a wide variety of species.

habitat corridor *See* habitat linkage.

habitat degradation The process whereby humans alter a natural ecosystem so much that it cannot support its characteristic species anymore.

habitat fragmentation The process whereby human activities reduce once large, unending wildernesses to several increasingly smaller and isolated ecosystem fragments.

habitat interior Habitat away from edges and associated altered environmental conditions. Also called core habitat.

habitat linkage Connection between protected areas that allows for dispersal and migration. Also known as habitat corridor, movement corridor, or wildlife corridor. Compare to stepping stone habitat.

habitat loss The outright destruction of natural habitats and ecosystems.

habitat matrix An area of unsuitable habitat that surrounds a suitable habitat patch.

habituate Slowly making an animal used to the presence of people, to give tourists a more intimate wildlife experience.

hard release A translocation strategy that involves releasing individuals without assistance. Compare to soft release.

head-starting A programme that raises threatened animals in captivity during their young, vulnerable stages before they are released into the wild.

healthy ecosystem A subjective term that describes a complex and adaptive ecosystem in which all the different ecosystem processes are intact and functioning normally.

heavy metals Metals, such as mercury, lead, and bismuth, with relatively high densities or atomic weights that are toxic in high amounts and often biomagnify in the environment.

herbarium (plural: herbaria) A collection of plant specimens with their provenance data (e.g. location, how collected, collector name), preserved for scientific study.

herbivore A species that gets its nutrients and energy from eating photosynthetic plants. Also called a primary consumer.

heterosis A level of genetic variation that improves individual evolutionary fitness.

heterozygous Condition where an individual received different two alleles of the same gene from their parents.

homozygous Condition where an individual received two identical alleles from each parent.

human-wildlife conflict Situations where humans are negatively impacted during their interaction with wildlife.

husbandry Techniques used in the care, cultivation, and breeding of plants and animals, often in captivity.

hybrid Offspring that results from mating between individuals from closely-related species.

hybrid vigour Hybrid offspring that are so strong in an evolutionary sense that they outcompete their parent species.

hydrocarbons Compounds of hydrogen and carbon molecules; one of the primary components of fossil fuels.

hydrological fracturing The environmentally destructive process where subterranean rock is forced open by pressurised liquid to release oil or gas inside. Poses several environmental and human health hazards.

immigration The act of moving into a new area with the aim of settling down there. Compare to emigration.

Important Bird and Biodiversity Area (IBA) A programme by BirdLife International that uses set criteria to identify areas that are globally important for the conservation of populations.

in situ conservation Protecting existing populations and ecosystems in the wild. Compare to ex situ conservation.

inbreeding depression Reduced offspring fitness following mating among closely-related individuals.

inbreeding Mating among closely-related individuals; includes self-fertilisation.

indicator species Sensitive species that are used for ecosystem monitoring and evaluate conservation actions.

indirect use value The value we gain from biodiversity — water filtration by wetlands, soil protection by plants, ecotourism — that does not involve harvesting or destroying the natural resource. Also known as public goods, or non-consumptive use value.

Industrial Revolution The period from about 1760 to around 1830 when rural societies became industrial and urban due, primarily, to a shift from homemade/handmade products to machine-powered, special purpose production of goods, especially in the agricultural and textiles sectors.

integrated conservation A conservation paradigm that focuses on the social and economic benefits of conservation. It is often associated with landscape-scale action and collaborations between a wide variety of stakeholders, including private business and local communities. Compare to fortress conservation.

integrated conservation and development project (ICDP) Conservation project that also provides for the economic needs and welfare of local people.

Intergovernmental Panel on Climate Change (IPCC) Leading scientists brought together by the UN to study the causes and implications of climate change.

Intergovernmental Science-Policy Platform on Biodiversity and Ecosystem Services (IPBES) Leading scientists brought together by the UN to study nature's contributions to people (NCP).

intermediate disturbance hypothesis A theory that predicts that intermediate levels of disturbance maximises biodiversity because it increases opportunities for a greater variety of species to live in an area.

intrinsic value Values attached to nature for its own sake, independent of human benefits.

introduction Creating a new population by moving individuals to suitable areas outside that species' historical range.

invasive species A species that causes ecological and/or economic harm to areas outside its native range.

island biogeography A model that predicts that more species live on larger islands than smaller islands; the model can be used to predict the impact of habitat loss on species extinctions, by viewing remaining habitat as an "island" in the "sea" of a degraded ecosystem.

International Union for Conservation of Nature and Natural Resources (IUCN) A major international conservation organisation who, among other tasks, maintains Red Lists of threatened species. Previously known as the World Conservation Union.

Key Biodiversity Area (KBA) An area deemed a conservation priority based on standardised criteria and thresholds that account for concentrations of threatened species and/or globally significant population aggregations.

keystone resource *See* limiting resource.

keystone species Species that constitute only a small proportion of their ecosystem's overall living biomass but have such disproportionately important roles that their disappearance would lead to drastic environmental changes.

Kyoto Protocol An international treaty adopted in Kyoto, Japan in December 1997 where governments committed to reduce greenhouse gas emissions that cause climate change.

land grabbing Contentious large-scale land acquisitions by foreign companies and individuals, generally to produce food and biofuels for their own people, with little to no benefit to the local community.

land reclamation The process of rehabilitating degraded land to a more productive state.

land sharing A conservation philosophy — based on the premise that humans and nature can coexist sustainably — proposing that food be produced in areas of low-yielding, wildlife-friendly agriculture on a larger land footprint. Compare to land sparing.

land sparing A conservation philosophy — based on the premise that humans and nature cannot coexist — proposing that areas set aside for intensive (and not necessarily nature-friendly) agriculture would leave more areas of untouched wilderness. Compare to land sharing.

Lazarus species Species that were believed to be extinct only to be miraculously discovered later. Essentially a metaphorical return from the dead, much like the biblical story of Lazarus.

leaching The loss of water-soluble nutrients from the soil by excessive irrigation or runoff.

legal title The right to land ownership that is recognised by the government.

light pollution Excessive and inappropriate artificial light that negatively impacts biodiversity.

limiting resource Any requirement that restricts the size or distribution of a population, and without which the population cannot survive. Also called keystone resource.

locally extinct A species that no longer exists in a place where it used to occur, but still exists elsewhere. Also called extirpated.

lumping A conservative taxonomic approach that prefers to combine two or more closely-related taxa into a single taxon. Compare to splitting.

management plan A formal document that describes how a protected area should be run to accomplish its goals and objectives.

mangrove swamp A type of tropical coastal wetland characterised by distinctive woody plants with aerial roots that can tolerate saltwater.

marine protected area (MPA) Protected areas specifically seeking to protect our oceanic and coastal environments.

market failure Misallocation of resources, which allow a small number of people or businesses to profit or benefit at the expense of the rest of society, who will bear much of these costs in the future.

mass extinction event Periods characterised by the widespread extinction of many species over a short period of time.

maximum sustainable yield The greatest number of individuals that can be harvested without detriment to a population.

megafauna Very large animals that typically weigh over 1,000 kg. Generally considered to be rhinoceros, hippopotamus, whales, giraffes, and elephants.

mesopredator release Situation where mid-sized predators (e.g. jackal) flourish in the absence of their natural enemies. Associated with lethal control of apex predators (e.g. leopards) that endanger livestock.

metapopulation Shifting populations linked by movements between them. In essence, a "population of populations".

microclimate A distinctive climate restricted in a small area that differs from the climate of the surrounding area.

microloans Very small loans offered to very poor borrowers that lack a credit record, collateral, or even access to banking accounts. These loans are meant to promote economic development and help the borrowers out of poverty by providing seed money for a business; the loan is paid back once the business is successful.

microplastics Plastic particles smaller than 1 mm (some are microscopic) that are either manufactured intentionally small or originate from the breakdown of larger pieces of plastic.

migratory (species) A species in which a significant proportion of individuals cyclically and predictably move from one area to another in search of seasonally available resources. Compare to nomadic species and resident species.

minimum dynamic area (MDA) The smallest area of suitable habitat required to sustain a minimum viable population.

minimum viable population (MVP) The smallest number of individuals necessary for a population to have a chance of long-term persistence.

mixed-use zoning *See* zoning

morphological definition of species Individuals that are distinct from other groups in their morphology, physiology, or biochemistry.

morphology The appearance of an organism, which includes external (e.g. shape, colour, size, structure, patterns) and internal features (i.e. anatomy).

morphospecies A species that is distinct based on their appearance that do not yet have a scientific name.

mountain-top extinction Climate change driven extinction of specialist species living on mountain tops which cannot disperse elsewhere without leaving their habitat.

movement corridor *See* habitat linkage.

mutations Changes in genes and chromosomes that give rise to genetic variation.

mutualistic relationship A biological (symbiotic) relationship where two species benefit each other. Compare to parasitism.

Nagoya Protocol The shortened name for the *Nagoya Protocol on Access to Genetic Resources and the Fair and Equitable Sharing of Benefits Arising from their Utilization to the Convention on Biological Diversity*, an international agreement through the UN to prevent biopiracy.

native species A species occurring in a place naturally, without the influence of people. Compare to invasive species.

natural history The observational study of animals, plants, and other aspects of biodiversity. Also used to refer to the ecology and other distinctive characters of a species.

natural resources Aspects of biodiversity that are valued by people.

natural selection Changes in a population in response to specific factors in the environment and sexual selection whereas those with the most adaptive traits are best able to survive.

naturalised An exotic species that is thoroughly integrated in their new environment.

naturalist A person who studies or is an expert in natural history.

nature deficit disorder A situation where spending less time in nature lead to behavioural problems in children that lasts through adulthood.

nature's contributions to people (NCP) All the positive (e.g. food provisioning) and negative (e.g. disease transmission) contributions of biodiversity to people's quality of life. Compare to ecosystem services.

neocolonialism The use of economic, political, cultural, or other pressures to gain control or influence over other countries or regions. *See also* land grabbing.

niche A multi-dimensional space that explains the role and position of a species in its environment. It includes essential resource limits, as well as the species' interactions with its biotic and abiotic environment.

niche model *See* species distribution model.

no-take zone An area where hunting, fishing, and collection of natural products are not allowed.

noise pollution Excessive and inappropriate man-made noise that negatively impacts biodiversity.

nomadic (species) A species in which a significant proportion of individuals have no fixed territories, and wander from place to place, in search of limited resources, with no fixed route. Compare to resident species.

non-consumptive use value The benefits gained from natural resources that are not collected, harvested, consumed, converted, or destroyed during use.

non-invasive techniques Research techniques that cause minimal disturbance to study individuals or study sites.

non-governmental organisation (NGO) A private organisation that acts to benefit society in some way; many conservation organisations are NGOs.

normative discipline A discipline that incorporates human values, not just facts, and uses scientific methods to understand those values and achieve its goals. Compare to scientific discipline.

nutrient pollution Excessive nutrients added to water bodies, causing excessive algae growth and eutrophication.

ocean acidification The decrease in the pH of a marine environment, caused by the excessive uptake of carbon dioxide (CO_2) from the atmosphere.

ocean deoxygenation *See* ocean suffocation.

ocean suffocation Warmer surface water absorbs less atmospheric oxygen; this combined with decreased circulation of dissolved oxygen to deeper waters, due to climate change, limits the available oxygen for marine fauna. Also known as ocean deoxygenation.

ocean warming Increased water temperatures in oceanic environments due to climate change.

old-growth forest A forest that has never been logged.

omnivore A species that obtain energy and nutrients by eating both plants and animals.

open-access resources Natural resources such as water, air, and fish populations that are freely used by many different groups of people. Also called common-pool resources.

option value The potential of an organism to provide a currently unknown economic benefit at some point in the future.

outbreeding The situation where individuals of different species or same species with different adaptations (perhaps from distant populations) mate to produce offspring.

outbreeding depression Lowered fitness that occasionally occurs when individuals of different species or of widely different populations mate and produce offspring. Lowered fitness is caused by inheritance of traits not well-suited to the current environment.

overexploitation *See* overharvesting.

overharvesting Harvesting of natural resources at rates faster than recovery, causes the natural resource's decline or loss. Also known as overexploitation.

ozone layer An area in the stratosphere, consisting of high ozone (O_3) concentrations, that shields humans and biodiversity from the sun's harmful ultraviolet (UV) rays.

PADDD Acronym for protected area downgrading, downsizing, and degazettement, the legal process through which protected areas become weaker and smaller, or their protection is eliminated completely.

palustrine ecosystems Freshwater ecosystems, such as some wetlands and bogs, characterised by non-flowing water.

paper park Parks that appear on official government lists, but are invisible on the ground, thus providing little contribution to conservation.

parasitic (relationship) A biological relationship between two organisms from which one species benefits while the other is negatively affected. Compare to mutualistic relationship.

Paris Agreement The world's first comprehensive agreement on climate change, aimed to hold global warming below 2°C through fossil fuel divestment and financing measures such as forest protection.

passive integrated transponder (PIT) tags Tracking tags with internal microchips that can be used to identify an animal or plant in which it is implanted.

pastoralist Nomadic livestock farmers who move their herds in search of fresh pasture and water.

payment for ecosystem services (PES) Markets that enable landowners to receive direct payments for protecting and restoring ecosystems and ecosystems services.

persecution Indiscriminate mistreatment or killing of a group of animals such as predators.

persistent organic pollutants (POP) Harmful organic pollutants that bioaccumulate because they are resistant to environmental degradation.

perverse subsidies Financial incentives governments provide to industries that result in environmentally destructive activities.

pesticide drift The process in which pesticides are being transported away from their source through air, along rivers, and even in groundwater.

phenological mismatch The disruption of timed aspects of a species' life cycle, such as migration and breeding, which may cause some populations to decline and others to increase in abundance. Often used when referring to the impacts of climate change. Also called trophic asynchrony.

phenotype An organism's morphology, anatomy, physiology, and biochemistry, as an expression of an individual's genotype.

phenotypic plasticity The ability of an organism to change its phenotype in response to changes in the environment.

photochemical smog Air pollution, visible to the naked eye, that forms when chemical pollutants in the atmosphere react with ultraviolet light from the sun.

photosynthesis The process through which plants (and some other organisms) convert sunlight energy into chemical energy, the "fuel" required to sustain life.

pioneer species The first species to colonise an area during the process of ecological succession.

plant blindness A common perception that animals take precedence above plants, the latter seen as the backdrop of the environment rather than the critical foundation of every natural community and food web on Earth.

polymorphic gene A gene that has multiple forms or alleles.

population A group of individuals of the same species that interact with one another. Compare to metapopulation.

population and habitat viability assessments (PHVA) Population viability assessments that also consider an ecosystem's ability to support viable wildlife populations.

population biology The study of population dynamics over time and space.

population bottleneck The phenomenon when small population size lead to the loss of rare alleles, and thus genetic diversity, from one generation to the next.

population health and the environment (PHE) Human development that integrates family planning and human health with biodiversity conservation to achieve better outcomes than with single-sector approaches.

population rescue A type of metapopulation where continuous movement of individuals from a source population prevents a sink population from going extinct.

population viability analysis (PVA) A risk assessment for a species or population that uses demographic data and mathematical methods to predict the likelihood of a population or species going extinct at some point in the future.

predator Animals, such as a sharks and lions, that hunt, kill, and eat other animals. Compare to carnivore.

prescribed burn A fire set deliberately to maintain a fire-adapted ecosystem and to avoid dangerous accumulation of fuel loads. Also called a controlled burn.

prey An animal that is hunted by a predator.

primary consumer *See* herbivore.

primary producer Green plants, algae, seaweeds, and other photosynthetic organisms that obtain their energy directly from the sun. Also known as autotrophs.

primary productivity *See* ecosystem productivity.

productive use value The value of natural resources that is sold at markets. Compare to consumptive use value.

protected area An area managed primarily for the maintenance of biodiversity.

public outreach Efforts, such as public talks, workshops, school visits, and guided walks, aimed at raising the general public's awareness and understanding of conservation activities.

Ramsar Convention on Wetlands An international agreement that recognises the ecological, scientific, economic, cultural, and recreational value of freshwater, estuarine, and coastal marine ecosystems.

range-restricted (species) A species that occurs in a geographically small area and nowhere else.

range-shift gap When a physical gap in suitable habitat prevents a species from dispersing from one place to another.

rapid biodiversity assessment (RAP) A biodiversity inventory compiled under tight deadlines to answer urgent questions and inform urgent decisions. Also known as a rapid assessment plan.

reconciliation ecology The science of establishing and maintaining areas to protect biodiversity where people live and work.

recruitment The increase in a population's number of reproducing individuals as immigrants arrive or young become old enough to reproduce. Described in plant ecology as the presence of seedlings or newly germinated individuals, not vegetative reproduction.

Red Data Books *See* Red Lists.

Red List criteria Quantitative measures developed to reflect a taxon's (or ecosystem's) risk of extinction.

Red Lists Detailed lists of threatened wildlife compiled by the IUCN and its affiliate organisations.

Reducing Emissions from Deforestation and Forest Degradation (REDD+) A UN programme that uses financial incentives to reduce greenhouse gas emissions due to ecosystem destruction.

reference site A control site (or ecosystem) that provide a practical target for restoration and can be used to quantitatively assess of the success of a restoration project.

refugee species Species pushed to live or persist in suboptimal habitat by threats present in more suitable areas.

reintroduction Releasing individuals into areas where they occurred in the past but where they no longer occur.

relict species A member of a once diverse and widespread group of species that has continued to survive while its close relatives have gone extinct.

remote sensing Obtaining ecosystem data without making physical contact (i.e. boots on the ground) with the observation site, using e.g. satellite images and aerial photographs.

resident species Species that do not depend on dispersal (migration or nomadism) for survival.

resilience The ability to rapidly recover after a disturbance event.

resistance The ability to maintain stable throughout and after a disturbance event.

restocking Increasing the size and genetic diversity of existing populations by releasing individuals that have been raised in captivity or that have been collected from other wild populations. Also referred to as augmentation.

restoration ecology The scientific study of restoring damaged ecosystems, communities, and populations.

resurrection biology *See* de-extinction.

Rio Summit *See* Earth Summit.

riparian zone The area directly next to a water feature, such as a riverine forest.

rivet-popper hypothesis Compares biodiversity to the rivets that hold the airliner together; just as an airliner can only lose so many rivets before it falls apart, so will the progressive loss of species systematically weaken ecosystem stability until it collapses. Compare to species redundancy hypothesis.

Sahel An ecological transition zone that stretches across the northern parts of Africa; separates the Sahara Desert (north) from tropical Africa (south).

scavenger An animal that feeds on dead plant material, animal carcasses, and items discarded by humans.

scientific discipline A discipline that embraces knowledge based on observable phenomena that can be verified by other researchers working under the same conditions. Compare to normative discipline.

scientific method Creation of new knowledge and the verification of existing knowledge through systematic observations and measurements.

sea level rise The increase in the volume of water in the world's oceans due to climate change, resulting in an increase in global mean sea level.

secondary consumer Predators and carnivores that eat other animals.

secondary poisoning Non-target individuals that are poisoned or killed when they come into contact with poisons, such as insecticides.

seed bank (1) A ex situ collection of seeds that is stored for conservation of genetic diversity; (2) a natural in situ collection of dormant seed present in the soil.

seed disperser An animal that moves plant seeds away from parent plants, thereby allowing seedlings to colonise new areas away from parent plants and siblings.

seed scarification The weakening or opening of a seed's coat, which can happen chemically, thermally, or mechanically during seed dispersal. Scarification is a prerequisite for germination of many seeds.

sensitivity analysis Exploratory analyses where key assumptions, computations, and/or input are systematically changed, and subsequent results compared with the original model to assess the effect of those changes on model output.

sentinel species A species used as an early warning system for environmental hazards because they are more sensitive to certain conditions than humans.

shifting baseline syndrome Judging ecosystem condition against reference points (baselines) which themselves represent significant changes from an even earlier state of the system.

shifting cultivation *See* slash-and-burn agriculture.

siltation The process by which water becomes turbid due to fine soil particles suspended in the water. Associated with erosion and runoff.

sink population A subpopulation, which is part of a metapopulation, that receives new individuals from a connected source population.

sixth extinction episode The current mass extinction event, caused by human activities.

slash-and-burn agriculture Traditional farming practice in which farmers prepare agricultural lands by clearing land for fuel wood followed by burning the remaining vegetation for fertilisation. Crops are then grown for a few years before the plot is abandoned and the process is repeated elsewhere. Also called shifting cultivation.

SLOSS debate A discussion framework that conservation biologists use to debate the relative advantages of a single large conservation areas over several small conservation areas.

smallholder farmer Farmers that own small plots of land and rely on family labour to grow subsistence crops.

Society for Conservation Biology (SCB) An international non-profit professional organisation with a mission to advance the science and practice of conserving the Earth's biological diversity.

soft release A translocation strategy that involves keeping individuals in an enclosed area at the release site for a period of time before release; it may also include some form of assistance after release to increase opportunities for success. Compare to hard release.

source population A subpopulation, which is part of a metapopulation, from which individuals are dispersed to other locations.

specialist species A species adapted to a restricted set of environmental conditions, or have very particular (e.g. dietary, temperature) needs. Compare to generalist species.

speciation The formation of a new species through evolution and genetic drift.

species distribution model (SDM) The process of using geospatial analysis to predict the distribution of species based the distribution of suitable environmental conditions.

species diversity The full variety of species from single-celled organisms, like bacteria, to larger multicellular organisms, like animals and everything in between.

species redundancy hypothesis Holds that ecosystem stability is best maintained by ensuring that there is redundancy in ecosystem functioning, accomplished by ensuring that each ecosystem has a variety of (seemingly redundant) species performing similar roles.

species richness The total number of species found at a location or in a community.

species-area relationship The prediction that large areas (islands, habitats) contain more species than smaller areas because large areas are better buffered from extinction events since they can maintain large enough populations to ensure long-term persistence.

splitting A liberal taxonomic approach that prefers to classify closely-related taxa as individual entities. Compare to lumping.

stepping stone habitat A special type of habitat linkage that facilitates dispersal along a patchwork of isolated habitat patches within a matrix of unsuitable or inhospitable habitat.

stochastic model A model where each iteration will result in a different outcome. Compare to deterministic model.

stochasticity Random variation that happens by chance.

subpopulation A subset of a larger population. Often used in reference to fragmented populations or metapopulations.

substitute species A common species used to fill persistent data gaps that affects the conservation management of an at-risk species. Compare to surrogate species.

succession Describes the gradual process during which ecosystems change after a disturbance; these changes can include changes to the species present, the soil chemistry, and microclimatic characteristics. More generally known as ecological succession,

surrogate species A common species, closely related to one or more species of concern, that is used to assess broader biodiversity patterns during conservation planning studies. Compare to substitute species.

surveys A catch-all term that describes methods to monitor aspect of biodiversity, such as population size or ecosystem health.

sustainable agricultural intensification *See* conservation agriculture.

sustainable development Economic activities that satisfies both present and future needs for resources and employment without compromising the natural world.

symbiotic relationship A biological relationship (e.g. parasitism) between two organisms. Obligate symbiosis describes a relationship where one species cannot survive without the other, while facultative symbiosis describes a relationship in which one species can live independent of the other. Compare to mutualistic relationship.

systematic conservation planning A structured approach to identifying conservation priorities by identifying the species or populations that lack protection and identifying the actions or areas that will best fill those protection gaps.

taxon (plural: taxa) A catch-all term describing biological units of classification, such as a single species or a group of related species. All monkeys fall under the order Primates, but they are also part of the class Mammalia. Species, orders, and classes are all distinct taxa, or taxonomic groups.

taxonomist Specialist scientist involved in the identification, classification, and naming of species.

thermal pollution The degradation of an environment by changing its temperature. Often used when referring to aquatic ecosystems.

thermal shock Rapid changes in water temperatures leading to excessive stress or damage to aquatic ecosystems.

threatened species A species that is classified as *Vulnerable, Endangered,* or *Extinct* according to the IUCN Red List criteria, and thus considered at risk of extinction if current conditions persist.

traditional ecological knowledge (TEK) Evolving knowledge acquired by indigenous and local peoples over hundreds or thousands of years through direct contact with the environment. Also called local ecological knowledge.

traditional people Self-sufficient human societies that have lived a rural, non-industrialised lifestyle for many generations, depend on the land and self-harvested natural resources for survival, and are not integrated into mainstream society. Sometimes also called traditional tribes, protecting their cultural practices and way of life has special status under international law.

tragedy of the commons The gradual loss of open-access resources because of unregulated use.

transfrontier conservation area (TFCA) A large region that crosses international boundaries and encompasses one or more protected areas as well as the surrounding multi-use areas, all managed as a single conservation unit.

translocation The capture, transport, and release of animals or plants from one location to another.

treaty *See* convention.

trophic asynchrony *See* phenological mismatch.

trophic cascade The situation where one keystone species' loss has rippling effects at other trophic levels.

trophic levels The different levels in a biological community (e.g. primary producer; herbivore; carnivore, decomposer), each sharing the same position in a food chain.

umbrella species A species whose protection indirectly benefits other species and ecosystem components with which they share their landscape.

UNESCO The United Nations Educational, Scientific and Cultural Organisation. Manages the list of World Heritage Sites.

United Nations An intergovernmental organisation that promotes international co-operation and facilitates international law and order.

United Nations Environmental Programme (UNEP) Coordinates the UN's environmental activities, including assisting developing countries in implementing sound environmental policies and practices.

urban heat island effect Occurs when absorbed sun energy from modified surfaces, such as asphalt roads, causes urban areas to be warmer than the surrounding natural environment.

urbanisation The increase in the proportion of people moving from rural areas to live in urban areas.

use value The direct and indirect benefits humans gain from biodiversity.

voluntary transaction The assumption that monetary transactions take place only when it benefits both parties involved. This principle frequently fails to account for harm to people not directly involved in the transaction and to society leading to market failure.

wilderness areas Large blocks of land that have been minimally affected by human activity, have a low human population density, and are not likely to be developed soon.

wildlife In the context of this textbook, the term refers to all the wild organisms on Earth, including but not restricted to animals, plants, fungi, and bacteria.

wildlife corridor *See* habitat linkage.

wildlife crossing A structure, such as an underpass tunnel, overpass or canopy bridge of fish ladder, that enables wildlife to safely disperse over human-made barriers.

wildlife trafficking The illegal trade of protected species and their body parts.

wilful ignorance Purposefully ignoring the environmental damage caused directly or indirectly by their activities.

World Bank An international financial institution that provides loans for development projects in developing countries.

World Heritage Site Natural and/or cultural areas of international significance recognised by the United Nations.

World Parks Congress Organised by the IUCN every 10 years, it is the largest gathering of organisations and individuals involved in protected areas management worldwide.

zoning A management method of dealing with conflicting demands on protected areas by setting aside designated areas where certain regulated human activities are permitted.

zoonotic disease Diseases, such as rabies, that can be transmitted from animals to humans. Compare to anthroponotic disease.

Index

www.ingramcontent.com/pod-product-compliance
Lightning Source LLC
LaVergne TN
LVHW070906300926
849162LV00012B/1865